System-on-Chip:
Next Generation Electronics

Edited by
Bashir M. Al-Hashimi

School of Electronics and Computer Science
University of Southampton, UK

The Institution of Electrical Engineers

Published by: The Institution of Electrical Engineers, London,
United Kingdom

The Institution of Electrical Engineers,
Michael Faraday House,
Six Hills Way, Stevenage,
Herts., SG1 2AY, United Kingdom

www.iee.org

British Library Cataloguing in Publication Data

System on chip : next generation electronics.-
(IEE circuits, devices and systems series)
1. Systems on a chip 2. Embedded computer systems
3. Microelectronics
I. Al-Hashimi, Bashir M. II. Institution of Electrical Engineers

621.3'95

ISBN-10: 0863415520
ISBN-13: 978-086341-552-4

Typeset in India by Newgen Imaging Systems (P) Ltd., Chennai, India
Printed in the UK by MPG Books Ltd., Bodmin, Cornwall

Learner
Services

Please return
on or before
the last date
stamped below

CITY COLLEGE
NORWICH

1 5 APR 2013

IEE CIRCUITS, DEVICES AND SYSTEMS SERIES 18

Series Editors: Professor Dr. R. Soin
Dr. D. Haigh
Professor Y. Sun

System-on-Chip: Next Generation Electronics

Other volumes in the Circuits, Devices and Systems series:

To May, Sara, Haneen and Zahara

Preface

System-on-chip (SoC) is widely believed to represent the next major market for microelectronics, and there is considerable interest world-wide in developing effective methods and tools to support the SoC paradigm. The work presented in this book grew out of a special issue 'Embedded Microelectronic Systems: Status and Trends', IEE Proceedings: Computers and Digital Techniques, April/June 2005.

Recently, a number of excellent books on SoC have been published, most of which have focused on a particular area of SoC research. The field of SoC is broad and expanding, and the prime objective of this book is to provide a timely and coherent account of the recent advances in some of the SoC key research areas in one volume. In order to achieve this, 25 international research groups were invited to contribute to the book. Each contribution has an up-to-date research survey highlighting the key achievements and future trends. To facilitate the understanding of the numerous research topics covered in the book, each chapter has some background covering the basic principles, and an extensive up-to-date list of references. To enhance the book's readability, the 25 chapters have been grouped into eight parts, each part examining a particular theme of SoC research in depth.

In general, complete SoC designs consist of hardware and software components, which are traditionally developed separately and combined at a later stage of the design. This, however, increases time-to-market and system cost which is in contraction with some of the SoC drivers. To address such difficulties and to cope with the continuing increased design complexity, new design methodologies that support SoCs are needed. Part I of the book contains six chapters (1 to 6) dealing with the systematic and concurrent design, analysis and optimisation of SoC-based embedded systems. Software plays a very important role in the design of SoC; Part II has three chapters devoted to embedded software characterisation (Chapter 7), retargetable compilation (Chapter 8) and power-aware software generation (Chapter 9).

Excessive power dissipation in SoC does not only limit their applications in portable devices, but also results in increased packaging and cooling costs. Managing the power issue is fundamental to successfully expending Moore's law. Until recently, dynamic power has been the dominant source of power consumption, however, leakage power is becoming a significant fraction of the total power in deep-submicron designs. Part III contains four chapters (9 to 13) describing effective techniques for reducing the dynamic and leakage power consumption. These techniques can be applied at various levels of the design hierarchy to allow designers to meet the challenging power constraints in SoC. There are some compelling advantages of employing reconfigurable devices in SoC in terms of speed, power, cost and time-to-market. Architectures and design methods for reconfigurable computing are discussed in Part IV of the book (Chapter 14).

Telecomm and multimedia applications require mixed-signal SoCs; Chapter 15 of Part V describes methods and tools that automate the process of modelling and generating analogue/RF cores for such SoCs. The International Technology Roadmap on Semiconductors (ITRS – http://public.itrs.net/) predicts that the use of clock-less designs will be increased in future SoCs to cope with timing issues. Chapter 16 of Part five is concerned with the synthesis and design automation of asynchronous systems. A key element in achieving functional design is the on-chip communication that interconnects the SoC cores. Bus-based interconnections provide the current SoC communication. However, SoCs complexity is increasing with the continuing scaling down of CMOS feature sizes. According to ITRS'03, an average SoC will contain >50 processing and memory blocks in 2008 and 100 such blocks in 2012. Consequently, it may not be viable to continue to effectively employ bus-based communication in future SoC. To address this concern, and improve performance of future SoCs, different interconnection technologies are being developed. Part VI (Chapters 17 and 18) is devoted to network-on-chip, a new interconnection technology where SoC cores communicate with each other by sending packets over an on-chip network.

Part VII of the book contains three chapters investigating functional design validation and verification, which are important factors that contribute to the ultimate costs of an SoC. Chapters 19 and 20 focus on simulation-based techniques that have been developed to validate complex hardware/software systems, whilst Chapter 21 considers formal verification as a way of verifying system correctness. The high level of integration is making the cost of testing SoC expensive, mainly due to the volume of test data and limited test access to embedded cores. The ITRS'03 predicts that if the current trends are maintained, by 2015 the cost of testing a transistor will approach or even exceed the cost of manufacturing. Therefore, low-cost design-for-test techniques for SoCs are required, which is the subject of the final part (Part VIII) of the book. This part has four chapters, test-resource partitioning (Chapter 22), multi-site testing (Chapter 23), on-chip timing measurement (Chapter 24) and yield and reliability (Chapter 25).

Book audience

It is the intention of this book to contain a diverse coverage of SoC main research themes, each theme is discussed in depth and therefore the book will appeal to broader readership. SoC is a popular PhD research topic and is appearing as part of the syllabus for both undergraduate and postgraduate Electronics and Computer Engineering courses at many universities, and I hope that this book will complement the research and teaching that is taking place in this area. Also, the book should serve as a valuable reference for designers and managers interested in various aspects of SoC design and test.

Acknowledgements

First and foremost I am indebted to all the contributing authors of this book. I acknowledge the financial support from the Engineering & Physical Sciences Research Council (EPSRC), UK, to my research activities in SoC which made this book possible. I like to thank all the researchers whom I have worked with and who have broadened my knowledge of the SoC field in particular, Dr. M. Schmitz, Dr. P. Rosinger, Dr. T. Gonciari and Dr. D. Wu. Finally, I wish to thank Sarah Kramer (Senior Commissioning Editor) and Phil Sergeant at the IEE, their assistance during this project have been invaluable and Geoff Merrett for his work on the book cover.

Bashir M. Al-Hashimi
Southampton, UK
June 2005

Contents

3   Analysis and optimisation of heterogeneous real-time embedded systems   **75**
Paul Pop, Petru Eles and Zebo Peng

13 Leakage power analysis and reduction for nano-scale circuits   **415**
Amit Agarwal, Saibal Mukhopadhyay, Chris H. Kim,
Arijit Raychowdhury and Kaushik Roy

PART IV – Reconfigurable computing   **449**

14 Reconfigurable computing: architectures and design methods   **451**
T.J. Todman, G.A. Constantinides, S.J.E. Wilton, O. Mencer, W. Luk and
P.Y.K. Cheung

25 Yield and reliability prediction for DSM circuits   **857**

Thomas S. Barnett and Adit D. Singh

Authors and Contributors

Part I System Level Design

Simon Künzli, Lothar Thiele,
Eckart Zitzler
Department Information
Technology and Electrical Engineering
Computer Engineering and
Networks Lab
ETH Zurich, Switzerland

Rafik Henia, Arne Hamann,
Marek Jersak, Razvan Racu,
Kai Richter, Rolf Ernst
Institute of Computer and
Communication Network Engineering
Technical University of Braunschweig
D-38106 Braunschweig, Germany

Paul Pop, Petru Eles, Zebo Peng
Computer and Information
Science Dept.
Linköping University SE-581 83
Sweden

Jaehwan John Lee
Department of Electrical and Computer
Engineering
Purdue School of Engineering and
Technology
Indiana Univ. - Purdue Univ.
Indianapolis
723 West Michigan St. Indianapolis
IN 46202, USA

Vincent John Mooney III
Centre for Research on Embedded
Systems and Technology
School of Electrical and Computer
Engineering
Georgia Institute of Technology
Georgia
USA

Axel Jantsch, Ingo Sander
Royal Institute of Technology
SB-100 44 Stockholm
Sweden

Prabhat Mishra
Department of Computer and
Information Science and Engineering
University of Florida
Gainesville
FL 32611
USA

Nikil Dutt
Center for Embedded Computer
Systems
University of California, Irvine
CA 92697
USA

Part II Embedded Software

Gunnar Braun
CoWare, Inc.
Technologiezentrum am Europaplatz
Dennewartstrasse 25-27
52068 Aachen
Germany

**Rainer Leupers, Manuel Hohenauer,
Kingshuk Karuri, Jianjiang Ceng,
Hanno Scharwaechter,
Heinrich Meyr, Gerd Ascheid**
Institute for Integrated Signal
Processing Systems
RWTH Aachen University of
Technology
Germany

Edward Lee, Stephen Neuendorffer
EECS Department
University of California at
Berkeley
Berkeley, CA, USA
and Xilinx Research Labs
San Jose, CA
USA

**J. Hu, G. Chen, M. Kandemir,
N. Vijaykrishnan**
Department of Computer Science and
Engineering
Pennsylvania State University
University Park, PA 16802
USA

Part III Power Reduction and Management

P. Marchal, F. Catthoor
IMEC
Kapeldreef75
B-3001 Leuven, Belgium

J.I. Gomez, D. Atienza
Universidad Complutense de Madrid
Spain

S. Mamagkakis
Democritus University of Thrace
Greece

Niraj K. Jha
Department of Electrical Engineering
Princeton University
Princeton, NJ 08544, USA

Afshin Abdollahi, Massoud Pedram
Dept. of Electrical Engineering
University of Southern California
Los Angeles, CA 90211
USA

**Amit Agarwal, Saibal
Mukhopadhyay, Chris H. Kim,
Arijit Raychowdhury,
Kaushik Roy**
School of Electrical and Computer
engineering
Purdue University
West Lafayette, Indiana 47907-1285
USA

Part IV Reconfigurable Computing

G. A. Constantinides, P. Y. K. Cheung
Department of Electrical and Electronic
Engineering
Imperial College, Exhibition Road
London SW7 2BT,
England, UK

T. J. Todman, O. Mencer, W. Luk
Department of computing
180 Queen's Gate
South Kensington Campus
Imperial College London
SW7 2AZ
England, UK

S. J. E. Wilton
Department of Electrical and Computer
Engineering
University of Toronto
10 King's College Rd.
Toronto, Ontario
Canada M5S 1A4

Part V Architectural Synthesis

Georges G. E. Gielen
Department of Electrical Engineering -
ESAT-MICAS
Katholieke Universiteit Leuven
Belgium

Danil Sokolov, Alex Yakovlev
School of Electrical, Electronic and
Computer Engineering
University of Newcastle upon Tyne
NE17RU, UK

Part VI Network-on-chip

Luca Benini, Davide Bertozzi
The Dip. di Elettronica, Informaticae
Sistemistica
Universit'a di Bologna
Bologna 40136, Italy

Manish Amade
Department of Electronics & Computer
Engineering
University of California at San Diego
USA

Aristides Efthymiou
School of Informatics
University of Edinburgh
Edinburgh, UK

Tomaz Felicijan,
Douglas Edwards
Department of Computer
Science
University of Manchester
Manchester, UK

Luciano Lavagno
Politecnico di Torino
Dipartimento di Elettronica
Corso Ducadegli Abruzzi 24
Torino 10129
Italy

Part VII Simulation and Verification

Ian G. Harris
Department of Computer Science
University of California Irvine
Irvine, CA 92697
USA

Sungjoo Yoo
SoC Research Center
System LSI Division
Samsung Electronics, Korea

A. Jerraya
System-Level Synthesis Group
TIMA Laboratories
France

Rolf Drechsler, Daniel Große
Institute of Computer Science
University of Bremen
28359 Bremen
Germany

Part VIII Manufacturing Test

Krishnendu Chakrabarty
Department of Electrical and Computer
Engineering
Duke University
Box 90291
Durham
NC 27708
USA

**Sandeep Kumar Goel, Erik Jan
Marinissen**
Philips Research Laboratories
IC Design - Digital
Design & Test
High Tech Campus 5
5656AE, Eindhoven
The Netherlands

Peter M. Levine, Gordon W. Robert
Microelectronics and Computer
Systems Laboratory
McGill University
3480 University Street
Montreal, Quebec
H3A 2A7, CANADA

Adit D. Singh
Department of Electrical and Computer
Engineering
Auburn University
Auburn, Alabama 36849, USA

Thomas S. Barnett
IBM Microelectronics
Essex Junction
Vermont 05452, USA

Part I
System-level design

Chapter 1

Multi-criteria decision making in embedded system design

Simon Künzli, Lothar Thiele and Eckart Zitzler

1.1 Introduction

Embedded systems are usually evaluated according to a large variety of criteria such as performance, cost, flexibility, power and energy consumption, size and weight. As these kinds of non-functional objectives are very often conflicting, there is no single optimal design but a variety of choices that represent different design trade-offs. As a result, a designer is not only interested in one implementation choice but in a well-chosen set that best explores these trade-offs.

In addition, embedded systems are often complex in that they consist of heterogeneous subcomponents such as dedicated processing units, application-specific instruction set processors, general-purpose computing units, memory structures and communication means like buses or networks. Therefore, the designer is faced with a huge design space.

Embedded systems are resource constrained because of tight cost bounds. Therefore, there is resource sharing on almost all levels of abstraction and resource types that makes it difficult for a designer to assess the quality of a design and the final effect of design choices. This combination of a huge design space on the one hand and the complexity in interactions on the other hand makes automatic or semi-automatic (interactive) methods for exploring different designs important.

Besides the above-mentioned multiple objectives in the design of embedded systems, there are tight constraints on the design time. One possibility to accommodate late design changes and a short time-to-market is to choose a very flexible design, close to a general-purpose computing system. On the other hand, this approach sacrifices almost all other quality criteria of a design. As a consequence, embedded systems are usually domain-specific and try to use the characteristics of the particular application domain in order to arrive at competitive implementations. In order to achieve acceptable design times though, there is a need for automatic or semi-automatic (interactive) exploration methods that take into account the application domain, the level of abstraction on which the exploration takes place and that can cope with conflicting criteria.

Following the usual hierarchical approach to embedded system design, there are several layers of abstraction on which design choices must be taken. Above the technology layer one may define the abstraction levels 'logic design and high-level synthesis', 'programmable architecture', 'software compilation', 'task level' and 'distributed operation'. These terms are explained in more detail in Section 1.2. Design space exploration takes place on all these layers and is a generic tool within the whole design trajectory of embedded systems.

A simplified view on the integration into an abstraction layer is shown in Figure 1.1. For example, if the layer of abstraction is the 'programmable architecture', then the generation of a new design point may involve the choice of a cache architecture. The estimation of non-functional properties may be concerned with the performance of task execution on the underlying processor architecture, the size of the cache or the total energy consumption. The estimation may either be done using analytic methods or by a suitable simulator by use of suitable input stimuli, e.g. memory access traces. In any case, properties of the sub-components (from logic design) are necessary, e.g. the relations among area, power consumption, structure and size of the cache. The generation of new design points has to satisfy various constraints e.g. in terms of feasible cache sizes or structures. The choice of a cache will then lead to refined constraints for the design of its sub-components (digital design layer).

Figure 1.1 makes also apparent the interplay between exploration on the one hand and estimation on the other. The methods and tools applied to the estimation of non-functional properties very much depend on the particular abstraction layer and the design objectives. For example, if the average timing behaviour is of concern, very often simulation-based approaches are used. On the other hand, worst-case timing usually requires analytic methods. Estimation is particularly difficult as only a limited knowledge about the properties of sub-components and the system environment in terms of input stimuli is available. For example, on the system-level, the sub-components to be used are not designed yet and the individual tasks of the application may not be fully specified. This chapter mainly focuses on the generation of new design points and the decision process that finally leads to a design decision, estimation will not be covered.

The purpose of the chapter is to review existing approaches to design space exploration of embedded systems and to describe a generic framework that is based on multi-objective decision making, black-box optimisation and randomised search strategies. The framework is based on the PISA (Platform and Programming language independent Interface for Search Algorithms) protocol that specifies a problem-independent interface between the search/selection strategies on the one hand and the domain-specific estimation and variation operators on the other. It resolves the current problem that state-of-the-art exploration and search strategies are not (easily)

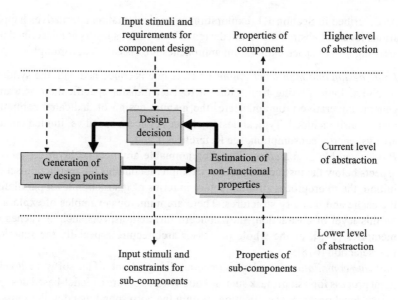

Figure 1.1   *Embedding of exploration in a hierarchical design trajectory for embedded systems*

accessible to solve the domain-specific exploration problems in embedded systems design. The main questions the chapter would like to answer can be phrased as follows. How can one apply efficient design space exploration to a new design problem in embedded systems design? How can one integrate a new estimation methodology into a complete design space exploration in a simple and efficient way?

In Section 1.2 existing approaches to design space exploration in embedded systems are reviewed and classified. Section 1.4 describes the proposed framework that is based on strategies that select promising design points (Section 1.4.1), implementation of domain-specific variation operators to determine a suitable neighbourhood of current set of design points (Section 1.4.2), and the implementation based on the PISA protocol (Section 1.4.3). A simple running example for cache exploration is used to illustrate the different steps. Finally, Section 1.4.4 shortly describes a more complex application, i.e. the system-level exploration of a stream processing architecture.

1.2   Approaches to design space exploration

There are a vast number of approaches available that make use of an automated or semi-automated design space exploration in embedded systems design. Therefore, only a representative subset will be discussed with an emphasis on the exploration strategies, whereas the different estimation methods for non-functional properties will not be discussed further.

As described in Section 1.1, exploration of implementation alternatives happens at various levels of abstraction in the design. These various layers are described next and existing design space exploration approaches are classified accordingly:

- *Logic design and high-level synthesis*: Here one is concerned with the synthesis of digital logic starting from either a register-transfer specification or a more general imperative program. Here, the manual design of dedicated computing units is also included. Typical design choices concern speed vs. implementation area vs. energy consumption, see References 1 and 2.
- *Programmable architecture*: The programmable architecture layer contains all aspects below the instruction set. For example, it contains the instruction set definition, the microprocessor architecture in terms of instruction level parallelism, the cache and memory structures. There are numerous examples of exploration on this level of abstraction; they concern different aspects such as caches and memories [3–5], or the whole processor architecture especially the functional unit selection [6–8].
- *Software compilation*: This layer concerns all ingredients of the software development process for a single task such as code synthesis from a model-based design or a high-level program specification. Within the corresponding compiler, possible exploration tasks are code size vs. execution speed vs. energy consumption. There are attempts to perform a cross-layer exploration with the underlying processor architecture, see References 9 and 10.
- *Task Level*: If the whole application is partitioned into tasks and threads, the task level refers to operating system issues like scheduling, memory management and arbitration of shared resources. Therefore, typical trade-offs in choosing the scheduling and arbitration methods are energy consumption vs. average case vs. worst case timing behaviour, e.g. Reference 11.
- *Distributed operation*: Finally, there exist applications that run on distributed resources. The corresponding layer contains the hardware aspects of distributed operation (such as the design of communication networks) as well as methods of distributed scheduling and arbitration. On this level of abstraction, which is sometimes called system level, one is interested in the composition of the whole system that consists of various computing and communication resources. System-level design not only refers to the structure of the system, but also involves the mapping of application to the architecture and the necessary (distributed) scheduling and arbitration methods. This highest level of abstraction seems to be especially suited for exploration methods, see e.g. results on the communication infrastructure [12,13], on distributed systems [14] or multiprocessor systems and systems-on-chip [15–19].

The above approaches combine several important aspects such as the integration of the exploration into the whole design process, the specific estimation method used to evaluate the properties of design points and finally the method that is used to perform the actual exploitation. Following the focus of the chapter, the existing approaches can be classified in a way that is orthogonal to the abstraction layers, namely the methods that are applied to perform the exploration itself. This way it becomes apparent that the exploration process is largely independent of the abstraction level. This property will be used later on in defining the new generic framework.

If only a single objective needs to be taken into account in optimisation, the design points are totally ordered by their objective value. Therefore, there is a single optimal design (if all have different objective values). The situation is different if multiple objectives are involved. In this case, design points are only partially ordered, i.e. there is a set of incomparable, optimal solutions. They reflect the trade-offs in the design. Optimality in this case is usually defined using the concept of Pareto-dominance: A design point dominates another one if it is equal or better in all criteria and strictly better in at least one. In a set of design points, those are called Pareto-optimal which are not dominated by any other.

Using this notion, available approaches to the exploration of design spaces can be characterised as follows.

1 *Exploration by hand*: The selection of design points is done by the designer himself. The major focus is on efficient estimation of the selected designs, e.g. Reference 16.

2 *Exhaustive search*: All design points in a specified region of the design parameters are evaluated. Very often, this approach is combined with local optimisation in one or several design parameters in order to reduce the size of the design space, see References 4 and 20.

3 *Reduction to a single objective*: For design space exploration with multiple conflicting criteria, there are several approaches available that reduce the problem to a set of single criterion problems. To this end, manual or exhaustive sampling is done in one (or several) directions of the search space and a constraint optimisation, e.g. iterative improvement or analytic methods, is done in the other, see References 2, 3, 8 and 12.

4 *Black-box randomised search*: The design space is sampled and searched via a black-box optimisation approach, i.e. new design points are generated based on the information gathered so far and by defining an appropriate neighbourhood function (variation operator). The properties of these new design points are estimated which increases the available information about the design space. Examples of sampling and search strategies used are Pareto Simulated Annealing [21] and Pareto Tabu Search [7,10], evolutionary multi-objective optimisation [13,14,18,22] or Monte Carlo methods improved by statistical estimation of bounds [1]. These black-box optimisations are often combined with local search methods that optimise certain design parameters or structures [11].

5 *Problem-dependent approaches*: In addition to the above classification, one can also find a close integration of the exploration with a problem-dependent characterisation of the design space. Several possibilities have been investigated so far.

- Use the parameter independence in order to prune the design space, e.g. References 17 and 23.
- Restrict the search to promising regions of design space, e.g. Reference 6.

- Investigate the structure of the Pareto-optimal set of design points, e.g. using hierarchical composition of sub-component exploration and filtering, e.g. References 5 and 15.
- Explicitly model the design space, use an appropriate abstraction, derive a formal characterisation by symbolic techniques and use pruning techniques, e.g. Reference 24.

Finally, usually an exhaustive search or a black-box randomised search is carried out for those parts of the optimisation that are inaccessible for tailored techniques.

From the above classification, one can state that most of the above approaches use randomised search techniques one way or the other, at least for the solution of sub-problems. This observation does not hold for the exploration by hand or the exhaustive search, but these methods are only feasible for small design spaces with a few choices of the design parameters. Even in the case of a reduction to a single objective or in the case of problem-dependent approaches, sub-optimisation tasks need to be solved, either single objective or multi-objective and randomised (black-box) search techniques are applied.

While constructing tools that perform design space exploration of embedded systems at a certain level of abstraction, the question arises, how to apply exploration to a new design problem. How does one connect the problem-specific parts of the exploration with a randomised black-box search engine? What is an appropriate interface between the generic and problem-dependent aspects? Which search strategy should one use? How can one achieve a simple implementation structure that leads to a reliable exploration tool? Section 1.4 is devoted to this problem.

The basis of the proposed solution is the protocol PISA, see Reference 25. It is tailored towards black-box randomised search algorithms and is characterised by the following properties. (1) The problem-specific and the generic parts of the exploration method are largely independent from each other, i.e. the generic search and selection should be treated as a black-box (separation of concerns). (2) The framework itself should not depend on the machine types, operating systems or programming languages used (portability). (3) The protocol and framework should be tailored towards a reliable exploration. The main components of the proposed framework in Figure 1.2 are a refinement of Figure 1.1. It shows the separation into the problem-specific variation and estimation part on the one hand and generic black-box search on the other.

1.3   A simple example: design space exploration of cache architectures

Before the PISA framework is described in more detail (in Section 1.4.3), this section introduces a simple example application that will be used throughout the remainder of this chapter for illustration purposes. Note, that it is not the purpose of the example to present any new results in cache optimisation.

The example problem to solve is to optimise the architecture of a cache for a predefined benchmark application. The solution space for the problem is restricted to

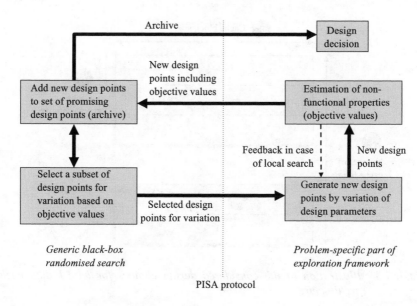

Figure 1.2   *Overview of the proposed framework for design space exploration based on the PISA protocol*

Table 1.1   *Parameters determining a cache architecture*

No.	Parameter	Range
1	No. of cache lines	2^k, with $k = 6 \ldots 14$
2	Block size	2^k bytes, with $k = 3 \ldots 7$
3	Associativity	2^k, with $k = 0 \ldots 5$
4	Replacement strategy	LRU or FIFO

L1 data caches only, i.e. the design choices include the cache size, the associativity level, the block size and the replacement strategy. The goal is to identify a cache architecture that (1) maximises the overall computing performance with respect to the benchmark under consideration and (2) minimises the chip area needed to implement the cache in silicon.

In Table 1.1, all parameters and possible values for the cache architecture are given. A design point is therefore determined by three integer values and a Boolean value. The integers denote the number of cache lines, the cache block size and the cache associativity; the Boolean value encodes the replacement strategy: 'false' denotes FIFO (first-in-first-out), 'true' denotes LRU (least recently used). Figure 1.3 graphically depicts the design parameters. The values for the number of cache lines,

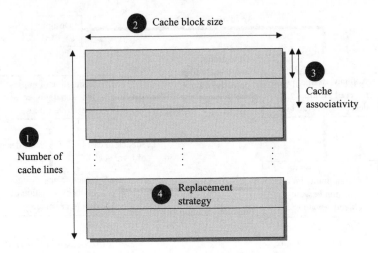

Figure 1.3   Illustration of the considered design choices for an L1 data cache architecture

block size and associativity have to be powers of 2, due to restrictions in the tools used for evaluation of the caches.

The first objective according to which the cache parameters are to be optimised is the CPI (cycles per instruction) achieved for a sample benchmark application and the second objective is the chip area needed to implement the cache on silicon. To estimate the corresponding objective values, two tools were used, namely `sim-outorder` of SimpleScalar [26] and CACTI [27] provided by Compaq. The first tool served to estimate the CPI for the benchmark compress95 running on the plain text version of the GNU public licence as application workload. The smaller the CPI for compress95 for a particular solution, the better is the solution for this objective. The second tool calculated an estimate for the silicon area needed to implement the cache. The smaller the area, the better is the cache for the area objective.

1.4   A general framework for design space exploration

As discussed in Section 1.2, the proposed general framework for design space exploration separates application-specific aspects from the optimisation strategy. The resulting two parts are implemented as independent processes communicating via text files, as will be detailed in Section 1.4.3. This concept (Figure 1.2) reflects the working principle of black-box randomised search algorithms.

Black-box methods are characterised by the fact that they do not make any assumptions about the objective functions, and in this sense they treat the design criteria as black-boxes which can contain arbitrarily complex functionalities. Initially, they create one or several designs at random, which are then evaluated with respect to the objective functions under consideration. Afterwards, the information about the already considered design(s) is used in order to generate one or several different designs that are then evaluated as well. This process is repeated until a certain number of iterations has been carried out or another stopping condition is fulfilled. The goal here is to exploit structural properties of the design space such that only a fraction of the design space needs to be sampled to identify optimal and nearly optimal solutions, respectively. This implies that different search space characteristics require different search strategies, and accordingly various black-box optimisers such as randomised local search, simulated annealing, evolutionary algorithms, etc. and variants thereof are available, see Reference 28.

Two principles form the basis for all randomised search algorithms: selection and variation. On the one hand, selection aims at focusing the search on promising regions of the search space as will be discussed in Section 1.4.1. This part is usually problem independent. On the other hand, variation means generating new designs by slightly modifying or combining previously generated ones. Although standard variation schemes exists – details can be found in Section 1.4.2 – the generation of new designs based on existing ones is strongly application dependent, similarly to the internal representation and the evaluation of designs.

1.4.1   Selection

The selection module implements two distinct phases: selection for variation and selection for survival. The former type of selection chooses the most promising designs from the set of previously generated designs that will be varied in order to create new designs. For practical reasons, though, not all of the generated designs will be kept in memory. While, e.g. simulated annealing and tabu search only store one solution in the working memory (in this case, selection for variation simply returns the single, stored solution), evolutionary algorithms operate on a population of solutions, which is usually of fixed size. As a consequence, another selection phase is necessary in order to decide which of the currently stored designs and the newly created ones will remain in the working memory. This phase is often called selection for survival or environmental selection, in analogy to the biological terminology used in the context of evolutionary algorithms.

1.4.1.1   Selection for variation

Selection for variation is usually implemented in a randomised fashion. One possibility to choose N out of M designs is to hold tournaments between two solutions that are picked at random from the working memory based on a uniform probability distribution. For each tournament, the better design is copied to a temporary set which is also denoted as a mating pool – again a term mainly used within the field of evolutionary computation. By repeating this procedure, several designs can be selected for variation, where high-quality designs are more likely to have one or multiple copies in the mating pool. This selection method is known as binary tournament selection; many alternative schemes exist as well (see Reference 29).

Most of these selection algorithms assume that the usefulness or quality of a solution is represented by a scalar value, the so-called fitness value. While fitness assignment is straight forward in the case of a single objective function, the situation

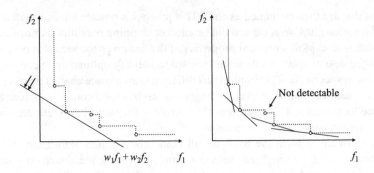

Figure 1.4   Illustration of the weighted-sum approach for two objectives. The left-hand side shows how a particular weight combination (w_1, w_2) uniquely identifies one Pareto-optimal design. The right-hand side demonstrates that not for all Pareto-optimal designs such a weight combination exists

is more complex in a multi-objective scenario. Here, one can distinguish between three conceptually different approaches.

- *Aggregation*: Traditionally several optimisation criteria are aggregated into a single objective by, e.g. summing up the distinct objective function values, where weight coefficients are used to control the influence of each criterion. The difficulty with this approach, though, is the appropriate setting of the weights. This usually requires more knowledge about the design space than is actually available. Furthermore, optimising a particular weight combination yields one Pareto-optimal solution. To obtain several optimal trade-off designs, multiple weight combinations need to be explored either in parallel or subsequently. Nevertheless, not necessarily all Pareto-optimal designs can be found as illustrated in Figure 1.4. The weighted-sum approach is only able to detect all solutions if the front of Pareto-optimal solutions is convex. Similar problems occur with many other aggregation methods, see Reference 30.
- *Objective switching*: The first papers using evolutionary algorithms to approximate the Pareto set suggested to switch between the different objectives during the selection step. For instance, Schaffer [31] divided selection for variation into n selection steps where n corresponds to the number of optimisation criteria; in the ith step, designs in the working memory were chosen according to their ith objective function value.
- *Dominance-based ranking*: Nowadays, most popular schemes use fitness assignments that directly make use of the dominance relation or extensions of it. By pairwise comparing all the designs in the working memory, different types of information can be extracted. The dominance rank gives the number of solutions by which a specific solution is dominated, the dominance count represents the number of designs that a particular design dominates and the dominance depth denotes the level of dominance when the set of designs is divided into non-overlapping non-dominated fronts (see Reference 28 for details).

These fitness assignment schemes can also be extended to handle design constraints. For dominance-based approaches, the dominance relation can be modified such that feasible solutions by definition dominate infeasible ones, while among infeasible designs the one with the lower constraint violation is superior – for feasible solutions, the definition of dominance remains unchanged [28]. An alternative is the penalty approach which can be used with all of the above schemes. Here, the overall constraint violation is calculated and summarised by a real value. This value is then added to the original fitness value (assuming that fitness is to be minimised), thereby, infeasible solutions are penalised.

Finally, another issue that is especially important in the presence of multiple objectives is maintaining diversity among the designs stored. If the goal is to identify a set of Pareto-optima, special techniques are necessary in order to prevent the search algorithm from converging to a single trade-off solution. Most modern multi-objective optimisers integrate some diversity preservation technique that estimates the density of solutions in the space defined by the objective functions. For instance, the density around a solution can be estimated by calculating the Euclidean distance to the next closest solution. This density information can then be incorporated into the fitness, e.g. by adding original fitness value and density estimate. Again, there is variety of different methods that cannot be discussed here in detail.

1.4.1.2 Selection for survival

When approximating the Pareto set, it is desirable not to lose promising designs due to random effects. Therefore, selection for survival is usually realised by a deterministic algorithm. Similar issues as with selection for variation come into play here; however, almost all search methods make sure that designs not dominated among those in the working memory are preferred over dominated ones with respect to environmental selection. If there are too many non-dominated solutions, then additional diversity information is used to further discriminate among these designs. Furthermore, as many randomised search algorithms only keep a single solution in the working memory, often a secondary memory, a so-called archive (see also Figure 1.2), is maintained that stores the current approximation of the Pareto set. For instance, PAES [32], a randomised local search method for multi-objective optimisation, checks for every generated design whether it should be added to the archive, i.e. whether it is dominated by any other archive member. If the design was inserted, dominated designs are removed. If the archive size is exceeded after insertion, a design with the highest density estimate is deleted.

A theoretical issue that has been investigated recently by different researchers [33,34] addresses the loss in quality per iteration. Optimally, the current set of designs represents the best Pareto set approximation among all solutions ever considered during the optimisation run – given the actual memory constraints. This goal is difficult to achieve in general, but Laumanns *et al.* [33] proposed an archiving method by which the loss can be bound and kept arbitrarily small by adjusting the memory usage accordingly.

1.4.1.3   Multi-objective optimisers

The above discussion could only touch the aspects involved in the design of the selection process for a multi-objective randomised search algorithm. In fact, a variety of methods exist [28], which over the time have become more complex [35]. In the evolutionary computation field, even a rapidly growing sub-discipline emerged focusing on the design of evolutionary algorithms for multiple criteria optimisation [36]. However, an application engineer who would like to carry out a design space exploration is not necessarily an expert in the optimisation field. He is rather interested in using state-of-the-art multi-objective optimisers. For this reason, the proposed design space exploration framework separates the general search strategy from the application-specific aspects such as variation. Thereby, it is possible to use precompiled search engines without any implementation effort.

1.4.2   Variation

In this subsection the application-specific part of the proposed design space exploration framework is described. In particular, the variation module encapsulates the representation of a design point and the variation operators, see also the overview in Figure 1.2. It is the purpose of this component in the design space exploration framework to generate suitable new design points from a given set of selected ones. Therefore, the variation is problem-specific to a large extent and provides a major opportunity in including domain knowledge.

1.4.2.1   Representation

A formal description of a design needs to be appropriately encoded in the optimisation algorithm. The main objectives for suitable design representations are as follows.

- The encoding should be designed in a way that enables an efficient generation of design points in an appropriate neighbourhood, see also the next subsection on variation operators.
- The representation should be able to encode all relevant design points of the design space. In particular, if the design space has been pruned using problem-dependent approaches, the chosen representation should reflect these constraints in a way that enables efficient variation for the determination of a neighbourhood.
- The design parameters should be independent of each other as much as possible in order to enable a suitable definition of variation operators.

A representation of a solution can, e.g., consist of real or integer values, or vectors thereof to encode clock speeds, memory size, cache size, etc. Bit vectors can be used to describe the allocation of different resources. Another class of representations could be the permutation of a vector with fix elements to represent, e.g., a certain task scheduling. Furthermore, variable length data structures such as trees or lists can be used for the representation of, e.g., graphs (see Reference 29 for an overview).

All parameters for the representation have to lie inside the problem specification that spans the design space of possible solutions. A solution parameter could therefore be, e.g., a real value in the range given in the specification, an integer value in a list of possible integers or a selected edge in a problem specification graph.

The cache example uses integer values to represent the number of cache lines in the solution cache, the block size and the associativity. The integer values are the actual cache parameters, such that these lie in the range specified in Table 1.1. The cache line replacement strategy is represented by a Boolean value.

1.4.2.2   Variation operators

The purpose of the variation operators is to determine new design points given a set of selected, previously evaluated, design points. There are several objectives for selecting appropriate variation operators.

- The variation operators operate on the design representation and generate a local neighbourhood of the selected design points. These new design points will be evaluated by the estimation, see Figure 1.1. Therefore, the construction of the variation operators is problem-dependent and a major possibility to include domain knowledge.
- The constructed neighbourhood should not contain infeasible design points, if possible.
- In the case of infeasible design points where non-functional properties are outside of given constraints, one may use a feedback loop shown in Figure 1.1 in order to correct.
- The variation operator may also involve problem-dependent local search (e.g. by optimising certain parameters or hidden optimisation criteria) in order to relieve the randomised search from optimisation tasks that can better be handled with domain knowledge.

In principle, different variation operators can be distinguished according to the number of solutions they operate on. Most randomised search algorithms generate a single new design point by applying a randomised operator to a known design point. For simulated annealing and randomised local search algorithms this operator is called the neighbourhood function, whereas for evolutionary algorithms this operator is denoted as the mutation operator. The term mutation will be used in the remainder of this section.

In the context of evolutionary algorithms there also exists a second type of variation, in addition to mutation. Since evolutionary algorithms maintain a population of solutions, it is possible to generate one or more new solutions based on two or more existing solutions. The existing designs selected for variation are often referred to as parents, whereas the newly generated designs are called children. The operator that generates ≥ 1 children based on ≥ 2 parents is denoted as recombination.

Mutation: The assumption behind mutation is that it is likely to find better solutions in the neighbourhood of good solutions. Therefore, mutation operators are usually designed in such a way that the probability of generating a specific solution decreases with increasing distance from the parent. There exist several approaches to implement mutation. It is, e.g., possible to always change exactly one parameter in the representation of a solution and keep all other parameters unchanged. A different mutation operator changes each of n parameters with probability $1/n$, which leads to the fact that one parameter is changed in expectation. This approach is also used in the cache example.

Changing a parameter means changing its value, i.e. flipping a bit in a binary representation, or choosing new parameter values according to some probability distribution for an integer- or real-valued representation. For representations based on permutations of vector elements the mutation operator changes the permutation by exchanging two elements. If the specification is based on lists of possible values, the mutation operator selects a new element according to some probability distribution.

In general, a mutation operator should on the one hand produce a new solution that is 'close' to the parent solution with a high probability, but on the other hand be able to produce any solution in the design space, although with very small probability. This is to prevent the algorithm from being stuck in a local optimum.

The cache example uses the following mutation operator. Each of the design parameters is mutated with probability 0.25 (as there are four different parameters). The change that is applied to each of the parameters is normally distributed, i.e., the value of a parameter is increased by a value that is normally distributed around 0 inside the ranges given in Table 1.1, e.g. the block size parameter change is normally distributed between -4 and $+4$. Note, that in the example changes of size 0 are also allowed, i.e. the parameter remains unchanged.

Recombination: Recombination takes two or more solutions as input and then generates new solutions that represent combinations of the parents. The idea behind recombination is to take advantage of the good properties of each of the parents to produce even better children. In analogy to the mutation operator, a good recombination vector should produce solutions that lie 'between' the parents either with respect to the parameter space or to the objective space.

For vectors in general, recombination of two parents can be accomplished by cutting both solutions at randomly chosen positions and rearranging the resulting pieces. For instance, one-point crossover creates a child by copying the first half from the first parent and the second half from the second parent. If the cut is made at every position, i.e. at each position randomly either the value from the first or the second parent is copied, the operator is called uniform recombination.

A further approach for real-valued parameters is to use the average of the two parents' parameter values, or some value between the parents' parameter values. A detailed overview of various recombination operators for different representation data structures can be found in Reference 29.

For the cache example uniform recombination was used, i.e. for each of the parameters, like cache block size, it was randomly decided from which parent solution the parameter for the first child solution should be used, while all unused parameters of the parent solutions are then used for the second child solution. See Figure 1.5 on the right-hand side for a graphical representation of uniform recombination.

Infeasible solutions: It can happen that after mutation or recombination a generated solution is not feasible, i.e. the solution represented by the parameters does not

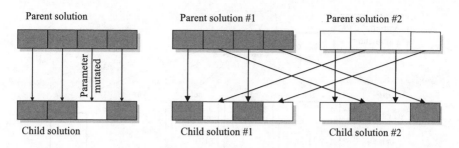

Figure 1.5   Variation operators used in the cache example: mutation (left), recombination (right)

describe a valid system. To solve this problem there are different possibilities. First, one could ensure that the variation operators do not create infeasible solutions by controlling the construction of new solutions, one can call this approach 'valid by construction'. Second, one could implement a repair method that turns constructed solutions that are infeasible into feasible ones by fixing the infeasible parameters. The third possibility is to introduce an additional constraint and to penalise infeasible designs in the way as described in Section 1.4.1. Finally, one can use the concept of penalty functions in order to guide the search away from areas with infeasible design points.

1.4.3   Implementation issues

In this section the protocol used in PISA is briefly introduced, see also Reference 25. It is the purpose of PISA to make state-of-the-art randomised search algorithms for multi-objective optimisation problems readily available. Therefore, for a new design space exploration task in embedded system design, one can concentrate on the problem-dependent aspects, where the domain knowledge comes in. The protocol has to be implemented by any design space exploration tool that would like to benefit from precompiled and ready-to-use search algorithms available at http://www.tik.ee.ethz.ch/pisa. Besides, the website also contains a set of application problems and benchmark applications for the development of new randomised search algorithms. The detailed protocol including file formats and data type definitions is given in Reference 25. In the protocol description, the application-specific part is called 'variator' and the search algorithm is denoted 'selector', according to Figure 1.6. The variator also contains the estimation of non-functional properties.

The details of the protocol have been designed with several objectives in mind.

- Small amounts of data that need to be communicated between the two different processes (selector and variator).
- The communicated data should be independent of the problem domain in order to enable a generic implementation of the selector process.
- Separation into problem-independent (selector) and problem-dependent (variator) processes.

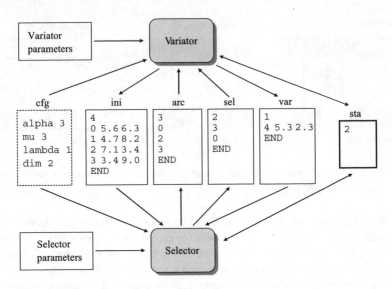

Figure 1.6   Communication between modules through text files as defined by the
PISA protocol. The files contain sample data

- The implementation of the protocol should be as much as possible independent of the programming languages, hardware platforms and operating systems. It should enable a reliable (delay-independent) execution of the design space exploration.

The protocol defines the sequence of actions performed by the selector and variator processes. The communication between the two processes is done by exchange of text files over a common file system. The handshake protocol is based on states and ensures that during the optimisation process only one module is active at any time. During the inactive period a process polls the state file for changes. Whenever a module reads a state that requires some action on its part, the operations are performed and the next state is set.

The core of the optimisation process consists of state 2 and state 3. In each iteration the selector chooses a set of parent individuals and passes them to the variator. The variator generates new child solutions on the basis of the parents, computes the objective function values of the new individuals and passes them back to the selector.

In addition to the core states two more states are necessary for normal operation. State 0 and state 1 trigger the initialisation of the variator and the selector, respectively. In state 0 the variator reads the necessary parameters. Then, the variator creates an initial population, determines the objective values of the individuals and passes the initial population to the selector. In state 1, the selector also reads the required parameters, then selects a sample of parent individuals and passes them to the variator.

The four states 0–3 provide the basic functionality of the PISA protocol. To add some flexibility the PISA protocol defines a few more states which are mainly used to terminate or reset both the variator process and the selector process. Table 1.2 gives

Table 1.2   States for the PISA protocol

State	Action	Next state
State 0	Variator reads parameters and creates initial solutions	State 1
State 1	Selector reads parameters and selects parent solutions	State 2
State 2	**Variator generates and evaluates new solutions**	**State 3**
State 3	**Selector selects solutions for variation**	**State 2**
State 4	Variator terminates	State 5
State 6	Selector terminates	State 7
State 8	Variator resets. (Getting ready to start in state 0)	State 9
State 10	Selector resets. (Getting ready to start in state 0)	State 11

The main states of the protocol are printed in bold face.

an overview of all defined states. The additional states 4–11 are not mandatory for a basic implementation of the protocol.

The data transfer between the two modules introduces some overhead compared to a traditional monolithic implementation. Thus, the amount of data exchange for each individual should be minimised. Since all representation-specific operators are located in the variator, the selector does not have to know the representation of the individuals. Therefore, it is sufficient to convey only the following data to the selector for each individual: an identifier and its objective vector. In return, the selector only needs to communicate the identifiers of the parent individuals to the variator. The proposed scheme allows to restrict the amount of data exchange between the two modules to a minimum.

For PISA-compliant search algorithms to work correctly, a designer has to ensure, that all objectives are to be 'minimised'. In addition the variator and selector have to agree on a few common parameters: (1) the population size α, (2) the number of parent solutions μ, (3) the number of child solutions λ and (4) the number of objectives dim. These parameters are specified in the parameter file with suffix cfg, an example file is shown in Figure 1.6.

The selector and the variator are normally implemented as two separate processes. These two processes can be located on different machines with possibly different operating systems. This complicates the implementation of a synchronisation method. Most common methods for interprocess communication are therefore not applicable.

In PISA, the synchronisation problem is solved using a common state variable which both modules can read and write. The two processes regularly read this state variable and perform the corresponding actions. If no action is required in a certain state, the respective process sleeps for a specified amount of time and then rereads the state variable. The state variable is an integer number stored to a text file with suffix sta. The protocol uses text files instead of, e.g. sockets, because file access is completely portable between different platforms and familiar to all programmers.

All other data transfers between the two processes besides the state are also performed using text files. The initial population is written by the variator to the file with suffix ini, the population is written by the selector to a file with suffix arc. In a text file with suffix sel the selector stores the parent solutions that are selected for variation. The newly generated solutions are passed from the variator to the selector through a file with suffix var. All text files for data transfer have to begin with the number of elements that follow and to end with the keyword END.

Once the receiving process has completely read a text file, it has to overwrite the file with 0, to indicate that it successfully read the data.

1.4.4   Application of PISA for design space exploration

For the cache example presented in Section 1.3, the variator part was written in Java. The mutation and recombination operator were implemented as described in Section 1.4.2, and the combination with a selector is PISA-compliant as described in Section 1.4.3. The selector was downloaded from the PISA website. The design space exploration for L1 data caches was performed using strength Pareto evolutionary algorithm (SPEA2), an evolutionary multi-objective optimiser described in Reference 35.

The solutions selected by SPEA2 for variation were pairwise recombined with probability 0.8 and the resulting solutions were then mutated with probability 0.8. Afterwards the generated solutions were added to the population and passed to the search algorithm for selection.

The design space with all solutions is shown in Figure 1.7. These design points have been generated using exhaustive search in order to compare the heuristic search with the Pareto front of optimal solutions. The front of non-dominated solutions found

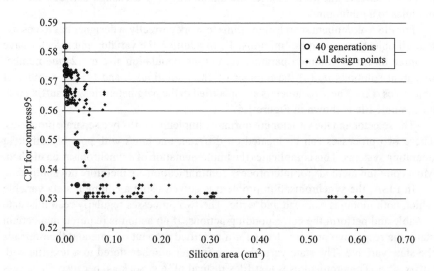

Figure 1.7   All 540 possible design points determined using exhaustive search and the design points found by the multi-objective search algorithm SPEA2 after 40 generations

Table 1.3 Details of ten non-dominated solutions for the simple example of a cache exploration found after a typical design space exploration run

No.	CPI	Area	Design parameters			
1	0.5774	0.001311	LRU	2^7 cache lines	Block size 8	d. m.
2	0.5743	0.001362	LRU	2^7 cache lines	Block size 8	2 sets
3	0.5622	0.022509	FIFO	2^8 cache lines	Block size 64	8 sets
4	0.5725	0.002344	LRU	2^7 cache lines	Block size 16	2 sets
5	0.5488	0.024018	LRU	2^{10} cache lines	Block size 32	8 sets
6	0.5319	0.027122	LRU	2^{10} cache lines	Block size 32	16 sets
7	0.5666	0.002898	LRU	2^6 cache lines	Block size 32	2 sets
8	0.5653	0.003629	FIFO	2^6 cache lines	Block size 64	d. m.
9	0.5307	0.044902	FIFO	2^{10} cache lines	Block size 64	8 sets
10	0.5626	0.004907	LRU	2^6 cache lines	Block size 64	2 sets

These solutions are marked with circles in Figure 1.7.

for the cache example with SPEA2 after a typical optimisation run with 40 generations for a population size of six solutions is marked with circles. The details of the solutions in the population after 40 generations are represented in Table 1.3.

Although the cache design space exploration problem is simple in nature, one can make some observations which also hold for more involved exploration problems. The two objectives, namely the minimisation of the silicon area and the minimisation of the CPI, are conflicting, resulting in an area vs. performance trade-off. This results in the fact that there is not a single optimal solution, but a front of Pareto-optimal solutions. All points on this front represent different promising designs, leaving the final choice for the design of the cache up to the designer's preference. Further, one can observe in Figure 1.7 that the evolutionary algorithm found solutions close to the Pareto-optimal front.

The reduction of the problem to a single objective optimisation problem, e.g. using a weighted-sum approach, is difficult already for this simple example, because it represents a true multi-objective problem. It is not at all clear how to relate area to performance, which would be needed for the weighted-sum approach.

As a more involved example, the design space exploration of complex stream processor architectures on the system level has been performed using the PISA framework. To this end, a variator process 'EXPO' has been implemented which is available on the website of PISA also (Figure 1.8). The representation of design points, the variation operators, the local search method to reduce the design space and the way in which infeasible design points are avoided or repaired are all specific to the application domain of stream processors. These methods are based on models for stream processing tasks, a specification of the workload generated by traffic flows and a description of the feasible space of architectures involving computation and communication resources.

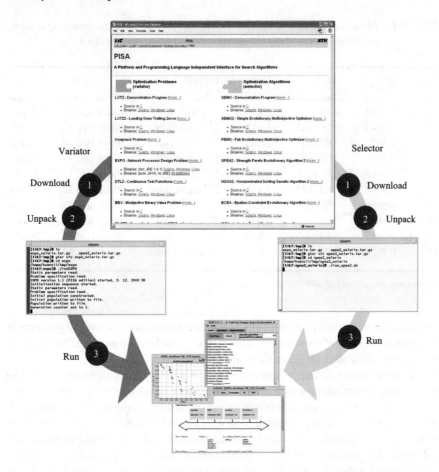

Figure 1.8   On the PISA website, many different ready-to-use evolutionary search algorithms can be downloaded. Additionally, a design space exploration tool is offered for download at the website. The only steps needed for a first design space exploration using the PISA framework are shown: (1) just download a variator, e.g. the exploration tool EXPO and one of the search algorithms on the right-hand side of the website, (2) unpack the tools and then (3) run them

the evaluation of a single design point, the tool makes use of a new method to estimate end-to-end packet delays and queueing memory, taking task scheduling policies and bus arbitration schemes into account. The method is analytical and is based on a high level of abstraction, where the goal is to quickly identify interesting architectures, which may then be subjected to a more detailed evaluation, e.g. using simulation. The approach used in EXPO and results are described in much more detail in Reference 22.

For the simple cache example the design space could have been explored using exhaustive search instead of employing evolutionary algorithms, which actually was done to determine all solutions shown in Figure 1.7. For larger design spaces, as the one explored with EXPO, exhaustive search is prohibitive, and only randomised search algorithms can be used. It has been shown in many studies (e.g. in Reference 28) that evolutionary algorithms perform better on multi-objective optimisation problems than do other simpler randomised search algorithms.

PISA enables the use of different evolutionary algorithms without having to change the implementation of the exploration tools. A recent study [37] has shown that for EXPO the quality of the approximation of the Pareto-optimal front may differ between different evolutionary algorithms. With a modular framework based on the PISA protocol it is possible to test the design space exploration performance of different randomised search algorithms to find the search algorithm most suitable to the exploration problem.

1.5 Conclusions

This chapter introduced a framework for design space exploration of embedded systems. It is characterised by (1) multiple optimisation criteria, (2) randomised search algorithms and (3) a software interface that clearly separates problem-dependent and problem-independent parts of an implementation. In particular, the interface PISA formally characterises this separation. It is implemented in a way that is independent of programming language used and the underlying operating system. As a result, it is easily possible to extend any existing method to estimate non-functional properties with an effective multi-objective search.

It should be pointed out, that effective automatic or semi-automatic (interactive) exploration needs deep knowledge about the specific optimisation target, i.e. the level of abstraction, the optimisation goals, efficient and accurate estimation methods. Nevertheless, the PISA framework separates the problem-dependent variation and estimation from the generic search and selection. Therefore, the user is relieved from dealing with the complex and critical selection mechanisms in multi-objective optimisation. On the other hand, his specific domain knowledge will be important when designing the variation operators that determine a promising local neighbourhood of a given search point.

Finally, it is common knowledge that the class of randomised search algorithms described in the chapter does not guarantee to find the optimal solutions. In addition, if there is domain knowledge available that allows problem-specific exploration methods to be applied, then there is little reason to use a generic approach. But usually, those analytic methods do not exist for complex optimisation scenarios as found in embedded system design.

Acknowledgement

The work described in this chapter was supported by the Swiss Government within the KTI program. The project was part of the MEDEA+ project SpeAC.

References

1 BRUNI, D., BOGLIOLO, A., and BENINI, L.: 'Statistical design space exploration for application-specific unit synthesis'. Proceedings of the 38th *Design
   automation conference*, (ACM Press, New York, 2001) pp. 641–646
2 CHANTRAPORNCHAI, C., SHA, E.-M., and HU, X.S.: 'Efficient acceptable design exploration based on module utility selection', *IEEE Transactions
   on Computer Aided Design of Integrated Circuits and Systems*, 2000, **19**(1),
   pp. 19–29
3 GHOSH, A., and GIVARGIS, T.: 'Analytical design space exploration of caches
   for embedded systems', in NORBERT, W., and DIEDERIK, V. (Eds): *Design,
   automation and test in Europe* conference and exhibition (DATE 03), Munich,
   Germany, (IEEE Press, Los Alamitos, CA, 2003) pp. 650–655
4 SHIUE, W.-T., and CHAKRABARTI, C.: 'Memory exploration for low power,
   embedded systems'. Proceedings of the 36th ACM/IEEE *Design automation* conference, New Orleans, Louisiana, (ACM Press, New York, 1999)
   pp. 140–145
5 SZYMANEK, R., CATTHOOR, F., and KUCHCINSKI, K.: 'Time-
   energy design space exploration for multi-layer memory architectures', in
   GEORGES, G., and JOAN, F. (Eds): Proceedings of the seventh ACM/IEEE
   Design, automation and test in Europe conference, Paris, France (ACM Press,
   New York, 2004) p. 10318
6 HEKSTRA, G., HEI, G.L., BINGLEY, P., and SIJSTERMANS, F.: 'Tri-
   Media CPU64 design space exploration', in KUEHLMANN, A., and
   CHASE, C., (Eds): 1999 IEEE International Conference on Computer
   Design, Austin, TX, USA (IEEE Press, Los Alamitos, CA, 1999)
   pp. 593–598
7 PALERMO, G., SILVANO, C., and ZACCARIA, V.: 'A flexible framework
   for fast multi-objective design space exploration of embedded systems', in
   JORGE, J.C., and ENRICO, M. (Eds): PATMOS 2003 – international workshop on power and timing modeling, vol. 2799 of lecture notes in computer
   science, Torino, Italy (Springer, Berlin, Germany, 2003) pp. 249–258
8 RAJAGOPAL, S., CAVALLARO, J., and RIXNER, S.: 'Design space exploration for real-time embedded stream processors', *IEEE Micro*, 2004, **24**(4),
   pp. 54–66
9 ZITZLER, E., TEICH, J., and BHATTACHARYYA, S.: 'Evolutionary algorithms for the synthesis of embedded software', *IEEE Transactions on VLSI
   Systems*, 2000, **8**(4), pp. 452–456
10 AGOSTA, G., PALERMO, G., and SILVANO, C.: 'Multi-objective co-
   exploration of source code transformations and design space architectures for
   low-power embedded systems', in HISHAM, H., ANDREA, O., ROGER, L.W.,
   and LORIE, M.L. (Eds): Proceedings of the 2004 ACM symposium on *Applied
   computing*, Nicosia, Cyprus (ACM Press, New York, 2004) pp. 891–896
11 BAMBHA, N.K., BHATTACHARYYA, S., TEICH, J., and ZITZLER, E.:
   'Hybrid search strategies for dynamic voltage scaling in embedded

multiprocessors'. Proceedings of the ninth international symposium on *Hardware/software codesign*, Copenhagen, Denmark (ACM Press, New York, 2001) pp. 243–248

12 LAHIRI, K., RAGHUNATHAN, A., and DEY, S.: 'Design space exploration for optimizing on-chip communication architectures', *IEEE Transactions on Computer-Aided Design of Integrated Circuits and Systems*, 2004, **23**(6), pp. 952–961

13 EISENRING, M., THIELE, L., and ZITZLER, E.: 'Handling conflicting criteria in embedded system design', *IEEE Design and Test of Computers*, 2000, **17**, pp. 51–59

14 ANLIKER, U., BEUTEL, J., and DYER, M. *et al.*: 'A systematic approach to the design of distributed wearable systems', *IEEE Transactions on Computers*, 2004, **53**(8), pp. 1017–1033

15 ABRAHAM, S., RAU, B.R., and SCHREIBER, R.: 'Fast design space exploration through validity and quality filtering of subsystem designs'. Technical Report HPL-2000-98, HP Labs Technical Reports, 1998

16 GAJSKI, D.D., VAHID, F., NARAYAN, S., and GONG, J.: 'System-level exploration with SpecSyn'. Proceedings of the 35th *Design automation* conference, San Francisco, California (ACM Press, New York, 1998) pp. 812–817

17 GIVARGIS, T., VAHID, F., and HENKEL, J.: 'System-level exploration for Pareto-optimal configurations in parameterized system-on-a-chip', *IEEE Transactions on Very Large Scale Integrated Systems*, 2002, **10**(4), pp. 416–422

18 BLICKLE, T., TEICH, J., and THIELE, L.: 'System-level synthesis using evolutionary algorithms', *Journal on Design Automation for Embedded Systems*, 1998, **3**, pp. 23–58

19 THIELE, L., CHAKRABORTY, S., GRIES, M., and KÜNZLI, S.: 'Design space exploration of network processor architectures', in CROWLEY, P., FRANKLIN, M.A., HADIMIOGLU, H., and ONUFRYK, P.Z. (Eds): 'Network processor design: issues and practices, vol. 1' (Morgan Kaufmann Publishers, San Francisco, CA, 2003), ch. 4, pp. 55–90

20 ZHUGE, Q., SHAO, Z., XIAO, B., and SHA, E.H.-M.: 'Design space minimization with timing and code size optimization for embedded DSP', in GUPTA, R., NAKAMURA, Y., ORAILOGLU, A., and CHOU, P.H. (Eds): Proceedings of the 1st IEEE/ACM/IFIP international conference on *Hardware/software codesign and system synthesis*, Newport Beach, CA, (ACM Press, New York, 2003) pp. 144–149

21 CZYZAK, P., and JASZKIEWICZ, A.: 'Pareto-simulated annealing – a metaheuristic technique for multi-objective combinatorial optimization', *Journal of Multi-Criteria Decision Analysis*, 1998, **7**, pp. 34–47

22 THIELE, L., CHAKRABORTY, S., GRIES, M., and KÜNZLI, S.: 'A framework for evaluating design tradeoffs in packet processing architectures'. Proceedings of the 39th *Design automation* conference (DAC), New Orleans, LA (ACM Press, New York, 2002) pp. 880–885

23   PALESI, M., and GIVARGIS, T.: 'Multi-objective design space explo-
      ration using genetic algorithms', in HENKEL, J., XIAOBO SHARON HU,
      GUPTA, R., and SRI PARAMESWARAN (Eds): Proceedings of the tenth inter-
      national symposium on *Hardware/software codesign*, Estes Park, Colorado,
      (ACM Press, New York, 2002) pp. 67–72

24   NEAM, S., SZTIPANOVITS, J., and KARSAI, G.: 'Design-space con-
      struction and exploration in platform-based design'. Technical Report ISIS-
      02-301, Vanderbilt University, Institute for Software Integrated Systems,
      2002

25   BLEULER, S., LAUMANNS, M., THIELE, L., and ZITZLER, E.:
      'PISA – a platform and programming language independent interface for
      search algorithms', in FONSECA, C.M., FLEMING, P.J., ZITZLER, E.,
      DEB, K., and THIELE, L. (Eds): *Evolutionary multi-criterion optimization*
      (EMO 2003), Lecture notes in computer science (Springer, Berlin, 2003)
      pp. 494–508

26   BURGER, D., and AUSTIN, T.M.: 'The simplescalar tool set, version 2.0',
      SIGARCH Computer Architecture News, 1997, **25** (3), pp. 13–25

27   SHIVAKUMAR, P., and JOUPPI, N.P.: 'Cacti 3.0: an integrated cache tim-
      ing, power and area model'. Technical Report WRL Research Report 2001/2,
      Compaq, Western Research Laboratory, August 2001

28   DEB, K.: 'Multi-objective optimization using evolutionary algorithms' (Wiley,
      Chichester, UK, 2001)

29   BÄCK, T., FOGEL, D.B., and MICHALEWICZ, Z. (Eds): 'Handbook of Evo-
      lutionary Computation' (IOP Publishing and Oxford University Press, Bristol,
      UK, 1997)

30   MIETTINEN, K.: 'Nonlinear Multiobjective Optimization' (Kluwer, Boston,
      1999)

31   SCHAFFER, J.D.: 'Multiple objective optimization with vector evalu-
      ated genetic algorithms', in GREFENSTETTE, J.J. (Ed.): Proceedings of
      an international conference on *Genetic algorithms and their applications*
      (Lawrence Erlbaum Associates, Mahwah, NJ, 1985) pp. 93–100

32   KNOWLES, J.D., and CORNE, D.W.: 'Approximating the non-dominated
      front using the Pareto Archived Evolution Strategy', *Evolutionary Computation*,
      2000, **8**(2), pp. 149–172

33   LAUMANNS, M., THIELE, L., DEB, K., and ZITZLER, E.: 'Combin-
      ing convergence and diversity in evolutionary multiobjective optimization',
      Evolutionary Computation, 2002, **10**(3), pp. 263–282

34   KNOWLES, J.D.: 'Local-search and hybrid evolutionary algorithms for pareto
      optimization'. PhD thesis, University of Reading, 2002

35   ZITZLER, E., LAUMANNS, M., and THIELE, L.: 'SPEA2: improving the
      strength pareto evolutionary algorithm for multiobjective optimization', in
      GIANNAKOGLOU, K. *et al.* (Eds): *Evolutionary methods for design, opti-
      misation and control with application to industrial problems* (EUROGEN
      2001) (International Center for Numerical Methods in Engineering (CIMNE),
      Barcelona, Spain, 2002) pp. 95–100

36 ZITZLER, E., DEB, E., THIELE, L., COELLO, C.A.C., and CORNE, D. (Eds): *Evolutionary multi-criterion optimization* (EMO 2001), Lecture notes in computer science, vol. 1993 (Springer, Berlin, Germany, 2001)

37 ZITZLER, E., and KÜNZLI, S.: 'Indicator-based selection in multiobjective search'. Proceedings of the eighth international conference on *Parallel problem solving from nature* (PPSN VIII), vol. 3242 of Lecture notes in computer science, Birmingham, UK (Springer, Berlin, Germany, 2004)

Chapter 2

System-level performance analysis – the SymTA/S approach

Rafik Henia, Arne Hamann, Marek Jersak, Razvan Racu,
Kai Richter and Rolf Ernst

2.1 Introduction

With increasing embedded system complexity, there is a trend towards heterogeneous, distributed architectures. Multiprocessor system-on-chip designs (MpSoCs) use complex on-chip networks to integrate multiple programmable processor cores, specialised memories and other intellectual property (IP) components on a single chip. MpSoCs have become the architecture of choice in industries such as network processing, consumer electronics and automotive systems. Their heterogeneity inevitably increases with IP integration and component specialisation, which designers use to optimise performance at low power consumption and competitive cost. Tomorrow's MpSoCs will be even more complex, and using IP library elements in a 'cut-and-paste' design style is the only way to reach the necessary design productivity.

Systems integration is becoming the major challenge in MpSoC design. Embedded software is increasingly important to reach the required productivity and flexibility. The complex hardware and software component interactions pose a serious threat to all kinds of performance pitfalls, including transient overloads, memory overflow, data loss and missed deadlines. The International Technology Roadmap for Semiconductors, 2003 Edition (http://public.itrs.net/Files/2003ITRS/Design2003.pdf) names system-level performance verification as one of the top three codesign issues.

Simulation is state-of-the-art in MpSoC performance verification. Tools from many suppliers support cycle-accurate cosimulation of a complete hardware and software system. The cosimulation times are extensive, but developers can use the same simulation environment, simulation patterns and benchmarks in both function and

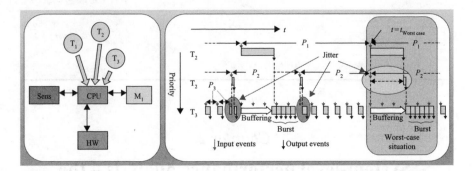

Figure 2.1   CPU subsystem

performance verification. Simulation-based performance verification, however, has conceptual disadvantages that become disabling as complexity increases.

MpSoC hardware and software component integration involves resource sharing that is based on operating systems and network protocols. Resource sharing results in a confusing variety of performance runtime dependencies. For example, Figure 2.1 shows a central processing unit (CPU) subsystem executing three processes. Although the operating system activates T_1, T_2 and T_3 strictly periodically (with periods P_1 P_2 and P_3, respectively), the resulting execution sequence is complex and leads to output bursts.

As Figure 2.1 shows, T_1 can delay several executions of T_3. After T_1 completes, T_3 – with its input buffers filled – temporarily runs in burst mode with the execution frequency limited only by the available processor performance. This leads to transient T_3 output burst, which is modulated by T_1's execution.

Figure 2.1 does not even include data-dependent process execution times, which are typical for software systems, and operating system overhead is neglected. Both effects further complicate the problem. Yet finding simulation patterns – or use cases – that lead to worst-case situations as highlighted in Figure 2.1 is already challenging.

Network arbitration introduces additional performance dependencies. Figure 2.2 shows an example. The arrows indicate performance dependencies between the CPU and digital signal processor (DSP) subsystems that the system function does not reflect. These dependencies can turn component or subsystem best-case performance into system worst-case performance – a so-called scheduling anomaly. Recall the T_3 bursts from Figure 2.1 and consider that T_3's execution time can vary from one execution to the next. There are two critical execution scenarios, called corner cases: The minimum execution time for T_3 corresponds to the maximum transient bus load, slowing down other components' communication, and vice versa.

The transient runtime effects shown in Figures 2.1 and 2.2 lead to complex system-level corner cases. The designer must provide a simulation pattern that reaches each corner case during simulation. Essentially, if all corner cases satisfy the given performance constraints, then the system is guaranteed to satisfy its constraints under all possible operation conditions. However, such corner cases are

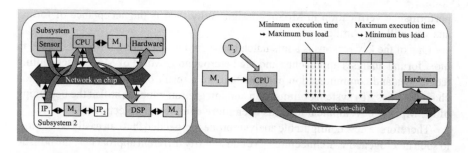

Figure 2.2   Scheduling anomaly

extremely difficult to find and debug, and it is even more difficult to find simulation patterns to cover them all. Reusing function verification patterns is not sufficient because they do not cover the complex non-functional performance dependencies that resource sharing introduces. Reusing component and subsystem verification patterns is not sufficient because they do not consider the complex component and subsystem interactions.

The system integrator might be able to develop additional simulation patterns, but only for simple systems in which the component behaviour is well understood. Manual corner-case identification and pattern selection is not practical for complex MpSoCs with layered software architectures, dynamic bus protocols and operating systems. In short, simulation-based approaches to MpSoC performance verification are about to run out of steam, and should essentially be enhanced by formal techniques that systematically reveal and cover corner cases.

Real-time systems research has addressed scheduling analysis for processors and buses for decades, and many popular scheduling analysis techniques are available. Examples include rate-monotonic scheduling and earliest deadline first [1], using both static and dynamic priorities; and time-slicing mechanisms like time division multiple access (TDMA) or round robin [2]. Some extensions have already found their way into commercial analysis tools, which are being established, e.g. in the automotive industry to analyse individual units that control the engine or parts of the electronic stability program.

The techniques rely on a simple yet powerful abstraction of task activation and communication. Instead of considering each event individually, as simulation does, formal scheduling analysis abstracts from individual events to event streams. The analysis requires only a few simple characteristics of event streams, such as an event period or a maximum jitter. From these parameters, the analysis systematically derives worst-case scheduling scenarios, and timing equations safely bound the worst-case process or communication response times.

It might be surprising that – up to now – only very few of these approaches have found their way into the SoC (system-on-chip) design community by means of tools. Regardless of the known limitations of simulation such as incomplete corner-case coverage and pattern generation, timed simulation is still the preferred means of performance verification in MpSoC design. Why then is the acceptance of formal analysis still very limited?

One of the key reasons is a mismatch between the scheduling models assumed in most formal analysis approaches and the heterogenous world of MpSoC scheduling techniques and communication patterns that are a result of (a) different application characteristics; (b) system optimisation and integration which is still at the beginning of the MpSoC development towards even more complex architectures.

Therefore, a new configurable analysis process is needed that can easily be adapted to such heterogeneous architectures. We can identify different approaches: the holistic approach that searches for techniques spanning several scheduling domains; and hierarchical approaches that integrate local analysis with a global flow-based analysis, either using new models or based on existing models and analysis techniques.

In the following section, the existing analysis approaches from the literature on real-time analysis are reviewed and key requirements for their application to MpSoC design are identified. In Section 2.3, the fundamentals and basic models of the SymTA/S technology are introduced. Section 2.4 surveys a large number of extensions that enable the analysis of complex applications. Section 2.5 shows how the overall analysis accuracy can be deliberately increased when designers specify few additional correlation information. Automatic optimisations using evolutionary algorithms is explained in Section 2.6, while Section 2.7 introduces the idea of sensitivity analysis. An experiment is carried out in Section 2.8 and conclusions are drawn in Section 2.9

2.2   Formal techniques in system performance analysis

Formal approaches to heterogeneous systems are rare. The 'holistic' approach [3,4] systematically extends the classical scheduling theory to distributed systems. However, because of the very large number of dependencies, the complexity of the equations underlying the analysis grows with system size and heterogeneity. In practice, the holistic approach is limited to those system configurations which simplify the equations, such as deterministic TDMA networks. However, there is, up to now, no general procedure to set-up and solve the holistic equations for arbitrary systems. This could explain why such holistic approaches are largely ignored by the SoC community even though there are many proposals for multiprocessor analysis in real-time computing.

Gresser [5] and Thiele [6] established a different view on scheduling analysis. The individual components or subsystems are seen as entities which interact, or communicate, via event streams. Mathematically speaking, the stream representations are used to capture the dependencies between the equations (or equations sets) that describe the individual component's timing. The difference to the holistic approach (that also captures the timing using system-level equations) is that the compositional models are well structured with respect to the architecture. This is considered a key benefit, since the structuring significantly helps designers to understand the complex dependencies in the system, and it enables a surprisingly simple solution. In the 'compositional' approach, an output event stream of one component turns into an input event stream of a connected component. Schedulability analysis, then, can be seen as a flow-analysis problem for event streams that, in principle, can be solved iteratively using event stream propagation.

Both approaches use a highly generalised event stream representation to tame the complexity of the event streams. Gresser uses a superpositional 'event vector system', which is then propagated using complex event dependency matrices. Thiele *et al.* use a more intuitive model. They use 'numerical' upper and lower bound event 'arrival curves' for event streams, and similar 'service curves' for execution modelling.

This generality, however, has its price. Since they introduced new stream models, both Thiele and Gresser had to develop new scheduling analysis algorithms for the local components that utilise these models; the host of existing work in real-time systems can not be reused. Furthermore, the new models are far less intuitive than the ones known from the classical real-time systems' research, e.g. the model of rate-monotonic scheduling with its periodic tasks and worst-case execution times. A system-level analysis should be simple and comprehensible, otherwise its acceptance is extremely doubtful.

The compositional idea is a good starting point for the following considerations. It uses some event stream representation to allow component-wise local analysis. The local analysis results are, then, propagated through the system to reach a global analysis result. We do not necessarily need to develop new local analysis techniques if we can benefit from the host of work in real-time scheduling analysis. For example, in Figure 2.1, even if input and output streams seem to have totally different characteristics, the number of T_3's output events can be easily bounded over a longer time interval. The bursts only occur temporarily, representing a transient overload within a generally periodic event stream. In other words, some key characteristics of the original periodic stream remain even in the presence of heavy distortion.

A key novelty of the SymTA/S approach is that it uses intuitive 'standard event models' (Section 2.3.2) from real-time systems' research rather than introducing new, complex stream representations. Periodic events or event streams with jitter and bursts [7] are examples of standard models that can be found in literature. The SymTA/S technology allows us to extract this information from a given schedule and automatically interface or adapt the event stream to the specific needs within these standard models, so that designers can safely apply existing subsystem techniques of choice without compromising global analysis.

2.3 The SymTA/S approach

SymTA/S [8] is a formal system-level performance and timing analysis tool for heterogeneous SoCs and distributed systems. The application model of SymTA/S is described in Section 2.3.1. The core of SymTA/S is a technique to couple local scheduling analysis algorithms using event streams [9,10]. Event streams describe the possible input/output (I/O) timing of tasks. Input and output event streams are described by standard event models which are introduced in detail in Section 2.3.2. The analysis composition using event streams is described in Section 2.3.3. A second key property of the SymTA/S compositional approach is the ability to adapt the possible timing of events in an event stream. The event stream adaptation concept is described in Section 2.3.4.

2.3.1 SymTA/S application model

A task is activated due to an activating event. Activating events can be generated in a multitude of ways, including expiration of a timer, external or internal interrupt and task chaining. Each task is assumed to have one input first in first out (FIFO). A task reads its activating data from its input FIFO and writes data into the input FIFO of a dependent task. A task may read its input data at any time during one execution. The data is therefore assumed to be available at the input during the whole execution of the task. SymTA/S also assumes that input data is removed from the input FIFO at the end of one execution.

A task needs to be mapped on to a 'computation' or 'communication' resource to execute. When multiple tasks share the same resource, then two or more tasks may request the resource at the same time. In order to arbitrate request conflicts, a resource is associated with a 'scheduler' which selects a task to which it grants the resource out of the set of active tasks according to some scheduling policy. Other active tasks have to wait. 'Scheduling analysis' calculates worst-case (sometimes also best-case) task response times, i.e. the time between task activation and task completion, for all tasks sharing a resource under the control of a scheduler. Scheduling analysis guarantees that all observable response times will fall into the calculated (best-case, worst-case) interval. Scheduling analysis is therefore conservative. A task is assumed to write its output data at the end of one execution. This assumption is standard in scheduling analysis.

Figure 2.3 shows an example of a system modelled with SymTA/S. The system consists of two resources each with two tasks mapped on it. R_1 and R_2 are both assumed to be priority scheduled. *Src*1 and *Src*2 are the sources of the external

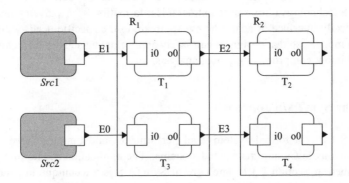

Figure 2.3   System modelled with SymTA/S

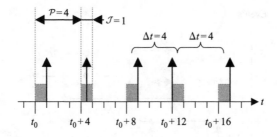

Figure 2.4   Example of an event stream that satisfies the event model ($\mathcal{P} = 4$, $\mathcal{J} = 1$)

activating events at the system inputs. The possible timing of activating events is captured by so-called 'event models', which are introduced in Section 2.3.2.

2.3.2   SymTA/S standard event models

Event models can be described by sets of parameters. For example, a 'periodic with jitter'event model has two parameters ($\mathcal{P}$, $\mathcal{J}$) and states that each event generally occurs periodically with period $\mathcal{P}$, but that it can jitter around its exact position within a jitter interval $\mathcal{J}$. Consider an example where ($\mathcal{P}$, $\mathcal{J}$) = (4, 1). This event model is visualised in Figure 2.4. Each gray box indicates a jitter interval of length $\mathcal{J} = 1$. The jitter intervals repeat with the event model period $\mathcal{P} = 4$. The figure additionally shows a sequence of events which satisfies the event model, since exactly one event falls within each jitter interval box, and no events occur outside the boxes.

An event model can also be expressed using two 'event functions' $\eta^{u}(\Delta t)$ and $\eta^{l}(\Delta t)$.

Definition 2.1 (Upper event function)   *The upper 'event function' $\eta^{u}(\Delta t)$ specifies the maximum number of events that can occur during any time interval of length Δt.*

Definition 2.2 (Lower event function)   *The lower 'event function' $\eta^{l}(\Delta t)$ specifies the minimum number of events that have to occur during any time interval of length Δt.*

Event functions are piecewise constant step functions with unit-height steps, each step corresponding to the occurrence of one event. Figure 2.5 shows the event functions for the event model ($\mathcal{P} = 4$, $\mathcal{J} = 1$). Note that at the points where the functions step, the smaller value is valid for the upper event function, while the larger value is valid for the lower event function (indicated by dark dots). For any time interval of length Δt, the actual number of events is bound by the upper and lower event functions. Event functions resemble arrival curves [11] which have been successfully used by Thiele *et al.* for compositional performance analysis of network processors [12]. In the following, the dependency of η^{u} and η^{l} on Δt is omitted for brevity.

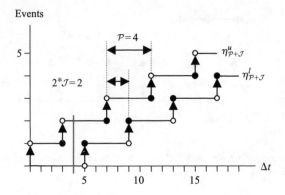

Figure 2.5   Upper and lower event functions for the event model ($\mathcal{P} = 4$, $\mathcal{J} = 1$)

A 'periodic with jitter'event model is described by the following event functions $\eta^u_{\mathcal{P}+\mathcal{J}}$ and $\eta^l_{\mathcal{P}+\mathcal{J}}$ [13].

$$\eta^u_{\mathcal{P}+\mathcal{J}} = \left\lceil \frac{\Delta t + \mathcal{J}}{\mathcal{P}} \right\rceil \tag{2.1}$$

$$\eta^l_{\mathcal{P}+\mathcal{J}} = \max\left(0, \left\lfloor \frac{\Delta t - \mathcal{J}}{\mathcal{P}} \right\rfloor \right) \tag{2.2}$$

To get a better feeling for event functions, imagine a sliding window of length Δt that is moved over the (infinite) length of an event stream. Consider $\Delta t = 4$ (grey vertical bar in Figure 2.5). The upper event function indicates that at most two events can be observed during any time interval of length $\Delta t = 4$. This corresponds, e.g. to a window position between $t_0 + 8.5$ and $t_0 + 12.5$ in Figure 2.4. The lower event function indicates that no events have to be observed during $\Delta t = 4$. This corresponds, e.g. to a window position between $t_0 + 12.5$ and $t_0 + 16.5$ in Figure 2.4.

In addition, 'distance' functions $\delta^{\min}(N \geq 2)$ and $\delta^{\max}(N \geq 2)$, are defined to return, respectively, the minimum and maximum distance between $N \geq 2$ consecutive events in an event stream.

Definition 2.3 (Minimum distance function)   *The minimum distance function* $\delta^{\min}(N \geq 2)$ *specifies the minimum distance between $N \geq 2$ consecutive events in an event stream.*

Definition 2.4 (Maximum distance function)   *The maximum distance function* $\delta^{\max}(N \geq 2)$ *specifies the maximum distance between $N \geq 2$ consecutive events in an event stream.*

For 'periodic with jitter' event models the following distance functions are obtained

$$\delta^{\min}(N \geq 2) = \max\{0, (N-1) \times \mathcal{P} - \mathcal{J}\} \tag{2.3}$$

$$\delta^{\max}(N \geq 2) = (N-1) \times \mathcal{P} + \mathcal{J} \tag{2.4}$$

For example, the minimum distance between two events in a 'periodic with jitter' event model with ($\mathcal{P} = 4$, $\mathcal{J} = 1$) is three time units, and the maximum distance between two events is five time units.

Periodic with jitter event models are well suited to describe generally periodic event streams, which often occur in control, communication and multimedia systems [14]. If the jitter is zero, then the event model is strictly periodic. If the jitter is larger than the period, then two or more events can occur at the same time, leading to bursts. To describe a 'bursty' event model, the 'periodic with jitter' event model can be extended with a $d_{\min}$ parameter that captures the minimum distance between events within a burst. A more detailed discussion can be found in Reference 13.

Additionally, 'sporadic' events are also common [14]. Sporadic event streams are modelled with the same set of parameters as periodic event streams. The difference is that for sporadic event streams, the lower event function $\eta^l(\Delta t)$ is always zero. The maximum distance function $\delta^{\max}(N \geq 2)$ approaches infinity for all values of N [13]. Note that 'jitter' and $d_{\min}$ parameters are also meaningful in sporadic event models, since they allow us to accurately capture sporadic transient load peaks.

Event models with this small set of parameters have several advantages. First, they are easily understood by a designer, since period, jitter, etc. are familiar event stream properties. Second, the corresponding event functions and distance functions can be evaluated quickly, which is important for scheduling analysis to run fast. Third, as will be shown in Section 2.3.3.2, compositional performance analysis requires the modelling of possible timing of output events for propagation to the next scheduling component. Event models as described above allow us to specify simple rules to obtain output event models (Section 2.3.3.1) that can be described with the same set of parameters as the activating event models. Therefore, there is no need to depart from these event models whatever the size and structure of the composed system (hence the term 'standard'). This makes the compositional performance analysis approach very general.

2.3.3 Analysis composition

In the compositional performance analysis methodology [13,14], local scheduling analysis and event model propagation are alternated, during system-level analysis. This requires the modelling of possible timing of output events for propagation to the next scheduling component. In the following, first the output event model calculation is explained. Then the compositional analysis approach is presented.

2.3.3.1   Output event model calculation

The SymTA/S standard event models allow us to specify simple rules to obtain output event models that can be described with the same set of parameters as the activating event models. The output event model period obviously equals the activation period. The difference between maximum and minimum response times (the response time jitter) is added to the activating event model jitter, yielding the output event model jitter (Equation (2.5)).

$$\mathcal{J}_{\text{out}} = \mathcal{J}_{\text{act}} + (t_{\text{resp,max}} - t_{\text{resp,min}}) \qquad (2.5)$$

Note that if the calculated output event model has a larger jitter than period, this information alone would indicate that an early output event could occur before a late previous output event, which obviously cannot be correct. In reality, output events cannot follow closer than the minimum response time of the producer task. This is indicated by the value of the 'minimum distance' parameter.

2.3.3.2   Analysis composition using standard event models

In the following, the compositional analysis approach is explained using the system example in Figure 2.3. Initially, only event models at the external system inputs are known. Since an activating event model is available for each task on R_1, a local scheduling analysis of this resource can be performed and output event models are calculated for T_1 and T_3 (Section 2.3.3.1). In the second phase, all output event models are propagated. The output event models become the activating event models for T_2 and T_4. Now, a local scheduling analysis of R_2 can be performed since all activating event models are known.

However, it is sometimes impossible to perform system-level scheduling analysis as explained above. This is shown in the system example in Figure 2.6.

Figure 2.6 shows a system consisting of two resources, R_1 and R_2, each with two tasks mapped on to it. Initially, only the activating event models of T_1 and T_3 are known. At this point the system cannot be analysed, because on every resource an activating event model for one task is missing, i.e. response times on R_1 need to be

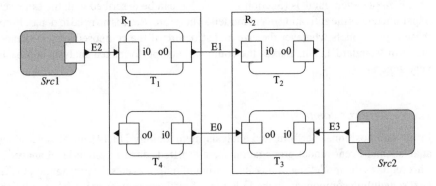

Figure 2.6   Example of a system with cyclic scheduling dependency

calculated to be able to analyse R_2. On the other hand, R_1 cannot be analysed before analysing R_2. This problem is called 'cyclic scheduling dependency'.

One solution to this problem is to initially propagate all external event models along all system paths until an initial activating event model is available for each task [15]. This approach is safe since on one hand scheduling cannot change an event model period. On the other hand, scheduling can only 'increase' an event model jitter [7]. Since a smaller jitter interval is contained in a larger jitter interval, the minimum initial jitter assumption is safe.

After propagating external event models, global system analysis can be performed. A global analysis step consists of two phases [13]. In the first phase local scheduling analysis is performed for each resource and output event models are calculated (Section 2.3.3.1). In the second phase, all output event models are propagated. It is then checked if the first phase has to be repeated because some activating event models are no longer up-to-date, meaning that a newly propagated output event model is different from the output event model that was propagated in the previous global analysis step. Analysis completes if either all event models are up-to-date after the propagation phase or if an abort condition, e.g. the violation of a timing constraint, has been reached.

2.3.4   Event stream adaptation

A key property of the SymTA/S compositional performance analysis approach is the ability to adapt the possible timing of events in an event stream (expressed through the adaptation of an event model [13]). There are several reasons to do this. It may be that a scheduler or a scheduling analysis for a particular component requires certain event stream properties. For example, rate-monotonic scheduling and analysis [1] require strictly periodic task activation. Alternatively, an integrated IP component may require certain event stream properties. External system outputs may also impose event model constraints, e.g. a minimum distance between output events or a maximum acceptable jitter. Such a constraint may be the result of a performance contract with an external subsystem [16]. Event stream adaptation can also be done for the sole purpose of 'traffic shaping' [13]. Traffic shaping can be used, e.g. to reduce transient load peaks, in order to obtain more regular system behaviour. Practically, event model 'adaptation' is distinguished from event model 'shaping' in SymTA/S [17]. Adaptation is required to satisfy an event model constraint, while shaping is voluntary to obtain more regular system behaviour. Two types of event adaptation functions (EAF) are currently implemented in SymTA/S: a 'periodic' EAF produces a periodic event stream from a 'periodic with jitter' input event stream. A d_{min}-EAF enforces a minimum distance between output events.

2.4   Complex embedded applications

Compositional performance analysis as described so far is not applicable to embedded applications with complex task dependencies. This is because it uses a simple activation model where the completion of one task directly leads to the activation of a dependent task. However, activation dependencies in realistic embedded applications are usually more complex. A consumer task may require a different amount of data per execution than produced by a producer task, leading to multi-rate systems. Task activation may also be conditional, leading to execution-rate intervals. Furthermore, a task may consume data from multiple task inputs. Then, task activation timing is a function of the possible arrival timing of all required input data. Tasks with multiple inputs also allow us to form cyclic dependencies (e.g. in a control loop).

In this section, the focus is on multiple inputs (both AND- and OR-activation) and functional cycles [18]. Multi-rate systems and conditional communication are not considered, since these features have not yet been incorporated into SymTA/S. The theoretical foundations can be found in Reference 19.

2.4.1   Basic thoughts

The activation function of a consumer task C with multiple inputs is a Boolean function of input events at the different task inputs. An imposed restriction is that activation must not be invalidated due to the arrival of additional tokens [20]. This means that negation is not allowed in the activation function. Consequently, the only acceptable Boolean operators are AND and OR, since an input is negated in all other commonly used Boolean operators (NOT, XOR, NAND, NOR).

In order to perform scheduling analysis on the resource to which task C is mapped, activating event functions for task C have to be calculated from all input event functions. In the following it is shown how to do this for AND- and OR-activation using standard event models (Section 2.3.2). An extended discussion covering event models in general can be found in Reference 19.

2.4.2   AND-activation

For a consumer task C with multiple inputs, AND-activation implies that C is activated after an input event has occurred at each input i. An example of an AND-activated task with three inputs is shown in Figure 2.7.

Note that AND-activation requires input data buffering, since at some inputs data may have to wait until data has arrived at all other inputs for one consumer

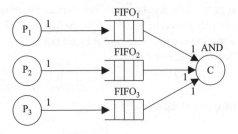

Figure 2.7   Example of an AND-activated task C

activation. The term 'AND-buffering' is used to refer to this source of buffering. The term 'token' [21] is used to refer to the amount of data required for one input event.

2.4.2.1   AND-activation period

To ensure bounded AND-buffer sizes, the period of all input event models must be the same. The period of the activating event model equals this period.

$$\mathcal{P}_i \stackrel{!}{=} \mathcal{P}_j; \quad i,j = 1 \ldots k \quad \Rightarrow$$
$$\mathcal{P}_{\text{AND}} = \mathcal{P}_i; \quad i = 1 \ldots k \tag{2.6}$$

2.4.2.2   AND-activation jitter

In order to obtain the AND-activation jitter, consider how often activation of the AND-activated task can occur during any time interval Δt. Obviously, during any time interval Δt, the port with the smallest minimum number of available tokens determines the minimum number of AND-activations. Likewise, the port with the smallest maximum number of available tokens determines the maximum number of AND-activations.

The number of available tokens at port i during a time interval Δt depends on both the number of tokens arriving during Δt, and on the number of tokens that arrived earlier, but did not yet lead to an activation because tokens at one or more other ports are still missing. This is illustrated in the following example. Assume that the task in Figure 2.7 receives tokens at each with the following 'periodic with jitter' input event models:

$$\mathcal{P}_1 = 4, \quad \mathcal{J}_1 = 0$$
$$\mathcal{P}_2 = 4, \quad \mathcal{J}_2 = 2$$
$$\mathcal{P}_3 = 4, \quad \mathcal{J}_3 = 3$$

Figure 2.8 shows a possible sequence of input events that adhere to these event models, and the resulting AND-activation events. The numbering of events in the figure indicates which events together lead to one activation of AND-activated task C. As can be seen, the minimum distance between two AND-activations (activations 3 and 4 in Figure 2.8) equals the minimum distance between two input events at input 3, which is the input with the largest input event model jitter. Likewise, the maximum distance between two AND-activations (activations 1 and 2 in Figure 2.8) equals the maximum distance between two input events at input 3. It is not possible to find a different sequence of input events leading to a smaller minimum or a larger maximum distance between two AND-activations. From this it results that the input with the largest input event jitter determines the activation jitter of the AND-activated task, i.e.

$$\mathcal{J}_{\text{AND}} = \max\{\mathcal{J}_i\}; \quad i = 1 \ldots k \tag{2.7}$$

This statement also remains true if the first set of input events does not arrive at the same time (as is the case in Figure 2.8). A proof is given in Reference 19. Calculation

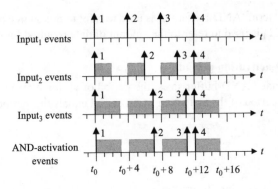

Figure 2.8   AND-activation timing example

of the worst-case delay and backlog at each input due to AND-buffering can also be found in Reference 19.

Note that in some cases it may be possible to calculate phases between the arrivals of corresponding tokens in more detail, e.g. through the use of inter-event stream contexts (Section 2.5.3). It may then be possible to calculate a tighter activating jitter if it can be shown that a certain input cannot (fully) influence the activation timing of an AND-activated task, because tokens at this input arrive relatively early. This is particularly important for the analysis of functional cycles (Section 2.4.4).

2.4.3   OR-activation

For a consumer task C with multiple inputs, OR-activation implies that C is activated each time an input event occurs at any input of C. Contrary to AND-activation, input event models are not restricted, and no OR-buffering is required, since a token at one input never has to wait for tokens to arrive at a different input in order to activate C. Of course, activation buffering is still required.

An example of an OR-activated task with two inputs is shown in Figure 2.9. Assume the following 'periodic with jitter' event models at the two inputs of task C:

$$P_1 = 4, \quad J_1 = 2$$
$$P_2 = 3, \quad J_2 = 2$$

The corresponding upper and lower input event functions are shown in Figure 2.10. Since each input event immediately leads to one activation of task C, the upper and lower activating event functions are constructed by adding the respective input event functions. The result is shown in Figure 2.11(a).

Recall a key requirement of compositional performance analysis, namely, that event streams are described in a form that can serve both as input for local scheduling analysis, and can be produced as an output of local scheduling analysis for propagation to the next analysis component (Section 2.3.3.2). Due to the irregularly spaced steps (visible in Figure 2.11(a)), the 'exact' activating event functions cannot be described by a 'periodic with jitter' event model, and thus cannot serve directly as input for

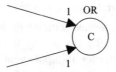

Figure 2.9   Example of an OR-activated task C

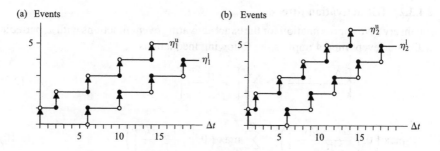

Figure 2.10   Upper and lower input event functions in the OR-example. (a) OR input 1 ($\mathcal{P}_1 = 4, \mathcal{J}_1 = 2$); (b) OR input 2 ($\mathcal{P}_2 = 3, \mathcal{J}_2 = 2$)

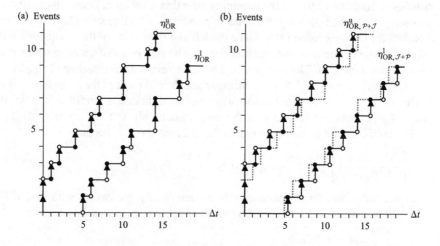

Figure 2.11   Upper and lower activating event functions in the OR-example. (a) Exact; (b) periodic with jitter approximation

local scheduling analysis. Furthermore, after local scheduling analysis a 'periodic with jitter' output event model has to be propagated to the next analysis component. An activation jitter is required in order to calculate an output jitter (Section 2.3.3.1). Therefore, conservative approximations for the exact activating event functions that can be described by a 'periodic with jitter' event model ($\mathcal{P}_{OR}, \mathcal{J}_{OR}$) need to be found. The intended result is shown in Figure 2.11 (the exact curves appear as dotted lines).

2.4.3.1 OR-activation period

The period of OR-activation is the least common multiple $\text{LCM}(P_i)$ of all input event model periods (the 'macro period'), divided by the sum of input events during the macro period assuming zero jitter for all input event streams.

$$\mathcal{P}_{\text{OR}} = \frac{\text{LCM}(P_i)}{\sum_{i=1}^{n} \text{LCM}(P_i)/P_i} = \frac{1}{\sum_{i=1}^{n}(1/P_i)} \tag{2.8}$$

2.4.3.2 OR-activation jitter

A conservative approximation for the exact activating event functions with a 'periodic with jitter' event model implies the following inequations.

$$\left\lceil \frac{\Delta t + \mathcal{J}_{\text{OR}}}{\mathcal{P}_{\text{OR}}} \right\rceil \geq \sum_{i=1}^{n} \left\lceil \frac{\Delta t + \mathcal{J}_i}{P_i} \right\rceil \tag{2.9}$$

$$\max\left(0, \left\lfloor \frac{\Delta t - \mathcal{J}_{\text{OR}}}{\mathcal{P}_{\text{OR}}} \right\rfloor\right) \leq \sum_{i=1}^{n} \max\left(0, \left\lfloor \frac{\Delta t - \mathcal{J}_i}{P_i} \right\rfloor\right) \tag{2.10}$$

In order to be as accurate as possible, the minimum jitter that satisfies in Equations (2.9) and (2.10) must be found. It can be shown that the minimum jitter that satisfies in Equation (2.9) and the minimum jitter that satisfies in Equation (2.10) are the same [19]. In the following, the upper approximation (in Equation (2.9)) is used to calculate the OR-activation jitter. Since the left and right sides of this inequation are only piecewise continuous, the inequation cannot be simply transformed to obtain the desired minimum jitter. The solution used here is to evaluate in Equation (2.9) piecewise for each interval $[\Delta t_j, \Delta t_{j+1}]$, during which the right side of the inequation has a constant value $k_j \in \mathbb{N}$. For each constant piece of the right side, a condition for a 'local jitter' $\mathcal{J}_{\text{OR,j}}$ is obtained that satisfies the inequation for all $\Delta t : \Delta t_j < \Delta t \leq \Delta t_{j+1}$.

For each constant piece of the right side, in Equation (2.9) becomes

$$\left\lceil \frac{\Delta t + \mathcal{J}_{\text{OR,j}}}{\mathcal{P}_{\text{OR}}} \right\rceil \geq k_j; \quad \Delta t_j < \Delta t \leq \Delta t_{j+1}, k_j \in \mathbb{N}$$

Since the left side of this inequation is monotonically increasing with Δt, it is sufficient to evaluate it for the smallest value of Δt, which approaches Δt_j, i.e.

$$\lim_{\epsilon \to +0} \left\lceil \frac{\Delta t_j + \epsilon + \mathcal{J}_{\text{OR,j}}}{\mathcal{P}_{\text{OR}}} \right\rceil \geq k_j; \quad k_j \in \mathbb{N}$$

$$\Leftrightarrow \quad \lim_{\epsilon \to +0} \frac{\Delta t_j + \epsilon + \mathcal{J}_{\text{OR,j}}}{\mathcal{P}_{\text{OR}}} > k_j - 1$$

$$\Leftrightarrow \quad \lim_{\epsilon \to +0} (\mathcal{J}_{\text{OR,j}} + \epsilon) > (k_j - 1) \times \mathcal{P}_{\text{OR}} - \Delta t_j$$

$$\Leftrightarrow \quad \mathcal{J}_{\text{OR,j}} \geq (k_j - 1) \times \mathcal{P}_{\text{OR}} - \Delta t_j \tag{2.11}$$

The global minimum jitter is then the smallest value which satisfies all local jitter conditions. As already said, $\eta_{\text{OR}}^{\text{u}}$ displays a pattern of distances between steps which

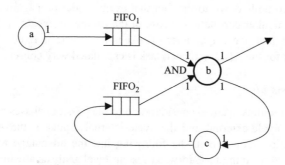

Figure 2.12   Example of a cyclic dependency

repeats periodically every macro period. Therefore, it is sufficient to perform the above calculation for one macro period. An algorithm can be found in Reference 22.

2.4.4   Cyclic task dependencies

Tasks with multiple inputs allow us to build cyclic dependencies. A typical application is a control loop, where one task represents the controller and the other represents a model of the controlled system. A task graph with a cycle is shown in Figure 2.12.

Tasks with multiple inputs in cycles are assumed to be AND-activated, and to produce one token at each output per execution. This implies that at least one initial token must be present inside the cycle to avoid deadlock [21], and that the number of tokens inside the cycles remains constant. Consequently, the period of the cycle-external event model determines the period of all cycle tasks. Finally, exactly one cycle task with one cycle-external and one cycle-internal input is assumed to exist in a cycle. All other cycle tasks only have cycle-internal inputs. These restrictions allow us to concisely discuss the main issues resulting from functional cycles. A much more general discussion can be found in Reference 19.

In Section 2.4.2 it was established that the activation jitter of an AND-activated task is bounded by the largest input jitter. As was the case for cyclic scheduling dependencies (Section 2.3.3.2), system analysis starts with an initial assumption about the cycle-internal jitter of the AND-activated task, since this value depends on the output jitter of that task, which has not been calculated yet. A conservative starting point is to initially assume zero internal jitter. Now analysis and event model propagation can be iterated around the cycle, hoping to find a fix-point.

However, if only one task along the cycle has a response time which is an interval, then after the first round of analysis and event model propagation the internal input jitter of the AND-activated task will be larger than the external input jitter. In the SymTA/S compositional performance analysis approach, this larger jitter will be propagated around the cycle again, resulting in an even larger jitter at the cycle-internal input of the AND-activated task (Section 2.3.3.2). It is obvious that the jitter appears unbounded if calculated this way.

The problem boils down to the fact that event model propagation as presented so far captures neither correlations between the timing of events in different event streams nor the fact that the number of tokens in a cycle is fixed. Therefore, the activation jitter for the AND-activated task is calculated very conservatively.

2.4.5   Analysis idea

Cycle analysis requires detailed consideration of the possible phases between tokens arriving at the cycle-external and the cycle-internal inputs of the AND-activated task. The solution proposed in the following has the advantage to require only minor modifications to the feed-forward system-level analysis already supported by SymTA/S. The idea is as follows: initially, the cycle-internal input is assumed not to increase the activation jitter of the AND-activated task. This allows to 'cut' the cycle-internal edge, rendering a feed-forward system which can be analysed as explained in Section 2.3.3.2. Then the time it takes a token to travel around the cycle is calculated, and the validity of the initial assumption is verified.

In the following, the idea is explained for cycles with one initial tokens. Assume an external 'periodic with jitter' event model with period $\mathcal{P}_{ext}$ and jitter $\mathcal{J}_{ext}$. Let t_{ff}^{min} and t_{ff}^{max} be, respectively, the minimum and maximum sum of worst-case response times of all tasks belonging to a cycle (the 'time around the cycle') as obtained through analysis of the corresponding feed-forward system. Let us further assume that after analysis of the corresponding feed-forward system, $t_{ff}^{max} \leq \mathcal{P}_{ext}$.

At system startup, the first token arriving at the cycle-external input will immediately activate the AND-concatenated task together with the initial token already waiting at the cycle-internal input. No further activation of the AND-activated task is possible until the next token becomes available at the cycle-internal input of that task. If feed-forward analysis was valid, then this will take between t_{ff}^{min} and t_{ff}^{max} time units.

The maximum distance between two consecutive external tokens is $\delta_{ext}^{max}(2) = \mathcal{P}_{ext} + \mathcal{J}_{ext}$ (Equation (2.4)). From $t_{ff}^{max} \leq \mathcal{P}_{ext}$ it follows that it is not possible that the '2nd' external token arriving as 'late' as possible after the '1st' external token has to wait for an internal token.

The '3rd' external token can arrive at most $\delta_{ext}^{max}(3) = 2 \times \mathcal{P}_{ext} + \mathcal{J}_{ext}$ after the '1st' external token. Therefore, if both the '2nd' and the '3rd' external tokens arrive as late as possible, then the '3rd' arrives $\mathcal{P}_{ext}$ after the '2nd'. From $t_{ff}^{max} \leq \mathcal{P}_{ext}$ it follows that the '3rd' external token arriving as 'late' as possible after the '1st' external token cannot wait for an internal token, even if the '2nd' external token also arrived as 'late' as possible. This argument can be extended to all further tokens. Thus, no external token arriving as late as possible has to wait for an internal token.

Activation of task b also cannot happen earlier than the arrival of an external token. Therefore, the activating event model of task b is conservatively captured by the external input event model (Equation (2.12)). This approach is therefore valid for a cycle with $M = 1$ initial token, for which $t_{ff}^{max} \leq \mathcal{P}_{ext}$.

$$\mathcal{P}_{act} = \mathcal{P}_{ext}; \quad \mathcal{J}_{act} = \mathcal{J}_{ext} \tag{2.12}$$

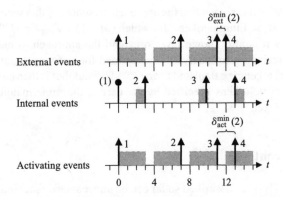

Figure 2.13 *Possible event sequence for the cycle example. Grey boxes indicate jitter intervals during which an event can occur. Note that line 2 displays the possible timing of internal events depending on the previous activating event, while lines 1 and 3 display the possible timing of events independent of previous events*

For example, assume that in the system in Figure 2.12 task b is activated externally with ($\mathcal{P}_{b,ext} = 4$, $\mathcal{J}_{b,ext} = 3$). Let us further assume that feed-forward analysis has determined the time around the cycle to be $[t_{ff}^{min}, t_{ff}^{max}] = [2,3]$, i.e. each internal input event follows between [2,3] time units after the previous activating event. Figure 2.13 shows a snapshot of a sequence of external, internal and activating events for task b (numbers indicate corresponding input events and the resulting activating event). The first internal event is due to the initial token. As can be seen, activating event timing can be described by the same event model as external input event timing. If on the other hand analysis of the corresponding feed-forward system determines $t_{ff}^{max} > \mathcal{P}_{ext}$, then this statement is no longer true, since for example the '3rd' internal event could occur later than the latest possible '3rd' external event.

In Figure 2.13 it can also be seen that an 'early' external token may have to wait for an internal token since two token arrivals at the cycle-internal input of task b cannot follow closer than t_{ff}^{min}, and thus

$$\delta_{act}^{min}(2) = \begin{cases} \delta_{ext}^{min}(2); & t_{ff}^{min} \leq \delta_{ext}^{min}(2) \\ t_{ff}^{min} & ; & t_{ff}^{min} > \delta_{ext}^{min}(2) \end{cases} \tag{2.13}$$

Effectively, if $t_{ff}^{min} > \delta_{ext}^{min}(2)$, then the cycle acts like a d_{min}-EAF with $d_{min} = t_{ff}^{min}$ (Section 2.3.4). This additional effect of the cycle does not 'require' a new scheduling analysis, since the possible activation timing is only tightened. All possible event timing in the tighter model is already included in the wider model. Therefore, the results in Equation (2.12) remain valid. However, it is worthwhile to perform scheduling analysis again with the tighter activating event model for the AND-concatenated task, since results will be more accurate.

In Reference 19 it is shown that the approach presented in this section is also valid for a cycle with $M > 1$ initial tokens, for which $(M - 1) \times \mathcal{P}_{ext} < t_{ff}^{max} \leq M \times \mathcal{P}_{ext}$. In Reference 19 it is also shown how to extend the approach to nested cycles. In SymTA/S, the feed-forward analysis is performed for every cycle, and the required number of initial tokens is calculated from t_{ff}^{max}. This number is then compared against the number of cycle-tokens specified by the user in the same manner as any other constraint is checked.

2.5 System contexts

Performance analysis as described so far can be unnecessarily pessimistic, because it ignores certain correlations between consecutive task activations or assumes a very pessimistic worst-case load distribution over time.

In SymTA/S, advanced performance analysis techniques taking correlations between successive computation or communication requests as well as correlated load distribution into account, have been added in order to yield tighter analysis bounds. Cases where such correlations have a large impact on system timing are especially difficult to simulate and, hence, are an ideal target for formal performance analysis. Such correlations are called 'system contexts'.

In Section 2.5.1, using an example of a hypothetical set-top box, the assumptions made by a typical performance analysis, called 'context-blind' analysis, are reviewed. Then, the analysis improvements that can be obtained when considering two different types of system contexts separately and also in combination are shown: 'intra event stream contexts', which consider correlations between successive computation or communication requests (Section 2.5.2), and 'inter event stream contexts', which consider possible phases between events in different event streams (Section 2.5.3). The combination of both system contexts is explained in Section 2.5.4.

2.5.1 Context-blind analysis

The SoC implementation of a hypothetical set-top box shown in Figure 2.14 is used as an example throughout this section. The set-top box can process Motion Pictures Expert Group-2 (MPEG-2) video streams arriving from the RF-module (*rf_video*) and sent via the bus (BUS) to the TV (*tv*). In addition, a decryption unit (*DECRYPTION*) allows us to decrypt encrypted video streams. The set-top box can additionally process IP traffic and download web-content via the bus (*ip*) to the hard-disk (*hd*).

The focus will be on worst-case response time calculation for the system bus. Assume 'static priority-based scheduling' on the bus. The priorities are assigned as follows: $enc > dec > ip$. MPEG-2 video frames are assumed to arrive periodically from the RF-module. The arrival period is normalised to 100. The core execution and communication times of the tasks are listed in Table 2.1.

The worst-case response time of ip, calculated by a context-blind analysis, is 170. As can be seen in Figure 2.15, even though a data dependency exists between

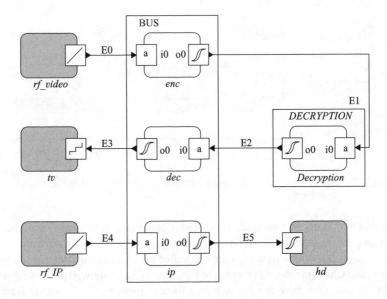

Figure 2.14 Hypothetical set-top-box system

Table 2.1 Core execution times

Task	CET
enc	[10,30]
dec	[10,30]
ip	[50,50]
Decryption	[40,40]

enc and *dec*, which may even out their simultaneous activation, a context-blind analysis assumes that in the worst-case all communication tasks are activated at the same instant. Furthermore, even though MPEG-2 frames may have different sizes depending on their type, a context-blind analysis assumes that every activation of *enc* and *dec* leads to a maximum transmission time of one MPEG-2 frame.

2.5.2 Intra event stream context

Context-blind analysis assumes that in the worst case, every scheduled task executes with its worst-case execution time for each activation. In reality, different events often activate different behaviours of a computation task with different worst-case execution time (WCET), or different bus loads for a communication task. Therefore, a lower maximum load (and a higher minimum load) can be determined for a sequence of successive activations of a higher-priority task if the types of the activating events

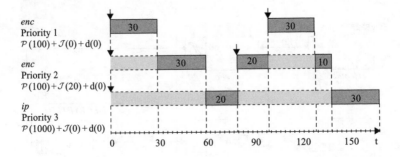

Figure 2.15   Worst-case response time calculation for 'ip without' contexts, using SymTA/S

are considered. This in turn leads to a shorter calculated worst-case response time (and a longer best-case response time) of lower-priority tasks. The correlation within a sequence of different activating events is called an 'intra event stream context'.

Mok and Chen introduced this idea in Reference 23 and showed promising results for MPEG-streams where the average load for a sequence of I-, P- and B-frames is much smaller than in a stream that consists only of large I-frames, which is assumed by a context-blind worst-case response time analysis. However, the periodic sequence of types of activating events was supposed to be completely known.

In reality, intra event stream contexts can be more complicated. If no complete information is available about the types of the activating events, it is no longer possible to apply Mok's and Chen's approach. Mok and Chen also do not clearly distinguish between different types of events on one hand, and different task behaviours, called 'modes' [24], on the other. However, this distinction is crucial for subsystem integration and compositional performance analysis. Different types of events are a property of the sender, while modes are a property of the receiver. Both can be specified separately from each other and later correlated. Furthermore, it may be possible to propagate intra event stream contexts along a chain of tasks. It is then possible to also correlate the modes of consecutive tasks.

In SymTA/S, intra event stream contexts are extended by allowing minimum and maximum conditions for the occurrence of a certain type of event in a sequence of a certain length n, in order to capture partial information about an event stream, n is an arbitrary integer value. A single worst-case and a single best-case sequence of events with length n can be determined from the available min- and max-conditions that can be used to calculate the worst- and best-case load due to any number of consecutive activations of the consumer task. In Reference 25, the static-priority preemptive response-time calculation is extended to exploit this idea.

In the following, this approach is applied to the set-top box example. Suppose that the video stream, sent from the RF to the bus, is encoded in one of several patterns of I-, P- and B-frames (IBBBBB, IBBPBB, IPBBBB...), or that several video streams are interleaved. Therefore, it is impossible to provide a fixed sequence of successive frame types in the video stream. However, it may be possible to determine min- and max-conditions for the occurrence of each frame type.

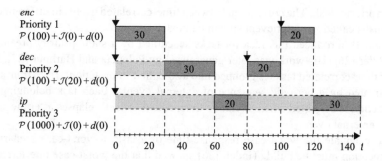

Figure 2.16 Worst-case response time calculation for *ip* considering *'intra'* contexts

The communication times of tasks *enc* and *dec* depend on the received frame type. I-frames have the largest size and lead to the longest execution time, P-frames have the middle size and B-frames have the smallest size. Therefore, the mode corresponding to the transmission of an I-frame has the largest communication time and the mode corresponding to the transmission of a B-frame has the lowest communication time.

Having both intra event stream context information and modes of the consumer tasks, a weight-sorted worst-case sequence of frame types with length n can be determined. The reader interested in knowing the algorithm to exploit min- and max-conditions is referred to Reference 25.

Now the worst-case load produced on the bus can be determined for l successive activations of *enc* and *dec*. This is performed, by iterating through the weight-sorted sequence starting from the first event and adding up loads until the worst-case load for l activations has been calculated. If l is bigger than n, the sequence length, the algorithm goes only through l mod n events and adds the resulting load to the load of the whole sequence multiplied by l div n.

In Figure 2.16, assuming that the worst-case sequence of frame types with length 2 is IP; and that the transmission time for an I-frame is 30 and for a P-frame is 20, the calculated worst-case response time of *ip*, when considering the available intra event stream context information, is shown. As can be seen, for both tasks *enc* and *dec*, the produced load on the bus due to a transmission of two successive MPEG-2 frames is smaller than in the context-blind case (see Figure 2.15). This leads to a reduction of the calculated worst-case response time of *ip*: 150 instead of 170.

2.5.3   Inter event stream context

Context-blind analysis assumes that all scheduled tasks sharing a resource are independent and that in the worst case all tasks are activated simultaneously. In reality, activating events are often time-correlated, which rules out simultaneous activation of all tasks. This in turn may lead to a lower maximum number (and higher minimum number) of interrupts of a lower-priority task through higher-priority tasks, resulting in a shorter worst-case response time (and longer best-case response time) of the lower-priority task. The correlation between time-correlated events in different event streams is called an 'inter event stream context'.

Tindell introduced this idea for tasks scheduled by a static priority preemptive scheduler [26]. His work was later generalised by Palencia and Harbour [27]. Each set of time-correlated tasks is grouped into a so-called 'transaction'. Each transaction is activated by a periodic sequence of external events. Each task belonging to a transaction is activated when a relative time, called 'offset', elapses after the arrival of the external event.

To calculate the worst-case response time of a task, a worst-case scenario for its execution must be build. Tindell [26] showed that the worst-case interference of a transaction on the response time of a task occurs at the 'critical instant' which corresponds to the most-delayed activation of a higher-priority task belonging to the transaction. The activation times of the analysed task and all higher-priority tasks have to happen as soon as possible after the critical instant.

Since all activation times of all higher-priority tasks belonging to a transaction are candidates for the critical instant of the transaction, the worst-case response time of a lower-priority task has to be calculated for all possible combinations of all critical instants of all transactions that contain higher-priority tasks, to find the absolute worst case.

In the following, Tindell's approach is applied to the set-top box example. Due to the data dependency among *enc*, *decryption* and *dec*, these tasks are time-correlated. The offset between the activations of *enc* and *decryption* corresponds to the execution time of *enc*. Based on this offset and the execution time of *decryption*, the offset between the activations of *enc* and *dec* can be calculated.

In order to show the analysis improvement due to inter event stream contexts in isolation, assume for now that all video-frames are I-frames. Figure 2.17 shows for the inter event stream context case the calculated worst-case response time of *ip* due to interrupts by *enc* and *dec*. As can be seen, a gap exists between successive executions of *enc* and *dec*. Since *ip* executes during this gaps, one interrupt less of *ip* is calculated (in this case through *enc*). This leads to a reduction of the calculated worst-case response time of *ip*: 140 instead of 170.

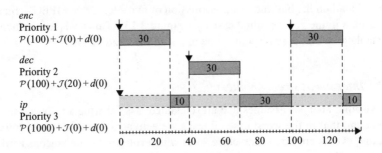

Figure 2.17   Worst-case response time calculation for 'ip' considering 'inter' contexts

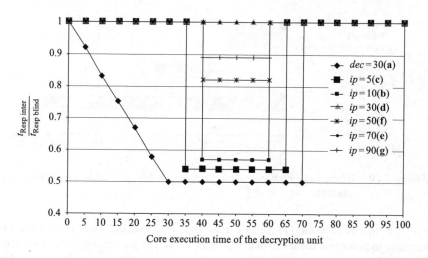

Figure 2.18   Improved worst-case response time calculation due to 'inter' contexts

In Figure 2.18, analysis improvements with inter event stream context information in relation to the context-blind case are shown as a function of the offset between *enc* and *dec*, which is equal to the execution time of the decryption unit.

Curve **a** shows the reduction of the calculated worst-case response time of *dec*. Depending on the offset, *dec* is either partially (offset value less than 30), completely (offset value more than 70) or not interrupted at all by *enc* (offset value between 30 and 70). The latter case yields a maximum reduction of 50 per cent.

Curves **b–g** show the reduction in the calculated worst-case response time of *ip* for different IP traffic sizes. The reduction is visible in the curves as dips. If no gaps exists between two successive executions of *enc* and *dec*, no worst-case response time reduction of *ip* can be obtained (offset value less than 30 or more than 70). If a gap exists, then sometimes one interrupt less of *ip* can be calculated (either through *enc* or *dec*), or there is no gain at all (curves **d** and **f**). Since the absolute gain that can be obtained equals the smaller worst-case execution time of *enc* and *dec*, the relative worst-case response time reduction is bigger for shorter IP traffic.

An important observation is that inter event stream context analysis reveals the dramatic influence that a small local change, e.g. the speed of the decryption unit reading data from the bus and writing the results back to the bus, can have on system performance, e.g. the worst-case transmission time of lower-priority IP traffic.

2.5.4   Combination of contexts

'Inter' event stream contexts allow us to calculate a tighter number of interrupts of a lower-priority task through higher-priority tasks. 'Intra' event stream contexts allow us to calculate a tighter load for a number of successive activations of a higher-priority task. The two types of contexts are orthogonal: the worst-case response time of a lower-priority task is reduced because fewer high-priority task activations

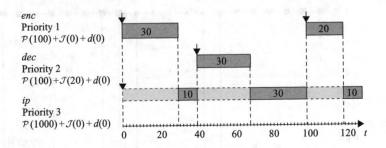

Figure 2.19   Worst-case response time calculation for 'ip' with 'combination' of contexts

can interrupt its execution during a certain time interval, and also because the time required to process a sequence of activations of each higher-priority task is reduced. Therefore, performance analysis can be further improved if it is possible to consider both types of contexts in combination. This is shown in Figure 2.19 for the worst-case response time calculation of ip: 130 instead of 170.

In Figure 2.20, analysis improvements considering both inter and intra event stream contexts in relation to the context-blind case are shown as a function of the offset between enc and dec. Curve **a** shows the reduction of the calculated worst-case response time of dec. Since dec is interrupted at most once by enc, and the worst-case load produced due to one activation of enc is the transmission time of one I-frame, no improvement is obtained through the context combination in comparison to curve **a** in Figure 2.18.

Curves **b**–**g** show the reduction of the calculated worst-case response time of ip for different IP traffic sizes. When comparing curves **b** and **c** (IP traffic sizes of 5 and 10) to curves **b** and **c** in Figure 2.18, it can be seen that no improvement is obtained through the context combination. This is due to the fact that ip is interrupted at most once by enc and at most once by dec. Therefore, as in case **a**, the calculated worst-case load produced by the video streams is the same no matter whether the available intra event stream context information is considered or not.

Curve **d** shows that for an IP traffic size of 30, no improvements are obtained through the context combination in comparison to the 'context-blind' case. This is due to the fact that for all offset-values, ip is interrupted exactly once by enc and exactly once by dec, and that the calculated worst-case load produced by the video streams due to one activation is the same no matter if intra event stream contexts are considered or not.

Curves **e** and **f** show that for IP traffic sizes of 50 and 70 improvements are obtained as a result of the context combination in comparison to both the intra and inter event stream context analysis. Since intra and inter event stream contexts are orthogonal, the reduction of the calculated worst-case response time of ip due to the intra event stream context is constant for all offset values. Since no reduction due to inter event stream context can be obtained for an offset value of 0 (equivalent to the inter event stream context-blind case), the reduction shown in the curve for this

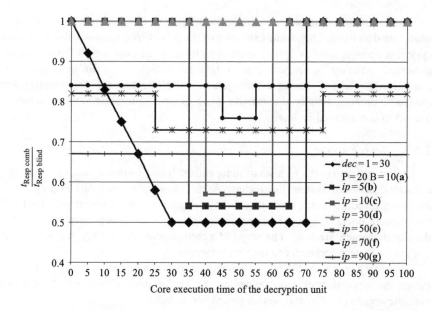

Figure 2.20   Analysis improvement due to the 'combination of intra and inter' contexts

offset value can only be a result of the intra event stream context. On the other hand, the additional reduction between the offset values 25 and 75 is obtained due to the inter event stream context.

Curve **g** shows that for an IP traffic size of 90, even though the inter event stream context leads to an improvement (see curve **g** in Figure 2.18), the improvement due to the intra event stream context dominates, since no dip exists in the curve, i.e. no additional improvements are obtained due to the context combination in comparison to the intra event stream context case.

This example shows that considering the combination of system contexts can yield considerably tighter performance analysis bounds compared with a context-blind analysis. Equally important, this example reveals the dramatic influence that a small local change can have on system performance. Systematically identifying such system-level influences of local changes is especially difficult using simulation due to the large number of implementations that would have to be synthesised and executed. On the other hand, formal performance analysis can systematically and quickly identify such corner cases. All these results took a couple of milliseconds to compute using SymTA/S.

2.6   Design space exploration for system optimisation

This section gives an overview of the compositional design space exploration framework used in SymTA/S which is based on evolutionary optimisation techniques.

First, system parameters which can be subject to optimisation, i.e. the search space, are described. Then some examples of metrics expressing desired or undesired system properties used as optimisation objectives in the exploration framework are presented. Afterwards, we will explain how the search space can be defined and dynamically modified during exploration in order to allow the designer to guide the search process. Finally, the iterative design space exploration loop performed in SymTA/S is explained in detail.

2.6.1   Search space

The entire system is seen as a set of independent 'chromosomes', each representing a distinct subset of system parameters. A chromosome carries variation operators necessary for combination with other chromosomes of its type. Currently in SymTA/S, the standard operator's mutation and crossover which are independently applied to the chromosomes, are used. The scope of a chromosome is arbitrary, it reaches from one single system parameter to the whole system.

The search space and the optimisation objectives can be multidimensional, which means that several system parameter can be explored simultaneously to optimise multiple objectives. Possible search parameter include:

- mapping of tasks onto different resources
- changing priorities on priority-scheduled resources
- changing time slot sizes and time slot order on TDMA or round robin scheduled resources
- changing the scheduling policy on a resource
- modifying resource speed
- traffic shaping

Traffic shaping is included in the search space because it increases the design space and allows us to find solutions which are not possible without traffic modulation. This shall be shown with a small example.

Consider the task set in Table 2.2 scheduled according to the static-priority pre-emptive policy. All tasks are activated periodically except T_0 which has a very large jitter leading to the simultaneous arrival of three activations in the worst case.

Two experiments are conducted. The first one with the original activating event models and the second one using a shaper at the input of T_0 extending the minimum

Table 2.2   Simple task set

Name	Activating event model	CET	Deadline
T_0	$\mathcal{P}(100) + \mathcal{J}(200) + d(10)$	4	8
T_1	$\mathcal{P}(100) + \mathcal{J}(0) + d(0)$	8	12
T_2	$\mathcal{P}(100) + \mathcal{J}(0) + d(0)$	5	21
T_3	$\mathcal{P}(100) + \mathcal{J}(0) + d(0)$	3	24

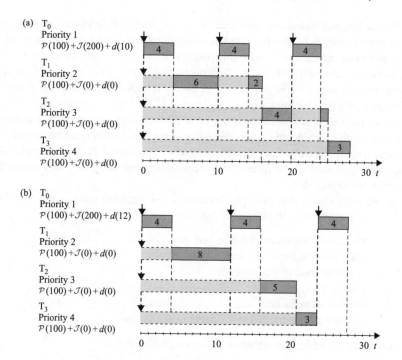

Figure 2.21   *WC scheduling scenarios $T_0 > T_1 > T_2 > T_3$. (a) minimum distance 10, (b) minimum distance 12*

distance to 12. In the first experiment no priority assignment leading to a system fulfilling all constraints is found. However, in the second experiment the priority assignment $T_0 > T_1 > T_2 > T_3$ leading to a working system is found. The reason for this is that extending the minimum distance of successive activations of T_0 relaxes the impact of the burst and leads to more freedom for the lower-priority tasks to execute. This results in less preemption and thus earlier completion for T_1, T_2 and T_3. Figures 2.21(a) and (b) visualise this effect by showing the worst-case scheduling scenarios for the priority assignment $T_0 > T_1 > T_2 > T_3$ with minimum distances 10 and 12.

Note that in the general case concerning distributed systems with complex performance dependencies, optimisation through traffic shaping is not applicable in such a straight forward manner. Nevertheless, traffic shaping can broaden considerably the solution-space by restricting event streams, leading to increased freedom on cross-related event streams.

2.6.2 Optimisation objectives

The SymTA/S exploration framework is capable of performing a multi-objective optimisation of several concurrent optimisation objectives, leading usually to the discovery of several 'Pareto-optima'.

Pareto-optimal solutions represent a certain trade-off between two or more optimisation objectives, leaving it to the designer to decide which solution to adopt. More precisely, given a set V of k-dimensional vectors $v \in \mathbb{R}^k$. A vector $v \in V$ dominates a vector $w \in V$ if for all elements $0 \leq i < k$ we have $v_i \leq w_i$ and for at least one element l we have $v_l < w_l$. A vector is called Pareto-optimal if it is not dominated by any other vector in V.

Optimisation objectives can be any kind of metric, defined on desired or undesired properties of the considered system. Note that some metrics only make sense in combination with constraints. Each design alternative considered during the exploration process is associated with a fitness vector containing one entry for every concurrent optimisation objective.

In the following some example optimisation objectives used in the SymTA/S exploration framework will be introduced using the following notation:

R maximum response time of a task or
       maximum end-to-end latency along a path
D deadline (task or end-to-end)
ω constant weight > 0
k number of tasks or
       number of constrained tasks/paths in the system

1 minimisation of the (weighted) sum of completion times

$$\sum_{i=1}^{k} \omega_i \times R_i$$

2 minimisation of the maximum lateness

$$\max(R_1 - D_1, \ldots, R_k - D_k)$$

3 maximisation of the minimum earliness

$$\min(D_1 - R_1, \ldots, D_k - R_k)$$

4 minimisation of the (weighted) average lateness

$$\sum_{i=1}^{k} \omega_i \times (R_i - D_i)$$

5 maximisation of the (weighted) average earliness

$$\sum_{i=1}^{k} \omega_i \times (D_i - R_i)$$

6 minimisation of end-to-end latencies
7 minimisation of jitters
8 minimisation of the sum of communication buffer sizes

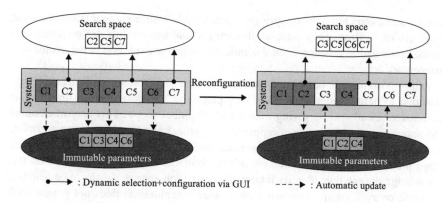

Figure 2.22   Search space definition

The choice of the metric for optimisation of a specific system is very important to obtain satisfying results. Example metrics 4 and 5, for instance, express the average timing behaviour of a system with regard to its timing constraints. They might mislead an evolutionary algorithm and prevent it from finding system configurations fulfilling all timing constraints, since met deadlines compensate linearly for missed deadlines. For systems with hard real-time constraints, metrics with higher penalties for missed deadline and fewer rewards for met deadlines can be more appropriate, since they lead to a more likely rejection of system configurations violating hard deadline constraints. The following example metric penalises violated deadlines in an exponential way and can be used to optimise the timing properties of a system with hard real-time constraints:

$$\sum_{i=0}^{k} c_i^{R_i - D_i}, \quad c_i > 1 \text{ constant}$$

2.6.3   Defining the search space and controlling the exploration

The designer defines the current search space, by selecting and configuring a set of chromosomes representing the desired search space. System parameters not included inside the selected chromosomes remain immutable during the exploration. Figure 2.22 shows this principle.

The set of chromosomes representing the search space serves as a blueprint for specific individuals (phenotypes) used during exploration. The variation operators (i.e. crossover and mutation) for these individuals are applied chromosome-wise.

The chromosomes are encoded and varied independently. There are two reasons why independent encoding and variation have been chosen. First, it is easier to establish a constructively correct encoding on a small subset of design decisions. Such an encoding scheme ensures that all chromosome values correspond to valid decisions such that any chromosome variation is constructively valid. This improves the optimisation process as it greatly reduces the effort of checking a generated design for validity. It allows using the analysis engine of SymTA/S which requires correct design parameters to apply analysis (e.g. sum of time slots no longer than the period, legal priority setting, etc.). Second, it is easy to add and remove design parameters to the optimisation process, even dynamically, which is exploited in the exploration framework.

Chromosomes can be defined arbitrarily as fine or coarse grain. This enables the designer to define the search space very precisely. The designer can limit certain parameters locally while giving others a more global scope. This way of defining the search space represents a compositional approach to optimisation and allows us to scale the search process. The designer can conduct several well-directed exploration steps providing insight into the system's performance dependencies. Based on this knowledge she can then identify interesting design sub-spaces, worthy to be searched in-depth or even completely. An a priori global exploration does not permit such a flexibility and neglects the structure of the design space, giving the designer no possibility to modify and select the exploration strategy. In the worst case, when the composition of the design space is unfavourable, this can lead to non-satisfying results with no possibility for the designer to intervene. In many approaches the only possibility for the designer in such a case consists in restarting the exploration, hoping for better results.

One important precondition for this approach to design space exploration is the dynamic configurability of the search space. The exploration framework allows the designer to redirect the exploration in a new direction without discarding already obtained results. She can for example downsize the search space by fixing parameters having the same values in (nearly) all obtained Pareto-optimal solutions, or expand it with parameters not yet considered. Note that this methodology is more flexible than separate local parameter optimisation and subsequent recombination.

2.6.4   Design space exploration loop

Figure 2.23 shows the design space exploration loop performed in the exploration framework [28]. The 'Optimisation controller' is the central element. It is connected to the 'Scheduling analysis' and to an 'Evolutionary optimiser'. The 'Scheduling analysis' checks the validity of a given system parameter set, that is represented by an individual, in the context of the overall heterogeneous system. The 'Evolutionary optimiser' is responsible for the problem-independent part of the optimisation problem, i.e. elimination of individuals and selection of interesting individuals for variation. Currently, SPEA2 (Strength Pareto Evolutionary Algorithm 2) [29] and FEMO (fair evolutionary multi-objective optimiser) [30] are used for this part. They are coupled via PISA (Platform and Programming Language Independent Interface for Search algorithms) [31], discussed in the previous chapter by Künzli.

Note that the selection and elimination strategy depends on the used multi-objective optimiser. For instance FEMO [30], eliminates all dominated individuals in every iteration and pursue a fair sampling strategy, i.e. each parent participates in the creation of the same number of offspring. This leads to a uniform search in the neighbourhood of elitist individuals.

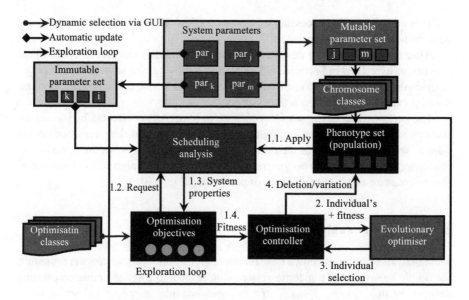

Figure 2.23   Design space exploration loop

The problem-specific part of the optimisation problem is coded in the chromosomes and their variation operators.

Before the exploration loop can be started the designer has to select the desired search space (see Section 2.6.1) and the optimisation objectives (see Section 2.6.2) she wants to optimise. The chromosomes representing the search space are included in the evolutionary optimisation, while all other system parameters remain immutable. After the designer has selected the search space and the optimisation task, SymTA/S is initialised with the immutable part of the system and the selected chromosomes are used as blueprints to create the initial population.

For each individual in the population the following is done:

- Step 1.1: The chromosomes of the considered individual are applied to the SymTA/S engine. This completes the system and it can be analysed.
- Step 1.2 + 1.3: Each optimisation objective requests the necessary system properties of the analysed system to calculate its fitness value.
- Step 1.4: The fitness values are communicated to the 'Optimisation controller'.

Once these four steps are performed for each individual inside the population the 'Optimisation controller' sends a list of all individuals and their fitness values to the 'Evolutionary optimiser' (step 2). Based on the fitness values the 'Evolutionary optimiser' creates two lists: a list of individuals which are to be deleted and a list of individuals selected for variation and sends them back to the 'Optimisation controller' (step 3). Based on the two lists the 'Optimisation controller' then manipulates the population, i.e. she deletes the according individuals and creates new offspring based on the individuals selected for variation and adds them to the population (step 4).

This completes the processing of one generation. The whole loop begins again with the new created population.

After each iteration the designer can choose to modify the search space. This consists, as explained in Section 2.6.3, in adding/removing chromosomes to/from the individuals. The re-evaluation of the fitness values is performed automatically and the next exploration iteration is then started.

The performance of the search procedure in SymTA/S is affected by the search strategy of the optimiser, the coding of the chromosomes and their variation operations as well as the choice of the optimisation objectives. As far as the optimiser is concerned, it is known that no general purpose optimisation algorithm exists that is able to optimise effectively all kinds of problems [32].

2.7   Sensitivity analysis

Most analysis techniques known from literature give a pure 'Yes/No' answer regarding the timing behaviour of a specific system with respect to a set of timing constraints defined for that system. Usually, the analyses consider a predefined set of input parameters and determine the response times, and thus, the schedulability of the system.

However, in a realistic system design process it is important to get more information with respect to the effects of parameter variations on system performance, as such variations are inevitable during implementation and integration. Capturing the bounds within which a parameter can be varied without violating the timing constraints offers more flexibility for the system designer and supports future changes. These bounds shows how 'sensitive' the system or system parts are to system configuration changes.

Liu and Layland [1] defined a maximum load bound on a resource that guarantees the schedulability of that resource when applying a rate-monotonic priority assignment scheme. The proposed algorithm is limited to specific system configurations: periodically activated tasks, tasks with deadlines at the end of their periods and tasks that do not share common resources (like semaphores) or that do not inter-communicate.

Later on, Lehoczky [33] extended this approach to systems with arbitrary task priority assignment. However, his approach does not go beyond the limitations mentioned above. Steve Vestal [34] proposed a fixed-priority sensitivity analysis for tasks with linear computation times and linear blocking time models. His approach is still limited to tasks with periodic activation patterns and deadlines equal to the period. Punnekkat [35] proposed an approach that uses a combination of a binary search algorithm and a slightly modified version of the response time schedulability tests proposed by Audsley and Tindell [7,36].

In the following is presented a brief overview about the sensitivity analysis algorithm and the analysis models and metrics used in SymTA/S. As already mentioned above, different approaches were proposed for the sensitivity analysis of different system parameters. However, these approaches can perform only single resource analysis as they are bounded by local constraints (tasks deadlines). Due to a fast increase of system complexity and heterogeneity, the current distributed systems usually have to satisfy global constraints rather than local ones. End-to-end deadlines or global buffer limits are an example of such constraints. Hence, the formal sensitivity analysis approaches used at resource level cannot be transformed and applied at system level, as this implies huge effort and less flexibility.

The sensitivity analysis framework used in SymTA/S combines a binary search technique and the compositional analysis model implemented in SymTA/S. As described in Section 2.3, SymTA/S couples the local scheduling analysis algorithms into a global analysis model.

Since deadlines are the major constraints in real-time systems it makes sense to measure the sensitivity of path latencies. As the latency of a path is determined by the response times of all tasks along that path, and the response time of a task directly depends on its core execution time, the following represent important metrics for the sensitivity analysis.

1. Maximum variation of the core execution time of a task without violating the system constraints or the system schedulability. If the system is not schedulable or constraints are violated then find the maximum value of the task core execution time that leads to a conforming system.
2. Minimum speed of a resource. The decrease of a resource speed directly affects the core execution times of all tasks mapped on that resource but also reduces the energy required by that resource. If the system is currently not schedulable or constraints are violated then find the minimum resource speed that determines a conforming system.

Variation of task execution/communication times: The search interval is determined by the current WCET value $t_{\text{core,max}}$ and the value corresponding to the maximum utilisation bound of the resource holding the analysed task. If the current utilisation of resource R is denoted by R_{load} and the maximum utilisation bound of resource R is denoted by $R_{\text{load,max}}$, then the search interval is determined by:

$$[t_{\text{core,max}}; t_{\text{core,max}} + \mathcal{P} \times (R_{\text{load,max}} - R_{\text{load}})]$$

where $\mathcal{P}$ represents the activation period in the case of periodic tasks or the minimum inter-arrival time in the case of sporadic tasks. If, for the current system configuration, the constraints are violated or the system is not schedulable then the search interval is determined by $[0; t_{\text{core,max}}]$.

The algorithm selects the interval middle value and verifies whether or not the constraints are satisfied for the configuration obtained by replacing the task WCET value with the selected value. If 'yes', then the upper half of the interval becomes the new search interval, otherwise the lower half of the interval is searched. The algorithm iterates until the size of the search interval becomes smaller than a specific predefined value (abort condition).

Variation of resource speed: The same algorithm is applied to find the minimum speed at which a resource can operate. If, for the current configuration, the constraints

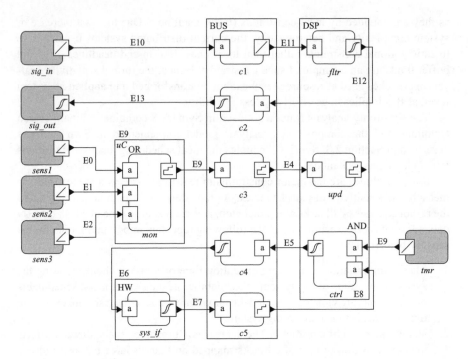

Figure 2.24   System-on-chip example

are satisfied and the system is schedulable then the search space is determined by $[R_{\text{speed,min}}; R_{\text{speed}}]$ where R_{speed} is the current speed factor (usually 1) and $R_{\text{speed,min}}$ is the speed factor corresponding to the maximum resource utilisation bound. Otherwise, the search space is $[R_{\text{speed}}; R_{\text{speed,max}}]$ where $R_{\text{speed,max}}$ is the speed factor corresponding to the maximum operational speed of that resource.

The ideal value for the maximum resource utilisation bound is 100 per cent. However, the experiments performed on different system models show that, for utilisation values above 98 per cent, the run-time of the sensitivity analysis algorithm drastically increases. This is due to an increase of the analysed period (busy period) in case of local analysis scheduling algorithms. Moreover, a resource load above 98 per cent is not realistic in practice due to variations of the system clock frequency or other distorting elements.

2.8   System-on-chip example

In this section, using SymTA/S, the techniques from the previous sections are applied to analyse the performance of a system-on-chip example shown in Figure 2.24.

The embedded system in Figure 2.24 represents a hypothetical SoC consisting of a micro-controller (uC), a 'DSP' and dedicated hardware ('HW'), all connected via an on-chip bus ('BUS'). 'DSP' and uC are equipped with local memory. The 'HW' acts

Table 2.3   Core execution and communication times

Computation task	C	Communication task	C
mon	[10,12]	*c1*	[8,8]
sys_if	[15,15]	*c2*	[4,4]
fltr	[12,15]	*c3*	[4,4]
upd	[5,5]	*c4*	[4,4]
ctrl	[20,23]	*c5*	[4,4]

Table 2.4   Event models at external system inputs

Input	s/p	$\mathcal{P}_{in}$	$\mathcal{J}_{in}$	$d_{min,in}$
sens1	*s*	1000	0	0
sens2	*s*	750	0	0
sens3	*s*	600	0	0
sig_in	*p*	60	0	0
tmr	*p*	70	0	0

as an interface to a physical system. It runs one task (*sys_if*) which issues actuator commands to the physical system and collects routine sensor readings. *sys_if* is controlled by task *ctrl*, which evaluates the sensor data and calculates the necessary actuator commands. *ctrl* is activated by a periodic timer ('*tmr*') and by the arrival of new sensor data (AND-activation in a cycle). Two initial tokens are assumed in the cycle.

The physical system is additionally monitored by three sensors (*sens1–sens3*), which produce data sporadically as a reaction to irregular system events. This data is registered by an OR-activated monitor task (*mon*) on the *uC*, which decides how to update the control algorithm. This information is sent to task *upd* on the 'DSP', which updates parameters into shared memory.

The DSP additionally executes a signal-processing task (*fltr*), which filters a stream of data arriving at input *sig_in*, and sends the processed data via output *sig_out*. All communication, except for shared-memory on the 'DSP', is carried out by communication tasks *c1–c5* over the on-chip 'BUS'. Core execution times for each task are shown in Table 2.3.

The event models in Table 2.4 are assumed at system inputs.

In order to function correctly, the system has to satisfy a set of path latency constraints (Table 2.5). Constraints 1 and 3 have been explicitly specified by the designer. The '2nd' constraint implicitly follows from the fact that the cycle contains two initial tokens. Constraint 3 is defined for causally dependent tokens [20]. Additionally, a maximum jitter constraint is imposed at output *sig_out* (Table 2.6).

Table 2.5   Path latency constraints

Constraint #	Path		Maximum latency
1	*sens1, sens2, sens3* → *upd*		70
2	*sig_in* → *sig_out*		60
3	cycle (*ctrl* → *ctrl*)		140

Table 2.6   Output jitter constraint

Constraint #	Output	Event model period	Event model jitter
4	*sig_out*	$\mathcal{P}_{sig\_out} = 60$	$\mathcal{J}_{sig\_out,max} = 18$

Table 2.7   Scheduling analysis results on uC

Task	s/p	Activating EM	r	s/p	Output EM
mon	s	$\mathcal{P}$ (250) $\mathcal{J}$ (500) d(0)	[10, 36]	s	$\mathcal{P}$ (250) $\mathcal{J}$ (526) d(10)

2.8.1   Analysis

Static priority scheduling is used both on the DSP and the BUS. The priorities on the BUS and DSP, respectively, are assigned as follows: $c1 > c2 > c3 > c4 > c5$ and $fltr > upd > ctrl$.

Performance analysis results were obtained using SymTA/S [8]. In the first step, SymTA/S performs OR-concatenation of the output event models of *sens1–sens3* and obtains the following 'sporadic' activating event model for task *mon*:

$$\mathcal{P}_{act} = \mathcal{P}_{OR} = 250, \ \mathcal{J}_{act} = \mathcal{J}_{OR} = 500$$

The large jitter is due to the fact that input events happening at the same time lead to a burst of up to three activations (no correlation between *sens1–sens3* is assumed). Since task 'mon' is the only task mapped onto *uC*, local scheduling analysis can now be performed for this resource, in order to calculate the minimum and maximum response times, as well as the output event model of task *mon*. The results of this analysis are shown in Table 2.7.

The worst-case response time of task *mon* increases compared to its worst-case core execution time, since later activations in a burst have to wait for the completion of the previous activations. The output jitter increases by the difference between

Table 2.8   Context-blind and sensitive analysis

Computation task	Resp$_{blind}$	Resp$_{sens}$	Communication tasks	Resp$_{blind}$	Resp$_{sens}$
mon	[10,36]	[10,36]	c1	[8,8]	[8,8]
sys_if	[15,17]	[15,15]	c2	[4,12]	[4,4]
fltr	[12,15]	[12,15]	c3	[4,16]	[8,12]
upd	[5,22]	[5,22]	c4	[4,28]	[8,20]
ctrl	[20,53]	[20,53]	c5	[4,32]	[8,32]

maximum and minimum core execution times compared to the activation jitter. The minimum distance between output events equals the minimum core execution time.

At this point, the rest of the system cannot be analysed, because on every resource activating event models for at least one task are missing. SymTA/S therefore generates a conservative starting-point by propagating all output event models along all paths until an initial activating event model is available for each task. SymTA/S then checks that the system cannot be overloaded in the long term. This calculation requires only activation periods and worst-case core execution times and thus can be done before response-time calculation.

System-level analysis can now be performed by iterating local scheduling analysis and event model propagation. SymTA/S determines that task *ctrl* belongs to a cycle, checks that AND-concatenation is selected and then proceeds to analyse the corresponding feed-forward system. SymTA/S executes until a fix-point for the whole system has been reached, and then compares the calculated performance values against performance constraints.

Table 2.8 shows the calculated response times of the computation and communication tasks with and without taking into account inter contexts. As can be observed, the exploitation of context information leads to much tighter response time intervals in the given example. This in turn reduces the calculated worst-case values for the constrained parameters. Table 2.9 shows that, in contrast to the inter context-blind analysis, all system constraints are satisfied when performance analysis takes inter context into account. In other words, a context-blind analysis would have discarded a solution which in reality is valid.

2.8.2   Optimisations

In this section, the architecture optimisation of the system-on-chip example is shown. Optimisation objectives are the four defined constraints. We try to minimise the latencies on paths 1–3 and the jitter at output *sig_out*.

In the first experiment the search space consists of the priority assignments on the BUS and the DSP. Table 2.10 shows the existing Pareto-optimal solutions. In the first two columns, tasks are ordered by priority, highest priority on the left. In the last four

Table 2.9   Constraint values for context-blind and sensitive analysis

#	Constraint	Inter context-blind	Inter context-sensitive
1	*sens1, sens2, sens3 → upd*	74	70
2	*sig_in → sig_out*	35	27
3	*cycle (ctrl → ctrl)*	130	120
4	$\mathcal{J}_{sig\_out,max} = 18$	11	3

Table 2.10   Pareto optimal solutions

#	Bus tasks	DSP tasks	Constrained values			
			1	2	3	4
1	*c1, c2, c3, c4, c5*	*upd, fltr, ctrl*	**55**	42	120	18
2	*c1, c2, c4, c3, c5*	*upd, fltr, ctrl*	59	42	112	18
3	*c2, c1, c4, c5, c3*	*upd, fltr, ctrl*	63	42	**96**	18
4	*c1, c2, c3, c4, c5*	*fltr, upd, ctrl*	70	**27**	120	**3**

columns, the actual value for all four constrained values is given. The best reached values for each constraint are emphasised.

As can be observed there are several possible solutions, each with its own advantages and disadvantages. We also observe that in each solution one constraint is only barely satisfied. A designer might want to find some alternative solutions where all constraints are fulfilled with a larger margin to the respective maximum values.

The search space is now extended by using a shaper at the output of task *mon*. It is making sense to perform traffic shaping at this location, because the OR-activation of *mon* can lead, in the worst-case scenario, to bursts at its output. That is, if all three 'sensors' trigger at the same time, *mon* will send three packets over the BUS with a distance of 10 time units, which is its minimum core execution time. This transient load peak affects the overall system performance in a negative way. A shaper is able to increase this minimum distance in order to weaken the global impact of the worst-case burst.

Table 2.11 shows Pareto-optimal solutions using a shaper at the output of *mon* extending the minimum distance of successive events to 12 time units, and thus weakening the global impact of the worst-case burst. The required buffer for this shaper is minimal, because at most one packet needs to be buffered at any time.

We observe that several new solutions are found. Not all best values for each constraint from the first attempt are reached, yet configurations 3 and 5 are interesting since they are more balanced regarding the constraints.

Table 2.11   *Pareto optimal solutions: shaper at mon output*

#	BUS tasks	DSP tasks	Constrained values			
			1	2	3	4
1	c2, c1, c3, c4, c5	upd, fltr, crtl	**59**	42	120	18
2	c1, c2, c4, c3, c5	upd, fltr, ctrl	63	42	112	18
3	c3, c2, c1, c4, c5	fltr, upd, ctrl	64	35	120	11
4	c2, c1, c5, c4, c3	upd, fltr, ctrl	67	42	**96**	18
5	c2, c3, c1, c5, c4	fltr, upd, ctrl	68	**31**	134	7

Table 2.12   *Sensitivity analysis of tasks WCETs*

Task	Current WCET	Max WCET	Slack
c1	8	8	0
c2	4	4	0
c3	4	7.65	3.65
c4	4	10.65	6.65
c5	4	22.5	18.5
upd	5	5	0
fltr	15	15	0
ctrl	23	30	7
sys_if	15	36	21
mon	12	15.66	3.66

2.8.3   Sensitivity analysis

This section presents the results of the sensitivity analysis algorithms described in Section 2.7 applied to system configuration #2 shown in Table 2.10. Table 2.12 shows the current WCET, the maximum WCET allowed as well as the free WCET slack obtained for the particular configuration.

The bar diagrams in Figure 2.25 show the system flexibility with respect to variations of tasks WCETs. It can be easily stated that the tasks and channels along the filter path ($c1$, $c2$, $fltr$) are very inflexible due to the jitter constraint defined at *sig_out*.

Table 2.13 presents the minimum resource speed factors that still guarantee that the system meets all its constraints. A particular observation can be made considering the speed of BUS. The results in Table 2.12 show that $c1$ and $c2$ are totally inflexible. However, Table 2.13 shows that the DSP can be speed-down with maximum

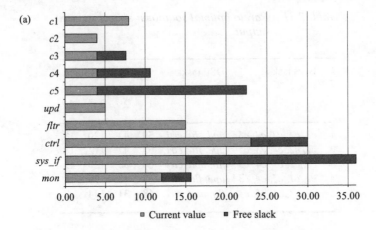

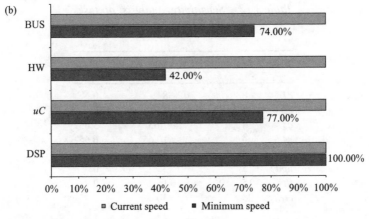

Figure 2.25   Sensitivity analysis results. (a) Tasks WCETS; (b) speed factors

26 per cent. By applying the sensitivity analysis for $c1$ and $c2$ only the WCET has been modified, the best-case execution time (BCET) remaining constant. Contrary, by changing the BUS speed both values, WCET and BCET, were changed. This led to a smaller jitter at channel output and to a higher flexibility for the BUS speed.

Figure 2.25(b) shows the results presented in Table 2.13. The timing constraints of filter-path ($c1, c2, fltr$) and system-reactive-path ($mon, c3, upd$) lead to rigid DSP properties with respect to later system changes.

2.9   Conclusions

The component integration step is critical in MpSoC design since it introduces complex component performance dependencies, many of which cannot be fully overseen by anyone in a design team. Finding simulation patterns covering all corner cases will soon become virtually impossible as MpSoCs grow in size and complexity, and

Table 2.13   Sensitivity analysis of resource
speed factors

Resource	Current factor	Min factor	Slack
HW	1	0.42	0.58
DSP	1	1	0
BUS	1	0.74	0.26
uC	1	0.77	0.23

performance verification is increasingly unreliable. In industry, there is an urgent need for systematic performance verification support in MpSoC design.

The majority of work in formal real-time analysis can be nicely applied to individual, local components or subsystems. However, the well-established view on scheduling analysis has been shown to be incompatible with the component integration style which is common practice in MpSoC design due to heavy component reuse. The recently adopted event stream view on component interactions represents a significant improvement for all kinds of system performance related issues.

First, the stream model elegantly illustrates the consequences of (a) resource sharing and (b) component integration, two of the main sources of complexity. This helps to identify previously unknown global performance dependencies, while tackling the scheduling problem itself locally where it can be overseen.

Second, the use of intuitive stream models such as periodic events, jitter, burst and sporadic streams, allows us to adopt existing local analysis and verification techniques. Essentially, SymTA/S provides automatic interfacing and adaptation among the most popular and practically used event stream models. In other words, SymTA/S is the enabling technology for the reuse of known local component design and verification techniques without compromising global analysis.

In this chapter, the basic ideas underlying the SymTA/S technology are presented. SymTA/S has a large variety of features that enable the analysis of complex embedded applications which can be found in practice. This includes multi-input tasks with complex activation functions, cyclic functional dependencies between tasks, systems with mutually exclusive execution modes and correlated task execution (intra and inter contexts). These powerful concepts make SymTA/S a unique performance analysis tool that verifies end-to-end deadlines, buffer over-/underflows, and transient overloads. SymTA/S eliminates key performance pitfalls and systematically guides the designer to likely sources of constraint violations.

And the analysis with SymTA/S is extremely fast (10 s for the system in Section 2.8, including optimisation). The turn-around times are within seconds. This opens the door to all sorts of explorations, which is absolutely necessary for system optimisation. SymTA/S uses genetic algorithms to automatically optimise systems with respect to multiple goals such as end-to-end latencies, cycles, buffer memory and others. Exploration is also useful for sensitivity analysis in order to determine slack and other popular measures of flexibility. This is specifically useful in systems which might experience later changes or modifications, a design scenario often found in industry. A large set of experiments demonstrates the application of SymTA/S and the usefulness of the results.

The SymTA/S technology was already applied in case studies in telecommunication, multimedia and automobile manufacturing projects. The cases had a very different focus. In one telecommunications project, a severe transient-fault system integration problem, that not even prototyping could solve, was resolved. In the multimedia case study, a complex two-stage dynamic memory scheduler was modelled and analysed to derive maximum response times for buffer sizing and priority assignment. In several automotive studies, it was shown how the technology enables a formal software certification procedure. The case studies have demonstrated the power and wide applicability of the event flow interfacing approach. The approach scales well to large, heterogeneous embedded systems including MpSoC. And the modularity allows us to customise SymTA/S libraries to specific industrial needs.

The SymTA/S approach can be used as a serious alternative or supplement to performance simulation. The unique technology allows comprehensive system integration and provides much more reliable performance analysis results at far less computation time.

References

1   LIU, C.L., and LAYLAND, J.W.: 'Scheduling algorithm for multiprogramming in a hard-real-time environment', *Journal of the ACM*, 1973, **20**

2   JENSEN, C.L.E., and TOKUDA, H.: 'A time-driven scheduling model for real-time operating systems'. Proceedings of the 6th IEEE *Real-time systems* symposium (RTSS1985), (IEEE CS Press, Los Alamitos, CA, 1985)

3   TINDELL, K., and CLARK, J.: 'Holistic schedulability analysis for distributed real-time systems', *Microprocessing and Microprogramming – Euromicro Journal (Special Issue on Parallel Embedded Real-Time Systems)*, 1994, **40**, pp. 117–134

4   GUTIERREZ, J.J., PALENCIA, J.C., and HARBOUR, M.G.: 'On the schedulability analysis for distributed hard real-time systems'. Proceedings of the ninth Euromicro workshop on *Real-time systems*, Toledo, Spain, 1997, pp. 136–143

5   GRESSER, K.: 'An event model for deadline verification of hard real-time systems'. Proceedings of the fifth Euromicro workshop on *Real-time systems*, Oulu, Finland, 1993, pp. 118–123

6   THIELE, L., CHAKRABORTY, S., and NAEDELE, M.: 'Real-time calculus for scheduling hard real-time systems'. Proceedings of the international symposium on *Circuits and systems* (ISCAS), Geneva, Switzerland, 2000

7   TINDELL, K.W.: 'An extendible approach for analysing fixed priority hard real-time systems', *Journal of Real-Time Systems*, 1994, **6**(2), pp. 133–152

8 HAMANN, A., HENIA, R., JERSAK, M., RACU, R., RICHTER, K., and ERNST, R.: 'SymTA/S – Symbolic timing analysis for systems'. http://www.symta.org/

9 RICHTER, K., and ERNST, R.: 'Event model interfaces for heterogeneous system analysis'. Proceedings of the *Design, automation and test in Europe* conference (DATE'02), Paris, France, March 2002

10 RICHTER, K., ZIEGENBEIN, D., JERSAK, M., and ERNST, R.: 'Model composition for scheduling analysis in platform design'. Proceedings of the *39th Design Automation Conference*, New Orleans, USA, June 2002

11 CRUZ, R.L.: 'A calculus for network delay', *IEEE Transactions on Information Theory*, 1991, **37**(1), pp. 114–141

12 THIELE, L., CHAKRABORTY, S., GRIES, M., and KÜNZLI, S.: 'Design space exploration of network processor architectures', in FRANKLIN, M., CROWLEY, P., HADIMIOGLU, H., and ONUFRYK, P. (Eds): 'Network processor design issues and practices, vol. 1' (Morgan Kaufmann, San Francisco, CA, October 2002) ch. 4, pp. 55–90

13 RICHTER, K., RACU, R., and ERNST, R.: 'Scheduling analysis integration for heterogeneous multiprocessor SoC'. Proceedings of the 24th international *Real-time systems* symposium (RTSS'03), Cancun, Mexico, December 2003

14 RICHTER, K., JERSAK, M., and ERNST, R.: 'A formal approach to MpSoC performance verification.' *IEEE Computer*, 2003, **36**(4), pp. 60–77

15 RICHTER, K.: 'Compositional scheduling analysis using standard event models'. PhD thesis, Technical University of Braunschweig, 2004

16 TINDELL, K., KOPETZ, H., WOLF, F., and ERNST, R.: 'Safe automotive software development'. Proceedings of the *Design, automation and test in Europe* (DATE'03), Munich, Germany, March 2003

17 Technical University of Braunschweig. 'SymTA/S – Symbolic Timing Analysis for Systems'. http://www.symta.org

18 JERSAK, M., and ERNST, R.: 'Enabling scheduling analysis of heterogeneous systems with multi-rate data dependencies and rate intervals'. Proceedings of the 40th *Design automation* conference, Anaheim, USA, June 2003

19 JERSAK, M.: 'Compositional performance analysis for complex embedded applications'. PhD thesis, Technical University of Braunschweig, 2004

20 ZIEGENBEIN, D.: 'A Compositional Approach to Embedded System Design'. PhD thesis, Technical University of Braunschweig, 2003

21 LEE, E.A., and MESSERSCHMITT, D.G.: 'Synchronous dataflow', *Proceedings of the IEEE*, 1987, **75**(9), pp. 1235–1245

22 HOURI, M.Y.: 'Task graph analysis with complex dependencies'. Master's thesis, Institute of Computer and Communication Networks Engineering, Technical University of Braunschweig, 2004

23 MOK, A., and CHEN, D.: 'A multiframe model for real-time tasks', *IEEE Transactions on Software Engineering*, 1997, **23**(10), pp. 635–645

24 ZIEGENBEIN, D., RICHTER, K., ERNST, R., THIELE, L., and TEICH, J.: 'SPI – A system model for heterogeneously specified embedded systems', *IEEE*

Transactions on Very Large Scale Integration (VLSI) Systems, 2002, **10**(4), pp. 379–389

25   JERSAK, M., HENIA, R., and ERNST, R.: 'Context-aware performance analysis for efficient embedded system design'. Proceedings of the *Design automation and test in Europe*, Paris, France, March 2004

26   TINDELL, K.W.: 'Adding time-offsets to schedulability analysis'. Technical Report YCS 221, University of York, 1994

27   PALENCIA, J.C., and HARBOUR, M.G.: ' Schedulablilty analysis for tasks with static and dynamic offsets'. Proceedings of the 19th IEEE *Real-time systems* symposium (RTSS98), Madrid, Spain, 1998.

28   HAMANN, A., JERSAK, M., RICHTER, K., and ERNST, R.: 'Design space exploration and system optimization with symTA/S – symbolic timing analysis for systems'. Proceedings of the 25th international *Real-time systems* symposium (RTSS'04), Lisbon, Portugal, December, 2004

29   ZITZLER, E., LAUMANNS, M., and THIELE, L.: 'SPEA2: Improving the strength Pareto evolutionary algorithm'. Technical Report 103, Gloriastrasse 35, CH-8092 Zurich, Switzerland, 2001

30   LAUMANNS, M. , THIELE, L., ZITZLER, E., WELZL, E., and DEB, K.: 'Running time analysis of multi-objective evolutionary algorithms on a simple discrete optimisation problem'. *In* Parallel Problem Solving From Nature — PPSN VII, 2002

31   BLEULER, S., LAUMANNS, M., THIELE, L., and ZITZLER, E.: 'PISA – a platform and programming language independent interface for search algorithms'. http://www.tik.ee.ethz.ch/pisa/

32   WOLPERT, D.H., and MACREADY, W.G.: 'No free lunch theorems for optimisation', *IEEE Transactions on Evolutionary Computation*, 1997, **1**(1), pp. 67–82

33   LEHOCZKY, J., SHA, L., and DING, Y.: 'The rate monotonic scheduling algorithm: exact characterization and average case behavior'. Proceedings of the *Real-time systems* symposium, Santa Monica, CA, USA, 1989, pp. 201–209

34   VESTAL, S.: 'Fixed-priority sensitivity analysis for linear compute time models', *IEEE Transactions on Software Engineering*, 1994, **20**(4), pp. 308–317

35   PUNNEKKAT, S., DAVIS, R., and BURNS, A.: 'Sensitivity analysis of real-time task sets', *ASIAN*, 1997, pp. 72–82

36   AUDSLEY, N.C., BURNS, A., RICHARDSON, M.F., TINDELL, K., and WELLINGS, A.J.: 'Applying new scheduling theory to static priority preemptive scheduling', *Journal of Real-Time Systems*, 1993, **8**(5), pp. 284–292

Chapter 3

Analysis and optimisation of heterogeneous real-time embedded systems

Paul Pop, Petru Eles and Zebo Peng

3.1 Introduction

Embedded real-time systems have to be designed such that they implement correctly the required functionality. In addition, they have to fulfil a wide range of competing constraints: development cost, unit cost, reliability, security, size, performance, power consumption, flexibility, time-to-market, maintainability, correctness, safety, etc. Very important for the correct functioning of such systems are their timing constraints: 'the correctness of the system behaviour depends not only on the logical results of the computations, but also on the physical instant at which these results are produced' [1].

Real-time systems have been classified as 'hard' real-time and 'soft' real-time systems [1]. Basically, hard real-time systems are systems where failing to meet a timing constraint can potentially have catastrophic consequences. For example, a brake-by-wire system in a car failing to react within a given time interval can result in a fatal accident. On the other hand, a multi-media system, which is a soft real-time system, can, under certain circumstances, tolerate a certain amount of delays resulting maybe in a patchier picture, without serious consequences besides some possible inconvenience to the user.

Many real-time applications, following physical, modularity or safety constraints, are implemented using 'distributed architectures'. Such systems are composed of several different types of hardware components, interconnected in a network. For such systems, the communication between the functions implemented on different nodes has an important impact on the overall system properties such as performance, cost, maintainability, etc.

The analysis and optimisation approaches presented are aimed towards heterogeneous distributed hard real-time systems that implement safety-critical applications where timing constraints are of utmost importance to the correct behaviour of the application.

The chapter is organised in ten sections. Section 3.2 presents the heterogeneous real-time embedded systems addressed, and the type of applications considered. Sections 3.3 and 3.4 introduce the current state-of-the-art on the analysis and optimisation of such systems. The rest of the chapter focuses in more detail on some techniques for multi-cluster systems. The hardware and software architecture of multi-clusters, together with the application model, are outlined in Section 3.5. Section 3.6 identifies partitioning and mapping and frame packing as design optimisation problems characteristic to multi-clusters. We present an analysis for multi-cluster systems in Section 3.7, and show, in Section 3.8, how this analysis can be used to drive the optimisation of the packing of application messages to frames. The last two sections present the experimental results of the frame packing optimisation and conclusions.

3.1.1   Automotive electronics

Although the discussion in this chapter is valid for several application areas, it is useful, for understanding the distributed embedded real-time systems evolution and design challenges, to exemplify the developments in a particular area.

If we take the example of automotive manufacturers, they were reluctant, until recently, to use computer controlled functions onboard vehicles. Today, this attitude has changed for several reasons. First, there is a constant market demand for increased vehicle performance, more functionality, less fuel consumption and less exhausts, all of these at lower costs. Then, from the manufacturers' side, there is a need for shorter time-to-market and reduced development and manufacturing costs. These, combined with the advancements of semiconductor technology, which is delivering ever-increasing performance at lower and lower costs, has led to the rapid increase in the number of electronically controlled functions onboard a vehicle [2].

The amount of electronic content in an average car in 1977 had a cost of $110. Currently, the cost is $1341, and it is expected that this figure will reach $1476 by the year 2005, continuing to increase because of the introduction of sophisticated electronics found until now only in high-end cars [3,4]. It is estimated that in 2006 the electronics inside a car will amount to 25 per cent of the total cost of the vehicle (35 per cent for the high-end models), a quarter of which will be due to semiconductors [3,5]. High-end vehicles currently have up to 100 microprocessors implementing and controlling various parts of their functionality. The total market for semiconductors in vehicles is predicted to grow from $8.9 billions in 1998 to $21 billion in 2005, amounting to 10 per cent of the total worldwide semiconductors market [2,3].

At the same time with the increased complexity, the type of functions implemented by embedded automotive electronics systems has also evolved. Thanks to the semiconductors revolution, in the late 1950s, electronic devices became small enough to be installed on board vehicles. In the 1960s the first analogue fuel injection system appeared, and in the 1970s analogue devices for controlling transmission, carburetor, and spark advance timing were developed. The oil crisis of the 1970s led to the demand of engine control devices that improved the efficiency of the engine, thus reducing fuel consumption. In this context, the first microprocessor-based injection control system appeared in 1976 in the United States. During the 1980s, more sophisticated systems began to appear, such as electronically controlled braking systems, dashboards, information and navigation systems, air conditioning systems, etc. In the 1990s, development and improvement have concentrated in the areas such as safety and convenience. Today, it is not uncommon to have highly critical functions like steering or braking implemented through electronic functionality only, without any mechanical backup, as is the case in drive-by-wire and brake-by-wire systems [6,7].

The complexity of electronics in modern vehicles is growing at a very high pace, and the constraints – in terms of functionality, performance, reliability, cost and time-to-market – are getting tighter. Therefore, the task of designing such systems is becoming increasingly important and difficult at the same time. New design techniques are needed, which are able to

- successfully manage the complexity of embedded systems
- meet the constraints imposed by the application domain
- shorten the time-to-market
- reduce development and manufacturing costs

The success of such new design methods depends on the availability of analysis and optimisation techniques, beyond those corresponding to the state-of-the-art, which are presented in the next section.

3.2 Heterogeneous real-time embedded systems

3.2.1 Heterogeneous hardware architecture

Currently, distributed real-time systems are implemented using architectures where each node is dedicated to the implementation of a single function or class of functions. The complete system can be, in general, composed of several networks, interconnected with each other (see Figure 3.1). Each network has its own communication

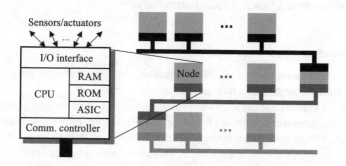

Figure 3.1   Distributed hard real-time systems

protocol, and inter-network communication is via a *gateway* which is a node connected to both networks. The architecture can contain several such networks, having different types of topologies.

A network is composed of several different types of hardware components, called 'nodes'. Typically, every node, also called 'electronic control unit' (ECU), has a communication controller, CPU, RAM, ROM and an I/O interface to sensors and actuators. Nodes can also have ASICs in order to accelerate parts of their functionality.

The microcontrollers used in a node and the type of network protocol employed are influenced by the nature of the functionality and the imposed real-time, fault-tolerance and power constraints. In the automotive electronics area, the functionality is typically divided in two classes, depending on the level of criticalness:

- *Body electronics* refers to the functionality that controls simple devices such as the lights, the mirrors, the windows, the dashboard. The constraints of the body electronic functions are determined by the reaction time of the human operator that is in the range of 100–200 ms. A typical body electronics system within a vehicle consists of a network of 10–20 nodes that are interconnected by a low bandwidth communication network such as LIN. A node is usually implemented using a single-chip 8-bit microcontroller (e.g. Motorol-a 68HC05 or Motorola 68HC11) with some hundred bytes of RAM and kilobytes of ROM, I/O points to connect sensors and to control actuators, and a simple network interface. Moreover, the memory size is growing by more than 25 per cent each year [6,8].

- *System electronics* are concerned with the control of vehicle functions that are related to the movement of the vehicle. Examples of system electronics applications are engine control, braking, suspension, vehicle dynamics control. The timing constraints of system electronic functions are in the range of a couple of milliseconds to 20 ms, requiring 16- or 32-bit microcontrollers (e.g. Motorola 68332) with about 16 kB of RAM and 256 kB of ROM. These microcontrollers have built-in communication controllers (e.g. the 68HC11 and 68HC12 automotive family of microcontrollers have an on-chip CAN controller), I/O to sensors and actuators, and are interconnected by high bandwidth networks [6,8].

Section 3.5 presents more details concerning the hardware and software architecture considered by our analysis and optimisation techniques.

3.2.2   Heterogeneous communication protocols

As the communications become a critical component, new protocols are needed that can cope with the high bandwidth and predictability required.

There are several communication protocols for real-time networks. Among the protocols that have been proposed for vehicle multiplexing, only the Controller Area Network (CAN) [9], the Local Interconnection Network (LIN) [10] and SAE's J1850 [11] are currently in use on a large scale. Moreover, only a few of them are suitable for safety-critical applications where predictability is mandatory [12]. Rushby [12] provides a survey and comparison of communication protocols for safety-critical embedded systems. Communication activities can be triggered either dynamically, in response to an event, or statically, at predetermined moments in time.

- Therefore, on one hand, there are protocols that schedule the messages statically based on the progression of time, for example, the SAFEbus [13] and SPIDER [14] protocols for the avionics industry, and the TTCAN [15] and Time-Triggered Protocol (TTP) [1] intended for the automotive industry.
- On the other hand, there are several communication protocols where message scheduling is performed dynamically, such as CAN used in a large number of application areas including automotive electronics, LonWorks [16] and Profibus [17] for real-time systems in general, etc. Out of these, CAN is the most well known and widespread event-driven communication protocol in the area of distributed embedded real-time systems.
- However, there is also a hybrid type of communication protocols, such as Byte-flight [18] introduced by BMW for automotive applications and the FlexRay protocol [19], that allows the sharing of the bus by event-driven and time-driven messages.

The time-triggered protocols have the advantage of simplicity and predictability, while event-triggered protocols are flexible and have low cost. Moreover, protocols such as TTP offer fault-tolerant services necessary in implementing safety-critical applications. However, it has been shown [20] that event-driven protocols such as CAN are also predictable, and fault-tolerant services can also be offered on top of protocols such as the TTCAN. A hybrid communication protocol such as FlexRay offers some of the advantages of both worlds.

3.2.3   Heterogeneous scheduling policies

The automotive suppliers will select, based on their own requirements, the scheduling policy to be used in their ECU. The main approaches to scheduling are

- *Static cyclic scheduling* algorithms are used to build, off-line, a schedule table with activation times for each process, such that the timing constraints of processes are satisfied.
- *Fixed priority scheduling* (FPS). In this scheduling approach each process has a fixed (static) priority which is computed off-line. The decision on which ready process to activate is taken on-line according to their priority.
- *Earliest deadline first* (EDF). In this case, that process will be activated which has the nearest deadline.

Typically, processes scheduled off-line using static cyclic scheduling are non-pre-emptable, while processes scheduled using techniques such as FPS and EDF are pre-emptable. Another important distinction is between the event- and time-triggered approaches.

- *Time-triggered*. In the time-triggered approach activities are initiated at prede-termined points in time. In a distributed time-triggered system it is assumed that the clocks of all nodes are synchronised to provide a global notion of time. Time-triggered systems are typically implemented using 'non-pre-emptive static cyclic scheduling', where the process activation or message communication is done based on a schedule table built off-line.

- *Event-triggered.* In the event-triggered approach activities happen when a significant change of state occurs. Event-triggered systems are typically implemented using 'pre-emptive fixed-priority-based scheduling', or 'earliest deadline first', where, as response to an event, the appropriate process is invoked to service it.

There has been a long debate in the real-time and embedded systems communities concerning the advantages of each approach [1,21,22]. Several aspects have been considered in favour of one or the other approach, such as flexibility, predictability, jitter control, processor utilisation and testability.

Lönn and Axelsson [23] have performed an interesting comparison of ET and TT approaches from a more industrial, in particular automotive perspective. The conclusion is that one has to choose the right approach, depending on the particularities of the application.

For certain applications, several scheduling approaches can be used together. Efficient implementation of new, highly sophisticated automotive applications, entails the use of time-triggered process sets together with event-triggered ones implemented on top of complex distributed architectures.

3.2.4   Distributed safety-critical applications

Considering the automotive industry, the way functionality has been distributed on an architecture has evolved over time. Initially, distributed real-time systems were implemented using architectures where each node is dedicated to the implementation of a single function or class of functions, allowing the system integrators to purchase nodes implementing required functions from different vendors, and to integrate them into their system [24]. There are several problems related to this restricted mapping of functionality:

- The number of such nodes in the architecture has exploded, reaching, for example, more than 100 in a high-end car, incurring heavy cost and performance penalties.
- The resulting solutions are sub-optimal in many aspects, and do not use the available resources efficiently in order to reduce costs. For example, it is not possible to move a function from one node to another node where there are enough available resources (e.g. memory, computation power).
- Emerging functionality, such as brake-by-wire in the automotive industry, is inherently distributed, and achieving an efficient fault-tolerant implementation is very difficult in the current setting.

This has created a huge pressure to reduce the number of nodes by integrating several functions in one node and, at the same time, to distribute certain functionality over several nodes (see Figure 3.2). Although an application is typically distributed over one single network, we begin to see applications that are distributed across

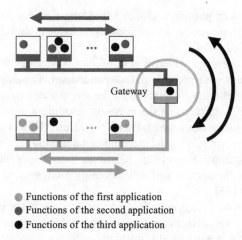

Functions of the first application
Functions of the second application
Functions of the third application

Figure 3.2   Distributed safety-critical applications

several networks. For example, in Figure 3.2, the third application, represented as black dots, is distributed over two networks.

This trend is driven by the need to further reduce costs, improve resource usage, but also by application constraints such as having to be physically close to particular sensors and actuators. Moreover, not only are these applications distributed across networks, but their functions can exchange critical information through the gateway nodes.

3.3   Schedulability analysis

There is a large quantity of research [1,25,26] related to scheduling and schedulability analysis, with results having been incorporated in analysis tools such as TimeWiz [27], RapidRMA [28], RTA-OSEK Planner [29] and Aires [30]. The tools determine if the timing constraints of the functionality are met, and support the designer in exploring several design scenarios, and help to design optimised implementations.

Typically, the timing analysis considers independent processes running on single processors. However, very often functionality consists of distributed processes that have data and control dependencies, exclusion constraints, etc. New schedulability analysis techniques are needed which can handle distributed applications, data and control dependencies, and accurately take into account the details of the communication protocols that have an important influence on the timing properties. Moreover, highly complex and safety-critical applications can in the future be distributed across several networks, and can use different, heterogeneous, scheduling policies.

Pre-emptive scheduling of independent processes with static priorities running on single-processor architectures has its roots in the work of Liu and Layland [31]. The approach has been later extended to accommodate more general computational models and has also been applied to distributed systems [32]. Several surveys on this topic have been published [25,26,33]. Static cyclic scheduling of a set of data dependent software processes on a multiprocessor architecture has also been intensively researched [1,34].

Lee *et al* [35] has proposed an earlier deadline first strategy for non-pre-emptive scheduling of process with possible data dependencies. Pre-emptive and non-pre-emptive static scheduling are combined in the cosynthesis environment proposed by Dave *et al* [36,37]. In many of the previous scheduling approaches researchers have assumed that processes are scheduled independently. However, processes can be sporadic or aperiodic, are seldom independent and normally they exhibit precedence and exclusion constraints. Knowledge regarding these dependencies can be used in order to improve the accuracy of schedulability analyses and the quality of the produced schedules [38].

It has been claimed [22] that static cyclic scheduling is the only approach that can provide efficient solutions to applications that exhibit data dependencies. However, advances in the area of fixed priority pre-emptive scheduling show that such applications can also be handled with other scheduling strategies [39].

One way of dealing with data dependencies between processes in the context of static priority-based scheduling has been indirectly addressed by the extensions proposed for the schedulability analysis of distributed systems through the use of the 'release jitter' [32]. Release jitter is the worst-case delay between the arrival of a process and its release (when it is placed in the ready-queue for the processor) and can include the communication delay due to the transmission of a message on the communication channel.

In References 32 and 40 time 'offset' relationships and 'phases', respectively, are used in order to model data dependencies. Offset and phase are similar concepts that express the existence of a fixed interval in time between the arrivals of sets of processes. The authors show that by introducing such concepts into the computational model, the pessimism of the analysis is significantly reduced when bounding the time behaviour of the system. The concept of 'dynamic offsets' has been later introduced and used to model data dependencies [41].

Currently, more and more real-time systems are used in physically distributed environments and have to be implemented on distributed architectures in order to meet reliability, functional, and performance constraints.

Researchers have often ignored or very much simplified the communication infrastructure. One typical approach is to consider communications as processes with a given execution time (depending on the amount of information exchanged) and to schedule them as any other process, without considering issues such as communication protocol, bus arbitration, packaging of messages, clock synchronisation, etc. [40].

Tindell and Clark [32] integrate processor and communication scheduling and provide a 'holistic' schedulability analysis in the context of distributed real-time systems. The validity of the analysis has been later confirmed in Reference 42.

In the case of a distributed system the response time of a process also depends on the communication delay due to messages. In Reference 32 the analysis for messages is done in a similar way as for processes: a message is seen as a non-pre-emptable process that is 'running' on a bus. The response time analyses for processes and messages are combined by realising that the 'jitter' (the delay between the 'arrival' of a process – the time when it becomes ready for execution – and the start of its execution) of a destination process depends on the 'communication delay' (the time it takes for a message to reach the destination process, from the moment it has been produced by the sender process) between sending and receiving a message. Several researchers have provided analyses that bound the communication delay for a given communication protocol:

- CAN protocol [20];
- time-division multiple access protocol [32];
- asynchronous transfer mode protocol [43];
- token ring protocol [44],
- fiber distributed data interface protocol [45].
- time-triggered protocol [46];
- FlexRay protocol [47].

Based on their own requirements, the suppliers choose one particular scheduling policy to be used. However, for certain applications, several scheduling approaches can be used together.

One approach to the design of such systems, is to allow ET and TT processes to share the same processor as well as static (TT) and dynamic (ET) communications to share the same bus. Bus sharing of TT and ET messages is supported by protocols which support both static and dynamic communication [19]. We have addressed the problem of timing analysis for such systems [47].

A fundamentally different architectural approach to heterogeneous TT/ET systems is that of heterogeneous multi-clusters, where each cluster can be either TT or ET. In a 'time-triggered cluster' processes and messages are scheduled according to a static cyclic policy, with the bus implementing a TDMA protocol, for example, the time-triggered protocol. On 'event-triggered clusters' the processes are scheduled according to a priority-based pre-emptive approach, while messages are transmitted using the priority-based CAN bus. In this context, we have proposed an approach to schedulability analysis for multi-cluster distributed embedded systems [48]. This analysis will be outlined in Section 3.7.

When several event-driven scheduling policies are used in a heterogeneous system, another approach [49] to the verification of timing properties is to couples the analysis of local scheduling strategies via an event interface model.

3.4 Design optimisation

3.4.1 *Traditional design methodology*

There are several methodologies for real-time embedded systems design. The aim of a design methodology is to coordinate the design tasks such that the time-to-market is minimised, the design constraints are satisfied and various parameters are optimised.

The main design tasks that have to be performed are described in the following sections.

3.4.1.1   Functional analysis and design

The functionality of the host system, into which the electronic system is embedded, is normally described using a formalism from that particular domain of application. For example, if the host system is a vehicle, then its functionality is described in terms of control algorithms using differential equations, which are modelling the behaviour of the vehicle and its environment. At the level of the embedded real-time system which controls the host system, the functionality is typically described as a set of functions, accepting certain inputs and producing some output values.

The typical automotive application is a control application. The controller reads inputs from sensors, and uses the actuators to control the physical environment (the vehicle). A controller can have several modes of operation, and can interact with other electronic functions, or with the driver through switches and instruments.

During the functional analysis and design stage, the desired functionality is specified, analysed and decomposed into sub-functions based on the experience of the designer. Several suppliers and manufacturers have started to use tools such as Statemate [50], Matlab/Simulink [51], ASCET/SD [52] and SystemBuild/ MatrixX [53] for describing the functionality, in order to eliminate the ambiguities and to avoid producing incomplete or incoherent specifications.

At the level of functional analysis the exploration is currently limited to evaluating several alternative control algorithms for solving the control problem. Once the functionality has been captured using tools such as Matlab/Simulink, useful explorations can involve simulations of executable specifications in order to determine the correctness of the behaviour, and to assess certain properties of chosen solutions.

3.4.1.2   Architecture selection and mapping

The architecture selection task decides what components to include in the hardware architecture and how these components are connected.

According to current practice, architecture selection is an *ad hoc* process, based on the experience of the designer and previous product versions.

The mapping task has to decide what part of the functionality should be implemented on which of the selected components.

The manufacturers integrate components from suppliers, and thus the design space is severely restricted in current practice, by the fact that the mapping of functionality to an ECU is fixed.

3.4.1.3   Software design and implementation

This is the phase in which the software is designed and the code is written.

The code for the functions is developed manually for efficiency reasons, and thus the exploration that would be allowed by automatic code generation is limited.

At this stage the correctness of the software is analysed through simulations, but there is no analysis of timing constraints, which is left for the scheduling and schedulability analysis stage.

3.4.1.4 Scheduling and schedulability analysis

Once the functions have been defined and the code has been written, the scheduling task is responsible for determining the execution strategy for the functions 'inside an ECU', such that the timing constraints are satisfied.

Simulation is extensively used to determine if the timing constraints are satisfied. However, simulations are very time consuming and provide no guarantees that the timing constraints are met.

In the context of static cyclic scheduling, deriving a schedule table is a complex design exploration problem. Static cyclic scheduling of a set of data-dependent software processes on a multiprocessor architecture has received a lot of attention [1,34]. Such research has been used in commercial tools such as TTP-Plan [54] which derives the static schedules for processes and messages in a time-triggered system using the time-triggered protocol for communication.

If fixed priority pre-emptive scheduling is used, exploration is used to determine how to allocate priorities to a set of distributed processes [55]. Their priority assignment heuristic is based on the schedulability analysis from Reference 32. For earliest deadline first the issue of distributing the global deadlines to local deadlines has to be addressed [56].

3.4.1.5 Integration

In this phase the manufacturer has to integrate the ECUs from different suppliers.

There is a lack of tools that can analyse the performance of the interacting functionality, and thus the manufacturer has to rely on simulation runs using the realistic environment of a prototype car. Detecting potential problems at such a late stage requires time-consuming extensive simulations. Moreover, once a problem is identified it takes a very long time to go through all the previous stages in order to fix it. This leads to large delays on the time-to-market.

In order to reduce the large simulation times, and to guarantee that potential violations of timing constraints are detected, manufacturers have started to use in-house analysis tools and commercially available tools such as Volcano Network Architect (for the CAN and LIN buses) [57].

Volcano makes inter-ECU communication transparent for the programmer. The programmer only deals with 'signals' that have to be sent and received, and the details of the network are hidden. Volcano provides basic API calls for manipulating signals. To achieve interoperability between ECUs from different suppliers, Volcano uses a 'publish/subscribe' model for defining the signal requirements. Published signals are made available to the system integrator by the suppliers, while subscribed signals are required as inputs to the ECU. The system integrator makes the publish/subscribe connections by creating a set of CAN frames, and creating a mapping between the data in frames and signals [58]. Volcano uses the analysis by Tindell *et al* [20] for bounding the communication delay of messages transmitted using the CAN bus.

3.4.1.6   Calibration, testing, verification

These are the final stages of the design process. Because not enough analysis, testing and verification has been done in earlier stages of the design, these stages tend to be very time consuming, and problems identified here lead to large delays in product delivery.

3.4.2   *Function architecture co-design and platform-based design*

New design methodologies are needed, which can handle the increasing complexity of heterogeneous systems, and their competing requirements in terms of performance, reliability, low power consumption, cost, time-to-market, etc. As the complexity of the systems continues to increase, the development time lengthens dramatically, and the manufacturing costs become prohibitively high. To cope with this complexity, it is necessary to reuse as much as possible at all levels of the design process, and to work at higher and higher abstraction levels.

'Function/architecture co-design' is a design methodology proposed in References 59 and 60, which addresses the design process at higher abstraction levels. Function/architecture co-design uses a top-down synthesis approach, where trade-offs are evaluated at a high level of abstraction. The main characteristic of this methodology is the use, at the same time with the top-down synthesis, of a bottom-up evaluation of design alternatives, without the need to perform a full synthesis of the design. The approach to obtaining accurate evaluations is to use an accurate modelling of the behaviour and architecture, and to develop analysis techniques that are able to derive estimates and to formally verify properties relative to a certain design alternative. The determined estimates and properties, together with user-specified constraints, are then used to drive the synthesis process.

Thus, several architectures are evaluated to determine if they are suited for the specified system functionality. There are two extremes in the degrees of freedom available for choosing an architecture. At one end, the architecture is already given, and no modifications are possible. At the other end of the spectrum, no constraints are imposed on the architecture selection, and the synthesis task has to determine, from scratch, the best architecture for the required functionality. These two situations are, however, not common in practice. Often, a 'hardware platform' is available, which can be 'parameterised' (e.g. size of memory, speed of the buses, etc.). In this case, the synthesis task is to derive the parameters of the platform architecture such that the functionality of the system is successfully implemented. Once an architecture is determined and/or parameterised, the function/architecture co-design continues with the mapping of functionality onto the instantiated architecture.

This methodology has been used in research tools such as Polis [61] and Metropolis [62], and has also led to commercial tools such as the Virtual Component Co-design (VCC) [63].

In order to reduce costs, especially in the case of a mass market product, the system architecture is usually reused, with some modifications, for several product lines. Such a common architecture is denoted by the term 'platform', and consequently the design tasks related to such an approach are grouped under the term 'platform-based design' [64]. The platform consists of a hardware infrastructure together with software components that will be used for several product versions, and will be shared with other product lines, in the hope to reduce costs and the time-to-market.

Keutzer *et al* [64] have proposed techniques for deriving such a platform for a given family of applications. Their approach can be used within any design methodology for determining a system platform that later on can be parameterised and instantiated to a desired system architecture.

Considering a given application or family of applications, the system platform has to be instantiated, deciding on certain parameters, and lower level details, in order to suit that particular application(s). The search for an architecture instance starts from a certain platform, and a given application. The application is mapped and compiled on an architecture instance, and the performance numbers are derived, typically using simulation. If the designer is not satisfied with the performance of the instantiated architecture, the process is repeated.

In the remainder of the chapter we will consider a platform consisting of event- and time-triggered clusters, using the CAN and TTP protocols for communication, respectively. We will discuss analysis and optimisation techniques for the configuration of the platform such that the given application is schedulable.

3.5 Multi-cluster systems

One class of heterogeneous real-time embedded systems is that of 'multi-cluster' systems. We consider architectures consisting of two clusters, one time-triggered and the other event-triggered, interconnected by gateways (see Figure 3.2):

- In a 'time-triggered cluster' (TTC) processes and messages are scheduled according to a static cyclic policy, with the bus implementing a TDMA protocol such as, e.g. the time-triggered protocol (TTP) [65].
- On 'event-triggered clusters' (ETC) the processes are scheduled according to a priority-based pre-emptive approach, while messages are transmitted using the priority-based CAN bus [9].

The next two sections present the hardware and software architecture of a two-cluster system, while Section 3.5.3 presents the application model used. Section 3.6 will introduce design problems characteristic for multi-cluster systems composed of time-triggered clusters interconnected with event-triggered clusters: the partitioning of functionality between the TT and ET clusters, the mapping of functionality to the nodes inside a cluster and the packing of application message to frames on the TTP and CAN buses. Then, Section 3.8 will present two optimisation strategies for the frame packing problem.

3.5.1   Hardware architecture

A 'cluster' is composed of nodes which share a broadcast communication channel. Let $\mathcal{N}_T$ ($\mathcal{N}_E$) be the set of nodes on the TTC (ETC). Every 'node' $\mathcal{N}_i \in \mathcal{N}_T \cup \mathcal{N}_E$ includes a communication controller and a CPU, along with other components. The gateways, connected to both types of clusters, have two communication controllers, for TTP and CAN. The communication controllers implement the protocol services, and run independently of the node's CPU. Communication with the CPU is performed through a 'Message Base Interface' (MBI); see Figure 3.5.

Communication between the nodes on a TTC is based on the TTP [65]. The TTP integrates all the services necessary for fault-tolerant real-time systems. The bus access scheme is time-division multiple access (TDMA), meaning that each node N_i on the TTC, including the gateway node, can transmit only during a predetermined time interval, the TDMA slot S_i. In such a slot, a node can send several messages packed in a frame. A sequence of slots corresponding to all the nodes in the architecture is called a TDMA round. A node can have only one slot in a TDMA round. Several TDMA rounds can be combined together in a cycle that is repeated periodically. The sequence and length of the slots are the same for all the TDMA rounds. However, the length and contents of the frames may differ.

The TDMA access scheme is imposed by a message descriptor list (MEDL) that is located in every TTP controller. The MEDL serves as a schedule table for the TTP controller which has to know when to send/receive a frame to/from the communication channel.

There are two types of frames in the TTP. The initialisation frames, or I-frames, which are needed for the initialisation of a node, and the normal frames, or N-frames, which are the data frames containing, in their data field, the application messages. A TTP data frame (Figure 3.3) consists of the following fields: start of frame bit (SOF), control field, a data field of up to 16 bytes containing one or more messages,

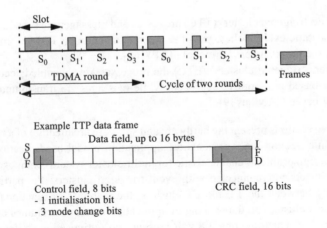

Figure 3.3   Time-triggered protocol

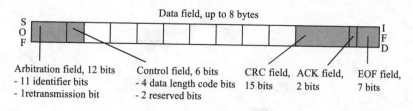

Figure 3.4   Controller area network data frame (CAN 2.0A)

and a cyclic redundancy check (CRC) field. Frames are delimited by the inter-frame delimiter (IDF, 3 bits).

For example, the data efficiency of a frame that carries 8 bytes of application data, i.e. the percentage of transmitted bits which are the actual data bits needed by the application, is 69.5 per cent (64 data bits transmitted in a 92-bit frame, without considering the details of a particular physical layer). Note that no identifier bits are necessary, as the TTP controllers know from their MEDL what frame to expect at a given point in time. In general, the protocol efficiency is in the range of 60–80 per cent [66].

On an ETC, the CAN [9] protocol is used for communication. The CAN bus is a priority bus that employs a collision avoidance mechanism, whereby the node that transmits the frame with the highest priority wins the contention. Frame priorities are unique and are encoded in the frame identifiers, which are the first bits to be transmitted on the bus.

In the case of CAN 2.0A, there are four frame types: data frame, remote frame, error frame and overload frame. We are interested in the composition of the data frame, depicted in Figure 3.4. A data frame contains seven fields: SOF, arbitration field that encodes the 11 bits frame identifier, a control field, a data field up to 8 bytes, a CRC field, an acknowledgement (ACK) field and an end of frame field (EOF).

In this case, for a frame that carries 8 bytes of application data, we will have an efficiency of 47.4 per cent [67]. The typical CAN protocol efficiency is in the range of 25–35 per cent [66].

3.5.2   Software architecture

A real-time kernel is responsible for activation of processes and transmission of messages on each node. On a TTC, the processes are activated based on the local schedule tables, and messages are transmitted according to the MEDL. On an ETC, we have a scheduler that decides on activation of ready processes and transmission of messages, based on their priorities.

In Figure 3.5 we illustrate our message passing mechanism. Here we concentrate on the communication between processes located on different clusters. We have previously presented the message passing within a TTC [68], and the infrastructure needed for communications in an ETC [20].

Let us consider the example in Figure 3.5, where we have an application consisting of four processes and four messages (depicted in Figure 3.5(b)) mapped on the two clusters in Figure 3.5(c). Processes P_1 and P_4 are mapped on node N_1 of the TTC,

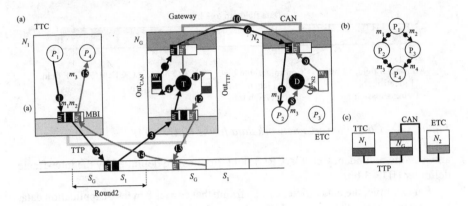

Figure 3.5   A message passing example

while P_2 and P_3 are mapped on node N_2 of the ETC. Process P_1 sends messages m_1 and m_2 to processes P_2 and P_3, respectively, while P_2 and P_3 send messages m_3 and m_4 to P_4. All messages have a size of one byte.

The transmission of messages from the TTC to the ETC takes place in the following way (see Figure 3.5). P_1, which is statically scheduled, is activated according to the schedule table, and when it finishes it calls the send kernel function in order to send m_1 and m_2, indicated in the figure by the number (1). Messages m_1 and m_2 have to be sent from node N_1 to node N_2. At a certain time, known from the schedule table, the kernel transfers m_1 and m_2 to the TTP controller by packing them into a frame in the MBI. Later on, the TTP controller knows from its MEDL when it has to take the frame from the MBI, in order to broadcast it on the bus. In our example, the timing information in the schedule table of the kernel and the MEDL is determined in such a way that the broadcasting of the frame is done in the slot S_1 of round 2 (2). The TTP controller of the gateway node N_G knows from its MEDL that it has to read a frame from slot S_1 of round 2 and to transfer it into its MBI (3). Invoked periodically, having the highest priority on node N_G, and with a period which guarantees that no messages are lost, the gateway process T copies messages m_1 and m_2 from the MBI to the TTP-to-CAN priority-ordered message queue Out$_{CAN}$ (4). Let us assume that on the ETC messages m_1 and m_2 are sent independently, one per frame. The highest priority frame in the queue, in our case the frame f_1 containing m_1, will tentatively be broadcast on the CAN bus (5). Whenever f_1 will be the highest priority frame on the CAN bus, it will successfully be broadcast and will be received by the interested nodes, in our case node N_2 (6). The CAN communication controller of node N_2 receiving f_1 will copy it in the transfer buffer between the controller and the CPU, and raise an interrupt which will activate a delivery process, responsible to activate the corresponding receiving process, in our case P_2, and hand over message m_1 that finally arrives at the destination (7).

Message m_3 (depicted in Figure 3.5 as a grey rectangle labelled 'm_3') sent by process P_2 from the ETC will be transmitted to process P_4 on the TTC. The transmission starts when P_2 calls its send function and enqueues m_3 in the priority-ordered

Out_{N_2} queue (8). When the frame f_3 containing m_3 has the highest priority on the bus, it will be removed from the queue (9) and broadcast on the CAN bus (10). Several messages can be packed into a frame in order to increase the efficiency of data transmission. For example, m_3 can wait in the queue until m_4 is produced by P_3, in order to be packed together with m_4 in a frame. When f_3 arrives at the gateway's CAN controller it raises an interrupt. Based on this interrupt, the gateway transfer process T is activated, and m_3 is unpacked from f_3 and placed in the Out_{TTP} FIFO queue (11). The gateway node N_G is only able to broadcast on the TTC in the slot S_G of the TDMA rounds circulating on the TTP bus. According to the MEDL of the gateway, a set of messages not exceeding $size_{S_G}$ of the data field of the frame travelling in slot S_G will be removed from the front of the Out_{TTP} queue in every round, and packed in the S_G slot (12). Once the frame is broadcast (13) it will arrive at node N_1 (14), where all the messages in the frame will be copied in the input buffers of the destination processes (15). Process P_4 is activated according to the schedule table, which has to be constructed such that it accounts for the worst-case communication delay of message m_3, bounded by the analysis in Section 3.7.1, and, thus, when P_4 starts executing it will find m_3 in its input buffer.

As part of our frame packing approach, we generate all the MEDLs on the TTC (i.e. the TT frames and the sequence of the TDMA slots), as well as the ET frames and their priorities on the ETC such that the global system is schedulable.

3.5.3 Application model

The functionality of the host system, into which the electronic system is embedded, is normally described using a formalism from that particular domain of application. For example, if the host system is a vehicle, then its functionality is described in terms of control algorithms using differential equations, which are modelling the behaviour of the vehicle and its environment. At the level of the embedded system which controls the host system, viewed as the system level for us, the functionality is typically described as a set of functions, accepting certain inputs and producing some output values.

There is a lot of research in the area of system modelling and specification, and an impressive number of representations have been proposed. Edward, [69] presents an overview, classification and comparison of different design representations and modelling approaches.

The scheduling and mapping design tasks deal with sets of interacting processes. A 'process' is a sequence of computations (corresponding to several building blocks in a programming language) which starts when all its inputs are available. When it finishes executing, the process produces its output values. Researchers have used, for example, 'dataflow process networks' (also called 'task graphs', or 'process graphs') [70] to describe interacting processes, and have represented them using directed acyclic graphs, where a node is a process and the directed arcs are dependencies between processes.

Thus, we model an application Γ as a set of process graphs $G_i \in \Gamma$ (see Figure 3.6). Nodes in the graph represent processes and arcs represent dependency between the

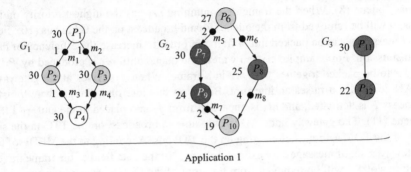

Application 1

Figure 3.6   Application model

connected processes. A 'process' is a sequence of computations (corresponding to several building blocks in a programming language) which starts when all its inputs are available. When it finishes executing, the process produces its output values. Processes can be pre-emptable or non-pre-emptable. 'Non-pre-emptable' processes are processes that cannot be interrupted during their execution, and are mapped on the TTC. 'Pre-emptable' processes can be can be interrupted during their execution, and are mapped on the ETC. For example, a higher priority process has to be activated to service an event, in this case, the lower priority process will be temporarily pre-empted until the higher priority process finishes its execution.

A process graph is polar, which means that there are two nodes, called source and sink, that conventionally represent the first and last process. If needed, these nodes are introduced as dummy processes so that all other nodes in the graph are successors of the source and predecessors of the sink, respectively.

The communication time between processes mapped on the same processor is considered to be part of the process worst-case execution time and is not modelled explicitly. Communication between processes mapped to different processors is performed by message passing over the buses and, if needed, through the gateway. Such message passing is modelled as a communication process inserted on the arc connecting the sender and the receiver process (the black dots in Figure 3.6).

Potential communication between processes in different applications is not part of the model. Technically, such a communication is implemented by the kernels based on asynchronous non-blocking send and receive primitives. Such messages are considered non-critical and are not affected by real-time constraints. Therefore, communications of this nature will not be addressed.

Each process P_i is mapped on a processor $\mathcal{M}(P_i)$ (mapping represented by hashing in Figure 3.6), and has a worst-case execution time C_i on that processor (depicted to the left of each node). The designer can provide manually such worst-case times, or tools can be used in order to determine the worst-case execution time of a piece of code on a given processor [71].

For each message we know its size (in bytes, indicated to its left), and its period, which is identical with that of the sender process. Processes and messages activated based on events also have a uniquely assigned priority, $priority_{Pi}$ for processes and $priority_{mi}$ for messages.

All processes and messages belonging to a process graph G_i have the same period $T_i = T_{Gi}$ which is the period of the process graph. A deadline D_{Gi} is imposed on each process graph G_i. Deadlines can also be placed locally on processes. Release times of some processes as well as multiple deadlines can be easily modelled by inserting dummy nodes between certain processes and the source or the sink node, respectively. These dummy nodes represent processes with a certain execution time but which are not allocated to any processing element.

3.6 Multi-cluster optimisation

Considering the types of applications and systems described in the previous section, and using the analysis outlined in Section 3.7, several design optimisation problems can be addressed.

In this section, we present problems which are characteristic to applications distributed across multi-cluster systems consisting of heterogeneous TT and ET networks:

- Section 3.6.1 briefly outlines the problem of partitioning the processes of an application into time- and event-triggered domains, and their mapping to the nodes of the clusters.
- Section 3.6.2 presents the problem of packing of messages to frames, which is of utmost importance in cost-sensitive embedded systems where resources, such as communication bandwidth, have to be fully utilised [58,72,73]. This problem will be discussed in more detail in Section 3.8.

The goal of these optimisation problems is to produce an implementation which meets all the timing constraints (i.e. the application is schedulable).

In order to drive our optimisation algorithms towards schedulable solutions, we characterise a given frame packing configuration using the degree of schedulability of the application. The 'degree of schedulability' [74] is calculated as:

$$\delta_\Gamma = \begin{cases} c_1 = \sum_{i=1}^{n} \max(0, r_i - D_i), & \text{if } c_1 > 0 \\ c_2 = \sum_{i=1}^{n} (r_i - D_i), & \text{if } c_1 = 0 \end{cases} \qquad (3.1)$$

where n is the number of processes in the application, r_i is the worst-case response time of a process P_i and D_i its deadline. The worst-case response times are calculated by the MultiClusterScheduling algorithm using the response time analysis presented in Section 3.7.

If the application is not schedulable, the term c_1 will be positive, and, in this case, the cost function is equal to c_1. However, if the process set is schedulable, $c_1 = 0$ and we use c_2 as a cost function, as it is able to differentiate between two alternatives, both leading to a schedulable process set. For a given set of optimisation parameters leading to a schedulable process set, a smaller c_2 means that we have improved the worst-case response times of the processes, so the application can potentially be implemented on a cheaper hardware architecture (with slower processors and/or buses). Improving the degree of schedulability can also lead to an improvement in the quality of control for control applications.

3.6.1   Partitioning and mapping

By partitioning, we denote the decision whether a certain process should be assigned to the TT or the ET domain (and, implicitly, to a TTC or an ETC, respectively). Mapping a process means assigning it to a particular node inside a cluster.

Very often, the partitioning decision is taken based on the experience and preferences of the designer, considering aspects such as the functionality implemented by the process, the hardness of the constraints, sensitivity to jitter, legacy constraints, etc. Let $\mathcal{P}$ be the set of processes in the application Γ. We denote with $\mathcal{P}_T \subseteq \mathcal{P}$ the subset of processes which the designer has assigned to the TT cluster, while $\mathcal{P}_E \subseteq \mathcal{P}$ contains processes which are assigned to the ET cluster.

Many processes, however, do not exhibit certain particular features or requirements which obviously lead to their implementation as TT or ET activities. The subset $\mathcal{P}^+ = \mathcal{P} \setminus (\mathcal{P}_T \cup \mathcal{P}_E)$ of processes could be assigned to any of the TT or ET domains. Decisions concerning the partitioning of this set of activities can lead to various trade-offs concerning, for example, the schedulability properties of the system, the amount of communication exchanged through the gateway, the size of the schedule tables, etc.

For part of the partitioned processes, the designer might have already decided their mapping. For example, certain processes, due to constraints such as having to be close to sensors/actuators, have to be physically located in a particular hardware unit. They represent the sets $\mathcal{P}_T^M \subseteq \mathcal{P}_T$ and $\mathcal{P}_E^M \subseteq \mathcal{P}_E$ of already mapped TT and ET processes, respectively. Consequently, we denote with $\mathcal{P}_T^* = \mathcal{P}_T \setminus \mathcal{P}_T^M$ the TT processes for which the mapping has not yet been decided, and similarly, with $\mathcal{P}_E^* = \mathcal{P}_E \setminus \mathcal{P}_E^M$ the unmapped ET processes. The set $\mathcal{P}^* = \mathcal{P}_T^* \cup \mathcal{P}_E^* \cup \mathcal{P}^+$ then represents all the unmapped processes in the application.

The mapping of messages is decided implicitly by the mapping of processes. Thus, a message exchanged between two processes on the TTC (ETC) will be mapped on the TTP bus (CAN bus) if these processes are allocated to different nodes. If the communication takes place between two clusters, two message instances will be created, one mapped on the TTP bus and one on the CAN bus. The first message is sent from the sender node to the gateway, while the second message is sent from the gateway to the receiving node.

Let us illustrate some of the issues related to partitioning in such a context. In the example presented in Figure 3.7 we have an application[1] with six processes, P_1 to P_6, and four nodes, N_1 and N_2 on the TTC, N_3 on the ETC and the gateway node N_G. The worst-case execution times on each node are given to the right of the application

---

[1] Communications are ignored for this example.

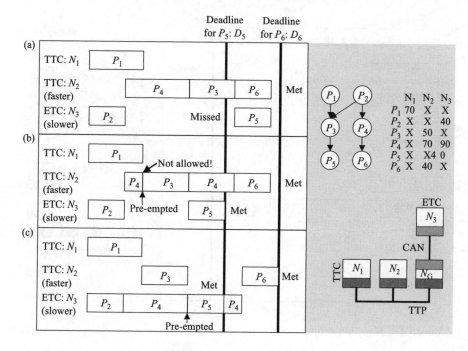

Figure 3.7   Partitioning example

graph. Note that N_2 is faster than N_3, and an 'X' in the table means that the process is not allowed to be mapped on that node. The mapping of P_1 is fixed on N_1, P_3 and P_6 are mapped on N_2, P_2 and P_5 are fixed on N_3, and we have to decide how to partition P_4 between the TT and ET domains. Let us also assume that process P_5 is the highest priority process on N_3. In addition, P_5 and P_6 have each a deadline, D_5 and D_6, respectively, as illustrated in the figure by thick vertical lines.

We can observe that although P_3 and P_4 do not have individual deadlines, their mapping and scheduling has a strong impact on their successors, P_5 and P_6, respectively, which are deadline constrained. Thus, we would like to map P_4 such that not only P_3 can start on time, but P_4 also starts soon enough to allow P_6 to meet its deadline.

As we can see from Figure 3.7(a), this is impossible to achieve by mapping P_4 on the TTC node N_2. It is interesting to observe that, if pre-emption would be allowed in the TT domain, as in Figure 3.7(b), both deadlines could be met. This, however, is impossible on the TTC where pre-emption is not allowed. Both deadlines can be met only if P_4 is mapped on the slower ETC node N_3, as depicted in Figure 3.7(c). In this case, although P_4 competes for the processor with P_5, due to the pre-emption of P_4 by the higher priority P_5, all deadlines are satisfied.

For a multi-cluster architecture the communication infrastructure has an important impact on the design and, in particular, the mapping decisions. Let us consider the example in Figure 3.8. We assume that P_1 is mapped on node N_1 and P_3 on node N_3

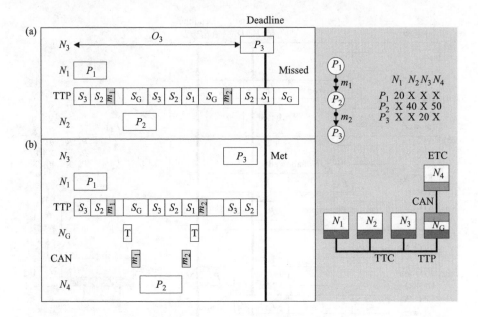

Figure 3.8   Mapping example

on the TTC, and we are interested to map process P_2. P_2 is allowed to be mapped on the TTC node N_2 or on the ETC node N_4, and its execution times are depicted in the table to the right of the application graph.

In order to meet the deadline, one would map P_2 on the node it executes fastest, N_2 on the TTC, see Figure 3.8(a). However, this will lead to a deadline miss due to the TTP slot configuration which introduces communication delays. The application will meet the deadline only if P_2 is mapped on the slower node, i.e. node N_4 in the case in Figure 3.8(b).[2] Not only is N_4 slower than N_2, but mapping P_2 on N_4 will place P_2 on a different cluster than P_1 and P_3, introducing extra communication delays through the gateway node. However, due to the actual communication configuration, the mapping alternative in Figure 3.8(b) is desirable.

Using the notation introduced, the partitioning and mapping problem can be described more exactly as follows. As an input we have an application Γ given as a set of process graphs and a two-cluster system consisting of a TT and an ET cluster. As introduced previously, $\mathcal{P}_T$ and $\mathcal{P}_E$ are the sets of processes already partitioned into TT and ET, respectively. Also, $\mathcal{P}_T^M \subseteq \mathcal{P}_T$ and $\mathcal{P}_E^M \subseteq \mathcal{P}_E$ are the sets of already mapped TT and ET processes. We are interested to find a partitioning for processes in $\mathcal{P}^+ = \mathcal{P} \backslash (\mathcal{P}_T \cup \mathcal{P}_E)$ and decide a mapping for processes in $\mathcal{P}^* = \mathcal{P}_T^* \cup \mathcal{P}_E^* \cup \mathcal{P}^+$, where $\mathcal{P}_T^* = \mathcal{P}_T \backslash \mathcal{P}_T^M$, and $\mathcal{P}_E^* = \mathcal{P}_E \backslash \mathcal{P}_E^M$ such that imposed

---

[2] Process T in Figure 3.8(b) executing on the gateway node N_G is responsible for transferring messages between the TTP and CAN controllers.

deadlines are guaranteed to be satisfied. We have highlighted a possible solution to this problem [75].

3.6.2 Frame packing

In both the TTP and CAN protocols messages are not sent independently, but several messages having similar timing properties are usually packed into frames. In many application areas, such as automotive electronics, messages range from one single bit (e.g. the state of a device) to a couple of bytes (e.g. vehicle speed, etc.). Transmitting such small messages one per frame would create a high communication overhead, which can cause long delays leading to an unschedulable system. For example, 65 bits have to be transmitted on CAN for delivering one single bit of application data. Moreover, a given frame configuration defines the exact behaviour of a node on the network, which is very important when integrating nodes from different suppliers.

Let us consider the motivational example in Figure 3.9, where we have the process graph from Figure 3.9(d) mapped on the two-cluster system from Figure 3.9(e): P_1 and P_4 are mapped on node N_1 from the TTC, while P_2 and P_3 are mapped on N_2 from ETC. The data field of the frames is represented with a black rectangle, while the other frame fields are depicted with a grey colour. We consider a physical implementation of the buses such that the frames will take the time indicated in the figure by the length of their rectangles. We are interested to find a frame configuration such that the application is schedulable.

In the system configuration of Figure 3.9(a) we consider that, on the TTP bus, the node N_1 transmits in the first slot (S_1) of the TDMA round, while the gateway transmits in the second slot (S_G). Process P_3 has a higher priority than process P_2, hence P_2 will be interrupted by P_3 when it receives message m_2. In such a setting, P_4 will miss its deadline, which is depicted as a thick vertical line in Figure 3.9. Changing the frame configuration as in Figure 3.9(b), so that messages m_1 and m_2 are packed into frame f_1 and slot S_G of the gateway comes first, processes P_2 and P_3 will receive m_1 and m_2 sooner and thus reduce the worst-case response time of the process graph, which is still larger than the deadline. In Figure 3.9(c), we also pack m_3 and m_4 into f_2. In such a situation, the sending of m_3 will have to be delayed until m_4 is queued by P_2. Nevertheless, the worst-case response time of the application is further reduced, which means that the deadline is met, thus the system is schedulable.

However, packing more messages will not necessarily reduce the worst-case response times further, as it might increase too much the worst-case response times of messages that have to wait for the frame to be assembled, this is the case with message m_3 in Figure 3.9(c).

This design optimisation problem can be formulated more exactly as follows. As input to the frame-packing problem we have an application Γ given as a set of process graphs mapped on an architecture consisting of a TTC and an ETC interconnected through a gateway. We consider that the partitioning and mapping of processes has been already decided.

We are interested to find a mapping of messages to frames (a frame packing configuration) denoted by a 4-tuple $\psi = \langle \alpha, \pi, \beta, \sigma \rangle$ such that the application Γ is

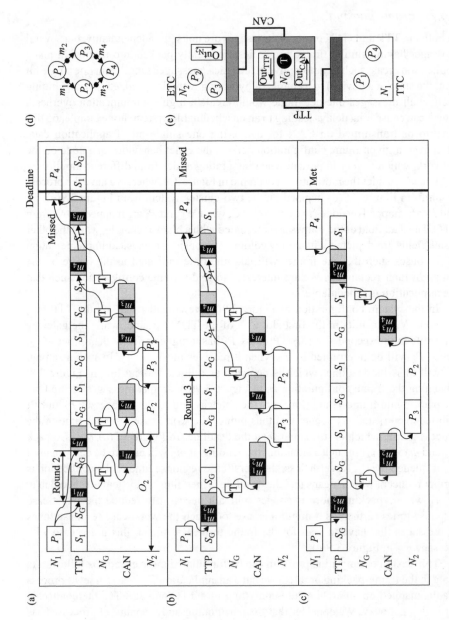

Figure 3.9 Frame packing optimisation example

schedulable. Once a schedulable system is found, we are interested to further improve the 'degree of schedulability' so the application can potentially be implemented on a cheaper hardware architecture (with slower buses and processors).

Determining a frame configuration ψ means deciding on:

- The mapping of application messages transmitted on the ETC to frames (the set of ETC frames α), and their relative priorities, π. Note that the ETC frames α have to include messages transmitted from an ETC node to a TTC node, messages transmitted inside the ETC cluster, and those messages transmitted from the TTC to the ETC.
- The mapping of messages transmitted on the TTC to frames, denoted by the set of TTC frames β, and the sequence σ of slots in a TDMA round. The slot sizes are determined based on the set β, and are calculated such that they can accommodate the largest frame sent in that particular slot. We consider that messages transmitted from the ETC to the TTC are not statically allocated to frames. Rather, we will dynamically pack messages originating from the ETC into the 'gateway frame', for which we have to decide the data field length (see Section 3.5.2).

Several details related to the schedulability analysis were omitted from the discussion of the example. These details will be discussed in the next section.

3.7 Multi-cluster analysis and scheduling

Once a partitioning and a mapping is decided, and a frame packing configuration is fixed, the processes and messages have to be scheduled. For the TTC this means building the schedule tables, while for the ETC the priorities of the ET processes have to be determined and their schedulability has to be analysed.

The analysis presented in this section works under the following assumptions:

- All the processes belonging to a process graph G have the same period T_G. However, process graphs can have different periods.
- The offsets are *static* (as opposed to *dynamic* [42]), and are smaller than the period.
- The deadlines are arbitrary, i.e. can be larger than the period.

The basic idea is that on the TTC an application is schedulable if it is possible to build a schedule table such that the timing requirements are satisfied.

On the ETC, the answer whether or not a system is schedulable is given by a 'schedulability analysis'. Thus, for the ETC we use a 'response time analysis', where the schedulability test consists of the comparison between the worst-case response time r_i of a process P_i and its deadline D_i. Response time analysis of data dependent processes with static priority pre-emptive scheduling has been proposed in [39,40,42], and has been also extended to consider the CAN protocol [20]. The authors use the concept of 'offset' in order to handle data dependencies. Thus, each process P_i is characterised by an offset O_i, measured from the start of the process graph, that indicates the earliest possible start time of P_i. Such an offset is, for example, O_2 in

Figure 3.9(a), as process P_2 cannot start before receiving m_1. The same is true for messages, their offset indicating the earliest possible transmission time. The response time analysis employed is presented in Section 3.7.1.

However, determining the schedulability of an application mapped on a multi-cluster system cannot be addressed separately for each type of cluster, since the inter-cluster communication creates a circular dependency: the static schedules determined for the TTC influence through their offsets the worst-case response times of the processes on the ETC, which in turn influence the schedule table construction on the TTC. In Figure 3.9(b) packing m_1 and m_2 in the same frame leads to equal offsets for P_2 and P_3. Because of this, P_3 will delay P_2 (which would not be the case if m_2 sent to P_3 would be scheduled in round 3, e.g.) and thus the placement of P_4 in the schedule table has to be accordingly delayed to guarantee the arrivals of m_3 and m_4.

In our analysis we consider the influence between the two clusters by making the following observations:

- The start time of process P_i in a schedule table on the TTC is its offset O_i.
- The worst-case response time r_i of a TT process is its worst-case execution time, i.e. $r_i = C_i$ (TT processes are not pre-emptable).
- The worst-case response times of the messages exchanged between two clusters have to be calculated according to the schedulability analysis described in Section 3.7.1.
- The offsets have to be set by a scheduling algorithm such that the precedence relationships are preserved. This means that, if process P_B depends on process P_A, the following condition must hold: $O_B \geq O_A + r_A$. Note that for the processes on a TTC which receive messages from the ETC this translates to setting the start times of the processes such that a process is not activated before the worst-case arrival time of the message from the ETC. In general, offsets on the TTC are set such that all the necessary messages are present at the process invocation.

The MultiClusterScheduling algorithm in Figure 3.10 receives as input the application Γ, the frame configuration ψ, and produces the offsets ϕ and worst-case response times ρ.

The algorithm sets initially all the offsets to 0 (line 1). Then, the worst-case response times are calculated using the ResponseTimeAnalysis function (line 4) using the analysis presented in Section 3.7.1. The fixed-point iterations that calculate the response times at line 3 will converge if processor and bus loads are smaller than 100 per cent [39]. Based on these worst-case response times, we determine new values ϕ^{new} for the offsets using a list scheduling algorithm (line 6). We now have a schedule table for the TTC and worst-case response times for the ETC, which are pessimistic. The following loop will reduce the pessimism of the worst-case response times.

The multi-cluster scheduling algorithm loops until the degree of schedulability δ_Γ of the application Γ cannot be further reduced (lines 8–20). In each loop iteration, we select a new offset from the set of ϕ^{new} offsets (line 10), and run the response time analysis (line 11) to see if the degree of schedulability has improved (line 12). If δ_Γ has not improved, we continue with the next offset in ϕ^{new}.

MultiClusterScheduling(Γ, $\mathcal{M}$, Ψ)
-- determines the set of offsets ϕ and worst-case response times ρ
1   **for each** $O_i \in \phi$ **do** $O_i = 0$ **end for** -- initially all offsets are zero
2   -- determine initial values for the worst-case response times
3   -- according to the analysis in Section 3.7.1
4   ρ = ResponseTimeAnalysis(Γ, $\mathcal{M}$, Ψ, ϕ)
5   -- determine new values for the offsets, based on the response times ρ
6   ϕ^{new} = ListScheduling(Γ, $\mathcal{M}$, Ψ, ρ)
7    $\theta_\Gamma = \infty$ -- consider the system unschedulable at first
8   **repeat** -- iteratively improve the degree of schedulability δ_Γ
9   **for each** $O_i^{new} \in \phi^{new}$ **do** -- for each newly calculated offset
10       $O_i^{old} = \phi.O_i$; $\phi.O_i = \phi^{new} O_i^{new}$ -- set the new offset, remember old
11   ρ^{new} = ResponseTimeAnalysis(Γ, $\mathcal{M}$, Ψ, Φ)
12   δ_Γ^{new} = SchedulabilityDegree(Γ, ρ)
13   **if** $\delta_\Gamma^{new} < \delta_\Gamma$ **then** -- the schedulability has improved
14   -- offsets are recalculated using ρ^{new}
15   ϕ^{new} = ListScheduling(Γ, $\mathcal{M}$, Ψ, ρ^{new})
16   **break** -- exit the for-each loop
17   **else** -- the schedulability has not improved
18         $\phi.O_i = O_i^{old}$ -- restore the old offset
19   **end for**
20   **until** θ_Γ has not changed
21   **return** ρ, ϕ, δ_Γ
end MultiClusterScheduling

Figure 3.10   *The* MulticlusterScheduling *algorithm*

When a new offset O_i^{new} leads to an improved δ_Γ we exit the for-each loop 9–19 that examines offsets from ϕ^{new}. The loop iteration 8–20 continues with a new set of offsets, determined by ListScheduling at line 15, based on the worst-case response times ρ^{new} corresponding to the previously accepted offset.

In the multi-cluster scheduling algorithm, the calculation of offsets is performed by the list scheduling algorithm presented in Figure 3.11. In each iteration, the algorithm visits the processes and messages in the ReadyList. A process or a message in the application is placed in the ReadyList if all its predecessors have been already scheduled. The list is ordered based on the priorities [76]. The algorithm terminates when all processes and messages have been visited.

In each loop iteration, the algorithm calculates the earliest time moment (*offset*) when the process or message $node_i$ can start (lines 5–7). There are four situations:

1   The visited node is an ET message. The message m_i is packed into its frame f (line 9), and the offset O_f of the frame is updated. The frame can only be transmitted after all the sender processes that pack messages in this frame have finished executing. The offset of message m_i packed to frame f is equal to the frame offset O_f.

ListScheduling(Γ, $\mathcal{M}$, Ψ, ρ) - - determines the set of offsets ϕ
1   *ReadyList* = source nodes of all process graphs in the application
2   **while** *ReadyList* $\neq \oslash$ **do**
3   *node$_i$* = Head(*ReadyList*)
4   offset = 0 - - determine the earliest time when an activity can start
5   **for each** direct predecessor *node$_j$* of *node$_i$* **do**
6   *offset* = max(*offset*, $O_j + r_j$)
7   **end for**
8   **if** *node$_i$* is a message m_i **then**
9   PackFrame(m_i, f) - - pack each ready message m into its frame f
10   O_f = max(O_f, *offset*) - - update the frame offset
11   **if** f is complete **then** - - the frame is complete for transmission
12   **if** $f \in \propto$ **then** - - f is an ET frame
13   - - the offset of messages is equal to the frame offset
14   **for each** $m_j \in f$ **do** $O_j = O_f$ **end for**
15   **else** - - f is a TT frame
16   <round, slot> = ScheduleTTFrame(f, *offset*, Ψ)
17   - - set the TT message offsets based on the round and slot
18   **for each** $m_j \in f$ **do** O_j = round* $T_T DMA + O_{slot}$ **end for**
19   **endif; endif**
20   **else** - - *node$_i$* is a process P_i
21   **if** $\mathcal{M}(P_i) \in \mathcal{N}_E$ **then** - - if process P_i is mapped on the ETC
22   O_i = offset – the ETC process can start immediately
23   **else** - - process P_i is mapped on the TTC
24   - - P_i has to wait also for the processor $\mathcal{M}(P_i)$ to become available
25   O_i = max(offset, ProcessorAvailable($\mathcal{M}(P_i)$))
26   **end if; end if;**
27   Update(ReadyList)
28   **end while**
29   **return** offsets
end ListScheduling

Figure 3.11   ListScheduling *algorithm*

2   The node is a TT message. In this case, when the frame is ready for transmission, it is scheduled using the ScheduleTTFrame function (presented in Figure 3.12), which returns the *round* and the *slot* where the frame has been placed (line 16 in Figure 3.11). In Figure 3.12, the round immediately following *offset* is the initial candidate to be considered (line 2). However, it can be too late to catch the allocated slot, in which case the next round is considered (line 4). For this candidate round, we have to check if the slot is not occupied by another frame. If so, the communication has to be delayed for another round (line 7). Once a frame has been scheduled, we can determine the offsets and worst-case response times

ScheduleTTFrame (f, *offset*, ψ)

-- returns the slot and the round assigned to frame f

1   *slot* = the slot assigned to the node sending f -- the frame slot

2   *round* = *offset* / T_{TDMA} -- the first round which could be a candidate

3   **if** *offset* – *round* * T_{TDMA} > O_{slot} **then** -- the *slot* is missed

4   *round* = *round* + 1 -- if yes, take the next round

5   **end if**

6   **while** *slot* is occupied **do**

7   *round* = *round* + 1

8   **end while**

9   **return** *round*, slot

end ScheduleTTFrame

Figure 3.12   Frame scheduling on the TTC

(Figure 3.11, line 18). For all the messages in the frame the offset is equal to the start of the slot in the TDMA round, and the worst-case response time is the slot length.

3   The algorithm visits a process P_i mapped on an ETC node. A process on the ETC can start as soon as its predecessors have finished and its inputs have arrived, hence $O_i = \textit{offset}$ (line 22). However, P_i might experience, later on, interference from higher priority processes.

4   Process P_i is mapped on a TTC node. In this case, besides waiting for the predecessors to finish executing, P_i will also have to wait for its processor $\mathcal{M}(P_i)$ to become available (line 25). The earliest time when the processor is available is returned by the ProcessorAvailable function.

Let us now turn the attention back to the multi-cluster scheduling algorithm in Figure 3.10. The algorithm stops when the δ_Γ of the application Γ is no longer improved, or when a limit imposed on the number of iterations has been reached. Since in a loop iteration we do not accept a solution with a larger δ_Γ, the algorithm will terminate when in a loop iteration we are no longer able to improve δ_Γ by modifying the offsets.

3.7.1   Schedulability analysis for the ETC

For the ETC we use a response time analysis. A 'response time analysis' has two steps. In the first step, the analysis derives the worst-case response time of each process (the time it takes from the moment is ready for execution, until it has finished executing). The second step compares the worst-case response time of each process to its deadline and, if the response times are smaller or equal to the deadlines, the system is schedulable. The analysis presented in this section is used in the ResponseTimeAnalysis function (line 4 of the algorithm in Figure 3.10).

Thus, the response time analysis [77] uses the following equation for determining the worst-case response time r_i of a process P_i:

$$r_i = C_i + \sum_{\forall P_j \in hp(P_i)} \left\lceil \frac{r_i}{T_j} \right\rceil C_j \qquad (3.2)$$

where C_i is the worst-case execution time of process P_i, T_j is the period of process P_j and $hp(P_i)$ denotes the set of processes that have a priority higher than the priority of P_i.

The summation term, representing the interference I_i of higher priority processes on P_i, increases monotonically in r_i, thus solutions can be found using a recurrence relation. Moreover, the recurrence relations that calculate the worst-case response time are guaranteed to converge if the processor utilisation is under 100 per cent.

The previously presented analysis assumes that the deadline of a process is smaller or equal to its period. This assumption has later been relaxed [32] to consider 'arbitrary deadlines' (i.e. deadlines can be larger than the period). Thus, the worst-case response time r_i of a process P_i becomes:

$$r_i = \max_{q=0,\, 1,\, 2\ldots} (J_i + w_i(q) - qT_i) \qquad (3.3)$$

where J_i is the jitter of process P_i (the worst-case delay between the arrival of a process and the start of its execution), q is the number of busy periods being examined and $w_i(q)$ is the width of the level-i busy period starting at time qT_i. The level-i busy period is defined as the maximum time a processor executes processes of priority greater than or equal to the priority of process P_i, and is calculated as [32]:

$$w_i(q) = (q+1)C_i + B_i + \sum_{\forall P_j \in hp(P_i)} \left\lceil \frac{w_i(q) + J_j}{T_j} \right\rceil C_j \qquad (3.4)$$

The pessimism of the previous analysis can be reduced by using the information related to the precedence relations between processes. The basic idea is to exclude certain worst-case scenarios, from the critical instant analysis, which are impossible due to precedence constraints.

Methods for schedulability analysis of data dependent processes with static priority pre-emptive scheduling have been proposed [39,40,41,42]. They use the concept of 'offset' (or 'phase'), in order to handle data dependencies. Tindell [39] shows that the pessimism of the analysis is reduced through the introduction of offsets. The offsets have to be determined by the designer.

In their analysis [39], the response time of a process P_i is:

$$r_i = \max_{q=0,\, 1,\, 2\ldots} \left(\max_{\forall P_j \in G} \left(w_i(q) + O_j + J_j - T_G \right.\right.$$
$$\left.\left. \left(q + \left\lceil \frac{O_j + J_j - O_i - J_i}{T_G} \right\rceil \right) - O_i \right) \right) \qquad (3.5)$$

where T_G the period of the process graph G, O_i and O_j are offsets of processes P_i and P_j, respectively, and J_i and J_j are the release jitters of P_i and P_j. In Equation (3.5), the level-i busy period starting at time qT_G is

$$w_i(q) = (q + 1)C - i + B_i + I_i \qquad (3.6)$$

In the previous equation, the blocking term B_i represents interference from lower priority processes that are in their critical section and cannot be interrupted, and C_i represents the worst-case execution time of process P_i. The last term captures the interference I_i from higher priority processes in the application, including higher priority processes from other process graphs. Tindell [39] presents the details of the interference calculation.

Although this analysis is exact (both necessary and sufficient), it is computationally infeasible to evaluate. Hence, Tindell [39] proposes a feasible but not exact analysis (sufficient but not necessary) for solving Equation (3.5). Our implementations use the feasible analysis provided in Tindell [39] for deriving the worst-case response time of a process P_i.

We are now interested to determine the worst-case response time of frames and the worst-case queuing delays experienced by a frame in a communication controller.

Regarding the worst-case response time of messages, we have extended the CAN analysis from messages [20] and applied it in the contest of frames on the CAN bus:

$$r_f = \max_{q=0, 1, 2\ldots} (J_f + W_f(q) + (1 + q)C_f) \qquad (3.7)$$

In the previous equation J_f is the jitter of frame f which in the worst case is equal to the largest worst-case response time $r_{S(m)}$ of a sender process $S(m)$ which sends message m packed into frame f:

$$J_f = \max_{\forall m \in f} (r_{S_m}) \qquad (3.8)$$

In Equation (3.7), W_f is the 'worst-case queuing delay' experienced by f at the communication controller, and is calculated as:

$$W_f(q) = w_f(q) - qT_f \qquad (3.9)$$

where q is the number of busy periods being examined, and $w_f(q)$ is the width of the level-f busy period starting at time qT_f.

Moreover, in Equation (3.7), C_f is the worst-case time it takes for a frame f to reach the destination controller. On CAN, C_f depends on the frame configuration and the size of the data field, s_f, while on TTP it is equal to the slot size in which f is transmitted.

The worst-case response time of message m packed into a frame f can be determined by observing that $r_m = r_f$.

The worst-case queueing delay for a frame (W_f in Equation (3.7)) is calculated differently for each type of queue:

1   The output queue of an ETC node, in which case $W_f^{N_i}$ represents the worst-case time a frame f has to spend in the Out_{N_i} queue on ETC node N_i. An example of such a frame is the one containing message m_3 in Figure 3.9(a), which is sent by

process P_2 from the ETC node N_2 to the gateway node N_G, and has to wait in the Out_{N2} queue.

2 The TTP-to-CAN queue of the gateway node, in which case W_f^{CAN} is the worst-case time a frame f has to spend in the Out_{CAN} queue of node N_G. In Figure 3.9(a), the frame containing m_1 is sent from the TTC node N_1 to the ETC node N_2, and has to wait in the Out_{CAN} queue of gateway node N_G before it is transmitted on the CAN bus.

3 The CAN-to-TTP queue of the gateway node, where W_f^{TTP} captures the time f has to spend in the Out_{TTP} queue node N_G. Such a situation is present in Figure 3.9(a), where the frame with m_3 is sent from the ETC node N_2 to the TTP node N_1 through the gateway node N_G where it has to wait in the Out_{TTP} queue before it is transmitted on the TTP bus, in the S_G slot of node N_G.

On the TTC, the synchronisation between processes and the TDMA bus configuration is solved through the proper synthesis of schedule tables, hence no output queues are needed. The frames sent from a TTC node to another TTC node are taken into account when determining the offsets, and are not involved directly in the ETC analysis.

The next sections show how the worst queueing delays are calculated for each of the previous three cases.

3.7.1.1   Worst-case queuing delays in the Out_{N_i} and Out_{CAN} queues

The analyses for $W_f^{N_i}$ and W_f^{CAN} are similar. Once f is the highest priority frame in the Out_{CAN} queue, it will be sent by the gateway's CAN controller as a regular CAN frame, therefore the same equation for w_f can be used:

$$w_f(q) = B_f + \sum_{\forall f_j \in hp(f)} \left\lceil \frac{w_f(q) + J_j}{T_j} \right\rceil C_j \qquad (3.10)$$

The intuition is that f has to wait, in the worst case, first for the largest lower priority frame that is just being transmitted (B_f) as well as for the higher priority $f_j \in hp(f)$ frames that have to be transmitted ahead of f (the second term). In the worst case, the time it takes for the largest lower priority frame $f_k \in p(f)$ to be transmitted to its destination is:

$$B_f = \max_{\forall f k \in lp(f)} (C_k) \qquad (3.11)$$

Note that in our case, $lp(f)$ and $hp(f)$ also include messages produced by the gateway node, transferred from the TTC to the ETC.

3.7.1.2   Worst-case queuing delay in the Out_{TTP} queue

The time a frame f has to spend in the Out_{TTP} queue in the worst case depends on the total size of messages queued ahead of f (Out_{TTP} is a FIFO queue), the size S_G of the data field of the frame fitting into the gateway slot responsible for carrying the CAN messages on the TTP bus, and the period T_{TDMA} with which this slot S_G is

MultiClusterConfiguration(Γ)
1   - - determine an initial partitioning and mapping $\mathcal{M}$,
2   - - and an initial frame configuration ψ^0
3    $<\mathcal{M}, \psi^0> = $ PartitioningAndMapping(Γ)
4   - - the frame packing optimization algorithm
5    $\psi = $ FramePackingOptimization($\Gamma, \mathcal{M}, \psi^0$)
6   - - test if the resulted configuration leads to a schedulable application
7   **if** MultiClusterScheduling($\Gamma, \mathcal{M}, \psi$) returns schedulable **then**
8   **return** $\mathcal{M}, \psi$
9   **else**
10   **return** unschedulable
11   **endif**
end MultiClusterConfiguration

Figure 3.13   The general frame packing strategy

circulating on the bus [46]:

$$w_f^{\text{TTP}}(q) = B_f + \left\lfloor \frac{(q+1)s_f + I_f(w_f(q))}{S_G} \right\rfloor T_{\text{TDMA}} \tag{3.12}$$

where I_f is the total size of the frames queued ahead of f. Those frames $f_j \in hp(f)$ are ahead of f, which have been sent from the ETC to the TTC, and have higher priority than f:

$$I_f(w) = \sum_{\forall f_j \in hp(f)} \left\lceil \frac{w_f + J_j}{T_j} \right\rceil s_j \tag{3.13}$$

where the frame jitter J_j is given by Equation (3.8).

The blocking term B_f is the time interval in which f cannot be transmitted because the slot S_G of the TDMA round has not arrived yet. In the worst case (i.e. the frame f has just missed the slot S_G), the frame has to wait an entire round T_{TDMA} for the slot S_G in the next TDMA round.

3.8   Frame-packing optimisation strategy

The general multi-cluster optimisation strategy is outlined in Figure 3.13. The MultiClusterConfiguration strategy has two steps:

1   In the first step, line 3, the application is partitioned on the TTC and ETC clusters, and processes are mapped to the nodes of the architecture using the PartitioningAndMapping function. The partitioning and mapping can be done with an optimisation heuristic [75]. As part of the partitioning and mapping process, an initial frame configuration $\psi^0 = \langle \alpha^0, \pi^0, \beta^0 \sigma^0 \rangle$ is derived. Messages exchanged by processes partitioned to the TTC will be mapped to TTC frames,

while messages exchanged on the ETC will be mapped to ETC frames. For each message sent from a TTC process to an ETC process, we create an additional message on the ETC, and we map this message to an ETC frame. The sequence σ^0 of slots for the TTC is decided by assigning in order nodes to the slots ($S_i = N_i$). One message is assigned per frame in the initial set β^0 of TTC frames. For the ETC, the frames in the set α^0 initially hold each one single message, and we calculate the message priorities Π^0 based on the deadlines of the receiver processes.

2   The frame packing optimisation, is performed as the second step (line 5 in Figure 3.13). The FramePackingOptimization function receives as input the application Γ, the mapping $\mathcal{M}$ of processes to resources and the initial frame configuration ψ^0, and it produces as output the optimised frame packing configuration ψ. Such an optimisation problem is NP complete [78], thus obtaining the optimal solution is not feasible. We present two frame packing optimisation strategies, one based on a simulated annealing approach, presented in Section 3.8.1, while the other, outlined in Section 3.8.2, is based on a greedy heuristic that uses intelligently the problem-specific knowledge in order to explore the design space.

If after these steps the application is unschedulable, we conclude that no satisfactory implementation could be found with the available amount of resources.

Testing if the application Γ is schedulable is done using the MultiClusterScheduling (MCS) algorithm (line 7 in Figure 3.13). The multi-cluster scheduling algorithm, presented in Figure 3.10, takes as input an application Γ, a mapping $\mathcal{M}$ and an initial frame configuration ψ^0, builds the TT schedule tables, sets the ET priorities for processes, and provides the global analysis.

3.8.1   Frame packing with simulated annealing

The first algorithm we have developed is based on a simulated annealing (SA) strategy [78], and is presented in Figure 3.14. The algorithm takes as input the application Γ, a mapping $\mathcal{M}$ and an initial frame configuration ψ^0, and determines the frame configuration ψ which leads to the best degree of schedulability δ_Γ (the smaller the value, the more schedulable the system, see Section 3.6).

Determining a frame configuration ψ means finding the set of ETC frames α and their relative priorities π, and the set of TTC frames B, including the sequence σ of slots in a TDMA round.

The main feature of a SA strategy is that it tries to escape from a local optimum by randomly selecting a new solution from the neighbours of the current solution. The new solution is accepted if it is an improved solution (lines 9–10 of the algorithm in Figure 3.14). However, a worse solution can also be accepted with a certain probability that depends on the deterioration of the cost function and on a control parameter called temperature (lines 12–13).

In Figure 3.14 we give a short description of this algorithm. An essential component of the algorithm is the generation of a new solution ψ_{new} starting from the current one ψ_{current}. The neighbours of the current solution ψ_{current} are obtained by

SimulatedAnnealing(Γ, $\mathcal{M}$, ψ^0)
1   -- given an application Γ finds out if it is schedulable and produces
2   -- the configuration $\langle \Psi, \pi, \beta, \sigma \rangle$ leading to the smallest δ_Γ
3   -- initial frame configuration
4    $\psi_{current} = \psi^0$
5   *temperature* = initial temperature *TI*
6   **repeat**
7   **for** $i = 1$ to temperature length *TL* **do**
8   generate randomly a neighboring solution ψ_{new} of $\psi_{current}$
9   δ = MultiClusterScheduling(Γ, $\mathcal{M}$, ψ_{new}) -
                MultiClusterScheduling(Γ, $\mathcal{M}$, $\psi_{current}$)
10   **if** $\delta < 0$ **then** $\psi_{current} = \psi_{new}$
11   **else**
12   generate q = Random (0, 1)
13   **if** $q < e^{-\delta/temperature}$ **then** $\psi_{current} = \psi_{new}$ **end if**
14   **end if**
15   **end for**
16   *temperature* = ε * temperature
17   **until** stopping criterion is met
18   **return** SchedulabilityTest(Γ, $\mathcal{M}$, ψ_{best}), solution ψ best
     corresponding to the best degree of schedulablity δ_Γ
end SimulatedAnnealing

Figure 3.14   The SimulatedAnnealing *algorithm*

performing transformations (called moves) on the current frame configuration $\psi_{current}$ (line 8). We consider the following moves:

- moving a message m from a frame f_1 to another frame f_2 (or moving m into a separate single-message frame);
- swapping the priorities of two frames in α;
- swapping two slots in the sequence σ of slots in a TDMA round.

For the implementation of this algorithm, the parameters *TI* (initial temperature), *TL* (temperature length), ε (cooling ratio) and the stopping criterion have to be determined. They define the 'cooling schedule' and have a decisive impact on the quality of the solutions and the CPU time consumed. We are interested to obtain values for *TI*, *TL* and ε that will guarantee the finding of good quality solutions in a short time.

We performed long runs of up to 48 h with the SA algorithm, for ten synthetic process graphs (two for each graph dimension of 80, 160, 240 320, 400, see Section 3.9) and the best ever solution produced has been considered as the optimum. Based on further experiments we have determined the parameters of the SA algorithm so that the optimisation time is reduced as much as possible but the near-optimal result is still produced. For example, for the graphs with 320 nodes, *TI* is 700, *TL* is 500 and

ε is 0.98. The algorithm stops if for three consecutive temperatures no new solution has been accepted.

3.8.2   Frame packing greedy heuristic

The OptimizeFramePacking greedy heuristic (Figure 3.15) constructs the solution by progressively selecting the best candidate in terms of the degree of schedulability.

We start by observing that all activities taking place in a multi-cluster system are ordered in time using the offset information, determined in the StaticScheduling function based on the worst-case response times known so far and the application structure (i.e. the dependencies in the process graph). Thus, our greedy heuristic outlined in Figure 3.15, starts with building two lists of messages ordered according to the ascending value of their offsets, one for the TTC, $messages_\beta$, and one for ETC, $messages_\alpha$. Our heuristic is to consider for packing in the same frame messages which are adjacent in the ordered lists. For example, let us consider that we have three messages, m_1 of 1 byte, m_2 of 2 bytes and m_3 of 3 bytes, and that messages are ordered as m_3, m_1, m_2 based on the offset information. Also, assume that our heuristic has suggested two frames, frame f_1 with a data field of 4 bytes, and f_2 with a data field of 2 bytes. The PackMessages function will start with m_3 and pack it in frame f_1. It continues with m_2, which is also packed into f_1, since there is space left for it. Finally, m_3 is packed in f_2, since there is no space left for it in f_1.

The algorithm tries to determine, using the for-each loops in Figure 3.15, the best frame configuration. The algorithm starts from the initial frame configuration ψ^0, and progressively determines the best change to the current configuration. The quality of a frame configuration is measured using the MultiClusterScheduling algorithm, which calculates the degree of schedulability δ_Γ (line 13). Once a configuration parameter has been fixed in the outer loops it is used by the inner loops:

- Lines 10–15: The innermost loops determine the best size S_α for the currently investigated frame f_α in the ETC frame configuration $\alpha_{current}$. Thus, several frame sizes are tried (line 11), each with a size returned by RecomendedSizes to see if it improves the current configuration. The RecomendedSizes($messages_\alpha$) list is built recognising that only messages adjacent in the $messages_\alpha$ list will be packed into the same frame. Sizes of frames are determined as a sum resulted from adding the sizes of combinations of adjacent messages, not exceeding 8 bytes. For the previous example, with m_1, m_2 and m_3, of 1, 2 and 3 bytes, respectively, the frame sizes recommended will be of 1, 2, 3, 4, and 6 bytes. A size of 5 bytes will not be recommended since there are no adjacent messages that can be summed together to obtain 5 bytes of data.
- Lines 9–16: This loop determines the best frame configuration α. This means deciding on how many frames to include in α (line 9), and which are the best sizes for them. In α there can be any number of frames, from one single frame to n_α frames (in which case each frame carries one single message). Once a configuration α_{best} or the ETC, minimising δ_Γ, has been determined (saved in line 16), the algorithm looks for the frame configuration β which will further improve δ_Γ.

OptimizeFramePacking$(\Gamma, \mathcal{M}, \psi^0)$ - - produces the frame configuration ψ leading to the smallest degree of schedulability δ_Γ

1 π^0 = HOPA - - the initial priorities π^0 are updated using the HOPA heuristic

2 - - build the message lists ordered ascending on their offsets

3 $messages_\beta$ = ordered list of n_β messages on the TTC; $messages_\alpha$ = ordered list of n_α messages on the ETC

4 **for each** $slot_i \in \sigma_{current}$ **do for each** $slot_j \in \sigma_{current} \wedge slot_i \neq slot_j$ **do** - - determine the best TTP slot sequence σ

5   Swap$(slot_i; slot_j)$ - - tentatively swap slots $slot_i$ with $slot_j$

6   **for each** $\beta_{current}$ with 1 to n_β frames **do** - - determine the best frame packing configuration β for the TTC

7   **for each** frame $f_\beta \in \beta_{current}$ **do for each** frame size $S_\beta \in$ RecomendedSizes$(message_\beta)$

8   **do** - - determine the best frame size for f_β

9          $\beta_{current}.f_\beta.S = S_\beta$

10   **for each** $\alpha_{current}$ with 1 to n_α frames **do** - - determine the best frame packing configuration α for the ETC

11   **for each** frame $f_\alpha \in \alpha_{current}$ **do for each** frame size $S_\alpha \in$ RecomendedSizes$(messages_\alpha)$

12   **do** - - determine the best frame size for f_α

13            $\alpha_{current}.f_\alpha.S = S_\alpha$

14            $\psi_{current} = < \alpha_{current}, \pi^0, \beta_{current}, \sigma_{current} >$; PackMessages$(\psi_{current}, messages_\beta \cup messages_\alpha)$

15   δ_Γ =MultiClusterScheduling$(\Gamma, \mathcal{M}, \psi_{current})$

16   **if** δ_Γ $(\psi_{current})$is best so far **then** $\psi_{best} = \psi_{current}$ **end if** - - remember the best configuration so far

17   **end for; end for; if** $\exists \psi_{best}$ **then** $\alpha_{current}.f_\alpha.S = \alpha_{best}.f_\alpha.S$ **end if** - - remember the best frame size for f_α

18   **end for; if** $\exists \psi_{best}$ **then** $\alpha_{current} = \alpha_{best}$ **end if** - - remember the best frame packing configuration α

19   **end for; end for; if** $\exists \psi_{best}$ **then** $\beta_{current}.f_\beta.S = \beta_{best}.f_\beta.S$ **end if** - - remember the best frame size for f_β

        end for; if $\exists \psi_{best}$ **then** β_{best} **end if** - - remember the best frame packing configuration β

      end for; if $\exists \psi_{best}$ **then**$\sigma_{current}.slot_i = \sigma_{current}.slot$ **end if;** - - remember the best slot sequence σ; **end for**

20 **return** SchedulabilityTest$(\Gamma, \mathcal{M}, \psi_{best})$, ψ_{best}

end OptimizeFramePacking

Figure 3.15 The OptimizeFramePacking algorithm

- Lines 7–17: The best size for a frame f_β is determined similarly to the size for a frame f_α.
- Lines 6–18: The best frame configuration β_{best} is determined. For each frame configuration β tried, the algorithm loops again through the innermost loops to see if there are better frame configurations α in the context of the current frame configuration $\beta_{current}$.
- Lines 4–19: After a β_{best} has been decided, the algorithm looks for a slot sequence σ starting with the first slot and tries to find the node which, when transmitting in this slot, will reduce δ_Γ. Different slot sequences are tried by swapping two slots within the TDMA round (line 5).

For the initial message priorities π^0 (initially, there is one message per frame) we use the 'heuristic optimised priority assignment' (HOPA) approach [55], where priorities in a distributed real-time system are determined, using knowledge of the factors that influence the timing behaviour, such that the degree of schedulability of the system is improved (line 1). The ETC message priorities set at the beginning of the algorithm are not changed by our greedy optimisation loops. The priority of a frame $f_\alpha \in \alpha$ is given by the message $m \in f_\alpha$ with the highest priority.

The algorithm continues in this fashion, recording the best ever ψ_{best} configurations obtained, in terms of δ_Γ, and thus the best solution ever is reported when the algorithm finishes.

3.9   Experimental results

For the evaluation of our frame-packing optimisation algorithms we first used process graphs generated for experimental purpose. We considered two-cluster architectures consisting of 2, 4, 6, 8 and 10 nodes, half on the TTC and the other half on the ETC, interconnected by a gateway. Forty processes were assigned to each node, resulting in applications of 80, 160, 240, 320 and 400 processes.

We generated both graphs with random structure and graphs based on more regular structures such as trees and groups of chains. We generated a random structure graph deciding for each pair of two processes if they should be connected or not. Two processes in the graph were connected with a certain probability (between 0.05 and 0.15, depending on the graph dimension) on the condition that the dependency would not introduce a loop in the graph. The width of the tree-like structures was controlled by the maximum number of direct successors a process can have in the tree (from 2 to 6), while the graphs consisting of groups of chains had 2 to 12 parallel chains of processes. Furthermore, the regular structures were modified by adding a number of 3 to 30 random cross-connections.

The mapping of the applications to the architecture has been done using a simple heuristic that tries to balance the utilisation of processors while minimising communication. Execution times and message lengths were assigned randomly using both uniform and exponential distribution within the 10–100 ms and 1–2 bytes ranges, respectively. For the communication channels we considered a transmission speed of 256 kbps and a length below 20 meters. All experiments were run on a SUN Ultra 10.

Table 3.1   Evaluation of the frame-packing optimisation algorithms

No. of processes	Straightforward solution (SP)			OptimizeFramePacking (OFP)			SimulatedAnnealing (SA)
	average (%)	max (%)	time (s)	average (%)	max (%)	time (s)	time (s)
80	2.42	17.89	0.09	0.40	1.59	4.35	235.95
160	16.10	42.28	0.22	2.28	8.32	12.09	732.40
240	40.49	126.4	0.54	6.59	21.80	49.62	2928.53
320	70.79	153.08	0.74	13.70	30.51	172.82	7585.34
400	97.37	244.31	0.95	31.62	95.42	248.30	22099.68

The first result concerns the ability of our heuristics to produce schedulable solutions. We have compared the degree of schedulability δ_Γ obtained from our OptimizeFramePacking (OFP) heuristic (Figure 3.15) with the near-optimal values obtained by SA (Figure 3.14). Obtaining solutions that have a better degree of schedulability means obtaining tighter worst-case response times, increasing the chances of meeting the deadlines.

Table 3.1 presents the average percentage deviation of the degree of schedulability produced by OFP from the near-optimal values obtained with SA. Together with OFP, a straightforward approach (SF) is presented. The SF approach does not consider frame packing, and thus each message is transmitted independently in a frame. Moreover, for SF we considered a TTC bus configuration consisting of a straightforward ascending order of allocation of the nodes to the TDMA slots; the slot lengths were selected to accommodate the largest message frame sent by the respective node, and the scheduling has been performed by the MultiClusterScheduling algorithm in Figure 3.10.

In Table 3.1 we have one row for each application dimension of 80–400 processes, and a header for each optimisation algorithm considered. For each of the SF and OFP algorithms we have three columns in the table. In the first column, we present the average percentage deviation of the algorithm from the results obtained by SA. The percentage deviation is calculated according to the formula:

$$\text{deviation} = \frac{\theta_\Gamma^{\text{approach}} - \theta_\Gamma^{\text{SA}}}{\theta_\Gamma^{\text{SA}}} 100 \tag{3.14}$$

The second column presents the maximum percentage deviation from the SA result, and the third column presents the average execution time of the algorithm, in seconds. For the SA algorithm we present only its average execution times.

Table 3.1 shows that when packing messages to frames, the degree of schedulability improves dramatically compared to the straightforward approach. The greedy heuristic OptimizeFramePacking performs well for all the graph dimensions, having, e.g., run-times which are on average under 50 for applications with 240 processes.

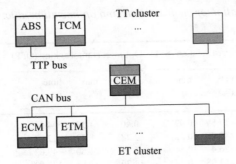

Figure 3.16   Hardware architecture for the cruise controller

When deciding on which heuristic to use for design space exploration or system synthesis, an important issue is the execution time. On average, our optimisation heuristics needed a couple of minutes to produce results, while the simulated annealing approach had an execution time of up to 6 h.

3.9.1   The vehicle cruise controller

A typical safety-critical application with hard real-time constraints, is a vehicle cruise controller (CC). We have considered a CC system derived from a requirement specification provided by the industry. The CC delivers the following functionality: it maintains a constant speed for speeds over 35 km/h and under 200 km/h, offers an interface (buttons) to increase or decrease the reference speed and is able to resume its operation at the previous reference speed. The CC operation is suspended when the driver presses the brake pedal.

The specification assumes that the CC will operate in an environment consisting of two clusters. There are four nodes which functionally interact with the CC system: the Anti-lock Braking System (ABS), the Transmission Control Module (TCM), the Engine Control Module (ECM) and the Electronic Throttle Module (ETM) (see Figure 3.16).

It has been decided to map the functionality (processes) of the CC over these four nodes. The ECM and ETM nodes have an 8-bit Motorola M68HC11 family CPU with 128 kbytes of memory, while the ABS and TCM are equipped with a 16-bit Motorola M68HC12 CPU and 256 kbytes of memory. The 16-bit CPUs are twice as fast than the 8-bit ones. The transmission speed of the communication channel is 256 kbps and the frequency of the TTP controller was chosen to be 20 MHz.

We have modelled the specification of the CC system using a set of 32 processes and 17 [72] where the mapping of processes to the nodes is also given. The period was chosen 250 ms, equal to the deadline.

In this setting, the straightforward approach SF produced an end-to-end worst-case response time of 320 ms, greater than the deadline, while both the OFP and SA heuristics produced a schedulable system with a worst-case response time of 172 ms.

This shows that the optimisation heuristic proposed, driven by our schedulability analysis, is able to identify that frame packing configuration which increases the schedulability degree of an application, allowing the developers to reduce the implementation cost of a system.

3.10 Conclusions

Heterogeneous distributed real-time systems are used in several application areas to implement increasingly complex applications that have tight timing constraints. The heterogeneity is manifested not only at the hardware and communication protocol levels, but also at the level of the scheduling policies used. In order to reduce costs and use the available resources more efficiently, the applications are distributed across several networks.

We have introduced the current state-of-the-art analysis and optimisation techniques available for such systems, and addressed in more detail a special class of heterogeneous distributed real-time embedded systems called multi-cluster systems.

We have presented an analysis for multi-cluster systems and outlined several characteristic design problems, related to the partitioning and mapping of functionality and the optimisation of the access to the communication infrastructure. An approach to schedulability-driven frame packing for the synthesis of multi-cluster systems was presented as an example of solving such a design optimisation problem. We have developed two optimisation heuristics for frame configuration synthesis which are able to determine frame configurations that lead to a schedulable system. We have shown that by considering the frame packing problem, we are able to synthesise schedulable hard real-time systems and to potentially reduce the overall cost of the architecture.

The main message of the presented research is that efficient analysis and optimisation methods are needed and can be developed for the efficient implementation of applications distributed over interconnected heterogeneous networks.

References

1   KOPETZ, H.: 'Real-time systems – design principles for distributed embedded applications' (Kluwer Academic Publishers, 1997)
2   KOPETZ, H.: 'Automotive electronics'. Proceedings of the 11th Euromicro conference on *Real-time systems*, 1999, pp. 132–140
3   HANSEN, P.: 'The Hansen report on automotive electronics' http://www. hansenreport.com/, July–August, 2002
4   LEEN, G., and HEFFERNAN, D.: 'Expanding automotive electronic systems', *Computer*, 2002, **35**(1), pp. 88–93
5   JOST, K.: 'From fly-by-wire to drive-by-wire'. *Automotive Engineering International*, 2001, http://www.sae.org/automeg/electronics09-2001

6   CHIODO, M.: 'Automotive electronics: a major application field for hardware-software co-design' in 'Hardware/software co-design' (Kluwer Academic Publishers, 1996), pp. 295–310

7   X-by-Wire Consortium, *X-By-Wire*: Safety related fault tolerant systems in vehicles, http://www.vmars.tuwien.ac.at/projects/xbywire/, 1998

8   KOPETZ, H.: 'Automotive electronics – present state and future prospects'. Proceedings of the 25th international symposium on *Fault-tolerant computing*, 1995

9   Robert   Bosch   GmbH: CAN   Specification,   Version   2.0, http://www.can.bosch.com/, 1991

10   Local interconnect network protocol specification, http://www.lin-subbus.org, 2003

11   SAE Vehicle Network for Multiplexing and Data Communications Standards Committee, SAE J1850 Standard, 1994

12   RUSHBY, J.: 'Bus architectures for safety-critical embedded systems'. *Lecture notes in computer science*, vol. 2211 (Springer Verlag, Heidelberg, 2001), pp. 306–323

13   HOYME, K., and DRISCOLL, K.: 'SAFEbus', *IEEE Aerospace and Electronic Systems Magazine*, 1992, **8**(3), pp. 34–39

14   MINER, P.S.: 'Analysis of the SPIDER fault-tolerance protocols'. Proceedings of the 5th NASA *Langley formal methods workshop*, 2000

15   INTERNATIONAL ORGANIZATION FOR STANDARDIZATION: 'Road vehicles – Controller area network (CAN) – Part 4: Time-triggered communication'. ISO/DIS 11898-4, 2002

16   Echelon: LonWorks: The   LonTalk   Protocol   Specification, http://www.echelon.com, 2003

17   PROFIBUS   INTERNATIONAL: *PROFIBUS   DP   Specification*, http://www.profibus.com/, 2003

18   BERWANGER, J., PELLER, M., and GRIESSBACH, R.: 'A new high performance data bus system for safety-related applications', http://www.byteflight.de, 2000

19   THE FLEXRAY GROUP: FlexRay Requirements Specification, Version 2.0.2, http://www.flexray-group.com/, 2002

20   TINDELL, K., BURNS, A., and WELLINGS, A.: 'Calculating CAN message response times', *Control Engineering Practice*, 1995, **3**(8), pp. 1163–1169

21   AUDSLEY, N., TINDELL, K., and BURNS, A.: 'The end of line for static cyclic scheduling?'. Proceedings of the Euromicro Workshop on *Real-time systems*, 1993, pp. 36–41

22   XU, J., and PARNAS, D.L.: 'On satisfying timing constraints in hard-real-time systems', *IEEE Transactions on Software Engineering*, 1993, **19**(1), pp. 132–146

23   LÖNN, H., and AXELSSON, J.: 'A comparison of fixed-priority and static cyclic scheduling for distributed automotive control applications'. Proceedings of the Euromicro conference on *Real-time systems*, 1999, pp. 142–149

24 EAST-EEA project, *ITEA Full Project Proposal*, http://www.itea-office.org, 2002

25 AUDSLEY, N., BURNS, A., DAVIS, R., TINDELL, K., and WELLINGS, A.: 'Fixed priority preemptive scheduling: an historical perspective', *Real-Time Systems*, 1995, **8**(2/3), pp. 173–198

26 BALARIN, F., LAVAGNO, L., MURTHY, P., and SANGIOVANNI-VINCENTELLI, A.: 'Scheduling for embedded real-time systems', *IEEE Design and Test of Computers*, January–March, 1998, **15**(1), 71–82

27 TimeWiz: http://www.timesys.com

28 RapidRMA: http://www.tripac.com

29 RTA-OSEK Planner, http://www.livedevices.com

30 Aires, http://kabru.eecs.umich.edu/aires/

31 LIU, C.L., and LAYLAND, J.W.: 'Scheduling algorithms for multiprogramming in a hard-real-time environment', *Journal of the ACM*, 1973, **20**(1), pp. 46–61

32 TINDELL, K., and CLARK, J.: 'Holistic schedulability analysis for distributed hard real-time systems', *Microprocessing & Microprogramming*, 1994, **50** (2–3), pp. 117–134

33 STANKOVIC, J.A., and RAMAMRITHAM, K.: 'Advances in real-time systems' (IEEE Computer Society Press, Washington, 1993)

34 XU, J., and PARNAS, D.L.: 'Priority scheduling versus pre-run-time scheduling', *Journal of Real Time Systems*, 2000, **18**(1), pp. 7–24

35 LEE, C., POTKONJAK, M., and WOLF, W.: 'Synthesis of hard real-time application specific systems', *Design Automation for Embedded Systems*, 1999, **4**(4), pp. 215–241

36 DAVE, B.P., and JHA, N.K.: 'COHRA: hardware-software cosynthesis of hierarchical heterogeneous distributed systems', *IEEE Transactions on CAD*, 1998 **17**(10), pp. 900–919

37 DAVE, B.P., LAKSHMINARAYANA, G., and JHA, N.J.: 'COSYN: Hardware-software co-synthesis of heterogeneous distributed embedded systems', *IEEE Transactions on VLSI Systems*, 1999, **7**(1), pp. 92–104

38 BURNS, A., and WELLINGS, A.: 'Real-time systems and programming languages' (Addison Wesley, Reading, MA, 2001)

39 TINDELL, K.: 'Adding time-offsets to schedulability analysis'. Department of Computer Science, University of York, Report No. YCS-94-221, 1994

40 YEN, T.Y., and WOLF, W.: 'Hardware-software co-synthesis of distributed embedded systems' (Kluwer Academic Publishers, 1997)

41 PALENCIA, J.C., and GONZÁLEZ HARBOUR, M: 'Exploiting precedence relations in the schedulability analysis of distributed real-time systems', Proceedings of the 20th IEEE *Real-time systems* symposium, 1999, pp. 328–339

42 PALENCIA, J.C., and GONZÁLEZ HARBOUR, M.: 'Schedulability analysis for tasks with static and dynamic offsets'. Proceedings of the 19th IEEE *Real-time systems* symposium, 1998, pp. 26–37

43 ERMEDAHL, H., HANSSON, H., and SJÖDIN, M.: 'Response-time guarantees in ATM Networks'. Proceedings of the *IEEE Real-time systems* symposium, 1997, pp. 274–284

44   STROSNIDER, J.K., and MARCHOK, T.E.: 'Responsive, deterministic IEEE 802.5 Token ring scheduling', *Journal of Real-Time Systems*, 1989, **1**(2), pp. 133–158

45   AGRAWAL, G., CHEN, B., ZHAO, W., and DAVARI, S.: 'Guaranteeing synchronous message deadlines with the token medium access control protocol'. *IEEE Transactions on Computers*, 1994, **43**(3), pp. 327–339

46   POP, P., ELES, P., and PENG, Z.: 'Schedulability-driven communication synthesis for time-triggered embedded systems', *Real-Time Systems Journal*, 2004, **26**(3), pp. 297–325

47   POP, T., ELES, P., and PENG, Z.: 'Holistic scheduling and analysis of mixed time/event-triggered distributed embedded systems'. Proceedings of the international symposium on *Hardware/software codesign*, 2002, pp. 187–192

48   POP, T., ELES, P., and PENG, Z.: 'Schedulability analysis and optimization for the synthesis of multi-cluster distributed embedded systems'. Proceedings of *Design automation and test in Europe* conference, 2003, pp. 184–189

49   RICHTER, K., JERSAK, M., and ERNST, R.: 'A formal approach to MpSoC performance verification', *Computer*, 2003, **36**(4), pp. 60–67

50   STATEMATE: http://www.ilogix.com

51   MATLAB/SIMULINK: http://www.mathworks.com

52   ASCET/SD: http://en.etasgroup.com/products/ascet_sd/

53   SYSTEMBUILD/MATRIXX: http://www.ni.com/matrixx

54   TTP-PLAN: http://www.tttech.com/

55   GUTIÉRREZ GARCÍA, J.J. and GONZÁLEZ HARBOUR, M.: 'Optimized priority assignment for tasks and messages in distributed Hard real-time systems'. Proceedings of the workshop on *Parallel and distributed real-Time systems*, 1995, pp. 124–132

56   JONSSON, J., and SHIN, K.G.: 'Robust adaptive metrics for deadline assignment in distributed hard real-time systems', *Real-Time Systems: The International Journal of Time-Critical Computing Systems*, 2002, **23**(3), pp. 239–271

57   VOLCANO NETWORK ANALYZER: http://www.volcanoautomotive.com/

58   RAJNAK, A., TINDELL, K., and CASPARSSON, L.: 'Volcano Communications Concept' (Volcano Communication Technologies AB, 1998)

59   KIENHUIS, B., DEPRETTERE, E., VISSERS, K., and VAN DER WOLF, P.: 'An approach for quantitative analysis of application-specific dataflow architectures'. Proceedings of the IEEE international conference on *Application-specific systems, architectures and processors*, 1997, pp. 338–349

60   TABBARA, B., TABBARA, A., and SANGIOVANNI-VINCENTELLI, A.: 'Function/architecture optimization and co-design of embedded systems' (Kluwer Academic Publishers, Boston, MA, 2000)

61   BALARIN, F. *et al.*: 'Hardware-software co-design of embedded systems: The POLIS approach' (Kluwer Academic Publishers, Boston, MA, 1997)

62   BALARIN, F., WATANABE, Y., HSIEH, H., LAVAGNO, L., PASERONE, C., and SANGIOVANNI-VINCENTELLI, A.: 'Metropolis: An integrated electronic system design environment', *Computer*, 2003, **36**(4), pp. 45–52

63   VIRTUAL COMPONENT CO-DESIGN: http://www.cadence.com/

64 KEUTZER, K., MALIK, S., and NEWTON, A.R.: 'System-level design: orthogonalization of concerns and platform-based design', *IEEE Transactions on Computer-Aided Design of Integrated Circuits and Systems*, 2000, **19**(12), pp. 1523–1543

65 TTTECH: TTP/C Specification Version 0.5, 1999, availabe at http://www.tttech.com

66 TTTECH: 'Comparison CAN–Byteflight–FlexRay–TTP/C'. Technical report, availabe at http://www.tttech.com

67 NOLTE, T., HANSSON, H., NORSTRÖM, C., and PUNNEKKAT, S.: 'Using bit-stuffing distributions in CAN analysis'. Proceedings of the IEEE/IEE *Real-time embedded systems* workshop, 2001

68 POP, P., ELES, P., and PENG, Z.: 'Scheduling with optimized communication for time triggered embedded systems'. Proceedings of the international workshop on *Hardware-software codesign*, 1999, pp. 178–182

69 EDWARDS, S.: 'Languages for digital embedded systems' (Kluwer Academic Publishers, Boston, MA, 2000)

70 LEE, E.A., and PARKS, T.M.: 'Dataflow process networks', *Proceedings of the IEEE*, 1995, **83**, pp. 773–801

71 PUSCHNER, P., and BURNS, A.: 'A review of worst-case execution-time analyses', *Real-Time Systems Journal*, 2000, **18**(2/3), pp. 115–128

72 POP, P.: 'Analysis and synthesis of communication-intensive heterogeneous real-time systems'. Linköping Studies in Science and Technology, PhD dissertation No. 833, 2003, available at http://www.ida.liu.se/~paupo/thesis

73 SANDSTRÖM, K., and NORSTRÖM, C.: 'Frame packing in real-time communication'. Proceedings of the international conference on *Real-time computing systems and applications*, 2000, pp. 399–403

74 POP, P., ELES, P., and PENG, Z.: 'Bus access optimization for distributed embedded systems based on schedulability analysis'. Proceedings of the *Design automation and test in Europe* conference, 2000, pp. 567–574

75 POP, P., ELES, P., PENG, Z., IZOSIMOV, V., HELLRING, M., and BRIDAL, O.: 'Design optimization of multi-cluster embedded systems for real-time applications'. Proceedings of *Design, automation and test in Europe* conference, 2004, pp. 1028–1033

76 ELES, P., DOBOLI, A., POP, P., and PENG, Z.: 'Scheduling with bus access optimization for distributed embedded systems', *IEEE Transactions on VLSI Systems*. 2000, **8**(5), pp. 472–491

77 AUDSLEY, N.C., BURNS, A., RICHARDSON, M.F., and WELLINGS, A.J.: 'Hard real-time scheduling: the deadline monotonic approach'. Proceedings of the 8th IEEE workshop on *Real-time operating systems and software*, 1991, pp. 127–132

78 REEVS, C.R.: 'Modern heuristic techniques for combinatorial problems' (Blackwell Scientific Publications, Oxford, 1993)

Chapter 4

Hardware/software partitioning of operating systems: focus on deadlock avoidance

Jaehwan John Lee and Vincent John Mooney III

4.1 Introduction

Primitive operating systems were first introduced in the 1960s in order to relieve programmers of common tasks such as those involving Input/Output (I/O). Gradually, scheduling and management of multiple jobs/programs became the purview of an Operating System (OS). Many fundamental advances, such as multithreading and multiprocessor support, have propelled both large companies and small to the forefront of software design.

Recent trends in chip design press the need for more advanced operating systems for System-on-a-Chip (SoC). However, unlike earlier trends where the focus was on scientific computing, today's SoC designs tend to be driven more by the needs of embedded computing. While it is hard to state exactly what constitutes embedded computing, it is safe to say that the needs of embedded computing form a superset of scientific computing. For example, real-time behaviour is critical in many embedded platforms due to close interaction with non-humans, e.g. rapidly moving mechanical parts. In fact, the Application-Specific Integrated Circuits (ASICs) preceding SoC did not integrate multiple processors with custom hardware, but instead were almost exclusively digital logic specialised to a particular task and hence very timing predictable and exact. Therefore, we predict that advances in operating systems for SoC focusing on Real-Time Operating System (RTOS) design provide a more natural evolution for chip design as well as being compatible with real-time systems.

Furthermore, thanks to the recent trends in the technologies of MultiProcessor SoC (MPSoC) and reconfigurable chips, many hardware Intellectual Property (IP) cores that implement software algorithms have been developed to speed up computation. However, efforts to fully exploit these innovative hardware IP cores have encountered many difficulties such as interfacing IP cores to a specific system, modifying IP cores to fulfil requirements of a system under consideration, porting device drivers and finally integrating both IP cores and software seamlessly. Much work of interfacing, modifying and/or porting IP cores and device drivers has relied on human resources. Hardware/software codesign frameworks can help reduce this burden on designers.

This chapter focuses on such research in the design of OS, especially RTOSes. We have implemented and upgraded the δ hardware/software RTOS/MPSoC design framework (shown in Figure 4.1). Since we have already described key aspects of our approach in References 1–5, in this chapter we first briefly explain the δ framework and then focus more on an exposition of deadlock issues. We believe deadlock issues are on the horizon due to the rapid evolution in MPSoC technology and the introduction of many innovative IP cores. We predict that future MPSoC designs will have hundreds of processors and resources (such as custom FFT hardware) all in a single chip; thus, systems will handle much more functionality, enabling a much higher level of concurrency and requiring many more deadlines to be satisfied. As a result, we predict there will be resource sharing problems among the many processors desiring the resources, which may result in deadlock more often than designers might realise.

The remainder of this chapter is organised as follows. Section 4.2 presents our target MPSoC architecture and then explains the δ hardware/software RTOS design framework version 2.0 including a description of two hardware RTOS components: a 'lock' cache and a dynamic memory allocator. Section 4.3 motivates deadlock issues and provides background about deadlock problems. Section 4.4 focuses on several new software/hardware solutions to such deadlock problems. Section 4.5 addresses experimental setup and shows various comparison results with applications that demonstrate how the δ framework could impact hardware/software partitioning in current and future RTOS/MPSoC designs. Finally, Section 4.6 concludes this chapter.

4.2   Hardware/software RTOS design

4.2.1   RTOS/MPSoC target

Figure 4.2 shows our primary target MPSoC consisting of multiple processing elements with L1 caches, a large L2 memory, and multiple hardware IP components with essential interfaces such as a memory controller, an arbiter and a bus system. The target also has a shared memory multiprocessor RTOS (Atalanta [6] developed at the Georgia Institute of Technology), which is small and configurable. The code of Atalanta RTOS version 0.3 resides in shared memory, and all processing elements (PEs) execute the same RTOS code and share kernel structures as well as the states of all processes and resources. Atalanta supports priority scheduling with priority inheritance as well as round-robin; task management such as task creation, suspension and resumption; various Inter Process Communication (IPC) primitives such as semaphores, mutexes, mailboxes, queues and events; memory management; and interrupts. As shown in Figure 4.2, hardware IP cores can be either integrated

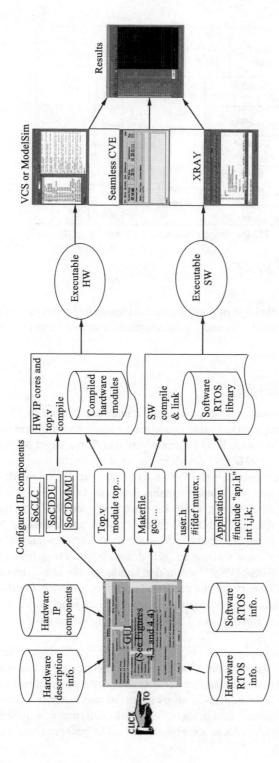

Figure 4.1   The 8 hardware/software RTOS design framework

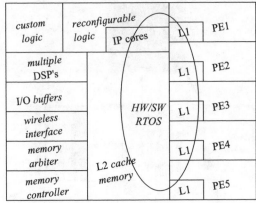

PE: Processing element, IP: Intellectual property

Figure 4.2   Future MPSoC

into the reconfigurable logic or implemented as custom logic. Besides, specialised IP cores such as DSP processors and wireless interface cores can also be integrated into the chip.

4.2.2   The δ framework

The δ hardware/software RTOS generation framework (shown in Figure 4.1) for MPSoC has been proposed to enable automatic generation of different mixes of pre-designed hardware/software RTOS components that fit the target MPSoC a user is designing so that RTOS/MPSoC designers can explore crucial decisions early in the design phase of their target product(s) [1–5]. Thus, the δ framework helps users explore which configuration is most suitable for users' target and application or set of applications. In other words, the δ framework is specifically designed to provide a solution to rapid RTOS/MPSoC (both hardware and software) design space exploration so that users can easily and quickly find a few optimal RTOS/MPSoC architectures that are most suitable to their design goals. The δ framework generates a configured RTOS/MPSoC design that is simulatable in a hardware/software cosimulation environment after the generated design is compiled. Hardware designs are described in a Hardware Description Language (HDL) such as Verilog. Software designs could be described in any language although we have only used C in our designs.

From the initial implementation [1, 2], we have extended the δ framework to include parameterised generators of hardware IP components (i.e. automatically configurable to fit a desired target architecture) as well as the generation of various types of bus systems. This section gives an overview of parameterised generators for a customised RTOS/MPSoC design including a bus configurator, a dynamic memory management unit generator and a custom 'lock' cache generator, and explains such

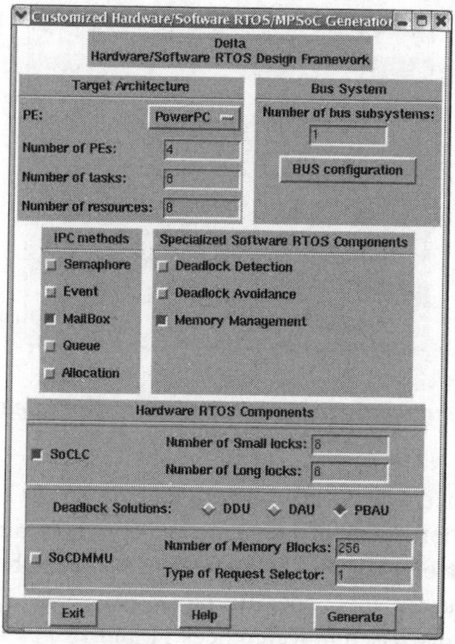

Customized Hardware/Software RTOS/MPSoC Generation

Delta
Hardware/Software RTOS Design Framework

Target Architecture

PE: PowerPC

Number of PEs: 4

Number of tasks: 8

Number of resources: 8

Bus System

Number of bus subsystems: 1

BUS configuration

IPC methods
- Semaphore
- Event
- MailBox
- Queue
- Allocation

Specialized Software RTOS Components
- Deadlock Detection
- Deadlock Avoidance
- Memory Management

Hardware RTOS Components

SoCLC
Number of Small locks: 8
Number of Long locks: 8

Deadlock Solutions: ◇ DDU ◇ DAU ◆ PBAU

SoCDMMU
Number of Memory Blocks: 256
Type of Request Selector: 1

Exit   Help   Generate

Figure 4.3   GUI of the δ framework

available IP components briefly. Many low-level details – e.g. details of the bus system generation – are not repeated in this chapter but instead are available in referenced works [1–22].

Figure 4.3 shows a Graphical User Interface (GUI) for the δ framework version 2.0, which now integrates four parameterised generators we have and generates an RTOS/MPSoC system. The GUI generates a top-level architecture file plus additional configuration files, used as input parameter files to generate specific hardware component files (i.e. modules) either using a dedicated generator or via Verilog PreProcessor (VPP [23]).

Here we summarise each generator briefly. For more information, please see specific references. When users want to create their own specific bus systems, by clicking 'Bus configuration' (shown at the top right of Figure 4.3), users can specify address and data bus widths as well as detailed bus topology for each subsystem in case a system has a hierarchical bus structure. After the appropriate inputs are entered, the tool will generate a user-specified bus system with the specified hierarchy. Further details about bus system generation are described in References 7 to 10.

At the bottom of Figure 4.3, there are several options for 'Hardware RTOS Components': the SoC Lock Cache (SoCLC), multiple deadlock detection/ avoidance solutions, and the SoC Dynamic Memory Management Unit (SoCDMMU). The details of these hardware RTOS components will be described in Section 4.2.3.

In addition to selecting hardware RTOS components, the δ framework version 2.0 can also manipulate the size and type of each RTOS component by use of input

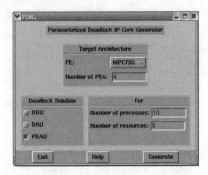

Figure 4.4   GUI for automatic generation of a hardware deadlock solution

parameters. For instance, when users want to include SoCLC, they specify the number of short locks (short locks are held for a very short time, e.g. 100 cycles or
less) and the number of long locks (equivalent to semaphores) according to the
expected requirements for their specific target (or goal). Detailed parameterised
SoCLC generation is discussed in References 11 and 18.

For deadlock hardware components, after a user selects either the Deadlock Detection Unit (DDU), the Deadlock Avoidance Unit (DAU) or the Parallel Banker's
Algorithm Unit (PBAU), the GUI tool shown in Figure 4.3 directs the generation
at the specified deadlock IP component with the designated type and specific size
according to the number of tasks and resources specified in the Target Architecture
window (see upper left of Figure 4.3) [7]. Figure 4.4 shows a separate deadlock
hardware solution generator [7]. The generation process is the same as we explained
above. In the screen shot shown in Figure 4.4, the user has selected four PEs all of
the same type, namely, MPC755.

For the SoCDMMU IP component, users specify the number of memory blocks
(available for dynamic allocation in the system) and several additional parameters,
and then the GUI tool generates a user-specified SoCDMMU. Details regarding
parameterised SoCDMMU generation are addressed in References 13 and 14.

We briefly describe our approach to HDL file generation in the following
example.

Example 4.1   Top-level architecture file generation in the δ framework
This example briefly describes a specific portion of the δ framework that generates a
top-level design file of a particular MPSoC with the SoCLC hardware IP component.
Figure 4.5 illustrates that the GUI tool (shown in Figure 4.3) generates a top-level
architecture file (i.e. a top-level file for the system being designed, where the top-level
file instantiates any number of additional modules needed in an overall hierarchy)
according to the description of a user-specified system with hardware IP components.
Let us assume that a user selects a system having three PEs and an SoCLC for eight
short locks and eight long locks. Then, the generation process starts with a description
of a system having an SoCLC (i.e. LockCache description) in the description library.

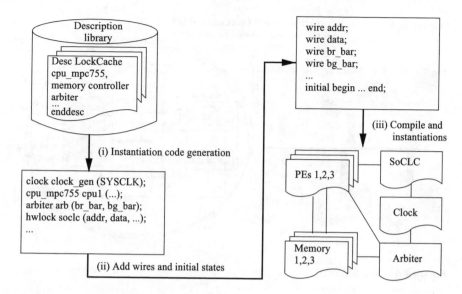

Figure 4.5   Top file generation in the δ framework

The LockCache description lists modules necessary to build a system containing an SoCLC, such as PEs, L2 memory, a memory controller, a bus arbiter, an interrupt controller and an SoCLC with the specified locks. The Verilog top file generator, which we call `Archi_gen`, writes all instantiation code for each module in the list of the LockCache description to a file. In case that a system contains multiple units of the same type of module (e.g. multiple PEs), `Archi_gen` also automatically includes multiple instantiation code of the same type IP with distinct identification numbers since some modules need to be instantiated multiple times. Then, `Archi_gen` writes necessary wires described in the LockCache description, and then writes initialisation routines necessary to execute simulation. Later by compiling `Top.v`, a specified target hardware architecture will be ready for exploration via standard simulation tools (in our case, Seamless CVE [37]) [2].   ∎

4.2.3   Hardware RTOS components

This subsection briefly summarises two available hardware IP components presented previously: SoCLC and SoCDMMU.

4.2.3.1   SoCLC

Synchronisation has always been a critical issue in multiprocessor systems. As multiprocessors execute a multitasking application on top of an RTOS, any important shared data structure, also called a Critical Section (CS), may be accessed for inter-process communication and synchronisation events occurring among the tasks/processors in the system.

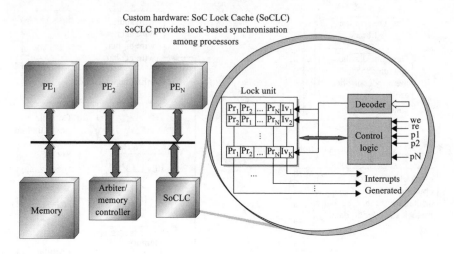

Figure 4.6   SoCLC

Previous work has shown that the SoCLC, which is a specialised custom hardware unit realising effective lock-based synchronisation for a multiprocessor shared-memory SoC as shown in Figure 4.6, reduces on-chip memory traffic, provides a fair and fast lock hand-off, simplifies software, increases the real-time predictability of the system and improves performance as well [15–18].

Akgul *et al.* [17] extended the SoCLC mechanism with a priority inheritance support implemented in hardware. Priority inheritance provides a higher level of real-time guarantees for synchronising application tasks. The authors present a solution to the priority inversion problem in the context of an MPSoC by integrating an Immediate Priority Ceiling Protocol (IPCP) [24] implemented in hardware. The approach also provides higher performance and better predictability for real-time applications running on an MPSoC.

Experimental results indicate that the SoCLC hardware mechanism with priority inheritance achieves a 75 percent speedup in lock delay (i.e. average time to access a lock during application execution [17]). The cost in terms of additional hardware area for the SoCLC with 128 locks supporting priority inheritance is approximately 10 000 NAND2 gates in TSMC .25μ chip fabrication technology [18].

4.2.3.2   SoCDMMU

The System-on-a-Chip Dynamic Memory Management Unit shown in Figure 4.7 is a hardware unit that allows a fast and deterministic way to dynamically allocate/de-allocate global (L2) memory among PEs [12]. The SoCDMMU is able to convert the PE address (virtual address) to a physical address. The memory mapped address or I/O port to which the SoCDMMU is mapped is used to send commands to the SoCDMMU (writing data to the port or memory-mapped location) and to receive the results of the command execution (reading from the port or memory-mapped location).

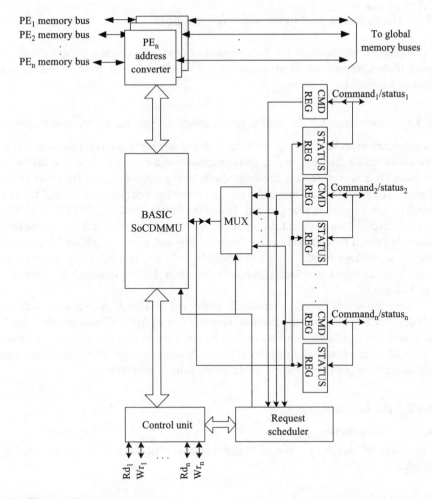

Figure 4.7   SoCDMMU

As shown in References 12 and 14, the SoCDMMU achieves a 4.4× over-all speedup in memory management during application transition time of several examples when compared with conventional memory allocation/deallocation tech-niques, i.e. `malloc()` and `free()`. The SoCDMMU is synthesisable and has been integrated into a system example including porting SoCDMMU functionality to an RTOS (so that the user can access SoCDMMU functionality using standard software memory management APIs) [12]. Also, the SoCDMMU-crossbar switch Generator (DX-Gt [13]) can configure and optimise the SoCDMMU and associated crossbar switch to suit a specific system (e.g. for a particular memory configuration and num-ber of PEs). In this way, DX-Gt automates the customisation and the generation of the hardware memory management functionalities with associated crossbar support.

4.3 Background and prior work for deadlock

In this section we motivate the development of deadlock-related software and hardware IP components and then introduce definitions and prior work related to our deadlock research.

4.3.1 Motivation for the design of deadlock-related hardware components

In most current embedded systems in use today, deadlock is not a critical issue due to the use of only a few (e.g. two or less) processors and a couple of custom hardware resources (e.g. direct memory access hardware plus a video decoder). However, in the coming years future chips may have five to twenty (or more) processors and ten to a hundred resources all in a single chip. This is the way we predict MPSoC will rapidly evolve. Even in the platform design area, Xilinx already has been able to include multiple PowerPC processors in the Virtex-II Pro and Virtex-IV FPGA [25]. Given current technology trends, we predict that MPSoC designers and users are going to start facing deadlock problems more and more often. That is, deadlock problems are on the horizon.

How can we efficiently and timely cope with deadlock problems in such an MPSoC? Although dynamic resource allocation in an MPSoC may produce deadlock problems, MPSoC architectures can be modified to provide efficient hardware solutions to deadlock. Before describing such solutions, we first introduce some definitions and our target system model in the following section.

4.3.2 Background

4.3.2.1 Definitions

Definitions of 'deadlock', 'livelock' and 'avoidance' in our context can be stated as follows:

Definition 4.1 *A system has a deadlock if and only if the system has a set of processes, each of which is blocked (e.g. preempted), waiting for requirements that can never be satisfied.*

Definition 4.2 *Livelock is a situation where a request for a resource is repeatedly denied and possibly never accepted because of the unavailability of the resource, resulting in a stalled process, while the resource is repeatedly made available for other process(es) which make progress.*

Definition 4.3 *Deadlock avoidance is a way of dealing with deadlock where resource usage is dynamically controlled not to reach deadlock (i.e. on the fly, resource usage is controlled to ensure that there can never be deadlock).*

In addition, we define two kinds of deadlock: request deadlock (R-dl) and grant deadlock (G-dl).

Definition 4.4 *For a given system, if a request from a process directly causes the system to have a deadlock at that moment, then we denote this case as* **request deadlock** *or* **R-dl** [7].

Definition 4.5 *For a given system, if the grant of a resource to a process directly causes the system to have a deadlock at that moment, then we denote this case as* **grant deadlock** *or* **G-dl** [7].

While a request deadlock (R-dl) example is described in Example 4.4 of Section 4.4.3, a grant deadlock (G-dl) example is described in the example presented in Section 4.5.6.1. Please note that we differentiate between R-dl and G-dl because our deadlock avoidance algorithm in Section 4.4.3 requires the distinction to be made. The distinction is required because some actions can only be taken for either R-dl or G-dl; e.g. for G-dl it turns out that perhaps deadlock can be avoided by granting the released resource to a lower priority process.

We now define 'single-instance resource' and 'multiple-instance resource'.

Definition 4.6 *A single-instance resource is a resource that services no more than one process at a time. That is, while the resource is processing a request from a process, all other processes requesting to use the resource must wait* [26].

Definition 4.7 *A multiple-instance resource is a resource that can service two or more processes at the same time, providing the same or similar functionality to all serviced processes* [26].

Example 4.2 An example of a multiple-instance resource
The SoC Dynamic Memory Management Unit dynamically allocates and deallocates segment(s) of global level two (L2) memory between PEs with very fast and deterministic time (e.g. four clock cycles) [12]. In a system having an SoCDMMU and 16 segments of global L2 memory, which can be considered as a 16-instance resource, rather than having each PE (or process) keep track of each segment, PEs request segment(s) from the SoCDMMU (which keeps track of the L2 memory). In this way, not only can the overhead of tracking segments for each PE be reduced but also interfaces between PEs and segments can be simplified because PEs request segment(s) from one place (i.e. the SoCDMMU). ∎

We also introduce the definitions of an 'H-safe sequence' and an 'H-safe state' used to clarify the Parallel Banker's Algorithm. Please note that the notion of 'safe' was first introduced by Dijkstra [27] and was later formalised into 'safe sequence', 'safe state' and 'unsafe state' by Habermann [28]. We refer to Habermann's 'safe sequence' as an 'H-safe sequence', to Habermann's 'safe state' as an 'H-safe state' and to Habermann's 'unsafe state' as an 'H-unsafe state' where the 'H' stands for Habermann.

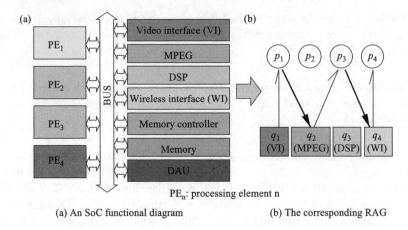

(a) An SoC functional diagram   (b) The corresponding RAG

Figure 4.8   A practical MPSoC realisation

Definition 4.8   *An H-safe sequence is an enumeration $p_1, p_2, \ldots, p_n$ of all the processes in the system, such that for each $i = 1, 2, \ldots, n$, the resources that p_i may request are a subset of the union of resources that are currently available and resources currently held by $p_1, p_2, \ldots, p_{i-1}$ [27,28].*

Theorem 4.1   *A system of processes and resources is in an H-safe state if and only if there exists an H-safe sequence $\{p_1, p_2, \ldots, p_n\}$. If there is no H-safe sequence, the system is in an H-unsafe state [28].*

4.3.2.2   System model in the view of deadlock

To address deadlock issues, we first show a modified MPSoC from Figure 4.2 in the following example:

Example 4.3   A future Request-Grant MPSoC

We introduce the device shown in Figure 4.8 as a particular MPSoC example. This MPSoC consists of four PEs and four resources: a Video and Image capturing interface (VI), an MPEG encoder/decoder, a DSP and a Wireless Interface (WI), which we refer to as q_1, q_2, q_3 and q_4, respectively, as shown in Figure 4.8(b). The MPSoC also contains memory, a memory controller and a DAU. In the figure, we assume that each PE has only one active process; i.e. each process p_1, p_2, p_3 and p_4, as shown in Figure 4.8(b), runs on PE1, PE2, PE3 and PE4, respectively. In the current state, resource q_1 is granted to process p_1, which in turn requests q_2. In the meantime, q_2 is granted to p_3, which requests q_4, while q_4 is granted to process p_4. The DAU in Figure 4.8 receives all requests and releases, decides whether or not the request or grant can cause a deadlock and then permits the request or grant only if no deadlock results.   ∎

We consider this kind of request-grant system with many resources and processes as shown in Figure 4.8 as our system model in the view of deadlock. Based on our system model, we first introduce prior work in deadlock research and then describe new approaches for such MPSoCs.

4.3.3 Prior work in deadlock research

4.3.3.1 Overview of prior deadlock research

Researchers have put tremendous efforts into deadlock research, three well-known areas of which are deadlock detection, prevention and avoidance [26, 27, 29, 30]. Among them, deadlock detection provides more freedom for a system since deadlock detection does not typically restrict the behaviour of a system, facilitating full concurrency. Deadlock detection, however, usually requires a recovery once a deadlock is detected. In contrast, deadlock prevention prevents a system from reaching deadlock by typically restraining request orders to resources in advance, implying restrictions on concurrency. One such method is the Priority Ceiling Protocol (PCP [24]), which is only a solution for a single processor system, though. Another method is the collective request method, which, however, tends to cause resource under-utilisation as well as process starvation [26]. Deadlock avoidance, by contrast, generally sits in-between; that is, deadlock avoidance normally gives more freedom with less restrictions than deadlock prevention [26]. As implemented in known algorithms, deadlock avoidance essentially requires knowledge about the maximum necessary resource requirements for all processes in a system, which unfortunately makes the implementation of deadlock avoidance difficult in real systems with dynamic workloads [27–30].

4.3.3.2 Deadlock detection

All software deadlock detection algorithms known to the authors to date have a run-time complexity of at least $O(m \times n)$, where m is the number of resources and n is the number of processes. In 1970, Shoshani *et al.* [29] proposed an $O(m \times n^2)$ run-time complexity detection algorithm, and about two years later, Holt [30] proposed an $O(m \times n)$ algorithm to detect a knot that tells whether deadlock exists or not. Both of the aforementioned algorithms (of Shoshani *et al.* and of Holt) are based on a Resource Allocation Graph (RAG) representation. Leibfried [31] proposed a method of describing a system state using an adjacency matrix representation and a corresponding scheme that detects deadlock with matrix multiplications but with a run-time complexity of $O(m^3)$. Kim and Koh [32] proposed an algorithm with $O(m \times n)$ time for 'detection preparation'; thus an overall time for detecting deadlock (starting from a system description that just came into existence, e.g. due to multiple grants and requests occurring within a particular time or clock cycle) of at least $O(m \times n)$.

4.3.3.3 Deadlock avoidance

A traditional well-known deadlock avoidance algorithm is the Banker's Algorithm (BA) [27]. The algorithm requires each process to declare the maximum requirement (claim) of each resource it will ever need. In general, traditional deadlock avoidance (i.e. based on some variant of BA) is more expensive than deadlock detection and may be impractical because of the following disadvantages: (1) an avoidance algorithm must be executed for every request prior to granting a resource, (2) deadlock avoidance tends to restrict resource utilisation, which may degrade normal system performance and (3) the maximum resource requirements (and thus requests) might not be known in advance [27, 33].

In 1990, Belik [34] proposed a deadlock avoidance technique in which a path matrix representation is used to detect a potential deadlock before the actual allocation of resources. However, Belik's method requires $O(m \times n)$ time complexity for updating the path matrix in releasing or allocating a resource and thus an overall complexity for avoiding deadlock of $O(m \times n)$, where m and n are the numbers of resources and processes, respectively. Furthermore, Belik did not mention any solution to livelock although livelock is a possible consequence of his deadlock avoidance algorithm.

4.4   New approaches to deadlock problems

In this section, we describe in detail deadlock related IP components, i.e. the Deadlock Detection hardware Unit (DDU), the DAU and the Parallel Banker's Algorithm Unit (PBAU).

4.4.1   Introduction

All of the algorithms referenced in Section 4.3 assume an execution paradigm of one instruction or operation at a time. With a custom hardware implementation of a deadlock algorithm, however, parallelism can be exploited.

Detection of deadlock is extremely important since any request for or grant of a resource might result in deadlock. Invoking software deadlock detection on every resource allocation event would typically cost too much computational power; thus, using a software implementation of deadlock detection and/or avoidance would perhaps be impractical in terms of the performance cost. A promising way of solving deadlock problems with small compute power is to implement deadlock detection and/or avoidance in hardware.

To handle this possibility, the DDU [19, 20], the DAU [21] utilising the DDU and the PBAU [22] have recently been proposed. These three hardware deadlock solutions improve the reliability and timeliness of applications running on an MPSoC under an RTOS. Of course, adding a centralised module on MPSoC may lead to a bottleneck. However, since resource allocation and deallocation are preferably managed by an OS (which already implies some level of centralised operation), adding hardware can potentially reduce the burden on software rather than becoming a bottleneck.

4.4.2   New deadlock detection methodology: the DDU

The DDU manipulates a simple Boolean representation of the types of each edge: the request edge of a process requesting a resource, the grant edge of a resource granted

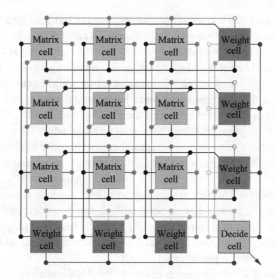

Figure 4.9   DDU architecture

to a process, or no activity (neither a request nor a grant) [7]. Since the DDU is implemented in a small amount of hardware, the designed deadlock detection unit hardly affects system performance (and potentially has no negative impact whatsoever) yet provides the basis for an enhanced deadlock detection methodology. The DDU has been proven to have a run-time complexity of $O(\min(m, n))$ using custom hardware [7].

The DDU consists of three parts as shown in Figure 4.9: matrix cells (part 1), weight cells (part 2) and a decide cell (part 3). Part 1 of the DDU is a realisation of the system state matrix M_{ij} (shown in Equation (4.1)) via use of an array of matrix cells that represents an array of α_{st} entries where $1 \leq s \leq m$ and $1 \leq t \leq n$. Since each matrix element α_{st} represents one of the following: $g_{s \to t}$ (a grant edge), $r_{t \to s}$ (a request edge) or 0_{st} (no edge) (i.e. α_{st} is ternary-valued), α_{st} can be minimally defined as a pair of two bits $\alpha_{st} = (\alpha_{st}^r, \alpha_{st}^g)$. If an entry α_{st} is a grant edge g, bit α_{st}^r is set to 0, and α_{st}^g is set to 1; if an entry α_{st} is a request edge r, bit α_{st}^r is set to 1, and α_{st}^g is set to 0; otherwise, both bits α_{st}^r and α_{st}^g are set to 0. Hence, an entry α_{st} can be only one of the following binary encodings: 01 (a grant edge), 10 (a request edge) or 00 (no edge).

$$M_{ij} = \begin{bmatrix} (\alpha_{11}^r, \alpha_{11}^g) & \cdots & (\alpha_{1t}^r, \alpha_{1t}^g) & \cdots & (\alpha_{1n}^r, \alpha_{1n}^g) \\ \vdots & \vdots & \vdots & \vdots & \vdots \\ (\alpha_{s1}^r, \alpha_{s1}^g) & \cdots & (\alpha_{st}^r, \alpha_{st}^g) & \cdots & (\alpha_{sn}^r, \alpha_{sn}^g) \\ \vdots & \vdots & \vdots & \vdots & \vdots \\ (\alpha_{m1}^r, \alpha_{m1}^g) & \cdots & (\alpha_{mt}^r, \alpha_{mt}^g) & \cdots & (\alpha_{mn}^r, \alpha_{mn}^g) \end{bmatrix} \qquad (4.1)$$

On top of the matrix, terminal edges (i.e. edges connected to a node with only incoming edges or only outgoing edges and thus provably not involved in deadlock) are iteratively found and removed to detect deadlock (i.e. edges are still remaining by the end of iterations). Such discoveries and removals of terminal edges are performed in part 2, which consists of two weight vectors: (1) a column weight vector W^c below the matrix cells and (2) a row weight vector W^r on the right-hand side of the array of matrix cells. Each element w_{c_t}, $1 \leq t \leq n$, in W^c is called a column weight cell, and each element w_{r_s}, $1 \leq s \leq m$, in W^r is called a row weight cell. Both w_{c_t} and w_{r_s} represent whether the corresponding node has terminal edges, non-terminal edges or neither. At the bottom right corner of the DDU is one decide cell (part 3) which calculates at each iteration whether there exist terminal edges (if none, all iterations are done) or whether there exist non-terminal edges (in order to check deadlock).

Figure 4.9 specifically illustrates the DDU for three processes and three resources. This DDU has nine matrix cells (3 × 3) for each edge element $(\alpha_{st}^r, \alpha_{st}^g)$ of M_{ij}, six weight cells (three for column processing and three for row processing), and one decide cell for deciding whether or not deadlock has been detected. The details of each cell are described in Reference 7. The area of Figure 4.9 mapped to a 0.3 μm standard cell library from AMIS [35] is 234 in units equivalent to minimum-sized two-input NAND gates in the library [19,20].

An RTOS/MPSoC system example with the DDU achieves approximately a 1400X speedup in deadlock detection time and a 46 percent speedup in application execution time over an RTOS/MPSoC system with a deadlock detection method in software (please see details in Section 4.5.5) [19,20].

4.4.3   New deadlock avoidance methodology: the DAU

The Deadlock Avoidance Unit, our new approach to deadlock avoidance, not only detects deadlock but also avoids possible deadlock within a few clock cycles and with a small amount of hardware. The DAU, if employed, tracks all requests and releases of resources and avoids deadlock by not allowing any grant or request that leads to a deadlock.

The disadvantages (1), (2) and (3) mentioned in Section 4.3.3.3 unfortunately make the implementation of deadlock avoidance difficult in real systems. Our novel DAU approach to mixing deadlock detection and avoidance (thus, not requiring advanced, a priori knowledge of resource requirements) contributes to easier adaptation of deadlock avoidance in an MPSoC by accommodating both maximum freedom (i.e. maximum concurrency of requests and grants depending on a particular execution trace) with the advantage of deadlock avoidance. Note that the DAU only supports systems with single-instance resources.

Algorithm 1 shows our deadlock avoidance approach. Rather than give an overview of Algorithm 1, we illustrate actual operation with Example 4.4.

Algorithm 1 Deadlock Avoidance Algorithm (DAA)

```
DAA (event) {
1   case (event) {
2       a request:
3           if the resource is available
```

```
 4              grant the resource to the requester
 5          else if the request would cause request deadlock (R-dl)
 6              if the priority of the requester greater than that of the owner
 7                  make the request be pending
 8                  ask the current owner of the resource to release the resource
 9              else
10                  ask the requester to give up resource(s)
11              end-if
12          else
13              make the request be pending
14          end-if
15          break

16   a release:
17          if any process is waiting for the released resource
18              if the grant of the resource would cause grant deadlock
19                  grant the resource to a lower priority process waiting
20              else
21                  grant the resource to the highest priority process waiting
22              end-if
23          else
24              make the resource become available
25          end-if
26 } end-case
}
```

Example 4.4 Avoidance of request deadlock

Figure 4.10 illustrates the DAU, implemented in Verilog HDL. The DAU consists of four parts: a DDU [19, 20], command registers, status registers and DAA logic (implementing Algorithm 1) with a finite state machine. The command registers receive request and grant commands from each PE. The processing results of the DAU are stored into status registers read by all PEs. The DAA logic mainly controls DAU behaviour, i.e. DAA logic interprets and executes commands (requests or releases) from PEs as well as returns processing results back to PEs via status registers.

We now show a sequence of requests and grants that would lead to R-dl as shown in Figure 4.11 and Table 4.1. In this example, we assume the following. (1) Process p_1 requires resources q_1 (VI) and q_2 (IDCT) to complete its job. (2) Process p_2 requires resources q_2 and q_3 (DSP). (3) Process p_3 requires resources q_3 and q_1.

The detailed sequence is shown in Table 4.1. At time t_1, process p_1 requests q_1. Then the DAU checks for the availability of the resource requested, i.e. the DAU checks if no other process either holds or is requesting the resource (line 3 of Algorithm 1). Since q_1 is available, q_1 is granted to p_1 immediately (line 4). Similarly, at time t_2, process p_2 requests and acquires q_2 (line 4), and, at time t_3, process p_3 requests and acquires q_3 (line 4). After that, at time t_4, process p_2 requests q_3; since

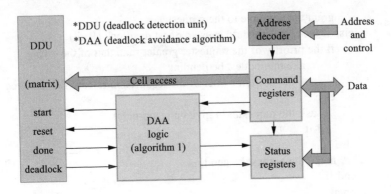

Figure 4.10   DAU architecture

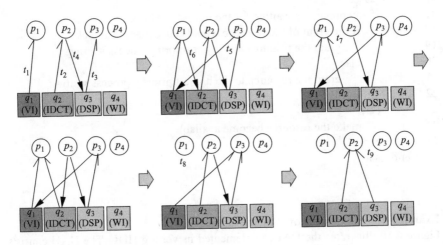

Figure 4.11   Events RAG (request deadlock)

q_3 is unavailable because it was already granted to p_3, the DAU checks the possibility of request deadlock (R-dl) (line 5). The request edge $p_2 \rightarrow q_3$ is temporarily written inside the DDU. Then, deadlock detection check is performed. Since the request does not cause R-dl (line 12), the request is valid and becomes pending (line 13). At time t_5, process p_3 requests q_1; since q_1 was already granted to p_1, and since the request does not cause R-dl, this request also becomes pending (lines 5, 12 and 13).

At time t_6, process p_1 requests q_2; since q_2 is unavailable, the DAU checks (via the DDU) whether the request would cause R-dl (line 5). Since at this time the request $p_1 \rightarrow q_2$ indeed will cause R-dl, the DAU identifies the potential R-dl. Thus, the DAU next compares the priority of p_1 with that of the current owner of q_2 (line 6). Since the priority of p_1 is higher than that of p_2 (i.e. the current owner of q_2), the DAU makes the request be pending for p_1 (line 7) and then asks p_2 to give up q_2 so that p_1 can proceed (line 8).

Table 4.1   A sequence of requests and grants that would lead to R-dl

Time	Events
t_0	The application starts
t_1	p_1 requests q_1; q_1 is granted to p_1
t_2	p_2 requests q_2; q_2 is granted to p_2
t_3	p_3 requests q_3; q_3 is granted to p_3
t_4	p_2 requests q_3, which becomes pending
t_5	p_3 requests q_1, which also becomes pending
t_6	p_1 requests q_2, which is about to lead to R-dl. However, the DAU detects the possibility of R-dl. Thus, the DAU asks p_2 to give up resource q_2
t_7	p_2 releases q_2, which is granted to p_1. A moment later, p_2 requests q_2 again
t_8	p_1 uses and releases q_1 and q_2. Then, while q_1 is granted to p_3, q_2 is granted to p_2
t_9	p_3 uses and releases q_1 and q_3, q_3 is granted to p_2
t_{10}	p_2 finishes its job, and the application ends

As a result, the DAU avoids the potential R-dl, and, at time t_7, p_2 gives up and releases q_2 (line 16). Then, since p_1 is waiting for q_2 (line 17), q_2 needs to be granted to p_1. However, there could be potential grant deadlock (G-dl) when any process is waiting for released resources (line 17); thus, the DAU checks potential G-dl before actually granting via use of the DDU (see References 7, 19 and 20). The DAU temporarily marks a grant of $q_2 \rightharpoonup p_1$ inside the DDU, and then to check potential G-dl, the DAU initiates the DDU to execute its deadlock detection algorithm. Since the temporary grant does not cause G-dl (line 20), it becomes a fixed grant; thus q_2 is granted to p_1 (line 21) (of course, p_2 has to request q_2 again at a later time in order for p_2 to continue making progress).

After using q_1 and q_2, p_1 releases q_1 and q_2 at time t_8. Then, while q_1 needs to be granted to p_3, q_2 needs to be granted to p_2. However, there could also be potential G-dl, the DAU again checks potential G-dl. Since a grant of $q_1 \rightharpoonup p_3$ does not cause G-dl (line 20), q_1 is safely granted to p_3 (line 21). Similarly, q_2 is granted to p_2 (lines 17, 20 and 21). Thus, p_3 uses q_1 and q_3 and at time t_9 releases q_1 and q_3; next q_3 is granted to p_2, which then uses q_2 and q_3 and finishes its job at time t_{10}. ∎

The DAU not only provides a solution to both deadlock and livelock but is also up to 312X faster than an equivalent software solution (please see details in Section 4.5.6) [21]. A more complete DAU description is available in References 7 and 21.

4.4.4   Parallel banker's algorithm

The DAU in Section 4.4.3 can only be used for systems exclusively with single-instance resources because the algorithm employed assumes all resources are single-instance resources. No easy way is known to extend the algorithm employed by the DAU to handle multiple-instance resources. However, a well-known solution exists for multiple-instance resources: the BA [27, 28]. Thus, we have devised the Parallel Banker's Algorithm Unit (PBAU), which can be used not only for a system with single-instance resources but also for a system with multiple-instance resources as well [22].

We now explain the main concept of our novel Parallel Banker's Algorithm (PBA) and its hardware implementation in the PBA Unit (PBAU). Algorithm 2 shows PBA for multiple-instance multiple-resource systems. PBA executes whenever a process is requesting resources and returns the status of whether the request is successfully granted or is rejected due to the possibility of deadlock. PBA decides if the system is still going to be sufficiently safe after the grant, i.e. if there exists at least one H-safe sequence of process executions after some allocation of resources that the process requested.

Before explaining the details of PBA, let us first introduce both notation used as shown in Table 4.2 and data structures used as shown in Table 4.3. In Table 4.2, array[], array[i][] and array[][j] mean 'all elements of the array', 'all elements of row i of the array', and 'all elements of column j of the array', respectively. In Table 4.3, Request[i][j] is a request for resource j from process i. If resource j is a single-instance resource, Request[i][j] is either '0' or '1'; otherwise, if resource j

Table 4.2   Notations for PBA

Notation	Explanation
p_i	A process
q_j	A resource
array[][] or array[]	All elements of the array
array[i][]	All elements of row i of the array
array[][j]	All elements of column j of the array

Table 4.3   Data structures for PBA

Name	Notation	Explanation
Request[i][j]	R_{ij}	Request from process i for resource j
Maximum[i][j]	X_{ij}	Maximum demand of process i for resource j
Available[j]	V_j	Current number of unused resource j
Allocation[i][j]	G_{ij}	Process i's current allocation of j
Need[i][j]	N_{ij}	Process i's potential for more j (Need[i][j] = Maximum[i][j] − Allocation[i][j])
Work[j]	W_j	A temporary storage (array) for Available[j]
Finish[i]	F_i	Whether process i may potentially complete successfully
Wait_count[i]	C_i	Wait count for process i; used to help break livelock

is a multiple-instance resource, Request[i][j] can take on values greater than one. Maximum[i][j] represents the maximum instance demand of process i for resource j. Available[j] indicates the number of available instances of resource j. Allocation[i][j] records the number of instances of resource j allocated to process i. Need[i][j] contains the number of additional instances of resource j that process i may need. Note that Need[i][j] = Maximum[i][j] – Allocation[i][j]. Work[] (i.e. Work[j] for all j) is a temporary storage for Available[] (i.e. Available[j] for all j). Finish[i] denotes whether or not process i can potentially complete successfully (we utilise the notion of an H-safe sequence to compute Finish[i]). Wait_count[i] is a counter for each process and is incremented by one each time a request is denied; proper use of Wait_count[i] can enable some potential livelock situations to be broken.

Algorithm 2 Parallel Banker's Algorithm (PBA)

PBA (Process p_i sends Request[i][] for resources) {
    STEP 1:
1   if ($\forall j$, (Request[i][j] $\leq$ Need[i][j])) /* $\forall$ means *for all* */
2   goto STEP 2
3   else deny p_i's request
    STEP 2:
4   if ($\forall j$, (Request[i][j] $\leq$ Available[j]))
5   goto STEP 3
6   else deny p_i's request, increment Wait_count[i] by one and return
    STEP 3: pretend to allocate requested resources
7      $\forall j$, Available[j] := Available[j] – Request[i][j]
8      $\forall j$, Allocation[i][j] := Allocation[i][j] + Request[i][j]
9      $\forall j$, Need[i][j] := Maximum[i][j] – Allocation[i][j]
    STEP 4: prepare for the H-safety check
10     $\forall j$, Work[j] := Available[j]
11     $\forall i$, Finish[i] := false
    STEP 5: H-safety check
12   Let *able-to-finish(i)* be ((Finish[i] == false) and ($\forall j$, Need[i][j] $\leq$ Work[j]))
13   Find all i such that *able-to-finish(i)*
14   if such i exists,
15         $\forall j$, Work[j] := Work[j] + $\Sigma_{i\ \text{such that }able\text{-}to\text{-}finish(i)}$ Allocation[i][j]
16         $\forall i$, if *able-to-finish(i)* then Finish[i] := true
17   repeat STEP 5
18   else (i.e. no such i exists) goto STEP 6 (end of iteration)
    STEP 6: H-safety decision
19   if ($\forall i$, (Finish[i] == true))
20   then pretended allocations anchor; p_i proceeds (i.e. H-safe)
21   else
22   restore the original state and deny p_i's request (i.e. H-unsafe)
}

Parallel Banker's Algorithm takes as input the maximum requirements of each process and guarantees that if the system began in an H-safe state, the system will always remain in an H-safe state. Tables (data structures or arrays) are maintained of available resources, maximum requirements, current allocations of resources and resources needed, as shown in Table 4.3. PBA uses these tables/matrices to determine whether the state of the system is either H-safe or H-unsafe. When resources are requested by a process, the tables are updated *pretending* the resources were allocated. If the tables will result in an H-safe state, then the request is actually granted; otherwise, the request is not granted, and the tables are returned to their previous states. Please note that possible livelock situations must be detected by alternate methods not part of PBA, e.g. via use of Wait_count[i].

Let us explain Algorithm 2 step by step. A process can request multiple resources at a time as well as multiple instances of each resource. In Step 1, when process i requests resources, PBA first checks if the request (i.e. Request[i][]) does not exceed the maximum claims (i.e. Need[i][]) for process i. If the request is within p_i's pre-declared claims, PBA continues to Step 2; otherwise, if the request is not within p_i's maximum claims, the request is denied with an error code.

In Step 2, PBA checks if there are sufficient available resources for this request. If sufficient resources exist, PBA continues to Step 3; otherwise, the request is denied.

In Step 3, it is pretended that the request could be fulfilled, and the tables are temporarily modified according to the request.

In Step 4, PBA prepares for the H-safety check, i.e. initialises variables Finish[] and Work[]. Work[] is used to search for processes that can finish their jobs by acquiring (if necessary) both resources currently Available[] and resources that will become available during the execution of an H-safe sequence (i.e. resources currently held by previous processes in an H-safe sequence, please see Definition 4.8).

At each iteration of Step 5, PBA tries to find processes that can finish their jobs by acquiring some or all resources available according to Work[] (please see the previous paragraph). If one or more such processes exist, PBA adds all resources that these processes hold to Work[], then declares these processes to be *able-to-finish* (i.e. Finish[i] := true for each process i), and finally repeats Step 5. On the other hand, if no such process exists – meaning either all processes became *able-to-finish* or no more processes can satisfy the comparison (i.e. Need[i][j] $\leq$ Work[j] for all j) – PBA moves to Step 6 to decide whether or not the pretended allocation state is H-safe.

In Step 6, if all processes have been declared to be *able-to-finish*, then the pretended allocation state is in an H-safe state (meaning there exists at least one identifiable H-safe sequence by which all processes can finish their jobs in the order of processes having been declared to be *able-to-finish* in the iterations of Step 5); thus, the requester can safely proceed. However, in Step 6, if there remain any processes unable to finish, the pretended allocation state may cause deadlock; thus, PBA denies the request, restores the original allocation state before the pretended allocation and also increases Wait_count[i] for the requester (process i).

The following example illustrates how PBA works in a simple yet general case.

Table 4.4   A resource allocation state

	Maximum		Allocation		Need		Available	
	q_1	q_2	q_1	q_2	q_1	q_2	q_1	q_2
p_1	3	2	1	1	2	1	1	1
p_2	2	1	1	0	1	1		
p_3	1	2	0	0	1	2		

Table 4.5   Initial resource allocation state for case (1)

	Maximum		Allocation		Need		Available	
	q_1	q_2	q_1	q_2	q_1	q_2	q_1	q_2
p_1	3	2	1	1	2	1	0	1
p_2	2	1	2	0	0	1		
p_3	1	2	0	0	1	2		

Example 4.5   Resource allocation controlled by PBA

Consider a system with three processes p_1, p_2 and p_3 as well as two resources q_1 and q_2, where q_1 has three instances and q_2 has two instances. Table 4.4 shows a possible current resource allocation status in the system as well as maximum resource requirements for each process. Notice that Need[i][j] = Max[i][j] − Allocation[i][j].

Currently one instance of q_1 and one instance of q_2 are given to p_1, and another instance of q_1 is given to p_2. Thus, only one instance of q_1 and one instance of q_2 are available. At this moment, let us consider two cases. (1) When p_2 requests one instance of q_1, will it be safely granted? (2) When p_1 requests one instance of q_2, will it be safely granted? First, considering case (1), let us pretend to grant q_1 to p_2; then the allocation table would be changed as shown in Table 4.5.

Now PBA checks if the resulting system stays in an H-safe state (please see Theorem 4.1). That is, there must exist an H-safe sequence even if all processes were to request up to their maximum needs after the pretended grant [27,28,30]. The following corresponds to Step 5 of PBA. From Table 4.5, if p_2 requests one more instance of q_2 (i.e. up to p_2's maximum claim), since q_2 is available, q_2 is going to be granted to p_2, which will enable p_2 to finish its job and release all resources. Then, the available resources will be two instances of q_1 and one instance of q_2 as shown in Table 4.6.

Next, p_1 can acquire these available resources, finish its job and release all resources; the available resources will be three instances of q_1 and two instances of q_2 as shown in Table 4.7.

Table 4.6   *Resource allocation state in case (1) after*
p₂ finishes

	Maximum		Allocation		Need		Available	
	q_1	q_2	q_1	q_2	q_1	q_2	q_1	q_2
p_1	3	2	1	1	2	1	2	1
p_2	2	1	0	0	2	1		
p_3	1	2	0	0	1	2		

Table 4.7   *Resource allocation state in case (1) after*
p₁ finishes

	Maximum		Allocation		Need		Available	
	q_1	q_2	q_1	q_2	q_1	q_2	q_1	q_2
p_1	3	2	0	0	3	2	3	2
p_2	2	1	0	0	2	1		
p_3	1	2	0	0	1	2		

Table 4.8   *A resource allocation state in case (2)*

	Maximum		Allocation		Need		Available	
	q_1	q_2	q_1	q_2	q_1	q_2	q_1	q_2
p_1	3	2	1	2	2	0	1	0
p_2	2	1	1	0	1	1		
p_3	1	2	0	0	1	2		

Similarly, p_3 can acquire these available resources and finally finish its job. As a result, an H-safe sequence exists in the order p_2, p_1 and p_3. That is, after the grant of one instance of q_1 to p_2, the system remains in an H-safe state.

Now considering case (2), let us pretend to grant one instance of q_2 to p_1; then the allocation table would be changed as shown in Table 4.8 (which is appropriately altered from Table 4.4). From this moment on, neither processes p_1, p_2 nor p_3 can acquire up to its declared maximum unless another process releases resources that the process holds. Thus, the system will not remain in any H-safe state. As a result, the algorithm will deny the request in case (2). ∎

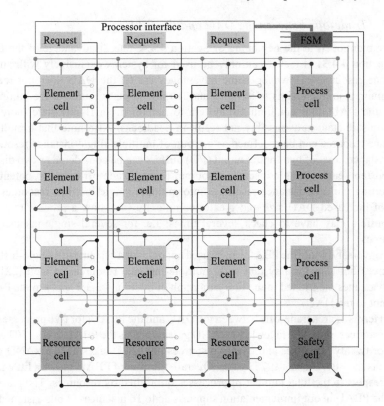

Figure 4.12 PBAU architecture

4.4.5 PBAU architecture

Figure 4.12 illustrates the PBAU architecture. The PBAU is composed of element cells, process cells, resource cells and a safety cell in addition to a Finite State Machine (FSM) and a processor interface.

The Processor Interface (PI) consists of command registers and status registers. PI receives and interprets commands (requests or releases) from processes as well as accomplishes simple jobs such as setting up the number of maximum claims and available resources as well as adjusting the number of allocated and available resources in response to a release of resources. PI also returns processing results back to PEs via status registers as well as activates the FSM in response to a request for resources from a process. Details of each cell are in Reference 7. For exact area estimation, instead of the PBAU shown in Figure 4.12, which is for three processes and three resources, we implemented a PBAU for five processes and five resources (as well as several larger PBAUs [7,22]). For a PBAU with five processes and five resources, when mapped to a 0.25 μm standard cell library from QualCore Logic [36], the resulting area is 1303 in units equivalent to minimum-sized two-input NAND gates in the library; similarly, for a PBAU with twenty processes and twenty resources, the area is 19753 NAND gates in the same technology.

4.4.6   Trade-offs between the DAU and the PBAU

As we mentioned in the beginning of Section 4.4.4, one disadvantage of the DAU (in Section 4.4.3) is that it can only be used for systems exclusively with single-instance resources. However, some advantages are (i) the DAU does not require maximum resource claims in advance and (ii) it restricts resource usage minimally since the DAU does not limit the system to remain in an H-safe state (note that an H-unsafe state may possibly not result in deadlock). Also note that a multiple-instance resource can be redefined as a group of multiple single-instance resources. For instance, a set of two Input/Output (I/O) buffers can be considered as two distinct I/O buffers, each a single-instance resource; thus, a trade-off exists in potentially converting multiple-instance resources into multiple single-instance resources at a cost of increased DAU size and greater complexity (e.g. now a process requiring I/O buffers may have to check several times, i.e. for several single-instance I/O resources).

Conversely, while the PBAU can be used not only for a system with single-instance resources but also for a system with multiple-instance resources as well [22], the disadvantages mentioned in the first paragraph of Section 4.3.3.3 apply to the PBAU (but not to the DAU).

In terms of area trade-offs between the DAU and the PBAU, for five processes and five resources, the DAU takes 1597 NAND gates whereas the PBAU takes 1303 gates, and for twenty processes and twenty resources, the DAU takes 15247 NAND gates whereas the PBAU takes 19753 gates. Overall, the area of PBAU grows a little faster with respect to the total number of processes multiplied by resources. The reason is that the PBAU in our implementation supports up to 16 instances (4 bits assigned) for each resource while the DAU supports only single-instance resources (2 bits assigned for each matrix element).

4.5   Experimentation and results

In this section, we first explain the detailed base MPSoC for experimentation and various configured RTOS/MPSoCs. Then, we demonstrate performance comparisons among the RTOS/MPSoC systems with applications.

4.5.1   Base MPSoC for experimentation

Prior to inclusion of any hardware RTOS components, all configured RTOS/MPSoC experimental simulations presented in this chapter have exactly the same base system consisting of four Motorola MPC755s and four resources as introduced in Section 4.3.2.2. We implemented most of the base system in Verilog HDL; however, please note that we did not implement the MPC755 in Verilog (PE cores and corresponding simulation models are typically provided by vendors, e.g. Seamless CVE [37] provides processor support packages). Each MPC755 has separate instruction and data L1 caches each of size 32 kB. Four resources available are a video interface (VI) device, a DSP, an IDCT unit and a wireless interface (WI) device.

Table 4.9   Configured RTOS/MPSoCs

System	Configured components on top of base pure software RTOS
RTOS1	Pure software RTOS with priority inheritance support in software (Section 4.2.1)
RTOS2	SoCLC with immediate priority ceiling protocol in hardware (Section 4.2.3.1)
RTOS3	SoCDMMU in hardware (Section 4.2.3.2)
RTOS4	Pure software RTOS with a software deadlock detection algorithm
RTOS5	DDU in hardware (Section 4.4.2)
RTOS6	DAA (i.e. Algorithm 1) in software (Section 4.4.3)
RTOS7	DAU in hardware (Section 4.4.3)
RTOS8	PBA (i.e. Algorithm 2) in software (Section 4.4.4)
RTOS9	PBAU in hardware (Section 4.4.5)

These four resources have timers, interrupt generators and input/output ports as necessary to support proper simulation. The base system also has a bus arbiter, a clock driver, a memory controller and 16 MB of shared memory. The master clock rate of the bus system is 10 ns (the minimum external clock period for MPC755 [38], which was designed in .22μ technology). Code for each MPC755 runs on an instruction-accurate (not cycle-accurate) MPC755 simulator provided by Seamless CVE [37].

The experimental simulations were carried out using Seamless Co-Verification Environment (CVE) [37] aided by Synopsys VCS [39] for Verilog HDL simulation and XRAY [40] for software debugging. We have used Atalanta RTOS version 0.3 [6], a shared-memory multiprocessor RTOS, introduced in Section 4.2.1.

4.5.2   Configured RTOS/MPSoCs for experimentation

Using the δ hardware/software RTOS design framework, we have specified various RTOS/MPSoC configurations as shown in Table 4.9. All RTOS/MPSoC configurations are generated primarily based on the base MPSoC described in the previous section.

4.5.3   Execution time comparison between RTOS1 and RTOS2

This section presents the performance comparison between SoCLC (please see Section 4.2.3.1 and References 15–18 for more detail) with priority inheritance in hardware versus the full software Atalanta RTOS with priority inheritance in software. In this comparison, the application used is an algorithmic model of a robot control application with an MPEG decoder. Five tasks in the application represent recognising objects, avoiding obstacles, moving, displaying robot trajectory and recording data. More details are described in Reference 18.

For performance comparison, lock delay and overall execution time for each architecture were measured. The first architecture does not include SoCLC and is named as the 'RTOS1' case; the second architecture includes SoCLC and is named

Table 4.10   Simulation results of the robot application

(Time in clock cycles)	RTOS1	RTOS2	Speedup[&]
Lock delay*	6701	3834	1.75×
Overall execution*	112170	78226	1.43×

* The time unit is a bus clock, and the values are averaged.
[&] The speedup is calculated according to the formula by Hennessy and Patterson [41].

Table 4.11   Execution Time of some SPLASH-2 benchmarks using glibc malloc() *and* free()

Benchmark	Total exe. time (cycles)	Memory management time (cycles)	Percentage of time used for memory management
LU	318307	31512	9.90
FFT	375988	101998	27.13
RADIX	694333	141491	20.38

as the 'RTOS2' case. As seen from Table 4.10, RTOS2 (the SoCLC with priority inheritance in hardware) achieves a 75 percent speedup (i.e. 1.75×) in lock delay and a 43 percent speedup (i.e. 1.43×) in overall execution time when compared to RTOS1 (Atalanta RTOS with priority inheritance in software).

4.5.4   Execution time comparison between RTOS1 and RTOS3

This section demonstrates a performance comparison between RTOS1 and RTOS3. For the performance comparison, several benchmarks taken from the SPLASH-2 application suite have been used: Blocked LU Decomposition (LU), Complex 1D FFT (FFT) and Integer Radix Sort (RADIX) [42,43].

Table 4.11 shows the execution time of the benchmarks in clock cycles and the total number of cycles consumed in memory management when the benchmarks use conventional memory allocation/deallocation techniques (glibc [44,45] malloc() and free()).

Table 4.12 shows the same information introduced in Table 4.11 but with the benchmarks using the SoCDMMU for memory allocation/deallocation. Also, Table 4.12 shows the reduction in memory management execution time due to using the SoCDMMU instead of using glibc malloc() and free() functions. This reduction in the memory management execution time yields speedups in the benchmark execution time. As we can see in Table 4.12, using the SoCDMMU tends to

Table 4.12   Execution time of some SPLASH-2 benchmarks using the SoCDMMU

Benchmark	Total time (cycles)	Memory mgmt. time (cycles)	Percentage of time used for memory mgmt.	Percentage of reduction in time used to manage memory	Percentage of reduction in benchmark exe. time
LU	288271	1476	0.51	95.31	9.44
FFT	276941	2951	1.07	97.10	26.34
RADIX	558347	5505	0.99	96.10	19.59

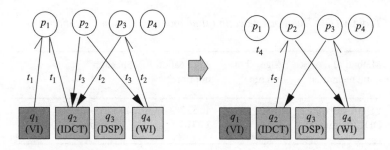

Figure 4.13   Events RAG

speed up application execution time and this speedup is almost equal to the percentage of time consumed by conventional software memory management techniques. For more details, please see Reference 14.

4.5.5   Execution time comparison between RTOS4 and RTOS5

In this experiment, we wanted to identify the performance difference in an application executing using the DDU versus a software deadlock detection algorithm. In RTOS5, the MPSoC has a DDU for five processes and five resources. We devised an application example inspired by the Jini lookup service system [46], in which client applications can request services through intermediate layers (i.e. lookup, discovery and admission). In this experiment, we invoked one process on each PE and prioritised all processes, p_1 being the highest and p_4 being the lowest. The video frame we use for the experiment is a test frame whose size is 64 by 64 pixels. The IDCT processing time of the test frame takes approximately 23 600 clock cycles.

We show a sequence of requests and grants that finally leads to a deadlock as shown in Figure 4.13 and Table 4.13. When the IDCT is released by p_1 at time t_4, the IDCT is granted to p_2 since p_2 has a higher priority than p_3. This last grant will lead to a deadlock in the SoC. More details are described in References 7 and 20.

With the above scenario, we measured both deadlock detection time Δ and application execution time from the application start (t_0) until the detection of a deadlock

Table 4.13   A sequence of requests and grants

Time	Number	Events
t_0	e_0	The application starts
t_1	e_1	p_1 requests IDCT and VI; IDCT and VI are granted to p_1 immediately
t_2	e_2	p_3 requests IDCT and WI; WI is granted to p_3 immediately
t_3	e_3	p_2 requests IDCT and WI. Both p_2 and p_3 wait IDCT
t_4	e_4	IDCT is released by p_1
t_5	e_5	IDCT is granted to p_2 since p_2 has a higher priority than p_3

Table 4.14   Deadlock detection time and application execution time

Method of implementation	Algorithm run time*	Application run time*	Speedup
PDDA software	1830	40 523	$\dfrac{40\,523 - 27\,714}{27\,714} = 46\%$
DDU(hardware)	1.3	27 714	

* The time unit is a bus clock, and the values are averaged.

in two cases: (1) on top of RTOS4 (a software parallel deadlock detection algorithm (PDDA)) and (2) RTOS5 (the DDU). Note that the RTOS initialisation time was excluded (i.e. the RTOS is assumed to be fully operational at time t_0). Table 4.14 shows that (1) in average the DDU achieved a $1408\times$ speedup over the PDDA in software and that (2) the DDU gave a 46 percent speedup in application execution time over the PDDA. The application invoked deadlock detection ten times. Note that a different case where deadlock does not occur so early would of course not show a 46 percent speedup, but instead would show a potentially far lower percentage speedup; nonetheless, for critical situations where early deadlock detection is crucial, our approach can help significantly.

4.5.6   Execution time comparison between RTOS6 and RTOS7

In this experiment, we wanted to identify the performance difference in an application executing on top of RTOS6 (DAA, i.e. Algorithm 1 in software) versus on top of RTOS7 (i.e. the MPSoC with a DAU for five processes and five resources).

4.5.6.1   Application example I

This application performs the same job briefly described in Section 4.5.5. We show a sequence of requests and grants that would lead to grant deadlock (G-dl) as shown in Figure 4.14 and Table 4.15.

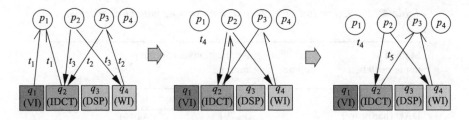

Figure 4.14   Events RAG (grant deadlock)

Table 4.15   A sequence of requests and grants that could lead to grant deadlock (G-dl)

Time	Events
t_0	The application starts
t_1	p_1 requests q_1 and q_2, which are granted to p_1 immediately
t_2	p_3 requests q_2 and q_4; only q_4 is granted to p_3 since q_2 is not available
t_3	p_2 also requests q_2 and q_4
t_4	q_1 and q_2 are released by p_1
t_5	Then, the DAU tries to grant q_2 to p_2 (see an arc from q_2 to p_2 in the middle graph of Figure 4.14) since p_2 has a priority higher than p_3. However, the DAU detects potential G-dl. Thus, the DAU grants q_2 to p_3, which does not lead to a deadlock
t_6	q_2 and q_4 are used and released by p_3
t_7	q_2 and q_4 are granted to p_2
t_8	p_2 finishes its job, and the application ends

Recall that there is no constraint on the ordering of the resource usage. That is, when a process requests a resource and the resource is available, it is granted immediately to the requesting process. At time t_1, process p_1, running on PE1, requests both VI and IDCT, which are then granted to p_1. After that, p_1 starts receiving a video stream through VI and performs some IDCT processing. At time t_2, process p_3, running on PE3, requests IDCT and WI to convert a frame to an image and to send the image through WI. However, only WI is granted to p_3 since IDCT is unavailable. At time t_3, p_2 running on PE2 also requests IDCT and WI, which are not available for p_2. When IDCT is released by p_1 at time t_4, IDCT would typically (assuming the DAU is not used) be granted to p_2 since p_2 has a priority higher than p_3; thus, the system would typically end up in deadlock (i.e. consider the right-side graph in Figure 4.13 where $p_2 \rightarrow q_2$ is changed to $q_2 \rightharpoonup p_2$). However, the DAU discovers the potential G-dl (see the middle graph in Figure 4.14) and then avoids the G-dl by granting IDCT to p_3 even though p_3 has a priority lower than p_2 (see the right-side graph in Figure 4.14). Then, p_3 uses and releases IDCT and WI at time t_6. After that, IDCT and WI are granted to p_2 at time t_7; p_2 then finishes its job at time t_8.

Table 4.16   Execution time comparison example with G-dl possible

Method of implementation	Algorithm runtime*	Application runtime*	Speedup
DAA in software	2188	47 704	$\dfrac{47\,704 - 34\,791}{34\,791} = 37\%$
DAU (hardware)	7	34 791	

* The time unit is a bus clock, and the values are averaged.

With the above scenario, we wanted to measure two figures, average execution time of deadlock avoidance algorithms and total execution time of the application in two cases: (1) on top of RTOS6 (using DAA, i.e. Algorithm 1 in software) versus (2) on top of RTOS7 (using the DAU in hardware).

4.5.6.2   Experimental result I

Table 4.16 shows that the DAU achieves a $312\times$ speedup in average algorithm execution time and gives a 37 percent speedup in application execution time over avoiding deadlock with DAA in software. Note that the application invoked deadlock avoidance 12 times (since every request and release results in deadlock avoidance algorithm invocation).

4.5.6.3   Application example II

We also carried out another experiment in the case of request deadlock (R-dl), which is already introduced in Example 4.4 of Section 4.4.3. This application also performs a job very similar to the job briefly described in Section 4.5.5, but in this execution trace the order of requests is different (see Table 4.1 in Section 4.4.3 for the trace; see Reference 7 for even more detail). We similarly measured two figures, average deadlock avoidance algorithm execution time and total execution time of the application in two cases: (1) on top of RTOS6 versus (2) RTOS7.

4.5.6.4   Experimental result II

Table 4.17 demonstrates that the DAU achieves a $294\times$ speedup in average algorithm execution time and gives a 44 percent speedup in application execution time over avoiding deadlock with DAA in software. Note that the application invoked deadlock avoidance 14 times.

4.5.7   *Execution time comparison between RTOS8 and RTOS9*

To measure the performance difference between RTOS8 (using the BA in software) and RTOS9 (exploiting the PBAU), we execute a sample robotic application which performs the following: recognising objects, avoiding obstacles and displaying trajectory requiring DSP processing; robot motion and data recording involving accessing IO buffers; and proper real-time operation (e.g. maintaining balance) of

Table 4.17   Execution time comparison example with R-dl possible

Method of implementation	Algorithm runtime*	Application runtime*	Speedup
DAA in software	2102	55 627	$\dfrac{55\,627 - 38\,508}{38\,508} = 44\%$
DAU (hardware)	7.14	38 508	

* The time unit is a bus clock, and the values are averaged.

Table 4.18   A sequence of requests and releases

Time	Events
t_0	The application starts, and the numbers of available resources in the system are set
$t_1 \sim t_5$	$p_1 \sim p_5$ set their maximum claims for resources as shown in Table 4.19
t_6	p_1 requests one instance of r_2
t_7	p_2 requests two instances of r_1
t_8	p_3 requests three instances of r_1 and two instances of r_3
t_9	p_4 requests two instances of r_1, one instance of r_2 and one instance of r_3
t_{10}	p_5 requests two instances of r_3
t_{11}	p_1 requests two instances of r_2 and one instance of r_3
t_{12}	p_5 requests one instance of r_1. So far, granting all requests results in a system in an H-safe state
t_{13}	p_5 again requests one more instance of r_1, which results in an H-unsafe state. Thus, this request is denied. The wait count for p_5 is increased
t_{14}	p_3 releases two instances of r_1 and two instances of r_3
t_{15}	p_3 initiates a faulty request (i.e. it requests five instances of r_1, r_2 and r_3, respectively), which of course is denied
t_{16}	p_5 again requests one more instance of r_1, which now results in an H-safe state. Thus, this request is granted. The wait count for p_5 is cleared
t_{17}	p_1 finishes its job and releases three instances of r_2 and one instance of r_3
t_{18}	p_2 releases two instances of r_1
t_{19}	p_3 releases one instance of r_1
t_{20}	p_4 releases two instances of r_1, one instance of r_2 and one instance of r_3
t_{21}	p_5 releases two instances of r_1 and two instances of r_3, the application ends

the robot demanding fast and deterministic allocation and deallocation of memory blocks. This application invokes a sequence of requests and releases. In a specific trace, the sequence has ten requests, six releases and five claim settings as shown in Table 4.18 with one faulty request that violates a pre-declared maximum claim (e.g. Request[i][j] > Need[i][j]) and one additional request that leads to an H-unsafe state. Please note that every command is processed by an avoidance algorithm (either PBAU or BA in software).

Table 4.19 Initial resource allocation state at time t_5

	Maximum			Allocation			Need			Available		
	q_1	q_2	q_3	q_1	q_2	q_3	q_1	q_2	q_3	q_1	q_2	q_3
p_1	7	5	3	0	0	0	7	5	3	10	5	7
p_2	3	2	2	0	0	0	3	2	2			
p_3	9	0	2	0	0	0	9	0	2			
p_4	2	2	2	0	0	0	2	2	2			
p_5	4	3	3	0	0	0	4	3	3			

Table 4.20 Resource allocation states

	Allocation			Need			Available		
	r_1	r_2	r_3	r_1	r_2	r_3	r_1	r_2	r_3
At time t_{10}									
p_1	0	1	0	7	4	3	3	3	2
p_2	2	0	0	1	2	2			
p_3	3	0	2	6	0	0			
p_4	2	1	1	0	1	1			
p_5	0	0	2	4	3	1			
At time t_{12}									
p_1	0	3	1	7	2	2	2	1	1
p_2	2	0	0	1	2	2			
p_3	3	0	2	6	0	0			
p_4	2	1	1	0	1	1			
p_5	1	0	2	3	3	1			

Detailed sequence explanation is as follows. There are five processes and three resources in the system. Table 4.19 shows the available resources and maximum claims of each process in the system at time t_5 ('Maximum' equals 'Need' currently).

Table 4.20 shows the resource allocation state at time t_{10} as processes are using resources. After two more requests, Table 4.20 shows the resource allocation state at time t_{12}. So far, all requests result in H-safe states. However, at time t_{13} in Table 4.18, when p_5 requests one additional instance of resource r_1, the system would result in an H-unsafe state if granted. Thus, PBAU rejects the request; Wait_count[5] (see Table 4.3 and Section 4.4.4) for p_5 is incremented, and p_5 needs to re-request r_1 later.

At time t_{14}, p_3 releases two instances of r_1 and two instances of r_3, and the resulting allocation state is shown in Table 4.21. At time t_{16}, p_5 re-requests one additional instance of resource r_1, and the request is granted as shown Table 4.21. Wait_count[5] for p_5 is cleared (set to zero).

Table 4.21   Resource allocation states

	Allocation			Need			Available		
	r_1	r_2	r_3	r_1	r_2	r_3	r_1	r_2	r_3
At time t_{14}									
p_1	0	3	1	7	2	2	4	1	3
p_2	2	0	0	1	2	2			
p_3	1	0	0	8	0	2			
p_4	2	1	1	0	1	1			
p_5	1	0	2	3	3	1			
At time t_{16}									
p_1	0	3	1	7	2	2	3	1	3
p_2	2	0	0	1	2	2			
p_3	1	0	0	8	0	2			
p_4	2	1	1	0	1	1			
p_5	2	0	2	2	3	1			

Table 4.22   Application execution time comparison for PBAU test

Method of implementation	Algorithm execution time	PBAU speedup	Application execution time	Application speedup
BA in software PBAU (hardware)	5398.4 3.32	$\dfrac{5398.4 - 3.32}{3.32} = 1625\times$	221 259 185 716	$\dfrac{221\,259 - 185\,716}{185\,716} = 19\%$

* The time unit is a clock cycle, and the values are averaged.

After time t_{16}, as time progresses, all processes finish their jobs and release allocated resources.

With the above scenario, summarised in Tables 4.18–4.21, we measure two figures, average deadlock avoidance algorithm execution time and total application execution time in two cases: (1) on top of RTOS8 (using BA in software) versus (2) on top of RTOS9 (exploiting the PBAU).

Table 4.22 shows that PBAU achieves approximately a $1600\times$ speedup in average deadlock avoidance algorithm execution time and gives a 19 percent speedup in application execution time over avoiding deadlock with BA in software. Please note that during the run-time of the application, each avoidance method (PBAU or BA in software) is invoked 22 times in both cases, respectively (since every request and every release invokes a deadlock avoidance calculation). Table 4.23 represents the average algorithm execution time distribution in terms of different types of commands.

Thus, while BA in software spends roughly 5400 clock cycles on average at each invocation in this experiment, PBAU only spends 3.32 clocks on average.

Table 4.23   Algorithm execution time comparison between PBAU vs. BA in software

	Set available	Set max claim	Request command	Release command	Faulty command
# of commands	1	5	9	6	1
BA in software	416	427	11 337	2270	560
PBAU (hardware)	1	1	6.5	1	2

* The time unit is a clock cycle, and the values are averaged if there are multiple commands of the same type. '#' denotes 'the number of'.

4.6   Conclusions

This chapter presents a methodology for hardware/software partitioning of operating systems among pre-designed hardware and software RTOS pieces. The δ hardware/software RTOS/MPSoC codesign framework has been used to configure and generate simulatable RTOS/MPSoC designs having both appropriate hardware and software interfaces for each specified system architecture. The δ framework is specifically designed to help RTOS/MPSoC designers more easily and quickly explore their design space with available hardware and software modules so that they can efficiently search and discover a few optimal solutions matched to the specifications and requirements of their design.

We have configured, generated and simulated various RTOS/MPSoC systems with available hardware/software RTOS components such as SoCLC, SoCDMMU, DDU, DAU, PBAU and equivalent software modules, respectively. From the simulations using Seamless CVE from Mentor Graphics, we show that our methodology is a viable approach to rapid hardware/software partitioning of OS. In addition, we demonstrated the following with experiments. (i) A system with the SoCLC shows a 75 percent speedup in lock delay and a 43 percent speedup in overall execution time when compared to a system implementing priority inheritance and lock handling in software. (ii) A system with the SoCDMMU reduced benchmark execution time by 9.44 percent or more as compared to a system without the SoCDMMU. (iii) An RTOS/MPSoC system with the DDU achieved approximately a 1400X speedup in deadlock detection time and a 46 percent speedup in application execution time over an RTOS/MPSoC system with a deadlock detection method in software. (iv) A system with the DAU reduced deadlock avoidance time by 99 percent (about 300X) and application execution time by 44 percent as compared to a system with a deadlock avoidance algorithm in software. (v) The PBAU achieved a roughly 1600X speedup in average deadlock avoidance algorithm execution time and a 19 percent speedup in application execution time over avoiding deadlock with a version of the Bankers Algorithm in software.

In summary, we present recent updates to the δ hardware software RTOS partitioning framework. We focus on the DAU and the PBAU, the first work known to the authors on hardware support for deadlock avoidance in MPSoC/RTOS hardware/software codesign.

Acknowledgements

This research is funded by NSF under INT-9973120, CCR-9984808 and CCR-0082164. We would like to acknowledge donations received from Denali, HP, Intel, QualCore, Mentor Graphics, National Semiconductor, Sun and Synopsys.

References

1 LEE, J., RYU, K., and MOONEY, V.: 'A framework for automatic generation of configuration files for a custom RTOS'. Proceedings of the international conference on *Engineering of reconfigurable systems and algorithms* (ERSA'02) Las Vegas, Nevada, 2002, pp. 31–37

2 LEE, J., MOONEY, V., DALEBY, A., INGSTROM, K., KLEVIN, T., and LINDH, L.: 'A comparison of the RTU hardware RTOS with a hardware/software RTOS'. Proceedings of the *Asia and South Pacific design automation* conference (ASPDAC 2003), Kitakyushu, Japan, January 2003, pp. 683–688

3 MOONEY, V., and BLOUGH, D.: 'A hardware-software realtime operating system framework for SOCs', *IEEE Design and Test of Computers*, November–December 2002, pp. 44–51

4 MOONEY, V.: 'Hardware/software partitioning of operating systems', in *Embedded software for SoC*, by JERRAYA, A., YOO, S., VERKEST, D., and WEHN, N. (Eds): (Kluwer Academic Publishers, Boston, MA, USA, 2003) pp. 187–206

5 LEE, J., and MOONEY, V.: 'Hardware/software partitioning of operating systems: Focus on deadlock detection and avoidance', *IEE Proceedings Computer and Digital Techniques*, March 2005, **152**(2), pp. 167–182

6 SUN, D., BLOUGH, D., and MOONEY, V.: 'Atalanta: A new multiprocessor RTOS kernel for system-on-a-chip applications'. Technical Report GIT-CC-02-19, College of Computing, Georgia Tech, Atlanta, GA, 2002, http://www.coc.gatech.edu/research/pubs.html

7 LEE, J.: 'Hardware/software deadlock avoidance for multiprocessor multiresource system-on-a-chip'. PhD thesis, School of ECE, Georgia Institute of Technology, Atlanta, GA, Fall 2004, http://etd.gatech.edu/theses/available/etd-11222004-083429/

8 RYU, K., and MOONEY, V.: 'Automated bus generation for multiprocessor SoC Design'. Proceedings of the *Design automation and test in Europe* conference (DATE'03), Munich, Germany, March 2003, pp. 282–287.

9 RYU, K., and MOONEY, V.: 'Automated bus generation for multiprocessor SoC design'. *IEEE Transactions on Computer-Aided Design of Integrated Circuits and Systems*, 2004, **23**(11), pp. 1531–1549

10   RYU, K.: 'Automatic generation of bus systems'. PhD thesis, School of ECE, Georgia Institute of Technology, Atlanta, GA, USA, Summer 2004, http://etd.gatech.edu/theses/available/etd-07122004-121258/

11   AKGUL, B., and MOONEY, V.: 'PARLAK: Parametrized Lock Cache generator'. Proceedings of the *Design automation and test in Europe* conference (DATE'03), Munich, Germany, March 2003, pp. 1138–1139

12   SHALAN, M., and MOONEY, V.: 'Hardware support for real-time embedded multiprocessor system-on-a-chip memory management'. Proceedings of the tenth international symposium on *Hardware/software codesign* (CODES'02), Ester Park, CO, May 2002, pp. 79–84

13   SHALAN, M., SHIN, E., and MOONEY, V.: 'DX-Gt: Memory management and crossbar switch generator for multiprocessor system-on-a-chip'. Proceedings of the 11th workshop on *Synthesis and system integration of mixed information technologies* (SASIMI'03), Hiroshima, Japan, April 2003, pp. 357–364

14   SHALAN, M.: 'Dynamic memory management for embedded real-time multiprocessor system-on-a-chip'. PhD thesis, School of ECE, Georgia Institute of Technology, Atlanta, GA, USA, Fall 2003, http://etd.gatech.edu/theses/available/etd-11252003-131621/

15   AKGUL, B., LEE, J., and MOONEY, V.: 'A System-on-a-Chip Lock Cache with task preemption support'. Proceedings of the international conference on *Compilers, architecture and synthesis for embedded systems* (CASES'01), Atlanta, CA, November 2001, pp. 149–157

16   AKGUL, B., and MOONEY, V.: 'The System-on-a-Chip Lock Cache', *Transactions on Design Automation for Embedded Systems*, 2002, 7(1–2), pp. 139–174

17   AKGUL, B., MOONEY, V., THANE, H., and KUACHAROEN, P.: 'Hardware support for priority inheritance'. Proceedings of the IEEE *Real-time systems symposium* (RTSS'03), Cancun, Mexico, December 2003, pp. 246–254

18   AKGUL, B.: 'The System-on-a-Chip Lock Cache'. PhD thesis, School of ECE, Georgia Institute of Technology, Atlanta, GA, USA, Spring 2004, http://etd.gatech.edu/theses/available/etd-04122004-165130/

19   SHIU, P., TAN, Y., and MOONEY, V.: 'A novel parallel deadlock detection algorithm and architecture'. Proceedings of the 9th international symposium on *Hardware/software codesign* (CODES'01), Copenhagen, Denmark, April 2001, pp. 30–36

20   LEE, J., and MOONEY, V.: 'An O(min(m,n)) parallel deadlock detection algorithm'. ACM Trans on Design automation of electronic systems (TODAES), July 2005, 10(3), pp. 573–586

21   LEE, J., and MOONEY, V.: 'A novel deadlock avoidance algorithm and its hardware implementation'. Proceedings of the 12th international conference on *Hardware/software codesign and system synthesis* (CODES/ISSS'04), Stockholm, Sweden, September 2004, pp. 200–205

22   LEE, J., and MOONEY, V.: 'A novel O(n) Parallel Banker's Algorithm for system-on-a-chip'. Proceedings of the *Asia and South Pacific design*

automation conference (ASPDAC 2005), Shanghai, China, January 2005, pp. 1304–1308

23 VERILOG PreProcessor: http://surefirev.com/vpp/

24 SHA, L., RAJKUMAR, R., and LEHOCZKY, J.: 'Priority inheritance protocols: An approach to real-time synchronization', *IEEE Transactions on Computers*, 1990, **39**(9), pp. 1175–1185

25 XILINX: http://www.xilinx.com/

26 MAEKAWA, M., OLDHOEFT, A., and OLDEHOEFT, R.: 'Operating systems – advanced concepts' (Benjamin/Cummings Publishing Co., Menlo Park, CA, 1987)

27 DIJKSTRA, E.: 'Cooperating sequential processes'. Technical Report EWD-123, Technological University, Eindhoven, The Netherlands, September 1965

28 HABERMANN, A.: 'Prevention of system deadlocks', *Communications of the ACM*, 1969, **12**(7), pp. 373–377, 385

29 SHOSHANI, A., and COFFMAN, E. Jr.: 'Detection, prevention and recovery from deadlocks in multiprocess, multiple resource systems'. Proceedings of the 4th annual Princeton conference on *Information sciences and system*, Princeton, NJ, March 1970

30 HOLT, R.: 'Some deadlock properties of computer systems', *ACM Computing Surveys*, 1972, **4**(3), pp. 179–196

31 LEIBFRIED, T. Jr.: 'A deadlock detection and recovery algorithm using the formalism of a directed graph matrix', *Operation Systems Review*, 1989, **23**(2), pp. 45–55

32 KIM, J., and KOH, K.: 'An O(1) time deadlock detection scheme in single unit and single request multiprocess system'. IEEE *TENCON* '91, vol. 2, August 1991, pp. 219–223

33 COFFMAN, E., ELPHICK, M., and SHOSHANI, A.: 'System deadlocks', *ACM Computing Surveys*, 1971, **3**(2), pp. 67–78

34 BELIK, F.: 'An efficient deadlock avoidance technique', *IEEE Transactions on Computers*, 1990, **39**(7), pp. 882–888

35 AMI SEMICONDUCTOR: http://www.amis.com

36 QualCore Logic: http://www.qualcorelogic.com/

37 MENTOR GRAPHICS: 'Hardware/Software Co-Verification: Seamless', http://www.mentor.com/seamless/

38 MPC755 RISC microprocessor hardware specification, http://www.freescale.com/files/32bit/doc/data_sheet/MPC755EC.pdf

39 SYNOPSYS, VCS Verilog Simulator, http://www.synopsys.com/products/simulation/simulation.html

40 MENTOR GRAPHICS: XRAY_Debugger, http://www.mentor.com/xray/

41 HENNESSY, J., and PATTERSON, D.: 'Computer architecture – a quantitative approach' (Morgan Kaufmann Publisher, Inc., San Francisco, CA, 1996)

42 STANFORD UNIVERSITY: 'Stanford Parallel Applications for Shared Memory (SPLASH)', http://www-flash.stanford.edu/apps/SPLASH/

43   WOO, S., OHARA, M., TORRIE, E., SINGH, J., and GUPTA, A.: 'The SPLASH-2 programs: characterization and methodological considerations'. Proceedings of the 22nd international symposium on *Computer architecture*, June 1995, pp. 24–36

44   THE FREE SOFTWARE FOUNDATION: The GNU project, the GCC compiler, http://gcc.gnu.org/

45   THE FREE SOFTWARE FOUNDATION: The GNU project, the GNU C library, http://www.gnu.org/software/libc/manual/

46   MORGAN, S.: 'Jini to the rescue', *IEEE Spectrum*, April 2000, **37**(4), pp. 44–49

Chapter 5

Models of computation in the design process

Axel Jantsch and Ingo Sander

5.1 Introduction

A system-on-chip (SoC) can integrate a large number of components such as microcontrollers, digital signal processors (DSPs), memories, custom hardware and reconfigurable hardware in the form of field programmable gate arrays (FPGAs) together with analogue-to-digital (A/D) and digital-to-analogue (D/A) converters on a single chip (Figure 5.1). The communication structures become ever more sophisticated consisting of several connected and segmented buses or even packet switched networks. In total there may be dozens or hundreds of such components on a single SoC. These architectures offer an enormous potential but they are heterogeneous and tremendously complex. This also applies to embedded software. Moreover, the overall system complexity grows faster than system size due to the component interaction. In fact, intra-system communication is becoming the dominant factor for design, validation and performance analysis. Consequently, issues of communication, synchronisation and concurrency must play a prominent role in all system design models and languages.

The design process for SoCs is complex and sophisticated. From abstract models for requirements definition and system specification more and more refined models are derived leading eventually to low level implementation models that describe the layout and the assembler code. Most of the models are generated and processed either fully automatically or with tool support. Once created models have to be verified to check their consistency and correctness.

Figure 5.2 depicts a few of the models typically generated and transformed during a design project. Different design tasks require different models. A system level feasibility study and performance analysis needs key performance properties of the architecture, components and functions but not a full behavioural model. Scheduling and schedulability analysis need abstract task graphs, timing requirements and an abstract model of the scheduler in the operating system. Synthesis and verification

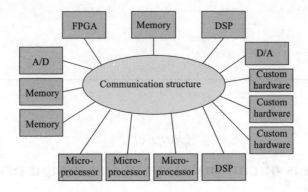

Figure 5.1   A possible system-on-a-chip architecture

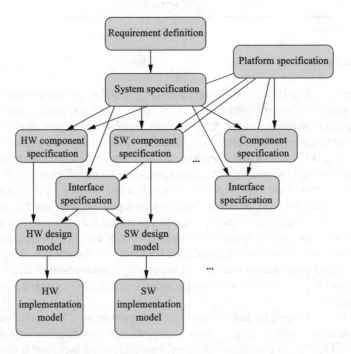

Figure 5.2   A SoC design process involves many models

tools need behavioural models at a proper abstraction level. Noise, EMC analysis, test pattern generators and many other tools have their own requirements on the models they use.

Since all design tasks put specific requirements on a model, we may ask, how strong the influence of a model of computation is on the potential and efficiency of design techniques. The answers are dependent on the specific design tasks and tools. We consider only a small selection of tasks, namely HW synthesis, simulation and formal verification. For models of computation for embedded software, see Part II, Chapter 7. Also, we cannot take all possible models into account, but we restrict the discussion to three specific MoC classes: untimed MoCs, synchronous MoCs, discrete and continuous time MoCs. They are distinguished by their abstraction of time and their synchronisation behaviour which will allow us to analyse design techniques with respect to these aspects. Other aspects such as data representation will not be covered.

In the next section we introduce the MoCs under consideration and review some of their important properties. In Section 5.3 we trace MoCs in different design phases and in Section 5.4, we discuss the importance of MoCs for synthesis, simulation and verification techniques.

5.2   Models of computation

We use the term 'Model of Computation' (MoC) to focus on issues of concurrency and time. Consequently, even though it has been defined in different ways by different authors (see for instance References 1–5), we use it to define the time representation and the semantics of communication and synchronisation between processes in a process network. Thus, a MoC defines how computation takes place in a structure of concurrent processes, hence giving semantics to such a structure [6,7]. These semantics can be used to formulate an abstract machine that is able to execute a model. 'Languages' are not computational models, but have underlying computational models. For instance the languages VHDL, Verilog and SystemC share the same discrete time, event driven computational model. On the other hand, languages can be used to support more than one computational model. In ForSyDe [8] the functional language Haskell [9] is used to express several models of computation. Libraries have been created for synchronous, untimed and discrete time models of computation. Standard ML has been used similarly [10]. SystemC has also been extended to support SDF (synchronous dataflow) and CSP (communicating sequential processes) MoCs in addition to its native discrete time MoC [11].

To choose the right model of computation is of utmost importance, since each MoC has certain properties. As an example consider a process network modelled as a discrete time system in SystemC. In the general case automatic tools will not be able to compute a static schedule for a single processor implementation, even if the process network would easily allow it. For this reason Patel and Shukla [11] have extended SystemC to support an SDF MoC. The same process network expressed as an SDF can then easily be statically scheduled by a tool.

Skillicorn and Talia [12] discuss models of computation for parallel architectures. Their community faces similar problems as those in design of embedded systems. In fact all typical parallel computer structures (SIMD, MIMD[1]) can be implemented on a SoC architecture. Recognising that programming of a large number of

---

[1] Flynn has classified typical parallel data structures in Reference 15 where SIMD is an abbreviation for Single Instruction, Multiple Data and MIMD for Multiple Instruction, Multiple Data.

communicating processors is an extremely complex task, they try to define properties for a suitable model of parallel computation. They emphasise that a model should hide most of the details (decomposition, mapping, communication, synchronisation) from programmers, if they are able to manage intellectually the creation of software. The exact structure of the program should be inserted by the translation process rather than by the programmer. Thus models should be as abstract as possible, which means that the parallelism has not even been made explicit in the program text. They point out that ad hoc compilation techniques cannot be expected to work on problems of this complexity, but advocate building software that is correct by construction rather than by verifying program properties after construction. Programs should be architecture independent to allow reuse. The model should support cost measures to guide the design process and should have guaranteed performance over a useful variety of architectures.

In the following sections, we present a number of important models of computations and give their key properties. Following References 1 and 7 we organise them according to their time abstraction. We distinguish between discrete time models, synchronous models where a cycle denotes an abstract notion of time, and untimed models. This is consistent with the tagged-signal model proposed by Lee and Sangiovanni-Vincentelli [2]. There each event has a time tag and different time tag structures result in different MoCs. For example, if the time tags correspond to real numbers we have a continuous time model; integer time tags result in discrete time models; time tags drawn from a partially ordered set result in an untimed MoC.

Models of computation can be organised along other criteria, e.g. along with the kinds of elements manipulated in a MoC which leads Paul and Thomas [3] to a grouping of MoCs for hardware artefacts, for software artefacts and for design artefacts. However, an organisation along properties that are not inherent is of limited use because it changes when MoCs are used in different ways.

A consequence of an organisation along the time abstraction is that all strictly sequential models such as finite state machines and sequential algorithms are not distinguished. All of them can serve for modelling individual processes, while the semantics of the MoC defines the process interaction and synchronisation.

5.2.1   *Continuous time models*

When time is represented by a continuous set, usually the real numbers, we talk of a continuous time MoC. Prominent examples of continuous time MoC instances are Simulink [13], VHDL-AMS and Modelica [14]. The behaviour is typically expressed as equations over real numbers. Simulators for continuous time MoCs are based on differential equation solvers that compute the behaviour of a model including arbitrary internal feedback loops.

Due to the need to solve differential equations, simulations of continuous time models are very slow. Hence, only small parts of a system are usually modelled with continuous time such as analogue and mixed signal components.

To be able to model and analyse a complete system that contains analogue components, mixed-signal languages and simulators such as VHDL-AMS have been developed. They allow us to model the pure digital parts in a discrete time MoC and the analogue parts in a continuous time MoC. This allows for complete system simulations with acceptable simulation performance. It is also a typical example where heterogeneous models based on multiple MoCs have clear benefits.

5.2.2 Discrete time models

Models where all events are associated with a time instant and the time is represented by a discrete set, such as the natural numbers, are called discrete time models.[2]

Discrete time models are often used for the simulation of hardware. Both VHDL [16] and Verilog [17] use a discrete time model for their simulation semantics. A simulator for discrete time MoCs is usually implemented with a global event queue that sorts occurring events. Discrete time models may have causality problems due to zero-delay in feedback loops, which are discussed in Section 5.2.4.

5.2.3 Synchronous models

In synchronous MoCs time is also represented by a discrete set, but the elementary time unit is not a physical unit but more abstract due to two abstraction mechanisms:

1 Each event occurs in a specific evaluation cycle (also called time slot or clock cycle). The occurrence of evaluation cycles is globally synchronised even for independent parts of the system. But the relative occurrence of events within the same evaluation cycle is not further specified. Thus, events within an evaluation cycle are only partially ordered as defined by causality and data dependences only.
2 Intermediate events that are not visible at the end of an elementary evaluation cycle are irrelevant and can be ignored.

In each evaluation cycle all processes evaluate once and all events occurring during this process are considered to occur simultaneously.

The synchronous assumption can be formulated according to Reference 19. The synchronous approach considers '*ideal* reactive systems that produce their outputs *synchronously* with their inputs, their reaction taking no observable time'. This implies that the computation of an output event is instantaneous. The synchronous assumption leads to a clean separation between computation and communication. A global clock triggers computations that are conceptually simultaneous and

---

[2] Sometimes this group of MoCs is denoted as 'discrete event MoC'. However, strictly speaking 'discrete event' and 'discrete time' are independent, orthogonal concepts. The first denotes a model where the set of the event values is a discrete set while the second denotes a model with time values drawn from a discrete set, e.g. integers. In contrast, 'continuous time' and 'continuous event' models use continuous sets for time and event values, respectively, e.g. the real numbers. All four combinations occur in practice: continuous time/continuous event models, continuous time/discrete event models, discrete time/continuous event models and discrete time/discrete event models. See for instance Reference 18 for a good coverage of discrete event models.

instantaneous. This assumption frees the designer from the modelling of complex communication mechanisms and provides a solid base for formal methods.

A synchronous design technique has been used in hardware design for clocked synchronous circuits. A circuit behaviour can be described deterministically independent of the detailed timing of gates by separating combinational blocks from each other with clocked registers. An implementation will have the same behaviour as the abstract circuit under the assumption that the combinational blocks are 'fast enough' and that the abstract circuit does not include zero-delay feedback loops.

The synchronous assumption implies a simple and formal communication model. Concurrent processes can easily be composed together. However, feedback loops with zero-delay may cause causality problems which are discussed next.

5.2.4 Feedback loops in discrete time and synchronous models

Discrete time models allow zero-delay computation; in perfectly synchronous models this is even a basic assumption. As a consequence, feedback loops may introduce inconsistent behaviour. In fact, feedback loops as illustrated in Figure 5.3 may have no solution, may have one solution or may have many solutions.

Figure 5.3(a) shows a system with a zero-delay feedback loop that does not have a stable solution. If the output of the Boolean AND function is `True` then the output of the NAND function is `False`. But this means that the output of the AND function has to be `False`, which is in contradiction to the starting point of the analysis. Starting with the value `False` on the output of AND does not lead to a stable solution either. Clearly there is no solution to this problem.

Figure 5.3(b) shows a system with a feedback loop with multiple solutions. Here the system is stable, if both AND functions have `False` or if both AND functions have `True` as their output value. Thus the system has two possible solutions.

Figure 5.3(c) shows a feedback loop with only one solution. Here the only solution is that both outputs are `True`.

It is crucial for the design of safety-critical systems that feedback loops with no solution as in Figure 5.3(a) are detected and eliminated, since they result in an oscillator. Also feedback loops with multiple solutions imply a risk for safety-critical systems, since they lead to non-determinism. Non-determinism may be acceptable, if it is detected and the designer is aware of its implications, but may have serious consequences, if it stays undetected.

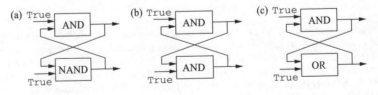

Figure 5.3   A feedback loop in a synchronous system. System (a) has no solutions, (b) has multiple solutions and (c) has a single solution

Since feedback loops in discrete time and synchronous models are of such importance there are several approaches which address this problem [6].

Microstep: In order to introduce an order between events that are produced and consumed in an event cycle, the concept of microsteps has been introduced into languages such as VHDL. VHDL distinguishes between two dimensions of time. The first one is given by a time unit, e.g. a picosecond, while the second is given by a number of delta-delays. A delta-delay is an infinitesimal small amount of time. An operation may take zero time units, but it takes at least one delta-delay. Delta-delays are used to order operations within the same time unit.

While this approach partly solves the zero-delay feedback problem, it introduces another problem since delta delays will never cause the advance of time measured in time units. Thus during an event cycle there may be an infinite amount of delta-delays. This would be the result, if Figure 5.3(a) would be implemented in VHDL, since each operation causes time to advance with one delta-delay. An advantage of the delta-delay is that simulation will reveal that the composite function oscillates. However, a VHDL simulation would not detect that Figure 5.3(b) has two solutions, since the simulation semantics of VHDL would assign an initial value for the output of the AND gates (False[3]) and thus would only give one stable solution, concealing the non-determinism from the designer. Another serious drawback of the microstep concept is that it leads to more complicated semantics, which complicates formal reasoning and synthesis.

Forbid zero-delays: The easiest way to cope with the zero-delay feedback problem is to forbid them. In case of Figures 5.3(a) and 5.3(b) this would mean the insertion of an extra delay function, e.g. after the upper AND function. Since a delay function has an initial value the systems will stabilise. Assuming an initial value of True, Figure 5.3(a) will stabilise in the current event cycle with the values False for the output of the NAND function and False for the value of the AND function. Figure 5.3(b) would stabilise with the output value True for both AND functions. A possible problem with this approach is that a stable system such as Figure 5.3(c) is rejected, since it contains a zero-delay feedback loop. This approach is adopted in the synchronous language Lustre [20] and in synchronous digital hardware design. When used in a synchronous MoC the resulting MoC variant is sometimes called 'clocked synchronous MoC' [1].

Unique fixed-point: The idea of this approach is that a system is seen as a set of equations for which one solution in the form of a fixed-point exists. There is a special value $\perp$ ('bottom') that allows it to give systems with no solution or many solutions a fixed-point solution. The advantage of this method is that the system can be regarded as a functional program, where formal analysis will show if the system has a unique solution. Also systems that have a stable feedback loop as in Figure 5.3(c) are accepted, while the systems of Figures 5.3(a) and 5.3(b) are rejected

---

[3] VHDL defines the data type Boolean by means of type Boolean is (False, True). At program start variables and signals take the leftmost value of their data type definitions; in case of the boolean data type the value False is used.

(the result will be the value ⊥ as the solution for the feedback loops). Naturally, the fixed-point approach demands more sophisticated semantics, but the theory is well understood [21]. Esterel has adopted this approach and the constructive semantics of Esterel are described in Reference 22.

Relation based: This approach allows the specification of systems as relations. Thus a system specification may have zero solutions, one solution or multiple solutions. Though an implementation of a system usually demands a unique solution, other solutions may be interesting for high-level specifications. The relation-based approach has been employed in the synchronous language Signal [23].

5.2.5   Untimed models

In untimed models there is no global notion of time. If one event does not depend directly or indirectly on another event, it is undefined if one event occurs at the same time as, earlier or later than the other event. Hence, the only ordering on the occurrence of events is determined by causal relationships. If one event depends on another event, it must occur after the other event.

5.2.5.1   Data flow process networks

Data flow process networks [24] are a special variant of Kahn process networks [25, 26]. In a Kahn process network processes communicate with each other via unbounded FIFO channels. Writing to these channels is 'non-blocking', i.e. they always succeed and do not stall the process, while reading from these channels is 'blocking', i.e. a process that reads from an empty channel will stall and can only continue when the channel contains sufficient data items ('tokens'). Processes in a Kahn process network are 'monotonic', which means that they only need partial information of the input stream to produce partial information of the output stream. Monotonicity allows parallelism, since a process does not need the whole input signal to start the computation of output events. Processes are not allowed to test an input channel for existence of tokens without consuming them. In a Kahn process network there is a total order of events inside a signal. However, there is no order relation between events in different signals. Thus Kahn process networks are only partially ordered which classifies them as an untimed model.

A data flow program is a directed graph consisting of nodes ('actors') that represent communication and arcs that represent ordered sequences ('streams') of events ('tokens') as illustrated in Figure 5.4. Empty circles denote nodes, arrows denote

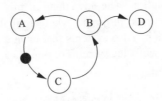

Figure 5.4   A data flow process network

streams and the filled circles denote tokens. Data flow networks can be hierarchical since a node can represent a data flow graph.

The execution of a data flow process is a sequence of 'firings' or 'evaluations'. For each firing tokens are consumed and tokens are produced. The number of tokens consumed and produced may vary for each firing and is defined in the 'firing rules' of a data flow actor.

Data flow process networks have been shown to be very valuable in digital signal processing applications. When implementing a data flow process network on a single processor, a sequence of firings, also called a 'schedule', has to be found. For general data flow models it is undecidable whether such a schedule exists because it depends on the input data.

Synchronous data flow (SDF) [27, 28] puts further restrictions on the data flow model, since it requires that a process consumes and produces a fixed number of tokens for each firing. With this restriction it can be tested efficiently, if a finite static schedule exists. If one exists it can be effectively computed. Figure 5.5 shows an SDF process network. The numbers on the arcs show how many tokens are produced and consumed during each firing. A possible schedule for the given SDF network is {A,A,C,C,B,D}.

SDF is an excellent example of a MoC that offers useful properties by restricting the expressive power. There exists a variety of different data flow models each representing a different trade-off between interesting formal properties and expressiveness. For an excellent overview see Reference 24.

5.2.5.2 Rendezvous-based models

A rendezvous-based model consists of concurrent sequential processes. Processes communicate with each other only at synchronisation points. In order to exchange information, processes must have reached this synchronisation point, otherwise they have to wait for each other. Each sequential process has its own set of time tags. Only at synchronisation points processes share the same tag. Thus there is a partial order of events in this model. The process algebra community uses rendezvous-based models. The CSP model of Hoare [29] and the CCS (Calculus of Communicating Systems) model of Milner [30, 31] are prominent examples. The language Ada [32] has a communication mechanism based on rendezvous.

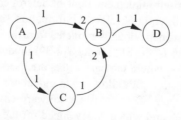

Figure 5.5   A synchronous data flow process network

5.2.6   Heterogeneous models of computation

A lot of effort has been spent to mix different models of computation. This approach has the advantage, that a suitable model of computation can be used for each part of the system. On the other hand, as the system model is based on several computational models, the semantics of the interaction of fundamentally different models has to be defined, which is no simple task. This even amplifies the validation problem, because the system model is not based on a single semantics. There is little hope that formal verification techniques can help and thus simulation remains the only means of validation. In addition, once a heterogeneous system model is specified, it is very difficult to optimise systems across different models of computation. In summary, while heterogeneous MoCs provide very general, flexible and useful simulation and modelling environment, cross-domain validation and optimisation will remain elusive for many years for any heterogeneous modelling approach. In the following an overview of related work on mixed models of computation is given.

In *charts [33] hierarchical finite state machines are embedded within a variety of concurrent models of computations. The idea is to decouple the concurrency model from the hierarchical FSM semantics. An advantage is that modular components, e.g. basic FSMs, can be designed separately and composed into a system with the model of computation that best fits to the application domain. It is also possible to express a state in an FSM by a process network of a specific model of computation. *charts has been used to describe hierarchical FSMs that are composed using data flow, discrete event and synchronous models of computations.

The composite dataflow [34] integrates data and control flow. Vectors and the conversion from scalar values to vectors and vice versa are integral parts of the model. This allows us to capture the timing effects of these conversions without resorting to a synchronous or discrete time MoC. Timing of processes is represented only to the level to determine if sufficient data are available to start a computation. In this way the effects of control and timing on dataflow processing are considered at the highest possible abstraction level because they only appear as data dependency problems. The model has been implemented to combine Matlab and SDL into an integrated system specification environment [35].

Internal representations like the system property intervals (SPI) model [36] and FunState [37] have been developed to integrate a heterogeneous system model into one abstract internal representation. The idea of the SPI model is to allow for global system analysis and system optimisation across language boundaries, in order to allow reliable and optimised implementations of heterogeneously specified embedded real-time systems. All synthesis relevant information, such as resource utilisation, communication and timing behaviour, is extracted from the input languages and transformed into the semantics of the SPI model. An SPI model is a set of parameterised communicating processes, where the parameters are used for the adaptation of different models of computation. SPI allows us to model non-determinism through the use of behavioural intervals. There exists a software environment for SPI that is called the SPI workbench and which is developed for the analysis and synthesis of heterogeneous systems.

The FunState representation refines the SPI model by adding the capability of explicitly modelling state information and thus allows the separation of data flow from control flow. The goal of FunState is not to provide a unifying specification, but it focuses only on specific design methods, in particular scheduling and validation. The internal FunState model shall reduce design complexity by representing only the properties of the system model relevant to these design methods.

The most well-known heterogeneous modelling and simulation framework is Ptolemy [6,38]. It allows us to integrate a wide range of different MoCs by defining the interaction rules of different MoC domains.

5.3 MoCs in the design flow

From the previous sections it is evident that different models fundamentally have different strengths and weaknesses. There is no single model that can satisfy all purposes and thus models of computation have to be chosen with care.

Let us revisit the discussed MoCs while considering the different design phases and the design flow. For the sake of simplicity we only identify five main design tasks as illustrated in Figure 5.6. Early on, the feasibility analysis requires detailed studies of critical issues that may concern performance, cost, power or any other functional

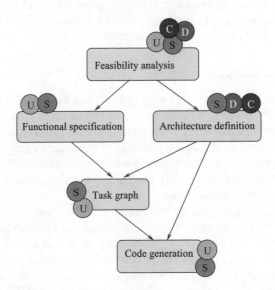

Figure 5.6   Suitability of MoCs in different design phases. 'C' stands for continuous time MoC; 'D' for discrete time MoC; 'S' for synchronous MoC; and 'U' for untimed MoC. More than one label on a design phase means, that all of the MoCs are required since no single MoC is sufficient by itself

or non-functional property. The functional specification determines the entire system functionality (at a high abstraction level) and constitutes the reference model for the implementation. Independent of the functional specification is the architecture specification, which may come with performance and functional models of processors, buses and other resources. The task graph breaks the functionality in concurrent activities (tasks), which are mapped onto architecture resources. Once resource binding and scheduling has been performed, the detailed implementation for the resources is created.

The essential difference of the four main computational models that we introduced in the previous section, is the representation of time. This feature alone weighs heavily with respect to their suitability for design tasks and development phases.

5.3.1   Continuous time models

Continuous time MoCs are mostly used to accurately model and analyse existing or prospective devices. They reflect detailed electrical and physical properties with high precision. Hence, they are ideal to study and model tiny entities in great detail but they are unsuitable to analyse and simulate large collections and complex systems due to the overwhelming amount of detail. They are usually not used to specify and constrain behaviour but may serve as reference models for the implementation. Thus, they are frequently used in feasibility studies, to analyse critical issues, and in architectural models to represent analogue or mixed signal components in the architecture. Analogue synthesis is still not well automated and hence continuous time models are rarely used as input to synthesis tools.

5.3.2   Discrete time models

The discrete time MoC constitutes a very general basis for modelling and simulation of almost arbitrary systems. With the proper elementary components it can serve to model digital hardware consisting of transistors and gates; systems-on-chip consisting of processors, memories, and buses; networks of computers, clients and servers; air traffic control systems; evolution of prey–predator populations; and much more [18]. In fact it is most popular and widely used in an enormous variety of engineering, economic and scientific applications.

However, it cannot be used for everything. In the context of hardware and software design the discrete time model has the drawback that a precise delay information cannot be synthesised. To provide a precise delay model for a piece of computation may be useful for simulation and may be appropriate for an existing component, but it hopelessly over-specifies the computation for synthesis. Assume a multiplication is defined to take 5 ns. Shall the synthesis tool try to get as close to this figure as possible? What deviation is acceptable? Or should it be interpreted as 'max 5 ns'? Different tools will give different answers to these questions and synthesis for different target technologies will yield very different results and none of them will match the simulation of the discrete time model. The situation becomes even worse, when a delta-delay based model is used. As we discussed in Section 5.2.4 the delta-delay model elegantly solves the problem of non-determinism for simulation, but it requires a mechanism for globally ordering the events. Essentially, a synthesis system had to synthesise a similar mechanism together with the target design, which is an unacceptable overhead.

These problems notwithstanding, synthesis systems for both hardware and software have been developed for languages based on time models. VHDL and Verilog-based tools are the most popular and successful examples. They have avoided these problems by ignoring the discrete time model and interpreting the specification according to a clocked synchronous model. Specific coding rules and assumptions allow the tool to identify a clock signal and infer latches or registers separating the combinatorial blocks. The drawbacks of this approach are that one has to follow special coding guidelines for synthesis, that specification and implementation may behave differently, and in general that the semantics of the language is complicated by distinguishing between a simulation and a synthesis semantics. The success of this approach illustrates that mixing different MoCs in the same language is practical. It also demonstrates the suitability of the clocked synchronous model for synthesis but underscores that the discrete time model is not synthesisable.

5.3.3 Synchronous models

The synchronous models represent a sensible compromise between untimed and discrete time models. Most of the timing details can be ignored but we can still use an abstract time unit, the evaluation or clock cycle, to reason about the timing behaviour. Therefore it has often a natural place as an intermediate model in the design process. Lower level synthesis may start from a synchronous model. Logic and RTL synthesis for hardware design and the compilation of synchronous languages for embedded software are prominent examples. The result of certain synthesis steps may also be represented as a synchronous description such as scheduling and behavioural synthesis.

It is debatable if a synchronous model is an appropriate starting point for higher level synthesis and design activities. It fairly strictly defines that activities occurring in the same evaluation cycle but in independent processes are simultaneous. This imposes an unnecessarily strong coupling between unrelated processes and may restrict early design and synthesis activities too much.

On the other hand in many systems timing properties are an integral part of the system functionality and are therefore an important part of a system specification model. Complex control structures typically require a fine control over the relative timing of events and activities. As single chip systems increase in complexity, this feature becomes more common. Already today there is hardly any SoC design that does not exhibit complex control.

Synchronous models constitute a very good compromise for dealing with time at an abstract level. While they avoid the nasty details of low-level timing problems, they allow us to represent and analyse timing relations. In essence the clock or evaluation cycle defines 'abstract time budgets' for each block. The time budgets turn into timing constraints for the implementation of these blocks. The abstract time budgets constrain the timing behaviour without over-constraining it. Potentially there is a high degree of flexibility in this approach if the evaluation cycles of a synchronous MoC are not considered as fixed-duration clock cycles but rather as abstract time budgets, which do not have to be of identical duration in different parts of the design. Their duration could also change from cycle to cycle if required. Re-timing techniques exploit this flexibility [39, 40].

This feature of offering an intermediate and flexible abstraction level of time makes synchronous MoCs suitable for a wide range of tasks as indicated in Figure 5.6.

5.3.4   Untimed models

Untimed models have an excellent track record in modelling, analysing and designing signal processing systems. They are invaluable in designing digital signal processing algorithms and analysing their key performance properties such as signal-to-noise ratio.

Furthermore, they have nice mathematical features, which facilitate certain synthesis tasks. The tedious scheduling problem for software implementations is well understood and efficiently solvable for synchronous data flow graphs. The same can be said for determining the right buffer sizes between processes, which is a necessary and critical task for hardware, software and mixed implementations. How well the individual processes can be compiled to hardware or software depends on the language used to describe them. The data flow process model does not restrict the choice of these languages and is therefore not responsible for their support. For what it is responsible, i.e. the communication between processes and their relative timing, it provides excellent support due to a carefully devised mathematical model.

5.3.5   Discussion

Figure 5.6 illustrates this discussion and indicates in which design phases the different MoCs are most suitable. Note that several MoCs placed on a design phase bubble means that in general a single MoC does not suffice for that phase but several or all of them may be required.

No single MoC serves all purposes equally well. The emphasis is on 'equally well' because all of them are sufficiently expressive and versatile to be used in a variety of contexts. However, their different focus makes them more or less suitable for specific tasks. For instance a discrete time, discrete event model can be used to model and simulate almost anything. But it is extremely inefficient to use it to simulate and analyse complex systems when detailed timing behaviour is irrelevant. This inefficiency concerns both tools and human designers. Simulation of a discrete time model takes orders of magnitude longer than simulation of an untimed model. Formal verification is orders of magnitude more efficient for perfectly synchronous models than for discrete time models. Human designers are significantly more productive in modelling and analysing a signal processing algorithm in an untimed model than in a synchronous or discrete time model. They are also much more productive to model a complex, distributed system when they have appropriate and high-level communication primitives available, than when they have to express all communication with unprotected shared variables and semaphores. Hardware engineers working on the RT level (synchronous MoC) design many more gates per day than their counterparts not using a synchronous design style. Analogue designers are even less productive in terms of designed transistors per day because they deal with the full range of details at the physical and electrical level. Unfortunately, good abstractions at a higher level have not been found yet for analogue design with the consequence that analogue design is less automated and less efficient than digital design.

MoCs impose different restrictions which, if selected carefully, can lead to significant improvements in design productivity and quality. A strict finite state machine model can never have unbounded memory requirements. This property is inherent in any FSM model and does not have to be proved for every specific design. The amount of memory required can be calculated by static analysis and no simulation is required. This is in contrast to models with dynamic memory allocation where it is in general impossible to prove an upper bound for the memory requirement and long simulations have to be used to obtain a high level of confidence that the memory requirements are indeed feasible. FSM models are restrictive but if a problem suits these restrictions, the gain in design productivity and product quality can be tremendous.

A similar example is synchronous dataflow. If a system can be naturally expressed as an SDF graph, it can be much more efficiently analysed, scheduled and designed than the same system modelled as a general dataflow graph.

As a general guideline we can state that 'the productivity of tools and designers is highest if the least expressive MoC is used that still can naturally be applied to the problem'.

Thus, all the different computational models have their place in the design flow. Moreover, several different MoCs have to be used in the same design model because different sub-systems have different requirements and characteristics. This leads naturally to heterogeneous MoCs which can either be delayed within one language or with several languages under a coordination framework as will be discussed below.

5.4 Design activities

Next we investigate specific design tasks and their relation to MoCs. We do not intend to present an exhaustive list of activities, but we hope to illustrate the strong connection and interdependence of design tasks and models on which they operate.

5.4.1 Synthesis

Today several automatic synthesis steps are part of the standard design flow of ASICs and SoCs. Register Transfer Level (RTL) synthesis, technology mapping, placement and routing, logic and FSM synthesis are among those. Other techniques that have been researched and developed but not successfully commercialised are high-level synthesis, system level partitioning, resource allocation and task mapping. We take a closer look at RTL and high-level synthesis because they are particularly enlightening examples.

```
PROCESS (clk, reset)
 BEGIN
  IF reset = '0' THEN
     state <= 0;
  ELSIF clk'event AND clk = '1' THEN
     state <= nextstate;
  END IF;
END PROCESS
```

Figure 5.7   A VHDL process encoding the P_reg block of Figure 5.8

5.4.1.1 RTL synthesis

Register transfer level synthesis takes as input an HDL (Hardware Description Language) model of a process, for instance written in VHDL or Verilog, and generates a netlist of gates that adheres to a synchronous design style. Since VHDL and Verilog are simulation not synthesis languages, some of their constructs cannot be synthesised. Every RTL synthesis tool defines a synthesisable subset of the input language.[4] This subset definition has two objectives. First, constructs that cannot be synthesised into HW are excluded. Obvious examples are file I/O operations and dynamic memory management. Second, typical and efficient HW structures are encoded in the language subset. Synthesis tools will identify FSMs, memories, registers and combinatorial logic in the source model and translate them efficiently onto corresponding HW structures. For example, VHDL processes have to be written in a specific style with only one clock signal such that the synthesis tool can extract a combinatorial netlist with registers at the outputs. Figure 5.7 shows a VHDL process that would be interpreted as a FSM state register by most synthesis tools. If two other combinatorial processes are provided and properly modelled, the tool would derive a FSM structure as shown in Figure 5.8. P_reg reacts to a reset signal to go into the initial state, and to a clock signal to make a state transition.

The definition of a synthesisable subset and the particular interpretation of synthesis tools lead to a divergence of simulation semantics and synthesis semantics. There are three main motivations for this.

1   Some language constructs are pure simulation devices and there is no reason why anybody would want to synthesise them. Examples are file access and assertions.
2   Some language constructs are too expensive to implement in hardware and the current state of the art suggests that they should not be synthesised. Examples are multi-dimensional arrays and dynamic memory allocation. When future engineers conclude that such constructs should also be available to hardware designers, these restrictions may disappear.

---

[4] There are different subsets imposed by different tools, but they are not very essential and concern mostly issues of user convenience and synthesis performance rather than the semantics. There exists even an IEEE standard for a synthesisable subset.

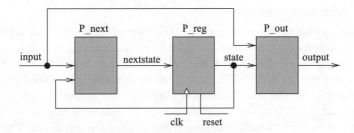

Figure 5.8 *A VHDL synthesis tool derives a state machine when the VHDL description contains three properly modelled processes. P_next is a combinatorial process defining the nextstate transition function. P_reg is a register storing the state. P_out models the output encoding function.*

3 The timing model of the simulation semantics is ill-suited for synthesis. The simulation semantics is based on a discrete time model and allows us to express delays in terms of nano- and picoseconds. In contrast the synthesised model is a clock synchronous MoC that simply cannot express physical time delays.

The last item interests us most because it shows that VHDL/Verilog-based simulation and synthesis use different models of computation, according to our scheme in Section 5.2. The simulation semantics is based on a discrete time MoC which is unsuitable for synthesis. Even if a delay of e.g. 2 ns could be accurately synthesised, it would over-constrain the following technology mapping, placement and routing steps and lead to a hopelessly inefficient implementation. Accurate synthesis of the delta-delay concept is even more elusive.

In contrast, the clocked synchronous MoC[5] allows us to separate synthesis of the behavior from timing issues. Since the clock structure and the scheduling of computations in clock cycles is already part of the input model, the RTL synthesis focuses on optimising the combinatorial blocks between registers. In an analysis step separate from synthesis the critical paths can be identified and the overall system performance can be assessed. Re-timing techniques, that move gates and combinatorial blocks across clock cycle boundaries, can shorten critical paths and increase overall performance. If all this proves insufficient the input model to RTL synthesis has to be modified.

In conclusion, for RTL synthesis a clocked synchronous MoC is the best choice because it reflects efficient hardware structures and allows for an effective separation of behavioural synthesis from timing optimisation. A lower level, discrete time MoC is entirely inadequate since it over-constrains the synthesis. Starting synthesis with a model based on a higher time abstraction, an untimed MoC, imposes fewer constraints on the synthesis process but consequently requires the synthesis task to include scheduling of operations as will be discussed next.

---

[5] Recall from Section 5.2.4 that a clocked synchronous MoC is a synchronous MoC variant where no feedback loops are allowed within the same clock cycle. Therefore the feedback loop in Figure 5.8 has to be broken by the P_reg register process.

5.4.1.2 High-level synthesis

High-level synthesis, later also called behavioural synthesis, as defined and researched heavily [41], includes the tasks of resource allocation, operation binding and operation scheduling. The input is an algorithm described in a sequential language such as C or as a VHDL process. 'Resource allocation' estimates the type and number of HW resources required to implement the algorithm, e.g. how many multipliers, adders, ALUs, etc. 'Operation binding' binds operations of the algorithm to allocated resources. 'Scheduling' assigns the operations to specific clock steps, thus determining when they will be executed. Figure 5.9 illustrates the scheduling procedure. From the algorithmic specification in Figure 5.9(a) the dataflow graph 5.9(b) is extracted to represent the data dependences. Figure 5.9(c) shows the scheduled dataflow graph by using the As-Soon-As-Possible (ASAP) scheduling principle.

The natural MoC for the input to high-level synthesis is an untimed MoC. Synchronous or discrete time MoCs are unsuitable because they both determine the execution time of individual operations, rendering the scheduling step superfluous. In fact the untimed model was the MoC chosen by all groups that developed high-level synthesis systems. This was either done by defining a dedicated language that could only express an untimed MoC, or by sub-setting a general purpose design language such as VHDL or Verilog. Resource allocation and operation binding concerns the

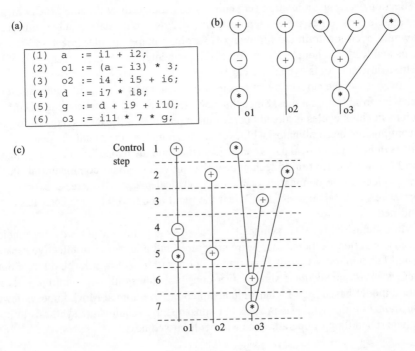

Figure 5.9 An algorithmic specification and its scheduled dataflow graph (a) algorithmic specification, (b) dataflow graph and (c) scheduled dataflow graph (from Reference 42)

refinement of computation. The abstraction level of the computation and the operators are not defined by the MoCs in Section 5.2. Thus, the untimed MoC is a suitable input to high-level synthesis independent of the kind of operations involved, simple adders and half-adders or highly complex processing elements.

5.4.1.3 Discussion

Other synthesis procedures also have their natural input and output MoC. Hence, each synthesis method has to be provided with input models that match its natural MoC, e.g. a clocked synchronous MoC for RTL and an untimed MoC for high-level synthesis. In practice this is accomplished in one of two ways. The obvious approach is to choose an input language that matches well with the natural MoC. If this is not desirable due to other constraints, a language subset and interpretation rules are established, that approximate the MoC required by the synthesis method. We call this technique the 'projection' of an MoC into a design language. It is illustrated in Figure 5.10.

Taking a step back we can contemplate the relation between synthesis methods and MoCs. They are mutually dependent and equally important. While it is in general correct that every synthesis method has 'natural MoCs' defining its input and output, we can also observe that the major synthesis steps follow naturally from the definition of the MoCs. For every significant difference between two MoCs we can formulate a synthesis step transforming one MoC into the other. On the other hand, the MoCs represent useful abstractions only if we can identify efficient synthesis methods that use them as input and output.

Our treatment of MoCs does not cover other relevant issues such as abstraction and refinement of computation and data types. We have focused foremost on time and therefore we could discuss the scheduling problem of high-level synthesis convincingly while we barely mentioned the allocation and binding tasks. We believe there are good arguments for using time as the primary criteria for categorising MoCs while other domains such as computation, communication and data lead to variants within

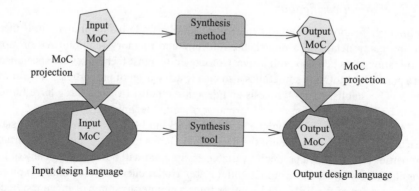

Figure 5.10   MoC projection into design languages

the same MoC. For a more thorough discussion of this question see Reference 43 or 7. For a further elaboration of domains and abstractions see Reference 44.

5.4.2   Simulation

All MoCs that we have discussed can be simulated. So the question that we have to ask is not, which MoC is suitable for simulation, but how efficiently a given MoC can be simulated. Also, we may want to distinguish different purposes of simulation and then we can ask if, for a given purpose, we should prefer one MoC to another.

It is obvious that discrete time MoC simulations are slower than synchronous MoC simulations which in turn are slower than untimed MoC simulations, because MoCs at lower abstraction levels require the computation of many more details. It has been reported that simulations of clock cycle true models, which correspond to our clocked synchronous MoCs, are 1–2 orders of magnitude faster than discrete event simulations, which correspond to our discrete time MoC [45]. Moving to an untimed MoC, e.g. functional or transaction level simulations, can further speed-up simulation by 1 to 2 orders of magnitude [45,46]. Higher abstraction in any of the domains time, data, computation and communication, improves simulation performance, but the time abstraction seems to be play a dominant role [47], because a higher time abstraction significantly reduces the number of events in a simulation uniformly in all parts of a model.

The disadvantage with abstract MoCs is the loss of accuracy. Detailed timing behaviour and the clock cycle period cannot be analysed in a synchronous MoC simulation. Transaction level models cannot unveil problems in the details of the interface and low-level protocol implementation. In an untimed MoC no timing related properties can be investigated and arithmetic overflow effects cannot be observed when using ideal, mathematical data types. Clearly, a trade-off between accuracy and simulation performance, as illustrated in Figure 5.11, demands that a design is simulated at various abstractions during a design project from specification to implementation.

5.4.3   Formal verification

Formal verification techniques experience a similar trade-off to simulation. If there are too many details in a model, the run-time and memory requirements of most verification tools become prohibitive. Consequently, most techniques are specialised on a particular MoC and sometimes also on a restricted set of properties. They follow the MoCs established by synthesis and design methods, because these have turned out to be useful MoCs for several formal verification techniques as well.

An example formal technique is model checking [48]. It requires a finite state machine (FSM) based model of the design and allows to express and verify various properties such as that a particular variable assignment will never occur in any of the possible states reachable from an initial state. Model checking essentially explores the state space of the FSM until it either finds a counter-example or it can prove the given property, e.g. by exploring the entire reachable state space.

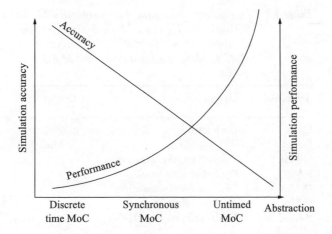

Figure 5.11   The trade-off between accuracy and simulation performance

Multiple, communicating FSMs can be handled but only by merging them into a single, flat FSM. This often leads to serious state space explosion problems. Due to clever algorithms and highly efficient data representations model checking can be applied to realistic designs and proves useful in practice.

The natural MoC for property checking is a synchronous MoC, just as for RTL synthesis, since it corresponds to a finite state machine and its evolution. Detailed timing information below the granularity of synchronous MoCs cannot be handled by model checking unless they are encoded in a way fitting into the MoC. On the other hand an untimed MoC would in principle be compatible with model checking but it would allow for infinitely many ways to merge multiple FSMs into a single one, thus magnifying the state space explosion problem even further.

Just as in the case of synthesis techniques, we can also observe that all formal verification techniques require specific MoCs as input descriptions. The basic principles, such as theorem proving, are often much more general but have to be specialised for a specific problem domain, and thus for a specific MoC, to make them useful in practice. Hence, a MoC serves by dramatically restricting the problem space and, if selected carefully, allows for efficient verification tools.

5.4.4   Summary

Table 5.1 summarises the discussed tasks and gives their respective MoCs.

As mentioned earlier, we have chosen to distinguish the MoCs according to their time abstraction. Therefore we can naturally analyse design tasks that have a strong relation to a particular time abstraction such as scheduling or cycle-true simulation. For an analysis of all other design tasks in a similar satisfactory way we would have to introduce MoC variants based on computation, data and communication abstractions.

Table 5.1   *Design activities with their respective MoCs*
          (U-MoC = Untimed MoC, S-MoC = Syn-
          chronous MOC, D-MoC = Discrete time
          MoC, C-MoC = Continuous time MoC)

Input MoC	Design task	Output MoC
U-MoC	High-level synthesis	S-MoC
S-MoC	RTL synthesis	D-MoC
U-MoC	Transaction level simulation	
S-MoC	Cycle-true simulation	
D-MoC	Discrete-event simulation	
C-MoC	Analogue simulation	
S-MoC	Model and property checking	

5.5   Conclusions

We have analysed the relation between some inherent properties of computational models and various design tasks and phases. Since this is an endeavour far beyond a single article we have taken time as our primary parameter and have defined four MoC classes based on the time abstraction: continuous time, discrete time, synchronous time and untimed MoC. This is justified because the chosen representation of time has a critical influence on synchronisation, communication and the overall system behaviour for systems described by communicating concurrent processes. For a more elaborate study that encompasses all design activities and phases we suggest to still use time abstraction as the primary criterion for defining MoCs but to use other abstractions and domains to introduce more MoC variants as suitable.

We have not carefully illuminated the relation between MoCs and design languages since it is an intricate one with many subtle connections and implications that requires a chapter of its own. For more, but not an exhaustive, elaboration of this issue see Reference 43.

The main targets of our study, complex, heterogeneous, embedded systems, require the use of all presented MoCs. But each MoC has a very specific place and role in the design process as illustrated by Figure 5.6 and Table 5.1. The usage of MoCs should be a conscious choice based on their inherent properties and the given objective and design task. Using them for the wrong purpose will lead to poor results that cannot be rectified by improving a synthesis or simulation algorithm.

But MoCs are not just predefined and given to us and we merely have to pick the right one. Rather, they have to be properly developed and defined for a particular purpose. This is a delicate task because we face a difficult trade-off. To simplify the overall design process and support tool interoperability we would like to have as few different MoCs as possible. However, if we aim at the best possible MoC for a specific task, we will have to integrate many, specialised MoCs in the design flow.

History shows, that the process of identifying, accepting and establishing MoCs is tedious and slow. The successful introduction of a new MoC is typically bound to a major paradigm change, such as the move from schematic entry design to RTL-based synthesis.

References

1 JANTSCH, A.: 'Models of embedded computation'. In RICHARD ZUROWSKI, Ed: 'Embedded Systems' (CRC Press, Boca Raton, FL, 2005), chapter 4

2 LEE, E.A., and SANGIOVANNI-VINCENTELLI, A.: 'A framework for comparing models of computation', *IEEE Transactions on Computer-Aided Design of Integrated Circuits and Systems*, 1998, **17**(12), pp. 1217–1229

3 PAUL, J.M., and THOMAS D.E.: 'Models of computation for systems-on-chip. In JERRAYA, A., and WOLF, W., Eds: 'Multiprocessor systems-on-chip' (Morgan Kaufman Publishers, San Mateo, CA, 2004) chapter 15

4 SAVAGE, J.E.: 'Models of computation, exploring the power of computing' (Addison Wesley, Reading, MA, 1998)

5 TAYLOR, R.G.: 'Models of computation and formal language' (Oxford University Press, New York, 1998)

6 EKER, J., JANNECK, J.W., LEE, E.A., *et al.*: 'Taming heterogeneity – the Ptolemy approach', *Proceedings of the IEEE*, 2003, **91**(1), pp. 127–144

7 JANTSCH, A.: 'Modeling embedded systems and SoCs – concurrency and time in models of computation'. *Systems on Silicon* (Morgan Kaufmann Publishers, San Mateo, CA, June 2003)

8 SANDER, I., and JANTSCH, A.: 'System modeling and transformational design refinement in ForSyDe', *IEEE Transactions on Computer-Aided Design of Integrated Circuits and Systems*, 2004, **23**(1), pp. 17–32

9 JONES, S.P.: 'Haskell 98 language and libraries' (Cambridge University Press, Cambridge, 2003)

10 MATHAIKUTTY, D., PATEL, H., and SHUKLA, S.: 'A functional programming framework of heterogeneous model of computations for system design'. Proceedings of the Forum on *Specification and design languages*, Lille, France, September 2004

11 PATEL, H.D., and SHUKLA, S.K.: 'SystemC Kernel extensions for heterogeneous system modeling' (Kluwer Academic Publishers, Boston/Dordrecht/London, June 2004)

12 SKILLICORN, D.B., and TALIA, D.: 'Models and languages for parallel computation', *ACM Computing Surveys*, 1998, **30**(2), pp. 123–169

13 DABNEY, J., and HARMAN, T.L.: 'Mastering SIMULINK 2' (Prentice Hall, New York, 1998)

14 ELMQVIST, H., MATTSSON, S.E., and OTTER, M.: 'Modelica – the new object-oriented modeling language'. Proceedings of the 12th European *Simulation* multiconference, June 1998

15   FLYNN M.J.: 'Some computer organisations and their effectiveness', *IEEE Transactions on Computers*, 1972, **C-21**(9), pp. 948–960

16   IEEE: *IEEE Standard VHDL Language Reference Manual* (IEEE, 2002)

17   IEEE: *IEEE Standard for Verilog Hardware Description Language* (IEEE, 2001)

18   CASSANDRAS, C.G.: 'Discrete event systems: Modeling and performance analysis' (Asken Associates, Boston, MA, 1993)

19   BENVENISTE, A., and BERRY, G.: 'The synchronous approach to reactive and real-time systems', *Proceedings of the IEEE*, 1991, **79**(9), pp. 1270–1282

20   HALBWACHS, N., CASPI, P., RAYMOND, P., and PILAUD, D.: 'The synchronous data flow programming language LUSTRE', *Proceedings of the IEEE*, 1991, **79**(9), pp. 1305–1320

21   WINSKEL, G.: 'The formal semantics of programming languages' (MIT Press, Cambridge, MA, 1993)

22   BERRY, G.: 'The constructive semantics of pure Esterel – draft version 3'. Technical report, INRIA, 06902 Sophia-Antipolis CDX, France, July 2, 1999

23   LE GUERNIC, P., GAUTIER, T., LE BORGNE, M., and LE MARIE, C.: 'Programming real-time applications with SIGNAL', *Proceedings of the IEEE*, 1991, **79**(9), pp. 1321–1335

24   LEE, E.A., and PARKS, T.M.: 'Dataflow process networks', *IEEE Proceedings*, 1995, **83**(5), pp. 773–799

25   KAHN, G.: 'The semantics of a simple language for parallel programming'. Proceedings of the *IFIP Congress 74*, Stockholm, Sweden, 1974 (North-Holland, Amsterdam, 1974)

26   KAHN, G., and MACQUEEN, D.B.: 'Coroutines and networks of parallel processes'. *IFIP '77* (North-Holland, Amsterdam, 1977)

27   LEE, E.A., and MESSERSCHMITT, D.G.: 'Static scheduling of synchronous data flow programs for digital signal processing', *IEEE Transactions on Computers*, 1987, **C-36**(1), pp. 24–35

28   LEE, E.A., and MESSERSCHMITT, D.G.: 'Synchronous data flow', *Proceedings of the IEEE*, 1987, **75**(9), pp. 1235–1245

29   HOARE, C.A.R.: 'Communicating sequential processes', *Communications of the ACM*, 1978, **21**(8), pp. 666–676

30   MILNER, R.: 'A calculus of communicating systems'. LNCS, 92 (Springer-Verlag, Heidelber, 1980)

31   MILNER, R.: 'Communication and concurrency' (Prentice Hall, New York, 1989)

32   BOOCH, G., and BRYAN, D.: 'Software engineering with Ada' (The Benjamin/ Cummings Publishing Company, Menlo Park, CA, 1994)

33   GIRAULT, A., LEE, B., and LEE, E.A.: 'Hierarchical finite state machines with multiple concurrency models', *IEEE Transactions on Computer-Aided Design of Integrated Circuits and Systems*, 1999, **18**(6), pp. 742–760

34   JANTSCH, A., and BJURÉUS, P.: 'Composite signal flow: A computational model combining events, sampled streams, and vectors'. Proceedings of the *Design and test Europe* conference (DATE), Paris, France, March 2000, pp. 154–160

35 BJURÉUS, P., and JANTSCH, A.: 'Modeling of mixed control and dataflow systems in MASCOT', *IEEE Transactions on Very Large Scale Integration (VLSI) Systems*, 2001, **9**(5), pp. 690–704

36 ZIEGENBEIN, D., RICHTER, K., ERNST, R., THIELE, L., and TEICH, J.: 'SPI – a system model for heterogeneously specified embedded systems', *IEEE Transactions on Very Large Scale Integration (VLSI) Systems*, 2002, **10**(4), pp. 379–389

37 STREHL, K., THIELE, L., GRIES, M., ZIEGENBEIN, D., ERNST, R., and TEICH, J.: 'FunState – an internal design representation for codesign', *IEEE Transactions on Very Large Scale Integration (VLSI) Systems*, 2001, **9**(4), pp. 524–544

38 LEE, E.A.: 'Overview of the ptolemy project'. Technical report UCB/ERL M03/25 (University of California, Berkeley, CA, July 2003)

39 ROSE, F., LEISERSON, C., and SAXE, J.: 'Optimizing synthesis circuitry by retiming'. Proceedings of the *Caltech* Conference on *VLSI*, 1983, pp. 41–67

40 WEHN, N., BIESENACK, J., LANGMAIER, T., *et al.*: 'Scheduling of behavioural VHDL by retiming techniques'. Proceedings *EuroDAC 94*, September 1994, pp. 546–551

41 GAJSKI, D., DUTT, N., WU, A., and LIN, S.: 'High level synthesis' (Kluwer Academic Publishers, New York, 1993)

42 ELES, P., KUCHCINSKI, K., and PENG, Z.: 'System synthesis with VHDL' (Kluwer Academic Publisher, New York, 1998)

43 JANTSCH, A., and SANDER, I.: 'System level specification and design languages', *IEE Proceedings on Computers and Digital Techniques*, 2005. Special issue on Electronic System Design; Invited paper

44 JANTSCH, A., KUMAR, S., and HEMANI, A.: 'The Rugby model: A framework for the study of modelling, analysis, and synthesis concepts in electronic systems'. Proceedings of *Design automation and test in Europe* (DATE), 1999

45 ROWSON J.A.: 'Hardware/software cosimulation'. Proceedings of the *Design automation* conference, 1994, pp. 439–440

46 WERNER, B., and MAGNUSSON, P.S.: 'A hybrid simulation approach enabling performance characterization of large software systems'. Proceedings of *MASCOTS*, 1997

47 KHOSRAVIPOUR, H.G.M., and GRIDLING, G.: 'Improving simulation efficiency by hierarchical abstraction transformations'. Proceedings of the Forum on *Design languages*, 1998

48 CLARKE, E.M., GRUMBERG, O., and PELED, D.A.: 'Model checking' (MIT Press, Cambridge, MA, 1999)

Chapter 6

Architecture description languages for programmable embedded systems

Prabhat Mishra and Nikil Dutt

6.1 Introduction

Embedded systems are everywhere – they run the computing devices hidden inside a vast array of everyday products and appliances such as cell phones, toys, handheld PDAs, cameras and microwave ovens. Cars are full of them, as are airplanes, satellites and advanced military and medical equipments. As applications grow increasingly complex, so do the complexities of the embedded computing devices. Figure 6.1 shows an example embedded system, consisting of programmable components including a processor core, coprocessors and memory subsystem. The programmable components are used to execute the application programs. Depending on the application domain, the embedded system can have application specific hardwares, interfaces, controllers and peripherals. The programmable components, consisting of a processor core, coprocessors and memory subsystem, are referred to as 'programmable embedded systems'. They are also referred to as 'programmable architectures'.

As embedded systems become ubiquitous, there is an urgent need to facilitate rapid design space exploration (DSE) of programmable architectures. This need for rapid DSE becomes even more critical given the dual pressures of shrinking time-to-market and ever-shrinking product lifetimes. Architecture Description Languages (ADLs) are used to perform early exploration, synthesis, test generation and validation of processor-based designs as shown in Figure 6.2. ADLs are used to specify programmable architectures. The specification can be used for generation of a software toolkit including the compiler, assembler, simulator and debugger. The application programs are compiled and simulated, and the feedback is used to modify the ADL specification with the goal of finding the best possible architecture for a given set

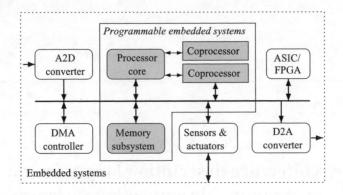

Figure 6.1   An example embedded system

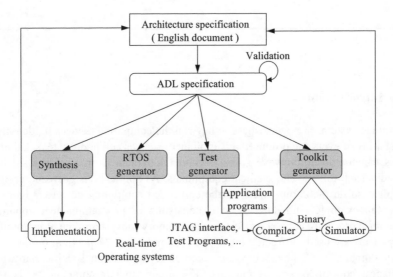

Figure 6.2   ADL-driven exploration, synthesis and validation of programmable architectures

of applications. The ADL specification can also be used for generating hardware prototypes under design constraints such as area, power and clock speed. Several researches have shown the usefulness of ADL-driven generation of functional test programs and test interfaces. The specification can also be used to generate device drivers for real-time operating systems (RTOS) [1].

Previously, researchers have surveyed architecture description languages for retargetable compilation [2], and systems-on-chip design [3]. Qin and Malik [2] surveyed the existing ADLs and compared the ADLs to highlight their relative strengths and weaknesses in the context of retargetable compilation. Tomiyama *et al.* [3] classified existing ADLs into four categories based on their main objectives: synthesis, compiler generation, simulator generation and validation. This chapter presents a comprehensive survey of existing ADLs and the accompanying methodologies for programmable embedded systems design.

The rest of the chapter is organised as follows: Section 6.2 describes how ADLs differ from other modelling languages. Section 6.3 surveys the existing ADLs. Section 6.4 presents the ADL-driven methodologies on software toolkit generation, hardware synthesis, exploration and validation of programmable embedded systems. This study forms the basis for comparing the relative merits and demerits of the existing ADLs in Section 6.5. Finally, Section 6.6 concludes this chapter with a discussion on expected features of future ADLs.

6.2   ADLs and other languages

The phrase 'architecture description language' has been used in the context of designing both software and hardware architectures. Software ADLs are used for representing and analysing software architectures [4,5]. They capture the behavioural specifications of the components and their interactions that comprises the software architecture. However, hardware ADLs capture the structure (hardware components and their connectivity) and the behaviour (instruction-set) of processor architectures. This chapter surveys hardware ADLs.

The concept of using machine description languages for specification of architectures has been around for a long time. Early ADLs such as ISPS [6] were used for simulation, evaluation and synthesis of computers and other digital systems. This chapter surveys contemporary ADLs.

How do ADLs differ from programming languages, hardware description languages, modelling languages and the like? This section attempts to answer this question. However, it is not always possible to answer the following question: given a language for describing an architecture, what are the criteria for deciding whether it is an ADL or not?

In principle, ADLs differ from programming languages because the latter bind all architectural abstractions to specific point solutions, whereas ADLs intentionally suppress or vary such binding. In practice, architecture is embodied and recoverable from code by reverse engineering methods. For example, it might be possible to analyse a piece of code written in C and figure out whether it corresponds to *Fetch* unit or not. Many languages provide architecture level views of the system. For example, C++ offers the ability to describe the structure of a processor by instantiating objects for the components of the architecture. However, C++ offers little or no architecture-level analytical capabilities. Therefore, it is difficult to describe architecture at a level of abstraction suitable for early analysis and exploration. More importantly, traditional programming languages are not a natural choice for describing architectures due to their inability in capturing hardware features such as parallelism and synchronisation.

Architecture description languages differ from modelling languages (such as UML) because the latter are more concerned with the behaviours of the whole

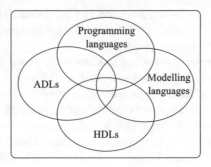

Figure 6.3   ADLs versus non-ADLs

rather than the parts, whereas ADLs concentrate on representation of components. In practice, many modelling languages allow the representation of cooperating components and can represent architectures reasonably well. However, the lack of an abstraction would make it harder to describe the instruction-set of the architecture.

Traditional Hardware Description Languages (HDLs), such as VHDL and Verilog, do not have sufficient abstraction to describe architectures and explore them at the system level. It is possible to perform reverse-engineering to extract the structure of the architecture from the HDL description. However, it is hard to extract the instruction-set behaviour of the architecture. In practice, some variants of HDLs work reasonably well as ADLs for specific classes of programmable architectures.

There is no clear line between ADLs and non-ADLs. In principle, programming languages, modelling languages and hardware description languages have aspects in common with ADLs, as shown in Figure 6.3. Languages can, however, be discriminated from one another according to how much architectural information they can capture and analyse. Languages that were born as ADLs show a clear advantage in this area over languages built for some other purpose and later co-opted to represent architectures. Section 6.5 will re-visit this issue in light of the survey results.

6.3   The ADL survey

Figure 6.4 shows the classification of ADLs based on two aspects: 'content' and 'objective'. The content-oriented classification is based on the nature of the information an ADL can capture, whereas the objective-oriented classification is based on the purpose of an ADL. Contemporary ADLs can be classified into six categories based on the objective: simulation-oriented, synthesis-oriented, test-oriented, compilation-oriented, validation-oriented and operating-system-oriented.

Architecture description languages can be classified into four categories based on the nature of the information: structural, behavioural, mixed and partial. The structural ADLs capture the structure in terms of architectural components and their

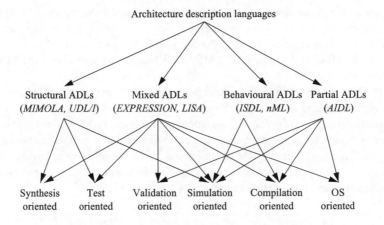

Figure 6.4   Taxonomy of ADLs

connectivity. The behavioural ADLs capture the instruction-set behaviour of the processor architecture. The mixed ADLs capture both structure and behaviour of the architecture. These ADLs capture complete description of the structure or behaviour or both. However, the partial ADLs capture specific information about the architecture for the intended task. For example, an ADL intended for interface synthesis does not require internal structure or behaviour of the processor.

Traditionally, structural ADLs are suitable for synthesis and test-generation. Similarly, behavioural ADLs are suitable for simulation and compilation. It is not always possible to establish a one-to-one correspondence between content and objective based classification. For example, depending on the nature and amount of information captured, partial ADLs can represent any one or more classes of the objective-based ADLs. This section presents the survey using content-based classification of ADLs.

6.3.1   Structural ADLs

Architecture description language designers consider two important aspects: level of abstraction versus generality. It is very difficult to find an abstraction to capture the features of different types of processors. A common way to obtain generality is to lower the abstraction level. Register transfer level (RT-level) is a popular abstraction level – low enough for detailed behaviour modelling of digital systems, and high enough to hide gate-level implementation details. Early ADLs were based on RT-level descriptions. This section briefly describes two structural ADLs: MIMOLA [7] and UDL/I [8].

6.3.1.1   MIMOLA

MIMOLA [7] is a structure-centric ADL developed at the University of Dortmund, Germany. It was originally proposed for micro-architecture design. One of the major advantages of MIMOLA is that the same description can be used for synthesis, simulation, test generation and compilation. A tool chain including the MSSH hardware synthesiser, the MSSQ code generator, the MSST self-test program compiler, the MSSB functional simulator and the MSSU RT-level simulator were developed based on the MIMOLA language [7]. MIMOLA has also been used by the RECORD [9] compiler.

MIMOLA description contains three parts: the algorithm to be compiled, the target processor model and additional linkage and transformation rules. The software part (algorithm description) describes application programs in a PASCAL-like syntax. The processor model describes micro-architecture in the form of a component netlist. The linkage information is used by the compiler in order to locate important modules such as program counter and instruction memory. The following code segment specifies the program counter and instruction memory locations [7]:

```
LOCATION_FOR_PROGRAMCOUNTER PCReg;
LOCATION_FOR_INSTRUCTIONS IM[0..1023];
```

The algorithmic part of MIMOLA is an extension of PASCAL. Unlike other high-level languages, it allows references to physical registers and memories. It also allows use of hardware components using procedure calls. For example, if the processor description contains a component named MAC, programmers can write the following code segment to use the multiply-accumulate operation performed by MAC:

```
res := MAC(x, y, z);
```

The processor is modelled as a net-list of component modules. MIMOLA permits modelling of arbitrary (programmable or non-programmable) hardware structures. Similar to VHDL, a number of predefined, primitive operators exist. The basic entities of MIMOLA hardware models are modules and connections. Each module is specified by its port interface and its behaviour. The following example shows the description of a multi-functional ALU module [7]:

```
MODULE ALU
    (IN inp1, inp2: (31:0);
    OUT outp: (31:0);
    IN ctrl: (1:0);
    )
CONBEGIN
        outp <- CASE ctrl OF
            0: inp1 + inp2 ;
            1: inp1 - inp2 ;
            2: inp1 AND inp2 ;
            3: inp1 ;
            END;
CONEND;
```

The CONBEGIN/CONEND construct includes a set of concurrent assignments. In the example a conditional assignment to output port *outp* is specified, which depends on the two-bit control input *ctrl*. The netlist structure is formed by connecting ports of module instances. For example, the following MIMOLA description connects two modules: *ALU* and accumulator *ACC*.

```
CONNECTIONS ALU.outp -> ACC.inp
            ACC.outp -> ALU.inp
```

The MSSQ code generator extracts instruction-set information from the module netlist. It uses two internal data structures: connection operation graph (COG) and instruction tree (I-tree). It is a very difficult task to extract the COG and I-trees even in the presence of linkage information due to the flexibility of an RT-level structural description. Extra constraints need to be imposed in order for the MSSQ code generator to work properly. The constraints limit the architecture scope of MSSQ to micro-programmable controllers, in which all control signals originate directly from the instruction word. The lack of an explicit description of processor pipelines or resource conflicts may result in poor code quality for some classes of VLIW or deeply pipelined processors.

6.3.1.2   UDL/I

Unified design language, UDL/I [8] is developed as a hardware description language for compiler generation in COACH ASIP design environment at Kyushu University, Japan. UDL/I is used for describing processors at an RT-level on a per-cycle basis. The instruction-set is automatically extracted from the UDL/I description [10], and then it is used for generation of a compiler and a simulator. COACH assumes simple RISC processors and does not explicitly support ILP or processor pipelines. The processor description is synthesisable with the UDL/I synthesis system [11]. The major advantage of the COACH system is that it requires a single description for synthesis, simulation and compilation. Designer needs to provide hints to locate important machine states such as program counter and register files. Due to difficulty in instruction-set extraction (ISE), ISE is not supported for VLIW and superscalar architectures.

Structural ADLs enable flexible and precise micro-architecture descriptions. The same description can be used for hardware synthesis, test generation, simulation and compilation. However, it is difficult to extract the instruction-set without restrictions on description style and target scope. Structural ADLs are more suitable for hardware generation than retargetable compilation.

6.3.2   Behavioural ADLs

The difficulty of instruction-set extraction can be avoided by abstracting behavioural information from the structural details. Behavioural ADLs explicitly specify the instruction semantics and ignore detailed hardware structures. Typically, there is a one-to-one correspondence between behavioural ADLs and instruction-set reference manual. This section briefly describes four behavioural ADLs: nML [12], ISDL [13], Valen-C [14] and CSDL [15,16].

6.3.2.1 nML

nML is an instruction-set oriented ADL proposed at Technical University of Berlin, Germany. nML has been used by code generators CBC [17] and CHESS [18], and instruction set simulators Sigh/Sim [19] and CHECKERS. Currently, CHESS/CHECKERS environment is used for automatic and efficient software compilation and instruction-set simulation [20].

nML developers recognised the fact that several instructions share common properties. The final nML description would be compact and simple if the common properties are exploited. Consequently, nML designers used a hierarchical scheme to describe instruction sets. The instructions are the topmost elements in the hierarchy. The intermediate elements of the hierarchy are partial instructions (PIs). The relationship between elements can be established using two composition rules: AND-rule and OR-rule. The AND-rule groups several PIs into a larger PI and the OR-rule enumerates a set of alternatives for one PI. Therefore, instruction definitions in nML can be in the form of an AND-OR tree. Each possible derivation of the tree corresponds to an actual instruction.

To achieve the goal of sharing instruction descriptions, the instruction set is enumerated by an attributed grammar [21]. Each element in the hierarchy has a few attributes. A non-leaf element's attribute values can be computed based on its children's attribute values. Attribute grammar is also adopted by other ADLs such as ISDL [13] and TDL [22]. The following nML description shows an example of instruction specification [12]:

```
op numeric_instruction(a:num_action, src:SRC, dst:DST)
action {
    temp_src = src;
    temp_dst = dst;
    a.action;
    dst = temp_dst;
}
op num_action = add | sub
op add()
action = {
    temp_dst = temp_dst + temp_src
}
```

The definition of `numeric_instruction` combines three PIs with the AND-rule: `num_action`, *SRC* and *DST*. The first PI, `num_action`, uses the

OR-rule to describe the valid options for actions: add or sub. The number of all possible derivations of numeric_instruction is the product of the size of num_action, *SRC* and *DST*. The common behaviour of all these options is defined in the action attribute of numeric_instruction. Each option for num_action should have its own action attribute defined as its specific behaviour, which is referred by the a.action line. For example, the above code segment has action description for add operation. Object code image and assembly syntax can also be specified in the same hierarchical manner.

nML also captures the structural information used by instruction-set architecture (ISA). For example, storage units should be declared since they are visible to the instruction-set. nML supports three types of storages: RAM, register and transitory storage. Transitory storage refers to machine states that are retained only for a limited number of cycles, e.g. values on buses and latches. Computations have no delay in nML timing model – only storage units have delay. Instruction delay slots are modelled by introducing storage units as pipeline registers. The result of the computation is propagated through the registers in the behaviour specification.

nML models constraints between operations by enumerating all valid combinations. The enumeration of valid cases can make nML descriptions lengthy. More complicated constraints, which often appear in DSPs with irregular instruction level parallelism (ILP) constraints or VLIW processors with multiple issue slots, are hard to model with nML. For example, nML cannot model the constraint that operation I1 cannot directly follow operation I0. nML explicitly supports several addressing modes. However, it implicitly assumes an architecture model which restricts its generality. As a result it is hard to model multi-cycle or pipelined units and multi-word instructions explicitly. A good critique of nML is given by Hantoog et al [23].

6.3.2.2 ISDL

Instruction Set Description Language (ISDL) was developed at MIT and used by the Aviv compiler [24] and GENSIM simulator generator [25]. The problem of constraint modelling is avoided by ISDL with explicit specification. ISDL is mainly targeted towards VLIW processors. Similar to nML, ISDL primarily describes the instruction-set of processor architectures. ISDL consists mainly of five sections: instruction word format, global definitions, storage resources, assembly syntax and constraints. It also contains an optimisation information section that can be used to provide certain architecture specific hints for the compiler to make better machine dependent code optimisations.

The instruction word format section defines fields of the instruction word. The instruction word is separated into multiple fields each containing one or more subfields. The global definition section describes four main types: tokens, non-terminals, split functions and macro definitions. Tokens are the primitive operands of instructions. For each token, assembly format and binary encoding information must be defined. An example token definition of a binary operand is:

```
Token X[0..1] X_R ival {yylval.ival = yytext[1] - '0';}
```

In this example, following the keyword Token is the assembly format of the operand. X_R is the symbolic name of the token used for reference. The ival is used to describe the value returned by the token. Finally, the last field describes the computation of the value. In this example, the assembly syntax allowed for the token X_R is X0 or X1, and the values returned are 0 or 1, respectively.

The value (last) field is used for behavioural definition and binary encoding assignment by non-terminals or instructions. Non-terminal is a mechanism provided to exploit commonalities among operations. The following code segment describes a non-terminal named XYSRC:

```
Non_Terminal ival XYSRC: X_D {$$ = 0;}   |
                         Y_D {$$ = Y_D + 1;};
```

The definition of XYSRC consists of the keyword Non_Terminal, the type of the returned value, a symbolic name as it appears in the assembly and an action that describes the possible token or non-terminal combinations and the return value associated with each of them. In this example, XYSRC refers to tokens X_D and Y_D as its two options. The second field (ival) describes the returned value type. It returns 0 for X_D or incremented value for Y_D.

Similar to nML, storage resources are the only structural information modelled by ISDL. The storage section lists all storage resources visible to the programmer. It lists the names and sizes of the memory, register files and special registers. This information is used by the compiler to determine the available resources and how they should be used.

The assembly syntax section is divided into fields corresponding to the separate operations that can be performed in parallel within a single instruction. For each field, a list of alternative operations can be described. Each operation description consists of a name, a list of tokens or non-terminals as parameters, a set of commands that manipulate the bitfields, RTL description, timing details and costs. RTL description captures the effect of the operation on the storage resources. Multiple costs are allowed including operation execution time, code size and costs due to resource conflicts. The timing model of ISDL describes when the various effects of the operation take place (e.g. because of pipelining).

In contrast to nML, which enumerates all valid combinations, ISDL defines invalid combinations in the form of Boolean expressions. This often leads to a simple constraint specification. It also enables ISDL to capture irregular ILP constraints. The following example shows how to describe the constraint that instruction I1 cannot directly follow instruction I0. The '[1]' indicates a time shift of one instruction fetch for the I0 instruction. The '~' is a symbol for NOT and '&' is for logical AND.

```
~(I1 *) & ([1] I0 *, *)
```

ISDL provides the means for compact and hierarchical instruction set specification. However, it may not be possible to describe instruction sets with multiple encoding formats using the simple tree-like instruction structure of ISDL.

6.3.2.3 Valen-C

Valen-C is an embedded software programming language proposed at Kyushu University, Japan [14,26]. Valen-C is an extended C language which supports explicit and exact bit-width for integer type declarations. A retargetable compiler (called Valen-CC) has been developed that accepts C or Valen-C programs as an input and generates the optimised assembly code. Although Valen-CC assumes simple RISC architectures, it has retargetability to some extent. The most interesting feature of Valen-CC is that the processor can have any datapath bit-width (e.g. 14 bits or 29 bits). The Valen-C system aims at optimisation of datapath width. The target processor description for Valen-CC includes the instruction set consisting of behaviour and assembly syntax of each instruction as well as the processor datapath width. Valen-CC does not explicitly support processor pipelines or ILP.

6.3.2.4 CSDL

Computer system description languages (CSDL) is a family of machine description languages developed for the Zephyr compiler infrastructure at the University of Virginia. CSDL consists of four languages: SLED [16], λ-RTL, CCL and PLUNGE. SLED describes assembly and binary representations of instructions [16], while λ-RTL describes the behaviour of instructions in the form of register transfers [15]. CCL specifies the convention of function calls [27]. PLUNGE is a graphical notation for specifying the pipeline structure.

Similar to ISDL, SLED (Specification Language for Encoding and Decoding) uses a hierarchical model for machine instruction. SLED models an instruction (binary representation) as a sequence of tokens, which are bit vectors. Tokens may represent whole instructions, as on RISC machines, or parts of instructions, as on CISC machines. Each class of token is declared with multiple fields. The construct patterns help to group the fields together and to bind them to binary values. The directive constructors help to connect the fields into instruction words. Similar to nML, SLED enumerates legal combinations of fields. There is neither a notion of hardware resources nor explicit constraint descriptions. Therefore, without significant extension, SLED is not suitable for use in VLIW instruction word description [2].

To reduce description effort, λ-RTL was developed. A λ-RTL description will be translated into register-transfer lists for the use of *vpo* (very portable optimiser). According to the developers [15], λ-RTL is a high order, strongly typed, polymorphic, pure functional language based largely on Standard ML [28]. It has many high-level language features such as name space (through the module and import directives) and function definition. Users can even introduce new semantics and precedence to basic operators.

In general, the behavioural languages have one feature in common: hierarchical instruction set description based on attribute grammar [21]. This feature simplifies the instruction set description by sharing the common components between operations. However, the capabilities of these models are limited due to the lack of detailed pipeline and timing information. It is not possible to generate cycle accurate simulators without certain assumptions regarding control behaviour. Due to lack of structural details, it is also not possible to perform resource-based scheduling using behavioural ADLs.

6.3.3   Mixed ADLs

Mixed languages captures both structural and behavioural details of the architecture. This section briefly describes five mixed ADLs: FlexWare, HMDES, TDL, EXPRESSION and LISA.

6.3.3.1   FlexWare

FlexWare is a CAD system for DSP or ASIP design [29]. The FlexWare system includes the CodeSyn code generator and the Insulin simulator. Both behaviour and structure are captured in the target processor description. The machine description for CodeSyn consists of three components: instruction set, available resources (and their classification) and an interconnect graph representing the datapath structure. The instruction set description is a list of generic processor macro instructions to execute each target processor instruction. The simulator uses a VHDL model of a generic parameterisable machine. The parameters include bit-width, number of registers, ALUs and so on. The application is translated from the user-defined target instruction set to the instruction set of the generic machine. Then, the code is executed on the generic machine.

6.3.3.2   HMDES

Machine description language HMDES was developed at the University of Illinois at Urbana-Champaign for the IMPACT research compiler [30]. C-like pre-processing capabilities such as file inclusion, macro-expansion and conditional inclusion are supported in HMDES. An HMDES description is the input to the MDES machine description system of the Trimaran compiler infrastructure, which contains IMPACT as well as the Elcor research compiler from HP Labs. The description is first pre-processed, then optimised and translated to a low-level representation file. A machine database reads the low-level files and supplies information for the compiler backend through a pre-defined query interface.

MDES captures both structure and behaviour of target processors. Information is broken down into sections such as format, resource usage, latency, operation and register. For example, the following code segment describes register and register file. It describes 64 registers. The register file describes the width of each register and other optional fields such as generic register type (virtual field), speculative, static and rotating registers. The value '1' implies speculative and '0' implies non-speculative.

```
SECTION Register {
  R0(); R1(); ... R63();
  'R[0]'(); ... 'R[63]'();
  ...
}

SECTION Register_ File {
  RF_i(width(32) virtual(i) speculative(1)
       static(R0...R63) rotating('R[0]'...'R[63]'));
  ...
}
```

MDES allows only a restricted retargetability of the cycle-accurate simulator to the HPL-PD processor family [31]. MDES permits description of memory systems, but limited to the traditional hierarchy, i.e. register files, caches and main memory.

6.3.3.3 TDL

Target description language TDL [22] was developed at Saarland University, Germany. The language is used in a retargetable postpass assembly-based code optimisation system called PROPAN [32]. A TDL description contains four sections: resource, instruction set, constraints and assembly format.

TDL offers a set of pre-defined resource types whose properties can be described by a pre-defined set of attributes. The pre-defined resource types comprise functional units, register sets, memories and caches. Attributes are available to describe the bit-width of registers, their default data type, the size of a memory, its access width and alignment restrictions. The designer can extend the domain of the pre-defined attributes and declare user-defined attributes if additional properties have to be taken into account.

Similar to behavioural languages, the instruction-set description of TDL is based on attribute grammar [21]. TDL supports VLIW architectures, so it distinguishes operation and instruction. The instruction-set section also contains definition of operation classes that groups operations for the ease of reference. TDL provides a non-terminal construct to capture common components between operations.

Similar to ISDL, TDL uses Boolean expressions for constraint modelling. A constraint definition includes a premise part followed by a rule part, separated by a colon. The following code segment describes constraints in TDL [22]:

```
op in {C0}: op.dst1 = op.src1;
op1 in {C1} & op2 in {C2}: !(op1 &&op2);
```

The first one enforces the first source operand to be identical to the destination operand for all operations of the operation class C0. The second rule prevents any operation of operation class C1 to be executed in parallel with an operation of operation class C2.

The assembly section deals with syntactic details of the assembly language such as instruction or operation delimiters, assembly directives and assembly expressions. TDL is assembly oriented and provides a generic modelling of irregular hardware constraints. TDL provides a well-organised formalism for VLIW DSP assembly code generation.

6.3.3.4   EXPRESSION

The above-mixed ADLs require explicit description of Reservation Tables (RT). Processors that contain complex pipelines, large amounts of parallelism and complex storage sub-systems, typically contain a large number of operations and resources (and hence RTs). Manual specification of RTs on a per-operation basis thus becomes cumbersome and error-prone. The manual specification of RTs (for each configuration) becomes impractical during rapid architectural exploration. The EXPRESSION ADL [33] describes a processor as a netlist of units and storages to automatically generate RTs based on the netlist [34]. Unlike MIMOLA, the netlist representation of EXPRESSION is coarse grain. It uses a higher level of abstraction similar to block-diagram level description in architecture manual.

EXPRESSION ADL was developed at University of California, Irvine. The ADL has been used by the retargetable compiler (EXPRESS [35]) and simulator (SIMPRESS [36]) generation framework. The framework also supports a graphical user interface (GUI) and can be used for design space exploration of programmable architectures consisting of processor cores, coprocessors and memories [37].

An EXPRESSION description is composed of two main sections: behaviour (instruction-set) and structure. The behaviour section has three subsections: operations, instruction and operation mappings. Similarly, the structure section consists of three subsections: components, pipeline/data-transfer paths and memory subsystem.

The operation subsection describes the instruction-set of the processor. Each operation of the processor is described in terms of its opcode and operands. The types and possible destinations of each operand are also specified. A useful feature of EXPRESSION is operation group that groups similar operations together for the ease of later reference. For example, the following code segment shows an operation group (alu_ops) containing two ALU operations: add and sub.

```
(OP_GROUP alu_ops
    (OPCODE add
        (OPERANDS (SRC1 reg) (SRC2 reg/imm) (DEST reg))
        (BEHAVIOR DEST = SRC1 + SRC2)
        ...
    )
    (OPCODE sub
        (OPERANDS (SRC1 reg) (SRC2 reg/imm) (DEST reg))
        (BEHAVIOR DEST = SRC1 - SRC2)
        ...
    )
)
```

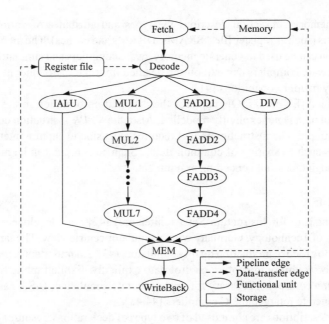

Figure 6.5 The DLX architecture

The instruction subsection captures the parallelism available in the architecture. Each instruction contains a list of slots (to be filled with operations), with each slot corresponding to a functional unit. The operation mapping subsection is used to specify the information needed by instruction selection and architecture-specific optimisations of the compiler. For example, it contains mapping between generic and target instructions.

The component subsection describes each RT-level component in the architecture. The components can be pipeline units, functional units, storage elements, ports and connections. For multi-cycle or pipelined units, the timing behaviour is also specified.

The pipeline/data-transfer path subsection describes the netlist of the processor. The 'pipeline path description' provides a mechanism to specify the units which comprise the pipeline stages, while the 'data-transfer path description' provides a mechanism for specifying the valid data-transfers. This information is used to both retarget the simulator, and to generate reservation tables needed by the scheduler [34]. An example path declaration for the DLX architecture [38] (Figure 6.5) is shown below. It describes that the processor has five pipeline stages. It also describes that the Execute stage has four parallel paths. Finally, it describes each path, e.g. it describes that the FADD path has four pipeline stages.

```
(PIPELINE Fetch Decode Execute MEM WriteBack)
(Execute (ALTERNATE IALU MULT FADD DIV))
(MULT (PIPELINE MUL1 MUL2 ... MUL7))
(FADD (PIPELINE FADD1 FADD2 FADD3 FADD4))
```

The memory subsection describes the types and attributes of various storage components (such as register files, SRAMs, DRAMs and caches). The memory netlist information can be used to generate memory aware compilers and simulators [39,40]. Memory aware compilers can exploit the detailed information to hide the latency of the lengthy memory operations [41].

In general, EXPRESSION captures the data path information in the processor. The control path is not explicitly modelled. Also, the VLIW instruction composition model is simple. The instruction model requires extension to capture inter-operation constraints such as sharing of common fields. Such constraints can be modelled by ISDL through cross-field encoding assignment.

6.3.3.5 LISA

LISA (Language for Instruction Set Architecture) [42] was developed at Aachen University of Technology, Germany with a simulator centric view. The language has been used to produce production quality simulators [43]. An important aspect of LISA language is its ability to capture control path explicitly. Explicit modelling of both datapath and control is necessary for cycle-accurate simulation. LISA has also been used to generate retargetable C compilers [44,45].

LISA descriptions are composed of two types of declarations: resource and operation. The resource declarations cover hardware resources such as registers, pipelines and memories. The pipeline model defines all possible pipeline paths that operations can go through. An example pipeline description for the architecture shown in Figure 6.5 is as follows:

```
PIPELINE int = {Fetch; Decode; IALU; MEM; WriteBack}
PIPELINE flt = {Fetch; Decode; FADD1; FADD2;
                FADD3; FADD4; MEM; WriteBack}
PIPELINE mul = {Fetch; Decode; MUL1; MUL2; MUL3; MUL4;
                MUL5; MUL6; MUL7; MEM; WriteBack}
PIPELINE div = {Fetch; Decode; DIV; MEM; WriteBack}
```

Operations are the basic objects in LISA. They represent the designer's view of the behaviour, the structure, and the instruction set of the programmable architecture. Operation definitions capture the description of different properties of the system such as operation behaviour, instruction set information and timing. These operation attributes are defined in several sections:

- The CODING section describes the binary image of the instruction word.
- The SYNTAX section describes the assembly syntax of instructions.
- The SEMANTICS section specifies the instruction-set semantics.
- The BEHAVIOR section describes components of the behavioural model.
- The ACTIVATION section describes the timing of other operations relative to the current operation.
- The DECLARE section contains local declarations of identifiers.

LISA exploits the commonality of similar operations by grouping them into one. The following code segment describes the decoding behaviour of two immediate-type (i_type) operations (ADDI and SUBI) in the DLX Decode stage. The complete behaviour of an operation can be obtained by combining its behaviour definitions in all the pipeline stages.

```
OPERATION i_type IN pipe_int.Decode {
    DECLARE {
        GROUP opcode={ADDI || SUBI}
        GROUP rs1, rd = {fix_register};
    }
    CODING {opcode rs1 rd immediate}
    SYNTAX {opcode rd '','' rs1 '','' immediate}
    BEHAVIOR { reg_a = rs1; imm = immediate; cond = 0;
    }
    ACTIVATION {opcode, writeback}
}
```

A language similar to LISA is RADL. RADL [46] was developed at Rockwell, Inc. as an extension of the LISA approach that focuses on explicit support of detailed pipeline behaviour to enable generation of production quality cycle-accurate and phase-accurate simulators.

6.3.4   Partial ADLs

The ADLs discussed so far capture a complete description of the processor's structure, behaviour or both. There are many description languages that capture partial information of the architecture needed to perform a specific task. This section describes two such ADLs.

AIDL is an ADL developed at the University of Tsukuba for design of high-performance superscalar processors [47]. It seems that AIDL does not aim at datapath optimisation but aims at validation of the pipeline behaviour such as data-forwarding and out-of-order completion. In AIDL, timing behaviour of the pipeline is described using interval temporal logic. AIDL does not support software toolkit generation. However, AIDL descriptions can be simulated using the AIDL simulator.

PEAS-I is a CAD system for ASIP design supporting automatic instruction set optimisation, compiler generation and instruction level simulator generation [48]. In the PEAS-I system, the GNU C compiler is used, and the machine description of GCC is automatically generated. Therefore, there exists no specific ADL in PEAS-I. Inputs to PEAS-I include an application program written in C and input data to the program. Then, the instruction set is automatically selected in such a way that the performance is maximised or the gate count is minimised. Based on the instruction set, GNU CC and an instruction level simulator are automatically retargeted.

6.4   ADL driven methodologies

The survey of ADLs is incomplete without a clear understanding of the supported methodologies. This section investigates the contribution of the contemporary ADLs in the following methodologies:

- software toolkit generation and exploration
- generation of hardware implementation
- top-down validation

6.4.1   *Software toolkit generation and exploration*

Embedded systems present a tremendous opportunity to customise designs by exploiting the application behaviour. Rapid exploration and evaluation of candidate architectures are necessary due to time-to-market pressure and short product lifetimes. ADLs are used to specify processor and memory architectures and generate software toolkit including compiler, simulator, assembler, profiler and debugger. Figure 6.6 shows a traditional ADL-based design space exploration flow. The application programs are compiled and simulated, and the feedback is used to modify the ADL specification with the goal of finding the best possible architecture for the given set of application programs under various design constraints such as area, power and performance.

An extensive body of recent work addresses ADL driven software toolkit generation and design space exploration of processor-based embedded systems, in both academia: ISDL [13], Valen-C [14], MIMOLA [9], LISA [42], nML [12], Sim-nML [49], EXPRESSION [33], and industry: ARC [50], Axys [51], RADL [46], Target [20], Tensilica [52], MDES [31].

One of the main purposes of an ADL is to support automatic generation of a high-quality software toolkit including at least an ILP compiler and a cycle-accurate

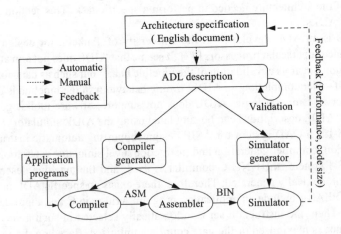

Figure 6.6   ADL driven design space exploration

simulator. However, such tools require detailed information about the processor, typically in a form that is not concise and easily specifiable. Therefore, it becomes necessary to develop procedures to automatically generate such tool-specific information from the ADL specification. For example, RTs are used in many ILP compilers to describe resource conflicts. However, manual description of RTs on a per-instruction basis is cumbersome and error-prone. Instead, it is easier to specify the pipeline and datapath resources in an abstract manner and generate RTs on a per-instruction basis [34].

This section describes some of the challenges in automatic generation of software tools (focusing on compilers and simulators) and surveys some of the approaches adopted by current tools.

6.4.1.1   Compilers

Traditionally, software for embedded systems was hand-tuned in assembly. With increasing complexity of embedded systems, it is no longer practical to develop software in assembly language or to optimise it manually except for critical sections of the code. Compilers which produce optimised machine specific code from a program specified in a high-level language (HLL) such as C/C++ and Java are necessary in order to produce efficient software within the time budget. Compilers for embedded systems have been the focus of several research efforts recently [53].

The compilation process can be broadly broken into two steps: analysis and synthesis [54]. During analysis, the program (in HLL) is converted into an intermediate representation (IR) that contains all the desired information such as control and data dependences. During synthesis, the IR is transformed and optimised in order to generate efficient target specific code. The synthesis step is more complex and typically includes the following phases: instruction selection, scheduling, resource allocation, code optimisations/transformations and code generation [55]. The effectiveness of each phase depends on the algorithms chosen and the target architecture. A further problem during the synthesis step is that the optimal ordering between these phases is highly dependent on the target architecture and the application program. As a result, traditionally, compilers have been painstakingly hand-tuned to a particular architecture (or architecture class) and application domain(s). However, stringent time-to-market constraints for SOC designs no longer make it feasible to manually generate compilers tuned to particular architectures. Automatic generation of an efficient compiler from an abstract description of the processor model becomes essential.

A promising approach to automatic compiler generation is the 'retargetable compiler' approach (see Part II embedded software, Chapter 8). A compiler is classified as retargetable if it can be adapted to generate code for different target processors with significant reuse of the compiler source code. Retargetability is typically achieved by providing target machine information (in an ADL) as input to the compiler along with the program corresponding to the application.

The complexity in retargeting the compiler depends on the range of target processors it supports and also on its optimising capability. Due to the growing amount of ILP features in modern processor architectures, the difference in quality of code generated by a naive code conversion process and an optimising ILP compiler can be enormous. Recent approaches on retargetable compilation have focused on developing optimisations/transformations that are 'retargetable' and capturing the machine-specific information needed by such optimisations in the ADL. The retargetable compilers can be classified into three broad categories, based on the type of the machine model accepted as input.

Architecture template based: Such compilers assume a limited architecture template which is parameterisable for customisation. The most common parameters include operation latencies, number of functional units, number of registers, etc. Architecture template based compilers have the advantage that both optimisations and the phase ordering between them can be manually tuned to produce highly efficient code for the limited architecture space. Examples of such compilers include the Valen-C compiler [14] and the GNU-based C/C++ compiler from Tensilica Inc. [52]. The Tensilica GNU-based C/C++ compiler is geared towards the Xtensa parameterisable processor architecture. One important feature of this system is the ability to add new instructions (described through an Instruction Extension Language) and automatically generate software tools tuned to the new instruction-set.

Explicit behavioural information based: Most compilers require a specification of the behaviour in order to retarget their transformations (e.g. instruction selection requires a description of the semantics of each operation). Explicit behavioural information based retargetable compilers require full information about the instruction-set as well as explicit resource conflict information. Examples include the AVIV [24] compiler using ISDL, CHESS [18] using nML and Elcor [31] using MDES. The AVIV retargetable code generator produces machine code, optimised for minimal size, for target processors with different instruction-set. It solves the phase ordering problem by performing a heuristic branch-and-bound step that performs resource allocation/assignment, operation grouping and scheduling concurrently. CHESS is a retargetable code generation environment for fixed-point DSP processors. CHESS performs instruction selection, register allocation and scheduling as separate phases (in that order). Elcor is a retargetable compilation environment for VLIW architectures with speculative execution. It implements a software pipelining algorithm (modulo scheduling) and register allocation for static and rotating register files.

Behavioural information generation based: Recognising that the architecture information needed by the compiler is not always in a form that may be well suited for other tools (such as synthesis) or does not permit concise specification, some research has focused on extraction of such information from a more amenable specification. Examples include the MSSQ and RECORD compiler using MIMOLA [9], retargetable C compiler based on LISA [44] and the EXPRESS compiler using EXPRESSION [33]. MSSQ translates Pascal-like HLL into microcode for micro-programmable controllers, while RECORD translates code written in a DSP-specific programming language, called data flow language (DFL), into machine code for the target DSP. The retargetable C compiler generation using LISA is based on reuse of a powerful C compiler platform with many built-in code optimisations and generation of mapping rules for code selection using the instruction semantics information [44]. The EXPRESS compiler tries to bridge the gap between explicit specification of all information (e.g. AVIV) and implicit specification requiring extraction of instruction-set (e.g. RECORD), by having a mixed behavioural/structural view of the processor.

6.4.1.2   Simulators

Simulators are critical components of the exploration and software design toolkit for the system designer. They can be used to perform diverse tasks such as verifying the functionality and/or timing behaviour of the system (including hardware and software), and generating quantitative measurements (e.g. power consumption) which can be used to aid the design process.

Simulation of the processor system can be performed at various abstraction levels. At the highest level of abstraction, a functional simulation of the processor can be performed by modelling only the instruction-set (IS). Such simulators are termed instruction-set simulators (ISS) or instruction-level simulators (ILS). At lower levels of abstraction are the cycle-accurate and phase-accurate simulation models that yield more detailed timing information. Simulators can be further classified based on whether they provide bit-accurate models, pin-accurate models, exact pipeline models and structural models of the processor.

Typically, simulators at higher levels of abstraction (e.g. ISS, ILS) are faster but gather less information as compared to those at lower levels of abstraction (e.g. cycle-accurate, phase-accurate). Retargetability (i.e. ability to simulate a wide variety of target processors) is especially important in the arena of embedded system design with emphasis on exploration and co-development of hardware and software. Simulators with limited retargetability are very fast but may not be useful in all aspects of the design process. Such simulators typically incorporate a fixed architecture template and allow only limited retargetability in the form of parameters such as number of registers and ALUs. Examples of such simulators are numerous in the industry and include the HPL-PD [31] simulator using the MDes ADL.

The model of simulation adopted has significant impact on the simulation speed and flexibility of the simulator. Based on the simulation model, simulators can be classified into three types: interpretive, compiled and mixed.

Interpretation based: Such simulators are based on an interpretive model of the processor's instruction-set. Interpretive simulators store the state of the target processor in host memory. It then follows a fetch, decode and execute model: instructions are fetched from memory, decoded and then executed in serial order. Advantages of this model include ease of implementation, flexibility and the ability to collect varied processor state information. However, it suffers from significant performance degradation as compared with the other approaches primarily due to the tremendous overhead in fetching, decoding and dispatching instructions. Almost all commercially available simulators are interpretive. Examples of research interpretive retargetable simulators include SIMPRESS [36] using EXPRESSION and GENSIM/XSIM [25] using ISDL.

Compilation based: Compilation-based approaches reduce the runtime overhead by translating each target instruction into a series of host machine instructions which manipulate the simulated machine state. Such translation can be done either at compile time (static compiled simulation) where the fetch–decode–dispatch overhead is completely eliminated, or at load time (dynamic compiled simulation) which amortises the overhead over repeated execution of code. Simulators based on the static compilation model are presented by Zhu and Gajski [56] and Pees *et al.* [43]. Examples of dynamic compiled code simulators include the Shade simulator [57] and the Embra simulator [58].

Interpretive + Compiled: Traditional interpretive simulation is flexible but slow. Instruction decoding is a time-consuming process in a software simulation. Compiled simulation performs compile time decoding of application programs to improve the simulation performance. However, all compiled simulators rely on the assumption that the complete program code is known before the simulation starts and is furthermore run-time static. Due to the restrictiveness of the compiled technique, interpretive simulators are typically used in embedded systems' design flow. Two recently proposed simulation techniques (JIT-CCS [59] and IS-CS [60]) combine the flexibility of interpretive simulation with the speed of the compiled simulation.

The 'just-in-time cache compiled simulation' (JIT-CCS) technique compiles an instruction during run-time, *just-in-time* before the instruction is going to be executed. Subsequently, the extracted information is stored in a simulation cache for direct reuse in a repeated execution of the program address. The simulator recognises if the program code of a previously executed address has changed and initiates a re-compilation. The 'instruction set compiled simulation' (IS-CS) technique performs time-consuming instruction decoding during compile time. In this case, an instruction is modified at run-time, the instruction is re-decoded prior to execution. It also uses an 'instruction abstraction' technique to generate aggressively optimised decoded instructions that further improve simulation performance [60,61].

6.4.2   Generation of hardware implementation

Recent approaches on ADL-based software toolkit generation enable performance driven exploration. The simulator produces profiling data and thus may answer questions regarding the instruction set, the performance of an algorithm and the required size of memory and registers. However, the required silicon area, clock frequency and power consumption can only be determined in conjunction with a synthesisable HDL model.

There are two major approaches in the literature for synthesisable HDL generation. The first one is a parameterised processor core-based approach. These cores are bound to a single processor template whose architecture and tools can be modified to a certain degree. The second approach is based on processor specification languages.

6.4.2.1 Processor template based

Examples of processor template-based approaches are Xtensa [52], Jazz [62] and PEAS [63]. Xtensa [52] is a scalable RISC processor core. Configuration options include the width of the register set, caches and memories. New functional units and instructions can be added using the Tensilica Instruction (TIE) language. A synthesisable hardware model along with software toolkit can be generated for this class of architectures. Improv's Jazz [62] processor is supported by a flexible design methodology to customise the computational resources and instruction set of the processor. It allows modifications of data width, number of registers, depth of hardware task queue and addition of custom functionality in Verilog. PEAS [63] is a GUI-based hardware/software codesign framework. It generates HDL code along with software toolkit. It has support for several architecture types and a library of configurable resources.

6.4.2.2 Specification language based

Figure 6.7 shows a typical framework of processor description language-driven HDL generation. Structure-centric ADLs such as MIMOLA are suitable for hardware generation. Some of the behavioural languages (such as ISDL and nML) are also used for hardware generation. For example, the HDL generator HGEN [25] uses ISDL description, and the synthesis tool GO [20] is based on nML. Itoh *et al.* [64] have proposed a micro-operation description-based synthesisable HDL generation.

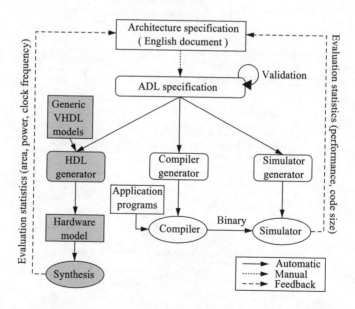

Figure 6.7   ADL-driven implementation generation and exploration

Mixed languages such as LISA and EXPRESSION capture both structure and behaviour of the processor. The synthesisable HDL generation approach based on LISA language [65] produces an HDL model of the architecture. The designer has the choice to generate a VHDL, Verilog or SystemC representation of the target architecture [66]. The HDL generation methodology presented by Mishra *et al.* [67] combines the advantages of the processor template-based environments and the language-based specifications using EXPRESSION ADL.

6.4.3   Top-down validation

Validation of microprocessors is one of the most complex and important tasks in the current System-on-Chip (SoC) design methodology. Figure 6.8 shows a traditional architecture validation flow. The architect prepares an informal specification of the microprocessor in the form of an English document. The logic designer implements the modules in the RTL. The 'RTL design' is validated using a combination of simulation techniques and formal methods. One of the most important problems in today's processor design validation is the lack of a golden reference model that can be used for verifying the design at different levels of abstraction. Thus, many existing validation techniques employ a 'bottom-up approach' to pipeline verification, where the functionality of an existing pipelined processor is, in essence, reverse-engineered from its RTL implementation.

Mishra [68] has presented an ADL-driven validation technique that is complementary to these bottom-up approaches. It leverages the system architects'

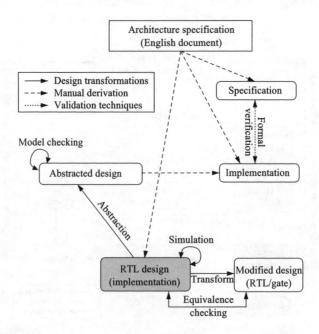

Figure 6.8   Traditional bottom-up validation flow

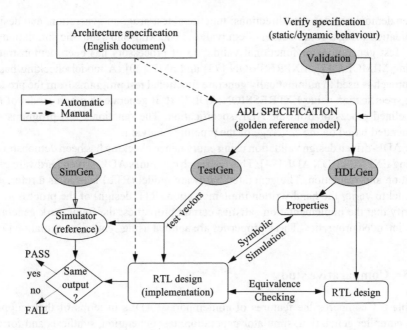

Figure 6.9   Top-down validation flow

knowledge about the behaviour of the programmable embedded systems through ADL constructs, thereby allowing a powerful 'top-down approach' to microprocessor validation. Figure 6.9 shows an ADL-driven top-down validation methodology. This methodology has two important steps: validation of ADL specification, and specification-driven validation of programmable architectures.

6.4.3.1   Validation of ADL specification

It is important to verify the ADL specification to ensure the correctness of the architecture specified and the generated software toolkit. Both static and dynamic behaviour need to be verified to ensure that the specified architecture is well formed. The static behaviour can be validated by analysing several static properties such as, connectedness, false pipeline and data-transfer paths and completeness using a graph-based model of the pipelined architecture [69,70].

The dynamic behaviour can be validated by analysing the instruction flow in the pipeline using a Finite State Machine (FSM) based model to verify several important architectural properties such as determinism and in-order execution in the presence of hazards and multiple exceptions [71,72].

6.4.3.2   Specification-driven validation

The validated ADL specification can be used as a golden reference model for top-down validation of programmable architectures. The top-down validation approach has been demonstrated in two directions: functional test program generation, and design validation using a combination of equivalence checking and symbolic simulation.

Test generation for functional validation of processors has been demonstrated using MIMOLA [7], EXPRESSION [73] and nML [20]. A model checking based approach is used to automatically generate functional test programs from the processor specification using EXPRESSION ADL [73]. It generates a graph model of the pipelined processor from the ADL specification. The functional test programs are generated based on the coverage of the pipeline behaviour.

ADL-driven design validation using equivalence checking has been demonstrated using EXPRESSION ADL [74]. This approach combines ADL-driven hardware generation and validation. The generated hardware model (RTL) is used as a reference model to verify the hand-written implementation (RTL design) of the processor. To verify that the implementation satisfies certain properties, the framework generates the intended properties. These properties are applied using symbolic simulation [74].

6.5   Comparative study

Table 6.1 compares the features of contemporary ADLs in terms of their support for compiler generation, simulator generation, test generation, synthesis and formal verification. Also, information captured by the ADLs is compared.

Since MIMOLA and UDL/I are originally HDLs, their descriptions are synthesisable and can be simulated using HDL simulators. MIMOLA appears to be successful for retargetable compilation for DSPs with irregular datapaths. However, since its abstraction level is rather low, MIMOLA is laborious to write. COACH (uses UDL/I) supports generation of both compilers and simulators. nML and ISDL support ILP compiler generation. However, due to the lack of structural information, it is not possible to automatically detect resource conflicts between instructions. MDES supports simulator generation only for the HPL-PD processor family. EXPRESSION has ability to automatically generate ILP compilers, reservation tables and cycle-accurate simulators. Furthermore, description of memory hierarchies is supported. LISA and RADL were originally designed for simulator generation. AIDL descriptions are executable on the AIDL simulator, and do not support compiler generation.

From the above comparison it is obvious that ADLs should capture both behaviour (instruction set) and structure (netlist) information in order to generate high-quality software toolkit automatically and efficiently. Behaviour information which is necessary for compiler generation should be explicitly specified for mainly two reasons. First, instruction-set extraction from netlists described in synthesis-oriented ADLs or HDLs does not seem to be applicable to a wide range of processors. Second, synthesis-oriented ADLs or HDLs are generally tedious to write for the purpose of DSE. Also, structure information is necessary not only to generate cycle-accurate simulators but also to generate ILP constraints which are necessary for high-quality ILP compiler generation.

ADLs designed for a specific domain (such as DSP or VLIW) or for a specific purpose (such as simulation or compilation) can be compact and it is possible to

Table 6.1   Comparison between different ADLs

	MIMOLA	UDL/I	nML	ISDL	HMDES	EXPRESSION	LISA	RADL	AIDL
Compiler generation	√	√	√	√	√	√	√		√
Simulator generation	√	√	√	√	√	√	√	√	△
Cycle-accurate simulation	√	△		√	√	√	√	√	△
Formal verification	√			△		△	√		
Hardware generation	√	√				√	√		
Test generation	√		√	√		√	√		
JTAG interface generation							√		
Structural information	√		√	√	△	√	√	√	
Memory information		√	√	√	√	√	√	√	

√ Supported.
△ Supported with restrictions.

automatically generate efficient (in terms of area, time and power) tools/hardwares. However, it is difficult to design an ADL for a wide variety of architectures to perform different tasks using the same specification. Generic ADLs require the support of powerful methodologies to generate high-quality results compared with domain-specific/task-specific ADLs.

6.6 Conclusions

In the past, an ADL was designed to serve a specific purpose. For example, MIMOLA and UDLI have features similar to a hardware description language and were used mainly for synthesis of processor architectures. Similarly, LISA and RADL were designed for simulation of processor architectures. Likewise, MDES and EXPRESSION were designed mainly for generating retargetable compilers.

The early ADLs were either structure-oriented (MIMOLA, UDL/I), or behaviour-oriented (nML, ISDL). As a result, each class of ADLs is suitable for specific tasks. For example, structure-oriented ADLs are suitable for hardware synthesis, and unfit for compiler generation. Similarly, behaviour-oriented ADLs are appropriate for generating compiler and simulator for instruction-set architectures, and unsuited for generating cycle-accurate simulator or hardware implementation of the architecture. The later ADLs (LISA and EXPRESSION) adopted the mixed approach where the language captures both structure and behaviour of the architecture.

At present, the existing ADLs are getting modified with the new features and methodologies to perform software toolkit generation, hardware generation, instruction-set synthesis, and test generation for validation of architectures. For example, nML is extended by Target Compiler Technologies [20] to perform hardware synthesis and test generation. Similarly, LISA language has been used for hardware generation [66,75], instruction encoding synthesis [76] and JTAG interface generation [77]. Likewise, EXPRESSION has been used for hardware generation [67], instruction-set synthesis [78], test generation [73,79] and specification validation [70,74].

The majority of the ADLs were designed mainly for processor architectures. MDES have features for specifying both processor and memory architectures. EXPRESSION allows specification of processor, memory and co-processor architectures [80]. Similarly, the language elements of LISA enable the description of processor, memory, peripherals and external interfaces [77,81].

In the future, the existing ADLs will go through changes in two dimensions. First, ADLs will specify not only processor, memory and co-processor architectures but also other components of the system-on-chip architectures including peripherals and external interfaces. Second, ADLs will be used for software toolkit generation, hardware synthesis, test generation, instruction-set synthesis, and validation of microprocessors. Furthermore, multiprocessor SoCs will be captured and various attendant tasks will be addressed. The tasks include support for formal analysis, generation of RTOS, exploration of communication architectures and support for interface synthesis. The emerging ADLs will have these features.

Acknowledgements

This work was partially supported by NSF grants CCR-0203813 and CCR-0205712. We thank Prof. Alex Nicolau and members of the ACES laboratory for their helpful comments and suggestions.

References

1 WANG, S., and MALIK, S.: 'Synthesizing operating system based device drivers in embedded systems'. Proceedings of international symposium on *Hardware/ software codesign and system synthesis (CODES+ISSS)*, 2003, pp. 37–44

2 QIN, W., and MALIK, S.: 'Architecture description languages for retargetable compilation', in 'The compiler design handbook: optimizations & machine code generation' (CRC Press, Boca Raton, FL, 2002)

3 TOMIYAMA, H., HALAMBI, A., GRUN, P., DUTT, N., and NICOLAU, A.: 'Architecture description languages for systems-on-chip design'. Proceedings of Asia Pacific conference on *Chip design language*, 1999, pp. 109–116

4 MEDVIDOVIC, N., and TAYLOR, R.: 'A framework for classifying and comparing architecture description languages'. In JAZAYERI, M., and SCHAUER, H. (Eds) Proceedings of *European software engineering conference (ESEC)* (Springer-Verlag, Berlin, 1997), pp. 60–76

5 PAUL CLEMENTS, C.: 'A survey of architecture description languages'. Proceedings of international workshop on *Software specification and design (IWSSD)*, 1996, pp. 16–25

6 BARBACCI, M.R.: 'Instruction set processor specifications (isps): The notation and its applications', *IEEE Transactions on Computers*, 1981, **C-30**(1): 24–40

7 LEUPERS, R., and MARWEDEL, P.: 'Retargetable code generation based on structural processor descriptions', *Design Automation for Embedded Systems*, 1998, 3(1), pp. 75–108

8 AKABOSHI, H.: '*A study on design support for computer architecture design*'. PhD thesis, Department of Information Systems, Kyushu University, Japan, 1996

9 LEUPERS, R., and MARWEDEL, P.: 'Retargetable generation of code selectors from HDL processor models'. Proceedings of *European design and test conference (EDTC)*, 1997, pp. 140–144

10 AKABOSHI, H., and YASUURA, H.: 'Behavior extraction of MPU from HDL description'. Proceedings of Asia Pacific conference on *Hardware description languages (APCHDL)*, 1994

11 http://pjro.metsa.astem.or.jp/udli. UDL/I Simulation/Synthesis Environment, 1997

12 FREERICKS, M.: The nML machine description formalism. Technical report TR SM-IMP/DIST/08, TU Berlin CS Dept., 1993

13   HADJIYIANNIS, G., HANONO, S., and DEVADAS, S.: 'ISDL: An instruction set description language for retargetability'. Proceedings of *Design automation conference (DAC)*, 1997, pp. 299–302

14   INOUE, A., TOMIYAMA, H., EKO, F., KANBARA, H., and YASUURA, H.: 'A programming language for processor based embedded systems'. Proceedings of Asia Pacific conference on *Hardware Description Languages (APCHDL)*, 1998, pp. 89–94

15   RAMSEY, N., and DAVIDSON, J.: 'Specifying instructions' semantics using λ-rt'. Technical report, University of Virginia, July 1999

16   RAMSEY, N., and FERNANDEZ, M.: 'Specifying representations of machine instructions'. Proceedings of *ACM TOPLAS*, May 1997

17   FAUTH, A., and KNOLL, A.: 'Automatic generation of DSP program development tools', Proceedings of international conference of *Acoustics, speech and signal processing (ICASSP)*, 1993, pp. 457–460

18   LANNEER, D., PRAET, J., KIFLI, A., SCHOOFS, K., GEURTS, W., THOEN, F., and GOOSSENS, G.: 'CHESS: Retargetable code generation for embedded DSP processors', in MARWEDEL, P., and GOOSSENS, G.: 'Code generation for embedded processors' (Kluwer Academic Publishers, Norwell, MA, 1995) pp. 85–102

19   LOHR, F., FAUTH, A., and FREERICKS, M.: 'Sigh/sim: An environment for retargetable instruction set simulation'. Technical report 1993/43, Department of Computer Science, Technical University of Berlin, Germany, 1993

20   http://www.retarget.com. Target Compiler Technologies

21   PAAKKI, J.: 'Attribute grammar paradigms – a high level methodology in language implementation', *ACM Computing Surveys*, 1995, 27(2): 196–256

22   KASTNER, D.: 'Tdl: A hardware and assembly description languages'. Technical report TDL 1.4, 2000, Saarland University, Germany

23   HARTOOG, M., ROWSON, J., REDDY, P., DESAI, S., DUNLOP, D., HARCOURT, E., and KHULLAR, N.: 'Generation of software tools from processor descriptions for hardware/software codesign'. Proceedings of *Design automation* conference *(DAC)*, 1997, pp. 303–306

24   HANONO, S., and DEVADAS, S.: 'Instruction selection, resource allocation, and scheduling in the AVIV retargetable code generator'. Proceedings of *Design automation* conference *(DAC)*, 1998, pp. 510–515

25   HADJIYIANNIS, G., RUSSO, P., and DEVADAS, S.: 'A methodology for accurate performance evaluation in architecture exploration'. Proceedings of *Design automation* conference *(DAC)*, 1999, pp. 927–932

26   INOUE, A., TOMIYAMA, H., OKUMA, H., KANBARA, H., and YASUURA, V.: 'Language and compiler for optimizing datapath widths of embedded systems', *IEICE Transactions of Fundamentals*, 1998, **E81-A(12)**: 2595–2604

27   BAILEY, M., and DAVIDSON, J.: 'A formal model and specification language for procedure calling conventions'. Proceedings of *Principle of programming languages (POPL)*, 1995, pp. 298–310

28 MILNER, R., TOFTE, M., and HARPER, R.: 'The definition of standard ML' (MIT Press, Cambridge, MA, 1989)

29 PAULIN, P., LIEM, C., MAY, T., and SUTARWALA, S.: 'FlexWare: A flexible firmware development environment for embedded systems'. Proceedings of Dagstuhl workshop on *Code generation for embedded processors*, 1994, pp. 67–84

30 GYLLENHAAL, J., RAU, B., and HWU, W.: 'Hmdes version 2.0 specification'. Technical report IMPACT-96-3, IMPACT Research Group, University of Illinois, Urbana. IL, 1996

31 The MDES User Manual. http://www.trimaran.org, 1997

32 KASTNER, D.: 'Retargetable postpass optimization by integer linear programming'. PhD thesis, 2000, Saarland University, Germany

33 HALAMBI, A., GRUN, P., GANESH, V., KHARE, A., DUTT, N., and NICOLAU, A.: 'EXPRESSION: A language for architecture exploration through compiler/simulator retargetability'. Proceedings of *Design automation and test in Europe (DATE)*, 1999, pp. 485–490

34 GRUN, P., HALAMBI, A., DUTT, N., and NICOLAU, A.: 'RTGEN: An algorithm for automatic generation of reservation tables from architectural descriptions'. Proceedings of international symposium on *System synthesis (ISSS)*, 1999, pp. 44–50

35 HALAMBI, A., SHRIVASTAVA, A., DUTT, N., and NICOLAU, A.: 'A customizable compiler framework for embedded systems'. Proceedings of *Software and compilers for embedded systems (SCOPES)*, 2001

36 KHARE, A., SAVOIU, N., HALAMBI, A., GRUN, P., DUTT, N., and NICOLAU, A.: 'V-SAT: A visual specification and analysis tool for system-on-chip exploration'. Proceedings of *EUROMICRO* conference, 1999, pp. 1196–1203

37 http://www.ics.uci.edu/~express. Exploration framework using EXPRESSION

38 HENNESSY, J., and PATTERSON, D.: 'Computer architecture: a quantitative approach' (Morgan Kaufmann Publishers Inc, San Mateo, CA, 1990)

39 MISHRA, P., MAMIDIPAKA, M., and DUTT, N.: 'Processor-memory co-exploration using an architecture description language', *ACM Transactions on embedded computing systems (TECS)*, 2004, 3(1), pp. 140–162

40 MISHRA, P., GRUN, P., DUTT, N., and NICOLAU, A.: 'Processor-memory co-exploration driven by an architectural description language'. Proceedings of international conference on *VLSI design*, 2001, pp. 70–75

41 GRUN, P., DUTT, N., and NICOLAU, A.: 'Memory aware compilation through accurate timing extraction'. Proceedings of *Design automation conference (DAC)*, 2000, pp. 316–321

42 ZIVOJNOVIC, V., PEES, S., and MEYR, H.: 'LISA – machine description language and generic machine model for HW/SW co-design'. IEEE workshop on *VLSI signal processing*, 1996, pp. 127–136

43 PEES, S., HOFFMANN, A., and MEYR, H.: 'Retargetable compiled simulation of embedded processors using a machine description language',

ACM Transactions on Design Automation of Electronic Systems, 2000, **5**(4), pp. 815–834

44   HOHENAUER, M., SCHARWAECHTER, H., KARURI, K. *et al.*: 'A methodology and tool suite for c compiler generation from adl processor models'. Proceedings of *Design automation and test in Europe (DATE)*, 2004, pp. 1276–1283

45   WAHLEN, O., HOHENAUER, M., LEUPERS, R., and MEYR, H.: 'Instruction scheduler generation for retragetable compilation'. *IEEE Design & Test of Computers*, 2003, **20**(1), pp. 34–41

46   SISKA, C.: 'A processor description language supporting retargetable multi-pipeline DSP program development tools'. Proceedings of international symposium on *System synthesis (ISSS)*, 1998, pp. 31–36

47   MORIMOTO, T., YAMAZAKI, K., NAKAMURA, H., BOKU, T., and NAKAZAWA, K.: 'Superscalar processor design with hardware description language aidl'. Proceedings of Asia Pacific conference on *Hardware description languages (APCHDL)*, 1994

48   SATO, J., ALOMARY, A., HONMA, Y., NAKATA, T., SHIOMI, A., HIKICHI, N., and IMAI, M.: 'Peas-i: A hardware/software codesign systems for ASIP development', *IEICE Transactions of Fundamentals*, 1994, **E77-A**(3): 483–491

49   RAJESH, V., and MOONA, R.: 'Processor modeling for hardware software codesign'. Proceedings of international conference on *VLSI design*, 1999, pp. 132–137

50   ARC CORES: http://www.arccores.com

51   http://www.axysdesign.com. Axys Design Automation

52   TENSILICA INC: http://www.tensilica.com

53   MARWEDEL, P., and GOOSSENS, G.: 'Code generation for embedded processors' (Kluwer Academic Publishers, 1995)

54   AHO A., SETHI R., and ULLMAN J.: 'Compilers: Principles, techniques and tools' (Addition-Wesley, Reading, MA, 1986)

55   MUCHNICK, S.S.: 'Advanced compiler design and implementation' (Morgan Kaufmann, San Francisco, CA, 1997)

56   ZHU, J., and GAJSKI, D.: 'A retargetable, ultra-fast, instruction set simulator'. Proceedings of *Design automation and test in Europe (DATE)*, 1999

57   CMELIK, R., and KEPPEL, D.: 'Shade: A fast instruction-set simulator for execution profiling', *ACM SIGMETRICS Performance Evaluation Review*, 1994, **22**(1), pp. 128–137

58   WITCHEL, E., and ROSENBLUM, M.: 'Embra: Fast and flexible machine simulation'. Proceedings of the ACM sigmetrics international conference on *Measurement and modeling of computer systems*, 1996, pp. 68–79

59   NOHL, A., BRAUN, G., SCHLIEBUSCH, O., LEUPERS, R., MEYR, H., and HOFFMANN, A.: 'A universal technique for fast and flexible instruction-set architecture simulation'. Proceedings of *Design automation conference (DAC)*, 2002, pp. 22–27

60 RESHADI, M., MISHRA, P., and DUTT, N.: 'Instruction set compiled simulation: A technique for fast and flexible instruction set simulation'. Proceedings of *Design automation* conference *(DAC)*, 2003, pp. 758–763

61 RESHADI, M., BANSAL, N., MISHRA, P., and DUTT, N.: 'An efficient retargetable framework for instruction-set simulation'. Proceedings of international symposium on *Hardware/software codesign and system synthesis (CODES+ISSS)*, 2003, pp. 13–18

62 IMPROV INC.: http://www.improvsys.com.

63 ITOH, M., HIGAKI, S., TAKEUCHI, Y., KITAJIMA, A., IMAI, M., SATO, J., and SHIOMI, A.: 'Peas-iii: An ASIP design environment'. Proceedings of international conference on *Computer design (ICCD)*, 2000

64 ITOH, M., TAKEUCHI, Y., IMAI, M., and SHIOMI, A.: 'Synthesizable HDL generation for pipelined processors from a micro-operation description', *IEICE Transactions of Fundamentals*, 2000, **E83-A**(3), pp. 344–400

65 SCHLIEBUSCH, O., CHATTOPADHYAY, A., STEINERT, M. *et al.*: 'RTL processor synthesis for architecture exploration and implementation'. Proceedings of *Design automation and test in Europe (DATE)*, 2004, pp. 156–160

66 SCHLIEBUSCH, O., HOFFMANN, A., NOHL, A., BRAUN, G., and MEYR, H.: 'Architecture implementation using the machine description language LISA'. Proceedings of Asia South Pacific design automation conference (ASPDAC)/international conference on *VLSI design*, 2002, pp. 239–244

67 MISHRA, P., KEJARIWAL, A., and DUTT, N.: 'Synthesis-driven exploration of pipelined embedded processors'. Proceedings of international conference on *VLSI design*, 2004

68 MISHRA, P.: 'Specification-driven validation of programmable embedded systems'. PhD thesis, University of California, Irvine, March 2004

69 MISHRA, P., TOMIYAMA, H., HALAMBI, A., GRUN, P., DUTT, N., and NICOLAU, A.: 'Automatic modeling and validation of pipeline specifications driven by an architecture description language'. Proceedings of Asia South Pacific design automation conference (ASPDAC)/international conference on *VLSI design*, 2002, pp. 458–463

70 MISHRA, P., and DUTT, N.: 'Automatic modeling and validation of pipeline specifications', *ACM Transactions on embedded computing systems (TECS)*, 2004, 3(1), pp. 114–139

71 MISHRA, P., TOMIYAMA, H., DUTT, N., and NICOLAU, A.: 'Automatic verification of in-order execution in microprocessors with fragmented pipelines and multicycle functional units'. Proceedings of *Design automation and test in Europe (DATE)*, 2002, pp. 36–43

72 MISHRA, P., DUTT, N., and TOMIYAMA, H.: 'Towards automatic validation of dynamic behavior in pipelined processor specifications', *Kluwer Design Automation for Embedded Systems(DAES)*, 2003, **8**(2–3), pp. 249–265

73 MISHRA, P., and DUTT, N.: 'Graph-based functional test program generation for pipelined processors'. Proceedings of *Design automation and test in Europe (DATE)*, 2004, pp. 182–187

74   MISHRA, P., DUTT, N., KRISHNAMURTHY, N., and ABADIR, M.: 'A top-down methodology for validation of microprocessors'. *IEEE Design & Test of Computers*, 2004, **21**(2), pp. 122–131

75   HOFFMANN, A., SCHLIEBUSCH, O., NOHL, A., BRAUN, G., WAHLEN, O., and MEYR, H.: 'A methodology for the design of application specific instruction set processors (asip) using the machine description language LISA'. Proceedings of international conference on *Computer-aided design (ICCAD)*, 2001, pp. 625–630

76   NOHL, A., GREIVE, V., BRAUN, G., HOFFMANN, A., LEUPERS, R., SCHLIEBUSCH, O., and MEYR, H.: 'Instruction encoding synthesis for architecture exploration using hierarchical processor models'. Proceedings of *Design automation conference (DAC)*, 2003, pp. 262–267

77   SCHLIEBUSCH, O., KAMMLER, D., CHATTOPADHYAY, A., LEUPERS, R., ASCHEID, G., and MEYR, H.: 'Automatic generation of JTAG interface and debug mechanism for ASIPs'. In *GSPx*, 2004

78   BISWAS, P., and DUTT, N.: 'Reducing code size for heterogeneous-connectivity-based VLIW DSPs through synthesis of instruction set extensions'. Proceedings of *Compilers, architectures, synthesis for embedded systems (CASES)*, 2003, pp. 104–112

79   MISHRA, P., and DUTT, N.: 'Functional coverage driven test generation for validation of pipelined processors'. Proceedings of *Design automation and test in Europe (DATE)*, 2005

80   MISHRA, P., DUTT, N., and NICOLAU, A.: 'Functional abstraction driven design space exploration of heterogeneous programmable architectures'. Proceedings of international symposium on *System synthesis (ISSS)*, 2001, pp. 256–261

81   BRAUN, G., NOHL, A., SHENG, W. *et al*: 'A novel approach for flexible and consistent ADL-driven ASIP design'. Proceedings of *Design automation conference (DAC)*, 2004, pp. 717–722

Part II

Embedded software

Part II

Embedded software

Chapter 7

Concurrent models of computation for embedded software

Edward Lee and Stephen Neuendorffer

7.1 Introduction

Embedded software has traditionally been thought of as 'software on small computers'. In this traditional view, the principal problem is resource limitations (small memory, small data word sizes and relatively slow clocks). The solutions emphasise efficiency; software is written at a very low level (in assembly code or C), operating systems with a rich suite of services are avoided and specialised computer architectures such as programmable DSPs and network processors are developed to provide hardware support for common operations. These solutions have defined the practice of embedded software design and development for the last 25 years or so.

Of course, thanks to the semiconductor industry's ability to follow Moore's law, the resource limitations of 25 years ago should have almost entirely evaporated. Why then has embedded software design and development changed so little? It may be because extreme competitive pressure in products based on embedded software, such as consumer electronics, rewards only the most efficient solutions. This argument is questionable, however, since there are many examples where functionality has proven more important than efficiency. We will argue that resource limitations are not the only defining factor for embedded software, and may not even be the principal factor now that the technology has improved so much.

Resource limitations are an issue to some degree with almost all software. So generic improvements in software engineering should, in theory, also help with embedded software. There are several hints, however, that embedded software is different in more fundamental ways. For one, object-oriented techniques such as inheritance, dynamic binding and polymorphism are rarely used in practice with embedded software development. In another example, processors used for embedded systems often avoid the memory hierarchy techniques that are used in general purpose processors to deliver large virtual memory spaces and faster execution using caches. In a third example, automated memory management, with allocation, deallocation and garbage collection, are largely avoided in embedded software. To be fair, there are some successful applications of these technologies in embedded software, such as the use of Java in cell phones, but their application remains limited and is largely confined to providing the services in embedded systems that are actually more akin with general purpose software applications (such as database services in cell phones).

There are further hints that the software solutions for embedded software may ultimately differ significantly from those for general purpose software. We point to four recent cases where fundamentally different software design techniques have been applied to embedded software. All four define concurrency models, component architectures and management of time-critical operations in ways that are significantly different from prevailing software engineering techniques. The first two are nesC with TinyOS [1,2], which was developed for programming very small programmable sensor nodes called 'motes', and Click [3,4], which was created to support the design of software-based network routers. These first two have an imperative flavour, and components interact principally through procedure calls. The next two are Simulink with Real-Time Workshop (from The MathWorks), which was created for embedded control software and is widely used in the automotive industry, and SCADE (from Esterel Technologies, see Reference 5), which was created for safety-critical embedded software and is used in avionics. These next two have a more declarative flavour, where components interact principally through messages rather than procedure calls. There are quite a few more examples that we will discuss below. The amount of experimentation with alternative models of computation for embedded software is yet a further indication that the prevailing software abstractions are inadequate.

Embedded systems are integrations of software and hardware where the software reacts to sensor data and issues commands to actuators. The physical system is an integral part of the design and the software must be conceptualised to operate in concert with that physical system. Physical systems are intrinsically concurrent and temporal. Actions and reactions happen simultaneously and over time, and the metric properties of time are an essential part of the behaviour of the system.

Software abstracts away time, replacing it with ordering. In the prevailing software abstraction, that of imperative languages such as C, C++ and Java, the 'order' of actions is defined by the program, but not by their 'timing'. This prevailing imperative abstraction is overlaid with another abstraction, that of threads or processes,[1] typically provided by the operating system, but occasionally by the language (as in Java).

We will argue that the lack of timing in the core abstraction is a flaw, from the perspective of embedded software, and that threads as a concurrency model are a poor match to embedded systems. They are mainly focused on providing an illusion of

---

[1] Threads are processes that can share data. The distinction between the two is not important in our discussion, so we will use the term 'threads' generically to refer to both.

concurrency in fundamentally sequential models, and they work well only for modest levels of concurrency or for highly decoupled systems that are sharing resources, where best-effort scheduling policies are sufficient.

Of the four cases cited above, not one uses threads as the concurrency model. Of the four, only one (Simulink) is explicit about timing. This may be a reflection of how difficult it is to be explicit about timing when the most basic notion of computation has abstracted time away. To be fair, the others do provide mechanisms to manage time-critical events. TinyOS and Click both provide access to hardware timers, but this access is largely orthogonal to the semantics. It is treated as an I/O interaction.

There are, of course, software abstractions that admit concurrency without resorting to threads. In functional languages (e.g. see Reference 6), programs are compositions of declarative relationships, not specifications of an order of operations. But although declarative techniques have been used in embedded software (e.g. Simulink and SCADE), functional languages have found almost no usage in embedded software. Thus, whether a language is imperative or declarative probably has little bearing on whether it is useful for embedded software.

Embedded software systems are generally held to a much higher reliability standard than general purpose software. Often, failures in the software can be life threatening (e.g. in avionics and military systems). We argue that the prevailing concurrency model based on threads does not achieve adequate reliability. In this prevailing model, interaction between threads is extremely difficult for humans to understand. The basic techniques for controlling this interaction use semaphores and mutual exclusion locks, methods that date back to the 1960s [7]. These techniques often lead to deadlock or livelock conditions, where all or part of a program cannot continue executing. In general purpose computing, this is inconvenient, and typically forces a restart of the program (or even a reboot of the machine). However, in embedded software, such errors can be far more than inconvenient. Moreover, software is often written without sufficient use of these interlock mechanisms, resulting in race conditions that yield non-deterministic program behaviour.

In practice, errors due to misuse (or no use) of semaphores and mutual exclusion locks are extremely difficult to detect by testing. Code can be exercised in deployed form for years before a design flaw appears. Static analysis techniques can help (e.g. Sun Microsystems' LockLint), but these methods are often thwarted by conservative approximations and/or false positives.

It can be argued that the unreliability of multi-threaded programs is due at least in part to inadequate software engineering processes. For example, better code reviews, better specifications, better compliance testing and better planning of the development process can help solve the problems. It is certainly true that these techniques can help. However, programs that use threads can be extremely difficult for programmers to understand. If a program is incomprehensible, then no amount of process improvement will make it reliable. For example, development schedule extensions are as likely to degrade the reliability of programs that are difficult to understand as they are to improve it.

Formal methods can help detect flaws in threaded programs, and in the process can improve the understanding that a designer has of the behaviour of a complex program.

But if the basic mechanisms fundamentally lead to programs that are difficult to understand, then these improvements will fall short of delivering reliable software.

All four of the cases cited above offer concurrency models that are much easier to understand than threads that interact via semaphores and mutual exclusion locks.

Simulink and SCADE are based on a synchronous abstraction, where components conceptually execute simultaneously, aligned with one or more interlocked clocks. SCADE relies on an abstraction where components appear to execute instantaneously, whereas Simulink is more explicit about the passage of time and supports definition of tasks that take time to execute and execute concurrently with other tasks. In both cases, every (correctly) compiled version of the program will execute identically, in that if it is given the same inputs, it will produce the same outputs. In particular, the execution does not depend on extraneous factors such as processor speed. Even this modest objective is often hard to achieve using threads directly.

TinyOS and Click offer concurrency models that are closer to the prevailing software abstractions, since they rely on procedure calls as the principle component interaction mechanism. However, neither model includes threads. The key consequence is that a programmer can rely on the atomicity of the execution of most program segments, and hence does not usually need to explicitly deal with mutual exclusion locks or semaphores. The result again is more comprehensible concurrent programs.

7.2   Concurrency and time

In embedded software, concurrency and time are essential aspects of a design. In this section, we outline the potential problems that software faces in dealing with these aspects.

Time is a relatively simple issue, conceptually, although delivering temporal semantics in software can be challenging. Time is about the ordering of events. Event x happens *before* event y, for example. But in embedded software, time also has a metric. That is, there is an amount of time between events x and y, and the amount of time may be an important part of the correctness of a system.

In software, it is straightforward to talk about the order of events, although in concurrent systems it can be difficult to control the order. For example, achieving a specified total ordering of events across concurrent threads implies interactions across those threads that can be extremely difficult to implement correctly. Research in distributed discrete-event simulation, for example, underscores the subtleties that can arise (e.g. see References 8 and 9).

It is less straightforward to talk about the metric nature of time. Typically, embedded processors have access to external devices called timers that can be used to measure the passage of time. Programs can poll for the current time, and they can set timers to trigger an interrupt at some time in the future. Using timers in this way implies immediately having to deal with concurrency issues. Interrupt service routines typically pre-empt currently executing software, and hence conceptually execute concurrently.

Concurrency in software is a challenging issue because the basic software abstraction is not concurrent. The basic abstraction in imperative languages is that the memory of the computer represents the current state of the system, and instructions transform that state. A program is a sequence of such transformations. The problem with concurrency is that from the perspective of a particular program, the state may change on its own at any time. For example, we could have a sequence of statements:

```
x = 5;
print x;
```

that results in printing the number '6' instead of '5'. This could occur, e.g., if after execution of the first statement an interrupt occurred, and the interrupt service routine modified the memory location where *x* was stored. Or it could occur if the computer is also executing a sequence of statements:

```
x = 6;
print x;
```

and a multitasking scheduler happens to interleave the executions of the instructions of the two sequences. Two such sequences of statements are said to be 'non-determinate' because, by themselves, these two sequences of statements do not specify a single behaviour. There is more than one behaviour that is consistent with the specification.

Non-determinism can be desirable in embedded software. Consider for example an embedded system that receives information at random times from two distinct sensors. Suppose that it is the job of the embedded software to fuse the data from these sensors so that their observations are both taken into account. The system as a whole will be non-determinate since its results will depend on the order in which information from the sensors is processed. Consider the following program fragment:

```
y = getSensorData();      // Block for data
x = 0.9 * x + 0.1 * y;    // Discounted average
print x;                  // Display the result
```

This fragment reads data from a sensor and calculates a running average using a discounting strategy, where older data has less effect on the average than newer data.

Suppose that our embedded system uses two threads, one for each sensor, where each thread executes the above sequence of statements repeatedly. The result of the execution will depend on the order in which data arrives from the sensors, so the program is non-determinate. However, it is also non-determinate in another way that was probably not intended. Suppose that the multitasking scheduler happens to execute the instructions from the two threads in interleaved order, as shown here:

```
y = getSensorData();      // From thread 1
y = getSensorData();      // From thread 2
x = 0.9 * x + 0.1 * y;    // From thread 1
x = 0.9 * x + 0.1 * y;    // From thread 2
print x;                  // From thread 1
print x;                  // From thread 2
```

The result is clearly not right. The sensor data read by thread 1 is ignored. The discounting is applied twice. The sensor data from thread 2 is counted twice. And the same (erroneous) result is printed twice.

A key capability for preventing such concurrency problems is 'atomicity'. A sequence of instructions is 'atomic' if during the execution of the sequence, no portion of the state that is visible to these instructions changes unless it is changed by the instructions themselves.

Atomicity is provided by programming languages and/or operating systems through 'mutual exclusion' mechanisms. These mechanisms depend on low-level support for an indivisible 'test and set'. Consider the following modification:

```
acquireLock();            // Block until acquired
y = getSensorData();      // Block for data
x = 0.9 * x + 0.1 * y;    // Discount old value
print x;                  // Display the result
releaseLock();            // Release the lock
```

The first statement calls an operating system primitive[2] that tests a shared, Boolean-valued variable, and if it is false, sets it to true and returns. If it is true, then it blocks, waiting until it becomes false. It is essential that between the time this primitive tests the variable and the time it sets it to true, that no other instruction in the system can access that variable. That is, the test and set occur as one operation, not as two. The last statement sets the variable to false.

Suppose we now build a system with two threads that each execute this sequence repeatedly to read from two sensors. The resulting system will not exhibit the problem above because the multitasking scheduler cannot interleave the executions of the statements. However, the program is still not correct. For example, it might occur that only one of the two threads ever acquires the lock, and so only one sensor is read. In this case, the program is not 'fair'. Suppose that the multitasking scheduler is forced to be fair, say by requiring it to yield to the other thread each time `releaseLock()` is called. The program is still not correct, because while one thread is waiting for sensor data, the other thread is blocked by the lock and will fail to notice new data on its sensor.

This seemingly trivial problem has become difficult. Rather than trying to fix it within the threading model of computation (we leave this an exercise), we will show that alternative models of computation make this problem easy.

Suppose that the program is given by the diagram in Figure 7.1.[3] Suppose that the semantics are those of Kahn process networks (PN) [10,11] augmented with a non-deterministic merge [12,13]. In that figure, the components (blocks) are called 'actors'. They have ports (shown by small triangles), with input ports pointing into

---

[2] Mutual exclusion locks may also be provided as part of a programming language. The 'synchronised' keyword in Java, e.g. performs the same function as our 'acquireLock' command.

[3] We give this program using a visual syntax to emphasise its concurrent semantics, and because visual syntaxes are commonly used for languages with similar semantics, such as SCADE and Simulink. But the visual syntax makes this no less a program.

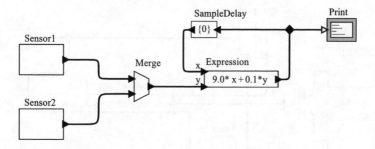

Figure 7.1   Process network realisation of the sensor fusion example

the blocks and output ports pointing out. Each actor encapsulates functionality that reads input values and produces output values.

In PN semantics, each actor executes continually in its own thread of control. The Sensor1 and Sensor2 actors will produce an output whenever the corresponding sensors have data (this could be done directly by the interrupt service routine, e.g.). The connections between actors represent sequences of data values. The Merge actor will non-deterministically interleave the two sequences at its input ports, preserving the order within each sequence, but yielding arbitrary ordering of data values across sequences. Suppose it is 'fair' in the sense that if a data value appears at one of the inputs, then it will 'eventually' appear at the output [14]. The remaining actors simply calculate the discounted average and display it. The SampleDelay actor provides an initial 'previous average' to work with (which prevents this program from deadlocking for lack of data at the input to the Expression actor). This program exhibits none of the difficulties encountered above with threads, and is both easy to write and easy to understand.

We can now focus on improving its functionality. Notice that the discounting average is not ideal because it does not take into account 'how old' the old data is. That is, there is no time metric. Old data is simply the data previously observed, and there is no measure of how long ago it was read. Suppose that instead of Kahn process networks semantics, we use 'discrete-event' (DE) semantics [8,15]. A modified diagram is shown in Figure 7.2. In that diagram, the meaning of a connection between actors is slightly different from the meaning of connections in Figure 7.1. In particular, the connection carries a sequence of data values as before, but each value has a 'time stamp'. The time stamps on any given sequence are non-decreasing. A data value with a time stamp is called an 'event'.

The Sensor1 and Sensor2 actors produce output events stamped with the time at which their respective interrupt service routines are executed. The merge actor is no longer non-deterministic. Its output is a chronological merge of the two input sequences.[4] The TimeGap actor produces on its output an event with the same time

---

[4] A minor detail is that we have to decide how to handle simultaneous input events. We could e.g. produce them both at the output with the one from the top input port preceding the one at the bottom input port. The semantics of simultaneous events is considered in detail in Reference 15.

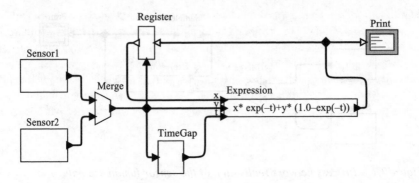

Figure 7.2   Discrete event realisation of an improved sensor fusion example

stamp as the input but whose value is the elapsed time between the current event and the previous event (or between the start of execution and the current event if this is the first event). The expression shown in the next actor calculates a better discounted average, one that takes into account the time elapsed. It implements an exponential forgetting function.

The Register actor in Figure 7.2 has somewhat interesting semantics. Its output is produced when it receives a trigger input on the bottom port. The value of the output is that of a 'previously observed' input (or a specified initial value if no input was previously observed). In particular, at any given time stamp, the value of the output does not depend on the value of the input, so this actor breaks what would otherwise be an unresolvable causality loop.

Even with such a simple problem, threaded concurrency is clearly inferior. PN offers a better concurrency model in that the program is easier to construct and to understand. The DE model is even better because it takes into account metric properties of time, which matter in this problem.

In real systems, the contrast between these approaches is even more dramatic. Consider the following two program fragments:

```
acquireLockA();
acquireLockB();
x = 5;
print x;
releaseLockB();
releaseLockA();
```

and

```
acquireLockB();
acquireLockA();
x = 5;
print x;
releaseLockA();
releaseLockB();
```

If these two programs are executed concurrently in two threads, they could deadlock. Suppose the multitasking scheduler executes the first statement from the first program followed by the first statement from the second program. At this point, the second statement of both programs will block! There is no way out of this. The programs have to be aborted and restarted.

Programmers who use threads have tantalising simple rules to avoid this problem. For example, 'always acquire locks in the same order' [16]. However, this rule is almost impossible to apply in practice because of the way programs are modularised. Any given program fragment is likely to call methods or procedures that are defined elsewhere, and those methods or procedures may acquire locks. Unless we examine the source code of every procedure we call, we cannot be sure that we have applied this rule.[5]

Deadlock can, of course, occur in PN and DE programs. If in Figure 7.1 we had omitted the SampleDelay actor, or in Figure 7.2 we had omitted the Register actor, the programs would not be able to execute. In both cases, the Expression actor requires new data at all of its input ports in order to execute, and that data would not be able to be provided without executing the Expression actor.

The rules for preventing deadlocks in PN and DE programs are much easier to apply than the rule for threads. For certain models of computation, whether deadlock occurs can be checked through static analysis of the program. This is true of the DE model used above for the improved sensor fusion problem, for example. So, not only was the model of computation more expressive in practice (i.e. it more readily expressed the behaviour we wanted), but it also had stronger formal properties that enabled static checks that prove the absence of certain flaws (deadlock, in this case).

We will next examine a few of the models of computation that have been used for embedded systems.

7.3 Imperative concurrent models

As mentioned above, TinyOS and Click have an imperative flavour. What this means is that when one component interacts with another, it gives a command, 'do this'. The command is implemented as a procedure call. Since these models of computation are also concurrent, we call them 'imperative concurrent' models of computation.

In contrast, when components in Simulink and SCADE interact, they simply offer data values, 'here is some data'. It is irrelevant to the component when (or even whether) the destination component reacts to the message. These models of computation have a declarative flavour, since instead of issuing commands, they declare relationships between components that share data. We call such models of computation 'declarative concurrent' models of computation.

We begin with the imperative concurrent models of computation.

---

[5] In principle, it might be possible to devise a programming language where the locks that are acquired by a procedure are part of the type signature of the procedure, much as in Java where the exceptions that are thrown by a procedure are part of its type signature. However, we know of no language that does this.

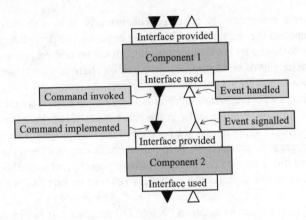

Figure 7.3   A representation of a nesC/TinyOS configuration

7.3.1   nesC/TinyOS

TinyOS is a specialised, small-footprint operating system for use on extremely resource-constrained computers, such as 8-bit microcontrollers with small amounts of memory [1]. It is typically used with nesC, a programming language that describes 'configurations', which are assemblies of TinyOS components [2].

A visual rendition of a two-component configuration is shown in Figure 7.3, where the visual notation is that used in Reference 2. The components are grey boxes with names. Each component has some number of interfaces, some of which it 'uses' and some of which it 'provides'. The interfaces it provides are put on top of the box and the interfaces it uses are put on the bottom. Each interface consists of a number of methods, shown as triangles. The filled triangles represent methods that are called 'commands' and the unfilled triangles represent 'event handlers'. Commands propagate downwards, whereas events propagate upwards.

After initialisation, computation typically begins with events. In Figure 7.3, Component 2 might be a thin wrapper for hardware, and the interrupt service routine associated with that hardware would call a procedure in Component 1 that would 'signal an event'. What it means to signal an event is that a procedure call is made upwards in the diagram via the connections between the unfilled triangles. Component 1 provides an event handler procedure. The event handler can signal an event to another component, passing the event up in the diagram. It can also call a command, downwards in the diagram. A component that provides an interface provides a procedure to implement a command.

Execution of an event handler triggered by an interrupt (and execution of any commands or other event handlers that it calls) may be pre-empted by another interrupt. This is the principle source of concurrency in the model. It is potentially problematic because event handler procedures may be in the middle of being executed when an interrupt occurs that causes them to begin execution again to handle a new event. Problems are averted through judicious use of the 'atomic' keyword in nesC. Code that is enclosed in an atomic block cannot be interrupted (this is implemented very efficiently by disabling interrupts in the hardware).

Clearly, however, in a real-time system, interrupts should not be disabled for extensive periods of time. In fact, nesC prohibits calling commands or signalling events from within an atomic block. Moreover, no mechanism is provided for an atomic test-and-set, so there is no mechanism besides the atomic keyword for implementing mutual exclusion. The system is a bit like a multithreaded system but with only one mutual exclusion lock. This makes it impossible for the mutual exclusion mechanism to cause deadlock.

Of course, this limited expressiveness means that event handlers cannot perform non-trivial concurrent computation. To regain expressiveness, TinyOS has tasks. An event handler may 'post a task'. Posted tasks are executed when the machine is idle (no interrupt service routines are being executed). A task may call commands through the interfaces it uses. It is not expected to signal events, however. Once task execution starts, it completes before any other task execution is started. That is, task execution is atomic with respect to other tasks. This greatly simplifies the concurrency model, because now variables or resources that are shared across tasks do not require mutual exclusion protocols to protect their accesses. Tasks may be pre-empted by event handlers, however, so some care must be exercised when shared data is accessed here to avoid race conditions. Interestingly, it is relatively easy to statically analyse a program for potential race conditions [2].

Consider the sensor fusion example from above. A configuration for this is sketched in Figure 7.4. The two sensors have interfaces called 'reading' that accept a command, a signal, an event. The command is used to configure the sensors. The event is signalled when an interrupt from the sensor hardware is handled. Each time such an event is signalled, the Fuser component records the sensor reading and posts a task to update the discounted average. The task will then invoke the command in the print interface of the Printer component to display the result. Because tasks execute atomically with respect to one another, in the order in which they are posted, the only tricky part of this implementation is in recording the sensor data. However, tasks in TinyOS can be passed arguments on the stack, so the sensor data can be recorded there. The management of concurrency becomes extremely simple in this example.

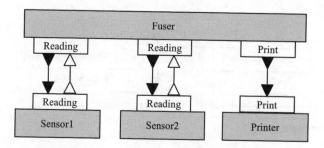

Figure 7.4   A sketch of the sensor fusion problem as a nesC/TinyOS configuration

In effect, in nesC/TinyOS, concurrency is much more disciplined than with threads. There is no arbitrary interleaving of code execution, there are no blocking operations to cause deadlock and there is a very simple mechanism for managing the one non-deterministic pre-emption that can be caused by interrupts. The price paid for this, however, is that applications must be divided into small, quickly executing procedures to maintain reactivity. Since tasks run to completion, a long-running task will starve all other tasks.

7.3.2   Click

Click was originally developed for designing software implementations of network routers on general purpose computers running Linux [3,4]. It has been recently adapted for designing software for specialised network processors [17], and has proven to offer effective abstractions for this style of embedded software, at least. The abstractions have a great deal of potential for any embedded software that deals with multiple converging asynchronous streams of stimuli.

As with nesC/TinyOS, in the Click model, connections between components represent method bindings. Click does not have the bidirectional interfaces of TinyOS, but it has its own twist that can be used to accomplish similar objectives. In Click, connections between ports can be 'push' or 'pull'. In a push connection, the method call originates with the source of the data. That is, the producer component calls the consumer component. In a pull connection, the method call originates with the consumer. That is, the consumer component calls the producer component to demand data. It is worth noting that there are middleware frameworks with similar push/pull semantics, such as the CORBA event service [18,19]. These, however, are aimed at distributed computation rather than at managing concurrency within a single CPU. Click executes in a single thread, and we will see that this simplifies the design of Click applications compared with what would be required by distributed models.

Figure 7.5 shows a Click model using the visual notation from Reference 4. Boxes again represent components, and ports are shown either as rectangles (for output ports) or triangles (for input ports). If a port is filled in black, then it is required to link to a push connection. If it is filled in white, then it is required to link to a pull connection.

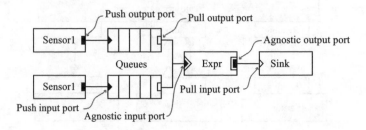

Figure 7.5   A representation of a Click program

If it has a double outline, then it is agnostic, and can be linked to either type of connection.

A component with just a push output port, like Sensor1 and Sensor2 in Figure 7.5, can function as a thin wrapper around hardware that will produce data. Conceptually, the component autonomously[6] initiates a reaction by pushing data on its output port, which means calling a method in a downstream component. That method in the downstream component may itself trigger further reactions by either pushing data to output ports or pulling data from input ports.

In the example shown in Figure 7.5, the components downstream of Sensor1 and Sensor2 are Queues. They have push inputs and pull outputs. When a method is called to push data into them, that method simply stores the data on a queue. When a method is called to pull data from their outputs, either a datum is provided or a null value is provided to indicate that no data are available.

Click runs in a single thread, so the push and pull methods of the queue component will be atomic with respect to one another. Thus, no special care needs to be taken to manage the fact that callers from the left and from the right will both access the same (queue) data structure.

Click maintains a task queue and executes tasks from this queue whenever the main loop becomes idle. Polling the sensors for data, for example, is accomplished by tasks that are always present on the task queue. In the example shown in Figure 7.5, the Sink component has a single pull input. This component would, e.g., have a task on the Click task queue that is repeatedly executed and pulls data from the input port. The upstream component, labelled Expr, has agnostic input and output ports. Because of the way it is connected, these ports will be used as pull ports. A pull from the Sink will cause the Expr component to pull data from the queues. Note that the Expr component can implement a scheduling strategy (such as round robin) to access the queues fairly. Generally, scheduling can be accomplished by components that have pull inputs and/or push outputs and that post tasks on the event queue.

It is easy to see how the example in Figure 7.5 could be adapted to implement the sensor fusion problem. Once again, the representation is simple and clear, with no particular difficulties due to concurrency. The primary mechanism for avoiding deadlock is the style that a pull should return null if no data are available. The danger of livelock is largely eliminated by avoiding feedback loops, although several interesting models include feedback loops that do not livelock because of the logic contained in components (see Reference 4, section 2.6). Data races do not occur accidentally because methods execute atomically. Nonetheless, on a coarser level, non-deterministic interactions like those in the sensor fusion example are easy to define. Indeed, these kinds of interactions are common in the application domain that Click targets, network routers.

---

[6] Currently, Click accomplishes this by repeatedly executing a task that polls the hardware, instead of an interrupt service routine, but it does not seem hard to adapt the model to leverage interrupt service routines, if desired.

7.3.3   Others

There are many other models with imperative concurrent semantics. Here, we briefly mention some that have been applied to the design of embedded systems.

7.3.3.1   Bluespec

In a Bluespec [20,21] model, components not only contain methods, but also 'activation rules' and 'execution constraints'. Each activation rule describes an atomic state update in the system, which can be performed whenever the associated execution constraints are satisfied. Bindings between methods enable complex state updates to be specified compositionally as a group of methods.

Conceptually, state updates in a Bluespec system occur sequentially. However, in some cases activation rules operate on independent portions of the system state, in which case they are called 'conflict free'. These 'conflict-free' rules represent parallelism in a system and can be executed concurrently. Bluespec discovers conflict-free rules through static program analysis and generates run-time scheduling logic.

Livelock in Bluespec models is prevented by a requirement that no method can cause itself to be executed through a sequence of method invocations. This requirement is guaranteed through static analysis of component compositions. Deadlock in Bluespec models cannot generally be avoided, since it is possible that a state in execution is reached where there are no activation rules whose execution constraints can be satisfied.

Bluespec has seen significant application in the specification of digital logic circuits [22,23]. Current compilers map a composition to a synchronous circuit that executes an activation rule in a single cycle. Methods are converted into combinational logic, which is guaranteed to be acyclic given the constraints on re-entrant methods. In each cycle every rule executes concurrently, but the results are gated so that only the state updates corresponding to a set of conflict-free rules are committed to the system state.

Bluespec models can also be synthesised directly into sequential software, which can be used to efficiently simulate synthesised digital logic systems. In software, it is more efficient to make scheduling decisions for activation rules initially and to only execute code corresponding to a single activation rule at a time. In comparison with direct simulation of synthesised digital logic, this technique offers significant speedup for many applications, since only committed activation rules are actually executed. Additionally, given coarse-grained activation rules, it seems possible to execute more than one rule in software concurrently.

7.3.3.2   Koala

Koala [24,25] is a model and language for components with procedure-call interfaces and a visual syntax with 'provides' and 'requires' ports that get connected. It has been proposed for use in the design of consumer electronics software specifically. As with nesC/TinyOS and Click, in a Koala model, connections between components represent method bindings. Communication occurs through method arguments and return values and the interaction between communicating components is primarily sequential. Koala allows components to contain arbitrary code and perhaps to encapsulate arbitrary operating system threads.

A 'configuration' is an interconnection of components plus configuration-specific code in something called a 'module'. To get hierarchy, the configuration can export its own requires and provides interfaces, and these can be mediated by the module. For example, the module can translate a particular provided method into a sequence of calls to provided methods of the components (e.g. to initialise all the components). The module is configuration specific, and is not itself a component, so it does not pollute the component library. The module can also provide services that are required by the components. For example, a component may require values for configuration parameters, and the module can provide those values. Partial evaluation is used to avoid introducing overhead in doing things this way.

Modules offer a much richer form of hierarchical abstraction than either nesC or Click provide. Modules are also used to implement primitive components, thus providing the leaf cells of the hierarchy.

Each 'requires' interface must be connected to either a module or a 'provides' interface (input port). A 'provides' interface, however, can be connected to zero or more 'requires' interfaces. An example is given in Reference 24 where components require a particular hardware interface (an I2C bus) that must be provided by a configuration. Operating system and scheduling services also interact with components through requires and provides interfaces. Thus, the language provides clean mechanisms for relating hardware requirements to software services.

A limited form of dynamic binding is provided in the form of 'switches', which work together with a module to direct procedure calls. These can be used at run time to direct a method call to one or another component. Switches can also be used with 'diversity interfaces' (see below), in which case, partial evaluation will likely lead to static binding and the elimination of some components from a configuration (components that are not used).

'Diversity' means one definition, multiple products. Koala's features support this well, particularly through its partial evaluation and static binding, which avoid the overhead often incurred by making components flexible. The authors compare the use of 'requires' interfaces to property lists in more conventional component architectures with set() and get() methods, and point out that set() and get() make it more difficult to optimise when properties are set at design time. Instead of 'providing' interfaces that must be filled in by the configuration (e.g. set()), Koala components have 'required' interfaces that the configuration must provide. These are called 'diversity interfaces'.

Koala components can provide 'optional interfaces' (fashioned after COM's query interface mechanism), which are automatically extended with an isPresent function, which the component is required to implement. For example the presence of an interface may depend on the hardware configuration. A component may also require an 'optional interface' (which is, to be sure, odd terminology), in which case the component can query for whether a configuration has a matching 'provides' interface.

The hierarchical structure, components with provides and requires interfaces and bindings concepts come from the architecture description language Darwin [26], but the modules and diversity schemes are new.

7.4   Declarative concurrent models

As mentioned above, Simulink and SCADE have a declarative flavour. The interactions between components are not 'imperative' in that one component does not 'tell the other what to do'. Instead, a 'program' is a declaration of the relationships among components. In this section, we examine a few of the models of computation with this character.

7.4.1   Simulink

Simulink was originally developed as a modelling environment, primarily for control systems. It is rooted in a continuous-time semantics, something that is intrinsically challenging for any software system to emulate. Software is intrinsically discrete, so an execution of a Simulink 'program' often amounts to approximating the specified behaviour using numerical integration techniques.

A Simulink 'program' is an interconnection of blocks where the connections are the 'variables', but the value of a variable is a function, not a single value. To complicate things, it is a function defined over a continuum. The Integrator block e.g., takes as input any function of the reals and produces as output the integral of that function. In general, any numerical representation in software of such a function and its integral is an approximation, where the value is represented at discrete points in the continuum. The Simulink execution engine (which is called a 'solver') chooses those discrete points using sometimes quite sophisticated methods.

Although initially Simulink focused on simulating continuous dynamics and providing excellent numerical integration, more recently it acquired a discrete capability. Semantically, discrete signals are piecewise-constant continuous-time signals. A piecewise constant signal changes value only at discrete points on the time line. Such signals are intrinsically easier for software, and more precise approximations are possible.

In addition to discrete signals, Simulink has discrete blocks. These have a 'sampleTime' parameter, which specifies the period of a periodic execution. Any output of a discrete block is a piecewise constant signal. Inputs are sampled at multiples of the sampleTime.

Certain arrangements of discrete blocks turn out to be particularly easy to execute. An interconnection of discrete blocks that all have the same sampleTime value, for example, can be efficiently compiled into embedded software. But even blocks with different sampleTime parameters can yield efficient models, when the sampleTime values are related by simple integer multiples.

Fortunately, in the design of control systems (and many other signal processing systems), there is a common design pattern where discrete blocks with harmonically

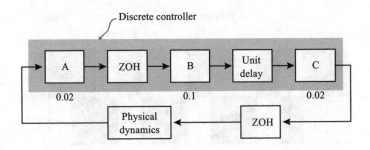

Figure 7.6   A representation of a Simulink program

related sampleTime values are commonly used to specify the software of embedded control systems.

Figure 7.6 shows schematically a typical Simulink model of a control system. There is a portion of the model that is a model of the physical dynamics of the system to be controlled. There is no need, usually, to compile that specification into embedded software. There is another portion of the model that represents a discrete controller. In this example, we have shown a controller that involves multiple values of the sampleTime parameter, shown as numbers below the discrete blocks. This controller is a specification for a program that we wish to execute in an embedded system.

Real-Time Workshop is a product from The MathWorks associated with Simulink. It takes models as shown in Figure 7.6 and generates code. Although it will generate code for any model, it is intended principally to be used only on the discrete controller, and indeed, this is where its strengths come through.

The discrete controller shown in Figure 7.6 has fast running components (with sampleTime values of 0.02 or 20 ms) and slow running components (with sample-Time values of 0.1 or 1/10 of a second). In such situations, it is not unusual for the slow running components to involve much heavier computational loads than the fast running components. It would not do to schedule these computations to execute atomically, as is done in TinyOS and Click (and SCADE, as discussed below). This would permit the slow running component to interfere with the responsivity (and time correctness) of the fast running components.

Simulink with Real-Time Workshop uses a clever technique to circumvent this problem. The technique exploits an underlying multitasking operating system with pre-emptive priority-driven multitasking. The slow running blocks are executed in a separate thread from the fast running blocks, as shown in Figure 7.7. The thread for the fast running blocks are given higher priority than that for the slow running blocks, ensuring that the slow running code cannot block the fast running code. So far, this just follows the principles of rate-monotonic scheduling [27].

But the situation is a bit more subtle than this, because data flows across the rate boundaries. Recall that Simulink signals have continuous-time semantics, and that discrete signals are piecewise constant. The slow running blocks should 'see' at their input a piecewise constant signal that changes values at the slow rate. To guarantee

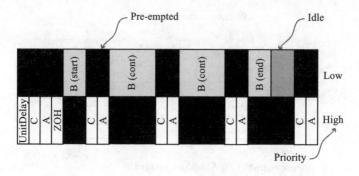

Figure 7.7   A simplified representation of a Simulink schedule

that, the model builder is required to put a zero-order hold (ZOH) block at the point of the rate conversion. Failure to do so will trigger an error message. Cleverly, the code for the ZOH runs at the rate of the slow block but at the priority of the fast block. This makes it completely unnecessary to do semaphore synchronisation when exchanging data across these threads.

When rate conversions go the other way, from slow blocks to fast blocks, the designer is required to put a UnitDelay block, as shown in Figure 7.6. This is because the execution of the slow block will typically stretch over several executions of the fast block, as shown in Figure 7.7.[7] To ensure determinacy, the updated output of the block must be delayed by the worst case, which will occur if the execution stretches over all executions of the fast block in one period of the slow block. The unit delay gives the software the slack it needs in order to be able to permit the execution of the slow block to stretch over several executions of the fast one. The UnitDelay executes at the rate of the slow block but at the priority of the fast block.

This same principle has been exploited in Giotto [28], which constrains the program to always obey this multirate semantics and provides (implicitly) a unit delay on every connection. In exchange for these constraints, Giotto achieves strong formal structure, which results in, among other things, an ability to perform schedulability analysis (the determination of whether the specified real-time behaviour can be achieved by the software).

The Simulink model does have some weaknesses, however. The sensor fusion problem that we posed earlier does not match its discrete multitasking model very well. While it would be straightfoward to construct a discrete multitasking model that polls the sensors at regular (harmonic) rates, reacting to stimulus from the sensors at random times does not fit the semantics very well. The merge shown in Figure 7.2 would be challenging to accomplish in Simulink, and it would not benefit much from the clever code generation techniques of Real-Time Workshop.

---

[7] This schedule is simplified, showing only the invocations of the methods associated with the blocks that produce outputs.

7.4.2 Discrete-event

In Figure 7.2, we gave a discrete-event model of an improved sensor fusion algorithm with an exponential forgetting function. Discrete-event modelling is widely used in electronic circuit design (VHDL and Verilog are discrete-event languages), in computer network modelling and simulation (e.g. OPNET Modeler[8] and Ns-2[9]), and in many other disciplines.

In discrete-event models, the components interact via signals that consist of 'events', which typically carry both a data payload and a time stamp. A straightforward execution of these models uses a centralised event queue, where events are sorted by time stamp, and a runtime scheduler dispatches events to be processed in chronological order. Compared with the Simulink/RTW model, there is much more flexibility in DE because discrete execution does not need to be periodic. This feature is exploited in the model of Figure 7.2, where the Merge block has no simple counterpart in Simulink.

A great deal of work has been done on efficient and distributed execution of such models, much of this work originating in either the so-called 'conservative' technique of Chandy and Misra [29] or the speculative execution methods of Jefferson [9]. Much less work has been done in adapting these models as an execution platform for embedded software, but there is some early work that bears a strong semantic resemblance to DE modelling techniques [30,31]. A significant challenge is to achieve the timed semantics efficiently while building on software abstractions that have abstracted away time.

7.4.3 Synchronous languages

SCADE [5] (Safety Critical Application Development Environment), a commercial product of Esterel Technologies, builds on the synchronous language Lustre [32], providing a graphical programming framework with Lustre semantics. Of the flagship synchronous languages, Esterel [33], Signal [34] and Lustre, Lustre is the simplest in many respects. All the synchronous languages have strong formal properties that yield quite effectively to formal verification techniques, but the simplicity of Lustre in large part accounts for SCADE achieving certification for use in safety critical embedded avionics software.[10]

The principle behind synchronous languages is simple, although the consequences are profound [35]. Execution follows 'ticks' of a global 'clock'. At each tick, each variable (represented visually by the wires that connect the blocks) may have a value (it can also be absent, having no value). Its value (or absence of value) is defined by functions associated with each block. That is, each block is a function from input values to output values. In Figure 7.8, the variables x and y at a particular tick are

---

[8] http://opnet.com/products/modeler/home.html
[9] http://www.isi.edu/nsnam/ns
[10] The SCADE tool has a code generator that produces C or ADA code that is compliant with the DO-178B Level A standard, which allows it to be used in critical avionics applications (see http://www.rtca.org).

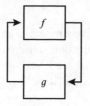

Figure 7.8   A simple feedback system illustrating the fixed point principles of synchronous languages

related by

$$x = f(y) \quad \text{and} \quad y = g(x)$$

The task of the compiler is to synthesise a program that, at each tick, solves these equations. Perhaps somewhat surprisingly, this turns out to be not difficult, well-founded and reasonably efficient.

An interesting issue with Lustre is that it supports multiple rates. That is, the master clock can be 'divided down' so that certain operations are performed on only some ticks of the clock. There is a well-developed formal 'clock calculus' that is used by the compiler to analyse systems with such multirate behaviour. Inconsistencies are detected by the compiler.

In SCADE, the functions associated with blocks can be defined using state machines. They can have behaviour that changes with each tick of the clock. This offers an expressive and semantically rich way to define systems, but most interestingly, it also offers opportunities for formal verification of dynamic behaviour. As long as the state machines have a finite number of states, then in principle, automated tools can explore the reachable state space to determine whether safety conditions can be guaranteed.

The non-deterministic merge of Figure 7.1 is not directly supported by Lustre. The synchronous language Signal [34] extends the principles of Lustre with a 'default' operator that supports such non-deterministic merge operations. The timed behaviour of Figure 7.2 is also not directly supported by Lustre, which does not associate any metric with the time between ticks. Without such a metric, the merging of sensor inputs in Figure 7.2 cannot be done deterministically. However, if these events are externally merged (e.g. in the interrupt service routines, which need to implement the appropriate mutual exclusion logic), then Lustre is capable of expressing the rest of the processing. The fact that there is no metric associated with the time between ticks means that Lustre programs can be designed to simply react to events, whenever they occur. This contrasts with Simulink, which has temporal semantics. Unlike Simulink, however, Lustre has no mechanisms for multitasking, and hence long running tasks will interfere with reactivity. A great deal of research has been done in recent years in 'desynchronising' synchronous languages, so we can expect in the future progress in this direction.

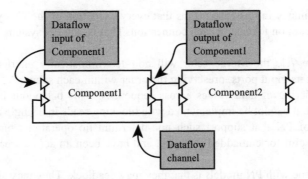

Figure 7.9   A diagram representing dataflow-oriented components

7.4.4   Dataflow

As with the other models of computation considered here, components in a dataflow model of computation also encapsulate internal state. However, instead of interacting through method calls, continuous-time signals, or synchronously defined variables, components interact through the asynchronous passing of data messages. Each message is called a 'token'. In this section, we will deal only with models where messages are guaranteed to be delivered in order and not lost. For these models it is common to interpret the sequence of tokens communicated from one port to another as a (possibly infinite) 'stream'. It is not uncommon to use visual representations for dataflow systems, as in Figure 7.9. In that figure, the wires represent streams, the blocks represent dataflow 'actors', and the triangles represent 'ports'. Input ports point into the block, and output ports point out. Feedback is supported by most variants of dataflow semantics, although when there is feedback, there is risk of deadlock. There are many variants of dataflow semantics. We consider a few of them here.

7.4.4.1   Kahn process networks

Figure 7.1, discussed above, has the semantics of Kahn process networks [10,11] augmented with a non-deterministic merge [12,13]. In PN semantics, each actor executes (possibly forever) in its own thread of control. The connections between actors represent streams of tokens. In Kahn/MacQueen semantics [10], the way that threads interact with the ports has a key constraint that guarantees determinacy. Specifically, a thread is not permitted to 'ask' an input port whether there are available tokens to read. It must simply read from the port, and if no tokens are available, the thread blocks until tokens become available. This behaviour is called 'blocking reads'. Correspondingly, when the thread produces an output token, it simply sends it to the output port and continues. It is not permitted to ask the output port whether there is room for the token, or whether the ultimate recipient of the token is ready to receive it. These simple rules turn out to sufficient to ensure that actors implement 'monotonic' functions over streams, which in turn guarantees determinacy

[36]. Determinacy in this case means that every execution of the PN system yields the same stream on tokens on each connection. That is, the PN system 'determines' the streams.

In Figure 7.1, the Merge actor will non-deterministically interleave the two sequences at its input ports, preserving the order within each sequence, but yielding arbitrary ordering of data values across sequences. This behaviour is not monotonic. In fact, it cannot be implemented with blocking reads in a single actor thread. Extensions of PN that support such non-deterministic operations turn out to be especially useful for embedded software, and have been an active area of research [12,13].

A key issue with PN models is that they may deadlock. They may also consume unbounded memory buffering tokens between actors. It turns out that it is undecidable whether a PN model deadlocks or executes in bounded memory. This means that no algorithm exists that can always answer these questions in finite time. Nonetheless, there are simple execution policies that guarantee that if a particular PN system can be executed without deadlock in bounded memory, then it will be executed without deadlock in bounded memory [37]. The undecidable problem is solved by a runtime policy, which does not need to solve the problem in bounded time. A practical implementation of this policy is available in the Ptolemy II system [38].

7.4.4.2   Dennis dataflow

In a distinct family of dataflow models of computation, instead of executing a (possibly infinite) thread, a component executes a sequence of distinct 'firings'. This style of dataflow model was introduced by Dennis in the 1970s [39], and was applied to the design of high-performance computer architectures for several years. Semantically, the sequence of firings, of course, can be considered to be a thread with a limited mechanism for storing state, so at a fundamental level, the distinction between PN and Dennis dataflow is not great [11]. But it turns out to be particularly convenient to formulate dataflow systems in terms of firings. A great deal of formal analysis of the system is enabled by this abstraction.

A firing is enabled by satisfaction of a 'firing rule'. The formal structure of firing rules has considerable bearing on the formal properties of the model as a whole [40]. Each firing reads a short sequence of input tokens and produces a short sequence of output tokens. The firing of a dataflow component might also update the internal state of a component, affecting the behavior of the component in future firings.

There are two common ways of implementing dataflow models. One possibility is to implement a centralised run-time scheduler that selects and executes actors whose firing rules are satisfied. A second possibility is to statically analyse the dataflow graph and construct a static, finite description of the schedule. The latter approach is preferable for embedded software, since the static analysis also yields execution time and memory usage information. However, for general dataflow models, it turns out to be undecidable whether such static schedules can be constructed [41]. A suite of decidable special cases of dataflow have been developed over the years, however, and some of these are quite promising for embedded software systems.

7.4.4.3 Decidable dataflow models

A simple special case of dataflow models restricts actors so that on each port, they produce and consume a fixed, pre-specified number of tokens. This model of computation has been called 'synchronous dataflow' (SDF) [42], but to avoid confusion with the (significantly different) synchronous languages (e.g. see Reference 32), it would perhaps better be called 'statically schedulable dataflow' (SSDF). Indeed, the key feature of this model of computation is that simple static analysis either yields a static schedule that is free of deadlock and consumes bounded memory, or proves that no such schedule exists [42].

Because of the constraint that actors produce and consume only fixed, pre-specified numbers of tokens on each firing, SSDF by itself cannot easily describe applications with data-dependent control structure. A number of extensions enrich the semantics in various ways.

Boolean dataflow [41,43] (BDF) and integer-controlled dataflow [44] (IDF) augment the model by permitting the number of tokens produced or consumed at a port to be symbolically represented by a variable. The value of this variable is permitted to change during execution, so data-dependent control flow can be represented. Static analysis can often still be performed, but in principle, it is undecidable whether a BDF or IDF program can execute without deadlock in bounded memory. Nonetheless, for many practical programs, static analysis often yields a proof that it can, and in the process also yields a 'quasi-static schedule', which is a finite representation of a schedule with data-dependent control flow.

The fact that BDF and IDF are undecidable formalisms, however, is inconvenient. Static analysis can fail to find a schedule even when such a schedule exists. Cyclo-static dataflow (CSDF) [45] offers slightly more expressiveness than SSDF by permitting the production and consumption rates at ports to vary periodically. SSDF can also be combined hierarchically with finite state machines (FSMs), and if the state transitions are constrained to occur only at certain disciplined times, the model remains decidable. This combination has been called heterochronous dataflow (HDF) [46]. Parameterised SSDF [47] offers similarly expressive variability of production and consumption rates while remaining within a decidable formalism. Most of these variants of dataflow are available in the Ptolemy II system [38] or in Ptolemy Classic [48].

7.4.5 PECOS

A final model that we consider shares a number of features with the previous, but also has some unique properties. In a PECOS [49–51] model, there are three types of components: 'active', 'event' and 'passive'. These components are composed hierarchically with the constraint that an active component must occur at the root of the tree. Active components are associated with an independent thread that is periodically activated. Event components are similar to active components, except they are triggered by aperiodic events occurring in the system. Event components are generally associated with sensors and actuators in the system and are triggered when a sensor has data or an actuator requires data. Passive components are executed by the element that contains them.

Connections between components represent a variable in shared memory that is read and written by the components connecting to it. Each passive component is specified by a single `execute()` method that reads the appropriate input variables and writes the correct output variables. The simplest PECOS model consists of an active component at the toplevel, containing only passive components. The entire execution occurs in the single thread, and consists of sequenced invocations of the `execute()` methods.

Active and event components are specified by a `synchronise()` method, in addition to the `execute()` method. In order to avoid data races, variables for communicating with active and event components are double buffered. The `synchronise()` method is executed by the component's container to copy the input and output variables. The `execute()` method that actually performs processing only accesses the variable copies.

7.5 Conclusions

The diversity and richness of semantic models for embedded software is impressive. This is clearly a lively area of research and experimentation, with many innovative ideas. It is striking that none of the concurrent models of computation considered in this chapter rely on threads as the principle concurrency mechanism. Yet prevailing industrial practice in embedded software often does, building the software by creating concurrent threads and using the mutual exclusion and semaphore mechanisms of a real-time operating system to manage concurrency issues. We argue that these mechanism are too difficult for designers to understand, and that, except in very simple systems, should not be used in raw form. At a minimum, a design pattern corresponding to a clean concurrent model of computation (such as process networks or synchronous component composition) is required to achieve truly reliable systems. But better than informal use of such design patterns is the use of languages or frameworks that enforce the pattern and provide proven implementations of the low-level details. We have outlined the key features of a few such promising languages and frameworks.

Acknowledgments

This chapter describes work that is part of the Ptolemy project, which is supported by the National Science Foundation (NSF award number CCR-00225610), and Chess (the Center for Hybrid and Embedded Software Systems), which receives support from NSF and the following companies: Aqilent, General Motors, Hewlett-Packard, Infineon, and Toyota.

References

1 HILL, J., SZEWCZYK, R., WOO, A., HOLLAR, S., CULLER, D., and PISTER, K.: 'System architecture directions for network sensors'. Proceedings of international conference on *Architectural support for programming languages and operating systems (ASPLOS)*, 2000

2 GAY, D., LEVIS, P., VON BEHREN, R., WELSH, M., BREWER, E., and CULLER, D.: 'The nesC language: A holistic approach to networked embedded systems'. Proceedings of *Programming language design and implementation (PLDI)*, 2003

3 KOHLER, E., MORRIS, R., CHEN, B., JANNOTTI, J., and KAASHOEK, M.F.: 'The Click modular router', *ACM Transactions on Computer Systems*, August 2000, **18**(3), pp. 263–297

4 KOHLER, E.: 'The click modular router'. PhD thesis, Massachusetts Institute of Technology, 2000

5 BERRY, G.: 'The effectiveness of synchronous languages for the development of safety-critical systems'. White paper, Esterel Technologies, 2003

6 HUDAK, P.: 'Conception, evolution, and application of functional programming languages', *ACM Computing Surveys*, September 1989, **21**(3), pp. 359–411

7 DIJKSTRA, E.: 'Cooperating sequential processes', in GENUYS, E.F., (Ed.): 'Programming languages' (Academic Press, New York, 1968)

8 CHANDY, K., and MISRA, J.: 'Asynchronous distributed simulation via a sequence of parallel computations', *Communications of the ACM*, 1981, **24**, pp. 198–206

9 JEFFERSON, D.: 'Virtual time', *ACM Transactions Programming Languages and Systems*, 1985, **7**, pp. 404–425

10 KAHN, G., and MACQUEEN, D.B.: 'Coroutines and networks of parallel processes', in GILCHRIST, B., (Ed.): 'Information processing' (North-Holland Publishing Co., Amsterdam, 1977)

11 LEE, E.A., and PARKS, T.M.: 'Dataflow process networks', *Proceedings of the IEEE*, 1995, **83**, pp. 773–801

12 ARVIND, and BROCK, J.D.: 'Resource managers in functional programming', *Journal of Parallel and Distributed Computing*, 1984, pp. 5–21

13 DE KOCK, E., ESSINK, G., SMITS, W., *et al.*: 'Yapi: Application modeling for signal processing systems'. Proceedings of the 37th *Design automation conference (DAC'2000)*, 2000, pp. 402–405

14 PANANGADEN, P., and SHANBHOGUE, V.: 'The expressive power of indeterminate dataflow primitives', *Information and Computation*, 1992, **98**, pp. 99–131

15 LEE, E.A.: 'Modeling concurrent real-time processes using discrete events', *Annals of Software Engineering*, 1999, **7**, pp. 25–45

16 LEA, D.: 'Concurrent programming in java: design principles and patterns' (Addison-Wesley, Reading, MA, 1997)

17 SHAH, N., PLISHKER, W., and KEUTZER, K.: 'Np-click: a programming model for the intel ixp1200', in CROWLEY, P., FRANKLIN, M.,

HADIMIOGLU, H., and ONUFRYK, P., (Eds): 'Network processor design: issues and practices' (Elsevier, Amsterdam, vol. 2, 2004) pp. 181–201

18   OBJECT MANAGEMENT GROUP (OMG): 'CORBA event service specification, version 1.1', 2001

19   MAFFEIS, S.: 'Adding group communication and fault-tolerance to CORBA'. Proceedings of the USENIX Conference on *Object-oriented technologies (COOTS)*, 1995, pp. 16–29

20   AUGUSTSSON, L., SCHWARTZ, J., and NIKHIL, R.: 'Bluespec language definition'. Technical report, Sandburst Corporation, Andover, MA, 2000

21   NIKHIL, R.: 'Bluespec SystemVerilog: efficient, correct RTL from high-level specifications'. Proceedings of the conference on *Methods and models for codesign (MEMOCODE)*, San Diego, CA, 2004

22   HOE, J.C., and ARVIND: 'Hardware synthesis from term rewriting systems'. Proceedings of the international conference on *Very large scale integration (VLSI)* (Kluwer, Dordrecht, 1999) pp. 595–619

23   HOE, J.C., and ARVIND: 'Synthesis of operation-centric hardware descriptions'. Proceedings of international conference on *Computer aided design (ICCAD)*, 2000

24   VAN OMMERING, R., VAN DER LINDEN, F., KRAMER, J., and MAGEE, J.: 'The Koala component model for consumer electronics software', *IEEE Computer*, 2000, **33**, pp. 78–85

25   VAN OMMERING, R.: 'Building product populations with software components'. Proceedings of the international conference on *Software Engineering (ICSE)*, 2002, pp. 255–265

26   MAGEE, J. *et al.*: 'Specifying distributed software architectures', in 'ESEC '95' (Springer-Verlag, Berlin, 1995) pp. 137–153

27   LIU, C.L., and LEYLAND, J.W.: 'Scheduling algorithms for multiprogramming in a hard real time environment', *Journal of the ACM*, 1973, **20**, pp. 46–61

28   HENZINGER, T.A., HOROWITZ, B., and KIRSCH, C.M.: 'Giotto: A timetriggered language for embedded programming', in: 'EMSOFT 2001', Tahoe City, CA, Volume LNCS 2211 (Springer-Verlag, Berlin, 2001)

29   CHANDY, K.M., and MISRA, J.: 'Parallel program design: a foundation' (Addison Wesley, Reading, MA, 1988)

30   LIU, J., and LEE, E.A.: 'Timed multitasking for real-time embedded software', *IEEE Control Systems Magazine*, 2003, pp. 65–75

31   GHOSAL, A., HENZINGER, T.A., KIRSCH, C.M., and SANVIDO, M.A.: 'Event-driven programming with logical execution times'. Seventh international workshop on *Hybrid systems: computation and control* (HSCC). Volume Lecture Notes in Computer Science 2993 (Springer-Verlag, Berlin, 2004) pp. 357–371

32   HALBWACHS, N., CASPI, P., RAYMOND, P., and PILAUD, D.: 'The synchronous data flow programming language lustre', *Proceedings of the IEEE*, 1991, **79**, pp. 1305–1319

33 BERRY, G., and GONTHIER, G.: 'The esterel synchronous programming language: Design, semantics, implementation', *Science of Computer Programming*, 1992, **19**, pp. 87–152

34 GUERNIC, P.L., GAUTHIER, T., BORGNE, M.L., and MAIRE, C.L.: 'Programming real-time applications with signal', *Proceedings of the IEEE*, 1991, **79**

35 BENVENISTE, A., and BERRY, G.: 'The synchronous approach to reactive and real-time systems', *Proceedings of the IEEE*, 1991, **79**, pp. 1270–1282

36 KAHN, G.: 'The semantics of a simple language for parallel programming'. Proceedings of the *IFIP Congress 74* (North-Holland Publishing Co., Amsterdam, 1974) pp. 471–475

37 PARKS, T.M.: 'Bounded scheduling of process networks'. PhD thesis, EECS Department, University of California, Berkeley, CA, 1995

38 DAVIS, J. *et al.*: 'Ptolemy II – heterogeneous concurrent modeling and design in Java'. Memo M01/12, UCB/ERL, EECS UC Berkeley, CA 94720, 2001

39 DENNIS, J.B.: 'First version of a dataflow procedure language'. In Programming Symposium: Proceedings, Colloque sur la Programmation. Number 19 in Lecture Notes in Computer Science (Springer, Berlin, 1974) pp. 362–376

40 LEE, E.A.: 'A denotational semantics for dataflow with firing'. Technical report UCB/ERL M97/3, Electronics Research Laboratory, Berkeley, CA 94720, 1997

41 BUCK, J.T.: 'Scheduling dynamic dataflow graphs with bounded memory using the token flow model'. PhD thesis, University of California, Berkeley, CA, 1993

42 LEE, E.A., and MESSERSCHMITT, D.G.: 'Synchronous data flow', *Proceedings of the IEEE*, 1987

43 BUCK, J.T., and LEE, E.A.: 'Scheduling dynamic dataflow graphs with bounded memory using the token flow model'. IEEE international conference on *Acoustics, speech, and signal processing (ICASSP)*, 1993, Vol. 1, pp. 429–432

44 BUCK, J.T.: 'Static scheduling and code generation from dynamic dataflow graphs with integer-valued control systems'. IEEE Asilomar conference on *Signals, systems, and computers*, Pacific Grove, CA, 1994

45 BILSEN, G., ENGELS, M., LAUWEREINS, R., and PEPERSTRAETE, J.A.: 'Static scheduling of multi-rate and cyclo-static dsp applications'. Workshop on *VLSI signal processing* (IEEE Press, Washington, 1994)

46 GIRAULT, A., LEE, B., and LEE, E.A.: 'Hierarchical finite state machines with multiple concurrency models', *IEEE Transactions on Computer-aided Design of Integrated Circuits and Systems*, 1999, **18**(6), pp.742–760

47 BHATTACHARYA, B., and BHATTACHARYYA, S.S.: 'Parameterized dataflow modeling of dsp systems'. Proceedings of the international conference on *Acoustics, speech, and signal processing (ICASSP)*, Istanbul, Turkey, 2000, pp. 1948–1951

48 BUCK, J.T., HA, S., LEE, E.A., and MESSERSCHMITT, D.G.: Ptolemy: A framework for simulating and prototyping heterogeneous systems, *International Journal of Computer Simulation*, special issue on 'Simulation Software Development', 1994, **4**, pp. 155–182

49   WINTER, M., GENßLER, T., CHRISTOPH, A. *et al.*: 'Components for embedded software : The PECOS approach'. Workshop on *Composition languages*, in Conjunction with 16th European conference on *Object-oriented programming (ECOOP)*, Malaga, Spain, 2002

50   GENßLER, T., CHRISTOPH, A., WINTER, M. *et al.*: 'Components for embedded software: the PECOS approach'. Proceedings of the international conference on *Compilers, architecture, and synthesis for embedded systems (CASES)*, October 2002, pp. 19–26

51   NIERSTRASZ, O., ARÉVALO, G., DUCASSE, S. *et al.*: 'A component model for field devices'. Proceedings of the international working conference on *Component deployment*, IFIP/ACM, 2002

Chapter 8

Retargetable compilers and architecture exploration for embedded processors

Rainer Leupers, Manuel Hohenauer, Kingshuk Karuri,
Gunnar Braun, Jianjiang Ceng, Hanno Scharwaechter,
Heinrich Meyr and Gerd Ascheid

8.1   Introduction

Compilers translate high-level language source code into machine-specific assembly code. For this task, any compiler uses a model of the target processor. This model captures the compiler-relevant machine resources, including the instruction set, register files and instruction scheduling constraints. While in traditional target-specific compilers this model is built-in (i.e. it is hard-coded and probably distributed within the compiler source code), a 'retargetable compiler' uses an external processor model as an additional input that can be edited without the need to modify the compiler source code itself (Figure 8.1). This concept provides retargetable compilers with high flexibility w.r.t. the target processor.

Retargetable compilers have been recognised as important tools in the context of embedded system-on-chip (SoC) design for several years. One reason is the trend towards increasing use of programmable processor cores as SoC platform building blocks, which provide the necessary flexibility for fast adoption e.g. of new media encoding or protocol standards and easy (software based) product upgrading and debugging. While assembly language used to be predominant in embedded processor programming for quite some time, the increasing complexity of embedded application code now makes the use of high-level languages like C and C++ just as inevitable as in desktop application programming.

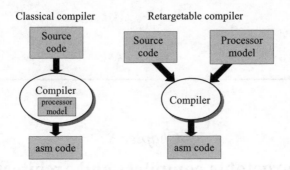

Figure 8.1   Classical vs. retargetable compiler

In contrast to desktop computers, embedded SoCs have to meet very high efficiency requirements in terms of MIPS per Watt, which makes the use of power-hungry, high-performance off-the-shelf processors from the desktop computer domain (together with their well-developed compiler technology) impossible for many applications. As a consequence, hundreds of different domain or even application-specific programmable processors have appeared in the semiconductor market, and this trend is expected to continue. Prominent examples include low-cost/low-energy microcontrollers (e.g. for wireless sensor networks), number-crunching digital signal processors (e.g. for audio and video codecs), as well as network processors (e.g. for internet traffic management).

All these devices demand their own programming environment, obviously including a high-level language (mostly ANSI C) compiler. This requires the capability of quickly designing compilers for new processors, or variations of existing ones, without the need to start from scratch each time. While compiler design traditionally has been considered a very tedious and manpower intensive task, contemporary retargetable compiler technology makes it possible to build operational (not heavily optimising) C compilers within a few weeks and more decent ones approximately within a single man-year. Naturally, the exact effort heavily depends on the complexity of the target processor, the required code optimisation and robustness level, and the engineering skills. However, compiler construction for new embedded processors is now certainly much more feasible than a decade ago. This permits us to employ compilers not only for application code development, but also for optimising an embedded processor architecture itself, leading to a true 'compiler/architecture codesign' technology that helps to avoid hardware–software mismatches long before silicon fabrication.

This chapter summarises the state-of-the-art in retargetable compilers for embedded processors and outlines their design and use by means of examples and case studies. In Section 8.2, we provide some compiler construction background needed to understand the different retargeting technologies. Section 8.3 gives an overview of some existing retargetable compiler systems. In Section 8.4, we describe how the above-mentioned 'compiler/architecture codesign' concept can be implemented in a processor architecture exploration environment. A detailed example of an industrial retargetable C compiler system is discussed in Section 8.5. Finally, Section 8.6 concludes and takes a look at potential future developments in the area.

8.2   Compiler construction background

The general structure of retargetable compilers follows that of well-proven classical compiler technology, which is described in textbooks such as [1–4]. First, there is a language 'frontend' for source code analysis. The frontend produces an 'intermediate representation' using which a number of machine-independent code optimisations are performed. Finally, the 'backend' translates the intermediate representation into assembly code, while performing additional machine-specific code optimisations.

8.2.1   Source language frontend

The standard organisation of a frontend comprises a 'scanner, a parser and a semantic analyser' (Figure 8.2). The scanner performs lexical analysis on the input source file, which is first considered just as a stream of ASCII characters. During lexical analysis, the scanner forms substrings of the input string to groups (represented by tokens), each of which corresponds to a primitive syntactic entity of the source language, e.g. identifiers, numerical constants, keywords or operators. These entities can be represented by regular expressions, for which in turn finite automata can be constructed and implemented that accept the formal languages generated by the regular expressions. Scanner implementation is strongly facilitated by tools like lex [5].

The scanner passes the tokenised input file to the parser, which performs syntax analysis w.r.t. the context-free grammar underlying the source language. The parser recognises syntax errors and, in case of a correct input, builds up a tree data structure that represents the syntactic structure of the input program.

Parsers can be constructed manually based on the LL(k) and LR(k) theory [2]. An LL(k) parser is a top-down parser, i.e. it tries to generate a derivation of the input program from the grammar start symbol according to the grammar rules. In each

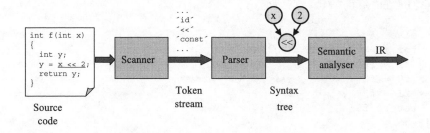

Figure 8.2   Source language frontend structure

step, it replaces a non-terminal by the right-hand side of a grammar rule. In order to decide which rule to apply out of possibly many alternatives, it uses a lookahead of k symbols on the input token stream. If the context-free grammar shows certain properties, this selection is unique, so that the parser can complete its job in linear time in the input size. The same also holds for LR(k) parsers which can process a broader range of context-free grammars, though. They work bottom-up, i.e. the input token stream is step-by-step reduced until finally reaching the start symbol. Instead of making a reduction step solely based on the knowledge of the k lookahead symbols, the parser additionally stores input symbols temporarily on a stack until enough symbols for an entire right-hand side of a grammar rule have been read. Due to this, the implementation of an LR(k) parser is less intuitive and requires some more effort than for LL(k).

Constructing LL(k) and LR(k) parsers manually provides some advantage in parsing speed. However, in most practical cases tools like yacc [5] (that generates a variant of LR(k) parsers) are employed for semi-automatic parser implementation.

Finally, the semantic analyser performs correctness checks not covered by syntax analysis, e.g. forward declaration of identifiers and type compatibility of operands. It also builds up a symbol table that stores identifier information and visibility scopes. In contrast to scanners and parsers, there are no widespread standard tools like lex and yacc for generating semantic analysers. Frequently, attribute grammars [1] are used, though, for capturing the semantic actions in a syntax-directed fashion, and special tools like ox [6] can extend lex and yacc to handle attribute grammars.

8.2.2 *Intermediate representation and optimisation*

In most cases, the output of the frontend is an intermediate representation (IR) of the source code that represents the input program as assembly-like, yet machine-independent low-level code. Three address code (Figures 8.3 and 8.4) is a common IR format.

There is no standard format for three address code, but usually all high-level control flow constructs and complex expressions are decomposed into simple statement sequences consisting of three-operand assignments and gotos. The IR generator inserts 'temporary variables' to store intermediate results of computations.

```
int fib (int m)
{ int f0 = 0, f1 = 1, f2, i;
  if (m <= 1) return m;
  else
  for (i = 2; i<=m; i++) {
    f2 = f0 + f1;
    f0 = f1;
    f1 = f2; }
  return f2;
}
```

Figure 8.3   Sample C source file fib.c (Fibonacci numbers)

```
int fib (int m_2)
{
 int f0_4, f1_5, f2_6, i_7, t1, t2, t3, t4, t6, t5;
          f0_4 = 0;
          f1_5 = 1;
          t1 = m_2 <= 1;
          if (t1) goto LL4;
          i_7 = 2;
          t2 = i_7 <= m_2;
          t6 = !t2;
          if (t6) goto LL1;
 LL3:     t5 = f0_4 + f1_5;
          f2_6 = t5;
          f0_4 = f1_5;
          f1_5 = f2_6;
 LL2:     t3 = i_7;
          t4 = t3 + 1;
          i_7 = t4;
          t2 = i_7 <= m_2;
          if (t2) goto LL3;
 LL1:     goto LL5;
 LL4:     return m_2;
 LL5:     return f2_6;
}
```

Figure 8.4   Three address code IR for source file fib.c. Temporary variable iden-
tifiers inserted by the frontend start with letter 't'. All local identifiers
have a unique numerical suffix. The particular IR format is generated
by the LANCE C frontend [7]

Three address code is a suitable format for performing different types of 'flow analysis', i.e. control and data flow analysis. Control flow analysis first identifies the basic block[1] structure of the IR and detects the possible control transfers between basic blocks. The results are captured in a control flow graph (CFG). Based on the CFG, more advanced control flow analyses can be performed, for example, in order to identify program loops. Figure 8.5 shows the CFG generated for the example from Figures 8.3 and 8.4.

Data flow analysis works on the statement level and determines interdependencies between computations. For instance, the data flow graph (DFG) from Figure 8.6 shows relations of the form 'statement X computes a value used as an argument in statement Y'.

---

[1] A basic block is a sequence of IR statements with unique control flow entry and exit points.

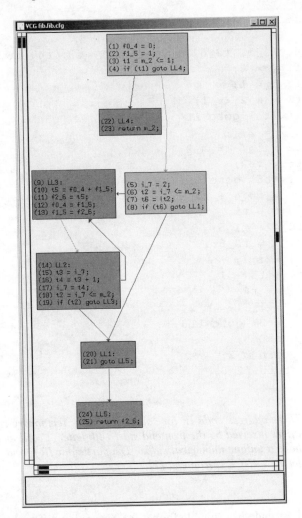

Figure 8.5   Control flow graph for fib.c

Both the CFG and the DFG form the basis for many code optimisation passes at the IR level. These include common subexpression elimination, jump optimisation, loop invariant code motion, dead code elimination and other 'Dragon Book' [1] techniques. Due to their target machine independence, these IR optimisations are generally considered complementary to machine code generation in the backend and are supposed to be useful 'on the average' for any type of target. However, care must be taken to select an appropriate IR optimisation sequence or script for each particular target, since certain (sometimes quite subtle) machine-dependencies do exist. For instance, common subexpression elimination removes redundant computations to save execution time and code size. At the assembly level, however, this effect might be over-compensated by the higher register pressure that increases

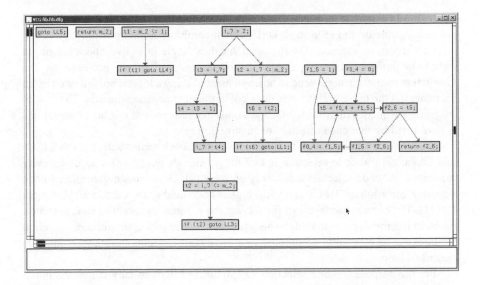

Figure 8.6   DFG for fib.c

the amount of spill code. Moreover, there are many interdependencies between the IR optimisations themselves. For instance, constant propagation generally creates new optimisation opportunities for constant folding, and vice versa, and dead code elimination is frequently required as a 'cleanup' phase in between other IR optimisations. A poor choice of IR optimisations can have a dramatic effect on final code quality. Thus, it is important that IR optimisations be organised in a modular fashion, so as to permit enabling and disabling of particular passes during fine-tuning of a new compiler.

8.2.3   Machine code generation

During this final compilation phase, the IR is mapped to target assembly code. Since for a given IR an infinite number of mappings as well as numerous constraints exist, this is clearly a complex optimisation problem. In fact, even many optimisation sub-problems in code generation are NP-hard, i.e. require exponential runtime for optimal solutions. As a divide-and-conquer approach, the backend is thus generally organised into three main phases: 'code selection', 'register allocation' and 'scheduling', which are implemented with a variety of heuristic algorithms. Dependent on the exact problem definition, all of these phases may be considered NP-hard, e.g. Reference 8 analyses the complexity of code generation for certain types of target machines. The rest of this subsection describes the different phases of machine code generation in detail.

8.2.3.1   Code selection

The IR of an application is usually constructed using primitive arithmetic, logical and comparison operations. The target architecture might combine several such primitive operations into a single instruction. A classical example is the multiply and accumulate (MAC) instruction found in most DSPs which combines a multiplication operation with a successive addition. On the other hand, a single primitive operation might have to be implemented using a sequence of machine instructions. For example, the C negation operation might need to be implemented using a logical not followed by an increment, if the target architecture does not implement negation directly. The task of mapping a sequence of primitive IR operations to a sequence of machine instructions is performed by the 'code selector' or 'instruction selector'.

In particular for target architectures with complex instruction sets, such as CISCs and DSPs, careful code selection is key for good code quality. Due to complexity reasons, most code selectors work only on trees [9], even though generalised code selection for arbitrary DFGs can yield higher code quality for certain architectures [10,11]. The computational effort for solving the NP-hard generalised code selection problem is normally considered too high in practice, whereas optimum code selection for trees can be efficiently accomplished using dynamic-programming techniques as described below.

For the purpose of code selection, the optimised IR is usually converted into a sequence of tree-shaped DFGs or Data Flow Trees (DFTs). Each instruction is represented as a 'tree-pattern' that can partially cover a sub-tree of a DFT and is associated with a cost. Figure 8.7 shows a set of instruction patterns for the Motorola 68 k CPU where each pattern has the same cost. As can be seen in Figure 8.8, the same DFT can be covered in multiple ways using the given instruction patterns. Using the cost metric for the machine instructions, the code selector aims at a minimum-cost covering of the DFTs by the given instruction patterns.

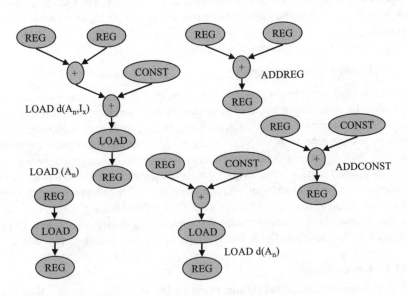

Figure 8.7   Five instruction patterns available for a Motorola 68 k CPU

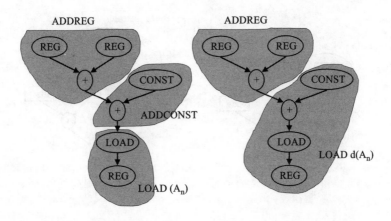

Figure 8.8   *Two possible coverings of a DFT using the instruction patterns from Figure 8.7*

```
Terminals: {MEM, +, -, *}
Non-terminals: {reg1, reg2}
Start symbol: reg1
```

Instruction Syntax	Cost	Rule
add reg1, reg2, reg1	2	reg1 → +(reg1, reg2)
add reg1, MEM, reg1	2	reg2 → +(reg1, MEM)
sub MEM, reg1, reg1	2	reg1 → −(MEM, reg1)
mul reg1, reg2, reg1	2	reg1 → *(reg1, reg2)
mul reg1, MEM, reg2	2	reg2 → *(reg1, MEM)
mac reg1, reg2, MEM	2	reg1 → +(*(reg1, reg2),MEM)
mov reg1, reg2	1	reg2 → reg1
mov reg2, reg1	1	reg1 → * reg2
load MEM, reg2	1	reg2 → MEM

Figure 8.9   *A context-free grammar representing different instruction patterns. The two non-terminals (reg1, reg2) in the grammar represent two different register classes. The right-hand side of each grammar rule is generated by pre-order traversal of the corresponding instruction pattern, the left-hand side refers to where the instruction produces its result*

Generally the code selection algorithms for DFTs use special context-free grammars to represent the instruction patterns. In such grammars, each rule is annotated with the cost of the corresponding pattern (Figure 8.9). The non-terminals in the grammar rules usually correspond to the different register classes or memory addressing mechanisms of the target architecture, and are placeholders for the results produced by the corresponding instructions. In each step of code selection, a sub-tree of the entire DFT is matched to a grammar rule, and the sub-tree is replaced by the non-terminal

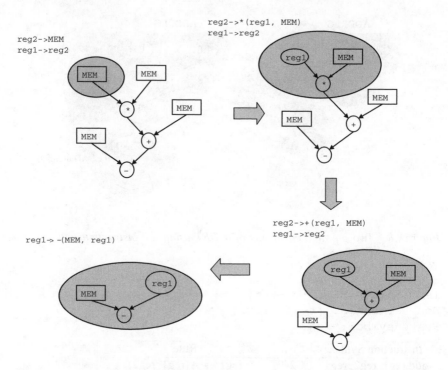

Figure 8.10   DFT covering using the grammar rules of Figure 8.9. In each step a sub-tree is replaced by the non-terminal(s) corresponding to one or more matching grammar rule(s)

in the corresponding rule (Figure 8.10). Since the context-free grammars representing instruction patterns are generally very ambiguous, each sub-tree, potentially, can be matched using different candidate rules. The optimum code selection, therefore, must take into account the costs of the different candidate rules to resolve these ambiguities.

The generalised dynamic-programming algorithm of code selection for DFTs applies a bottom-up technique to enumerate the costs of different possible covers using the available grammar rules. It calculates the minimum-cost match at each node of the DFT for each non-terminal of the grammar (Figure 8.11). After the costs of different matches are computed, the tree is traversed in a top-down fashion selecting the minimum-cost non-terminal (and the corresponding instruction pattern) for each node.

8.2.3.2   Register allocation

The code selector generally assumes that there are an infinite number of 'virtual registers' to hold temporary values. Subsequent to code selection, the register allocator decides which of these temporaries are to be kept in machine registers to ensure efficient access. Careful register allocation is key for target machines with RISC-like

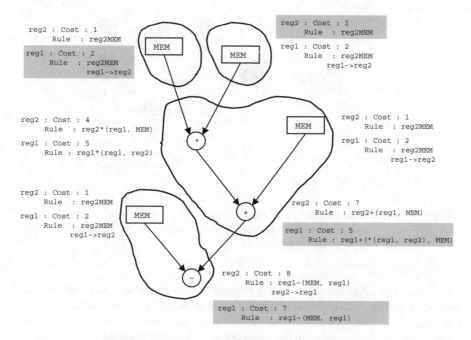

Figure 8.11   *Dynamic programming technique for code selection on the DFT of Figure 8.10 using the grammar rules of Figure 8.9. For each interme- diate node of the DFT, the minimum cost match for each non-terminal is calculated in a bottom-up fashion. The optimum cover consists of the rules in the grey boxes*

loadstore-architectures and large register files. Frequently, there are many more simul- taneously live variables than physically available machine registers. In such cases, the register allocator inserts 'spill code' to temporarily store register variables to main memory. Obviously, spill code needs to be minimised in order to optimise program performance and code size.

Many register allocators use a graph colouring approach [12,13] to accomplish this task. Although graph colouring is an NP-complete problem, a linear-time approx- imation algorithm exists that produces fairly good results. A register allocator first creates an 'interference graph' by examining the CFG and DFG of an application. Each node of the interference graph represents a temporary value. An edge between two nodes indicates that the corresponding temporaries are simultaneously 'alive' and therefore, cannot be assigned to the same register (Figure 8.12). If the target processor has K registers, then colouring the graph with K colours is a valid register assignment. Spills are generated when K-colouring of the graph is not possible.

8.2.3.3   Instruction scheduling

For target processors with instruction 'pipeline hazards' and/or 'instruction-level parallelism' (ILP), such as VLIW machines, instruction scheduling is an absolute

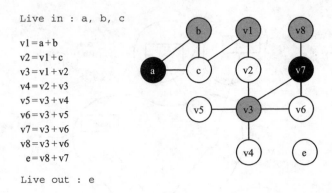

```
Live in : a, b, c

v1 = a + b
v2 = v1 + c
v3 = v1 + v2
v4 = v2 + v3
v5 = v3 + v4
v6 = v3 + v5
v7 = v3 + v6
v8 = v3 + v6
 e = v8 + v7

Live out : e
```

Figure 8.12   A code fragment and the corresponding interference graph. Since the graph is three-colourable, a valid register assignment with only three registers can be generated

necessity. For pipelined architectures without any hardware interlocks or data forwarding mechanism, instruction scheduling is required to ensure correct program semantics. For processors with high ILP, instruction scheduling is necessary to exploit the fine-grained parallelism in the applications. Like other backend phases, optimal scheduling is an intractable problem. However, there exist a number of powerful scheduling heuristics, such as list scheduling and trace scheduling [3].

Instruction scheduling is usually employed to resolve the (potentially) conflicting accesses to the same hardware resource(s) by a pair of instructions. The scheduler must take into account the structural, data or control hazards that usually result from pipelining. The schedulability of any two instructions, X and Y, under such hazards can usually be described using 'latency' and 'reservation' tables.

For the instruction X, the latency table specifies the number of cycles X takes to produce its result. This value places a lower bound on the number of instructions that Y, if it uses the result of X, must wait before it starts executing. In such cases, the scheduler must insert the required number of No-Operations (NOPs) (or other instructions that are not dependent on X) between X and Y to ensure correct program semantics. The reservation table specifies the number of hardware resources used by an instruction and the durations of such usages. The scheduler must ensure that X and Y do not make conflicting accesses to any hardware resource during execution. For example, if hardware multiplication takes three cycles to complete, then the scheduler must always keep two multiplication instructions at least three cycles apart (again by inserting NOPs or other non-multiplication instructions).

Scheduling can be done locally (for each basic block) or globally (across basic blocks). Most of the local schedulers employ the list scheduling technique. In contrast, the global scheduling techniques (such as trace scheduling, percolation scheduling, etc.) are more varied and out of scope of the current discussion.

In the list scheduling approach, the DFG of a basic block is topologically sorted on the basis of data-dependencies. The 'independent nodes' of the topologically sorted

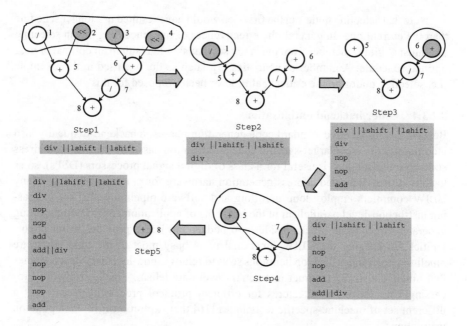

Figure 8.13   List scheduling applied to a DFG for a four-slot VLIW machine. For each step of the scheduling algorithm, nodes in the ready set are encircled, and the nodes finally selected by the algorithm are shown in grey colour. The partially scheduled code after each step is shown in grey boxes. Each node has been given a number for a better understanding of the explanation provided in the text

graph are entered into a 'ready' (for scheduling) set. Then, one member of the ready set is removed along with the edges emanating from it and is scheduled. The process is then recursively applied on the resulting graph.

An example of list scheduling for a four-slot VLIW machine, with three arithmetic logic units (ALUs) and one division unit, is presented in Figure 8.13. The division operation has a latency of four and all other operations have a latency of one. Initially, the ready set for the topologically sorted DFG consists of four nodes: two left shift and two division operations (nodes 1, 2, 3, 4). The left shift operations (nodes 2 and 3) can be scheduled simultaneously in two of the three available ALUs. However, only one division can be scheduled in the available division unit. The tie between the two division operations is broken by selecting node 4 which lies on the 'critical path' of the graph.

In the next scheduling step, only node 1 is available for scheduling. Although node 6 has no predecessor in the graph after removal of nodes 3 and 4, it can only be scheduled in the fifth cycle, i.e. after the division operation of node 3 produces its result. The scheduler schedules node 6 in step 3 after inserting two NOPs to take care of the latency of the division operation. The last two steps schedule the rest of the graph as shown in the figure.

Note that selecting node 1 in the first step would have resulted in a longer schedule than the current one. In general, the effectiveness of list scheduling lies in selecting the 'most eligible' member from the ready set that should result in an optimal or near optimal schedule. A number of heuristics [3], such as the one used in this example (i.e. selecting nodes on the critical path), have been proposed for this task.

8.2.3.4 Other backend optimisations

Besides the above three standard code generation phases, a backend frequently also incorporates different target-specific code optimisation passes. For instance, address code optimisation [14] is useful for a class of digital signal processors (DSPs), so as to fully utilise dedicated address generation hardware for pointer arithmetic. Many VLIW compilers employ loop unrolling and software pipelining [15] for increasing instruction-level parallelism in the hot spots of application code. Loop unrolling generates larger basic blocks inside loop bodies and hence provides better opportunities for keeping the VLIW functional units busy most of the time. Software pipelining rearranges the loop iterations so as to remove intra-loop data-dependencies that otherwise would obstruct instruction-level parallelism. Finally, Network Processing Unit (NPU) architectures for efficient protocol processing require yet a different set of machine-specific techniques [16] that exploit bit-level manipulation instructions.

The separation of the backend into multiple phases is frequently needed to achieve sufficient compilation speed but tends to compromise code quality due to interdependencies between the phases. In particular, this holds for irregular 'non-RISC' instruction sets, where the phase interdependencies are sometimes very tight. Although there have been attempts to solve the code generation problem in its entirety, e.g. based on integer linear programming [17], such 'phase-coupled' code generation techniques are still far from widespread use in real-word compilers.

8.3 Approaches to retargetable compilation

From the above discussions it is obvious that compiler retargeting mainly requires adaptations of the backend, even though IR optimisation issues certainly should not be neglected. In order to provide a retargetable compiler with a processor model, as sketched in Figure 8.1, a formal machine description language is required. For this purpose, dozens of different approaches exist. These can be classified w.r.t. the intended target processor class (e.g. RISC vs. VLIW) and the modelling abstraction level, e.g. purely behavioural, compiler-oriented vs. more structural, architecture-oriented modelling styles.

'Behavioural modelling languages' make the task of retargeting easier, because they explicitly capture compiler-related information about the target machine, i.e. instruction set, register architecture and scheduling constraints. On the other hand, they usually require good understanding of compiler technology. In contrast, 'architectural modelling languages' follow a more hardware design oriented approach and describe the target machine in more detail. This is convenient for users not much familiar with compiler technology. However, automatic retargeting gets more difficult, because a 'compiler view' needs to be extracted from the architecture model, while eliminating unnecessary details.

In the following, we will briefly discuss a few representative examples of retargetable compiler systems. For a comprehensive overview of existing systems see Reference 18.

8.3.1 MIMOLA

MIMOLA denotes both a mixed programming and hardware description language (HDL) and a hardware design system. As the MIMOLA HDL serves multiple purposes, e.g. register-transfer level (RTL) simulation and synthesis, the retargetable compiler MSSQ [19,20] within the MIMOLA design system follows the above-mentioned architecture-oriented approach. The target processor is described as an RTL netlist, consisting of components and interconnect. Figure 8.14 gives an example of such an RTL model.

Since the HDL model comprises all RTL information about the target machine's controller and data path, it is clear that all information relevant for the compiler backend of MSSQ is present, too. However, this information is only implicitly available, and consequently the lookup of this information is more complicated than in a behavioural model.

MSSQ compiles an 'extended subset' of the PASCAL programming language directly into binary machine code. Due to its early introduction, MSSQ employs only few advanced code optimisation techniques (e.g. there is no graph-based global register allocation), but performs the source-to-architecture mapping in a straightforward fashion, on a statement-by-statement basis. Each statement is represented by a DFG, for which an isomorphic subgraph is searched in the target data path. If this matching fails, the DFG is partitioned into simpler components, for which graph matching is invoked recursively.

In spite of this simple approach, MSSQ is capable of exploiting instruction-level parallelism in VLIW-like architectures very well, due to the use of a flexible instruction scheduler. However, code quality is generally not acceptable in case of complex instruction sets and load/store data paths. In addition, it shows comparatively high compilation times, due to the need for exhaustive graph matching.

The MIMOLA approach shows very high flexibility in compiler retargeting, since in principle any target processor can be represented as an RTL HDL model. In addition, it avoids the need to consistently maintain multiple different models of the same machine for different design phases, e.g. simulation and synthesis, as all phases can use the same 'golden' reference model. MSSQ demonstrates that retargetable compilation is possible with such unified models, even though it does not well handle architectures with complex instruction pipelining constraints (which is a limitation of the tool, though, rather than of the approach itself). The disadvantage, however, is that the comparatively detailed modelling level makes it more difficult to develop the

```
MODULE SimpleProcessor (IN inp:(7:0); OUT outp:(7:0)); STRUCTURE
IS TYPE InstrFormat = FIELDS        -- 21-bit horizontal instruction word
          imm:         (20:13);
          RAMadr:      (12:5);
          RAMctr:       (4);
          mux:         (3:2);
          alu:         (1:0);
        END;
      Byte = (7:0);  Bit = (0);  -- scalar types
PARTS                             -- instantiate behavioral modules
 IM: MODULE InstrROM (IN adr: Byte;  OUT ins: InstrFormat);
     VAR storage: ARRAY[0..255] OF InstrFormat;
     BEGIN ins <- storage[adr]; END;
 PC, REG: MODULE Reg8bit (IN data: Byte; OUT outp: Byte);
          VAR R: Byte;
          BEGIN R := data; outp <- R; END;
 PCIncr: MODULE IncrementByte (IN data: Byte; OUT inc: Byte);
        BEGIN outp <- INCR data; END;
 RAM: MODULE Memory (IN data, adr: Byte; OUT outp: Byte; FCT c: Bit);
      VAR storage: ARRAY[0..255] OF Byte;
      BEGIN
      CASE c OF: 0: NOLOAD storage; 1: storage[adr] := data; END;
      outp <- storage[adr];
      END;
 ALU: MODULE AddSub (IN d0, d1: Byte; OUT outp: Byte; FCT c: (1:0));
      BEGIN              -- "%" denotes binary numbers
      outp <- CASE c OF %00: d0 + d1; %01: d0 - d1; %1x: d0; END;
      END;
 MUX: MODULE Mux3x8 (IN d0,d1,d2: Byte; OUT outp: Byte; FCT c: (1:0));
      BEGIN outp <- CASE c OF 0: d0;  1: d1; ELSE: d2; END; END;
CONNECTIONS
 -- controller:                    -- data path:
 PC.outp        -> IM.adr;         IM.ins.imm      -> MUX.d0;
 PC.outp        -> PCIncr.data;    inp        ->  MUX.d1;  -- primary input
 PCIncr.outp    -> PC.data;        RAM.outp   ->  MUX.d2;
 IM.ins.RAMadr -> RAM.adr;         MUX.outp   ->  ALU.d1;
 IM.ins.RAMctr -> RAM.c;           ALU.outp   ->  REG.data;
 IM.ins.alu    -> ALU.C;           REG.outp   ->  ALU.d0;
 IM.ins.mux    -> MUX.c;           REG.outp   ->  outp;      -- primary output
END; -- STRUCTURE LOCATION_FOR_PROGRAMCOUNTER  PC;
LOCATION_FOR_INSTRUCTIONS IM; END; -- STRUCTURE
```

Figure 8.14   MIMOLA HDL model of a simple processor

model and to understand its interaction with the retargetable compiler, since e.g. the instruction set is 'hidden' inside the model.

Some of the limitations have been removed in RECORD, another MIMOLA HDL based retargetable compiler that comprises dedicated code optimisations for DSPs. In order to optimise compilation speed, RECORD uses an 'instruction set extraction technique' [21] that bridges the gap between RTL models and behavioural processor models. Key ideas of MSSQ, e.g. the representation of scheduling constraints by binary partial instructions, have also been adopted in the CHESS compiler [22,23], one of the first commercial tool offerings in that area. In the Expression compiler [24], the concept of structural architecture modelling has been further refined to increase the reuse opportunities for model components.

8.3.2   GNU C compiler

The widespread GNU C compiler gcc [25] can be retargeted by means of a machine description file that captures the compiler view of a target processor in a behavioural fashion. In contrast to MIMOLA, this file format is heterogeneous and solely designed for compiler retargeting. The gcc compiler is organised into a fixed number of different passes. The frontend generates a three address code like IR. There are multiple built-in 'Dragon Book' IR optimisation passes, and the backend is driven by a specification of instruction patterns, register classes and scheduler tables. In addition, retargeting gcc requires C code specification of numerous support functions, macros and parameters.

The gcc compiler is robust and well-supported, it includes multiple source language frontends, and it has been ported to dozens of different target machines, including typical embedded processor architectures like ARM, ARC, MIPS and Xtensa. However, it is very complex and hard to customise. It is primarily designed for 'compiler-friendly' 32-bit RISC-like loadstore architectures. While porting to more irregular architectures, such as DSPs, the gcc compiler can be ported to non-RISC processor architectures, too, this generally results in huge retargeting effort and/or insufficient code quality.

8.3.3   Little C compiler

Like gcc, retargeting the 'little C compiler' lcc [26,27] is enabled via a machine description file. In contrast to gcc, lcc is a 'lightweight' compiler that comes with much less source code and only a few built-in optimisations, and hence lcc can be used to design compilers for certain architectures very quickly. The preferred range of target processors is similar to that of gcc, with some further restrictions on irregular architectures, though.

In order to retarget lcc, the designer has to specify the available machine registers, as well as the translation of C operations (or IR operations, respectively) to machine instructions by means of 'mapping rules'. The following excerpt from lcc's Sparc machine description file [27] exemplifies two typical mapping rules:

```
addr: ADDP4(reg,reg) "%%%0+%%%1"
reg:  INDIRI1(addr) "ldsb [%0],%%%c\n"
```

The first line instructs the code selector how to cover address computations ('addr') that consist of adding two 4-byte pointers ('ADDP4') stored in registers ('reg'). The string '%%%0+%%%1' denotes the assembly code to be emitted, where '%0' and '%1' serve as placeholders for the register numbers to be filled later by the register allocator (and '%%' simply emits the register identifier symbol '%'). Since 'addr' is only used in context with memory accesses, here only a substring without assembly mnemonics is generated.

The second line shows the covering of a 1-byte signed integer load from memory ('INDIRI1'), which can be implemented by assembly mnemonic 'ldsb', followed by arguments referring to the load address ('%0', returned from the 'addr' mapping rule) and the destination register ('%c').

By specifying such mapping rules for all C/IR operations plus around 20 relatively short C support functions, lcc can be retargeted quite efficiently. However, lcc is very limited in the context of non-RISC embedded processor architectures. For instance, it is impossible to model certain irregular register architectures (as e.g. in DSPs) and there is no instruction scheduler, which is a major limitation for targets with instruction-level parallelism. Therefore, lcc has not found wide use in code generation for embedded processors so far.

8.3.4   CoSy

The CoSy system from ACE [28] is a retargetable compiler for multiple source languages, including C and C++. Like gcc, it includes several Dragon Book optimisations, but shows a more modular, extensible software architecture, which permits to add IR optimisation passes through well-defined interfaces.

For retargeting, CoSy comprises a backend generator that is driven by the CGD 'machine description format'. Similar to gcc and lcc, this format is full-custom and only designed for use in compilation. Hence, retargeting CoSy requires significant compiler know-how, particularly w.r.t. code selection and scheduling. Although it generates the backend automatically from the CGD specification, including standard algorithms for code selection, register allocation and scheduling, the designer has to fully understand the IR-to-assembly mapping and how the architecture constrains the instruction scheduler.

The CGD format follows the classical backend organisation. It includes mapping rules, a register specification, as well as scheduler tables. The register specification is a straightforward listing of the different register classes and their availability for the register allocator (Figure 8.15).

Mapping rules are the key element of CGD (Figure 8.16). Each rule describes the assembly code to be emitted for a certain C/IR operation, depending on matching conditions and cost metric attributes. Similar to gcc and lcc, the register allocator later replaces symbolic registers with physical registers in the generated code.

Mapping rules also contain a link to the CGD scheduler description. By means of the keywords 'PRODUCER' and 'CONSUMER', the instructions can be classified into groups, so as to make the scheduler description more compact. For instance, arithmetic instructions performed on a certain ALU generally have the same latency values. In the scheduler description itself (Figure 8.17), the latencies for pairs of instruction groups are listed as a table of numerical values. As explained in Section 8.2.3, these values instruct the scheduler to arrange instructions a minimum amount of cycles apart from each other. Different types of inter-instruction dependencies are permitted, here the keyword 'TRUE' denotes data dependency.[2]

---

[2] Data dependencies are sometimes called 'true', since they are induced by the source program itself. Hence, they cannot be removed by the compiler. In contrast, there are 'false', or anti-dependencies that are only introduced by code generation via reuse of registers for different variables. The compiler should aim at minimising the amount of false dependencies, in order to maximise the instruction scheduling freedom.

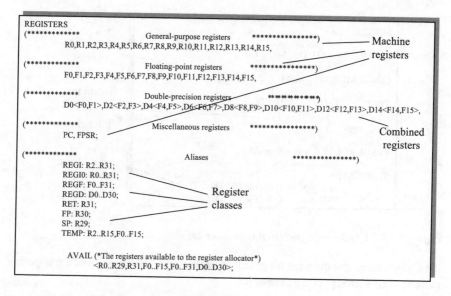

Figure 8.15   CGD specification of processor registers

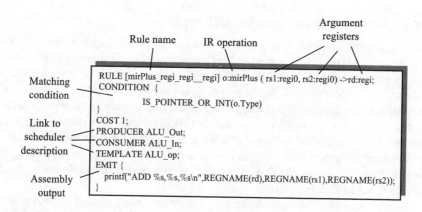

Figure 8.16   CGD specification of mapping rules

Via the 'TEMPLATE' keyword, a reservation table entry is referenced. The '&' symbol separates the resource use of an instruction group over the different cycles during its processing in the pipeline. For instance, in the last line of Figure 8.17, instruction group 'MUL_op' occupies resource 'EX_mul' for two subsequent cycles.

The CGD processor modelling formalism makes CoSy a quite versatile retargetable compiler. Case studies for RISC architectures show that the code quality produced by CoSy compilers is comparable to that of gcc. However, the complexity of

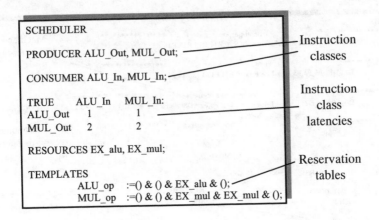

Figure 8.17   CGD specification of scheduler tables

the CoSy system and the need for compiler background knowledge make retargeting more tedious than e.g. in the case of lcc.

8.4   Processor architecture exploration

8.4.1   *Methodology and tools for ASIP design*

As pointed out in Section 8.1, one of the major applications of retargetable compilers in SoC design is to support the design and programming of application-specific instruction set processors (ASIPs). ASIPs receive increasing attention in both academia and industry due to their optimal flexibility/efficiency compromise [29]. The process of evaluating and refining an initial architecture model step-by-step to optimise the architecture for a given application is commonly called architecture exploration. Given that the ASIP application software is written in a high-level language like C, it is obvious that compilers play a major role in architecture exploration. Moreover, in order to permit frequent changes of the architecture during the exploration phase, compilers have to be retargetable.

Today's most widespread architecture exploration methodology is sketched in Figure 8.18. It is an iterative approach that requires multiple re-mapping of the application code to the target architecture. In each iteration, the usual software development tool chain (C compiler, assembler, linker) is used for this mapping. Since exploration is performed with a virtual prototype of the architecture, an instruction set simulator together with a profiler are used to measure the efficiency and cost of the current architecture w.r.t. the given (range of) applications, e.g. in terms of performance and area requirements.

We say that hardware (processor architecture and instruction set) and software (application code) 'match', if the hardware meets the performance and cost goals, and there is no over- or under-utilisation of HW resources. For instance, if the HW is not capable of executing the 'hot spots' of the application code under the given timing constraints, e.g. due to insufficient function units, too much spill code or too many pipeline stalls, then more resources need to be provided. On the other hand, if many function units are idle most of the time or half of the register file remains unused, this indicates an under-utilisation. Fine-grained profiling tools make such data available to the processor designer. However, it is still a highly creative process to determine the exact source of bottlenecks (application code, C compiler, processor instruction set or microarchitecture) and to remove them by corresponding modifications, while simultaneously overlooking their potential side effects.

If the HW/SW match is initially not satisfactory, the ASIP architecture is further optimised, dependent on the bottlenecks detected during simulation and profiling. This optimisation naturally requires HW design knowledge, and may comprise e.g. addition of application-specific custom machine instructions, varying register file sizes, modifying the pipeline architecture, adding more function units to the data path or simply removing unused instructions. The exact consequences of such modifications are hard to predict, so that usually multiple iterations are required in order to arrive at an optimal ASIP architecture that can be handed over to synthesis and fabrication.

With the research foundations of this methodology laid in the 1980s and 1990s (see Reference 18 for a summary of early tools), several commercial offerings are available now in the EDA (electronic design automation) industry, and more and more start-up companies are entering the market in that area. While ASIPs offer many advantages over off-the-shelf processor cores (e.g. higher efficiency, reduced royalty payments and better product differentiation), a major obstacle is still the potentially costly design and verification process, particularly concerning the SW tools shown in Figure 8.18. In order to minimise these costs and to make the exploration loop efficient, all approaches to processor architecture exploration aim at automating the retargeting of these tools as much as possible. In addition, a link to HW design has

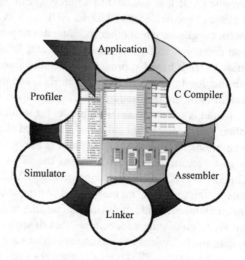

Figure 8.18   Processor architecture exploration loop

to be available in order to accurately estimate area, cycle time and power consumption of a new ASIP. In most cases this is enabled by automatic HDL generation capabilities for processor models, which provide a direct entry to gate-true estimations via traditional synthesis flows and tools.

One of the most prominent examples of an industrial ASIP is the Tensilica Xtensa processor [30]. It provides a basic RISC core that can be extended and customised by adding new machine instructions and adjusting parameters, e.g. for the memory and register file sizes. Software development tools and an HDL synthesis model can be automatically generated. Application programming is supported via the gcc compiler and a more optimising in-house C compiler variant. The Tensilica Xtensa, together with its design environment, completely implement the exploration methodology from Figure 8.18. On the other hand, the use of a largely predefined RISC core as the basic component poses limitations on the flexibility and the permissible design space. An important new entry to the ASIP market is Stretch [31]. Their configurable S5000 processor is based on the Xtensa core, but includes an embedded field-programmable gate array (FPGA) for processor customisation. While FPGA vendors have combined processors and configurable logic on a single chip for some time, the S5000 'instruction set extension fabric' is optimised for implementation of custom instructions, thus providing a closer coupling between processor and FPGA. In this way, the ASIP becomes purely SW-configurable and field-programmable, which reduces the design effort, yet at the expense of reduced flexibility.

8.4.2   ADL-based approach

More flexibility is offered by the tool suite from Target Compiler Technologies [23] that focuses on the design of ASIPs for signal processing applications. In this approach, the target processor can be freely defined by the user in the nML architecture description language (ADL). In contrast to a purely compiler-specific machine model, such as in the case of gcc or CoSy's CGD, an ADL such as nML also captures information relevant for the generation of other software development tools, e.g. simulator, assembler and debugger, and hence covers a greater level of detail. On the other hand, in contrast to HDL based approaches to retargetable compilation, such as MIMOLA, the abstraction level is still higher than RTL and usually allows for a concise explicit modelling of the instruction set. The transition to RTL only takes place once the ADL model is refined to an HDL model for synthesis.

LISATek is another ASIP design tool suite that originated at Aachen University [32]. It has first been produced by LISATek Inc. and is now available as a part of CoWare's SoC design tool suite [33]. LISATek uses the LISA 2.0 (Language for Instruction Set Architectures) ADL for processor modelling. A LISA model captures the processor resources like registers, memories and instruction pipelines, as well as the machine's instruction-set architecture (ISA). The ISA model is composed of 'operations' (Figure 8.19), consisting of 'sections' that describe the binary coding, timing, assembly syntax and behaviour of machine operations at different abstraction levels. In an instruction-accurate model (typically used for early architecture exploration), no pipeline information is present, and each operation corresponds to one

```
OPERATION ADD IN pipe.EX {
  // declarations
  DECLARE {
   INSTANCE writeback;
   GROUP src1, dst = { reg };
   GROUP src2 = { reg || imm };}

  // assembly syntax
  SYNTAX { "addc" dst "," src1 "," src2 }

  // binary encoding
  CODING { 0b0101 dst src1 src2 }

  // behavior (C code)
  BEHAVIOR {
   u32 op1, op2, result, carry;
   if (forward) {
    op1 = PIPELINE_REGISTER(pipe,EX/WB),result;}
   else {
    op1 = PIPELINE_REGISTER(pipe,DC/EX).op1;}
   result = op1 + op2;
   carry = compute_carry(op1, op2, result);
   PIPELINE_REGISTER(EX/WB).result = result;
   PIPELINE_REGISTER(EX/WB).carry = carry; }

  // pipeline timing
  ACTIVATION { writeback, carry_update }
}
```

*Figure 8.19 LISA operation example: execute stage of an ADD instruction in a
cycle-true model with forwarding hardware modelling*

instruction. In a more fine-grained, cycle-accurate model, each operation represents
a single pipeline stage of one instruction. LISATek permits the generation of software development tools (compiler, simulator, assembler, linker, debugger, etc.) from
a LISA model, and embeds all tools into an integrated GUI environment for application and architecture profiling. In addition, it supports the translation of LISA models
to synthesisable VHDL and Verilog RTL models. Figure 8.20 shows the intended
ASIP design flow with LISATek. In addition to an implementation of the exploration
loop from Figure 8.18, the flow also comprises the synthesis path via HDL models,
which enables back-annotation of gate-level hardware metrics.

In Reference 34 it has been exemplified how the LISATek architecture exploration methodology can be used to optimise the performance of an ASIP for an IPv6
security application. In this case study, the goal was to enhance a given processor

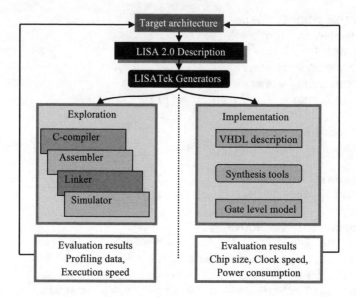

Figure 8.20   LISATek based ASIP design flow

architecture (MIPS32) by means of dedicated machine instructions and a microarchitecture for fast execution of the compute-intensive Blowfish encryption algorithm (Figure 8.21) in IPsec. Based on initial application C code profiling, hot spots were identified that provided first hints on appropriate custom instructions. The custom instructions were implemented as a co-processor (Figure 8.22) that communicates with the MIPS main processor via shared memory. The co-processor instructions were accessed from the C compiler generated from the LISA model via compiler intrinsics. This approach was feasible due to the small number of custom instructions required, which can be easily utilised with small modifications of the initial Blowfish C source code. LISATek-generated instruction-set simulators embedded into a SystemC based co-simulation environment were used to evaluate candidate instructions and to optimise the co-processor's pipeline microarchitecture on a cycle-accurate level.

Finally, the architecture implementation path via LISA-to-VHDL model translation and gate-level synthesis was used for further architecture fine-tuning. The net result was a 5× speedup of Blowfish execution over the original MIPS at the expense of an additional co-processor area of 22 k gates. This case study demonstrates that ASIPs can provide excellent efficiency combined with IP reuse opportunities for similar applications from the same domain. Simultaneously, the iterative, profiling based exploration methodology permits us to achieve such results quickly, i.e. typically within a few man-weeks.

The capability of modelling the ISA behaviour in arbitrary C/C++ code makes LISA very flexible w.r.t. different target architectures and enables the generation of high-speed ISA simulators based on the JITCC technology [35]. As in the MIMOLA

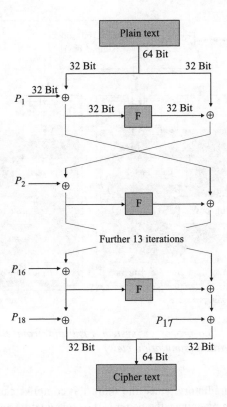

Figure 8.21   *Blowfish encryption algorithm for IPsec: P_i denotes a 32-bit subkey,*
F denotes the core subroutine consisting of substitutions and add/xor
operations

and Target approaches, LISATek follows the 'single golden model' paradigm, i.e. only
one ADL model (or automatically generated variants of it) is used throughout the
design flow in order to avoid consistency problems and to guarantee 'correct-by-
construction' software tools during architecture exploration. Under this paradigm, the
construction of retargetable compilers is a challenging problem, since in contrast to
special-purpose languages like CGD the ADL model is not tailored towards compiler
support only. Instead, similar to MIMOLA/MSSQ (see Section 8.3.1), the compiler-
relevant information needs to be extracted with special techniques. This is discussed
in more detail in the next section.

8.5 C compiler retargeting in the LISATek platform

8.5.1 Concept

The design goals for the retargetable C compiler within the LISATek environment
were to achieve high flexibility and good code quality at the same time. Normally,

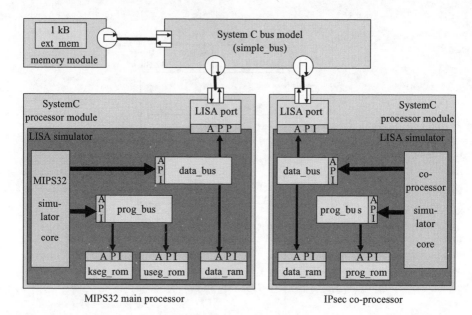

Figure 8.22   MIPS32/co-processor system resulting from Blowfish architecture exploration (simulation view)

these goals are contradictory, since the more the compiler can exploit knowledge of the range of target machines, the better is the code quality, and vice versa. In fact, this inherent trade-off has been a major obstacle for the successful introduction of retargetable compilers for quite some time.

However, a closer look reveals that this only holds for 'push-button' approaches to retargetable compilers, where the compiler is expected to be retargeted fully automatically once the ADL model is available. If compiler retargeting follows a more pragmatic user-guided approach (naturally at the cost of a slightly longer design time), then one can escape from the above dilemma. In the case of the LISA ADL, an additional constraint is the unrestricted use of C/C++ for operation behaviour descriptions. Due to the need for flexibility and high simulation speed, it is impossible to sacrifice this description vehicle. On the other hand, this makes it very difficult to automatically derive the compiler semantics of operations, due to large syntactic variances in operation descriptions. In addition, hardware-oriented languages like ADLs do not at all contain certain types of compiler-related information, such as C type bit widths, function calling conventions, etc., which makes an interactive, GUI-based retargeting environment useful, anyway.

In order to maximise the reuse of existing, well-tried compiler technology and to achieve robustness for real-life applications, the LISATek C compiler builds on the CoSy system (Section 8.3.4) as a backbone. Since CoSy is capable of generating the major backend components (code selector, register allocator, scheduler) automatically, it is sufficient to generate the corresponding CGD fragments (see

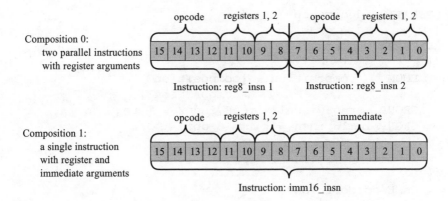

Figure 8.23   Two instruction encoding formats (compositions)

Figures 8.15–8.17) from a LISA model in order to implement an entire retargetable compiler tool chain.

8.5.2   Register allocator and scheduler

Out of the three backend components, the register allocator is the easiest one to retarget, since the register information is explicit in the ADL model. As shown in Figure 8.15, essentially only a list of register names is required, which can be largely copied from the resource declaration in the LISA model. Special cases (e.g. combined registers, aliases, special-purpose registers such as the stack pointer) can be covered by a few one-time user interactions in the GUI.

As explained in Section 8.2.3, generation of the instruction scheduler is driven by two types of tables: latency tables and reservation tables. Both are only implicit in the ADL model. Reservation tables model inter-instruction conflicts. Similar to the MSSQ compiler (Section 8.3.1), it is assumed that all such conflicts are represented by instruction encoding conflicts.[3] Therefore, reservation tables can be generated by examining the instruction encoding formats in a LISA model.

Figures 8.23 and 8.24 exemplify the approach for two possible instruction formats or 'compositions'. Composition 0 is VLIW-like and allows to encode two parallel 8-bit instructions. In composition 1, the entire 16 instruction bits are required due to an 8-bit immediate constant that needs to be encoded. In the corresponding LISA model, these two formats are modelled by means of a switch/case language construct.

The consequences of this instruction format for the scheduler are that instructions that fit into one of the 8-bit slots of composition 0 can be scheduled in either of the two, while an instruction that requires an immediate operand blocks other

---

[3]   This means that parallel scheduling of instructions with conflicting resource usage is already prohibited by the instruction encoding itself. Architectures for which this assumption is not valid appear to be rare in practice, and if necessary there are still simple workarounds via user interaction, e.g. through manual addition of artificial resources to the generated reservation tables.

```
OPERATION decode_op
{
 DECLARE
 {
   ENUM     composition = {composition0,
                            composition1};
   GROUP   reg8_insn1,  reg8_insn2 = { reg8_op } ;
   GROUP imm16_insn = { imm16_op};
 }
 SWITCH(compositions)
 {
   CASE composition0:
   {
    CODING AT (progam_counter)
        { insn_reg = = reg8_insn1 | | reg8_insn2 }
    SYNTAX { reg8_insn1 " , " reg_insn2}
   }
   CASE composition1:
   {
    CODING AT (progam_counter)
        { insn_reg = = imm16_insn }
    SYNTAX { imm16_insn }
   }
 }
}
```

Figure 8.24   LISA model fragment for instruction format from Figure 8.23

instructions from being scheduled in parallel. The scheduler generator analyses these constraints and constructs 'virtual resources' to represent the inter-instruction conflicts. Naturally, this concept can be generalised to handle more complex, realistic cases for wider VLIW instruction formats. Finally, a reservation table in CGD format (see Figure 8.17) is emitted for further processing with CoSy.[4]

The second class of scheduler tables, latency tables, depends on the resource access of instructions as they run through the different instruction pipeline stages. In LISA, cycle-accurate instruction timing is described via 'activation sections' inside the LISA operations. One operation can invoke the simulation of other operations downstream in the pipeline during subsequent cycles, e.g. an instruction fetch stage would typically be followed by a decode stage, and so forth. This explicit modelling of pipeline stages makes it possible to analyse the reads and writes to registers at

---

[4] We also generate a custom scheduler as an optional bypass of the CoSy scheduler. The custom scheduler achieves better scheduling results for certain architectures [36].

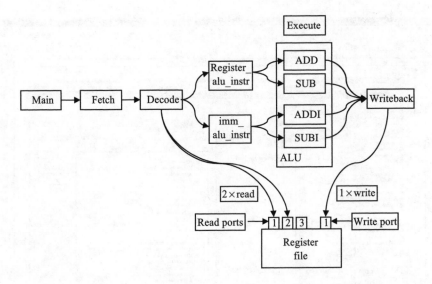

Figure 8.25   Register file accesses of an instruction during its processing over different pipeline stages

a cycle-true level. In turn, this information permits to extract the different types of latencies, e.g. due to a data dependency.

An example is shown in Figure 8.25, where there is a typical four-stage pipeline (fetch, decode, execute, writeback) for a load/store architecture with a central register file.[5] By tracing the operation activation chain, one can see that a given instruction makes two read accesses in stage 'decode' and a write access in stage 'writeback'. For arithmetic instructions executed on the ALU, for instance, this implies a latency value of 2 (cycles) in the case of a data dependency. This information can again be translated into CGD scheduler latency tables (Figure 8.17). The current version of the scheduler generator, however, is not capable of automatically analysing forwarding/bypassing hardware, which is frequently used to minimise latencies due to pipelining. Hence, the fine-tuning of latency tables is performed via user interaction in the compiler retargeting GUI, which allows the user to add specific knowledge in order to override potentially too conservative scheduling constraints, so as to improve code quality.

8.5.3   Code selector

As sketched in Section 8.2.3, retargeting the code selector requires specification of instruction patterns (or mapping rules) used to cover a DFG representation of the IR. Since it is difficult to extract instruction semantics from an arbitrary

---

[5] The first stage in any LISA model is the 'main' operation that is called for every new simulation cycle, similar to the built-in semantics of the 'main' function in ANSI C.

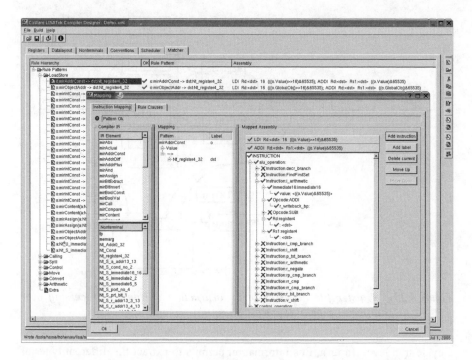

Figure 8.26   GUI for interactive compiler retargeting (code selector view)

C/C++ specification as in the LISA behaviour models, this part of backend retargeting is least automated. Instead, the GUI offers the designer a mapping dialogue (Figure 8.26, see also Reference 37) that allows for a manual specification of mapping rules. This dialog enables the 'drag-and-drop' composition of mapping rules, based on (a) the IR operations needed to be covered for a minimal operational compiler and (b) the available LISA operations. In the example from Figure 8.26, an address computation at the IR level is implemented with two target-specific instructions (LDI and ADDI) at the assembly level.

Although significant manual retargeting effort is required with this approach, it is much more comfortable than working with a plain compiler generator such as CoSy (Section 8.3.4), since the GUI hides many compiler internals from the user and takes the underlying LISA processor model explicitly into account, e.g. concerning the correct assembly syntax of instructions. Moreover, it ensures very high flexibility w.r.t. different target processor classes, and the user gets immediate feedback on consistency and completeness of the mapping specification.

The major drawback, however, of the above approach is a potential 'model consistency' problem, since the LISA model is essentially overlayed with a (partially independent) code selector specification. In order to eliminate this problem, yet retaining flexibility, the LISA language has recently been enhanced with 'semantic sections' [38]. These describe the behaviour of operations from a pure compiler

```
OPERATION ADD {
 DECLARE {
  GROUP src1, dst = { reg };
  GROUP src2 = { reg || imm };}
SYNTAX { "add" dst "," src1 "," src2 }
CODING { 0b0000 src1 src2 dst }
BEHAVIOR {
  dst = src1 + src2;
  if (((src1 < 0) && (src2 < 0)) ||
      ((src1 > 0) && (src2 > 0) &&
      (dst < 0)) ||
      ((src1 > 0) && (src2 < 0) &&
      (src1 > -src2)) ||
      (src1 < 0) && (src2 > 0) &&
      (-src1 < src2)))
  { carry = 1; }}}
```

Figure 8.27   Modelling of an add operation in LISA with carry flag generation as a side effect

```
OPERATION ADD {
 DECLARE {
  GROUP src1, dst = { reg };
  GROUP src2 = { reg || imm };}
SYNTAX { "add" dst "," src1 "," src2 }
CODING { 0b0000 src1 src2 dst }
SEMANTICS { _ADDI[_C] ( src1, src2 ) -> dst; }}

OPERATION reg {
 DECLARE {
  LABEL index; }
SYNTAX { "R" index=#U4 }
ODING { index=0bxxxx }
SEMANTICS { _REGI(R[index])<0..31> }}
```

Figure 8.28   Compiler semantics modelling of the add operation from Figure 8.27 and a micro-operation for register file access (micro-operation '_REGI')

perspective and in a canonical fashion. In this way, semantic sections eliminate syntactic variances and abstract from details such as internal pipeline register accesses or certain side effects that are only important for synthesis and simulation.

Figure 8.27 shows a LISA code fragment for an ADD instruction that generates a carry flag. The core operation ('dst = src1 + src2') could be analysed easily in this example, but in reality more C code lines might be required to capture this behaviour precisely in a pipelined model. The carry flag computation is modelled with a separate if-statement, but the detailed modelling style might obviously vary. On the other hand, the compiler only needs to know that (a) the operation adds two registers and (b) it generates a carry, independent of the concrete implementation.

The corresponding semantics model (Figure 8.28) makes this information explicit. Semantics models rely on a small set of precisely defined 'micro-operations' ('_ADDI' for 'integer add' in this example) and capture compiler-relevant side effects with special attributes (e.g. '_C' for carry generation). This is feasible, since the meaning of generating a carry flag (and similar for other flags like zero or sign) in instructions like ADD does not vary between different target processors.

Frequently, there is no one-to-one correspondence between IR operations (compiler dependent) and micro-operations (processor dependent). Therefore, the code selector generator that works with the semantic sections must be capable of implementing complex IR patterns by sequences of micro-operations. For instance, it might be needed to implement a 32-bit ADD on a 16-bit processor by a sequence of an ADD followed by an ADD-with-carry. For this 'lowering', the code selector generator relies on an extensible default library of 'transformation rules'. Vice versa, some LISA operations may have complex semantics (e.g. a DSP-like multiply accumulate) that cover multiple IR operations at a time. These complex instructions are normally not needed for an operational compiler but should be utilised in order to optimise code quality. Therefore, the code selector generator analyses the LISA processor model for such instructions and automatically emits mapping rules for them.

The use of semantic sections in LISA enables a much higher degree of automation in code selector retargeting, since the user only has to provide the semantics per LISA operation, while mapping rule generation is completely automated (except for user interactions possibly required to extend the transformation rule library for a new target processor).

The semantics approach eliminates the above-mentioned model consistency problem at the expense of introducing a potential 'redundancy' problem. This redundancy is due to the co-existence of separate behaviour (C/C++) and semantics (micro-operations) descriptions. The user has to ensure that behaviour and semantics do not contradict. However, this redundancy is easily to deal with in practice, since behaviour and semantics are local to each single LISA operation. Moreover, as outlined in Reference 39, co-existence of both descriptions can even be avoided in some cases, since one can generate the behaviour from the semantics for certain applications.

8.5.4   Results

The retargetable LISATek C compiler has been applied to numerous different processor architectures, including RISC, VLIW and network processors. Most importantly, it has been possible to generate compilers for all architectures with limited effort, the order of some man-weeks, dependent on the processor complexity. This indicates

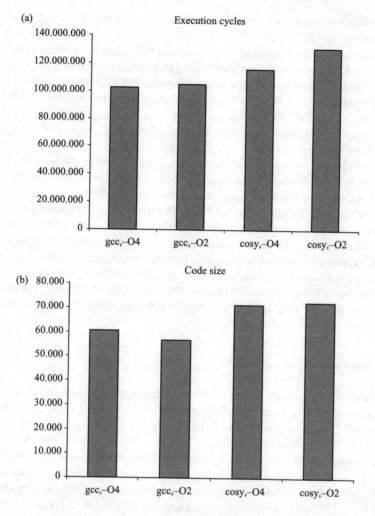

Figure 8.29 Comparison between the gcc compiler and the CoSy-based LISATek C compiler: (a) execution cycles and (b) code size

that the semi-automatic approach outlined in Section 8.5 works for a large variety of processor architectures commonly found in the domain of embedded systems.

While this flexibility is a must for retargetable compilers, code quality is an equally important goal. Experimental results confirm that the code quality is generally acceptable. Figure 8.29 shows a comparison between the gcc compiler (Section 8.3.2) and the CoSy based LISATek C compiler for a MIPS32 core and some benchmark programs. The latter one is an 'out-of-the-box' compiler that was designed within two man-weeks, while the gcc compiler due to its wide use most likely incorporates significantly more manpower. On average, the LISATek compiler shows an overhead of 10 per cent in performance and 17 per cent in code size. With specific compiler optimisations added to the generated backend, this gap could certainly be further narrowed.

Further results for a different target (Infineon PP32 network processor) show that the LISATek compiler generates better code (40 per cent in performance, 10 per cent in code size) than a retargeted lcc compiler (Section 8.3.3), due to more built-in code optimisation techniques in the CoSy platform. Another data point is the ST200 VLIW processor, where the LISATek compiler has been compared to the ST Multiflow, a heavily optimising target-specific compiler. In this case, the measured overhead has been 73 per cent in performance and 90 per cent in code size, which is acceptable for an 'out-of-the-box' compiler that was designed with at least an order of magnitude less time than the Multiflow. Closing this code quality gap would require adding special optimisation techniques, e.g. in order to utilise predicated instructions, which are currently ignored during automatic compiler retargeting. Additional optimisation techniques are also expected to be required for highly irregular DSP architectures, where the classical backend techniques (Section 8.2.3) tend to produce unsatisfactory results. From our experience we conclude that such irregular architectures can hardly be handled in a completely retargetable fashion, but will mostly require custom optimisation engines for highest code quality. The LISATek/CoSy approach enables this by means of a modular, extensible compiler software architecture, naturally at the expense of an increased design effort.

8.6   Summary and outlook

Motivated by the growing use of ASIPs in embedded SoCs, retargetable compilers have made their way from academic research to EDA industry and application by system and semiconductor houses. While still being far from perfect, they increase design productivity and help to obtain better quality of results. The flexibility of today's retargetable compilers for embedded systems can be considered satisfactory, but more research is required on how to make code optimisation more retargetable.

We envision a pragmatic solution where optimisation techniques are coarsely classified w.r.t. different target processor families, e.g. RISCs, DSPs, NPUs and VLIWs, each of which shows typical hardware characteristics and optimisation requirements. For instance, software pipelining and utilisation of SIMD (single instruction multiple data) instructions are mostly useful for VLIW architectures, while DSPs require address code optimisation and a closer coupling of different backend phases. Based on a target processor classification given by the user w.r.t. the above categories, an appropriate subset of optimisation techniques would be selected, each of which is retargetable only within its family of processors.

Apart from this, we expect a growing research interest in the following areas of compiler-related EDA technology.

Compilation for low power and energy: Low power and/or low-energy consumption have become primary design goals for embedded systems. As such systems are more and more dominated by software executed by programmable embedded processors, it is obvious that compilers may also play an important role, since they control the code efficiency. At first glance, it appears that program energy minimisation is identical to performance optimisation, assuming that power consumption is approximately constant over the execution time. However, this is only a rule-of-thumb, and the use of fine-grained instruction-level energy models [40,41] shows that there can be a trade-off between the two optimisation goals, which can be explored with special code generation techniques. The effect is somewhat limited, though, when neglecting the memory subsystem, which is a major source of energy consumption in SoCs. More optimisation potential is offered by exploitation of SoC (scratch-pad) memories, which can be treated as entirely compiler-controlled, energy efficient caches. Dedicated compiler techniques, such as [42,43], are required to ensure an optimum use of scratchpads for program code and/or data segments. See Part II, Chapter 9 for more details on 'software power optimisation'.

Source-level code optimisation: In spite of powerful optimising code transformations at the IR or assembly level, the resulting code can be only as efficient as the source code passed to the compiler. For a given application algorithm, an infinite number of C code implementations exist, possibly each resulting in different code quality after compilation. For instance, downloadable reference C implementations of new algorithms are mostly optimised for readability rather than performance, and high-level design tools that generate C as an output format usually do not pay much attention to code quality. This motivates the need for code optimisations at the source level, e.g. C-to-C transformations, that complement the optimisations performed by the compiler, while retaining the program semantics. Moreover, such C-to-C transformations are inherently retargetable, since the entire compiler is used as a backend in this case. Techniques like References 44–46 exploit the implementation space at the source level to significantly optimise code quality for certain applications, while tools like PowerEscape [47] focus on efficient exploitation of the memory hierarchy in order to minimise power consumption of C programs.

Complex application-specific machine instructions: Recent results in ASIP design automation indicate that a high performance gain is best achieved with complex application-specific instructions that go well beyond the classical custom instructions like multiply accumulate for DSPs. While there are approaches to synthesising such custom instructions based on application code analysis [48–50], the interaction with compilers is not yet well understood. In particular, tedious manual rewriting of the source code is still required in many cases to make the compiler aware of new instructions. This slows down the ASIP design process considerably, and in an ideal environment the compiler would automatically exploit custom instruction set extensions. This will require generalisation of classical code selection techniques to cover more complex constructs like directed acyclic graphs or even entire program loops.

References

1   AHO, A.V., SETHI, R., and ULLMAN, J.D.: 'Compilers – principles, techniques, and tools' (Addison-Wesley, Reading, MA, 1986)
2   APPEL, A.W.: 'Modern compiler implementation in C' (Cambridge University Press, Cambridge, 1998)

3   MUCHNIK, S.S.: 'Advanced compiler design & implementation' (Morgan Kaufmann Publishers, San Francisco, CA, 1997)
4   WILHELM, R., and MAURER, D.: 'Compiler design' (Addison-Wesley, Reading, MA, 1995)
5   MASON, T., and BROWN, D.: 'Lex & yacc' (O'Reilly & Associates, 1991)
6   BISCHOFF, K.M.: 'Design, implementation, use, and evaluation of Ox: An attribute-grammar compiling system based on Yacc, Lex, and C'. Technical Report 92-31, Department of Computer Science, Iowa State University, 1992
7   LEUPERS, R., WAHLEN, O., HOHENAUER, M., KOGEL, T., and MARWEDEL, P.: 'An executable intermediate representation for retargetable compilation and high-level code optimization'. International workshop on *Systems, architectures, modeling, and simulation* (SAMOS), Samos (Greece), July 2003
8   AHO, A., JOHNSON, S., and ULLMAN, J.: 'Code generation for expressions with common subexpressions', *Journal of the ACM* 1977, **24**(1)
9   FRASER, C.W., HANSON, D.R., and PROEBSTING, T.A.: 'Engineering a simple, efficient code generator', *ACM Letters on Programming Languages and Systems*, 1992, **1**(3)
10 BASHFORD, S., and LEUPERS, R.: 'Constraint driven code selection for fixed-point DSPs'. 36th *Design automation* conference (DAC), 1999
11 ERTL, M.A.: 'Optimal code selection in DAGs'. ACM symposium on *Principles of programming languages (POPL)*, 1999
12 CHOW, F., and HENNESSY, J.: 'Register Allocation by Priority-Based Coloring', *SIGPLAN Notices*, 1984, **19**(6)
13 BRIGGS, P.: 'Register allocation via graph coloring'. Ph.D. thesis, Department of Computer Science, Rice University, Houston/Texas, 1992
14 LIAO, S., DEVADAS, S., KEUTZER, K., TJIANG, S., and WANG, A.: 'Storage assignment to decrease code size'. ACM SIGPLAN conference on *Programming language design and implementation (PLDI)*, 1995
15 LAM, M.: 'Software pipelining: an effective scheduling technique for VLIW machines'. ACM SIGPLAN conference on *Programming language design and implementation (PLDI)*, 1988
16 WAGNER, J., and LEUPERS, R.: 'C compiler design for a network processor', *IEEE Transactions on CAD*, 2001, **20**(11)
17 WILSON, T., GREWAL, G., HALLEY, B., and BANERJI, D.: 'An integrated approach to retargetable code generation'. Seventh international symposium on *High-level synthesis (HLSS)*, 1994
18 LEUPERS, R., and MARWEDEL, P.: 'Retargetable compiler technology for embedded systems – tools and applications' (Kluwer Academic Publishers, ISBN 0-7923-7578-5, 2001)
19 MARWEDEL, P., and NOWAK, L.: 'Verification of hardware descriptions by retargetable code generation'. 26th *Design automation* conference, 1989
20 LEUPERS, R., and MARWEDEL, P.: 'Retargetable code generation based on structural processor descriptions', in 'Design Automation for Embedded Systems', Vol. 3, no. 1 (Kluwer Academic Publishers, 1998)

21 LEUPERS, R., and MARWEDEL, P.: 'A BDD-based frontend for retargetable compilers'. European *Design & test* conference (ED & TC), 1995
22 VAN PRAET, J., LANNEER, D., GOOSSENS, G., GEURTS, W., and DE MAN, H.: 'A graph based processor model for retargetable code generation'. European *Design and test* conference (ED & TC), 1996
23 Target Compiler Technologies: http://www.retarget.com.
24 MISHRA, P., DUTT, N., and NICOLAU, A.: 'Functional abstraction driven design space exploration of heterogenous programmable architectures'. International symposium on *System synthesis (ISSS)*, 2001
25 Free Software Foundation/EGCS: http://gcc.gnu.org
26 FRASER, C., and HANSON, D.: 'A retargetable C compiler: design and implementation' (Benjamin/Cummings, 1995)
27 FRASER, C., and HANSON, D.: LCC home page, http://www.cs.princeton.edu/software/lcc
28 Associated Compiler Experts: http://www.ace.nl
29 ORAIOGLU, A., and VEIDENBAUM, A.: 'Application specific microprocessors (guest editors' introduction)', *IEEE Design & Test Magazine*, Jan/Feb 2003
30 Tensilica Inc.: http://www.tensilica.com
31 Stretch Inc.: http://www.stretchinc.com
32 CoWare Inc.: http://www.coware.com
33 HOFFMANN, A., MEYR, H., and LEUPERS, R.: 'Architecture exploration for embedded processors with LISA' (Kluwer Academic Publishers, ISBN 1-4020-7338-0, 2002)
34 SCHARWAECHTER, H., KAMMLER, D., WIEFERINK, A. *et al.*: 'ASIP architecture exploration for efficient IPSec encryption: a case study'. International workshop on *Software and compilers for embedded systems (SCOPES)*, 2004
35 NOHL, A., BRAUN, G., SCHLIEBUSCH, O., HOFFMANN, A., LEUPERS, R., and MEYR, H.: 'A universal technique for fast and flexible instruction set simulation'. 39th *Design automation* conference (DAC), New Orleans (USA), 2002
36 WAHLEN, O., HOHENAUER, M., BRAUN, G. *et al.*: 'Extraction of efficient instruction schedulers from cycle-true processor models'. Seventh international workshop on *Software and compilers for embedded systems (SCOPES)*, 2003
37 HOHENAUER, M., WAHLEN, O., KARURI, K. *et al.*: 'A methodology and tool suite for C compiler generation from ADL processor models'. *Design automation & test in Europe* (DATE), Paris (France), 2004
38 CENG, J., SHENG, W., HOHENAUER, M. *et al.*: 'Modeling instruction semantics in ADL processor descriptions for C compiler retargeting'. International workshop on *Systems, architectures, modeling, and simulation (SA-MOS)*, 2004
39 BRAUN, G., NOHL, A., SHENG, W. *et al.*: 'A novel approach for flexible and consistent ADL-driven ASIP design'. 41st *Design automation* conference (DAC), 2004

40   LEE, M., TIWARI, V., MALIK, S., and FUJITA, M.: 'Power analysis and minimization techniques for embedded DSP software', *IEEE Trans. on VLSI Systems*, 1997, **5**(2)

41   STEINKE, S., KNAUER, M., WEHMEYER, L., and MARWEDEL, P.: 'An accurate and fine grain instruction-level energy model supporting software optimizations'. *PATMOS*, 2001

42   STEINKE, S., GRUNWALD, N., WEHMEYER, L., BANAKAR, R., BALAKRISHNAN, M., and MARWEDEL, P.: 'Reducing energy consumption by dynamic copying of instructions onto onchip memory'. *ISSS*, 2002

43   KANDEMIR, M., IRWIN, M.J., CHEN, G., and KOLCU, I.: 'Banked scratch-pad memory management for reducing leakage energy consumption'. *ICCAD*, 2004

44   FALK, H., and MARWEDEL, P.: 'Control flow driven splitting of loop nests at the source code level'. *Design automation and test in Europe* (DATE), 2003

45   LIEM, C., PAULIN, P., and JERRAYA, A.: 'Address calculation for retargetable compilation and exploration of instruction-set architectures'. 33rd *Design automation* conference (DAC), 1996

46   FRANKE, B., and O'BOYLE, M.: 'Compiler transformation of pointers to explicit array accesses in DSP applications'. International conference on *Compiler construction (CC)*, 2001

47   PowerEscape Inc.: http://www.powerescape.com

48   SUN, F., RAVI, S. *et al.*: 'Synthesis of custom processors based on extensible platforms'. *ICCAD*, 2002

49   GOODWIN, D., and PETKOV, D.: 'Automatic generation of application specific processors'. *CASES*, 2003

50   ATASU, K., POZZI, L., and IENNE, P.: 'Automatic application-specific instruction-set extensions under microarchitectural constraints'. *DAC*, 2003

Chapter 9

Software power optimisation

J. Hu, G. Chen, M. Kandemir and N. Vijaykrishnan

9.1 Introduction

Over the last decade, numerous software techniques have been proposed to improve the power efficiency of the underlining systems (Figure 9.1). These techniques differ from each other in various ways. Some of them employ compiler techniques to reduce the dynamic power consumption of processors by utilising voltage scaling, while some of the others focus on reducing static power consumption for memories by tuning the operating system (OS) or Java Virtual Machine (JVM). On the other hand, many of these techniques also share similarities. For example, there are a large number of power optimisation techniques targeting at memories, though they achieve their goals via different alternatives. There are also many techniques that improve power efficiency by controlling power modes of different components in the system. In short, two software power optimisation techniques could be similar from one aspect, but they could also be totally different if compared from another aspect. To better understand current software power optimisation techniques, we need to study them from different perspectives. In general, current techniques can be divided into different categories from four perspectives: power mode utilisation, operating level, power type and targeted components. Figure 9.2 gives a summary of these four perspectives.

Different operating modes (power modes/energy modes) are usually available in a power-aware system. Each mode consumes a different amount of power. This provision is available in processors, memory, disks and other peripherals. While these power-saving modes are extremely useful during idle periods, one has to pay a cost of exit latency for these hardware entities to transition back to the operational (active) state once the idle period is over. Software power optimisation techniques usually try to utilise as few hardware components as possible without paying performance penalties, and transition the rest into a power-conserving operating mode. Depending on how they utilise the operating modes, these techniques can be divided

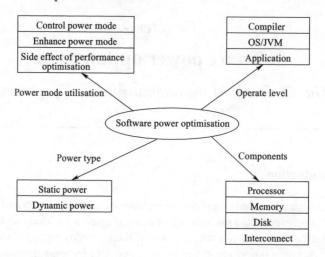

| Software |
| Architecture |
| Circuit |
| Process |

Figure 9.1   Power optimisation levels

Control power mode		Compiler
Enhance power mode		OS/JVM
Side effect of performance optimisation		Application

Power mode utilisation   Operate level

Software power optimisation

Power type   Components

Static power		Processor
Dynamic power		Memory
		Disk
		Interconnect

Figure 9.2   Dividing software power optimisation techniques from four perspectives

into three groups: control power modes, enhance power modes and side effect of performance optimisation. Some techniques achieve power savings by controlling the power modes without transforming the original program. That is, other than the additional power control instructions, the original program is not modified. For example, Hsu and Kremer [1] proposed a compiler algorithm to optimise programs for power usage using dynamic voltage scaling. This algorithm identifies program regions where the CPU can be slowed down with negligible performance loss, and inserts power control instructions for such regions to save power. Some techniques take a more aggressive step in utilising the power modes. They transform the original program so that the existing power modes can be utilised more efficiently. Zhang *et al.* [2] identified that significant additional power savings can be obtained when loop optimisation techniques such as loop fusion or loop fission are applied before inserting power control instructions into the original program. Power savings can also be obtained as a side-effect of performance-oriented optimisations (without utilising power modes). Kandemir *et al.* [3] evaluated power impact of several popular loop optimisation techniques that are used for performance optimisation. They found that most performance-oriented optimisations reduce the overall power consumption even though they tend to increase the data-path energy.

By looking at what level they operate on, current software power optimisation techniques can be put into three different categories: compiler, OS/JVM and application. Compiler techniques gather information from programs through program analysis tool, and then improve the power efficiency by transforming the program or inserting power mode control instructions at appropriate points in the program [1,2,4–6]. For example, the compiler can identify the last use of the instructions and places the corresponding cache lines that contain them into a low leakage mode to save power [2]. In another example, the compiler identifies the number of functional units to activate for different loops in the application, and then applies a leakage control mechanism to the unused units based on such compiler analysis [5]. OS/JVM techniques monitor the runtime behaviour of the applications and make decisions dynamically based on fixed or adaptive strategies for controlling the system power modes. For example, by tuning the frequency of invoking garbage collection based on object allocation and garbage creation rate, JVM can turn off more unused memory banks and save a significant amount of power consumption [7]. In a system where disks consume a large portion of total power, OS can predict session lengths based on previous disk accesses and shutdown disks between sessions adaptively to save power [8]. Both compiler and OS/JVM techniques are general in the sense that they can be applied to any application in the system. In comparison, there are also some techniques that target specific applications. For example, Rabiner and Chandrakasan [9] developed a power-efficient motion estimation algorithm used in power-constrained devices. Such application-specific techniques usually achieve better power savings than a general technique, but their applicability is limited to a small group of applications.

The main sources of power consumption in current complementary metal oxide semiconductor (CMOS)-technology-based circuits are dynamic power and static power (also called leakage power) [10]. Dynamic power is consumed whenever a component is utilised. In contrast, static power is consumed as long as power supply is maintained and is independent of component activity. Software power optimisation techniques can be classified into two groups based on whether a technique optimises dynamic power consumption or static power consumption. Dynamic voltage scaling [1,6] is recognised as one of the most effective techniques for reducing dynamic power consumption. It exploits the fact that a major portion of power of CMOS circuitry scales quadratically with the supply voltage. As a result, lowering the supply voltage can significantly reduce power dissipation. In Reference 9, the dynamic power consumption of the portable device is reduced by offloading some portion of the work to a high-powered base-station. Data locality optimisation techniques such as loop tiling reduces dynamic power consumption through the reduction of memory accesses and the associated power consumption. While static power consumption used to be an insignificant part of the overall power consumption, due to the continued shrinking of transistors and associated technology changes, static power consumption has become an important portion of overall power consumption. In fact, it is shown to be the dominant part of the chip power budget for 0.10 μm technology and below [11]. Most software power optimisation techniques reduce static power consumption by putting idle components into 'off' or 'sleep' mode in which static power consumption is much lower than in active mode. For example, a compiler can select a suitable instruction per cycle (IPC) for each loop, and then turn off unused or not frequently used integer

Table 9.1   Nine techniques and their categories

Technique	Power mode utilisation	Operating level	Power type	Targeted components
Hsu and Kremer [1]	Control power mode	Compiler	Dynamic	Processor
Xie *et al.* [6]	Control power mode	Compiler	Dynamic	Processor
Delaluz *et al.* [4]	Enhance power mode	Compiler	Static	Memory
Chen *et al.* [7]	Enhance power mode	OS/JVM	Static	Memory
Lu and Micheli [8]	Control power mode	OS/JVM	Dynamic	Disk
Zhang *et al.* [2]	Enhance power mode	Compiler	Static	Memory
Rabiner and Chandrakasan [9]	Side effect	Application	Dynamic	Processor
Kim *et al.* [5]	Enhance power mode	Compiler	Static	Processor
Kandemir *et al.* [3]	Side effect	Compiler	Dynamic	Memory and processor

ALUs to save static energy [5]. In a JVM environment, unused memory banks can be turned off to save static energy, and such benefits can be increased by carefully selecting an appropriate garbage collection frequency such that the idle time of unused memory banks is increased.

Software power optimisation techniques can also be categorised by the components they are applied to. Among the various components in a system, the major power consumers are memory, processor, disk and interconnect. There is much prior research focusing on memory [2,4,7], processor [1,5,6,9], disk [8] and interconnect [12]. Table 9.1 gives some related work on software power optimisations and their categories according to Figure 9.2.

In the following sections, we will present examples of software power optimisations. First, we will show how algorithmic decisions influence energy efficiency. Then, we will show how traditional compiler optimisations can affect power consumption. The rest of the chapter focuses on the use of software to control hardware configuration, power mode settings and enhancement of energy savings through power mode setting. These sections include examples from related work of other researchers in literature for pedagogic reasons.

9.2   Influence of algorithms on energy efficiency

Arguably, application-level energy optimisation has the most significant impact on system energy efficiency. Different algorithm implementations can have huge differences in the way in which computation, communication and data transferring are carried out in a given computing platform, leading to vastly different power/energy consumption. There are two possible directions for application-level energy optimisation. The first one is to develop a different algorithm to perform the exactly same task with much higher energy efficiency. For example, the computational complexity of different sorting algorithms will result in different energy consumption even though they achieve the same functionality. When these algorithms were used to sort 100 integers on SimplePower, an energy estimation tool, quick sort over bubble sort and heap sort reduced the datapath energy consumption by 83 per cent and 30 per cent [13]. In this case, energy reduction directly stems as a side-effect of performance optimizations.

The second option is to trade computational accuracy and throughput for energy savings. The second option is highly desirable where the energy is a major resource constraint and the computational accuracy can be sacrificed to prolong the system lifetime limited by energy source, e.g. battery-powered embedded systems for signal and image processing. Energy–quality (E–Q) scalability is used in Sinha *et al.*'s work [14] as a metric to evaluate the energy scalability of algorithms in terms of computational accuracy and energy consumption. E–Q performance is the quality of computation achieved for a given energy constraint. A better E–Q performance approaches to the maximum computational quality at less energy consumption. A good example to show the difference of algorithm E–Q performance is how to compute a simple power series as follows [14],

$$f(x) = 1 + k_1 x + k_2 x^2 + \cdots + k_N x^N$$

A straightforward algorithm to compute this power series uses an N-iteration loop where each term of $f(x)$ is evaluated in a single iteration and the computation proceeds from left to right in the expression. If the energy consumption for each loop iteration is assumed the same and k_i's values are similar, for a given $x = 2$, the computation till $N - 1$ iteration consumes $(N - 1)/N$ of the total energy of the loop while only achieving 50 per cent of the computation accuracy since $2^N / f(2) \approx 1/2$ [14].

However, if we change the computation in the reverse way, the first iteration computes $k_{N-1} x^{N-1}$, the first couple of iterations will approach the maximum accuracy at the minimum energy consumption (e.g. $2/N$ of the total energy, omitting other overheads). Such a computation delivers much better E–Q performance. The observation here is that the ordering of this power series evaluation to achieve higher E–Q performance is determined by the value of x. If $x > 1.0$, the reverse ordering should be selected. Otherwise, the original ordering performs better.

The implication of designing such an energy-scalable algorithm is the potential that the computation can be controlled to achieve significant energy reduction while guaranteeing acceptable accuracy. Figure 9.3 shows quality (Q) as a function of energy (E) where the transformed algorithm exhibits a much better E–Q scalability than the original one. To achieve the same computational quality, the transformed algorithm consumes much less energy than the original algorithm does since $E \ll E'$.

To design energy-scalable algorithms such as finite impulse response (FIR) filtering for digital signal processing (DSP) and discrete cosine transform (DCT) for image decoding in energy-constrained embedded systems, we can reorder the multiply-accumulate (MAC) operations in such a way that components with large weights will be first evaluated to approach the maximum accuracy (quality) as soon as possible. In case of energy shortage, the remaining computation component (e.g. MACs) with

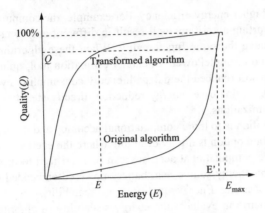

Figure 9.3   The E–Q performance of different algorithms

small weights can be omitted for energy savings. Such an algorithm also provides the system with the adaptability and flexibility to dynamically adjust the quality of services (e.g. computation, communication) based on the remaining power level of the battery.

9.3   Impact of compiler code optimisation

Many performance optimisations (such as optimal register allocation, redundant subexpression elimination and loop strength reduction) performed by modern compilers can also reduce energy consumption. For example, a compiler can improve the performance of a program by storing a frequently used value in a register instead of the main memory. Since accessing the register consumes less energy than accessing the main memory, this optimisation also reduces energy consumption. However, not all performance-oriented optimisations are beneficial from the energy perspective. An example of this case is compiler-guided speculative cache prefetching [15]. That is, the compiler predicts the data that will be used in the near further and inserts the instructions into certain points of the program to load these data into the cache in advance of their actual use. This optimisation reduces the overall execution time by overlapping the cache fetching time with the program execution time. However, such overlapping does not hide the energy consumed for fetching the data, and, consequently, it does not necessarily reduce the overall energy consumption.

Being one of the most important forms of compiler optimisations, loop transformations are widely used by the compiler, especially in array-intensive code, to improve the quality of the code, e.g. code transformations to improve data cache locality [16] for performance enhancement. In the rest of this section, we introduce some typical loop nest optimisations, i.e. linear loop transformations, tiling, unrolling, fusion and fission, and discuss their impact on power consumption.

Linear loop transformations attempt to improve cache performance, instruction scheduling and iteration-level parallelism by modifying the traversal order of the iteration space of the loop nest. The simplest form of loop transformation, called loop interchange, can improve data locality (cache utilisation) by changing the order of the loops. From the power consumption point of view, by applying this transformation we can expect a reduction in the total memory power due to better utilisation of the cache. For the power consumed in other parts of the system, if loop transformations can result in complex loop bounds and array subscript expressions that can potentially increase the power consumed in the core datapath.

Another important technique used to improve cache performance is blocking, or tiling. When it is used for cache locality, arrays that are too big to fit in the cache are broken up into smaller pieces (to fit in the cache). When we consider power, potential benefits from tiling depend on the changes in power dissipation induced by the optimisation on different system components. We can expect a decrease in power consumed in memory, due to better data reuse [17]. On the other hand, in the tiled code, we traverse the same iteration space of the original code using twice as many loops (in the most general case); this entails extra branch-control operations and macro calls. These extra computations might increase the power dissipation in the core.

Loop unrolling reduces the trip count of a given loop by putting more work inside the nest with the aim of reducing the number of memory accesses and promoting the register reuse. From the power point of view, fewer accesses to the memory means less power dissipation. In addition, we can also expect a reduction in the power consumed in the register file and data buses.

Loop fusion combines two loops into one loop. Since it improves data reuse, it can reduce the power consumed in the memory system. And, if used with scalar replacement, it can eliminate a number of memory references again reducing the memory system power. Also, since it reduces the number of loop nests, it eliminates lots of branch instructions that would otherwise contribute to a significant percentage of the core power.

In loop fission, the compiler breaks down a loop into two (or more) separate loops. When done solely for optimising the iteration scheduling, this transformation can increase power consumption in memory system and elsewhere due to an increase in the number of loop nests and a decrease in temporal locality. However, combining with low power mode available in cache or memory systems, loop fission has the potential to improve the opportunity for power mode control.

9.4 Software-controlled voltage scaling

Dynamic voltage scaling (DVS) is one of the most effective power reduction techniques. Lowering the supply voltage can reduce power dissipation since CMOS circuitry power quadratically scales with the supply voltage. Reduction of supply voltages also helps to reduce leakage power. See Part III, Chapter 11 for more information on DVS enabled systems.

Voltage scaling has been applied extensively in real-time software. Many of these applications are deadline-oriented in their execution. While completing an application before a specified deadline, there are no significant advantages to completing them early before the deadline. Software-oriented voltage scaling techniques exploit this property. If there is enough slack for a given task to be completed before its deadline, the supply voltage of the system is reduced to prolong the duration of the task to reduce the slack. An effective DVS algorithm is one that intelligently determines when to adjust the current frequency–voltage setting (scaling points) and to which frequency–voltage setting (scaling factors), so that considerable savings in energy can be achieved while the required performance is still delivered. Designing a good DVS algorithm is not an easy task. The overheads of transitions to and from different frequency–voltage settings may wipe out the benefits of DVS.

In addition to setting the voltage of a task, the available slack of different tasks can be varied by the software to enhance the overall effectiveness of power savings. For example, scheduling techniques can order the execution of the different tasks to maximise the opportunities for energy savings. Consider the tasks with deadlines and durations given in Table 9.2 to illustrate potential benefits from voltage scaling.

In the normal case without voltage scaling the tasks A and B would have completed within 7 units of time. In order to reduce the power, we can prolong the execution of B from 2 units of time to 5 units of time. This can be achieved by reducing the voltage of the system when operating on task B. In a practical system, it may be difficult to exactly make use of the entire slack due to the overheads of transitioning from one supply voltage to another. If the energy overhead due to the voltage transitioning to lower voltage and back is higher, it is not worthwhile to perform the transition.

Next, let us consider the scheduling of C and D. If C is scheduled as soon as it arrives at time unit 10, then neither task C nor task D can be operated at a lower voltage. However, if task C is not scheduled after task D, then task D can be prolonged from 3 units of time to 4 units of time. This illustrates a case where scheduling can provide energy savings.

Running as slowly as possible will minimise the CPU energy consumption but increase the overall system energy consumption, because the power-consuming non-CPU devices stay on longer. Consequently, when applying such voltage scaling approaches it is important to consider the impact on the rest of the system. Hsu and Kremer [1] proposed a compiler algorithm that addressed the aforementioned design issues. The idea is to identify the program regions in which the CPU is

Table 9.2   Example schedule

Task	Arrival time	Duration	Deadline
A	0	5	10
B	0	2	10
C	10	2	17
D	11	3	15

mostly idle due to memory stalls, and slow them down for energy reduction. On an architecture that allows the overlap of the CPU and memory activities, such slow-down will not result in serious performance degradation. As a result, it alleviates the situation where non-CPU systems components become dominant in the total system energy usage.

Voltage scaling approaches have also been applied in the context of on-chip multiprocessors. In Reference 18, the compiler takes advantage of heterogeneity in parallel execution between the loads of different processors and assigns different voltages/frequencies to different processors, if doing so reduces energy consumption without increasing overall execution cycles significantly. For example, consider a loop that is parallelised across four processors such that processors P_1, P_2, P_3 and P_4 have different numbers of instructions to execute. Consider that the loop is par-allelised in such a fashion that they have no dependencies in the four parallel code segments and that all processors need to synchronise before proceeding to execute the rest of the code. In such a case, the processors can be assigned a voltage to stretch the execution of lighter loads to match that of the processor having the heaviest load of instructions to execute. Given that the compiler can estimate the workload statically, it can also be used to set the voltage levels for the different processors. Since switching from one voltage level to another incurs performance and energy overheads, the software can also be used to enhance voltage reuse. For example, if processor P_1 is assigned a voltage V_1 when executing loop 1, the compiler can attempt to assign a parallel code segment such that P_1 can execute loop 2 fragment with the same voltage V_1.

9.5 Compiler-directed resource reconfiguration

Conventional computer systems are often designed with respect to the worst case in order to achieve high performance in all cases. Such a design philosophy is an overkill for most applications. For instance, a register file in an 8-wide superscalar microprocessor will require 16 read ports (assuming maximum two source operands per instruction) and 8 write ports to avoid any structural hazards in the worst case. Due to the large number of read/write ports, the register file will not only incur huge area overhead, but also significantly increase its dynamic and leakage power consumption. However, a majority of source operands (up to 66 per cent reported in Reference 19) are obtained from the bypassing network rather than the register file. Such an obser-vation can be utilised to design an energy-efficient register file with a significantly reduced number of read ports while only incurring a negligible performance loss [19]. Throughout the processor chip, most components/logic blocks are designed in such a way and put unnecessary burdens on transistor/die area budget and power budget. From power-saving perspective, reconfiguring the major processor blocks, or called resources such as register file, functional unit pool and caches to varying requirements of the application performance behaviour by using least resources is a very effective approach. Notice that reconfiguration has both performance and power overhead.

There are at least two possible ways to perform this dynamic resources reconfiguration: dynamic schemes and compiler schemes. Dynamic schemes monitor the application behaviour at runtime and use prediction schemes to direct resource reconfiguration when dynamic behaviour changes are detected. On the other hand, compiler schemes analyse the application program code and identify the discrepancy in resource demands between code segments. Then the compiler decides whether it is beneficial (e.g. for power savings) to reconfigure resources at that point. If so, a special instruction is inserted at that point by the compiler to perform dynamic reconfiguration at runtime. In general, dynamic schemes are more straightforward in implementation and can be accurate in detecting program behaviour changes with additional hardware. However, they need a profiling/learning period before a reconfiguration can be decided, which may delay an early reconfiguration. Further, a dynamic scheme will need additional time or multiple reconfigurations to reach a possible optimal configuration, which may further reduce the possible benefits. In contrast, the execution of compiler-inserted reconfiguration instructions reconfigures specific processor resources. The quality of the decision is determined by how well the compiler analyses the code. The insertion and execution of these special instructions incur code size and performance overhead. Compiler schemes also require instruction set architecture (ISA) extension to include this reconfiguration instruction. In this section, we will focus on compiler schemes for dynamic resource reconfiguration to reduce the power consumption in processors.

The granularity of reconfigurable resources can vary from the whole processor datapath pipeline, to macro blocks such as instruction issue queue, register file and caches, or to blocks such as functional units. For example, if the compiler finds out the instruction level parallelism of a particular large code segment is very low due to data dependences, the compiler can insert a special instruction right before the code segment to either convert the datapath from out-of-order issue mode to in-order issue mode, or shut off one or more clusters in a clustered datapath design, or shut off a portion of the instruction issue queue and functional units to achieve power savings. For signal or image processing applications mostly dealing with short-width values, the compiler can reconfigure the functional units by shutting off the high-bit portion not used and schedule short-value operations to those functional units. This is especially useful in VLIW (very long instruction word) architectures. The granularity of resource reconfiguration varies from applications, to function calls within a particular application, to loops within a function and even to much smaller code structures such as basic blocks. Figure 9.4 shows the points where compiler-directed resource reconfiguration can be invoked at different granularity.

In the rest of this section, we describe Hu *et al.*'s work [20] to show how the compiler can direct dynamic data cache reconfiguration at the granularity of loop boundaries for power saving. Caches are critical components in processors that form a memory hierarchy to bridge the increasing speed gap between the CPU and main memory (DRAM). Modern microprocessors, for both low-end embedded systems and high-end servers, dedicate a large portion of transistor budget and die area to cache memories. Consequently, on-chip caches contribute a significant share of the total processor power consumption, e.g. the instruction cache and data cache consume

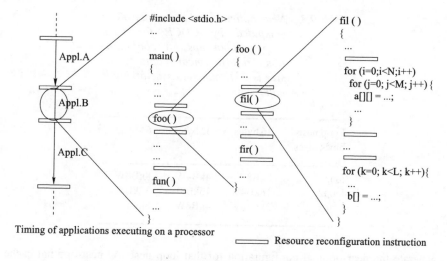

Timing of applications executing on a processor

Resource reconfiguration instruction

Figure 9.4   Compiler-directed resource reconfiguration at different granularities

43 per cent of the power budget in StrongARM SA-110 [21], and 22 per cent of the processor power in PowerPC [22]. Obviously, optimising power efficiency in caches is of first-class importance in designing low power systems.

In array-intensive code such as for signal and image processing applications, loop nests dominate the performance and power behaviour. Since each loop nest may manipulate different arrays in different ways, the cache behaviour of loop nests, which is determined by the footprints of the data manipulated by the loop nest, might be very different from each other. A conventional fixed-size data cache cannot adapt to the cache demands of different loop nests in array-intensive code to achieve power efficiency. Larger cache (in both cache size and associativity) reduces cache conflict misses, the main source of cache misses in array code, leading to better performance (assuming no impact on the clock cycle time). However, large caches increase the overall power consumption due to the large per access power consumption. On the other side, smaller cache than the one demanded by the loop nest will significantly increase cache conflict misses and cause accesses to the lower-level large memory. Besides the considerable performance loss in small cache configuration, the power overhead due to accessing the lower level caches or main memory may also be huge. The best solution is to find the right data cache configuration for each loop nest and reconfigure the data cache at loop nest boundary for both higher performance and higher power efficiency. Hu *et al.*'s work [20] provides such a solution, called compiler-directed cache polymorphism (CDCP).

The key idea in CDCP is to identify the reuse space of each array for a given loop nest and then to determine a near-optimal cache configuration that realises the data reuses into data localities. CDCP first simulates the interferences (conflicts) in reuse spaces among arrays and calculate an initial cache configuration based on these interference statistics and array sizes, followed by an optimising process to

Table 9.3   *Near-optimal cache configurations computed by CDCP for each loop nest in aps. A configuration '2kB8W' means data cache size is 2 kB and set associativity is eight ways*

Loop nests/ block sizes	16 bytes	32 bytes	64 bytes
loop1	2kB8W	4kB4W	8kB8W
loop2	16kB4W	16kB8W	32kB8W
loop3	2kB16W	4kB8W	8kB8W

generate the near-optimal configuration for that loop nest. A 'near-optimal cache configuration' is informally defined as the one with the smallest capacity and associativity that approximates the locality that can be obtained using a very large and fully associative cache. Intuitively, any increase in either cache size or associativity over this configuration would not produce any significant improvement. At code generation stage, the compiler uses the near-optimal cache configuration computed for each loop nest to insert a special resource reconfiguration instruction, if necessary, just before the first instruction of that loop. Therefore, with the near-optimal cache configuration, each loop nest can maximise its performance as well as keep energy consumption in control.

An application, *aps*, from Perfect Club for Mesoscale Hydro Model is used here to show how CDCP works. *aps* consists of three loop nest (loop1, loop2 and loop3), which are the kernel code, and manipulates 17 arrays. Given different data cache block sizes (e.g. 16 bytes, 32 bytes or 64 bytes), the near-optimal data cache configurations computed by CDCP for each loop nest are given in Table 9.3. These configurations are then passed to machine-dependent optimisation phases and code-generating stage to direct the compiler to insert data cache reconfiguration instructions (supported by ISA extensions).

The CDCP assumes a underlying reconfigurable cache supporting the dynamic reconfiguration at runtime, which is controlled by reconfiguration instructions setting a specific cache reconfiguration register. The address-bits adjustment for different cache reconfigurations is assumed to be taken care of by the reconfigurable mechanism implemented. As an example, Figure 9.5 shows a reconfigurable cache, where the cache has a total size of 32 kB, block size of 32 B, and set associativity of eight ways. Each cache way is logically divided into eight subarrays supporting fine-grain reconfiguration. One cache reconfiguration register (CRR) is used to control cache access with a particular configuration and can be updated by the special cache reconfiguration instructions inserted by the compiler to perform cache reconfiguration. The output of this reconfiguration register (CRR) controls which cache ways (way selection portion, 8 bits) and which logical subarrays (subarray selection

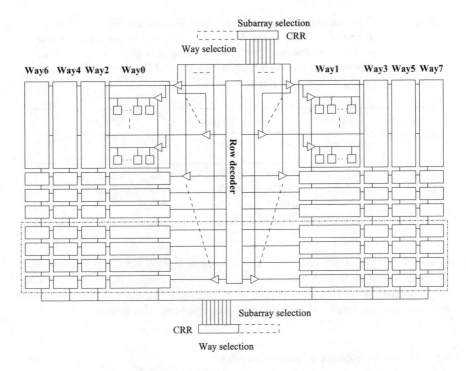

Figure 9.5   A reconfigurable cache supports compiler-directed cache polymorphism (CDCP). This figure only shows the reconfiguration control signals for way selection and subarray selection. The peripheral circuitry is omitted from this logical view

portion, 8 bits) within each way are active (normal mode and ready for access) or inactive (not accessible due to low-power control mode, either Gated_Vdd or DVS). For example, before the execution of loop2 in *aps*, the data cache is reconfigured to 16kB8W (32-byte block size) by the reconfiguration instruction setting all 8 bits (to 1's) in the way selection portion of CRR and only setting the high 4 bits (to 1's) in the subarray selection portion of CRR such that the lower four logical subarrays within each way are disabled and put into inactive mode as shown by the dash-dot line rectangle in Figure 9.5.

How about the performance using the near-optimal cache configuration for each loop nest when compared to some optimal (best performance achievable) configurations? With the same example application *aps*, Table 9.4 shows the data cache miss rate comparisons for each of its three loop nests under different cache block sizes. From this table, we can conclude that this CDCP is very effective in locating the near-optimal cache configuration, especially at moderate cache block sizes, e.g. 32 bytes and 64 bytes. Referring to Figure 9.5, in the execution of *aps*, data cache is reconfigured to 4kB4W for loop1 leading to a dynamic power reduction of up to 87.5 per cent, then reconfigured to 16kB8W for loop2 implying a power

Table 9.4   Data cache performance (miss rate) comparison for each loop nest
in aps. For each cache block size, Optimal is the cache performance using the configuration for best performance and the CDCP
column gives the cache performance using the near-optimal cache
configurations computed by CDCP

Loop nests/ block sizes	16 bytes		32 bytes		32 Bytes	
	Optimal	CDCP	Optimal	CDCP	Optimal	CDCP
loop1	0.0183	0.0183	0.0092	0.0109	0.0047	0.0053
loop2	0.0330	0.0339	0.0165	0.0185	0.0082	0.0088
loop3	0.0413	0.0588	0.0207	0.0207	0.0103	0.0103

reduction of up to 50 per cent, and reconfigured to 4kB8W for loop3 with potential
power saving of up to 87.5 per cent. These numbers are potential power reductions
only considering the data array part in the cache. Performance and power overhead
of cache reconfiguration is likely negligible in array-intensive code.

9.6   Software control of power modes

The software can be used to control the transition of the power mode of the different components based on their anticipated usage during the execution of a program.
Most components support multiple modes for power savings. For example, the processor can be operated at different voltages, the disk can be either spinning or shut
down or the memory can be in various sleep modes.

9.6.1   Memory mode control

Figure 9.6 shows possible power modes for a memory module and transitions between
modes. There are five operating modes: 'active', 'standby', 'napping', 'power-down',
and 'disabled'. Each mode is characterised by its 'power consumption' and the time
that it takes to transition back to the active mode ('resynchronisation time'). The resynchronisation times in cycles are shown along with arrows (assuming a negligible cost ε
for transitioning to a lower power mode).

We will utilise the approach presented by Delaluz et al. [4] to illustrate the use
of software control of power modes. Their approach uses a compiler to issue commands to control the power mode of DRAM modules. The application behaviour is
statically analysed to detect idleness of memory modules for selective power down.
This approach can be considered conservative since memory modules will not be
transitioned to low power modes unless one is absolutely sure that a module will not
be referenced for a while. However, its advantage is that there are no performance
overheads due to resynchronisation. The goal of this compiler-directed mechanism

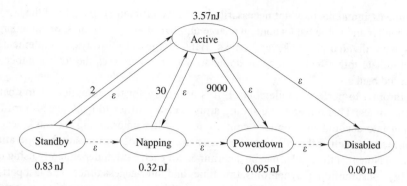

Figure 9.6   Power modes utilised

is to detect idle periods for each memory module, and to transition it into a lower power mode without paying any resynchronisation costs. Consequently, if the inter-access time is T, and the resynchronisation time is T_p (assuming less than T), then the compiler would transition the module into a lower energy mode (with a unit time energy of E_p) for the initial $T - T_p$ period (which would consume a total $\lfloor T - T_p \rfloor E_p$ energy), activate the module to bring it back to the active mode at the end of this period following which the module will resynchronise before it is accessed again (consuming $T_p E_a$ energy during transition assuming that E_a is the unit time energy for active mode as well as during the transition period). As a result, the total energy consumption with this transitioning would be $[T - T_p]E_p + T_p E_a$ without any resynchronisation overheads, while the consumption would have been $T E_a$ if there had been no transitioning. The compiler can evaluate all possible choices based on the mode energy, corresponding resynchronisation times and inter-access time, to select the best choice. Note that the compiler can select different low power modes for different idle periods of the same module depending on the duration of each idle period. When the inter-access time is ∞ (i.e. there is no next access), the module can be put into disabled mode.

The compiler-directed approach uses clustering to group the related (similar life-time access patterns) array variables together so that they can be placed in the same memory modules. This increases the likelihood of transitioning a memory module to a lower energy mode. On the other hand, placing variables that are accessed at different points of the execution in the same module would result in its longer residence in the active mode.

The default allocation of variables is in program declared order. Declaration order of array variables may have nothing to do with their access profiles and life times. Consequently, this order rarely leads to opportunities for effective use of low power operating modes. The strategy here is to analyse the program and determine the arrays with similar access behaviour and use this information to modify the declaration order of array variables so that those with similar access behaviour are declared consecutively. Note that this approach requires minimum modifications to the source code. The disadvantage is that depending on the array and bank sizes, the resulting module assignments may not necessarily be energy-efficient, especially if the arrays are smaller and some banks contain a large number of array variables, or some large arrays are divided across several banks. Bank alignment of arrays is implemented to eliminate this effect as long as doing so does not increases the total number of required banks.

In order to perform mode control, it is necessary for the compiler to find bank access times. This requires translating array access profiles to bank access profiles taking into account the memory configuration. After determining the bank access profile and detecting the idle slots, for each bank we can determine suitable operating modes. Note that the modes can be determined for each bank independently using the energy consumption, resynchronisation times and inter-access times. The last part of the compilation is to insert suitable mode transitioning instructions in the program code. Since most of the optimisations are on array-based applications, it is reasonable to choose the number of loop iterations as the basic unit for measuring time, requiring that all times be converted to iteration counts.

9.6.2   Controlling cache leakage power modes

While optimising dynamic power consumption is crucial in designing future micro-processors, controlling leakage power is arising as the next big challenge in processor design due to the continuous scaling down of the transistor threshold voltages and increasing number of integrated on-chip transistors [23]. See Part III, Chapter 13 for more information on leakage power in nano-scale circuits. There are various power modes that a CMOS circuit can be placed in to reduce leakage power [24]. The software can be used to determine when to make the power mode transitions.

The major component influencing leakage is subthreshold leakage current that is an exponential function of supply voltage V and threshold voltage $(-V_{th})$. Equation (9.1) [11] shows how subthreshold leakage current depends on threshold voltage and supply voltage:

$$I_{sub} = K_1 W e^{-V_{th}/nV_\theta}(1 - e^{-V/V_\theta}) \tag{9.1}$$

K_1 and n are experimentally derived, W is the gate width and V_θ in the exponents is the thermal voltage. At room temperature, V_θ is about 25 mV; it increases linearly as temperature increases. If I_{sub} grows enough to build up heat, V_θ will also start to rise, further increasing I_{sub} and possibly causing thermal runaway. Equation (9.1) suggests two ways to reduce I_{sub}. First, we could turn off the supply voltage – i.e. set V to zero so that the factor in parentheses also becomes zero. Second, we could increase the threshold voltage, which – because it appears as a negative exponent – can have a dramatic effect in even small increments.

Similar to the strategy described for memory mode transitions in the previous section, a compiler can be used to analyse the usage of the different components and set the supply voltage and threshold voltage to reduce leakage power in the idle state. We will consider two examples of leakage power mode savings using compiler analysis.

First, we will consider the compiler-directed instruction cache leakage optimisation proposed in Reference 2. In this approach, the compiler identifies the last use of the instructions and inserts special leakage control instruction to place the corresponding cache line that contains them into a low leakage mode. A compiler-based leakage optimisation strategy makes sense in a VLIW environment where the compiler has control of instruction execution order and is possible to significantly optimise instruction cache leakage energy.

Compiler strategy for cache leakage optimisation requires low-level circuit support, e.g. the dynamic scaling of the supply voltage in this case. As supply voltage to the cache cells reduces, the leakage current reduces significantly due to short-channel effects. The choice of the supply voltage influences whether the data is retained or not. When the normal supply voltage of 1.0 V is reduced below 0.3 V (for a 0.07 μm process), the data in the cells are no longer retained. Thus, 0.3 V supply voltage can be selected for the state-preserving leakage control mode. However, if state preservation is not a consideration, the supply voltage can be switched to 0 V to gain more reduction in energy. Consequently, there are two leakage power states for each cache line, a normal operational mode with high leakage current and a low leakage state in idle mode. The low leakage state can either be state-preserving or destroying based on application needs.

Compiler strategies for instruction cache leakage control can control the leakage state of each cache line through appropriate ISA support. Intuitively, two compiler strategies can be developed for turning off instruction cache lines. The first approach, called the 'conservative strategy', does not turn off an instruction cache line unless it knows for sure that the current instruction that resides in that line is dead (i.e. will not be referenced for the remaining part of the execution). The second approach is called the 'optimistic strategy' and turns off a cache line even if the current instruction instance in it is not dead yet. This might be a viable option if there is a large gap between two successive visits to the cache line. In such a case, the instructions need to be fetched from a lower level memory hierarchy, if the contents of a turned off cache line are required again.

Other questions that arise in the design of the leakage control mechanism are: what application characteristic should the compiler extract for directing these two leakage control modes? What granularity should the compiler insert the special leakage control instructions at? For example, the conservative strategy is based on determining the last usage of instructions. In other words, it tries to detect the use of the last instance of each instruction. Once this last use is detected, the corresponding cache line can be turned off. While it is possible to turn off the cache line immediately after the last use, such a strategy would not be very effective, because it would result in significant code expansion due to the large number of turn-off instructions inserted in the code. Also, such frequent turn-off instructions themselves would consume considerable dynamic energy. Obviously, instruction level analysis would not work for this leakage control. Notice that program behaviour, in general, is characterised by some high-level structures such as loops and function calls and those structures also produce program phases and phase changes during execution. Consequently, loop structures in program code were exploited for invoking leakage control in many research works in this area as is the case here.

9.6.2.1 Conservative strategy

Conservative strategy captures loop structures in program code and turns off instructions at the 'loop granularity level'. More specifically, when the execution exits a loop and it is for sure that this loop will not be visited again, the compiler turns off the cache lines that hold the instructions belonging to the loop (including those in the loop body as well as those used to implement the loop control code itself). While ideally one would want to issue turn-offs only for the cache lines that hold the instructions in the loop, identifying these cache lines is costly (i.e. it either requires some type of circuit support which itself would consume energy, or a software support which would be very slow). As a result, a more practical implementation is to turn off all cache lines when the execution exits loops. While this has the drawback of turning off the cache lines that hold the instructions sandwiched between the inner and outer loops of a nest (and lead to reactivation costs for such lines), the impact of this in practice is not too much as typically there are not many instructions sandwiched between nested loops. Turning off all cache lines also eliminates the complexity of selectively turning off a cache line in a set-associative cache and problems associated with multiple instructions from different loops being stored in a single cache line.

The idea behind the conservative strategy is illustrated in Figure 9.7 for a case that contains three loops, two of which are nested within the third one. Assume that once the outer loop is exited, this fragment is not visited again during execution. Here, Loop Body-I and Loop Body-II refer to the loop bodies of the inner loops. In the conservative strategy, when execution exits Loop Body-I (i.e. the loop that contains it), the compiler cannot turn off the cache lines occupied by it, because, there is an outer loop that will re-visit this loop body; in other words, the instructions in Loop Body-I are not dead yet. The same argument holds when execution exits Loop Body-II.

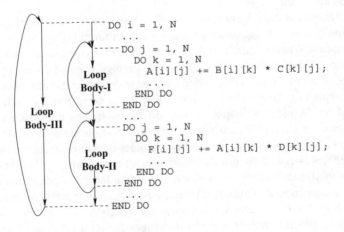

Figure 9.7   A code fragment that contains three loops

However, when execution exits the outer loop, the conservative strategy turns off all the cache lines that hold the instructions of this code fragment (i.e. all instructions in all three loops). As mentioned above, the leakage control instruction, in fact, turns off all the cache lines for implementation efficiency. It is clear that this strategy may not perform well if there is a large outermost loop that encloses a majority of the instructions in the code. In such cases, the cache lines occupied by the said instructions will not be turned off until the outermost loop finishes its execution. And, when this occurs, it might be too late to save any leakage energy. For example, if there is an outermost loop (in the application being optimised) that encloses the entire code (e.g. a convergence test in some array applications), this strategy will not generate very good results as it will have little opportunity to turn off cache lines.

9.6.2.2   Optimistic strategy

The optimistic strategy tries to remedy this drawback of the conservative scheme by turning off the cache lines optimistically. What we mean by optimism here is that the cache lines are turned off even if we know that the corresponding instruction instance(s) will be visited again, but the hope is that the gap (in cycles) between successive executions of a given instruction is large enough so that significant amounts of energy can be saved. Obviously, an important question here is how to make sure that at compile time (i.e. statically) there will be a large gap between successive executions of the same instruction. Here, as in the conservative case, the compiler works on a loop granularity. When execution exits a loop, the instructions in the loop body are turned off if either that loop will not be visited again (as in the conservative case) or the loop will be re-visited but there will be execution of another loop between the last and the next visit. Returning to the code fragment in Figure 9.7, when the execution exits Loop Body-I, the instructions in it can be turned off. This is because before Loop Body-I is visited again, the execution should proceed with another loop (the one with Loop Body-II), and the compiler optimistically assumes that this latter loop will take long time to finish. Similarly, the instructions in Loop Body-II can be turned off when the execution exits Loop Body-II.

Obviously, this strategy is more aggressive (in turning off the cache lines) than the conservative strategy. The downside is that in each iteration of the outer loop in Figure 9.7, it needs to re-activate the cache lines that hold Loop Body-I and Loop Body-II. The energy overhead of such a re-activation depends on the leakage saving mode employed. Also, since each reactivation incurs a performance penalty, the overall execution time impact due to the optimistic strategy can be expected to be much higher than that due to the conservative strategy. This will be particularly felt when the state-destroying leakage control mechanism is employed (since it takes longer to come back to normal operation from state-destroying than state-preserving mode due to the L_2 cache latency). If there are code fragments that are not enclosed by any loops, each such fragment can be treated as if it is within a loop which iterates only once. It should be emphasised, however, that both strategies turn off cache lines considering the instruction execution patterns. Since a typical cache line can accommodate several instructions during the course of execution, turning off a cache line may later lead to re-activation of the same cache line if another instruction wants to use it. This re-activation has both performance and energy penalty which should also be accounted for. Note that as the instruction cache gets bigger this problem will be of less importance. This is because in a larger cache we can expect less cache line sharing among instructions.

9.6.3   Controlling functional unit leakage mode

In this part, we show another example of a technique presented by Kim *et al.* [5] to demonstrate the interaction between the compiler and the architecture in affecting power control mechanisms that exploit application characteristics. In this case, the compiler attempts to exploit the idleness in functional units of wide-issue VLIW architectures. The basis of the work stems from the observation that there is an inherent variation in the maximum number of instructions that can be executed per cycle when executing an application. Not all parts of a given VLIW program can issue the maximum IPC. Due to data dependences and other reasons, it is typical that many VLIW cycles can execute fewer operations than that can be accommodated by the physical resources.

The approach proposed by Kim *et al.* [5] focuses on nest-intensive applications and identifies a suitable IPC assignment for basic blocks in the loop. After assigning an IPC to each loop, the next step is scheduling the code being optimised considering these IPCs. The scheduling strategy is adaptive as it changes the maximum IPC during the course of execution. In this step, the control flow graph (CFG) is traversed basic block by basic block, and for each block we schedule its operations using the IPC assigned to it. We also insert functional unit turn-on and turn-off instructions at the beginning and end of the associated loop containing the basic block considering its assigned IPC and the available integer ALUs in the architecture. When we set the IPC to k', the $k - k'$ integer ALUs are shut off (assuming that the machine contains a maximum of k integer ALUs). When (e.g. in executing the next nest in the code) we increase the IPC from k' to k'' where $k' < k'' < k$, we need to turn on $k'' - k'$ integer ALUs. As compared to dynamic IPC regulation strategies, the compilation approach incurs much less overhead as IPCs are determined at compile time. The only runtime overhead is the execution of turn-on/off instructions and extra execution cycles and energy due to increased schedule length.

This energy-saving strategy can, in general, increase execution time as it is not always possible to find a suitable IPC (lower than what the underlying machine can sustain) for each loop without increasing its schedule length. Increasing the schedule length also has an impact on energy consumption as all active functional units consume leakage energy during the extra cycles (coming from the reduced IPC). For the proposed strategy to reduce overall energy, the energy increase due to extra execution cycles should not offset the energy gains obtained through shutting off some integer ALUs. To illustrate this last point, let us consider Figure 9.8. This figure shows a schematic description of the original schedule and modified (optimised) schedule. As compared to the original schedule, the optimised schedule reduces the width

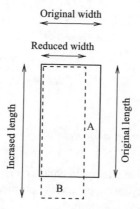

Figure 9.8   Original and modified schedules

(by restricting the maximum IPC) and increases the length (by increasing the number of cycles). By width of a schedule, we mean the number of integer ALUs used in executing the code. Let A denote the product of the total number of additional functional units that remain active in the original schedule as compared to the optimised schedule and the number of cycles of execution in the original schedule. Let B denote the product of the number of functional units active in the optimised schedule and the extra cycles in the optimised schedule. Let A' and B' denote the corresponding leakage energy expended due to slots in A and B. Clearly, this strategy is beneficial only if $A' > B'$ and the number of extra cycles is within a tolerable limit.

9.7   Enhancing power mode energy savings through data/code restructuring

In this section, we use two examples borrowed from Delaluz *et al.*'s work [25] and Zhang *et al.*'s work [2] to demonstrate how the compiler optimisations can be utilised to improve energy efficiency in a partitioned memory architecture with low power modes and to improve the instruction cache leakage savings from the compiler-directed strategies discussed in the previous section.

9.7.1   Compiler optimisation for partitioned memory architectures

In a partitioned off-chip memory (DRAM) architecture that consists of multiple memory banks, each of which can be operated in a number of operating modes (power modes) and the power modes can be controlled using either the compiler (as observed earlier in this chapter) or runtime system. Another important issue for a compilation approach to minimise the memory energy consumption is to restructure code and data such that as many memory banks as possible can be put into a low-power operating mode without impacting execution time (performance). This subsection introduces

```
DO i = 1, N
  {a[i], b[i]}
END DO
DO i = 1, N
  {c[i], d[i], e[i]}
END DO
DO i = 1, N
  {e[i], f[i]}
END DO
```

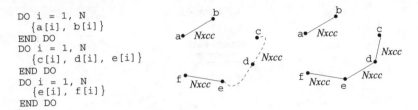

Figure 9.9   An example of array allocation

the compilation techniques that attempt to address this issue using a set of loop (iteration space) and memory layout (data space) optimisations: array allocation, loop fission, loop splitting and array renaming.

The basic idea behind the array allocation (data placement) algorithm is to place the arrays with the same (or similar) access patterns into the same (set of) bank(s). This is reasonable as such arrays are usually used (or not used) at the same time during which the memory bank(s) that hold them can be turned on (or off). By clustering arrays with similar access patterns together, the compiler optimisation can improve the opportunity for power mode control in such a multi-banked memory, leading to higher energy efficiency of the system.

As an example, let us consider the program fragment shown in Figure 9.9 whose array relation graph (ARG) with and without hyper-edge (spanning multiple nodes) is shown next to it. In ARG, the nodes represent the arrays declared in the program and the weight of an edge (or hyper-edge) represents the number of times (in cycles) two (or multiple) arrays that are incident on the edge are accessed in the same nested loop. Note that there are two edges, (a,b) and (e,f), and one hyper-edge (c,d,e) (shown in the first graph as a dashed path). This hyper-edge is then transformed into two (normal) edges (c,d) and (d,e) (as shown in the last graph), all with the same edge weight $Nxcc$ assuming that each nest has a cycle count of cc. Two obvious paths in this last graph are a-b and c-d-e-f. Assuming that the arrays are of the same size and a memory bank can hold two arrays, the compiler places a and b into the first bank, c and d into the second bank and e and f into the third.

Loop fission takes a nested loop that contains multiple statements in it and creates multiple nested loops each with a subset of the original statements. Loop fission helps to improve the effectiveness of the array allocation module by allowing a finer-granular control over the allocation of arrays. For instance, in the example shown in Figure 9.10 (assuming that the arrays are of the same size and each memory bank can hold at most two arrays), with the original nest, the two banks that contain the four arrays should be in the active mode throughout the entire execution. After the loop fission, on the other hand, only a single bank needs to be in the active mode during the execution of each loop (assuming that the original array allocation places a and b into one bank, and c and d into the other). The other bank can be put into a low power mode, thereby saving energy.

Loop splitting divides the index set of a nested loop into two or more disjoint parts. It was originally developed to break some data dependences by placing the

```
                                        DO i = 1, N
                                          {a[i], b[i]}
     DO i = 1, N                        END DO
       {a[i], b[i]}        ⟶           DO i = 1, N
       {c[i], d[i]}                       {c[i], d[i]}
     END DO                             END DO
```

Figure 9.10   Loop fission to improve array allocation

```
                                        DO i = 1, N/2
                                          {a[i], b[i]}
     DO i = 1, N                        END DO
       {a[i], b[i]}        ⟶           DO i = N/2 + 1, N
     END DO                               {a[i], b[i]}
                                        END DO
```

Figure 9.11   Loop splitting to improve array allocation

```
     DO i = 1, N                        DO i = 1, N
       {a[i], c[i]}                       {a[i], c[i]}
     END DO                             END DO
       ....                               ....
     DO i = 1, N           ⟶           DO i = 1, N
       {b[i], c[i]}                       {a[i], c[i]}
     END DO                             END DO
```

Figure 9.12   Array renaming to improve array allocation

source iteration of the dependence and the target iteration into separate loops. Note that since this optimisation does not change the execution order of loop iterations, it is always legal. Loop splitting has an important use in energy optimisation for large arrays that span multiple banks. For the example shown in Figure 9.11, assume that each of a and b spans two memory banks (i.e. total four memory banks are needed). During the execution of the original loop (the left one), all four banks should be in the active state. After the loop splitting, however, only two memory banks need to be in the active mode during the execution of each loop.

Array renaming is an optimisation that exploits the result of live variable analysis [26] to reuse the same memory space for storing multiple array variables whose lifetimes are disjoint. The code fragment in Figure 9.12 shows an example case with two arrays of disjoint lifetimes (a and b). That is, we assume that after the first nest, the array a is not needed, hence its memory space can be reused for some other array (in this case, the array b). Supposing that the arrays are of the same size and each bank can hold two arrays, in the original code (the left one), the array allocation module might place a and c into the same bank and b into another bank. In this case, during the execution of the first nest, only the first bank will be in the active mode and during the execution of the second nest, two banks will be in the active mode. On the other hand, if we reuse the same space for a and b (as shown on the right in the code), only one memory bank will be used during both the nests.

9.7.2   Impact of compiler optimisations on instruction cache leakage control

As discussed in the early part of this section, any compiler optimisations (especially those targeting at improving data locality) can modify the instruction execution order (sequence) dramatically leading to a significantly different energy picture. Such a characterisation of the impact of such optimisations on the effectiveness of the proposed mechanism can be fed-back to compiler writers, leading to better (e.g. energy-aware) compilation strategies. This subsection focuses on the impact of loop fission on compiler-directed instruction cache leakage management introduced in the previous section.

The loop fission transformation cuts the body of a for-loop statement into two parts. The first statement of the second part specifies the cut position. It is generally used for enhancing iteration-level parallelism (by placing statements with dependence sources into one loop and the ones with dependence sinks into the other), for improving instruction cache behaviour (by breaking a very large loop body into smaller, manageable sub-bodies with better instruction cache locality), and even for improving data cache locality (by separating the statements that access arrays that would create conflicts in the data cache). Different optimising compilers can employ this transformation for one or more of these reasons.

As an example, let us consider the fragment shown in Figure 9.13(a). If we distribute the outermost loop over the two groups of statements (denoted Body-I and Body-II in the figure), we obtain the fragment depicted in Figure 9.13(b). Figures 9.13(c) and (d), on the other hand, illustrate how the instructions in the fragments in Figures 9.13(a) and (b), respectively, would map to the instruction cache. The figure is used only for illustrative purposes and masks the details. In Figures 9.13(c) and (d), Header is the loop control code. Note that in the distributed version, Header is duplicated. Now, let us try to understand how this optimisation would influence the effectiveness of leakage optimisation strategies. First, let us focus

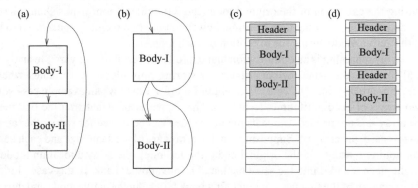

Figure 9.13   (a) A code fragment with a loop, (b) the distributed version of (a), (c) the instruction cache layout for (a), (d) the instruction cache layout for (d)

on Figure 9.13(c). During execution all three blocks (Header, Body-I and Body-II) need to be accessed very frequently, and there will be little opportunity (or energy benefit) in placing the cache lines in question into leakage control mode. If we consider the picture in Figure 9.13(d), on the other hand, when we are executing the first loop only the first Header and Body-I need to be activated. The second Header and Body-II can be kept in a leakage saving mode. Similarly, when we move to the second loop, during execution, only the second Header and Body-II need to be activated. Therefore, at any given time, the distributed alternative leads to the activation of fewer cache lines. However, the number of cache lines occupied by the code is one part of the big picture. Since we are focusing on the leakage energy consumption, we also need to consider the execution time. If, in this code fragment, data cache locality is a problem, then the first alternative (without distribution) might have shorter execution time if loop distribution destroys data cache locality. Consequently, although the alternative in Figure 9.13(d) will occupy fewer cache lines at a given time, it will keep those cache lines in the active mode for a longer duration of time. Consequently, there is a tradeoff here between the number of cache lines occupied and the time duration during which they are active.

9.8 Conclusions

Software can have significant impact on system energy consumption. Energy-efficiency is one of the critical criteria when designing or evaluating software for embedded systems with limited energy supply. There are many optimisation techniques that can improve energy-efficiency of the software. This chapter has shown various examples of software that can be used to control power modes, enhance power modes, reconfigure resources for energy-efficiency. In addition, we also showed examples of how algorithm modifications and traditional compiler optimisations influence power consumption.

Acknowledgement

This chapter includes results from prior research work performed by many of the graduate students of our research group.

References

1 HSU, C.-H., and KREMER, U.: 'The design, implementation, and evaluation of a compiler algorithm for CPU energy reduction'. Proceedings of the ACM SIGPLAN conference on *Programming language design and implementation*, San Diego, CA, June 2003

2 ZHANG, W., HU, J.S., DEGALAHAL, V., KANDEMIR, M., VIJAYKRISHNAN, N., and IRWIN, M.J.: 'Compiler-directed instruction cache

leakage optimization'. Proceedings of the 35th annual international symposium on *Microarchitecture*, Istanbul, Turkey, November 2002

3 KANDEMIR, M., VIJAYKRISHNAN, N., IRWIN, M.J., and YE, W.: 'Influence of compiler optimizations on system power'. Proceedings of the 37th *Design automation* conference, Los Angeles, CA, June 5–9, 2000

4 DELALUZ, V., KANDEMIR, M., VIJAYKRISHNAN, N., SIVASUBRAMANIAM, A., and IRWIN, M.J.: 'DRAM energy management using software and hardware directed power mode control'. Proceedings of the 7th international conference on *High performance computer architecture*, Monterrey, Mexico, January 2001

5 KIM, H.S., VIJAYKRISHNAN, N., KANDEMIR, M., and IRWIN, M.J.: 'Adapting instruction level parallelism for optimizing leakage in VLIW architectures'. Proceedings of workshop on *Languages, compilers, and tools for embedded systems*, San Diego, CA, June 11–13, 2003

6 XIE, F., MARTONOSI, M., and MALIK, S.: 'Compile-time dynamic voltage scaling settings: opportunities and limits'. Proceedings of the ACM SIGPLAN conference on *Programming language design and implementation*, San Diego, CA, June 2003

7 CHEN, G., KANDEMIR, M., VIJAYKRISHNAN, N., IRWIN, M.J., and WOLCZKO, M.: 'Adaptive garbage collection for battery-operated environments'. Proceedings the 2nd *USENIX Java virtual machine research and technology* symposium, San Francisco, CA, 1–2 August, 2002, pp. 1–12

8 LU, Y.-H., and DE MICHELI, G.: 'Adaptive hard disk power management on personal computers'. Proceedings of the IEEE Great Lakes symposium on *VLSI*, Ypsilanti, MI, March 4–6, 1999, pp. 50–53

9 RABINER, W., and CHANDRAKASAN, A.P.: 'Network driven motion estimation for portable video terminals', *IEEE Transactions on Circuits and Systems for Video Technology*, 1997, 7(4), pp. 644–653

10 BENINI, L., and DE Micheli G.: 'System-level power optimization: techniques and tools', *ACM Transaction on Design Automation of Electronic Systems*, 2000, 5(2), pp. 115–192

11 CHANDRAKASAN, A., BOWHILL, W.J., and FOX, F.: 'Design of high-performance microprocessor circuits' (IEEE Press, New York, 2001)

12 KIM, E.J., YUM, K.H., LINK, G., *et al.*: 'Energy optimization techniques in cluster interconnects'. Proceedings of the international symposium on *Low power electronics and design (ISLPED'03)*, Seoul, Korea, August 2003

13 VIJAYKRISHNAN, N., KANDEMIR, M., IRWIN, M.J., KIM, H., YE, W., and DUARTE, D.: 'Evaluating integrated hardware–software optimizations using a unified energy estimation framework', *IEEE Transactions on Computers*, 2003, 52(1), pp. 59–76

14 SINHA, A., WANG, A., and CHANDRAKASAN, A.P.: 'Algorithmic transforms for efficient energy scalable computation'. Proceedings of the international symposium on *Low power electronics and design (ISLPED)*, Rapallo, Italy, 12–15 October, 2000, pp. 62–73

15 MOWRY, T.C., LAM, M.S., and GUPTA, A.: 'Design and evaluation of a compiler algorithm for prefetching'. Proceedings of the fifth international conference on *Architectural support for programming languages and operating systems ASPLOS-5*, ACM Press, Boston, MA, 1992, pp. 62–73

16 WOLF, M.E., and LAM, M.: 'A data locality optimizing algorithm'. Proceedings of the SIGPLAN '91 conference on *Programming language design and implementation*, Toronto, Canada, June 1991, pp. 30–44

17 CATTHOOR, F., FRANSSEN, F., WUYTACK, S., NACHTERGAELE, L., and DEMAN, H.: 'Global communication and memory optimizing transformations for low power signal processing systems'. Proceedings of the IEEE workshop on *VLSI signal processing*, 1994, pp. 178–187

18 KADAYIF, I., KOLCU, I., KANDEMIR, M., VIJAYKRISHNAN, N., and IRWIN, M.J.: 'Exploiting processor workload heterogeneity for reducing energy consumption in chip multiprocessor'. Proceedings of the 7th *Design automation and test in Europe* conference *(DATE '04)*, Paris, France, February 2004

19 PARK, I., POWELL, M.D., and VIJAYKUMAR, T.N.: 'Reducing register ports for higher speed and lower energy'. Proceedings of the 35th annual ACM/IEEE international symposium on *Microarchitecture*, Istanbul, Turkey, 2002, pp. 171–182

20 HU, J.S., KANDEMIR, M., VIJAYKRISHNAN, N., and IRWIN, M.J.: 'Analyzing data reuse for cache reconfiguration', *ACM Transactions on Embedded Computing Systems (TECS)* (accepted for publication 2005)

21 MONTANARO, J. *et al.*: 'A 160-MHz, 32-b, 0.5-W CMOS RISC microprocessor', *Digital Technical Journal, Digital Equipment Corporation*, 1997, 9(1), pp. 49–62

22 BECHADE, R., FLAKER, R., KAUFFMANN, B., *et al.*: 'A 32b 66MHz 1.8W microprocessor'. Proceedings of the IEEE international *Solid-state circuits* conference, 1994, 16–18 February, pp. 208–209

23 KIM, N., AUSTIN, T., BLAAUW, D., *et al.*: 'Leakage current: Moore's law meets static power', *IEEE Computer Special Issue on Power- and Temperature-Aware Computing*, 2003, 36(12), pp. 68–75

24 DUARTE, D., FAI, Y.-T., VIJAYKRISHNAN, N., and IRWIN, M.J.: 'Evaluating run-time techniques for leakage power reduction'. Proceedings of the 7th Asia and South Pacific *Design automation* conference and 15th international conference on *VLSI design (VLSI design/ASPDAC '02)*, Bangalore, India, 7–11 January, 2002

25 DELALUZ, V., KANDEMIR, M., VIJAYKRISHNAN, N., and IRWIN, M.J.: 'Energy-oriented compiler optimizations for partitioned memory architectures'. International conference on *Compilers, architecture, and synthesis for embedded systems*, San Jose, CA, November, 2000, pp. 138–147

26 MUCHNICK, S.S.: 'Advanced compiler design implementation' (Morgan Kaufmann Publishers, San Francisco, CA, 1997)

Part III

Power reduction and management

Chapter 10

Power-efficient data management for dynamic applications

P. Marchal, J.I. Gomez, D. Atienza, S. Mamagkakis and F. Catthoor

10.1 The design challenges of media-rich services

Business analysts forecast a 250 billion dollar market for media-rich, mobile wireless terminals [1]. These systems require an enormous computational performance (40 GOPS[1]). Even though current PCs offer this performance requirement, they consume too much power (10–100 W). Mobile devices should consume at least two or three orders of magnitude less power [2]. Furthermore, they should be cheap to successfully penetrate the consumer market. Consequently and in spite of the design issues, the engineering and manufacturing costs need to be reduced. Industry strongly believes that platforms are a potential way to meet the above challenges.

10.1.1 The era of platform-based design

A platform is a fixed micro-architecture together with a programming environment that minimises mask-making costs and is flexible enough to work for a set of applications [3]. The production volumes can then remain high over an extended chip lifetime.

Given the strong energy constraints, we must choose the flavour of these platforms. Since power is cubic to the processing frequency, parallelism is an effective way to reduce power and energy consumption. Then, multiple simple processors are preferred to one complex speculative and out-of-order processor. In the right application domain, we can get better performance and spend less energy. Besides parallelism, heterogeneity is an alternative way to decrease the energy cost. For instance,

---

[1] Giga operations per second.

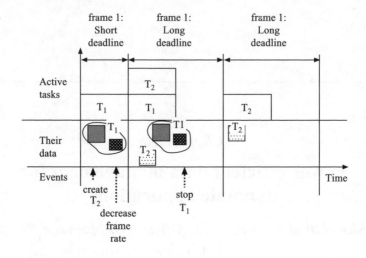

Figure 10.1   Characteristics of our application domain

the TI OMAP platform combines a RISC processor with a digital signal processor (DSP). The RISC is more energy-efficient for the input/output processing and simple control-dominated applications. The DSP, on the other hand, provides the computational performance for audio and video processing, while keeping the energy cost bounded.

Taking a look at the market, it is clear that heterogeneous multi-processor platforms are conquering the world of low power embedded systems: ST Nomadik [4], Philips Nexperia [5], TI OMAP [6].

10.1.2   The desire for media-rich services

Platforms perfectly support the next wave of media-rich, wireless applications, bound to flood the multi-billion dollar consumer market. Typical applications are media-players such the MPEG4 IM1 player.

We summarise the most important characteristics of the application domain in Figure 10.1:

- *Multi-threaded*: The systems contain multiple tasks which can execute in parallel. The tasks can either be independent or dependent. In Figure 10.1, the system contains two parallel tasks (T_1 and T_2).
- *A closed system*: The entire set of possible tasks is known at design time (i.e. we know the source code of every task to be executed in the system). However, the start time of each task and the exact instances of a task being executed at a precise instant, is only known at run time. User interaction and data-dependent conditions introduce non-deterministic behaviour in the system, making it impossible to accurately predict which tasks will be executed in parallel. We assume that no tasks can be downloaded on the system (such as e.g. Java applets or other software agents). For our example, this entails that no other types of tasks but T_1

and T_2 can occur at runtime. Conceptually, it is feasible to extend our framework and methodology for open systems, but we leave this for future work.

- *Time-constraints*: Tasks within multi-media applications are usually bound to time-constraints. The most common deadline is the frame-rate (see above). To have a fluid video display the tasks of a thread-frame have to finish within its deadline. We indicate the deadline imposed by the frame-rate on our application in Figure 10.1. In the first frame, we use a high frame-rate, i.e. a tight deadline for T_1. Thereafter, a user event relaxes the frame-rate. In the remainder of this text, we mainly focus on the frame-rate, despite the fact that other deadlines will in practice also occur.
- *Tasks are control/data flow graphs*: Each task is a control/data flow graph. Hence, parts of a task may be conditionally executed. As a result, which data and how frequently it is accessed may significantly vary at runtime. We take as a premise that at the start of each task we know how much memory it needs. The memory space can be used for the static data or as a heap for runtime allocated data. For instance, we assume that T_1 requires two data structures whereas T_2 only needs one.
- *Data-dominated*: The tasks are data-dominated. As a result, the energy of the data memory architecture dominates the system cost. On multi-media systems, this assumption is particularly true after the cost of the instruction memory hierarchy is optimised (e.g. with References 7 and 8). The data memory cost is then usually the remaining energy bottleneck. Consequently, optimising the data memory is the top priority, even if it afterwards slightly increases the processing energy consumption.

In the next subsection, we discuss the main challenges to integrate these applications on an embedded platform.

10.1.3 Memories rule power and performance

The memory system is an important contributor to the performance and power consumption of embedded software, particularly for multi-media applications [9,10]. The most well-known technique for improving the performance of the memory subsystem is introducing a layered memory architecture. Large memories used to store multi-media data have long access times. Therefore, they are too slow to supply data at a sufficient rate to the processing elements. As a result, the processing elements stall, thereby wasting time and energy. To improve the performance and reduce the energy cost, designers create a layered memory hierarchy. Each layer contains smaller memories to buffer the data that is frequently accessed by the processor.

In this work, we focus on how to exploit a layered memory architecture. Particularly, we optimise the available bandwidth to the multiple memories/banks of each layer. This problem consists of detecting a data assignment and instruction schedule that satisfy all time-constraints while minimising the energy consumption. Despite the many techniques already exist for this problem (Section 10.3), this approach improves the bandwidth within a basic block and assume that the memories are accessed by a 'single thread'. Moreover, these approaches require that access pattern of the application can be analysed at design-time. Unfortunately, in our application domain multiple-threads often share memory resources. Furthermore, the user determines which threads are running. As a consequence, we can only characterise the access pattern at runtime. We will show in Sections 10.4.1.1, 10.4.2.1 and 10.5.1 that existing techniques break down under these circumstances, resulting in energy and performance loss.

We will overview the techniques which we have developed to overcome the above limitations. We have investigated on the one hand design-time techniques for globally optimising the memory bandwidth, even across the tasks' boundaries (Sections 10.4.1.2 and 10.4.1.3 for the shared layer and section 10.4.2.2 for the local layer). On the other hand, we have developed a combined design-time approach for dealing with the dynamic behaviour (Section 10.5). It makes runtime decisions based on an extensive design time analysis phase. Finally, we present how these runtime decisions can be energy-efficiently implemented at runtime (Section 10.6). Before introducing our approach, we explain the memory architecture targeted throughout this text (Section 10.2) and the related work more in detail (Section 10.3).

10.2   Target architecture

During our research, we concentrate on a generic target architecture (Figure 10.2). Different processing tiles contain multiple processing elements that share a local

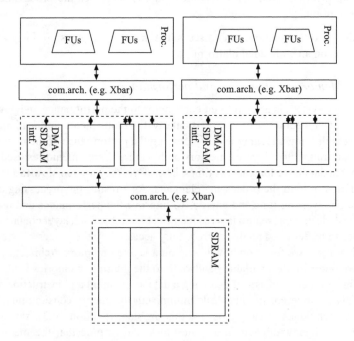

Figure 10.2   Target architecture for bandwidth optimisation

memory layer. The processing elements within the same tile are closely synchronised. A processing tile could be, for instance, a very long instruction word (VLIW) or a simple RISC processor (like on a TI OMAP). The local memory layer on a processing tile may comprise multiple scratch pad memories/banks. Again this closely resembles ST LX [11] or TI C6X [6] DSPs where up to eight memories are included in the local layer. We do not directly exploit cache memories, but focus on scratch pad memories. These software-controlled memories do not require complex tag-decoding logic [12]. Therefore, they have a lower energy cost per access compared to caches, and also reduce the indeterminacy of the system. To further reduce the global energy cost, we assume that they are heterogeneous: they can have different sizes, different number of ports and access time.

Furthermore, the processing tiles share an off-chip SDRAM (like on the TI OMAP or Philips Nexperia). We include the SDRAM in our overall target architecture, because it may consume up to 30 per cent of the system energy cost of a commercial hand-held device [13].[2]

We integrate a cross-bar as communication architecture both between the processing elements and the local layer as between the local layers and the shared SDRAM. Although a cross-bar is not the most energy-efficient architecture, its energy cost is currently limited to only 10 per cent of the global data transfer cost.[3] With this configuration, the communication architecture does not have any impact on the achievable performance, but the ports of the different memory layers (local memories and SDRAM) become potential bandwidth bottlenecks.

10.3   Surveying memory bandwidth optimisation

Memory bandwidth optimisation is a widely researched topic. Most of the related work is focused on improving the bandwidth of a single memory layer. This layer can either consist of multiple SRAMs or a large SDRAM memory (Figure 10.2). We discern two methods which are commonly applied/combined to optimise the memory bandwidth: data layout transformations and instruction scheduling techniques. Data layout transformations may comprise several optimisations: from deciding the optimal address in memory (and thus, the optimal memory module) for each application variable to the array elements relative order in memory. Variables lifetime can also be exploited to fully exploit the available memory space. Instruction scheduling tries to find a memory access ordering that optimises a specific cost. The goal is usually to increase the system's performance. Only a few methods exchange the performance gains for energy savings. In the next subsections, we outline them for SDRAMs and for the local memory layer.

---

[2] This percentage is for a complete system including speakers, LCD, etc.

[3] In principle, a more scalable communication architecture could be programmed or synthesised (such as e.g. Referene 14) However, research on advanced communication architectures falls outside the scope of this text.

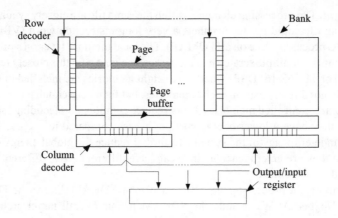

Figure 10.3   Multi-banked SDRAM architecture

10.3.1   SDRAM bandwidth

SDRAMs are mostly used for storing large multi-media data. The access time and energy cost of an SDRAM heavily depend on how it is used. In general, an SDRAM consists of several banks (Figure 10.3). Fetching or storing data in an SDRAM involves three memory operations. An activation operation decodes the row address, selects the appropriate bank and moves a page/row to the page buffer of the corresponding buffer. After a page is opened, a read/write operation moves data to/from the output pins of the SDRAM. Only one bank can use the pins at a time. When the next read/write accesses hit in the same page, the memory controller does not need to activate the page again (a 'page-hit'). The application can read these data elements at a lower access latency and lower energy cost. However, when another page is needed (a 'page-miss'), precharging the bank is needed first. Only thereafter the new page can be activated and the data can be read. Note that pages from different banks can be opened simultaneously. We can then interleave the access among banks in order to minimise the number of page-misses. The less page-misses occur, the better the performance and energy consumption of the SDRAM become. Most methods below focus on transforming the application such that page-misses are avoided.

10.3.1.1   Data layout transformations and data assignment techniques

For a fixed access schedule, the layout of the data in a memory bank defines how many page-misses occur (Figure 10.4). To illustrate this, we map the scalars a,b,c,d,e,f in two different ways onto the pages of an SDRAM bank. If a memory operation accesses an open page, a page-hit occurs (H). If, on the other hand, the next operation reads/writes to another page, a page-miss happens (M). For example, in the first layout, an access to c after one to a results in a page-hit, while an access to e after one to a causes a page-miss. Given the presented access sequence, four page-misses occur in the left layout. If we change the data layout, we can reduce the number of page-misses. For example, when we move e to the first page and b to the second

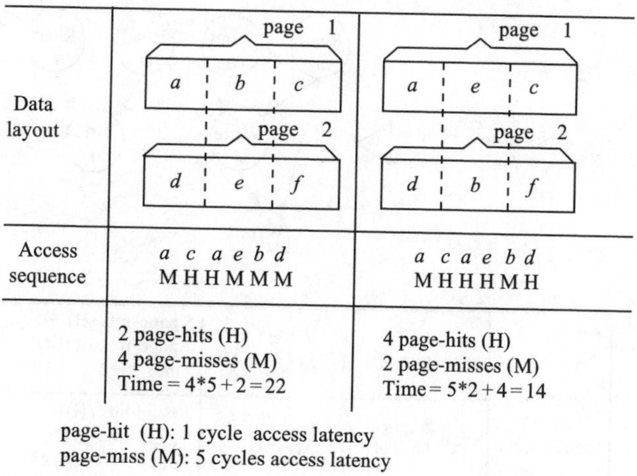

Data layout	page 1 a | b | c page 2 d | e | f	page 1 a | e | c page 2 d | b | f
Access sequence	$a\ \ c\ \ a\ \ e\ \ b\ \ d$ M H H M M M	$a\ \ c\ \ a\ \ e\ \ b\ \ d$ M H H H M H
	2 page-hits (H) 4 page-misses (M) Time = 4*5 + 2 = 22	4 page-hits (H) 2 page-misses (M) Time = 5*2 + 4 = 14

page-hit (H): 1 cycle access latency
page-miss (M): 5 cycles access latency

Figure 10.4   Different data layouts impact the number of page-misses

one, only two page-misses remain (Figure 10.4 – right). Furthermore, it reduces the execution time from 22 to 14 cycles. Since the data layout has such a large impact on the performance, several authors have proposed techniques to optimise it. Rixner *et al.* [18] partitioned arrays into tiles, each fitting into a single page. The tiles are derived such that the number of transitions between the tiles, and thus the number of page-misses, is minimised. Reference 16 proposes to layout the scalar variables inside the program text, reducing the overall page-misses.

In contrast with the older DRAM architectures, most SDRAMs nowadays have more than one bank. For example, the Rambus' SDRAMs have up to 32 banks. Multiple banks provide an alternative way to eliminate page-misses. For instance, Reference 17 distributes data with a high temporal affinity over different banks such that page-misses are avoided. Their optimisations rely on the fact that the temporal affinity in a single-threaded application is analysable at design time.

Thus, despite data assignment techniques exist for limiting the page-miss penalty, they are restricted to single-threaded, design-time analysable tasks. As we will motivate in Section 10.5.1, these techniques break down for dynamic multi-threaded applications.

10.3.1.2   Memory access reordering techniques

The access order also influences the number of page-misses. Consider the code of the basic block shown in Figure 10.5(a). Its data flow graph is also shown. After data dependence analysis, several memory access schedules are feasible (just two of them are shown). However, as depicted in the figure, the choice impacts the number of page-misses (and thus, the performance and energy consumption). We assume the left data layout in Figure 10.4. The data dependence analysis reveals that read accesses to a, b, d and e may be performed in any order. To hide the multiplication latencies,

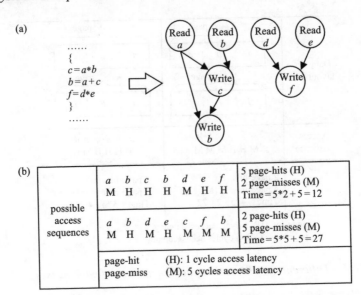

Figure 10.5   Access order impacts the number of page-misses

we may opt to generate the top schedule of Figure 10.5(b). It causes five page-misses out of seven accesses; reordering the accesses as shown in the second option helps to reduce the number of page-misses to just two. The potential performance improvement derived from the first schedule will very likely become a time penalty because of the extra page-misses. Moreover, the total energy consumption will be significantly larger.

As we will see in Section 10.4.1.1 for multi-threaded contexts, memory accesses from different concurrent tasks interfere with each other. As a consequence, task scheduling is a higher level way of changing the final access ordering.

We may classify the existing work in this area in two main threads: hardware approaches (trying to reorder accesses through smart memory controllers) or by software approaches (that rely on the compiler to perform code transformations and instruction scheduling optimisations). Several authors [18,19] propose hardware controllers to reorder the accesses. Typically, they buffer and classify memory access operations according to their type (precharge, row memory accesses and column memory accesses) and according to the accessed bank and the row. The hardware logic of the memory manager selects from this classified set which operation to execute first. Since we focus in low power design, we strive to simplify the hardware to the bare minimum and put the complexity of our designs as much as possible in the design-time preparation phase (subsection 10.5.2). In this way, we avoid the extra hardware which increases the energy consumption of all memory accesses.

Several software approaches have been presented too. Panda *et al.* [20] expose the special access modes of SDRAM memories to the compiler. As a result, their scheduler can hide the access latency to the SDRAMs. The work was started in the context of system synthesis, but later on extended to VLIW compilers [21]. Finally, Lyuh and Kim [22] combine the scheduling technique of Panda *et al.* [20] with the memory energy model of Delaluz *et al.* [23] for reducing the static SDRAM energy.

The above existing techniques rely on the fact that the access pattern can be analysed at design time for single-threaded applications. This is not the case in our application domain. Dynamism and data-dependent control flow in modern applications makes quasi unpredictable the final access pattern of a single thread. Matters become more complex in the multi-thread context: memory accesses from different threads are interleaved. Currently, no techniques analyse the access pattern across threads. Moreover, the dynamic behaviour of some multi-threaded applications further complicates the problem: the active task-set (set of tasks executing simultaneously) is only known at runtime. Therefore, it is impossible to predict the inter-tasks memory access interactions at design-time, since we cannot even know which tasks will be executed in parallel!

10.3.2 Bandwidth to the local memory layer

Complementary to the SDRAM layer, memory bandwidth optimisation has also been researched for the local memory layer. Their optimisation objective is mostly reducing memory area/energy while guaranteeing performance. In this subsection, we discern again techniques which only change the data assignment and the ones which combine it with instruction scheduling.

10.3.2.1 Data layout based techniques

In the synthesis community, many techniques were developed for synthesising a memory architecture which provides sufficient memory bandwidth, but is energy or area efficient too (e.g. References 24,25). They generate a memory architecture and decide on the data to memory assignment in a single step. As a consequence, this makes them not directly applicable for predefined memory architectures (such as on ASIPs or DSPs).

Modern DSPs usually have a local memory layer which consists of multiple SRAM memories. Let us consider the architecture in Figure 10.6 as an example. It has two single ported memory banks (X,Y) which can be read in parallel. Most compilers would model this memory layer as a monolithical memory with multiple ports. Under this assumption, data layout has almost no influence in the potential performance. The compiler will schedule in parallel as many memory operations as it can as load/store units exist on the architecture. Since the compiler is not bank-aware, it will even schedule accesses to the same memory bank in parallel. Although this simplifies the instruction scheduling, special hardware at runtime needs to serialise the parallel accesses to the same memory resource. The DSP is then stalled and performance is lost. Several authors therefore expose the local memory architecture to the linker.

A conscious data layout may help to alleviate this problem. Saghir *et al.* [26] maximise the performance by carefully distributing the data across the different memories. In this way, they ensure that as many accesses as possible can be executed in parallel.

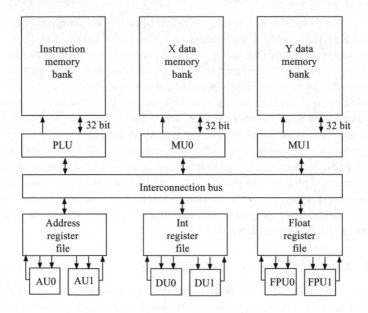

Figure 10.6   A VLIW architecture borrowed from Reference 26

Some very recent approaches [27] dynamically reallocate data in the local layer. At compile time they perform a life analysis of the task and insert instructionc to dynamically copy code segements and variables onto the scratchpad at runtime. They report energy reductions up to 34 per cent compared to static allocation. However, their technique cannot efficiently handle dynamic applications. A static analysis of a dynamic application cannot reaveal which data will be accessed at any point of the task. Furthermore, multi-tasked environments, where the local layer is shared between several tasks, are not considered at all.

10.3.2.2   Access order

The memory accesses order has also an important impact on the performance. and energy consumption. Indeed, changing the access ordering may allow us to achieve the same performance with a more energy-efficient memory system. Let us see how with an example (Figure 10.7). The application has one small array (B) and a bigger one (A). The memory system may consist of several instances of any of the modules in the 'Memory library'. Considering the access schedule 1, the architecture 1 is the most energy-efficient that enables the schedule. However, the dual port memory has an important impact on the energy consumption of this architecture. By rescheduling the memory accesses of the inner loop, we can eliminate the need for this memory (architecture 2 in Figure 10.7). It retains the same performance, but both data structures can now be mapped in a single port memory, thereby reducing the energy cost from 0.23 mJ to 0.13 mJ. From this example, it is clear that data layout and access scheduling are very effective in lowering the architecture cost. Because both

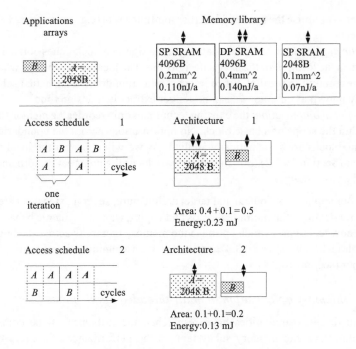

Figure 10.7 Reducing the memory cost with access ordering

techniques are so closely coupled, several authors propose to optimise memory layout and access ordering together.

An example is Reference 28. It optimises the memory bandwidth in a separate step before compilation, thereby outputting a (partial) data assignment which constrains the final instruction scheduling. It guarantees that enough memory bandwidth exists to meet the deadline, while remaining as energy-efficient as possible. This technique optimises the storage bandwidth within a basic block for memories with a uniform access time of the application. An extension to this work, Reference 29 indicates how this technique, initially developed for a system synthesis, can also be used on existing processor architectures. But in both cases, they only reorder the memory accesses within the scope of a basic block.

More global optimisation techniques can further improve the performance. In the past, several authors have proposed techniques to globally schedule instructions to parallelise code [30], but they do not consider how to optimise the memory bandwidth. Verhaegh *et al.* [31] define an operation schedule which reduces the number of memory ports. However, they do not take into account which data structures are accessed or how they are mapped onto the memory.

In summary, the main limitations of the above bandwidth optimisation techniques for both the local and the shared memory layer are:

1   *Single-threaded applications*: they optimise the memory bandwidth for a single task at a time. As a result, we cannot directly use them in our context, since we

want to optimise the bandwidth across multiple tasks (e.g. on the shared SDRAM layer).

2 *Static applications*: the above data layout/assignment techniques obtain information on the locality for the data at design time. The locality depends on which tasks are executing in parallel. Since in our application domain the actual schedule is only known at runtime, we can no longer extract it at design time.

3 *No global optimisation*: the existing techniques only reorder the memory accesses within the scope of a basic block. No optimisations across the boundaries of the basic blocks are systematically applied. As we will show for the local memory layer (Section 10.4.2), this significantly reduces the potential performance gains and energy savings.

For our application domain and target architecture, several extensions are clearly needed for dealing with multiple threads and coping with the dynamic behaviour. We discuss now techniques which optimise the memory hierarchy across the boundaries of a single task (subsection 10.3.3) and overview the techniques for managing dynamic behaviour (section 10.3.4).

10.3.3 *Memory optimisation in multi-threaded applications*

Different design communities have researched the influence of the communication architecture and memory subsystem on the performance of a multi-threaded application.

A large body of research exists in the high-performance computing domain on parallelising applications while reducing the communication cost (e.g. the SUIF-project [32] and the Paradigm compiler [33]). However, they target an architecture which is very different from ours. For example, they rely on complex hardware to guarantee data coherency and consistency, which may come at an important energy penalty. Furthermore, their techniques only work for statically analysable code and cannot cope with runtime variations which are typically present in modern multimedia applications. These limitations render this prior-art not directly applicable to our context.

Also in the embedded system's context many authors have studied multi-threaded applications. Li and Wolf [34] proposed a top-down hierarchical approach, compiling code on a heterogeneous multi-processor. The main disadvantage of this approach is that they use a synchronous data-flow model. It covers only a limited application domain and is not sufficient for our target domain.

Finally, since the middle of the last decennium, multi-processors systems have been widely researched in the system-level design community. Most techniques explore how to combine IP-blocks such that the system cost (albeit performance, energy or area) is reduced. In this context, the ordering and assignment of the tasks to the processing elements plays an important role in the system's performance (see Reference 35 for an overview). However, in recent years, the energy consumption has become an important bottleneck too. When energy is considered at all in task scheduling, the focus has been on the processing cores. Schmitz *et al.* [36] present a complete methodology to map multi-tasked and multimode applications onto heterogeneous multi-processor platforms. They focus on task and communication mapping, tasks scheduling and dynamic voltage scheduling. However, their model does not specifically include the memory system. They incorporate communication costs between tasks, but this information is not enough to efficiently optimise the memory hierarchy.

Unfortunately, only limited research exists in reducing the energy cost of the memory system. References 37 and 38 describe both a heuristic which does allocation, assignment, scheduling of multiple task-graphs. References 39 and 40 compile a task-graph on a given heterogeneous architecture. They explicitly model the memory system, interconnect and processing elements. The algorithm answers the following question: is it better to distribute the data (at a higher communication cost) or to keep data local (at a higher local memory cost). The above approaches use a naive memory architecture model and hardly incorporate the real behaviour of the interconnections and memories.

Recently, Reference 41 has discussed how many processors are required to execute code as energy-efficiently as possible. The task interaction is empirically accounted for (based on simulation), but this is not a scalable approach. Patel *et al.* [42] partitions the data space of a linked binary. Each part is then mapped onto a memory bank. It selects the partition which optimises the energy cost compared to a dual port memory. The performance of each partition is not accurately estimated since the technique does not account for memory stalls.

We identify the following limitations to the above techniques:

1 they target an architecture which is either too different from ours or is not detailed enough. As a result, we cannot reuse them to optimise the interaction between parallel executing tasks;
2 their program model is too limited for our application domain in which dynamic behaviour plays an important role too.

In the next section, we review the current techniques for coping with the dynamic behaviour.

10.3.4 Runtime memory management

Dynamic applications are slowly becoming desirable in the context of embedded systems. The unpredictability generated by the dynamism entails the usage of runtime policies for effective optimisations. These policies must be implemented efficiently to minimise the resulting overhead (Figure 10.8).

As indicated in the previous sections, for memory bandwidth optimisation, the policy making consists of scheduling the tasks (or their instructions) and (re)distributing their data across the available memories (step 1). To efficiently implement these decisions, we need to manage the memory space at runtime (step 2). In this section, we overview both the runtime decision taking and implementation techniques.

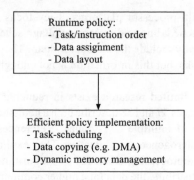

Figure 10.8   Runtime memory optimisation decomposed in two problems: decision taking and implementation

10.3.4.1 Runtime policies

In the context of embedded systems, only a few techniques decide where to store the data at runtime taking the memory architecture into account. For instance, References 12 and 43 decide at design time for each call-site to malloc/new to which memory the data should be assigned. They base their decision on simple criteria: object co-location to avoid conflict misses, object size and access frequency. Nearly no work has been done on memory-aware task scheduling for dynamic multi-media applications. One of the only contributions in this area is Reference 44. There, an OS scheduler directs the power mode transitions of the SDRAM modules, but performs no access scheduling or bank assignment.

10.3.4.2 Enforcing runtime policies

Dynamic memory management is a well-known problem. It has been widely researched in the context of general purpose computing. The main reason is that high-level programming languages intensively allocate data on the heap at runtime. For example, every time in C++ a new object is created, the 'new' function dynamically allocates memory space on the heap. Since applications allocate many differently sized data structures, the heap space easily becomes fragmented. This significantly reduces the available memory space and increases the allocation overhead. Several dynamic memory managers have been proposed for reducing fragmentation (References 45 and 46 give an overview). An important technique to eliminate fragmentation is adapting the dynamic memory manager to the allocation requests of the applications. Kiem-Phong [47] splits the available memory in pools. Every pool is then managed by a separate dynamic memory manager, which deals with a subset of the allocation requests. Usually, the subsets consist of the allocation requests with a similar size.

The authors of Reference 48 present a deterministic hardware-based dynamic memory manager. The memory is hierarchically managed. Each processor has its own memory pool which is controlled by the realtime operating system (RTOS).

Whenever the space in this pool is too limited, the processor allocates more memory from the shared memory pool. The shared pool is split in fixed sized blocks to simplify its management. The result is a memory manager which has very fast memory (de)allocations times.

In the context of multi-processors, the most scalable and fastest memory managers use a combination of private heaps combined with a shared pool [45,49]. These memory managers avoid typical multi-processor allocation problems such as blow-up of the required memory space, false sharing of cache-lines and contention of threads accessing the shared memory. However, they are unaware of the memory architecture and are complementary to our work. As we will show in Section 10.6, we reuse the above techniques to manage the memory space at runtime, but we have to carefully control their allocation overhead.

We conclude from the above that:

1   no decision techniques cope with the underlying memory architecture, albeit a multi-banked SDRAM or the local memory layer;
2   no current runtime decision techniques optimise the memory bandwidth;
3   despite dynamic memory management being a well-researched problem, limited support is available to integrate these decisions inside the code.

10.4   Memory bandwidth optimisation for platform-based design

Many techniques optimise the memory bandwidth for a single thread (Section 10.3), but they break down when applied to multi-threaded applications. In this section, we illustrate how parallel accesses from different processing elements either to the shared memory (subsection 10.4.1) or the local memory layer (subsection 10.4.2) degrade the system's performance and increase its energy consumption. For each of them, we will introduce techniques to mitigate the problems. Our techniques exploit data assignment and scheduling to optimise the behaviour of the memories.

10.4.1   The shared layer

As motivated, the shared layer usually is based on a multi-banked memory, such as an SDRAM. In this subsection, we first explain with an example why existing techniques break down (subsection 10.4.1.1). Then, we present how to overcome these limitations with data assignment and task scheduling (Sections 10.4.1.2–10.4.1.3).

10.4.1.1   Multi-threading causes extra page-misses

Over the past years, several techniques have been proposed to eliminate page-misses inside a single thread (Section 10.3.1), but they cannot cope the ones caused by parallel threads. A small example explains why (Figure 10.9). It consists of task 1 and task 2, running on a different processor tile and accessing data stored in the shared SDRAM memory. As explained above, the more page-misses occur on the SDRAM, the more energy is consumed.[4] One way to minimise the number of page-misses is to carefully assign the tasks' data to the banks of the SDRAM. Current techniques optimise the assignment of a single task at a time. In the case of our example, they generate layout A. If both tasks are sequentially executed, it results in only one page-miss for task1 (see sequential schedule). Also for task 2, only three page-misses occur, because its data (b and c) are distributed across the two banks (see again the sequential schedule).

As soon as both tasks execute in parallel while using layout A, extra delays and many more misses occur, because the SDRAM interleaves accesses from both tasks. For example, task 1 fetches a while task 2 reads simultaneously from b. With its single memory port, the SDRAM cannot access both data structures in parallel. Its interface has to serialise them, delaying the access to b with one cycle. Furthermore, every other access to a or b results in a page-miss, because they are stored on different pages in the same bank. The extra page-misses augment the energy cost and further delay the execution.

Interacting tasks on shared resources thus cause more delays and generate extra page-misses. Currently, no techniques can avoid this, because they optimise the data layout within a single task.

10.4.1.2   Optimising the data assignment across the tasks' boundaries

We have proposed a technique for reducing page-misses across the tasks' boundaries [50]. It stores frequently accessed data structures with high access locality in separate banks as much as possible. To identify these data structures, we have developed a heuristic parameter called selfishness.[5] At design-time, each task is analysed and profiled independently. Every relevant data structure of each task is characterised with a 'selfishness factor'. A data structure's selfishness is the average time between accesses (tba) divided by the average time between page-misses (tbm). It is a measure of spatial locality of the data structure; we finally weight it with the data structure's importance by multiplying it by the number of accesses to the data structure.

At runtime, when we know which tasks will co-occur in time, we decide the assignment of the alive data. We have implemented a greedy algorithm that assigns data to the banks by decreasing order of selfishness. The higher the selfishness becomes, the more important it is to store the data in a separate bank. Our algorithm distributes the data across banks such that the 'selfishness' of the banks is balanced. The selfishness of a bank is the sum of the selfishness of all data structures in the bank.

Consider the example in Figure 10.9. The sequential schedule gives us the required information for each data structure. a and b both have the same spatial locality, because the time between misses equals the entire duration of the task and the time between accesses is similar. However, because b is less frequently accessed than a, its selfishness is slightly lower. The selfishness of c is much lower than both a and

---

[4] For the clarity of our example, we only focus on their energy penalty, i.e. no performance penalty due to page-misses.
[5] For details of how to measure selfishness we refer to Reference 50.

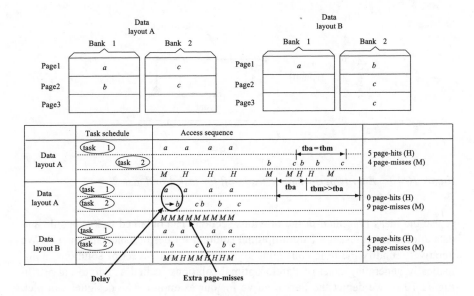

Figure 10.9 Interleaved accesses from different tasks cause page-misses and extra stalls

b, since for every access a page-miss occurs and it is less accessed. Therefore, when we schedule both tasks in parallel, our algorithm first separates the most selfish data structures a and b and then stores c with b (since the bank containing b is less selfish than the one with a). This corresponds with layout B in the figure. As a result, only five page-misses remain and the energy cost is significantly reduced compared with a naive layout: layout A just places one data structure after the other. This results in one page-miss per access.

10.4.1.3 Task ordering to trade-off energy/performance

Besides data assignment, the task order also heavily impacts the system's energy and performance. For instance, if we execute task 1 and task 2 sequentially, four page-misses occur. This is the most energy-efficient solution, but takes the longest time to execute. In contrast, when we execute both tasks in parallel, the execution time becomes shorter, but five page-misses occur, thus the energy cost increases.

Generally, by changing the task order we can trade-off the energy/performance of the system. We have developed an algorithm to schedule a set of tasks in such a way that the energy consumption is minimised while meeting a pre-fixed time deadline. For a given application, we first define the most likely combination of tasks that will happen at runtime (we call these combinations 'scenarios'. See Section 10.5). For each scenario and a specific time-constraint, we explore different schedule possibilities, trying to find the most energy-efficient. Once the relative schedule is defined, we must allocate the data of the tasks. For that purpose, we

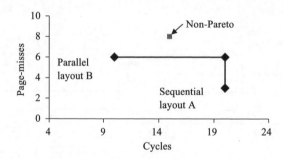

Figure 10.10   Energy/performance trade-off for task 1 and task 2

re-use the ideas presented in the previous section. The obtained task schedule and data allocation represent a Pareto-optimal solution for performance and energy. We iteratively modify the time-constraint in order to generate a set of solutions, automatically generating a set of Pareto-optimal solutions (called a 'Pareto curve'. In Figure 10.10 we depict the Pareto curve for our example). The designer can pick the operating point which best fits his needs from the generated trade-off points. All the details of this joined task schedule/data assignment technique are shown in Reference 51.

Memory aware task scheduling may be be also beneficial for performance. Assuming that it is always better to distribute the data across banks to reduce the number of page-misses, a conservative task schedule increases the assignment freedom. When the ratio 'number of data structures : number of banks' becomes high, insufficient banks are available to separate all energy critical data structures from each other. Data allocation alone does not suffice to decrease the number of page-misses. In such a situation, task scheduling is a good way to enlarge the freedom of the allocation process. Generally, sequential schedules result in the lowest energy consumption, but they have the worst execution time. In general, the trend is clear: the lower the execution time (scheduling more tasks in parallel), the higher the energy consumption. Of course, some schedules will not follow this tendency. Sometimes, too much parallelism is bad even for performance!! (if the number of page-misses increases too much and the applications are memory bounded, an aggressive parallel task schedule could increase the total execution time).

In Figure 10.11 time and SDRAM energy consumption values are shown for four different schedules of the same task-set. Schedule *D* corresponds to a sequential scheduling: as expected, the longest execution time with the lowest energy consumption. Full parallel schedule is shown in part *A* of the figure. *B* and *C* are intermediate schedules, that trade-off performance and energy consumption. As well as illustrating our point, Figure 10.11 also points out that the execution time of a task cannot be estimated independent of the other tasks running in parallel with it. The time penalty raised from sharing the SDRAM between several concurrent tasks reaches up to 300 per cent for *CMP* (in schedule *A*), compared to its execution time without other tasks executing on the platform.

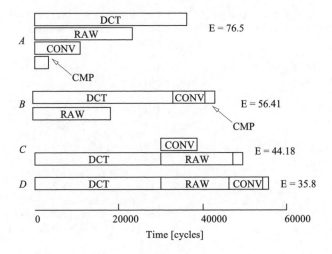

Figure 10.11 Scheduling outputs for four tasks

10.4.2 The local memory layer

10.4.2.1 Access conflicts reduce the system's performance

As indicated in Section 10.3.2, existing techniques only optimise the memory bandwidth in the scope of a basic-block. As a result, a large room for improvement remains. We illustrate this with a small example that consists of three data-dominated loops (see code in Figure 10.12 – left) which are executed on a platform that consists of three single-port memories: two 4 kB ones (0.11 nJ/access) and a 2 kB one (0.06 nJ/access).

Because the applications are data-dominated, the duration of the memory access schedule determines the performance of the loops. We assume that the remaining operations can be performed in parallel with the memory accesses or take only limited time. We will use the example presented in Figure 10.12 – left to study the influence of memory access schedule together with data assignment in the resulting performance and energy consumption.

Current compilers are not aware of the final data to memory assignment. During instruction scheduling, most compilers simply assume that any memory operation finishes after n-cycles. When the executed operation takes longer than presumed, the entire processor is stalled. As a result, often a large difference exists between the expected and the effective performance of the processor. We use a typical modulo scheduler [52] to generate our memory access scheduling. Note that a modulo scheduler may schedule read/write operations from the same instruction in the same cycle. This is the case in our examples: the write operation belongs to the iteration i while the read operation comes from iteration $i + 1$. Modulo scheduling may be applied when there are no data-dependent conditions in the loop body. The scheduler generates a memory access schedule for the inner-loops of 460 cycles (Figure 10.13(a)). However, the actual performance varies between 540 and 740 cycles. The schedule takes longer than expected because the processor has to serialise the accesses to D

```
int A[301],int B[100];int D[100]        int A[300],int B[100];int D[100]
int C[100]; int U[2];                   int C[100]; int U[2];
int i,j;

for(i=0;i<100;i++);// loop 1            for(int i=0; i<100; i++) // loop 2
 A[i+1]=A[i]+1;                            D[i]=C[i]+B[i];

                                        for(i=0; i<2; i++){ // loop 1&3
                                          for(int j=0; j<40; j++){
                                          D[j] = D[j-1]+ D[j];
for(i=0;i<100;i++) // loop 2              A[40*i+j]=A[40*i+j-1] + 1;
 D[i]=C[i]+B[i];                          }
                                          U[i] = D[39];
for(i=0;i<2;i++){ // loop 3             }
 for(j=0;j<40;j++) // loop 31           // remainder of loop 1
  D[j]=D[j-1]+D[j];                     for (int i=0; i<20; i++)
 U[i]=D[39];                              A[i+80] = A[i-1+80] + 1;
}
```

Figure 10.12   Motivational example: original code (left), code after fusion (right)

in loop 31. Extra stalls occur depending on whether the linker has assigned the C, B and/or D to the same memory.

Because how the linker assigns the data to the memories has such a large impact on the performance of the system, it is better to optimise the data assignment and the memory schedule together [28]. Our technique imposes restrictions on the assignment such that the energy is optimised, but still guarantees that the time-budget is met. The assignment constraints are modelled with a conflict graph (e.g. Figure 10.14 – left). The nodes correspond to the data structures of the application. An edge between two data structures indicates that we need to store them in different memories. Hence, the corresponding accesses to these data structures can be executed in parallel.

The assignment-constraints imposed by the conflict graph prevents certain accesses from happening in parallel. This restricts the feasible schedules to those respecting the access restrictions. Thus, the conflict graph links data assignment and instruction scheduling: for a specific conflict graph we can determine the fastest schedule possible when using the least energy consuming data assignment solution. For instance, the edge between A and C (Figure 10.14 – left) forces us to store both data structures in different memories. The fastest schedule for this conflict graph takes 540 cycles (Figure 10.13(b)). It consumes 64.4 nJ,[6] because the conflict edges force us to store both A and B in large memory (see complete assignment in Figure 10.14 – left). Note that the energy consumed in the memories only depends on the conflict graph, not on the access schedule. For a fixed conflict graph, the energy dissipated in the memories will remain the same even with different instruction schedules.

We can decrease the energy cost of the above assignment by reducing the number of conflicts. After eliminating the edges between B–D, C–D and B–C, the small

---

[6] We compute the energy consumption as follows: $\sum_{\forall m \in M} \sum_{\forall ds \in m} \mathrm{NrAccess}(ds) E^m_{access}$

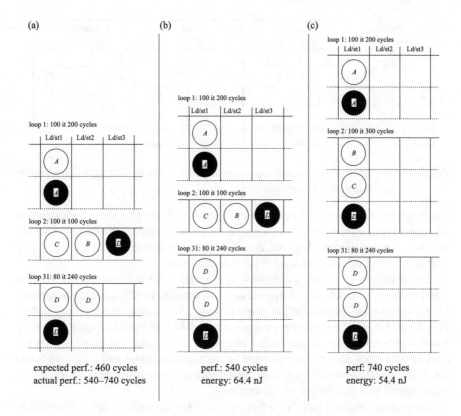

(a)

loop 1: 100 it 200 cycles

| Ld/st1 | Ld/st2 | Ld/st3 |

(b)

loop 1: 100 it 200 cycles

| Ld/st1 | Ld/st2 | Ld/st3 |

(c)

loop 1: 100 it 200 cycles

| Ld/st1 | Ld/st2 | Ld/st3 |

expected perf.: 460 cycles
actual perf.: 540–740 cycles

perf.: 540 cycles
energy: 64.4 nJ

perf: 740 cycles
energy: 54.4 nJ

*Figure 10.13   Empty issue slots in the memory access schedule of the inner-loops:
(a) existing compiler, (b) with fastest partial data assignment, (c) with
most energy-efficient partial assignment*

data structures B, D and C can be assigned in the smallest and most energy-efficient memory (Figure 10.14 – right). The energy consumption is then 54.4 nJ instead of the original 64.4 nJ. Fewer conflicts also imply that fewer memory accesses can execute in parallel. The fastest feasible schedule now takes 740 cycles (Figure 10.13(c)). The energy savings thus come at a performance loss.

However, many memory access slots remain empty (check again Figure 10.13). This is mainly due to: (1) inter-iteration dependencies. For instance, the initiation interval of loop 1 is 2, because A depends on itself. Hence, only 30 per cent of the available memory slots are used; (2) we do not use power hungry multi-port memories. Consequently, we cannot schedule operations that access the same data in parallel. For example, in loop 31 we cannot execute the accesses to D in parallel.

10.4.2.2   Global schedule: loop morphing

With more global optimisations, such as loop fusion [53], we can further compact the application's schedule. However, existing fusion techniques can only overlap

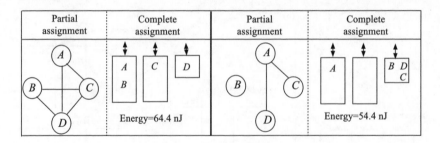

Figure 10.14   Partial assignment expressed with a conflict graphs: (left) fast; (right) more energy-efficient

few iterations. Consider the loop nests in Figure 10.16. We also show their iteration domains (each dot corresponds to a single iteration). Traditional techniques would only fuse four out of nine iterations, due to the different dimensions of the loop nests. This restriction limits the achievable performance gains.

Let us apply loop fusion to our previous example (code in Figure 10.12). We first select one conflict graph, which prevents certain accesses from happening in parallel. Let us consider the fastest conflict graph (Figure 10.14 – left). If we choose to fuse loop 1 and loop 2, the resulting schedule would take 440 cycles (Figure 10.15(a)). Another option would be to fuse loop 1 and loop 3. In this case, the loop nests are not conformable (they do not have the same number of dimensions and their loop limits are different). However, current loop fusion techniques will just fuse 40 iterations, resulting in a schedule length of 460 cycles.

Loop morphing is a technique that enables loop fusion beyond conformability limits (details can be found in Reference 54). Once we have decided which loops to fuse, our algorithm gradually tries to make them as similar as possible. Loop splitting and strip mining are iteratively applied to obtain conformable loop nests. Figure 10.17 shows the different steps to fully fuse the loops shown in Figure 10.16. First, we apply strip-mining to l2 to fit the number of dimensions of the other loop. The resulting loop nest is split to avoid the if-condition in the body of the loop. A new loop nest, with just one iteration, is generated (l3). We now transform the longest loop l1 such that it has the same length as l2. This transformation is accomplished through loop splitting. After that, we have two loop nests perfectly conformable, that we can easily fuse (Phase 2 of Figure 10.17). Note that eight iterations have been fused, instead of only four in Figure 10.16.

In our example in Figure 10.12, we first split loop 1. Two loops are generated: one with the first 80 iterations and a second one with the last 20 iterations. Strip mining is applied to the first of these two new loops. This way we obtain a two-level nested loop, similar to loop 3. Fusion is now straightforward. Finally, the 20 remainder iterations from loop 1 may be considered for subsequent fusion decisions. The final code after morphing is shown in Figure 10.12 – right. Always using the fastest conflict graph, loop morphing enables a schedule length of only 380 cycles (Figure 10.15(c)).

From the above, we conclude that for a given conflict graph we may decide which are the best loops to fuse, and use morphing to maximise the iterations fused. However,

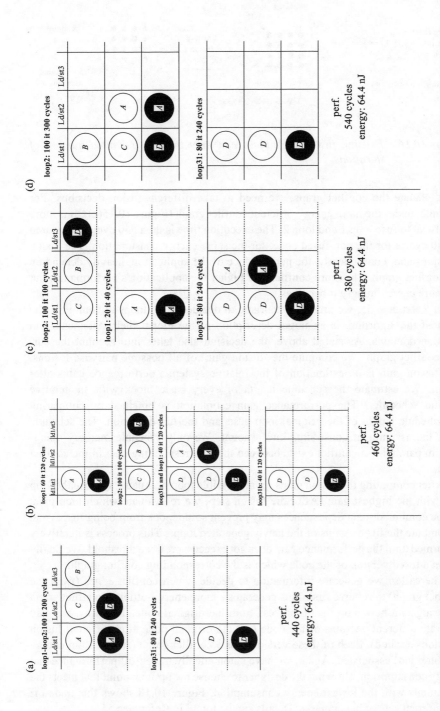

Figure 10.15 Loop fusion fills the issue-slots: (a)–(b) existing fusion techniques for the fastest partial data assignment, (c) fusion combined with loop splitting and strip mining, (d) best fusion for the energy-efficient partial assignment

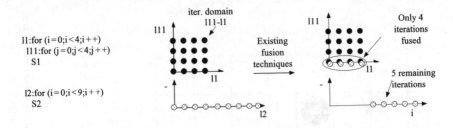

```
l1:for (i=0;i<4;i++)
 l11:for (j=0;j<4;j++)
  S1

l2:for (i=0;i<9;i++)
  S2
```

Figure 10.16   *Existing fusion techniques can only overlap a limited number of iterations*

if we change the conflict graph, we need to take different fusion decisions. For example, under the more energy-efficient conflict graph (Figure 10.15(c)), it is more beneficial to fuse loop 1 and loop 2. The execution time is then 540 cycles compared to 740 cycles for the non-fused code, for the same energy consumption. The fusion decisions and, consequently, the performance of the application, heavily depend on the conflict graph. The more conflicts the higher the application's performance, but the more energy hungry it becomes.

In Reference 55, we present a heuristic to decide which loops to combine. The input of the algorithm is an initial description of the loops, their statements and iteration domains. As stated above, the decision also takes into account the current conflict graph. We compute the 'fusion gain' of all possible pairwise fusions. The fusion gain is an estimation of the relative system's performance gains after fusion. We estimate the schedule length of every basic block with an iterative modulo scheduler. The performance gain estimation is obtained by comparing the schedule length of the original loop nests and the fused version. The schedule takes the assignment constraints into account. We only schedule memory operations in parallel if a conflict exists between their corresponding data in the conflict graph.

After computing the fusion gains for all possible loop nest pairs, we fuse the loop pair with the highest gain. After the fusion step, we re-evaluate which loop pairs can be combined (data dependences may prevent some loops from being fused) and re-compute the fusion gains of the newly generated loops. This process is iteratively performed until the performance gain does not exceed a prefixed threshold. We finally obtain a fused version of the code which is the corresponding conflict graph.

Thereafter, we generate information to decide which conflict edges (from the conflict graph) to remove first. This generates a more energy-efficient conflict graph that triggers a new loop fusion process from the original code. This way, we may generate different versions of the code (i.e. the original code after different fusion decisions applied). Each of these versions has a conflict graph (and thus, an energy consumption) associated. Again, we have automatically traded-off performance and energy consumption, allowing the designer to choose the optimal point that meets the constraints with the lowest energy consumption. Figure 10.18 shows this trade-off for different sets of benchmarks. Details can be found in Reference 55.

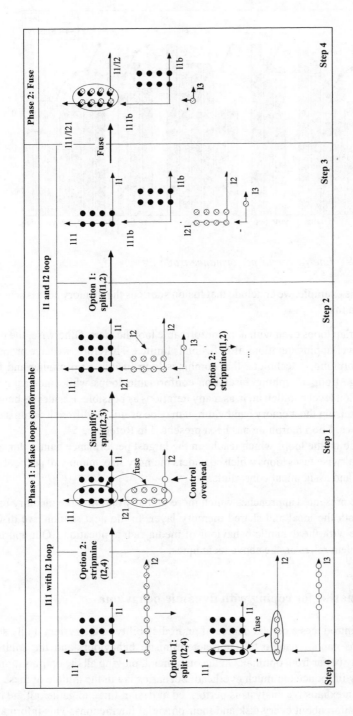

Figure 10.17 Morphing heuristic applied on an example

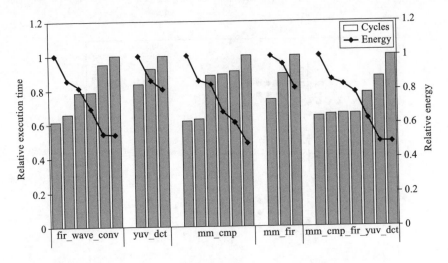

Figure 10.18   Energy versus performance trade-off

From this example, we conclude that fusion shortens the memory access schedule on condition that:

1   We overlap loops even with non-conformable loop headers. Otherwise, the number of overlapping iterations after fusion is limited. Therefore, we have proposed loop morphing, a technique that combines loop fusion, strip mining and loop splitting. Loop morphing fuses non-conformable loops while increasing the instruction level parallelism in as many iterations as possible. Besides its benefits for optimising the memory bandwidth, it may be useful for different optimisation objectives. Loop morphing has been presented in Reference 54.

2   We combine the loops which results in the largest performance gains. Our technique pairwise fuses loops which considers memory size, number of ports, access latency and assignment constraints.

We have presented approaches which more globally optimise the memory bandwidth for both the local and shared memory layer. In the next section, we discuss how to cope with the dynamic behaviour of media-rich applications. Our approach for this problem relies on the above techniques.

10.5   Scenarios for coping with dynamic behaviour

Dynamism introduces a certain degree of unpredictability in the system. Fully static optimisations cannot handle this non-deterministic behaviour, and the solutions obtained may be far from optimal. On the other hand, moving all the optimisations to runtime may introduce too much overhead. Scenarios lay in the middle of these two extremes. An exhaustive analysis is performed at design time, to gather all the relevant information about every task and their potential interactions. This information

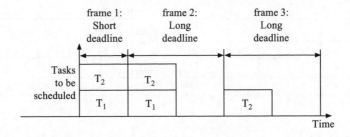

Figure 10.19   Dynamically created tasks with their deadline

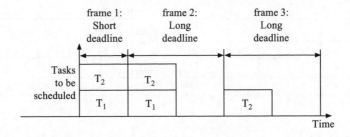

Figure 10.20   Our runtime task schedule/SDRAM assignment solution

will be later used at runtime to quickly take a decision. In this section, we will first examine the impact of this dynamism, focusing in the resulting energy consumption. Later, we will briefly explain the scenario approach.

10.5.1   Energy constraints demands for runtime decisions

Due to the dynamic behaviour of our application domain, it is more energy-efficient to assign the data and schedule the tasks at runtime. An example in the context of the SDRAM layer explains why (Figure 10.19). At the start of each frame, the user executes task 1 and/or task 2. They are the same tasks as in Section 10.4.1. We thus only know at runtime which tasks execute and which data they require. Also, note that the deadline varies from frame to frame. For example, at the start of frame 2, the user lowers the video quality, reducing the frame-rate by half. The system has then twice more time for each frame.

The optimal task order/data assignment decisions vary from frame to frame (see Figure 10.20). For example, to satisfy the short deadline in frame 1, we have to schedule both tasks in parallel. We obtain the least number of page-misses using layout B from Figure 10.9. However, in frame 2, only task 2 is active. As indicated in Figure 10.9, layout A is then more energy-efficient.

Finally, in frame 3, both tasks are started again. However, since the frame-rate is lower now, we can execute them sequentially and eliminate most page-misses with

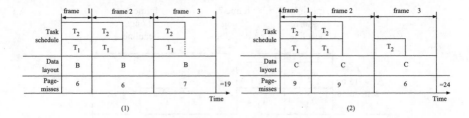

*Figure 10.21   State-of-the-art task schedule/SDRAM assignment: (1) design-time
and (2) operating system based solution*

layout A. So, in each frame, different scheduling/assignment decisions are optimal
for energy and we can only take these decisions at runtime.

Current approaches always generate more page-misses. Design-time techniques
only select one task schedule and layout (Figure 10.21(1)). This single design has to
meet the deadline for the worst-case load, i.e. task 1 and task 2 executed within the
short deadline (frame 1). The most energy-efficient design for this load is executing
both tasks in parallel and using layout B (Figure 10.9). This operating point is not
optimal for both frame 2 and frame 3. It results in seven more page-misses than the
above approach. A pure design-time technique is thus not energy-efficient.

Furthermore, also current runtime approaches are far from optimal. A typical
OS does not account for the specific behaviour of SDRAMs. As long as enough
processors are available, it schedules all tasks in parallel and assigns the data to the
first available free space (Figure 10.21(2)). By storing all the data in a single bank
and scheduling the tasks in parallel, 12 more page-misses occur than in the optimal
case. Again this solution is not energy-efficient.

These results indicate the potential benefits for a runtime technique which considers the SDRAM behaviour and can generate the solutions of Figure 10.20. Since it
should make complex task scheduling/data assignment at runtime, the main difficulty
is restricting its overhead. The local memory layer requires a similar approach, but
we restrict ourselves to the shared SDRAM layer.

10.5.2   *Our scenario-based approach*

We have proposed a mixed design-time/runtime approach for coping with the
dynamic behaviour (Section 10.6). The philosophy behind it is to take most schedul-
ing/assignment decisions at design-time for all frequently occurring task sets. In this
way, we can reuse our bandwidth optimisation techniques for multi-threaded applications and at the same time limit the runtime complexity. Only for the more seldomly
occurring task-sets a pure runtime decision is taken as a backup solution. In the next
paragraphs, we explain the main steps of our methodology (Figure 10.22).

10.5.2.1   Scenario identification

Fixing as many decisions as possible at design time comes at the risk of ignoring the actual behaviour and generating worst-case designs. For example, consider

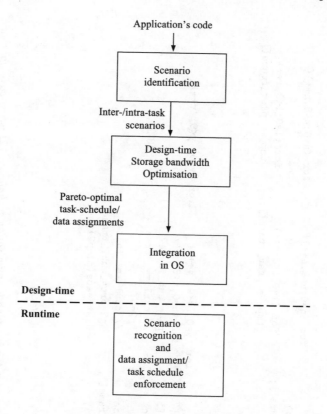

Figure 10.22   *A scenario-based design-time/runtime approach*

the code in Figure 10.23. Even though parts of the code are conditionally executed (e.g. mode and ctrl-conditions in the second loop nest), design-time techniques assume that both branches are executed, optimising thus the design for the worst-case load. As a consequence, we heavily over-estimate the required bandwidth and usually generate an over-dimensioned and energy-inefficient system.

To prevent this energy-loss, we try to capture the dynamic behaviour with scenarios. First, we analyse which tasks-sets often co-occur at runtime. We call them 'inter-task scenarios'. A similar but more restrictive concept is used by Reference 56. Second, we narrow down the data-dependent behaviour inside the tasks with 'intra-task scenarios'. An intra-task scenario is an execution path through the task (or a combination of execution paths) for different data-dependent parameters [57,58]. Both the inter- and intra-task scenarios should be manually extracted by the designer (using profiling). For example, in the code of Figure 10.23, we derive two intra-task scenarios, one for mode equals true and another one for mode equals false. Even though research outside IMEC is ongoing into how to identify scenarios [59,60], a more automated approach is still needed.

```
// scenario 1: mode=true                    // scenario 2: mode=false
for (y=0; y<9; y++){                         for (y=0; y<9; y++){
  ...                                          ...
  for (n=0;n<8;n++){                            for (n=0;n<8;n++){
    ...                                            ...
    for (l=0;l<8;l++)                              for (l=0;l<8;l++)
      tmp += prev_frame[];                           tmp += prev_frame[];
  ...                                          ...
}                                            }

for (y=0; y<9; y++){                         for (y=0; y<9; y++){
  for (n=0;n<8;n++){                            ...
    p1 = sub_frame2[];                       }
    // ctrl == true
    p2=0;                                                    (3-2)
    // ctrl == false
    p2=prev_sub2_frame[];
    dist+=abs(p1-p2);
  }
  ...
}

                (3-1)
```

```
for (y=0; y<9; y++){                         for (y=0; y<9; y++){
  ...                                          ...
  for (n=0;n<8;n++){                            for (n=0;n<8;n++){
    ...                                            ...
    for (l=0;l<8;l++)                              for (l=0;l<8;l++)
      tmp += prev_frame[];                           tmp += prev_frame[];
  ...                                          ...
}                                            }

for (y=0; y<9; y++){                         for (y=0; y<9; y++){
  if(mode)                                      // mode == true
  // mode == true                               for (n=0;n<8;n++){
  for (n=0;n<8;n++){                              p1 = sub_frame2[];
    p1 = sub_frame2[];                            // ctrl == true
    if (ctrl)                                     p2=0;
      p2=0;                                        // ctrl == false
    else                                           p2=prev_sub2_frame[];
    // ctrl == false                               dist+=abs(p1-p2);
      p2=prev_sub2_frame[];                      }
    dist+=abs(p1-p2);                            ...
  }                                            }
  ...
}                                                            (2)

                (1)
```

Figure 10.23 (1) data-dependent conditions as a limiting factor for bandwidth optimisation techniques, (2) worst-case approach, (3) scenario-based approach

After identifying the scenarios, we can represent each one with a data-flow graph on which we can easily apply our design-time techniques.

10.5.2.2 Storage bandwidth optimisation at design time

In the second step, we optimise the storage bandwidth of each scenario. Our design-time techniques generate for each scenario a set of task ordering/data assignment solutions. Each solution optimises the energy cost for a given time-budget. From this set, we only retain the Pareto-optimal solutions. For example, for our example's scenario in which task1 and task2 are active (Section 10.5.1), we would generate the Pareto curve of Figure 10.10. Finally, we integrate the Pareto set of each scenario into the OS. We provide more details in Reference 61.

10.5.2.3 Runtime phase

Then, at runtime, after identifying which scenario is activated and which is its deadline, we simply select the best prestored operating point on the Pareto curve and enforce its task ordering and data assignment decisions. For example, for frame 1 of our example (Section 10.5.1), our runtime mechanism then selects the leftmost operating point, scheduling both tasks in parallel with layout B. In contrast, for frame 2 with the relaxed deadline, it implements the rightmost operating point. If the scenario was not analysed at design time, we use a back-up solution. For example, we simply use an existing dynamic memory (DM) manager for assigning the data. Note that our approach leverages current design-time techniques, but requires that scenarios can be identified inside the application. Obviously, this partly restricts the applicability of our technique.[7]

10.6 DM management

The memory space available at runtime to our applications can be located in any physical memory of the system. It is managed with the help of a DM manager. DM management basically consists of two separate tasks, i.e. allocation and deallocation. Allocation searches for a memory block big enough to satisfy the request of a given application and deallocation returns this block to the available memory of the system in order to be reused later. In real applications, memory blocks with various sizes are requested and returned in a random order, creating 'holes' among used blocks (Figure 10.24). These holes are known as memory fragmentation [46]. We talk about 'internal fragmentation' when the wasted space is inside allocated partitions. Otherwise, we will name it external fragmentation. Internal fragmentation happens when allocated memory is larger than requested memory and not being used

---

[7] Another approach could be to start from existing OSs and make them account for the energy cost of the underlying memory architecture. All decisions are then made at runtime without design-time preparation. Even though we did not investigate this, we expect that such an approach causes too much energy overhead and violates more deadlines, but more research is still needed.

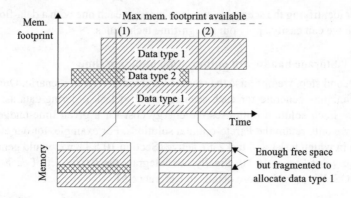

Figure 10.24   Fragmentation in DM management

(e.g. a free block of 5 kB is used for a request of 4 kB, hence 1 kB is wasted). If we suffer from external fragmentation, total memory space exists to satisfy the request, but it is not contiguous (e.g. if a request asks for 6 kB and there are several non-contiguous free blocks of 3 kB, the memory request cannot be satisfied). The DM manager has to take care of fragmentation issues.

We have classified all the important design options in different orthogonal decision trees, which can compose the design space of DM management. Based on these orthogonal trees, we can construct custom DM manager from its basic building elements. The most critical parts of the design search space are overviewed in the next paragraphs. For a complete description about the design space and how it can be used to build custom DM managers see References 62 and 63. Then, we analyse the most commonly used DM managers with the use of the design space. Finally, we explain our approach and why it is more energy-efficient than other state-of-the-art approaches to construct DM managers.

10.6.1   A brief summary of the DM management design space

The most important part is trying to prevent fragmentation. In Figure 10.25, we illustrate the most commonly used technique to prevent memory fragmentation. This technique assigns memory blocks to different memory segments (also called pools), which are then accessed by one pointer array that keeps track of the initial position of the pool with free blocks. The goal is to try to anticipate the size of the memory requests of the application. If every memory request is met with a memory block with the same size, then no memory space inside the block gets wasted. This means less internal fragmentation and a quick allocation of the requested block. For example, if an application usually employs blocks of 8 kB and 1024 kB, 50 per cent of its total memory allocation requests, its DM management will be simplified significantly if two memory pools of these sizes exist, then just two accesses are needed to return one block (i.e. one access to the pointer array and another one to update the first memory block pointer available to the next one).

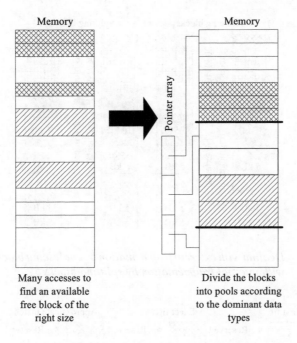

Many accesses to
find an available
free block of the
right size

Divide the blocks
into pools according
to the dominant data
types

Figure 10.25   Assigning the memory blocks to multiple pools

The second most important part of the DM management design space is about trying to deal with fragmentation. This means using defragmentation functions (i.e. coalescing and splitting memory blocks – see also Reference 64). On the left, we can see the coalescing of memory blocks, which is the way to deal with external fragmentation. If a requested size (e.g. 20 kB) is bigger than the size of the available memory blocks (e.g. 10 kB), then it is possible to coalesce the two smaller blocks. On the right, the splitting of a memory block is shown, which is the way to deal with internal fragmentation. If a requested size (e.g. 4 kB), is smaller than the size of the available memory block (e.g. 10 kB), then it is possible to split it into two smaller blocks. In this way, the remaining 6 kB are not wasted on internal fragmentation and can be re-used for a later request.

Finally, the third most important part of the design space is about trying to select the correct block to fulfill the memory request. This is done by the 'fit policies', some of which can be seen in Figure 10.27. On the left, it is shown how a 'first fit' policy works. This policy satisfies the memory request with the first block that it finds, that is not reserved and has enough or more space than the request. Needless to say, this policy is fast, but produces big amounts of internal fragmentation since blocks of large sizes can be used for allocation requests that are small. In the middle, the 'exact fit' policy is depicted. This policy will not stop looking for a block, unless it finds a block with the same size as the request. This policy eliminates internal fragmentation, but it is very slow due to the large amount of blocks that are needed to find the best candidate if the list of free blocks is extensive. Finally, on the right side, we can

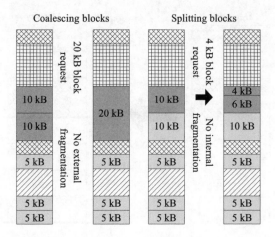

Figure 10.26　Dealing with external fragmentation by coalescing blocks and dealing with internal fragmentation by splitting blocks

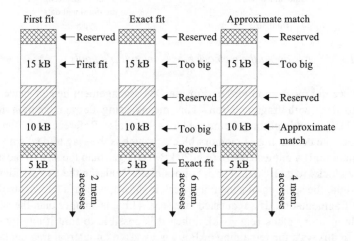

Figure 10.27　Fit policies example for a 5 kB block allocation

see the 'approximate match' policy. This policy satisfies the request according to a parameter, defined by the designer, which is a threshold that states how precisely it should look for a suitable block. In the case of Figure 10.27, this parameter states that the assigned block cannot be bigger than twice the size of the request. This is a more balanced approach than the two previous ones, neither wasting too much memory space nor slowing down the DM manager too much. However, the two previous options are also found in the literature in extreme cases where one of the two metrics (i.e. minimisation of memory space or performance) is much more important than the other one.

10.6.2 Existing memory managers focus on different parts of the design space

All the DM managers include these previous decisions of the design space in one way or the other in their designs. In the following we describe the main types of state-of-the-art DM managers.

First of all, one of the most popular DM managers, namely the Lea Allocator [64], is designed to optimise memory footprint by eliminating fragmentation, while preserving a reasonable speed. It is very frequently used in Linux-based systems. More specifically, it uses a very complex pool architecture, which prevents memory fragmentation and speeds up DM management. Then, it tries to defragment as much as possible, thus reducing even more the memory fragmentation, but slowing down DM management in a very significant percentage. Finally, it uses a combination of the previous best fit and first fit policies, which in total does not improve the speed a lot, but manages to preserve a reasonable low fragmentation level in a general context. The problem with the Lea Allocator, in the context of energy-efficiency, is that it uses far too many memory accesses trying to defragment. All these memory accesses cause the power consumption to increase extensively.

Second, another very popular type of DM manager is the one which uses many simple fixed-sized pools to allocate memory. This style is used by the Kingsley Allocator [46]. These allocators are very fast, but they deal poorly with fragmentation, thus use a big memory footprint. More specifically, they use a more straightforward definition of fixed-sized pools where only one size can fit in each pool perfectly and many lists of different sizes are pre-allocated during the initialisation of the DM managers, which try to prevent as much fragmentation as possible by placing the memory requests in the correct pools but can result in a large portion of memory wastage if all the memory pools of different sizes are not used. However, the truth is that the main goal of these kind of managers is quick-allocations and de-allocations and not limiting memory footprint consumption. As a result, since they use fixed-sized memory pools, they rarely (or even never) coalesce and split memory blocks thus ignoring defragmentation. This makes fragmentation usually even worse, but makes the DM manager even quicker. Finally, they also use a combination of best fit and first fit policies, which in total do not affect much speed, but unfortunately maintain the very high fragmentation level. The problem with this kind of allocator, in the context of energy-efficiency, is that high fragmentation affects the energy per access, because we have to assign data with more fragmentation to bigger, more power-hungry memories.

Finally, several custom DM managers exist to satisfy the needs of specific types of applications and their memory requests. An example is the Obstacks [46] custom DM manager, which is used to optimise a stack-like behaviour. Obstacks uses variable-sized pools, with no defragmentation support and an exact fit policy. This allocator is very fast and works well when many consecutive small-sized block allocations occur and finally one deallocation can be used to deallocate all the memory blocks at the same time (i.e. stack-like behaviour). Any other behaviour or big-sized requests make the fragmentation grow extensively, thus reducing its energy-efficiency because bigger memories are needed to store the data (as explained in the previous paragraph). The main limitation of these custom allocators is that they rely on a specific dynamic behaviour to work well and are manually designed and optimised. Unfortunately, no systematic methodology exists to create a custom DM manager from scratch to match a specific dynamic behaviour of an application.

10.6.3 Our approach to create a low power custom DM managers

Our approach consists of the study of the complete design search space. Also, we profile extensively applications to define its dynamic behaviour and its dominant dynamic data allocation sizes. The combination of these two elements produces a systematic methodology to create custom DM managers. In contrast with the previous approaches we address the whole design space rather than focusing on small subsets that may work well for a certain number of applications and a concrete metric (e.g. performance as Kingsley-based managers). Thus, we create custom DM managers with reduced power consumption, because we try at the same time to have few memory accesses and keep a low fragmentation level. The key point of our energy-efficient approach is the study of the trade-offs between low memory accesses to improve speed and low memory fragmentation to reduce the use of memory footprint (for more details see References 62 and 63). Moreover, all our custom DM managers are implemented in the middleware on top of the OS, hence platform-independent. This means that they do not require any hardware changes, as opposed to Reference 48.

As the basic template used in our energy-efficient DM managers, using the profiling information obtained in each concrete application, at least one fixed-sized pool is assigned to each of the sizes of the dominant data types. The sizes that are good candidates are those that imply at least 20 per cent of the total amount of allocation requests. This approach prevents most of the memory fragmentation and speeds up the DM manager by a significant factor (i.e. between 10 and 50 per cent) compared to those using only general-pools with many sizes. Then, we defragment only when this is absolutely needed to carefully balance the overhead of memory accesses required in the coalescing and splitting mechanisms. Mostly, we defragment when we are close to the high watermark of our memory space to try to avoid the allocation of the next requested memory block in a larger memory. Finally, a first fit policy is used only for dominant data types, for which we have already provided pools with their corresponding block size, thus it is equivalent to exact fit. The remaining data types must make concessions and sacrifice some memory footprint using an 'approximate fit' policy for preserving a certain degree of performance, because the number of memory accesses explodes if the DM manager searches the pools exhaustively to find the exact fit, as we indicated in the previous section.

We illustrate our technique with a real-life illustrative example: the Deficit Round Robin (DRR) application taken from the NetBench benchmarking suite [65]. It is a buffering and scheduling algorithm implemented in many wireless network routers today. Using the DRR algorithm the router tries to accomplish a fair scheduling by allowing the same amount of data to be passed and sent from each internal queue. It requires the use of DM because the input can vary enormously depending on the network traffic. The DRR algorithm has three types of dynamic data: the Internet packets, the packet headers and the list that accommodates the internal queues. It works in three phases when it is receiving packets. First, it checks the packet header, second it traverses the list of internal queues and, finally, it traverses the correct internal queue and stores the Internet packet in a first-in first-out (FIFO) order. Then, it forwards the packets in three additional phases. First, it traverses the list of internal queues, second, it picks up the first Internet packet and, finally, checks the packet header to forward it.

After analysing the DM behaviour and allocation sizes of this example with our approach, we decide to create two fixed-sized memory pools for the maintenance fields of the list of internal queues and the headers of the packets. Then, a general pool using approximate fit is used to store the internal variable-sized packets coming from the network that need to be forwarded. This choice was based on the dynamic sub-phases found with our approach in the algorithm, which indicate that the list of internal queues has 85 per cent of the total accesses on average and that this list is small enough to fit inside a small memory pool of 4 kB (or 8 kB in the worst case), which can increase significantly the locality when all the blocks are placed together and reduce enormously the complexity of DM management compared to other DM managers (i.e. 40 per cent less energy than Lea and 45 per cent less memory footprint than Kingsley-based DM managers).

Other experimental results in real-life embedded applications from the multi-media and wireless network domain show a significant reduction (up to 60 per cent) in power consumption with the use of our approach [66]. Also, when other factors are really limiting (e.g. memory footprint), trade-offs can be made to achieve the desired results (i.e. reduction of 60 per cent on average in memory footprint [62]) by an exhaustive exploration of the design space using our plug-and-play approach (see Reference 63 for more details). Finally, runtime behaviour profiling is embedded within our custom DM managers, so that real-time performance restrictions can be observed and deadlines met, using trade-offs between power consumption and performance [63].

10.7 Conclusions

In dynamic multi-threaded applications dealing with dynamic data and tasks is crucial. The memory bandwidth is both an issue at the shared SDRAM memory as well as on the local memory layer.

Modern multi-media applications contain multiple tasks and/or benefit from task parallelisation. However, tasks running on multiple processors can access the same memory in parallel. This causes access conflicts that delay the system and increase its energy cost. Since existing techniques also optimise the memory bandwidth inside a single task, they cannot cope with inter-task conflicts. A need thus exists for techniques that optimise the memory bandwidth across the tasks' boundaries. We have therefore introduced several task-ordering/data assignment techniques for both the local memory layer as the SDRAM.

Multi-media applications consist of multiple tasks which are started/stopped at runtime due to user events. Also the tasks themselves have become data-dependent. We have shown that design-time nor runtime techniques can effectively deal with this dynamic behaviour. Therefore, we have introduced a novel scenario-based memory bandwidth approach. It combines the best of the design-time and runtime techniques.

Finally, runtime memory management requires an efficient management of the free space. We have introduced a methodology which customises the DM managers for this purpose.

In future work, we want to automatically extract the inter-thread frame scenarios form the user's behaviour.

References

1 Semicon. www.semicon.org
2 DE MAN, H.: 'On nanoscale integration and gigascale complexity in the post.com world'. Proceedings of the ACM/IEEE *Design automation and test in Europe* conference, Paris, France, 2002
3 VINCENTELLI, A., and MARTIN, G.: 'A vision for embedded systems: platform-based design and software', *IEEE Design and Test Special of Computers*, 2001, **18**(6), pp. 23–33
4 ST NOMADIK: www.st.com/stonline/prodpres/dedicate/proc/proc.htm
5 Philips.: www.semiconductors.philips.com/platforms/nexperia
6 Texas Instruments. www.ti.com
7 VANDER AA, T., JAYAPALA, M., BARAT, F. *et al.*: 'Instruction buffering exploration for low energy vliws with instruction clusters'. Proceedings of the *ASP-DAC*, Yokohama, Japan, January 2004, pp. x–x
8 JAYAPALA, M., BARAT, F., OPDEBEECK, P., CATTHOOR, F., DECONINCK, G., and CORPORAAL, H.: 'A low energy clustered instruction memory hierarchy for long instruction word processors'. Proceedings of the *PATMOS*, September 2002, pp. 258–267
9 CATTHOOR *et al.*: 'Custom memory management methodology – exploration of memory organisation for embedded multimedia system design' (Kluwer Academic Publishers, Boston MA, 1998)
10 WOLF, W., and KANDEMIR, M.: 'Memory system optimization of embedded software', *Proceedings of the IEEE*, 2003, **91**(1), pp. 165–182
11 FARABOSCHI, P., BROWN, G., and FISCHER J.: 'Lx: a technology platform for customizable VLIW embedded processing'. Proceedings of the international symposium on *Computer architectures*, 2000, pp. 203–213
12 AVISSAR, O., BARUA, R., and STEWART D.: 'Heterogeneous memory management for embedded systems'. Proceedings of ACM international conference on *Compilers architecture and synthesis for embedded systems*, Atlanta, USA, September 2001, pp. 34–43
13 VIREDAZ, M., and WALLACHA, D.: 'Power evaluation of a handheld computer', *IEEE Micro*, 2003, **23**(1), pp. 66–74

14 RABAEY, J.: 'Silicon architectures for wireless systems: Part 2 configurable processors'. Presented as a tutorial hot chips conference, Stanford, USA, 2001

15 HETTIARATCHI, S., and CHEUNG, P.: 'Mesh partitioning approach to energy efficient data layout'. Proceedings of the ACM/IEEE *Design and automation and test in Europe* conference, Munich, Germany, March 2003, pp. 11076–11081

16 CHOI, Y., and KIM, T.: 'Memory layout techniques for variables utilizing efficient DRAM access modes'. Proceedings of the ACM/IEEE *Design automation* conference, San Diego, CA, December 2003, pp. 881–886

17 CHANG, H., and LIN, Y.: 'Array allocation taking into account SDRAM characteristics'. Proceedings of the ACM/IEEE Asia South Pacific *Design automation* conference, Yokohama, Japan, 2000, pp. 447–502

18 RIXNER, S., DALLY, W., KAPASI, U., MATTSON, P., and OWENS, J.: 'Memory access scheduling'. International symposium on *Computer architectures*, 2000, pp. 128–138

19 GHARSALLI, F., MEFTALI, S., ROUSSEAU, F., and JERRAYA, A.: 'Automatic generation of embedded memory wrapper for multiprocessor SoC'. Proceedings of the ACM/IEEE *Design automation* conference, New Orleans, USA, June 2002, pp. 596–601

20 PANDA, P., DUTT, N., and NICOLAU, A.: 'Exploiting off-chip memory access modes in high-level synthesis'. Proceedings of the ACM/IEEE international conference on *Computer aided design*, San Jose, CA, October 1997, pp. 333–340

21 GRUN, P., DUTT, N., and NICOLAU, A.: 'Memory aware compilation through timing extraction'. Proceedings of the 37th ACM/IEEE *Design automation* conference, San Jose, CA, June 2001, pp. 316–321

22 LYUH, C., and KIM, T.: 'Memory access scheduling and binding considering energy minimization in multi-bank memory systems'. Proceedings of the 41st ACM/IEEE *Design automation* conference, San Diego, CA, 2004, pp. 81–86

23 DELALUZ, V., KANDEMIR, M., VIJAYKRISHNAN, N., SIVASUBRAMANIAM, A., and IRWIN, M.: 'Hardware and software techniques for controlling dram power modes', *IEEE Transactions on Computers*, 2001, **50**(11), pp. 1154–1173

24 SCHMIT, H., and THOMAS, D.: 'Synthesis of application-specific memory designs', *IEEE Transactions on VLSI Systems*, 1997, **5**(1), pp. 101–111

25 HUANG, C., RAVI, S., RAGHUNATHAN, A., and JHA, N.: 'High-level synthesis of distributed logic-memory architectures'. Proceedings of the international conference on *Computer aided design* San Jose, CA, 2002, pp. 564–571

26 SAGHIR, M., CHOW, P., and LEE, C.: 'Exploiting dual data banks in digital signal processors'. Proceedings of the *ASPLOS*, June 1997, pp. 234–243

27 VERMA, M., WEHMEYER, L., and MARWEDEL, P.: 'Dynamic overlay of scratched memory for energy minimization'. Proceedings of the ACM/IEEE international symposium on *System synthesis*, Stockholm, Sweden, September 2004, pp. 104–110

28   WUYTACK, S., CATTHOOR, F., DE JONG, G., and DE MAN, H.: 'Minimizing the required memory bandwidth in VLSI system realizations', *IEEE Transactions on VLSI Systems*, 1999, **7**(4), pp. 433–441

29   CATTHOOR, F. *et al.*: 'Data access and storage management for embedded programmable processors' (Kluwer Academic Publishers, Boston, MA, 2002)

30   LAMPORT, L.: 'The parallel execution of do-loops', *Communications of ACM*, 1974, **17**(2), pp. 83–93

31   VERHAEGH, W., AARTS, E., VAN GORP, P., AND LIPPENS, P.: 'A two-stage solution approach for multidimensional periodic scheduling', *IEEE Transactions on Computer Aided Design of Integrated Circuits and Systems*, **10**(10), 2001, pp. 1185–1199

32   HALL, M., ANDERSON, M., AMARASINGHE, S., MURPHY, B., LIAO, B., BUIGNON, E., and LAM, M.: 'Maximizing multiprocessor performance with SUIF', *IEEE Computer*, 1996, **29**(12), pp. 84–89

33   BANERJEE, P., CHANDY, J., GUPTA, M. *et al.*: 'Overview of the PARADIGM compiler for distributed memory message-passing multicomputers', *IEEE Computer*, 1995, **28**(10), pp. 37–37

34   LI, Y., and WOLF, W.: 'Hierachical scheduling and allocation of multirate systems on heterogeneous multiprocessors'. Proceedings of the ACM/IEEE *Design automation and test in Europe* conference, Munich, Germany, 1997, pp. 134–139

35   EL-REWINI, H., HESHAM H. ALI, and LEWIS, T.: 'Task scheduling in multiprocessing systems', *IEEE Computer*, 1995, **28**(12), pp. 27–37

36   SCHMITZ, M.T., AL-HASHIMI, B.M., and ELES, P.: 'Cosynthesis of energy multimode embedded systems with consideration of mode-execution probabilities', *IEEE Transactions on Computer Aided Design of Integrated Circuits and Systems*, 2005, **24**(2), pp. 153–169

37   DAVE, B., LAKSSHMINARAYANA, G., and JHA, N.: 'COSYN: hardware-software co-synthesis of heterogeneous distributed embedded systems', *IEEE Transactions on VLSI Systems*, 1999, **7**(1), pp. 92–104

38   MEFTALI, S., GHARSALLI, F., ROUSSEAU, F., and JERRAYA, A.: 'An optimal memory allocation for application-specific multiprocessor system-on-chip'. Proceedings of the ACM/IEEE *Design automation and test in Europe* conference, Paris, France, September 2001, pp. 19–24

39   MADSEN, J., and JORGENSEN, P.: 'Embedded system synthesis under memory constraints'. Proceedings of the *Codes*, Rome, Italy, May 1999, pp. 188–192

40   SZYMANEK, R., and KUCHCINSKI, R.: 'A constructive algorithm for memory-aware task assignment and scheduling'. Proceedings of the *Codes*, April 2001, pp. 147–152

41   KANDEMIR, M., ZHANG, W., and KARAKOY, M.: 'Runtime code parallelization for on-chip multiprocessors'. Proceedings of the ACM/IEEE *Design automation and test in Europe* conference, Munich, Germany, 2003, pp. 10510–10515

42   PATEL, K., MACII, E., and PONCINO, M.: 'Synthesis of partitioned shared memory architectures for energy-efficient multi-processor SoC'. Proceedings of

the ACM/IEEE *Design automation and test in Europe* conference, Paris, France, 2004, pp. 700–701

43 TOMAR, S., KIM, S. VIJAYKRISHNAN, N., KANDEMIR, M., and IRWIN, M.: 'Use of local memory for efficient java execution'. Proceedings of the SVM/IEEE international conference on *Computer aided design*, San Jose, CA, 2001, pp. x–x

44 DELALUZ, V., SIVASUBRAMANIAM, A., KANDEMIR, M., VIJAYKRISHNAN, N., and IRWIN, M.: 'Scheduler-based DRAM energy management'. Proceedings of the 39th ACM/IEEE *Design and test in Europe* conference, New Orleans, USA, 2002, pp. 697–702

45 BERGER, E., MCKINLEY, K., BLUMOFE, R., and WILSON, P.: 'Hoard: a scalable memory allocator for multithreaded applications'. Proceedings of the 8th *ASPLOS*, November 1998, pp. 117–128

46 WILSON, P.R. *et al.*: 'Dynamic storage allocation: a survey and critical review DDR'. Technical report, Department of Computer Science, University of Texas, 1995

47 KIEM-PHONG VO: 'Vmalloc: a general and efficient memory allocator'. *Software Practice and Experience*, 1996, (26), pp. 1–18

48 SHALAN, M., and MOONEY, V. III: 'A dynamic memory management unit for embedded real-time systems-on-a-chip'. Proceedings of the ACM international conference on *Compilers, architecture and synthesis for embedded systems*, San Jose, CA, November 2000, pp. 180–186

49 VEE, V., and HU, W.: 'A scalable and efficient storage allocator on shared-memory multiprocessors', in International symposium on *Parallel architectures, algorithms and networks*, June 1999, pp. 230–235

50 MARCHAL, P., BRUNI, D., GOMEZ, J.I., BENINI, L., PINUEL, L., CATTHOOR, F., and CORPORAAL, H.: 'SDRAM-energy-aware memory allocation for dynamic multi-media applications on multi-processor platforms'. Proceedings of the ACM/IEEE *Design automation and test in Europe* conference, Munich, Germany, March 2003, pp. 10516–10523

51 GOMEZ, J.I., MARCHAL, P., BRUNI, D., BENINI, L., PRIETO, M., CATTHOOR, F., and CORPORAAL, H.: 'Scenario-based SDRAM-energy-aware scheduling for dynamic multi-media applications on multi-processor platforms'. Workshop on *Application specific processors (in conj. with MICRO)*, 2002

52 RAU, B.: 'Iterative modulo scheduling'. Technical report, HP Labs, 1995

53 WOLF, M.: 'Improving locality and parallelism in nested loops'. Technical report, CSL-TR-92-538, Stanford University, CA, USA, September 1992

54 MARCHAL, P., GOMEZ, J.I., PINUEL, L., VERDOOLAEGE, S., and CATTHOOR, F.: 'Loop morphing to optimise the memory bandwidth'. Proceedings of the *IEEE ASAP*, Galveston, TX, September 2004

55 MARCHAL, P., GOMEZ, J.I., PINUEL, L., VERDOOLAEGE, S., and CATTHOOR, F.: 'Optimizing the memory bandwidth with loop fusion'.

Proceedings of the ACM/IEEE international symposium on *System synthesis*, Stockholm, Sweden, September 2004

56   LEIJTEN, J.A.: 'Real-time constrained reconfigurable communication between embedded processors'. PhD thesis, Technishe Universiteit Eindhoven, November 1998

57   YANG, P.: 'Pareto-optimization based run-time task scheduling for embedded systems'. PhD thesis, Catholic University Leuven, 2004

58   YANG, P., MARCHAL, P., WONG, C. *et al.*: 'Multi-processor systems on chip', chapter: 'Cost-efficient mapping of dynamic concurrent tasks in embedded real-time multimedia systems' (Elsevier Science and Technology, 2004) pp. 46–58

59   POPLAVKO, P., PASTRNAK, M., BASTEN, T. VAN MEERBERGEN, J., and DE WITH, P.: 'Mapping of an MPEG-4 shape-texture decoder onto an on-chip multiprocessor'. PRORISC 2003, 14th workshop on *Circuits, systems and signal processing*, November 2003, pp. 139–147

60   PASTRNAK, M., POPLAVKO, P., DE WITH, P., and VAN MEERBERGEN, J.: 'Resource estimation for MPEG-4 video object shape-texture decoding on multiprocessor network-on-chip'. PROGRESS 2003, 4th seminar on *Embedded systems*, October 2003, pp. 185–193

61   MARCHAL, P., GOMEZ, J., BRUNI, D., BENINI, L., PINUEL, L., and CATTHOOR, F.: 'Integrated task-scheduling and data-assignment to enable SDRAM power/performance trade-offs in dynamic applications', *IEEE Design and Test of Computers*, 2004, **11**(1), pp. 1–12

62   ATIENZA, D., MAMAGKAKIS, S., CATTHOOR, F., MENDIAS, J.M., and SOUDRIS, M.: 'Dynamic memory management design methodology for reduced memory footprint in multimedia and wireless network applications'. Proceedings of the *Design automation and test in Europe* (DATE' 04) (IEEE Press, Paris, France, February 2004)

63   ATIENZA, D., MAMAGKAKIS, S., MENDIAS, J.M. *et al.*: 'Efficient system-level prototyping of power-aware dynamic memory managers for embedded systems', *Integration – The VLSI journal – Special Issue on Low Power Design*, October 2004, **38**(1)

64   LEA, D.: 'The lea 2.7.2 dynamic memory allocator', 2002, http://gee.cs. oswego.edu/dl/

65   NETBENCH: http//www.veritest.com/benchmarks/netbench

66   MAMAGKAKIS, S., ATIENZA, D., CATTHOOR, F., SOUDRIS, D., and MENDIAS, J.M.: 'Custom design of multi-level dynamic memory management subsystem for embedded systems'. Proceedings of *Signal processing symposium (SiPS)*, IEEE Signal Processing Society and IEEE Circuits and Systems Society Austin, Texas, USA, October 2004, pp. 170–175

Chapter 11

Low power system scheduling, synthesis and displays

Niraj K. Jha

11.1 Introduction

Low power system synthesis, specifically, system-on-a-chip (SoC) synthesis and hardware–software co-synthesis of distributed embedded systems, has attracted much attention. The system may be wireless and may have both quality of service (QoS) and real-time constraints. We will discuss such synthesis techniques in this chapter.

A significant fraction of the software and resource usage of a handheld computer system is devoted to its graphical user interface (GUI). GUIs are direct users of the display. They enable users to interact with the software. Since displays often are the greatest energy-consumers in such systems, it is important to optimise GUIs for energy. In addition, dynamic voltage scaling (DVS) and dynamic power management (DPM) techniques have traditionally been applied to computation and I/O intensive tasks. However, many modern applications are interactive in nature, necessitating new DVS/DPM techniques that take into account the user's perspective. Display power can be directly targeted by optimising the power of its various components.

We discuss low power system scheduling, synthesis and interactive systems in Sections 11.2, 11.3 and 11.4, respectively. We point out some open problems and conclude in Section 11.5.

11.2 Low power system scheduling

In this section, we discuss various low power distributed system scheduling techniques based on DVS and adaptive body biasing (ABB).

Whenever both DVS and DPM [1] are available for a processor, it is known that it is always advantageous to exploit DVS first.

The circuit delay depends on supply voltage V_{dd} and threshold voltage V_t as follows [2]:

$$\text{delay} = k \times V_{dd}/(V_{dd} - V_t)^\alpha \tag{11.1}$$

where k and α are constants, $1 < \alpha \leq 2$. Thus, delay increases as V_{dd} decreases. The switching power consumption (which currently dominates power consumption in complementary metal oxide semiconductor (CMOS) technology) is given by

$$P_{\text{switch}} = NC_L V_{dd}^2 f \tag{11.2}$$

where N is the switching activity, C_L the load capacitance and f the frequency. $N \times C_L$ is referred to as the switched capacitance. Note that f is inversely proportional to circuit delay. Hence, $P_{\text{switch}} \propto V_{dd}(V_{dd} - V_t)^\alpha$.

Another component of power consumption is short-circuit power that is incurred during logic transition when both the nMOS and pMOS networks of a CMOS gate conduct for a short duration. Together, switching and short-circuit power are referred to as dynamic power.

As technology feature size continues to scale, a third component of power, leakage power, is accounting for larger fractions of system power and will rival switching power by 2008/9. It is currently due to subthreshold leakage current, I_{sub}, as well as drain-body junction leakage current, I_j, and source-body junction leakage current, I_b, and is given by [3]:

$$P_{\text{leakage}} = I_s \left(\frac{W}{L}\right) V_{dd} e^{-V_t/nV_T} + | V_{bs} | (I_j + I_b) \tag{11.3}$$

where I_s and n are technology parameters, W and L are device geometries, V_T is the thermal voltage and V_{bs} is the body bias voltage. The first term corresponds to subthreshold leakage power (given by $I_{\text{sub}} \times V_{dd}$) and the second term to junction leakage power.

Table 11.1 shows the expected break-up between dynamic and leakage power in future technologies [4].

The input specification of real-time distributed systems is frequently given in terms of a set of task graphs. A task graph is a directed acyclic graph in which a node is associated with a coarse-grained task (e.g. discrete cosine transform) with a worst-case execution time (if the task can be run on more than one type

Table 11.1   Power consumption break-up for future technologies

Technology	0.07 μm	0.05 μm	0.035 μm
Dynamic power	78%	56%	33%
Leakage power	22%	44%	67%

of processing element, then the corresponding set of worst-case execution times is available) and an edge is associated with the amount of data transferred between tasks (or the worst-case communication times on the communication links the edge can be assigned to). The period associated with a task graph indicates the time interval between its successive executions. A hard deadline, by which time the task must complete execution, is given for each sink node and some intermediate nodes. An arrival time, by which time the task may begin execution, is given for each source node and possibly some intermediate nodes. A multi-rate system consists of multiple task graphs with different periods. The least common multiple of all the periods is called the hyperperiod. It is known that scheduling in the hyperperiod gives a valid schedule [5].

In addition to periodic task graphs, the system may also contain aperiodic tasks. An aperiodic task is invoked at any time and may have a hard or soft deadline. In case of soft deadlines, only the response time of the task needs to be minimised. For aperiodic tasks, generally a minimum inter-instance arrival time is specified.

Example 11.1   Figure 11.1 shows an embedded system specification consisting of two task graphs, each with the same period. The worst-case execution times of tasks T_1–T_7 at maximum voltage on a particular processor are shown next to them. The worst-case communication time for edge E_1 on a particular communication link is also shown. The communication times on the other edges are not shown since their corresponding edges are assumed to be mapped to the same processing element (PE). In distributed computing, it is traditionally assumed that intra-PE task communication takes negligible time compared to inter-PE task communication.   ∎

In addition to task graphs, a system synthesis algorithm requires information on the resource library. This library may consist of various PEs, such as general-purpose processors, dynamically reconfigurable field-programmable gate arrays (FPGAs) and application-specific integrated circuits (ASICs), in addition to communication links

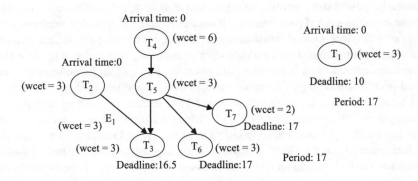

Figure 11.1   Periodic task graphs

and memory modules. The system may be implemented on a printed circuit board or a SoC.

11.2.1   DVS for distributed systems

Dynamic voltage scaling techniques for single processors have a rich history (see the survey article [6]). However, in this section, we concentrate on DVS techniques for distributed systems, which consist of multiple PEs connected with an interconnection network.

In Reference 7, a power-conscious algorithm is given for jointly scheduling multi-rate periodic task graphs with hard deadlines and aperiodic tasks with hard or soft deadlines. Periodic tasks are first scheduled statically and room made in the schedule for hard aperiodic tasks. Soft aperiodic tasks are scheduled dynamically with an on-line scheduler. It exploits the concepts of slack stealing and resource reclaiming to minimise the response times of aperiodic tasks. It uses DPM for parts of the schedule where DVS is not applicable.

In Reference 8, a list scheduling technique chooses the best two supply voltages for each task in a task graph specification. It uses dynamic recalculation of energy-sensitive task priorities for this purpose.

In Reference 9, a hybrid global/local search optimisation framework is given for DVS. Performance is a constraint under which an attempt is made to find the optimum voltage level for all the tasks that need to be executed. The schedule of tasks on different processors is assumed to be known *a priori*. The power consumed by the DVS hardware and the time to switch between voltages are also taken into account. A genetic algorithm is used for global search (coupled with a technique called simulated heating) and hill climbing, and Monte Carlo techniques for local search.

In Reference 10, a power-aware scheduling algorithm is presented for mission-critical applications. It satisfies min/max timing constraints and maximum power constraint. In addition, it also tries to meet minimum power constraints in order to make full use of free power (e.g. solar power) or to control power jitter.

Motivated by the work in Reference 11, the work in Reference 12 proposes a DVS scheme based on a static-priority list scheduling algorithm. It does effective slack allocation based on critical path analysis. It locates the critical path that minimises the ratio of total slack to total worst-case execution time on the path. It then extends the critical path such that the available slack is uniformly distributed. The algorithm terminates when no path can be extended anymore. It is optimal for non-pre-emptive fixed-priority scheduling on a single processor, but not for distributed systems. It does not target scenarios in which various tasks have different switching activities or heterogeneous distributed architectures in which different voltage-scalable processors have different voltage scaling characteristics. These limitations are overcome in Reference 13, in which simulated annealing is used to optimise task priorities and a fast slack allocation method is used based on the concept of energy gradients. An energy gradient is the negative of the derivative of energy with respect to the execution time of a task for a given supply voltage. For a single processor, allocating

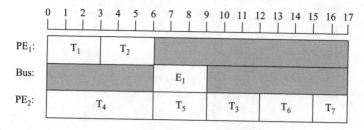

Figure 11.2   Initial schedule

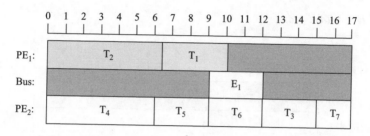

Figure 11.3   Optimised schedule

slack to the set of extensible tasks with the highest energy gradient is optimal (these are tasks for which an extension of the execution time does not lead to any deadline violations). Although no longer optimal for distributed systems, this remains a very effective heuristic.

Example 11.2   Consider the two task graphs in Figure 11.1 again. Assuming a distributed system consisting of two PEs connected by a bus, one can generate an initial schedule, as shown in Figure 11.2. This schedule has been derived based on a priority assigned to each task, which is an inverse of its latest possible start time that does not violate any deadlines. The schedule assumes worst-case execution and communication times for $V_{dd} = 1.8$ V. It also assumes that both PEs have communication buffers to allow processing and communication to go on in parallel. One critical path is (T_1, T_2, E_1, T_3, T_6, T_7). No voltage scaling is possible for this schedule since all scheduled events are on critical paths. However, if we change the execution order of tasks, as shown in Figure 11.3, then the critical path gets broken, allowing more flexibility in the schedule that can be exploited for voltage scaling for tasks T_1 and T_2. This results in a 40 per cent decrease in overall power consumption.   ∎

In Reference 14, a scheduling algorithm is presented for real-time distributed embedded systems that combines DVS and ABB to jointly optimise dynamic and leakage power. It derives an analytical expression to obtain the optimal supply voltage and body bias voltage for a given clock frequency. Based on this expression, it computes the optimal energy consumption for a given clock frequency and analyses the

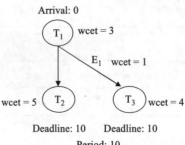

Figure 11.4   Task graph

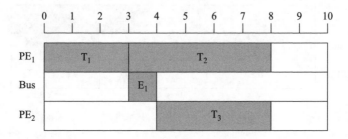

Figure 11.5   Schedule at nominal supply and threshold voltages

trade-off between energy consumption and execution time for a set of interdependent tasks under real-time constraints. It shows that in future CMOS technologies, using DVS+ABB, will be much more effective for reducing distributed system power than using DVS alone. A similar problem is also solved in Reference 15 using non-linear programming and mixed integer linear programming.

Example 11.3   Consider the task graph shown in Figure 11.4. Suppose the power consumptions of tasks T_1, T_2 and T_3 are c_1, c_2 and c_3, respectively. Figure 11.5 shows a valid schedule on a distributed system consisting of two PEs linked by a bus. Suppose initially for PE_1, $V_{dd} = 2.0$ V and $V_{th} = 0.6$ V whereas for PE_2, $V_{dd} = 1.5$ V and $V_{th} = 0.4$ V. Figure 11.6 shows a new schedule if the execution times of tasks T_1, T_2 and T_3 are extended to 4, 6 and 5 time units, respectively. Since the execution time of T_1 is extended from 3 to 4 time units, the speed of processor PE_1 can be scaled by a ratio of $\frac{4}{3}$.

Consider the 0.07 μm technology, in which dynamic power constitutes 78 per cent of total power consumption and leakage power 22 per cent (see Table 11.1). To scale the frequency of task T_1 by $\frac{4}{3}$, we can reduce the V_{dd} of PE_1 from 2.0 V to 1.55 V (assuming $\alpha = 1.4$ in Equation (11.1)). The dynamic power of T_1 reduces from $0.78c_1$ to $0.35c_1$ and the leakage power from $0.22c_1$ to $0.17c_1$, based on Equations (11.2) and (11.3), respectively. Thus, the overall power of T_1 reduces to $0.52c_1$. Similarly,

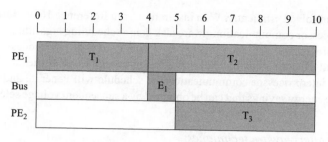

Figure 11.6   Schedule at scaled supply and threshold voltages

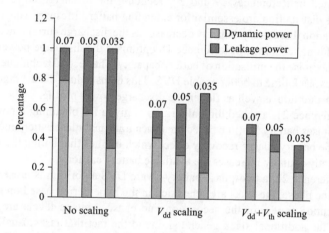

Figure 11.7   Power reduction for different technologies

the power consumption of T_2 and T_3 is reduced to $0.65c_2$ and $0.58c_3$, respectively. This results in an average power reduction of 42 per cent compared to no voltage scaling, assuming $c_1 = c_2 = c_3$.

Another way to approach the above problem is through combined supply and threshold voltage scaling. To scale the frequency of PE1 by $\frac{4}{3}$ for T_1, its V_{th} can be changed from 0.6 V to 0.64 V and V_{dd} from 2.0 V to 1.62 V. This reduces the dynamic power of T_1 to $0.38c_1$ and leakage power to $0.06c_1$ for a total of $0.44c_1$. Similarly, the power for T_2 and T_3 is reduced to $0.56c_2$ and $0.51c_3$, respectively. Thus, a greater overall reduction of 50 per cent is obtained, assuming $c_1 = c_2 = c_3$.

Figure 11.7 presents the normalised power consumption for the three technologies. It can be seen that V_{dd} scaling alone becomes much less effective than combined $V_{dd} + V_{th}$ scaling. ∎

In Reference 16, a two-phase method is presented that integrates task scheduling, ordering and voltage selection. In the first phase, task scheduling and ordering are performed to maximise the opportunities for voltage selection-based energy reduction in the second phase. Voltage selection is formulated as an integer programming problem and solved efficiently. With inspiration from Reference 16, in Reference 17, an interprocessor communication-aware task graph scheduling algorithm is presented for multi-processors. It reduces energy by reducing overall interprocessor communication and executing certain cycles at a lower voltage level. It tries to foresee the amount of interprocessor communication the schedule will generate and takes into account the energy savings that can be obtained by a subsequent voltage selection step.

11.2.2   Battery-aware techniques

In Reference 18, two battery-aware static scheduling techniques are presented. As suggested in References 19 and 20, reducing the discharge current level and shaping its distribution are essential for extending battery lifespan. This is based on the observation that battery capacity decreases as the discharge current increases. The first scheduling technique in Reference 18 optimises the discharge power profile in order to maximise the utilisation of battery capacity. The second technique efficiently re-allocates slack time to better enable DVS. This helps reduce the average discharge power consumption as well as flatten the discharge power profile.

In Reference 21, a scheduling algorithm is given for obtaining a power profile that maximises battery lifetime based on an accurate analytical battery model. It also exploits the battery charge recovery effect which implies that letting the battery rest intermittently actually increases the available battery capacity.

In Reference 22, a two-phase battery-aware DVS algorithm is presented. In the first off-line phase, the tasks are scheduled in the hyperperiod based on their worst-case execution times. In the second on-line phase, voltage levels are reassigned based on the additional slack generated due to the fact that tasks usually take less than their worst-case execution time. The procedure is applicable to multi-processor environments.

It is common for a system to contain multiple batteries. These batteries are typically discharged serially and completely. In Reference 23, it is shown that round-robin scheduling, in which batteries are discharged in a round-robin fashion, leads to longer battery lifetimes than serial scheduling.

An excellent survey of battery-aware techniques, which includes battery modelling and battery-efficient system design, can be found in Reference 24.

11.2.3   Power optimisation of communication links

Dynamic voltage scaling for communication links is a very new area of research. With increasing demands on system bandwidth, interconnection networks are becoming energy/power limited as well. Interconnection network fabrics were historically used in high-end multi-processor systems. Now they have proliferated to a wide range of systems – clusters, terabit Internet routers, server blades and on-chip networks. In some systems, they consume up to 40 per cent of the system power budget.

Components of a DVS link are shown in Figure 11.8. A transmitter converts digital binary signals to electrical signals. A receiver converts electrical signals back to binary signals. The signalling channel is modelled as a transmission line. The clock

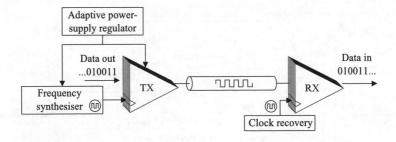

Figure 11.8   A DVS link

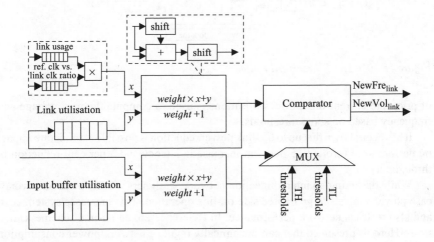

Figure 11.9   Hardware implementation of history-based DVS

recovery block compensates for delay through the signalling channel. The adaptive power supply regulator tracks the link frequency, regulates the voltage and feeds the voltage to multiple links. The frequency synthesiser generates the frequency. It is important to take into account the time and energy required to make voltage/frequency transitions in a DVS link.

In Reference 25, a history-based DVS method is presented to dynamically adjust the voltage and frequency of communication links to minimise their power consumption. In this method, each router predicts future communication traffic based on past link and input buffer utilisation. Since the method only relies on local traffic information, the hardware required in each router to implement DVS is quite simple, as shown in Figure 11.9. This hardware is placed at each output port of a router. It controls the multiple links of that port. Link utilisation is measured using a counter that counts the total number of cycles used to relay flits in a histroy interval (flit stands for a flow control unit and is a fixed-sized segment of a packet that is to be transmitted). A second counter determines the ratio between the router and link clock periods. A multiplier combines the two counter outputs to compute the link utilisation. A weighted average

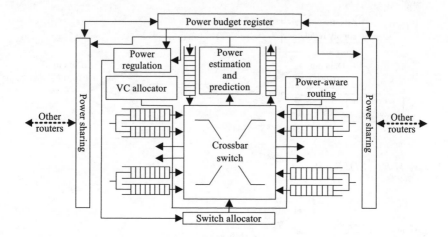

Figure 11.10   The PowerHerd router

of prior and current link/buffer utilisations is used to compute the new voltage and frequency based on some thresholds.

It has been shown that up to $6\times$ link power reduction is possible at moderate impact on performance (around 15 per cent increase in latency and 3 per cent reduction in throughput).

While designing an interconnection network, designers usually perform a worst-case power analysis to guarantee safe on-line operation. This increases system cost, and also restricts network performance. In Reference 26, an on-line scheme, called PowerHerd, is presented that can dynamically regulate network power consumption and guarantee that network peak power constraints are never exceeded. It is a distributed scheme. Each router maintains a local power budget, controls its local power consumption and exchanges spare power resources with its neighbouring routers to optimise network performance. Armed with PowerHerd, the designers can concentrate on the average-case power consumption, rather than the worst-case, allowing the use of more powerful interconnection networks in the system.

PowerHerd is composed of mechanisms for power estimation, prediction, sharing, regulation and routing built into each router, as shown in Figure 11.10. Initially, the user-defined global power budget, P_{GPB}, is divided evenly and stored in the power budget register at each router. This forms the local power constraint, P_{LPB}^{i}, at each router i. Using actual activity, each router estimates its power consumption at runtime, which is used to predict future activity and power consumption. Based on the estimate, each router decides if it has spare power to share with neighbouring routers, updating the power register accordingly. In each cycle, the power regulation mechanism dynamically throttles the switch allocator to keep the router power within the allocated power budget. Finally, the routing protocol steers network traffic towards routers with excess power budgets.

In Reference 27, a framework is discussed that can estimate power consumption of switch fabrics in network routers. It models node switches, internal buffers and interconnect wires within the fabric. This framework is used to analyse four switch fabric architectures. The framework is suitable for system architecture exploration based on high-performance network router design.

In Reference 28, a DPM method is presented for interconnection networks in which network links can be turned on and off based on network utilisation. This is done in a distributed fashion.

Leakage power modelling of networks has been addressed in Reference 29. It is shown that router buffers are the main candidates for leakage power optimisation. Leakage reduction policies are shown to reduce buffer leakage power by up to $30\times$.

11.3 Low power system synthesis

In this section, we survey various system synthesis techniques for low power SoCs and distributed embedded systems.

The key steps in system synthesis are allocation, assignment, scheduling and performance evaluation. Allocation determines the number of each type of PE and communication link in the system architecture. Assignment chooses a PE (link) to execute each task (communication) upon. Scheduling determines the time of execution of each task and communication. Performance evaluation involves computing the price, speed and power of the system architecture.

Three popular algorithms for system synthesis are iterative improvement, constructive and genetic. An iterative improvement algorithm starts with an initial solution and makes changes to it iteratively in order to improve its quality. A greedy approach would be to apply various available optimisation moves to the current system architecture and accept the one that improves the architecture quality the most. This is continued until no more improvement is possible. However, such an approach is likely to trap the algorithm in local minima. This problem can be alleviated by variable-depth iterative improvement which makes it possible to back-out of local minima.

Constructive algorithms incrementally synthesise the architecture by following a fixed set of rules. Since they are fast, they are effective when the problem size is large and only a constrained exploration of the design space is possible. However, they too can easily get trapped in local minima.

Genetic algorithms work on a pool of solutions (system architectures in this case) that evolve through multiple generations. Genetic operators, such as crossover and mutation, exchange information between solutions and introduce randomised local changes in them to try to improve solution quality. The highest-quality solutions are then selected for the next generation. Such algorithms excel at multi-objective optimisation and can easily get out of local minima.

11.3.1 Low power SoC synthesis

The method in Reference 30 considers an SoC with a fixed allocation of one processor, ASIC, instruction cache, data cache and main memory. As a case study, an MPEG-2 encoder is chosen to investigate the impact of different hardware/software partitions of the input specification between the processor and ASIC, and different system configurations such as cache size, cache line size, cache associativity and main memory size, on power dissipation of the SoC. The extension of this method in Reference 31 assumes the same SoC architecture, however, with the hardware fixed. The software is changed through various high-level transformations. This impacts cache and memory parameters. It investigates overall system energy. Another extension of this approach is given in Reference 32.

In Reference 33, the allocation of the SoC architecture is not fixed beforehand, and is hence not limited to a single processor and ASIC. It describes a tool called MOCSYN which synthesises real-time heterogeneous single-chip hardware/software architectures using an adaptive multi-objective genetic algorithm. It starts with a system specification consisting of multiple periodic task graphs as well as a database of core and SoC characteristics. The database consists of the worst-case execution times and average/peak power consumption of each task on each core on which the task can possibly run. Each core has a width, height, maximum clock frequency, variable indicating whether or not its communication is buffered and energy consumption per cycle dedicated to communication. In addition, information on core price is also available. A single system synthesis run produces multiple SoC designs which trade-off system price, power and area under real-time constraints. It assumes asynchronous communication between synchronous cores on the SoC and determines the best clock frequency for each core. It produces a hierarchical bus structure which balances ease of layout with the reduction of bus contention. It also performs floorplanning in the inner loop of system synthesis to accurately estimate global communication delays and energy, as well as clock network energy.

Figure 11.11 gives an overview of MOCSYN. First, the clock frequency is determined for each core, and some initial solutions obtained. In the outer cluster loop, the allocation is fixed. In the inner architecture loop, the task assignments are refined through various generations of the genetic algorithm. Then link priorities are assigned so that cores that communicate a lot may be placed close to each other in the block placement step. Links are re-prioritised based on more accurate global wiring information obtained from block placement. A hierarchical bus structure is obtained that trades off potential bus contention for ease of routing. Thereafter, the tasks and communication events are scheduled. Then another allocation is chosen and the above process repeated. The best architectures seen in any generation are given as outputs, each better than another in at least one aspect, e.g. system price or power.

In Reference 34, a functional partitioning method is given for synthesising low power real-time distributed embedded systems whose constituent nodes are SoCs. The input specification, given as a set of task graphs, is partitioned and each segment implemented as an SoC. It merges functional partitioning and SoC synthesis into a unified method. The genetic algorithm presented in Reference 33 is extended for this purpose.

In Reference 35, the target SoC architecture consists of one processor, instruction/data cache, main memory and several ASICs and peripherals. It shows the importance of adequate adaptation between core and interface parameters to minimise

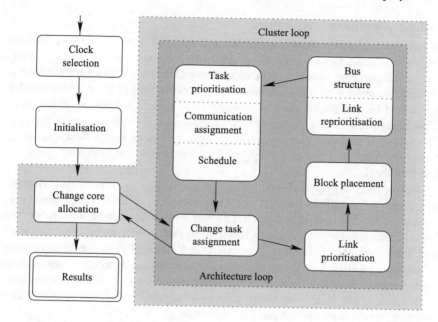

Figure 11.11   Overview of MOCSYN

power consumption. Cache parameters and configurations of cache buses have a significant effect in this respect.

In Reference 36, DVS is integrated into SoC synthesis. The target architecture consists of one processor and instruction/data cache. The input specification is assumed to consist of independent periodic tasks. The technique also addresses selection of the best processor core and determination of cache size and configuration to best enable DVS.

11.3.2   Low power distributed system synthesis

In a distributed system, PEs are not limited to a single chip. However, system synthesis still consists of solving the allocation, assignment, scheduling and performance evaluation problems. We discuss various types of distributed system synthesis algorithms next.

11.3.2.1   Iterative improvement algorithm

In Reference 37, the first work to integrate DVS into system synthesis is presented. Independent periodic tasks are assumed which are mapped to multiple processors connected by a bus. In addition to voltage, the switched capacitance is also implicitly optimised. Resource allocation is done using a gradient-driven iterative improvement heuristic. Tasks are also iteratively assigned to allocated processors based on an objective function. Additional load balancing is attempted in a post-processing step.

11.3.2.2   Constructive algorithms

In Reference 38, a constructive algorithm, called COSYN, is described which starts with a set of multi-rate periodic task graphs with real-time constraints and produces a price and power optimised distributed system architecture. It uses a combination of pre-emptive and non-pre-emptive scheduling. However, use of pre-emptive scheduling should be avoided as much as possible since it increases power consumption. It performs task clustering before system synthesis to make synthesis more tractable. This allows it to tackle very large task graphs (with more than a thousand tasks). It uses the concept of task graph pipelining to handle task graphs in which period is less than the deadline.

For medium- to large-scale embedded systems, such as telecom transport systems, task graphs are usually hierarchical, i.e. each node in an upper-level task graph may correspond to a full-fledged task graph at a lower level. If flat, non-hierarchical system architectures are derived from such hierarchical task graphs, many communication and processing bottlenecks may be created. In Reference 39, a constructive algorithm, called COHRA, is given to synthesise hierarchical distributed architectures from hierarchical or non-hierarchical real-time periodic task graphs. A hierarchical architecture is obtained by composing lower-level sub-architectures. COHRA also optimises power consumption and fault tolerance.

11.3.2.3   Genetic algorithms

In Reference 40, a genetic algorithm, called MOGAC, is used to synthesise real-time heterogeneous distributed architectures from multi-rate real-time periodic task graph specifications. It optimises both price and power. Genetic algorithms excel at such multi-objective optimisation. The number and type of PEs are not fixed *a priori*. Genetic algorithms allow solutions to cooperatively share information with each other, exploring the set of solutions that can only be improved in one way by being degraded in another (the Pareto-optimal set). MOGAC uses heuristics to allow multi-rate systems to be scheduled in reasonable time even when the periods are very different and possibly co-prime.

In Reference 41, a genetic algorithm called COWLS targets embedded systems consisting of servers and low power clients which communicate with each other through a channel of limited bandwidth, e.g. a wireless link. Clients may be mobile. It simultaneously optimises the price of the client–server system, power consumption of the client and response times of tasks with only soft deadlines, while meeting all the hard deadlines. It produces numerous solutions which trade off architectural features such as price, power and response time.

In Reference 42, a genetic algorithm targeting distributed systems consisting of processors and dynamically reconfigurable FPGAs is presented. It is based on a two-dimensional, multi-rate cyclic scheduling algorithm for such FPGAs. It determines task priorities based on real-time constraints and FPGA reconfiguration overhead information, and schedules tasks based on resource utilisation and reconfiguration condition in time and space. It optimises both system price and power. FPGA reconfiguration power is shown to consume a significant fraction of FPGA power.

A genetic algorithm is also used in Reference 43 to incorporate DVS into an energy minimisation technique for distributed embedded systems. It takes the power variations of tasks into account while performing DVS. An off-line voltage scaling heuristic is proposed which is fast enough to be used in system synthesis, starting from real-time periodic task graphs. Both task assignment and scheduling are done with the help of a genetic algorithm. This method has been extended in Reference 44.

11.3.2.4 Joint energy optimisation of processors and communication links

High-speed serial network interfaces are being used to connect processors and peripherals in distributed embedded systems. The fact that many such interfaces support multiple data rates can be exploited to perform power/performance trade-offs between communication and computation on processors that employ DVS. In Reference 45, a speed selection methodology is presented for globally optimising the energy consumption of embedded networked systems.

Earlier, it was mentioned that DVS is also possible in communication links. Naturally, performing simultaneous DVS in the processors and communication links in a distributed system can yield greater power savings than performing DVS in the processors alone. Such a method was presented in Reference 46. This obviously poses extra design challenges. The scheduling algorithm not only needs to consider real-time constraints, but also the underlying flow control techniques. The available system slack now has to be efficiently distributed among both processors and links, and communications may need to be scheduled through multiple hops.

11.3.2.5 QoS driven system synthesis

Quality of service is an important consideration in designing systems for real-time multi-media and wireless communication applications. In Reference 47, a DVS technique for partitioning a set of applications among multiple processors is given which minimises system energy while satisfying individual QoS requirements. QoS is a function of the required resources, such as bandwidth, CPU time and buffer space. The applications are assumed to be independent, have the same arrival times and no deadline constraints.

11.3.3 Low-energy network-on-chip architectures

Network-on-chip (NoC) architectures have recently received a lot of attention [48]. For more details, see Part IV of this book. They have regular tiles with a router embedded in each tile. The tiles may contain a general-purpose processor, digital signal processor, memory, etc. Inter-tile communication is based on routing of packets. In Reference 49, a method is given for mapping intellectual property cores of an SoC onto the tiles of a generic NoC architecture in order to minimise total communication energy. Performance constraints are satisfied through bandwidth reservation. While this is an indirect method of mapping tasks and communications to NoCs, in Reference 50 an energy-aware static scheduling algorithm is presented that directly assigns and schedules both tasks and communications on an NoC architecture under real-time constraints.

11.3.4 Task graph generation

The starting point for most algorithms and tools for SoC and distributed system synthesis is a set of task graphs. However, the embedded system applications are often written in C, and no task graphs are available for them. Manually deriving task graphs from C is a tedious and error-prone endeavour. To bridge this gap, a task graph extraction tool is described in Reference 51 that automatically extracts a set of task graphs from embedded applications written in C. It is available for download.

A popular random task graph generator, called TGFF, is presented in Reference 52. It can generate both independent tasks as well sets of task graphs. It creates a complete description of a scheduling problem instance. This includes attributes for processors, communication links, tasks and inter-task communication. The user is allowed to control correlations between attributes in a parametrised fashion.

Another source of realistic task graphs is the E3S suite [53]. It is based on EEMBC benchmarks. The E3S suite has data for 17 processors, including AMD ElanSC520, Analog Devices 21065L, Motorola MPC555 and Texas Instruments TMS320C6203. These processors are characterised based on the measured execution times of 47 tasks, power numbers derived from processor datasheets and additional information, such as die sizes and prices. E3S also contains communication resources modelling a number of different buses, e.g. CAN, IEEE1394, PCI, USB 2.0 and VME. There is one task set for each of the following application suites: automotive/industrial, consumer, networking, office automation and telecommunications.

11.4 Low power interactive systems

Power consumption is a major concern in mobile computing, e.g. with laptops and handheld computers. On such systems, a significant fraction of software is interactive, not compute-intensive. It has been shown in Reference 54 that over 90 per cent of the time and energy in such systems may be spent waiting for user input. Such idle times obviously offer significant opportunities for DVS and DPM. At the same time, since displays tend to consume a significant fraction of total system power in such systems and GUIs mediate between the user and the system, characterising and optimising the GUI energy is also an important new concern. Finally, display-related power can be directly targeted by optimising its various components. In this section, we discuss recent work in these areas.

11.4.1 Energy characterisation of GUIs

Comprehensive energy characterisation methodologies and experimental results have been presented in Reference 55. User interfaces consisted of an average of 48 per cent of the application code even a decade ago (early 1990s) [56]. Since modern user interfaces are mostly graphical, this only increases its fraction of source code and resource usage. In Reference 55, three different GUI platforms (Qt, Microsoft Windows and the X Window system) are characterised on three handheld computers (two versions of HP/Compaq iPAQ and a Sharp Zarus). A similar energy characterisation is possible

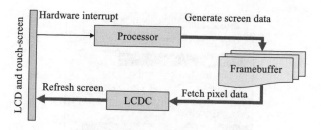

Figure 11.12   Hardware perspective

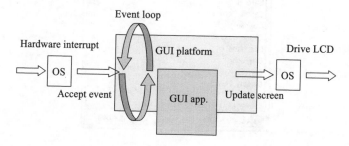

Figure 11.13   Software perspective

for laptops. The characterisation shows that the GUIs are energy-expensive, their different features consume drastically different amounts of energy and different GUI platforms display significant variations in their energy consumption.

The energy consumption of GUIs can be characterised from three different perspectives [55]: hardware, software and application. Figure 11.12 shows the hardware perspective. It consists of an LCD controller (LCDC). A framebuffer is implemented in main memory and stores data for a full screen. For a screen change, the processor generates new data for the changed screen pixels and stores them in the framebuffer. This implies higher energy consumption with increased temporal changes in the screen. To maintain screen data, the LCDC must sequentially read screen data from the framebuffer and refresh the LCD pixels. This implies higher energy consumption with increased spatial changes in the screen. The display itself consists of LCD power circuitry, front light and an LCD.

Figure 11.13 shows the software perspective. The OS handles user-generated interrupts. It produces events for the GUI platform, which are delivered to the GUI application through an event loop. The application instructs the platform on how the GUI should change. The platform coordinates GUIs of various applications, determines how the screen changes, generates new screen pixel data and then calls OS services to update the screen.

From the application perspective, a GUI consumes energy through user–GUI interaction sessions, in which the user locates the application, starts it, interacts with it and closes it. This typically consists of a series of window operations.

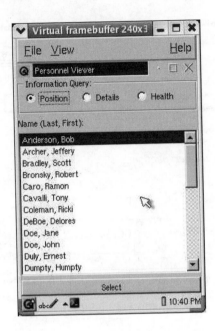

Figure 11.14   Personnel viewer: original

11.4.2   Energy-efficient GUI design

Traditional software power optimisation techniques usually reduce system energy during its busy time. This has limited utility in interactive systems where the system waits for the user input most of its time. Since the display must be on during system wait time, an effective way to reduce system energy is to improve user productivity. Studies from the field of psychology have shown that reading speed depends on the GUI layout and conciseness in addition to visibility. The cognitive process is governed by the Hick–Hyman Law [57,58]. Based on this law, a GUI should present as few choices as possible, e.g. split menus. The motor speed of humans is governed by the Fitts Law [59]. Based on this law, a GUI should utilise as much screen area as possible for widgets to be hit. Widgets that need to be hit sequentially should be placed next to each other.

In Reference 60, various techniques for energy-efficient GUI design are given that exploit the above laws. These techniques include low-energy colours, reduced screen changes, hot keys, user input cache, content placement, paged display and quick buttons. These techniques reduce the energy consumed in the display, storage, CPU and buses.

Example 11.4   Figures 11.14 and 11.15 show two implementations for a GUI for a personnel viewer: original and energy-efficient. The personnel viewer enables a user to scroll through a list of names and display information about selected individuals. Radio buttons at the top allow a user to select among position, affiliation

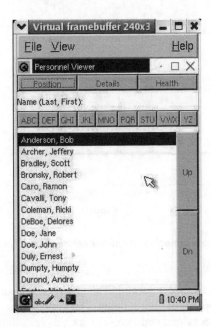

Figure 11.15   Personnel viewer: energy-efficient

and health queries. The original GUI requires two button presses to display information: radio button at the top to select the data and the pushbutton at the bottom to process the selection. The energy-efficient version replaces this with a single set of pushbuttons at the top. It also replaces the scrollbar with up–down buttons and alphabetic index tabs, and uses a low power colour scheme. This leads to an average improvement in performance by 35 per cent and in system energy by 45 per cent.   ∎

11.4.3   DVS/DPM for interactive systems

It was mentioned earlier that because of large system wait times (when the system waits for user input), there are major opportunities for DVS/DPM in interactive systems. One of the key problems in DVS/DPM is resource usage prediction. Most existing resource usage prediction techniques may work well for computation and I/O intensive tasks, but not for interactive tasks. Since the GUI takes care of system–user interaction, it often has *a priori* knowledge of how the system and user interact at a given moment. In Reference 54, techniques are given to use this knowledge and the above-mentioned psychological laws to predict user delays. Such delay predictions can be combined with DVS/DPM for aggressive power optimisation with little sacrifice in user productivity or satisfaction.

Example 11.5   Figure 11.16 shows the power consumption profile when some computation is performed using a Qtopia Calculator on the Sharp Zaurus SL-5500 PDA.

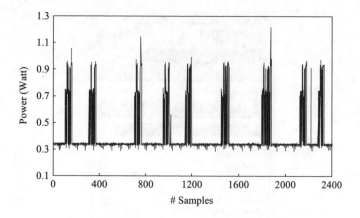

Figure 11.16   Power consumption of calculator on Sharp Zaurus

Power is sampled 400 times per second. The power peaks correspond to responses of the calculator to tappings of GUI buttons. The system waits for user inputs in the valleys while the OS does maintenance jobs such as handling timer interrupts and scheduling that account for the minor spikes in the valley. This is representative of interactive systems. An analysis of this application shows that 90 per cent of system energy and time is used up waiting for user input. The waiting periods are typically longer than 500 ms. Thus, the system can be put in a very low power mode. However, since it may take more than 150 ms to bring the system back to the normal mode from a very low power mode and the human perceptual threshold is between 50 ms and 100 ms, using the strategy of waking up the system only on next user input may lead to degradation in user productivity and be annoying to the user. Thus, DVS/DPM techniques for interactive systems need to predict when the system begins waiting for user input and how long is the wait. In Reference 54, this is done based on modelling the GUIs with their state transition diagrams and using psychological laws that model human perceptual, cognitive and motor delays to make the above predictions. ∎

11.4.4   Display-related power reduction

As shown earlier, an LCD system is composed of an LCD panel, framebuffer memory, LCD and framebuffer controller, and a backlight inverter and lamp, which are all heavy power-consumers. In Reference 61, techniques are given to minimise the energy of each of these components without causing any appreciable degradation in display quality. These techniques include variable-duty-ratio refresh, dynamic-colour-depth control and backlight luminance dimming with brightness compensation or contrast enhancement.

In Reference 62, several software-only techniques are introduced for LCD DPM. The first technique decreases the dot clock frequency that is used in processor-display communication, thus decreasing the display refresh rate as a side-effect. The second technique exploits the liquid crystal properties in the display to disable the LCD controller with minimum flicker. The third technique is based on reducing the image luminance. The last technique reduces backlight luminance based on environmental conditions.

In Reference 63, framebuffer compression methodologies are given for reducing display power. They are based on run-length encoding for on-the-fly compression, with a negligible impact on resources and time. When there are frequent partial framebuffer updates, an adaptive and incremental re-compression technique maintains its efficiency. These techniques reduce the activity of the framebuffer and associated buses during sweep operations.

In Reference 64, a concurrent brightness and contrast scaling technique is presented for backlit TFT-LCDs. It reduces backlight illumination while retaining image fidelity through preservation of the image contrast.

11.5 Chapter summary and open problems

Given that when both are applicable to processors in a distributed system, DVS is better than DPM, purely DPM-based system scheduling may not be that useful an area to pursue. This is specially true because many existing processors already have DVS capability. Of course, DPM will continue to be useful for other parts of the system which do not have DVS capability. Also, DVS cannot always get rid of all the idle slots in the system schedule. Thus, a combined DVS+DPM approach is preferable, applying DVS before DPM. In addition, targeting both switching and leakage power at the system level requires more research, e.g. by including ABB techniques in scheduling.

Battery-aware DVS+DPM approaches need further investigation. It is known that reducing the workload in a battery-operated system for a period of time leads to a recovery effect, which results in an increase in battery capacity [65]. More work needs to be done to exploit this effect in system scheduling.

Since interconnection networks are consuming increasingly larger fractions of system power, more DVS/DPM techniques for such networks are urgently needed. Power is also related to temperature. System-level thermal management will emerge as an important issue in the future. This will need to take both the network and PE tiles into account.

Low power system synthesis is also not a mature area. Currently, most such algorithms assume that the average power consumption of each task on each type of PE it can run on has been given. This is done because using more sophisticated power estimation techniques for processors, FPGAs and ASICs in the inner loop of system synthesis is currently not feasible. This points to the need for fast, yet relatively accurate, power estimation techniques, such as high-level macromodels, to drive system synthesis. Although some progress has been made in this direction for ASICs, high-level energy or power macromodels for processors and FPGAs deserve more attention. After obtaining such macromodels, the next step will be to integrate them in the inner loop of system synthesis.

Finally, when more sophisticated power-aware and battery-aware distributed system scheduling algorithms become available, they also need to be integrated in the inner loop of system synthesis.

References

1   BENINI, L., BOGLIOLO, A., and DE MICHELI, G.: 'A survey of design techniques for system-level dynamic power management', *IEEE Transactions on VLSI Systems*, 2000, **8**(3), pp. 299–316

2   BOWMAN, K.A., AUSTIN, L.B., EBLE, J.C., TANG, X., and MEINDL, J.D.: 'A physical alpha-power-law MOSFET model', *IEEE Journal of Solid-State Circuits*, 1999, **34**, pp. 1410–1414

3   KESHAVARZI, A., MA, S., NARENDRA, S., BLOECHEL, B., MISTRY, K., GHANI, T., BORKAR, S., and DE, V.: 'Effectiveness of reverse body bias for leakage control in scaled dual-Vt CMOS ICs'. Proceedings of the international symposium on *Low power electronics and design*, August 2001, pp. 207–212

4   DUARTE, D., VIJAYKRISHNAN, N., IRWIN, M.J., KIM, H.-S., and MCFARLAND, G.: 'Impact of scaling on the effectiveness of dynamic power reduction schemes'. Proceedings of the international confence on *Computer design*, September 2002, pp. 382–387

5   LAWLER, E.L., and MARTEL, C.U.: 'Scheduling periodically occurring tasks on multiple processors', *Information Processing Letters*, 1981, **7**, pp. 9–12

6   JHA, N.K.: 'Low power system scheduling and synthesis'. IEEE international conference on *Computer-aided design*, November 2001

7   LUO, J., and JHA, N.K.: 'Power-conscious joint scheduling of periodic task graphs and aperiodic tasks in distributed real-time embedded systems'. Proceedings of international conference on *Computer-aided design*, November 2000

8   GRUIAN, F., and KUCHCINSKI, K.: 'LEneS: task scheduling for low-energy systems using variable supply voltage processors'. Proceedings of the *Asia and South Pacific design automation* conference, January 2001, pp. 449–455

9   BAMBHA, N.K., BHATTACHARYA, S.S., TEICH, J., and ZITZLER, E.: 'Hybrid global/local search strategies for dynamic voltage scaling in embedded multiprocessors'. Proceedings of the international workshop on *HW/SW co-design*, March 2001, pp. 243–248

10   LIU, J., CHOU, P.H., BAGHERZADEH, N., and KURDAHI, F.: 'Power-aware scheduling under timing constraints for mission-critical embedded systems'. Proceedings of the *Design automation* conference, June 2001, pp. 840–845

11   YAO, F., DEMERS, A., and SHENKER, S.: 'A scheduling model for reduced CPU energy'. Proceedings of the annual symposium on *Foundations computer science*, 1995, pp. 374–381

12   LUO, J., and JHA, N.K.: 'Static and dynamic variable voltage scheduling algorithms for real-time heterogeneous distributed embedded systems'. Proceedings of international conference on *VLSI design*, January 2002

13 LUO, J., and JHA, N.K.: 'Power-profile driven variable voltage scaling for heterogeneous distributed real-time embedded systems'. Proceedings of international conference on *VLSI Design*, January 2003

14 YAN, L., LUO, J., and JHA, N.K.: 'Combined dynamic voltage scaling and adaptive body biasing for heterogeneous distributed real-time embedded systems'. Proceedings of international conference on *Computer-aided design*, November 2003

15 ANDREI, A., SCHMITZ, M., ELES, P., PENG, Z., and AL-HASHIMI, B.M.: 'Overhead-conscious voltage selection for dynamic and leakage energy reduction of time-constrained systems'. Proceedings of *Design automation and test in Europe* conference, February 2004, pp. 518–524

16 ZHANG, Y., HU, X., and CHEN, D.Z.: 'Task scheduling and voltage selection for energy minimization'. Proceedings of *Design automation* conference, June 2002, pp. 183–188

17 VARATKAR, G., and MARCULESCU, R.: 'Communication-aware task scheduling and voltage selection for total system energy minimization'. Proceedings of international conference on *Computer-aided design*, November 2003

18 LUO, J., and JHA, N.K.: 'Battery-aware static scheduling for distributed real-time embedded systems'. Proceedings of the *Design automation* conference, June 2001, pp. 444–449

19 PEDRAM, M., and WU, Q.: 'Design considerations for battery-powered electronics'. Proceedings of *Design automation* conference, June 1999

20 MARTIN, T., and SIEWIOREK, D.: 'The impact of battery capacity and memory bandwidth on CPU speed-setting: a case study'. Proceedings of international symposium on *Low power electronics and design*, August 1999, pp. 200–205

21 RAKHMATOV, D., VRUDHULA, S.B.K., and CHAKRABARTI, C.: 'Battery-conscious task sequencing for portable devices including voltage/clock scaling'. Proceedings of *Design automation* conference, June 2002, pp. 189–194

22 CHAKRABARTI, C., and AHMED, J.: 'A dynamic task scheduling algorithm for battery powered DVS systems'. Proceedings of international symposium on *Circuits and systems*, May 2004

23 BENINI, L., CASTELLI, G., MACII, A., MACII, E., PONCINO, M., and SCARSI, R.: 'Extending lifetime of portable systems by battery scheduling'. Proceedings of *Design automation and test in Europe* conference, March 2001, pp. 197–201

24 LAHIRI, K., RAGHUNATHAN, A., DEY, S., and PANIGRAHI, D.: 'Battery-driven system design: a new frontier in low power design'. Proceedings of the *Asia and South Pacific design automation* conference, January 2002

25 SHANG, L., PEH, L.-S., and JHA, N.K.: 'Dynamic voltage scaling with links for power optimization of interconnection networks'. Proceedings of the *High-performance computer architecture symposium*, February 2003

26 SHANG, L., PEH, L.-S., and JHA, N.K.: 'PowerHerd: dynamic satisfaction of peak power constraints in interconnection networks'. Proceedings of international conference on *Supercomputing*, June 2003

27   YE, T.T., BENINI, L., and DE MICHELI, G.: 'Analysis of power consumption on switch fabrics in network routers'. Proceedings of *Design automation* conference, June 2002, pp. 524–530

28   SOTERIOU, V., and PEH, L.-S.: 'Dynamic power management for power optimization of interconnection networks using on/off links'. Proceedings of international symposium on *High performance interconnects (hot interconnects)*, August 2003

29   CHEN, X., and PEH, L.-S.: 'Leakage power modeling and optimization in interconnection networks'. Proceedings of international symposium on *Low power and electronics design*, August 2003

30   HENKEL, J., and LI, Y.: 'Energy-conscious HW/SW-partitioning of embedded systems: a case study of an MPEG-2 encoder'. Proceedings of international Workshop on *HW/SW co-design*, March 1998, pp. 23–27

31   LI, Y., and HENKEL, J.: 'A framework for estimating and minimizing energy dissipation of embedded HW/SW systems'. Proceedings of the *Design automation* conference, June 1998, pp. 188–193

32   HENKEL, J.: 'A low power hardware/software partitioning approach for core-based embedded systems'. Proceedings of the *Design automation* conference, June 1999, pp. 122–127

33   DICK, R.P., and JHA, N.K. 'MOCSYN: multiobjective core-based single-chip system synthesis'. Proceedings of the *Design automation and test in Europe* conference, February 1999

34   FEI, Y. and JHA, N.K.: 'Functional partitioning for low power distributed systems of systems-on-a-chip'. Proceedings of the international conference on *VLSI design*, January 2002

35   GIVARGIS, T.D., HENKEL, J., and VAHID, F.: 'Interface and cache power exploration for core-based embedded system design'. Proceedings of the international conference on *Computer-aided design*, November 1999, pp. 270–273

36   HONG, I., KIROVSKI, D., QU, G., POTKONJAK, M., and SRIVASTAVA, M.B.: 'Power optimization of variable voltage core-based systems', *IEEE Transactions on Computer-Aided Design*, 1999, **18**(12), pp. 1702–1714

37   KIROVSKI, D., and POTKONJAK, M. 'System-level synthesis of low power hard real-time systems'. Proceedings of the *Design automation* conference, June 1997, pp. 697–702

38   DAVE, B., LAKSHMINARAYANA, G., and JHA, N.K.' COSYN: hardware-software co-synthesis of embedded systems', *IEEE Transactions on VLSI Systems*, 1999, **7**, pp. 92–104

39   DAVE, B., and JHA, N.K.: 'COHRA: hardware-software co-synthesis of hierarchical heterogeneous distributed embedded systems', *IEEE Transactions on Computer-Aided Design*, 1998, 17

40   DICK, R.P., and JHA, N.K.: 'MOGAC: a multiobjective genetic algorithm for the hardware-software co-synthesis of distributed embedded systems', *IEEE Transactions on Computer-Aided Design*, 1998, 17

41   DICK, R.P., and JHA, N.K.: 'COWLS: hardware-software co-synthesis of wireless low-power distributed embedded client-server systems', *IEEE Transactions Computer-Aided Design*, 2004, **23**(1), pp. 2–16

42   SHANG, L., and JHA, N.K.: 'Hardware-software co-synthesis of low power real-time distributed embedded systems using dynamically reconfigurable FPGAs'. Proceedings of the international conference on *VLSI design*, January 2002

43   SCHMITZ, M., and AL-HASHIMI, B.M.: 'Considering power variations of DVS processing elements for energy minimization in distributed systems'. Proceedings of the international symposium on *System synthesis*, November 2001

44   SCHMITZ, M.T., AL-HASHIMI, B.M., and ELES, P.: 'Energy-efficient mapping and scheduling for DVS enabled distributed embedded systems'. Proceedings of *Design automation and test in Europe* conference, February 2002, pp. 514–521

45   LIU, J., CHOU, P.H., and BAGHERZADEH, N.: 'Communication speed selection for embedded systems with networked voltage-scalable processors'. Proceedings of the international symposium on *Hardware-software co-design*, May 2002, pp. 169–174

46   LUO, J., PEH, L.-S., and JHA, N.K.: 'Simultaneous dynamic voltage scaling of processors and communication links in real-time distributed embedded systems'. Proceedings of *Design automation and test in Europe* conference, March 2003

47   QU, G., and POTKONJAK, M.: 'Energy minimization with quality of service'. Proceedings of the international symposium on *Low power electronics and design*, August 2000, pp. 43–49

48   BENINI, L., and DE MICHELI, G.: 'Networks on chips: a new SoC paradigm', *IEEE Computer*, 2002, **35**, pp. 70–78

49   HU, J., and MARCULESCU, R.: 'Exploiting the routing flexibility for energy/performance aware mapping of regular NoC architectures'. Proceedings of *Design, automation and test in Europe* conference, March 2003

50   HU, J., and MARCULESCU, R.: 'Energy-aware communication and task scheduling for network-on-chip architectures under real-time constraints'. Proceedings of *Design automation and test in Europe* conference, March 2004

51   VALLERIO, K., and JHA, N.K.: 'Task graph extraction for embedded system synthesis'. Proceedings of the international conference on *VLSI design*, January 2003. Tool available for download at http://www.princeton.edu/~vallerio/tools.html

52   DICK, R.P., RHODES, D.L., and WOLF, W.: 'TGFF: task graphs for free'. Proceedings of the international workshop on *Hardware-software co-design*, March 1998, pp. 97–101

53   E3S: the embedded system synthesis benchmarks suite, http://www.ece.northwestern.edu/~dickrp

54   ZHONG, L., and JHA, N.K.: 'Dynamic power optimization of interactive systems'. Proceedings of the international conference on *VLSI design*, January 2004

55   ZHONG, L., and JHA, N.K.: 'Graphical user interface energy characterization for handheld computers'. Proceedings of the international conference on *Compilers, architecture and synthesis for embedded systems*, November 2003

56   MYERS, B.A., and ROSSON, M.B.: 'Survey of user interface programming'. Proceedings of ACM conference on *Human factors in computing systems*, May 1992, pp. 195–202

57   HICK, W.E.: 'On the rate of gain of information', *Quarterly Journal of Experimental Psychology*, 1952, **4**, pp. 11–36

58   HYMAN, R.: 'Stimulus information as a determinant of reaction time', *Journal of Experimental Psychology*, 1953, **45**, pp. 188–196

59   FITTS, P.M.: 'The information capacity of the human motor system in controlling the amplitude of movement', *Journal of Experimental Psychology*, 1954, **47**(6), pp. 381–391

60   VALLERIO, K., ZHONG, L., and JHA, N.K.: 'Energy-efficient graphical user interface design'. Proceedings of international conference on *Pervasive computing and communications*, June 2004

61   CHOI, I., SHIM, H., and CHANG, N.: 'Low-power color TFT LCD display for handheld embedded systems'. Proceedings of international symposium on *Low power electronics and design*, August 2002, pp. 112–117

62   GATTI, F., ACQUAVIVA, A., BENINI, L., and RICCO, B.: 'Low power control techniques for TFT LCD displays'. Proceedings of international conference on *Compilers, architecture, and synthesis for embedded systems*, October 2002, pp. 218–224

63   SHIM, H., CHANG, N., and PEDRAM, M.: 'A compressed frame buffer to reduce display power consumption in mobile systems'. Proceedings of the *Asia and South Pacific design automation* conference, January 2004, pp. 819–824

64   CHENG, W.-C., HOU, Y., and PEDRAM, M.: 'Power minimization in a backlit TFT-LCD display by concurrent brightness and contrast scaling'. Proceedings of *Design automation and test in Europe* conference, March 2004

65   LINDEN, D.: 'Handbook of Batteries and Fuel Cells' (McGraw-Hill, New York, 1984)

Chapter 12

Power minimisation techniques at the RT-level and below

Afshin Abdollahi and Massoud Pedram

12.1 Introduction

A dichotomy exists in the design of modern microelectronic systems: they must be low power and high performance, simultaneously. This dichotomy largely arises from the use of these systems in battery-operated portable (wearable) platforms. Accordingly, the goal of low power design for battery-powered electronics is to extend the battery service life while meeting performance requirements. Unless optimisations are applied at different levels, the capabilities of future portable systems will be severely limited by the weight of the batteries required for an acceptable duration of service. In fixed, power-rich platforms, the packaging cost and power density/reliability issues associated with high power and high performance systems also force designers to look for ways to reduce power consumption. Thus, reducing power dissipation is a design goal even for non-portable devices since excessive power dissipation results in increased packaging and cooling costs as well as potential reliability problems. Meanwhile, following Moore's Law, integrated circuit densities and operating speeds have continued to go up in unabated fashion. The result is that chips are becoming larger, faster and more complex and because of this, consuming increasing amounts of power.

These increases in power pose new and difficult challenges for integrated circuit designers. While the initial response to increasing levels of power consumption was to reduce the supply voltage, it quickly became apparent that this approach was insufficient. Designers subsequently began to focus on advanced design tools and methodologies to address the myriad of power issues. Complicating designers' attempts to deal with these issues are the complexities – logical, physical and electrical – of contemporary IC designs and the design flows required to build them.

The established front-end approach to designing for lower power is to estimate and analyse power consumption at the register transfer level (RTL), and to modify the design accordingly. In the best case, only the RTL within given functional blocks is modified, and the blocks re-synthesised. The process is re-iterated until the desired results are achieved. Sometimes, though, the desired power consumption reductions may be achieved only by modifying the overall design architecture. Modifications at this level affect not only power consumption, but also other performance metrics, and may indeed greatly affect the cost of the chip. Thus, such modifications require re-evaluation and re-verification of the entire design. The architectural optimisation techniques, however, fall outside the coverage of the present chapter.

This chapter reviews a number of representative RTL design automation techniques that focus on low power design. It should be of interest to designers of power efficient devices, integrated circuit (IC) design engineering managers and electronic design automation (EDA) managers and engineers. More precisely, it covers techniques for sequential logic synthesis, RTL power management, multiple voltage design and leakage power minimisation and control techniques. Interested readers can find wide-ranging information on various aspects of low power design in References 1–3.

12.2   Multiple-voltage design

Using different voltages in different parts of a chip may reduce the global energy consumption of a design at a rather small cost in terms of algorithmic and/or architectural modifications. The key observation is that the minimum energy consumption in a circuit is achieved if all circuits paths are timing-critical (there is no positive slack in the circuit). A common voltage scaling technique is thus to operate all the gates on non-critical timing paths of the circuit at a reduced supply voltage. Gates/modules that are part of the critical paths are powered at the maximum allowed voltage, thus, avoiding any delay increase; the power consumed by the modules that are not on the critical paths, on the other hand, is minimised because of the reduced supply voltage. Using different power supply voltages on the same chip of circuitry requires the use of level shifters at the boundaries of the various modules (a level converter is needed between the output of a gate powered by a low V_{DD} and the input of a gate powered by a high V_{DD}, i.e. for a step-up change.) Figure 12.1 depicts a typical level converter design. Notice that if a gate that is supplied with $V_{DD,L}$ drives a fanout gate at $V_{DD,H}$, transistors N_1 and N_2 receive inputs at reduced supply and the cross-coupled positive channel metal oxide semiconductor (PMOS) transistors do the level conversion. Level converters are obviously not needed for a step-down change in voltage. Overhead of level converters can be mitigated by doing conversions at register boundaries and embedding the level conversion inside the flip flops (see Reference 4 for details).

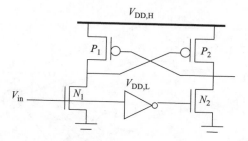

Figure 12.1   A typical level-converter design

A polynomial time algorithm for multiple-voltage scheduling of performance-constrained non-pipelined designs is presented by Raje and Sarrafzadeh [5]. The idea is to establish a supply voltage level for each of the operations in a data flow graph, thereby, fixing the latency of that operation. The goal is then to minimise the total power dissipation while satisfying the system timing constraints. Power minimisation is in turn accomplished by ensuring that each operation will be executed using the minimum possible supply voltage. The proposed algorithm is composed of a loop where, in each iteration, slacks of nodes in the acyclic data flow graph are calculated. Then, nodes with the maximum slack are assigned to lower voltages in such a way that timing constraints are not violated. The algorithm stops when no positive slack exists in the data flow graph. Notice that this algorithm assumes that the Pareto-optimal voltage vs. delay curve is identical for all computational elements in the data flow graph. Without this assumption, there is no guarantee that this algorithm will produce an optimal design.

In Reference 6, the problem is addressed for combinational circuits, where only two supply voltages are allowed. A depth-first search is used to determine those computational elements, which can be operated at low supply voltage without violating the circuit timing constraints. A computational element is allowed to operate at $V_{DD,L}$ only if all its successors are operating at $V_{DD,L}$. For example, Figure 12.2(a) demonstrates a clustered voltage scaling (CVS) solution in which each circuit path starts with $V_{DD,H}$ and switches to $V_{DD,L}$ when delay slack is available. The timing-critical path is shown with thick line segments. Here grey-coloured cells are running at $V_{DD,L}$. Level conversion (if necessary) is done in the flip flops at the end of the circuit paths. An extension to this approach is proposed in Reference 7, which is based on the observation that by optimising the insertion points of level converters, one can increase the number of gates using $V_{DD,L}$ without increasing the number of level converters. This leads to higher power savings. For example, in the CVS solution depicted in Figure 12.2(a), assume that the path delay from flip-flop FF_3 to gate G_2 is much longer than that of the path from FF_1 to G_2. In addition, assume that if we apply $V_{DD,L}$ to G_2, then the path from FF_3 to FF_5 through G_2 will miss its target combinational delay, i.e. G_2 must be assigned a supply level of $V_{DD,H}$. With the CVS approach, it immediately follows that G_3 must be assigned $V_{DD,H}$ although a potentially large positive slack remains in the path from FF_1 to G_2. The situation is the same

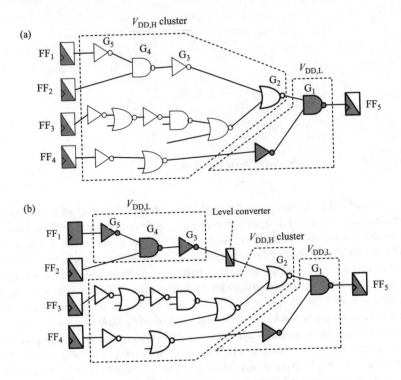

Figure 12.2   Examples of (a) CVS solution, (b) ECVS solution

for G_4 and G_5. Consequently, the CVS approach can miss opportunities for applying $V_{DD,L}$ to some gates in the circuit. If the insertion point of the level converter LC_1 is allowed to move up to the interface between G_3 and G_2, the gates G_3 through G_5 can be assigned a supply of $V_{DD,L}$, as depicted in Figure 12.2(b). The structure shown there is one that can be obtained by the extended CVS (ECVS) algorithm. Both CVS and ECVS assign the appropriate power supply to the gates by traversing the circuit from the primary outputs to the primary inputs in a topological order. ECVS allows a $V_{DD,L}$-driven gate to feed a $V_{DD,H}$-driven gate along with the insertion of a dedicated level converter.

Chen *et al.* [8] proposed an approach for voltage assignment in combinational logic circuits. First, a lower bound on dynamic power consumption is determined by exploiting the available slacks and the value of the dual-supply voltages that may be used in solving the problem of minimising dynamic power consumption of the circuit. Next, a heuristic algorithm is proposed for solving the voltage-assignment problem, where the values of the low and the high supply voltages are either specified by the user or fixed to the estimated ones.

Manzak and Chakrabarti [9] present resource- and latency-constrained scheduling algorithms to minimise power/energy consumption when the resources operate at multiple voltages. The proposed algorithms are based on efficient distribution of slack among the nodes in the data flow graph. The distribution procedure tries to implement the minimum energy relation derived using the Lagrange multiplier method in an iterative fashion.

An important phase in the design flow of multiple-voltage systems is that of assigning the most convenient supply voltage, selected from a fixed number of values, to each operation in the control-data flow graph (CDFG). The problem is to assign the supply voltages and to schedule the tasks so as to minimise the power dissipation under throughput/resource constraints. An effective solution has been proposed by Chang and Pedram [10]. The technique is based on dynamic programming and requires the availability of accurate timing and power models for the macro-modules in a RTL library. A preliminary characterisation procedure must then be run to determine an energy-delay curve for each module in the library and for all possible supply-voltage assignments. The points on the curve represent various voltage assignment solutions with different trade-offs between the performance and the energy consumption of the cell. Each set of curves is stored in the RTL library, ready to be invoked by the cost function that guides the multiple supply-voltage scheduling algorithm. We provide a brief description of the method for the simple case of CDFGs with a tree structure. The algorithm consists of two phases: first, a set of possible power-delay trade-offs at the root of the tree is calculated; then, a specific macro-module is selected for each node in such a way that the scheduled CDFG meets the required timing constraints. To compute the set of possible solutions, a power-delay curve at each node of the tree (proceeding from the inputs to the output of the CDFG) is computed; such a curve represents the power-delay trade-offs that can be obtained by selecting different instances of the macro-modules, and the necessary level shifters, within the subtree rooted at each specific node. The computation of the power-delay curves is carried out recursively, until the root of the CDFG is reached. Given the power-delay curve at the root node, that is, the set of trade-offs the user can choose from, a recursive preorder traversal of the tree is performed, starting from the root node, with the purpose of selecting which module alternative should be used at each node of the CDFG. Upon completion, all the operations are fully scheduled; therefore, the CDFG is ready for the resource-allocation step.

More recently, a level-converter free approach is proposed in Reference 11 where the authors try to eliminate the overhead imposed by level converters by suggesting a voltage scaling technique without utilising level converters. The basic initiative is to impose some constraints on the voltage differences between adjacent gates with different supply voltages based on the observation that there will be no static current if the supply voltage of a driver gate is higher than the subtraction of the threshold voltage of a PMOS from the supply voltage of a driven gate. Murugavel and Ranganathan [12] proposed behavioural-level power optimisation algorithms that use voltage and frequency scaling. In this work, the operators in a data flow graph are scheduled in the modules of the given architecture, by applying voltage and frequency scaling techniques to the modules of the architecture such that the power consumed by the modules is minimised. The global optimal selection of voltages and frequencies for the modules is determined through the use of an auction-theoretic model and a game-theoretic solution. The authors present a resource-constrained scheduling algorithm, which is based on applying the Nash equilibrium function to the game-theoretic formulation.

12.3 Dynamic voltage scaling and Razor logic

The dependence of both performance and power dissipation on supply voltage results in a trade-off in circuit design. High supply voltage results in high performance while low supply voltage makes an energy-efficient design. Dynamic voltage scaling (DVS) [13] is a powerful technique to reduce circuit energy dissipation in which the application or operating system identifies periods of low processor utilisation that can tolerate reduced frequency which allows reduction in the supply voltage. Since dynamic power scales quadratically with supply voltage, DVS significantly reduces energy consumption with a limited impact on system performance [14].

Several factors determine the voltage required to reliably operate a circuit in a given frequency. The supply voltage must be sufficiently high to fully evaluate the critical path in a single clock cycle (i.e. critical voltage). To ensure that the circuit operates correctly even in the worst-case operating environment some voltage margins are added to the critical voltage (e.g. process margin due to manufacturing variations, ambient margins to compensate high temperatures and noise margins due to uncertainty in supply and signal voltage levels).

To ensure correct operation under all possible variations, a conservative supply voltage is typically selected using corner analysis. Hence, margins are added to the critical voltage to account for uncertainty in the circuit models and to account for the worst-case combination of variations. However, such a worst-case combination of variations may be highly improbable; hence this approach is overly conservative.

In some approaches the delay of an embedded inverter chain is used as a prediction of the critical path delay of the circuit and the supply voltage is tuned during processor operation to meet a predetermined delay through the inverter-chain [15]. This approach to DVS allows dynamic adjustment of the operating voltage to account for global variations in supply voltage drop, temperature fluctuation and process variations. However, it cannot account for local variations, such as local supply voltage drops, intra-die process variations and cross-coupled noise, and therefore requires the addition of some margins to the critical voltage. Also, the delay of an inverter chain does not scale with voltage and temperature in the same way as the delays of the critical paths of the actual design, which can contain complex gates and pass-transistor logic, which again requires extra voltage margins.

Ernst *et al.* [16] proposed a different approach to DVS, referred to as Razor logic, which is based on dynamic detection and correction of speed path failures in digital designs. The basic idea is to tune the supply voltage by monitoring the error rate during operation, which eliminates the need for voltage margins that are necessary for 'always-correct' circuit operation in conventional DVS. In Razor logic, the operation at sub-critical supply voltages does not constitute a 'failure', but instead represents a trade-off between the power dissipation penalties incurred from error correction vs. the additional power savings obtained from operating at a lower supply voltage.

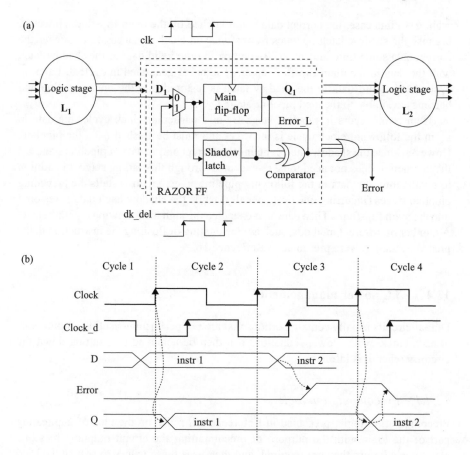

Figure 12.3   Illustration of Razor logic and DVS. (a) Pipeline augmented with Razor latches, (b) control lines for RAZOR flip-flops

The Razor logic based DVS utilises a combination of circuit and architectural techniques for low-cost error detection and correction of delay failures. Each flip-flop in the critical path is augmented with a 'shadow' latch which is controlled using a delayed clock. The operating voltage is constrained such that the worst-case delay meets the shadow latch setup time, even though the main flip-flop could fail. By comparing the values latched by the flip-flop and the shadow latch, a timing error in the main flip-flop can be detected. The value in the shadow latch, which is guaranteed to be correct, is subsequently utilised to correct the delay failure.

This concept is illustrated in Figure 12.3(a) for a pipeline stage. The operation of a Razor flip-flop is shown in Figure 12.3(b). In clock cycle 1, the combinational logic L_1 meets the setup time by the rising edge of the clock and both the main flip-flop and the shadow latch will latch the correct data. In this case, the error signal at the output of the XOR gate remains low and the operation of the pipeline is unaltered. In cycle 2, the combinational logic delay exceeds the intended delay due to sub-critical voltage scaling. In this case, the correct data is not latched by the main flip-flop. However, because the shadow latch operates from a delayed clock, it successfully latches the correct data some time in cycle 3. By comparing the valid data of the shadow latch with the data in the main flip-flop, an error signal is generated in cycle 3. Later, in cycle 4, the valid data in the shadow latch is restored into the main flip-flop and becomes available to the next pipeline stage L_2.

If an error occurs in pipeline stage L_1 in a particular clock cycle, the data in L_2 in the following clock cycle is incorrect and must be flushed from the pipeline. However, since the shadow latch contains the correct output data of pipeline stage L_1, the instruction does not need to be re-executed through this failing stage. In addition to invalidating the data in the following pipeline stage, an error stalls the preceding pipeline stages (incurring one cycle penalty) while the shadow latch data is restored into the main flip-flops. Then data is re-executed through the following pipeline stage. A number of different methods, such as clock gating or flushing the instruction in the preceding stages, were presented in Reference 16.

12.4   RTL power management

Digital circuits usually contain portions that are not performing useful computations at each clock cycle. Power reductions can then be achieved by shutting down the circuitry when it is idle.

12.4.1   *Precomputation logic*

Precomputation logic, presented in Reference 17, relies on the idea of duplicating part of the logic with the purpose of precomputing the circuit output values one clock cycle before they are required, and then uses these values to reduce the total amount of switching in the circuit during the next clock cycle. In fact, knowing the output values one clock cycle in advance allows the original logic to be turned off during the next time frame, thus eliminating any charging and discharging of the internal capacitances. Obviously, the size of the logic that pre-calculates the output values must be kept under control since its contribution to the total power balance may offset the savings achieved by blocking the switching inside the original circuit. Several variants to the basic architecture can then be devised to address this issue. In particular, sometimes it may be convenient to resort to partial, rather than global, shutdown, i.e. to select for power management only a (possibly small) subset of the circuit inputs.

The synthesis algorithm presented in Reference 17 suffers from the limitation that if a logic function is dependent on the values of several inputs for a large fraction of the applied input combinations, then no reduction in switching activity can be obtained. Monteiro *et al.* [18] focused on a particular sequential precomputation architecture where the precomputation logic is a function of all of the input variables. The authors call this architecture the 'complete input-disabling architecture'. It is shown that the complete input-disabling architecture can reduce power dissipation for a larger

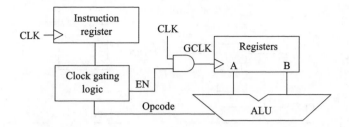

Figure 12.4   Clock gating logic for ALU in a typical processor microarchitecture with negative-edge triggered flip-flops

class of sequential circuits compared to the subset input-disabling architecture. The authors present an algorithm to synthesise precomputation logic for the complete input-disabling architecture.

12.4.2   Clock gating

Another approach to RT and gate-level dynamic power management, known as gated clocks [19–21], provides a way to selectively stop the clock, and thus, force the original circuit to make no transition, whenever the computation that is to be carried out at the next clock cycle is redundant. In other words, the clock signal is disabled according to the idle conditions of the logic network. For reactive circuits, the number of clock cycles in which the design is idle in some wait states is usually large. Therefore, avoiding the power waste corresponding to such states may be significant.

The logic for the clock management is automatically synthesised from the Boolean function that represents the idle conditions of the circuit (cf. Figure 12.4.) It may well be the case that considering all such conditions results in additional circuitry that is too large and too power-consuming. It may then be necessary to synthesise a simplified function, which dissipates the minimum possible power and stops the clock with maximum efficiency. The use of gated clocks has the drawback that the logic implementing the clock-gating mechanism is functionally redundant, and this may create major difficulties in testing and verification. The design of highly testable-gated clock circuits is discussed in Reference 22.

Another difficulty with clock gating is that one must stop hazards/glitches on EN signal from corrupting the clock signal to the register sets. This can be accomplished by introducing a transparent negative latch between EN and the AND gate as shown in Figure 12.5.

12.4.3   Computational kernels

Sequential circuits may have an extremely large number of reachable states, but during normal operation, these circuits tend to visit only a relatively small subset of the reachable states. A similar situation occurs at the primary outputs; while the circuit walks through the most probable states, only a few distinct patterns are generated at

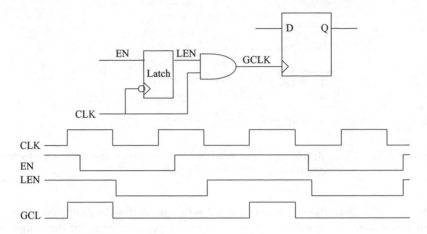

Figure 12.5   *Clock is disabled when EN = 0. Furthermore, a hazard on EN will be stopped from reaching GCLK*

the combinational outputs of the circuit. Many researchers have proposed approaches for synthesising a circuit that is fast and power-efficient under typical input stimuli, but continues to operate correctly even when uncommon input stimuli are applied to the circuit.

Reference 23 presents a power optimisation technique by exploiting the concept of computational kernel of a sequential circuit, which is a highly simplified logic block that imitates the steady-state behaviour of the original specification. This block is smaller, faster and less power-consuming than the circuit from which it is extracted and can replace the original network for a large fraction of the operation time.

The p-order computational kernel of an FSM is defined with respect to a given probability threshold p and includes the subset of the states, S_p, of the original FSM whose steady-state occupation probabilities are larger than p. The combinational kernel also includes the subset of states, R_p, where for each state in R_p there is an edge from a state in S_p to that state. As an example, consider the simple FSM shown in Figure 12.6(a) in which the input and output values are omitted for the sake of simplicity and the states are annotated with the steady-state occupation probabilities calculated through Markovian analysis of the corresponding state transition graph (STG). If we specify a probability threshold of $p = 0.25$, then the computational kernel of the FSM is depicted in Figure 12.6(b). States in black represent set S_p, while states in grey represent R_p. The kernel probability is $\text{Prob}(S_p) = 0.29 + 0.25 + 0.32 = 0.86$.

Given a sequential circuit with the standard topology depicted in Figure 12.7(a), the paradigm for improving its quality with respect to a given cost function (e.g. power dissipation, latency) is based on the architecture shown in Figure 12.7(b).

The basic elements of the architecture are: the combinational portion of the original circuit (block CL), the computational kernel (block K), the selector function (block S), the double state flip-flops (DSFFs) and the output multiplexers (MUX).

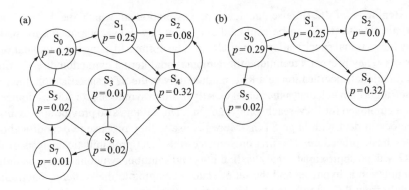

Figure 12.6   (a) Moore-type FSM and (b) its 0.25-order computational kernel

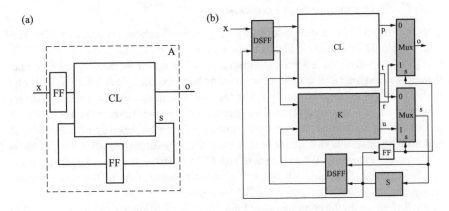

Figure 12.7   Illustration of computational kernel utilisation. (a) Baseline architecture, (b) Kernel-based optimised architecture

The computational kernel can be seen as a 'dense' implementation of the circuit from which it has been extracted. In other terms, K implements the core functions of the original circuit, and because of its reduced complexity, it usually implements such functions in a faster and more efficient way. The purpose of selector function S is that of deciding what logic block, between CL and K, will provide the output value and the next-state in the following clock cycle. To take a decision, S examines the values of the next-state outputs at clock cycle n. If the output and next-state values in cycle $n + 1$ can be computed by the kernel K, then S takes on the value 1. Otherwise, it takes on the value 0. The value of S is fed to a flip-flop, whose output is connected to the MUXs that select which block produces the output and the next-state. The optimised implementation is functionally equivalent to the original one. Computational kernels are a generalisation of the precomputation architecture from combinational and pipelined sequential circuits to finite state machines. Benini *et al.* [23] proposed an algorithm for generating the computational kernel of a FSM by iterative simplification of the original network by redundancy removal.

Benini *et al.* [24] raise the level of abstraction at which the kernel-based optimisation strategy can be exploited and show how RTL components for which only a functional specification is available can be optimised using the computational kernels. They present a technique for computational kernel extraction directly from the functional specification of a RTL module. Given the STG specification, the proposed algorithm calculates the kernel exactly through symbolic procedures similar to those employed for FSM reachability analysis. The authors also provide approximate methods to deal with large STGs. More precisely, they propose two modifications to the basic procedure. The first one replaces the exact probabilistic analysis of the STG with an approximate analysis. In the second solution, symbolic state probability computation is bypassed and the set of states belonging to the kernel is determined directly from RTL simulation traces of a given (random or user-provided) stream.

12.4.4   State machine decomposition

Decomposition of finite state machines for low power has been proposed by Monteiro and Oliveira [25]. The basic idea is to decompose the STG of a FSM into two STGs that jointly produce the equivalent input–output behaviour as the original machine. Power is saved because, except for transitions between the two sub-FSMs, only one of the sub-FSM needs to be clocked. The technique follows a standard decomposition structure. The states are partitioned by searching for a small subset of states with high probability of transitions among these states and a low probability of transitions to and from other states. This subset of states will then constitute a small sub-FSM that is active most of the time. When the small sub-FSM is active, the other larger sub-FSM can be disabled. Consequently, power is saved because most of the time only the smaller, more power-efficient, sub-FSM is clocked.

In Reference 26 the combinational logic block is partitioned (e.g. to CL_1 and CL_2) and the active part is decided based on the encoding of the present state. The states selected for one of the sub-FSM (i.e. M_1) are all encoded in such a way that the enable signal is always on for CL_1 while it is off for CL_2. Conversely, for all states in the other sub-FSM (i.e. M_2), the enable signal is always off for CL_1 while it is on for CL_2. Consequently, for all transitions within M_1, only CL_1 will be active and vice-versa.

Consider as an example $dk27$ FSM from the MCNC benchmark set, depicted in Figure 12.8. Assume that the input signal values, 0 and 1, occur with equal probabilities. The steady-state probabilities which are shown next to the states in this figure have been computed accordingly. Suppose we partition the FSM into two submachines M_1 and M_2 along the dotted line. Then around 40 per cent of the transitions occur in submachine M_1, 40 per cent of the transitions occur in submachine M_2 and 20 per cent of the transitions occur between submachines M_1 and M_2. Now suppose that the FSM is synthesised as two individual combinational circuits for submachines M_1 and M_2. Then we can turn off the combinational circuit for submachine M_2 when transitions occur within submachine M_1. Similarly, we can turn off the combinational circuit for submachine M_1 when transitions occur within submachine M_2. The states are partitioned such that the probability of transitions within any sub-FSM is maximised and the estimated overhead is minimised.

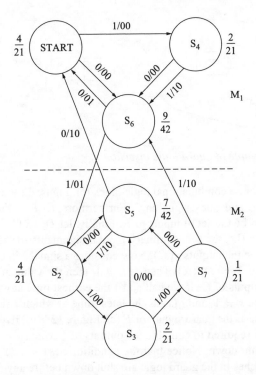

Figure 12.8 Example of an FSM (dk27) that may be decomposed into two sub-FSMs such that one sub-FSM can be shut off when the other is active and vice versa

These methods for FSM decomposition can be considered as extensions of the gated-clock for FSM self-loops approach proposed in Reference 27. In FSM decomposition the cluster of states that are selected for one of the sub-FSMs can be considered as a 'super-state' and then transitions between states in this cluster can be seen as self-loops on this 'super-state'.

12.4.5 Guarded evaluation

Guarded evaluation [28] is the last RT and gate-level shutdown technique we review in this section. The distinctive feature of this solution is that, unlike precomputation and gated clocks, it does not require one to synthesise additional logic to implement the shutdown mechanism; instead, it exploits existing signals in the original circuit. The approach is based on placing some guard logic, consisting of transparent latches with an enable signal, at the inputs of each block of the circuit that needs to be power-managed. When the block must execute some useful computation in a clock cycle, the enable signal makes the latches transparent. Otherwise, the latches retain their previous states, thus, blocking any transition within the logic block.

Guarded evaluation provides a systematic approach to identify where transparent latches must be placed within the circuit and by which signals they must be controlled.

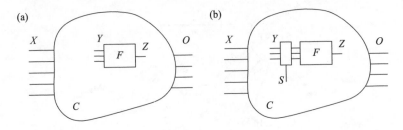

Figure 12.9   Example of guard logic insertion

For example, let C be a combinational logic block (cf. Figure 12.9(a)), X be the set of primary inputs to C and z be a signal in C. Furthermore, let F be the portion of logic that drives z and Y be the set of inputs to F. Finally, let $D_Z(X)$ be the observability don't-care set for z (i.e. the set of primary input assignments for which the value of z does not influence the outputs of C). Now consider a signal s in C which logically implies $D_Z(X)$, i.e. $s \Rightarrow D_Z(X)$. Then, if $s = 1$, then the value of z is not required to compute the outputs of C. If we call $t_e(Y)$ the earliest time at which any input to F can switch when $s = 1$, and $t_l(s)$ as the latest time at which s settles to one, then signal s can be used as the guard signal for F (cf. Figure 12.9(b)) if $t_l(s) < t_e(Y)$. This is because z is not required to compute the outputs of C when $s = 1$, and therefore, block F can be shut down. Notice that the condition $t_l(s) < t_e(Y)$ guarantees that the transparent latches in the guard logic are shut down before any of the inputs to F makes a transition.

This technique, referred to as pure guarded evaluation, has the desirable property that when applied, no changes in the original combinational circuitry are needed. On the other hand, if some resynthesis and restructuring of the original logic is allowed, a larger number of logic shutdown opportunities may become available.

12.5   Sequential logic synthesis for low power

Power can be minimised by appropriate synthesis of logic. The goal in this case is to minimise the so-called switched capacitance of the circuit by low power driven logic minimisation techniques.

12.5.1   State assignment

State encoding/assignment, as a crucial step in the synthesis of the controller circuitry, has been extensively studied. Roy and Prasad [29] were the first to address the problem of reducing switching activity of input state lines of the next state logic, during the state assignment, formulating it as a Minimum Weighted Hamming Distance problem [29]. Olson *et al.* [30] used a linear combination of switching activity of the next state lines and the number of literals as the cost function. Tsui *et al.* [31] used simulated annealing as a search strategy to find a low power state encoding that accounts for

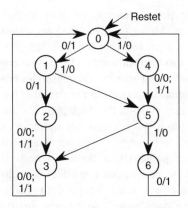

Figure 12.10   Excess-3 Converter state transition graph

both the switching activity of the next state lines and switched capacitance of the next state and output logic.

For example, consider the state transition graph for a BCD to Excess-3 Converter depicted in Figure 12.10. Assume that the transition probabilities of the thicker edges in this figure are more than those of the thin edges. The key idea behind all of the low power state assignment techniques is to assign minimum Hamming distance codes to the state pairs that have large inter-state transition probabilities. For example, the coding, $S_0 = 000$, $S_1 = 001$, $S_2 = 011$, $S_3 = 010$, $S_4 = 100$, $S_5 = 101$, $S_6 = 111$, $S_7 = 110$ fulfills this requirement.

Wu *et al.* [32] proposed the idea of realising a low power FSM by using T flip-flops. The authors showed that use of T flip-flops results in a natural clock gating and may result in reduced next state logic complexity. However, that work was mostly focused on BCD counters which have cyclic behaviour. The cyclic behaviour of counters results in a significant reduction of combinational logic complexity and, hence, lowers power consumption. Reference 33 introduces a mathematical framework for cycle representation of Markov processes and based on that, proposes solutions to the low power state assignment problem. The authors first identify the most probable cycles in the FSM and encode the states on these cycles with Grey codes. The objective function is to minimise the Weighted Hamming Distance. This reference also teaches how a combination of T and D flip-flops as state registers can be used to achieve a low power realisation of a FSM.

12.5.2   Retiming

Retiming is to reposition the registers in a design to improve the area and performance of the circuit without modifying its input–output behaviour. The technique was initially proposed by Leiserson and Saxe [34]. This technique changes the location of registers in the design in order to achieve one of the following goals: (1) minimise the clock period; (2) minimising the number of registers or (3) minimise the number of registers for a target clock period.

Minimising dynamic power for synchronous sequential digital designs is addressed in the literature. Monteiro *et al.* [35] presented heuristics to minimise the switching activity in a pipelined sequential circuit. Their approach is based on the fact that registers have to be positioned on the output edges of the computational elements that have high switching activity. The reason for power savings is that in this case the output of a register switches only at the arrival of the clock signal as opposed to potentially switching many times in the clock period. Consider the simple example of a logic gate belonging to a synchronous circuit and a capacitive load driven by the output gate. In CMOS technology, the power dissipated by gate is proportional to the product of the switching activity of the output node of the gate and the output load. At the output of gate some spurious transitions (i.e. glitches) may occur, which can result in a significant power waste. Suppose a register is inserted between the output of the gate and the capacitive load, in the new circuit the output of the register can make, at most, one transition per clock cycle. In fact, the gate output may have many redundant transitions but they are all filtered out by the register; hence, these logic hazards do not propagate to the output load.

The heuristic retiming technique of Reference 35 applies to a synchronous network with pipeline structure. The basic idea is to select a set of candidate gates in the circuit such that if registers are placed at their outputs, the total switching activity of the network gets minimised. The selection of the gates is driven by two factors: the amount of glitching that occurs at the output of each gate and the probability that such glitching propagates to the gates located in the transitive fanout. Registers are initially placed at the primary inputs of the circuit, and backward retiming (which consists of moving one register from all gate inputs to the output) is applied until all the candidate gates have received a register on their outputs. Then, registers that belong to paths not containing any of the candidate gates are repositioned, with the objective of minimising both the delay and the total number of registers in the circuit. This last retiming phase does not affect the registers that have been already placed at the outputs of the previously selected gates. In Reference 36, fixed-phase retiming is proposed to reduce dynamic power consumption. The edge-triggered circuit is first transformed to a two-phase level-clocked circuit, by replacing each edge-triggered flip-flop by two latches. Using the resulting level-clocked circuit, the latches of one phase are kept fixed, while the latches belonging to the other phase are moved onto wires with high switching activity and loading capacitance.

Fixed-phase retiming is best illustrated by the example shown below. Figure 12.11(a) shows a section of a pipelined circuit with edge-triggered flip-flops. The numbers on the edges represent the potential reduction in power dissipation when an edge-triggered flip-flop is present on that edge, assuming that the rest of the circuit remains unchanged. Negative values of power reduction indicate an increase in power dissipation when a flip-flop is placed on an edge. This reduction in power dissipation can be achieved if the edge has a high glitching-capacitance product [3]. After replacing each edge-triggered flip-flop by two back-to-back level-clocked latches, the resulting circuit is fixed-phase retimed to obtain the circuit in Figure 12.11(b).

Assuming a non-overlapping two-phase clocking scheme $\pi = \langle \phi_0 = 4, \gamma_0 = 1, \phi_1 = 4, \gamma_1 = 1 \rangle$ such as the one shown in Figure 12.11(c), power dissipation can

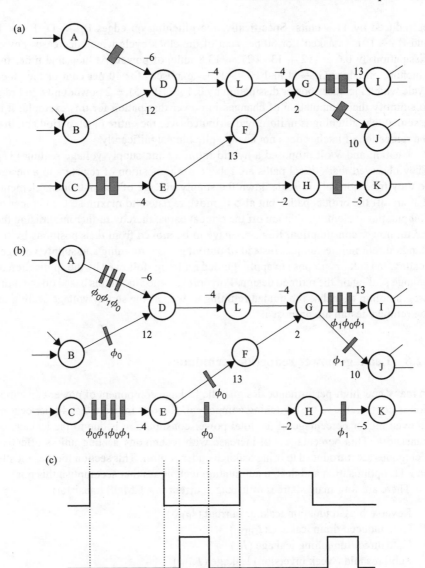

Figure 12.11 Illustration of fixed-phase retiming. (a) Initial edge-triggered circuit, (b) fixed-phase retimed circuit, (c) a two-phase clocking scheme $\pi = \langle \phi_0 = 4, \gamma_0 = 1, \phi_1 = 4, \gamma_1 = 1 \rangle$

be reduced by 11.8 units. Specifically, the glitching on edges B → D, E → F and E → H is 'masked' for 60 per cent of the clock cycle which decreases power dissipation by $0.6 \times (12 + 13 - 2) = 13.8$ units of power. At the same time, the glitching on edges G → J and H → K is 'exposed' for 40 per cent of the clock cycle which increases power dissipation by $0.4 \times (10 - 5) = 2$ power units. In order to simplify the computation of changes in power dissipation for this example, it is assumed that glitching is uniformly distributed over the entire clock period and that the relocation of latches does not change glitching significantly.

Chabini and Wolf proposed a hybrid retiming and supply voltage scaling [37]. They observed that critical paths are related to the position of registers in a design so they tried not only to scale down the supply voltage of computational elements that are off the critical paths, but also to move registers to maximise the number of computational elements that are off the critical paths, thereby further minimising the circuit power consumption. Registers have to be moved from their positions by the standard retiming technique. Instead of unifying basic retiming and supply voltages scaling, the authors propose to apply 'guided retiming' followed by the application of voltage scaling on the retimed design. Polynomial time algorithms based on dynamic programming to realise the guided retiming as well as the supply voltage scaling on the retimed design are proposed.

12.6   Leakage power reduction techniques

In many new high-performance designs, the leakage component of power consumption is comparable to the switching component. Reports indicate that 40 per cent or an even higher percentage of the total power consumption is due to the leakage of transistors. This percentage will increase with technology scaling unless effective techniques are introduced to bring leakage under control. This section focuses mostly on RTL optimisation and design automation techniques that accomplish this goal.

There are four main sources of leakage current in a CMOS transistor:

1   Reverse-biased junction leakage current (I_{REV})
2   Gate induced drain leakage (I_{GIDL})
3   Gate direct-tunnelling leakage (I_G)
4   Subthreshold (weak inversion) leakage (I_{SUB})

Let I_{OFF} denote the leakage of an OFF transistor ($V_{GS} = 0$ V for an NMOS device which results in $I_G = 0$).

$$I_{OFF} = I_{REV} + I_{GIDL} + I_{SUB}$$

Components, I_{REV} and I_{GIDL} are maximised when $V_{DB} = V_{DD}$. Similarly, for short-channel devices, I_{SUB} increases with V_{DB} because of the DIBL effect. Note the I_G is not a component of the OFF current, since the transistor gate must be at a high potential with respect to the source and substrate for this current to flow. An effective approach to overcome the gate leakage currents while maintaining excellent gate control is to replace the currently used silicon dioxide gate insulator with high-K dielectric material such as TiO_2 and Ta_2O_5. Use of the high-K dielectric will allow a less aggressive gate dielectric thickness reduction while maintaining the required gate overdrive at low supply voltages [38]. High-K gate dielectrics are expected to be introduced in 2006 [39]. Therefore, it is reasonable to ignore the I_G component of leakage. Among the three components of I_{OFF}, I_{SUB} is the dominant component. Hence, most leakage reduction techniques focus on I_{SUB}. For more details on the different leakage mechanisms, see Chapter 13.

12.6.1   Power gating and multi-threshold CMOS

The most obvious way of reducing the leakage power dissipation of a VLSI circuit in the STANDBY state is to turn off its supply voltage. This can be done by using one PMOS transistor and one NMOS transistor in series with the transistors of each logic block to create a virtual ground and a virtual power supply as depicted in Figure 12.12. In practice, only one transistor is necessary. Because of the lower on-resistance, NMOS transistors are usually used.

In the ACTIVE state, the sleep transistor is on. Therefore, the circuit functions as usual. In the STANDBY state, the transistor is turned off, which disconnects the gate from the ground. To lower the leakage, the threshold voltage of the sleep transistor must be large. Otherwise, the sleep transistor will have a high leakage current, which will make the power gating less effective. Additional savings may be achieved if the width of the sleep transistor is smaller than the combined width of the transistors in the pull-down network. In practice, Dual V_T CMOS or multi-threshold CMOS (MTCMOS) is used for power gating [40,41]. In these technologies there are several types of transistors with different V_T values. Transistors with a low V_T are used to implement the logic, while high-V_T devices are used as sleep transistors.

To guarantee the proper functionality of the circuit, the sleep transistor has to be carefully sized to decrease its voltage drop while it is on. The voltage drop on the sleep transistor decreases the effective supply voltage of the logic gate. Also, it increases the threshold of the pull-down transistors due to the body effect. This increases

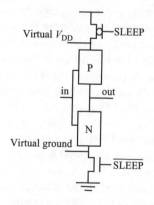

Figure 12.12   Power gating circuit

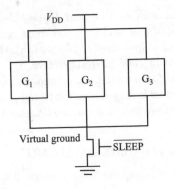

Figure 12.13   Using one sleep transistor for several gates

the high-to-low transition delay of the circuit. This problem can be solved by using a large sleep transistor. On the other hand, using a large sleep transistor increases the area overhead and the dynamic power consumed for turning the transistor on and off. Note that because of this dynamic power consumption, it is not possible to save power for short idle periods. There is a minimum duration of the idle time below which power saving is impossible. Increasing the size of the sleep transistors increases this minimum duration.

Since using one transistor for each logic gate results in a large area and power overhead, one transistor may be used for each group of gates as depicted in Figure 12.13. Notice that the size of the sleep transistor in this figure ought to be larger than the one used in Figure 12.12. To find the optimum size of the sleep transistor, it is necessary to find the vector that causes the worst case delay in the circuit. This requires simulating the circuit under all possible input values, a task that is not possible for large circuits.

In Reference 41, Kao and Chandrakasan describe a method to decrease the size of sleep transistors based on the mutual exclusion principle. In their method, the authors first size the sleep transistors to achieve delay degradation less than a given percentage for each gate. Notice that this guarantees that the total delay of the circuit will be degraded by less than the given percentage. In fact the actual degradation can be as much as 50 per cent smaller. The reason for this is that NMOS sleep transistors degrade only the high-to-low transitions and at each cycle only half of the gates switch from high to low. If two gates switch at different times (i.e. their switching windows are non-overlapping), then their corresponding sleep transistors can be shared.

Although sleep transistors can be used to disconnect logic gates from ground, using them to disconnect flip-flops from ground or supply voltage results in the loss of data. Hyo-Sig Won *et al.* [42] solve this problem by using high-threshold transistors for the inverters that hold data and low-threshold transistors for other parts of flip-flops. In the sleep mode, the low-threshold transistors are disconnected from the ground, but the two inverters that hold data stay connected to the ground. Since high-threshold transistors have been used in the inverters, their leakage is small. Other possibilities for saving data when MTCMOS is applied to a sequential circuit are to utilise leakage-feedback gates and flip-flops [43] or balloon latches [44].

12.6.2   Multiple-threshold cells

Multiple-threshold voltages have been available on many CMOS processes for a number of years. Multiple-threshold CMOS circuit, which has both high- and low-threshold transistors in a single chip, can be used to deal with the leakage problem. The high-threshold transistors can suppress the subthreshold leakage current, while the low-threshold transistors are used to achieve the high performance. Since the standby power is much larger for low V_T transistors compared to the high V_T ones, usage is limited to using low V_T transistors on timing-critical paths, with insertion rates of the order of 20 per cent or less. Since T_{ox} and L_{gate} are the same for high and low V_T transistors, low V_T insertion does not adversely impact the active power component or the design size. Drawbacks are that variation due to doping is uncorrelated between the high- and low-threshold transistors and extra mask steps incur a process cost.

The technology used for fabricating circuits can restrict the manner in which transistors can be mixed. For example, it may not be possible to use different threshold voltages for transistors in a stack due to their proximity. Furthermore, to simplify the design process and computer-aided design (CAD) algorithms, one may wish to restrict the way transistors are mixed. For example, when transistors of the same type are used in a logic cell, the size of multi-threshold cell library is only twice that of the original (single threshold) cell library. This reduces the library development time as well as the complexity and run time of CAD algorithms and tools that use the library.

In general, one expects that the leakage saving increases as the freedom to mix low and high V_T devices in a logic cell is increased. However, the percentage improvement is usually minor. Compared to the case of using logic cells with the same type of transistors (i.e. low-threshold or high-threshold) everywhere, Reference 45 reports an average of only 5 per cent additional leakage savings by using logic cells with the same type of transistors in a transistor stack.

Although using two threshold voltages instead of one significantly decreases the leakage current in a circuit, using more than two threshold voltages marginally improves the result [46]. This is true even when the threshold values are optimised to minimise the leakage for a given circuit. Thus, in many designs, only two threshold voltages are used.

12.6.3   Minimum leakage vector (MLV) method

The leakage current of a logic gate is a strong function of its input values. The reason is that the input values affect the number of OFF transistors in the NMOS and PMOS networks of a logic gate.

Table 12.1 shows the leakage current of a two-input NAND gate built in a 0.18 μm CMOS technology with a 0.2 V threshold voltage and a 1.5 V supply voltage. Input A is the one closer to the output of the gate.

The minimum leakage current of the gate corresponds to the case when both its inputs are zero. In this case, both NMOS transistors in the NMOS network are off, while both PMOS transistors are on. The effective resistance between the supply and the ground is the resistance of two OFF NMOS transistors in series. This is the maximum possible resistance. If one of the inputs is zero and the other is one,

Table 12.1   The leakage values of
a NAND gate

Inputs		Output	Leakage
A	B	O	current (nA)
0	0	1	23.06
0	1	0	51.42
1	0	0	47.15
1	1	0	82.94

the effective resistance will be the same as the resistance of one OFF NMOS transistor. This is clearly smaller than the previous case. If both inputs are one, both NMOS transistors will be on. On the other hand, the PMOS transistors will be off. The effective resistance in this case is the resistance of two OFF PMOS transistors in parallel. Clearly, this resistance is smaller than the other cases.

In the NAND gate of Table 12.1 the maximum leakage is about three times higher than the minimum leakage. Note that there is a small difference between the leakage current of the $A = 0$, $B = 1$ vector and the $A = 1$, $B = 0$ vector due to the body effect. The phenomenon whereby the leakage current through a stack of two or more OFF transistors is significantly smaller than a single device leakage is called the 'stack effect'. Other logic gates exhibit a similar leakage current behaviour with respect to the applied input pattern. As a result, the leakage current of a circuit is a strong function of its input values. It is possible to achieve a moderate reduction in leakage using this technique, but the reduction is not as high as the one achieved by the power gating method. On the other hand, the MLV method does not suffer from many of the shortcomings of the other methods. In particular,

1   No modification in the process technology is required.
2   No change in the internal logic gates of the circuit is necessary.
3   There is no reduction in voltage swing.
4   Technology scaling does not have a negative effect on its effectiveness or its overhead. In fact, the stack effect becomes stronger with technology scaling as DIBL worsens.

The first three facts make it very easy to use this method in existing designs. This technique is also referred to as input vector control (IVC) [47]. The problem of finding MLV for an arbitrary circuit is NP-complete [48] for which a number of heuristics have been proposed including a random simulation based approach presented in Reference 47. Bobba and Hajj [48] used a constraint graph to solve the problem for circuits with only a small number of inputs. An explicit branch and bound enumeration technique is described in Reference 49. For large circuits, bounds on the minimum and maximum leakage values were obtained by using heuristics. Abdollahi *et al.* [50] formulated the problem of determining the MLV using a series of Boolean satisfiability problems and solved accordingly. The authors report between 10 per cent and 55 per cent reduction in the leakage by using the MLV technique. Note that the saving is defined as $(1 - \text{Leakage}_{MLV}/\text{Leakage}_{AVG}) \times 100$, where Leakage_{MLV} is the leakage when the minimum leakage vector drives the circuit whereas Leakage_{AVG} is the expected leakage current under an arbitrary input combination (this is used because the input value prior to entering the sleep mode is unknown).

Lee and Blaauw [51] used the combination of MLV and dual-V_T assignment for leakage power reduction. They observe that within the performance constraints, it is more effective to switch off a high-V_T transistor than a low-V_T one. Naidu *et al.* [52] proposed an integer linear programming (ILP) model for circuits composed of NAND or AOI gates, which obtains the MLV. Gao and Hayes [53] proposed an ILP model for finding MLV, called the virtual gate or VG-ILP model. Virtual gates are cells that are added to the given circuit to facilitate model formulation, but have no impact on the functionality of the original circuit. The leakage current is viewed as a pseudo-Boolean function of the inputs, which is subsequently linearised. The authors resort to ILP to obtain the input MLV using linearised leakage current functions. They also propose a fast, heuristic technique for MLV calculation, which selectively relaxes variables of the ILP model, leading to a mixed-integer linear programming (MLP) model.

12.6.4 *Increasing the transistor channel lengths*

Active leakage of CMOS gates can be reduced by increasing their transistor channel lengths [54]. This is because there is a V_T roll-off due to the short channel effect (SCE). Therefore, different threshold voltages can be achieved by using different channel lengths. The longer transistor lengths used to achieve high-threshold transistors tend to increase the gate capacitance, which has a negative impact on the performance and dynamic power dissipation. Compared with multiple-threshold voltages, long channel insertion has similar or lower process cost, taken as the size increase rather than the mask cost. It results in lower process complexity. In addition, the different channel lengths track each other over process variation. This technique can be applied in a greedy manner to an existing design to limit the leakage currents [55]. A potential penalty is that the dynamic power dissipation of the up-sized gate is increased proportional to the effective channel length increase. In general, circuit power dissipation may not be saved unless the activity factor of the affected gates is low. Therefore, the activity factor must be taken into account when choosing gates whose transistor lengths are to be increased.

12.6.5 *Transistor sizing with simultaneous threshold and supply voltage assignment*

Increasing the threshold voltage of a transistor reduces the leakage current exponentially, but it has a marginal effect on the dynamic power dissipation. On the other hand, reducing the width of a transistor reduces both leakage and dynamic power, but at a linear rate only. Nguyen *et al.* [56] report an average 60 per cent and 75 per cent reduction in the total power dissipation by using sizing alone and sizing combined with V_T assignment, respectively. The combination of the technique with dual Vdd assignment resulted in only a marginal improvement, probably because of the optimisation algorithm used by the authors. Combining the three optimisations is currently an active area of research and will enable synthesising lower power circuits in the near future.

12.7　Conclusions

Several key elements emerge as enablers for an effective low power design methodology. The first is the availability of accurate, comprehensive power models. The second is the existence of fast, easy to use high-level estimation and design exploration tools for analysis and optimisation during the design creation process, while the third is the existence of highly accurate, high-capacity verification tools for tape-out power verification. As befitting a first-order concern, successfully managing the various power-related design issues will require that power be addressed at all phases and in all aspects of design, especially during the earliest design and planning activities. Advanced power tools will play central roles in these efforts.

An RTL design methodology supported by the appropriate design automation tools is one of the most effective methods of designing complex chips for lower power dissipation. Moreover, this methodology drastically reduces the risk of not meeting often harsh power constraints by the early identification of power hogs or hot spots, and enabling the analysis and selection of alternative solutions. Such methodologies have already been adopted by designers of complex chips and constitute the state-of-the-art in designing complex, high-performance, yet low power, designs.

This chapter reviewed a number of RTL techniques for low power design of VLSI circuits targeting both dynamic and leakage components of power dissipation in CMOS VLSI circuits. A more detailed review of techniques for low power design of VLSI circuits and systems can be found in many references, including Reference 1.

References

1　PEDRAM, M., and RABAEY, J. (Eds): 'Power aware design methodologies' (Kluwer Academic Publishers, Boston, MA 2002)
2　MACII, E. (Ed.): 'Ultra low-power electronics and design' (Kluwer Academic Publishers, Boston, MA 2004)
3　PIGUET, C. (Ed.): 'Low power electronics design' (CRC Press, Newman 2004)
4　HAMADA, M., TAKAHASHI, M., ARAKIDA, H. *et al.*: 'A top-down low power design technique using clustered voltage scaling with variable supply-voltage scheme'. Proceedings of the IEEE *Custom integrated circuits* conference (CICC'98), (IEEE press, Piscataway, NJ, USA, May 1998, pp. 495–498
5　RAJE, S., and SARRAFZADEH, M.: 'Variable voltage scheduling'. Proceedings of the international workshop on *Low power design*, August 1995, pp. 9–14

6 USAMI, K., and HOROWITZ, M.: 'Clustered voltage scaling technique for low-power design'. Proceedings of the international workshop on *Low power design*, 1995, pp. 3–8

7 USAMI, K., IGARASHI, M., MINAMI, F. *et al.*: 'Automated low-power technique exploiting multiple supply voltages applied to a media processor', *IEEE Journal of Solid-State Circuits*, 1998, **33**(3), pp. 463–472

8 CHEN, C., SRIVASTAVA, A., and SARRAFZADEH, M.: 'On gate level power optimization using dual supply voltages', *IEEE Transactions on VLSI Systems*, 2001, **9**, pp. 616–629

9 MANZAK, A., and CHAKRABARTI, C.: 'A low power scheduling scheme with resources operating at multiple voltages', *IEEE Transactions on VLSI Systems*, 2002, **10**(1), pp. 6–14

10 CHANG, J.M., and PEDRAM, M.: 'Energy minimization using multiple supply voltages', *IEEE Transactions VLSI Systems*, 1997, **5**(4), pp. 436–443

11 YEH, Y.-J., KUO, S.-Y., and JOU, J.-Y.: 'Converter-free multiple-voltage scaling techniques for low-power CMOS digital design', *IEEE Transactions Computer-Aided Design*, 2001, 20, pp. 172–176.

12 MURUGAVEL, A.K., and RANGANATHAN, N.: 'Game theoretic modeling of voltage and frequency scaling during behavioral synthesis'. Proceedings of *VLSI design*, 2004, pp. 670–673

13 MUDGE, T.: 'Power: A first class design constraint', *Computer*, 2001, **34**(4), pp. 52–57

14 PERING, T., BURD, T., and BRODERSEN, R.: 'The simulation and evaluation of dynamic voltage scaling algorithms'. Proceedings of international symposium on *Low power electronics and design 1998*, June 1998, pp. 76–81

15 GONZALEZ, R., GORDON, B., and HOROWITZ, M.: 'Supply and threshold voltage scaling for low power CMOS', *IEEE Journal of Solid-State Circuits*, 1997, **32**(8)

16 ERNST, D., NAM SUNG KIM, SHIDHARTHA DAS, *et al*: a low-power pipeline based on circuit-level timing speculation'. Proceedings of the 36th Annual international symposium *Microarchitecture (MICRO-36)*, IEEE Computer Society Press, 2003, pp. 7–18

17 ALIDINA, M., MONTEIRO, J., DEVADAS, S., GHOSH, A., and PAPAEFTHYMIOU, M.: 'Precomputation-based sequential logic optimization for low power', *IEEE Transactions on VLSI Systems*, 1994, **2**(4), pp. 426–436

18 MONTEIRO, J., DEVADAS, S., and GHOSH, A.: 'Sequential logic optimization For low power using input-disabling', *IEEE Transactions on Computer-Aided Design*, 1998, **17**(3), pp. 279–284

19 BENINI, L., SIEGEL, P., and DE MICHELI, G.:, 'Automatic synthesis of gated clocks for power reduction in sequential circuits', *IEEE Design Test Computer Magazine*, 1994, **11**(4), pp. 32–40

20 BENINI, L., and DE MICHELI, G.: 'Transformation and synthesis of FSM's for low power gated clock implementation', *IEEE Transactions on Computer-Aided Design*, 1996, **15**(6), pp. 630–643

21   BENINI, L., DE MICHELI, G., MACII, E., PONCINO, M., and SCARSI, R.: 'Symbolic synthesis of clock-gating logic for power optimization of control-oriented synchronous networks'. Proceedings of the *European design and test* conference, Paris, France, March 1997, pp. 514–520

22   BENINI, L., FAVALLI, M., and DE MICHELI, G.: 'Design for testability of gated-clock FSM's'. Proceedings of the *European design and test* conference, Paris, France, March 1996, pp. 589–596

23   BENINI, L., DE MICHELI, G., LIOY, A., MACII, E., ODASSO, G., and PONCINO, M.: 'Synthesis of power-managed sequential components based on computational kernel extraction', *IEEE Transactions on Computer-Aided Design,* 2001, **20**(9), pp. 1118–1131

24   BENINI, L., DE MICHELI, G., MACII, E., ODASSO, G., and PONCINO, M.: 'Kernel-based power optimization of RTL components: exact and approximate extraction algorithms'. Proceedings of *Design automation* conference, 1999, pp. 247–252

25   MONTEIRO, J., and OLIVEIRA, A.: 'Finite state machine decomposition for low power'. Proceedings of *Design automation* conference, June 1998, pp. 758–763

26   CHOW, S-H., HO, Y-C., and HWANG, T.: 'Low power realization of finite state machines a decomposition approach', *ACM Transactions on Design Automation of Electronic Systems,* 1996, **1**(3), pp. 315–340

27   BENINI, L., SIEGEL, P., and DE MICHELI, G.: 'Automatic synthesis of low-power gated-clock finite-state machines', *IEEE Transactions on Computer-Aided Design,* 1996, **15**(6), pp. 630–643

28   TIWARI, V., MALIK, S., and ASHAR, P.: 'Guarded evaluation: pushing power management to logic synthesis/design'. Proceedings of the ACM/IEEE international symposium on *Low power design,* Dana Point, CA, ACM press, New York, NY, USA, April 1995, pp. 221–226.

29   ROY, K., and PRASAD, S.: 'Syclop: synthesis of CMOS logic for low-power application'. Proceedings of international conference on *Computer design,* IEEE press, Piscataway, NJ, USA, October 1992, pp. 464–467

30   OLSON, E., and KANG, S.M.: 'Low-power state assignment for finite state machines'. Proceedings of international workshop on *Low power design,* ACM press, New York, NY, USA, April 1994, pp. 63–68

31   TSUI, C.Y., PEDRAM, M., and DESPAIN, A.M.: 'Low-power state assignment targeting two- and multilevel logic implementation', *IEEE Transactions on Computer-Aided Design,* 1998, **17**(12), pp. 1281–1291

32   WU, X., WEI, J., WU, Q., and PEDRAM, M.: 'Low-power design of sequential circuits using a quasi-synchronous derived clock', *International Journal of Electronics,* Taylor & Francis Publishing Group, 2001, **88**(6), pp. 635–643

33   IRANLI, A., REZVANI, P., and PEDRAM, M.: 'Low power synthesis of finite state machines with mixed D and T flip-flops'. Proceedings of Asia and South Pacific *Design automation* conference, IEEE press, Piscataway, NJ, USA, January 2003, pp. 803–808

34 LEISERSON, C.E., and SAXE, J.B.: 'Optimizing synchronous systems', *Journal of VLSI Computer Systems,* 1983, **1**(1), pp. 41–67

35 MONTEIRO, J., DEVADAS, S., and GHOSH, A.: 'Retiming sequential circuits for low power'. Proceedings of the international conference on *Computer-aided design*, Santa Clara, CA, November 1993, pp. 398–402

36 LALGUDI, K.N., and PAPAEFTHYMIOU, M.: 'Fixed-phase retiming for low power'. Proceedings of the international symposium on *Low-power electronics and design*, IEEE press, Piscataway, NJ, USA, August 1996, pp. 259–264

37 CHABINI, N., and WOLF, W.: 'Reducing dynamic power consumption in synchronous sequential digital designs using retiming and supply voltage scaling', *IEEE Transactions on VLSI Systems*, 2004, **12**(6), pp. 573–589

38 BORKAR, S.: 'Design challenges of technology scaling', *IEEE Micro*, 1999, **19**(4), pp. 23–29

39 Semiconductor Industry Association: International Technology Roadmap for semiconductors, 2003 edition, http://public.itrs.net/

40 MUTOH, S., DOUSEKI, T., MATSUYA, Y., AOKI, T., SHIGEMATSU, S., and YAMADA, J.: '1-V power supply high-speed digital circuit technology with multithreshold CMOS', *IEEE Journal of Solid-State Circuits*, 1995, **30**(8), pp. 847–854

41 KAO, J.T., and CHANDRAKASAN, A.P.: 'Dual-threshold voltage techniques for low-power digital circuits', *IEEE Journal of Solid-State Circuits*, 2000, **35**, pp. 1009–1018

42 HYO-SIG WON, KYO-SUU KIM, KWAU-OK JEONG, KI-TAE PARK, KYU-MYUNG CHOI, and JEONG-TAEK KONG: 'An MTCMOS design methodology and its application to mobile computing'. Proceedings of the international symposium on *Low power electronics and design*, ACM press, New York, NY, USA, August 2003

43 KAO, J.T., and CHANDRAKASAN, A.: 'MTCMOS sequential circuits'. Proceedings of the ESSCIRC, IEEE press, Piscataway, NJ, USA, 2001, pp. 332–339

44 SHIGEMATSU, S., MUTOH, S., MATSUYA, Y., TANABE, Y., and YAMADA, J.: 'A 1-V high-speed MTCMOS circuit scheme for power-down application circuits', *IEEE Journal of Solid State Circuits*, 1997, **32**, pp. 861–870

45 WEI, L., ROY, K., YE, Y., and DE, V.: 'Mixed-Vth (MVT) CMOS circuit design methodology for low power applications'. Proceedings of the *Design automation conference*, IEEE press, Piscataway, NJ, USA, June 1999, pp. 430–435

46 SRIVASTAVA, A.: 'Simultaneous Vt selection and assignment for leakage optimization'. Proceedings of the international symposium on *Low power electronics and design*, ACM press, New York, NY, USA, August 2003, pp. 146–151

47 HALTER, J., and NAJM, F.: 'A gate-level leakage power reduction method for ultra low power CMOS circuits'. IEEE *Custom integrated circuits* conference, IEEE press, Piscataway, NJ, USA, 1997, pp. 475–478

48 BOBBA, S., and HAJJ, I.N.: 'Maximum leakage power estimation for CMOS circuits'. Proceedings of IEEE Alessandro Volta memorial international workshop on *Low power design*, Como, Italy, March 4–5, 1999, pp. 116–124

49   JOHNSON, M., SOMASEKHAR, D., and ROY, K.: 'Models and algorithms for bounds in CMOS circuits'. *IEEE Transactions on Computer Aided Design of Integrated Circuits and Systems*, 1999, **18**(6), pp. 714–725

50   ABDOLLAHI, A., FALLAH F., and PEDRAM, M.:'Leakage current reduction in CMOS VLSI circuits by input vector control', *IEEE Transactions on Very Large Scale Integration (VLSI) Systems*, 2004, **12**(2), pp. 140–154

51   LEE, D., and BLAAUW, D.: 'Static leakage reduction through simultaneous threshold voltage and state assignment'. Proceedings of the *Design automation conference*, IEEE press, Piscataway, NJ, USA, June 2003, pp. 191–194

52   NAIDU, S., and JACOBS, E.: 'Minimizing standby leakage power in static CMOS circuits'. Proceedings of the *Design automation and test in Europe*, 2001, pp. 370–376

53   GAO, F., and HAYES, J.P.: 'Exact and heuristic approaches to input vector control for leakage reduction'. Proceedings of the international conference on *Computer-aided design*, November 2004, pp. 527–532

54   WEI, L., ROY, K., and DE, V.: 'Low voltage low power CMOS design techniques for deep submicron ICs'. Proceedings of the thirteenth international conference on *VLSI design*, IEEE press, Piscataway, NJ, USA, 2000, pp. 24–29

55   CLARK, L.T., PATEL, R., and BEATY, T.S.: 'Managing standby and active mode leakage power in deep sub-micron design'. Proceedings of the international symposium on *Low power electronics and design*, ACM press, New York, NY, USA, August 2004, pp. 274–279

56   NGUYEN, D., DAVARE, A., ORSHANSKY, M., CHINNERY, D., THOMPSON, B., and KEUTZER, K.: 'Minimization of dynamic and static power through joint assignment of threshold voltages and sizing optimization'. Proceedings of the international symposium on *Low power electronics and design*, ACM press, New York, NY, USA, August 2003, pp. 158–163.

Chapter 13

Leakage power analysis and reduction for nano-scale circuits

Amit Agarwal, Saibal Mukhopadhyay, Chris H. Kim,
Arijit Raychowdhury and Kaushik Roy

13.1 Introduction

Complementary metal oxide semiconductor (CMOS) devices have been scaled down aggressively in each technology generation to achieve higher integration density and performance [1–3]. With technology scaling, the supply voltage needs to be scaled down to reduce the dynamic power and maintain reliability [3]. However, this requires the scaling of the device threshold voltage (V_{th}) to maintain a reasonable gate over drive [4]. The threshold voltage (V_{th}) scaling and the V_{th} reduction due to short-channel effects (SCEs) (like drain-induced-barrier-lowering (DIBL), V_{th}-roll off) [4–6], result in an exponential increase in the subthreshold current (Figure 13.1). To control the short-channel effect and to increase the transistor drive strength, oxide thickness needs to be scaled down in each technology generation. The aggressive scaling of oxide thickness results in a high direct tunnelling current through the gate insulator of the transistor [4,5] (Figure 13.1). On the other hand, scaled devices require the use of the higher substrate doping density and the application of the 'halo' profiles (implant of high doping region near the source and drain junctions of the channel) to reduce the depletion region width of the source–substrate and drain–substrate junctions [4,5]. A lower depletion region width helps to control the short-channel effect. The high doping density near the source–substrate and drain–substrate junctions cause significantly large BTBT current through these junctions under high reversed bias [4,5] (Figure 13.1). Hence, with technology scaling, each

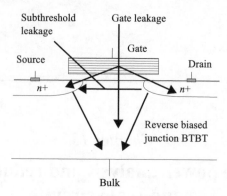

Figure 13.1   Major leakage components in a transistor

of these leakage components increases drastically, resulting in increase in the total leakage current [3].

In the nanometre regime, a significant portion of the total power consumption in high-performance digital circuits is due to leakage currents. Because high-performance systems are constrained to a predefined power budget, the leakage power reduces the available power, impacting performance. It also contributes to the power consumption during standby operation, reducing battery life [7]. Hence, techniques are necessary to reduce leakage power while maintaining the high performance. Moreover, as described above different components of leakage are becoming important with technology scaling. This increase in different leakage components has two major implications in leakage estimation and low power logic design. First, this results in a dramatic increase of the total leakage [7]. But, more importantly, each of the leakage components becomes equally important in nano-scaled devices [5,8]. Hence, the relative magnitudes of the leakage components play a major role in low-leakage logic design [9]. Each leakage reduction technique needs re-evaluation in scaled technologies where subthreshold conduction is not the only leakage mechanism. The new low power circuit techniques are required to reduce total leakage in high-performance nano-scale circuits. Furthermore, the magnitudes of each of these components depend strongly on the device geometry (namely, channel length, oxide thickness and transistor width), the doping profiles and temperature [4,5]. Hence, accurate estimation of the total leakage current is extremely important for designing low power CMOS circuits in the nanometre regime.

In this chapter, different leakage components and their impact on logic circuits is described. The basic physical mechanisms of the major leakage components, namely, the junction BTBT, the subthreshold and the gate leakage, are discussed. The leakage models based on (1) the device geometry, (2) the two-dimensional (2D) doping profile of the device and (3) the operating temperature are described. Different integrated circuit techniques to reduce overall leakage in both logic and cache memories are shown. A spectrum of circuit techniques including dual V_{th}, forward/reverse bias, dynamically varying the V_{th} during run time, sleep transistor, natural stacking, are reviewed. Based on these techniques, different leakage-tolerant schemes for logic and memories are summarised.

13.2 Different components of leakage current

In this section, the basic physical mechanisms of the major leakage components, namely, the junction BTBT, the subthreshold and the gate leakage, are discussed. The discussion is principally based on the NMOS transistors. However, it is equally applicable to PMOS transistors also. The discussion is based on the nano-scale devices with 2D non-uniform channel ('super halo' channel doping) and source/drain (S/D) doping profiles. A schematic of the device structure (symmetric about the middle of the channel) is shown in Figure 13.2 [10].

In nano-scale devices to control the threshold voltage and the short-channel effect, the 2D non-uniform doping profile is used. In the vertical direction (direction perpendicular to channel, direction Y in Figure 13.2), the doping is lower at the surface (to reduce threshold voltage) but increases at a depth below the surface (to control short-channel effect). This is known as 'retrograde' profile. In the lateral direction, the doping is designed to be higher near the source–substrate and drain–substrate junctions. A higher doping prevents the penetration of the source and drain depletion region within the channel, thereby reducing the short-channel effects. The patches of the higher doping regions near the source and drain junctions, are known as 'halo' regions and this type of laterally non-uniform profile is called the 'halo' profile. In nano-scale transistors both the 'retrograde' and 'halo' types of profiles are used together, resulting in a 2D non-uniform doping profile ('super halo'). The described non-uniform doping has a strong impact on the leakage components in nano-scale MOSFETS. The next three sections describe the basic physical mechanisms that govern different leakage components.

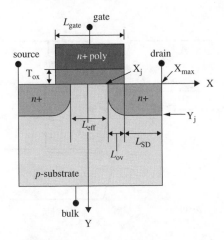

Figure 13.2   Device structure

13.2.1   *Junction band-to-band tunnelling current (I_{BTBT})*

A high electric field across a reverse biased p–n junction causes significant current to
flow through the junction due to tunnelling of electrons from the valence band of the
p-region to the conduction band of the n-region (Figure 13.3) [5]. Tunnelling occurs
when the total voltage drop across the junction (applied reverse bias (V_{app}) + built-in
voltage (ψ_{bi})) is more than the band-gap (Σ_g). The tunnelling current density through
a silicon p–n junction is given by [5]:

$$J_{b-b} = A \frac{E V_{app}}{\Sigma_g^{1/2}} \exp\left(-B \frac{\Sigma_g^{3/2}}{E}\right)$$

$$A = \frac{\sqrt{2m^*}q^3}{4\pi^3\hbar^2} \quad \text{and} \quad B = \frac{4\sqrt{2m^*}}{3q\hbar}$$

$$(13.1)$$

where m^* is the effective mass of electron, E is the electric field at the junction,
q is the electronic charge and $\hbar$ is the reduced Plank's constant. In an NMOSFET
when the drain and/or the source is biased at a potential higher than that of the
substrate, a significant BTBT current flows through the drain–substrate and/or the
source–substrate junctions. The total BTBT current in the MOSFET is the sum of the
currents flowing through two junctions and is given by:

$$I_{BTBT} = w_{eff} \int_l J_{b-b}(x, y)dl \Big|_{drain}$$
$$+ w_{eff} \int_l J_{b-b}(x, y)dl \Big|_{source}$$

$$(13.2)$$

$l :=$ Junction line $\equiv$

solution of the equation $Na(x, y) = Nd(x, y)$

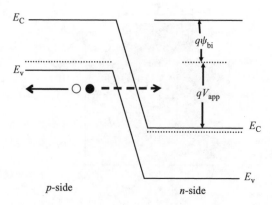

*Figure 13.3   Physical picture of valence band electron tunnelling in a reversed bias
p–n junction*

where $J_{b-b}(x, y)$ is the current density at a point (x, y) at the junction. Hence, the total junction BTBT current in a MOSFET can be obtained by considering the current from both junctions. It can be observed from Equation (13.1) that, the junction BTBT current is a strong function of the electric field across the drain–substrate (not source–substrate) junction. The electric field across a reverse-bias $p–n$ junction can be given by (assuming abrupt and step junctions):

$$\xi_{jn} = \sqrt{\frac{2q N_a N_d (V_{app} + V_{bi})}{\varepsilon_{si}(N_a + N_d)}} \tag{13.3}$$

where N_a and N_d are the doping in the p and n side, respectively, V_{bi} is built-in potential across the junction and V_{app} is the applied reverse bias. Although not explicitly shown in Equation (13.3), the junction electric field is also a strong function of the abruptness (how fast doping changes from p-type to n-type across the junction) of the $p–n$ junction. For same doping density and applied reverse bias, increasing the abruptness increases the junction electric field. The built-in potential also depends (weakly) on the doping at the two junctions and is given by:

$$V_{biside} = \frac{kT}{q} \ln\left(\frac{N_{aside} N_{dside}}{n_i^2}\right); \text{ assuming non-degenerate doping} \tag{13.4}$$

Using Equations (13.1)–(13.4) and considering the non-uniformity of the channel and the source–drain doping profile analytical expressions for the junction tunnelling current can be obtained as shown in Reference 11. From Equations (13.1) and (13.3) it can be observed that, application of a higher reverse bias across the junction results in an exponential increase in the junction tunnelling current. Hence, application of a reverse substrate bias ($V_{bs} < 0$) in the 'off-state' of an NMOS ($V_{gs} = 0$, $V_{ds} = V_{DD}$) exponentially increases the junction tunnelling current. Moreover, Equation (13.3) also suggests that a higher junction doping also increases the junction electric field. Hence, use of the highly doped 'halo' regions near the drain–substrate (or source–substrate) junction increases the junction tunnelling current. Hence, increasing the strength of the 'halo' doping has a negative impact on the junction tunnelling leakage. Figure 13.4 shows the variation of the junction tunnelling current with reverse substrate bias in predictive NMOS devices with $L_{eff} = 25$ nm ($V_{DD} = 0.7$ V) and 50 nm ($V_{DD} = 0.9$ V) and different doping profiles. It can be observed that, the tunnelling current increases with the application of reverse bias. Moreover, due to higher 'halo' doping and more abrupt junction, tunnelling current in the 25 nm device is significantly higher than that in the 50 nm device. Increasing the strength of the 'halo' doping also results in a higher junction tunnelling current as show in Figure 13.5 (for a predictive 25 nm device).

13.2.2 Subthreshold current (I_{ds})

In the 'off' state of a device ($V_{gs} < V_{th}$) the current flowing from the drain to the source of a transistor is known as the subthreshold current. The subthreshold current

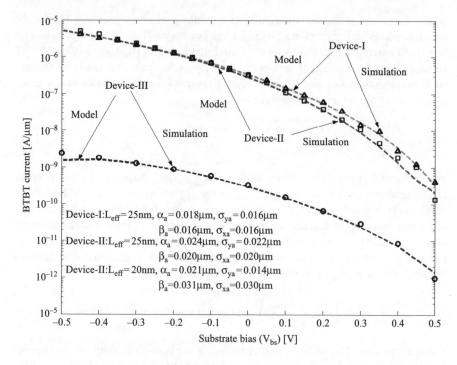

Figure 13.4   *Variation of BTBT current with substrate bias for NMOS transistor with $L_{eff} = 25\ nm$ ($V_{DD} = 0.7\ V$) and $Leff = 50\ nm$ ($V_{DD} = 0.9\ V$). Analytical results obtained with the model from [14] and simulation results are obtained using MEDICI device simulator*

flowing through a transistor is given by [5],

$$I_{sub} = \frac{w_{eff}}{L_{eff}} \mu \sqrt{\frac{q\varepsilon_{si}N_{cheff}}{2\Phi_s}} v_T^2 \exp\left(\frac{V_{gs} - V_{th}}{nv_T}\right)\left(1 - \exp\left(\frac{-V_{ds}}{v_T}\right)\right) \quad (13.5)$$

where N_{cheff} is the effective channel doping, Φ_s is the surface potential, n is the subthreshold swing and v_T is the thermal voltage given by kT/q. From Equation (13.5) it can be observed that the subthreshold leakage depends exponentially on the threshold voltage (V_{th}) of the device. Hence, the factors that modify the transistor threshold voltage have a strong impact on the subthreshold current. In the subthreshold region the inversion charge is negligible in the channel and hence the total charge in the channel principally consists of the depletion charge [5]. The depletion charge depends on the effective doping density (N_{cheff}) in the channel. In long-channel transistor theory, the threshold voltage is defined as the gate voltage at which the surface potential (i.e. the amount of bending of the conduction band in the silicon substrate) reaches

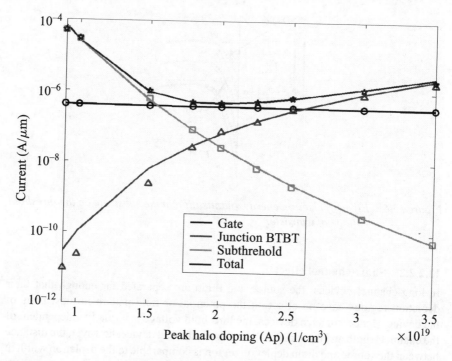

Figure 13.5   Variation of different leakage components with 'halo' doping strength for NMOS transistor with $L_{eff} = 25\,nm$ ($V_{DD} = 0.7\,V$)

$2\Phi_B = 2(kT/q)\ln(N_{cheff}/n_i)$ (Figure 13.6) [12]. Hence, the threshold voltage in a long-channel transistor is given by [4,5]:

$$V_{th} = V_{FB} + 2\Phi_B + \frac{\sqrt{2\varepsilon_{si}q\,N_{cheff}(2\Phi_B)}}{C_{ox}} \tag{13.6}$$

where V_{FB} is the flat-band potential. However, in a short-channel transistor the threshold voltage is affected by the following factors.

13.2.2.1 Body effect

Reverse biasing well-to-source junction of a MOSFET transistor widens the bulk depletion region and increases the threshold voltage [4,5]. The effect of body bias can be considered in the threshold voltage equation as [4,5]:

$$V_{th} = V_{FB} + 2\phi_B + \frac{\sqrt{2\varepsilon_{si}q\,N_{cheff}(2\phi_B - V_{BS})}}{C_{ox}} \tag{13.7}$$

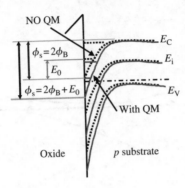

Figure 13.6   Effect of electron energy quantisation in the substrate on the threshold voltage of transistor

13.2.2.2   Short-channel effect

In long-channel devices, the source and drain are separated far enough that their depletion regions have no effect on the potential or field pattern in most parts of the device. Hence, for such devices, the threshold voltage is virtually independent of the channel length and drain bias [4]. In a short-channel device, however, the distance between the source and drain depletion region is comparable to the depletion width in the vertical direction. As a result, source–drain depletion width has a more pronounced effect on potential profiles and field patterns [4,5]. The source and drain depletion regions now penetrate more into the channel length, resulting in part of the channel being already depleted. Thus, gate voltage has to invert less bulk charge to turn a transistor on. This is known as short-channel effect, which reduces the threshold voltage of a transistor [5]. The fraction of the depletion charge supplied by the source–drain depletion region (ΔQ_B) to the total depletion charge (Q_B) required at the threshold condition is a measure of the short-channel effect. As the channel length is reduced or the drain bias is increased (which increases the drain depletion width) ΔQ_B increases, thereby reducing the threshold voltage [5]. The reduction of V_{th} with the reduction of the channel length is known as V_{th}-roll off. The reduction of V_{th} with the increase in the drain bias is known as drain-induced barrier lowering (DIBL) [4,5]. The short-channel effect can also be described as the fact that, for same gate voltages the amount of band bending in the Si–SiO$_2$ interface is more in a short-channel device compared to a long-channel device. Using this explanation the short-channel effect in V_{th} can be modelled as a lowering in the surface potential (Φ_S) required to obtain the threshold condition (say $\Delta\Phi_S$) from its long-channel zero bias value ($\Delta\Phi_{S0} = 2v_T \ln(N_{cheff}/n_i)$). Using the charge sharing model (Figure 13.7) and following the procedure given in References 13–15, the V_{th} of a short-channel transistor can be expressed as:

$$V_{th} = V_{FB} + (\Phi_{s0} - \Delta\Phi_s) + \gamma\sqrt{\Phi_{s0} - V_{bs}}\left(1 - \lambda\frac{X_d}{L_{eff}}\right) \tag{13.8}$$

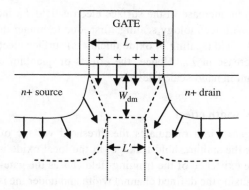

Figure 13.7   Diagram for charge sharing model explaining the reduction of V_{th} due to the source–drain depletion regions [5]

where $\gamma = \sqrt{2q\varepsilon_{si}N_{cheff}}/C_{ox}$ is the body factor, $X_d = \sqrt{2\varepsilon_{si}/qN_{cheff}}\sqrt{\Phi_{s0} - V_{bs}}$ is the depletion layer thickness and λ is a fitting parameter (≈ 1). The reduction in the surface potential due to short-channel effect ($\Delta\Phi_s$) can be modelled as:

$$\Delta\Phi_s = \frac{\upsilon_T \ln(N_{sdeff}N_{cheff}/n_i^2) - \Phi_{s0} + 0.5V_{ds}}{[\cosh(L_{eff}/\sqrt{(\varepsilon_{si}t_{ox}X_d)/(\eta\varepsilon_{sio2})})]} \tag{13.9}$$

where N_{sdeff} is the effective source/drain doping and η is another fitting parameter which is usually close to one [13]. It can be observed that increasing the drain bias (V_{DS}) or reducing the channel length increases $\Delta\Phi_S$ thereby reducing the threshold voltage. From the above discussion it can also be observed that, the short-channel effect can be reduced by reducing the drain and source depletion width. In a nano-scaled short-channel transistor this is achieved by adding the highly doped 'halo' implants near the source–substrate and the drain–substrate junction.

13.2.2.3   Quantum-confinement effect

In scaled devices, due to high electric field at the surface (ξ_s) and high substrate doping, the quantisation of inversion-layer electron energy modulates V_{th}. Quantum-mechanical behaviour of the electrons increases V_{th}, thereby reducing the subthreshold current, since more band bending is required to populate the lowest sub-band, which is at an energy higher than the bottom of the conduction band (Figure 13.6). When ξ_s is higher than 10^6 V/cm, electrons occupy only the lowest sub-band. In that case, the quantisation effect can be modelled as an increase in threshold voltage by an amount ΔV_{QM}, given by [5]:

$$\Delta V_{QM} = \left(1 + \frac{3t_{ox}}{X_d}\right)\left(\frac{E_0}{q} - \frac{kT}{q}\ln\left(\frac{8\pi q m_d\xi_s}{h^2 N_C}\right)\right) \tag{13.10}$$

where E_0 is the lowest sub-band energy given by [5] $E_o = [(3hq_s\xi_s/4\sqrt{2m_x})3/4]^{2/3}$, N_C is the effective conduction band density of states, m_x is the quantisation effective mass of electron and m_d is the density of states effective mass of electron. It can be observed that with an increase in the surface electric field E_0 increases and hence ΔV_{QM} increases. With technology scaling, first, due to higher doping and second, due to higher oxide field (scaling of oxide thickness) surface electric field increases resulting in an increase in E_0. Hence, the effects of quantum correction become extremely important in nano-scaled devices.

13.2.2.4 Narrow-width effect

The decrease in gate width modulates the threshold voltage of a transistor, and thereby modulates the subthreshold leakage. In the local oxide isolation (LOCOS) gate MOSFET, the existence of the fringing field causes the gate-induced depletion region to spread outside the defined channel width and under the isolations as shown in Figure 13.8(a). This results in an increase of the total depletion charge in the bulk region above its expected value. This effect becomes more substantial as the channel width decreases, and the depletion region underneath the fringing field is comparable to the classical depletion formed by the vertical field. This results in increase of threshold voltage due to the narrow-width effect [15,16]. This narrow-width effect can be modelled as an increase in the V_{th} by the amount given by [15]:

$$\Delta V_{NWE} = \frac{\pi q N_{cheff} X_d^2}{2 C_{ox} w_{eff}} = 3\pi \frac{t_{ox}}{w_{eff}} \phi_s \qquad (13.11)$$

A more accurate model of this effect can be found in Reference 16. A more complex effect is observed in trench isolation devices, known as the inverse-narrow-width effect. In the case of trench isolation devices, the depletion layer cannot spread under the oxide isolation [see Figure 13.8(b)]. Hence, the total depletion charge in the bulk does not increase, thereby eliminating the increase in the threshold voltage. On the other hand, due to the 2D field-induced edge fringing effect at the gate edge, formation of an inversion layer at the edges occurs at a lower voltage than the voltage required at the centre. Moreover, the overall gate capacitance now includes the sidewall capacitance (C_F) due to overlap of the gate with the isolation oxide. This increases the overall gate capacitance from C_{ox} to $(C_{ox} + C_F)$ in Equation (13.6) [17]. Hence, the threshold voltage reduces as shown in Figure 13.9 [17].

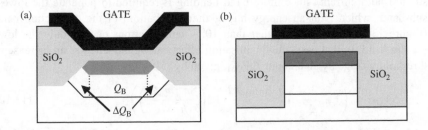

Figure 13.8   Two types of device structures and associated inversion – depletion layer (a) LOCOS gate MOSFET, (b) Trench isolated MOSFET

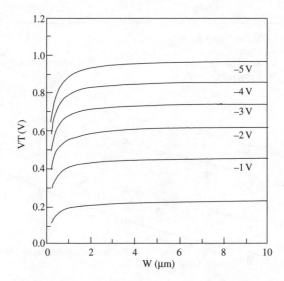

Figure 13.9   Variation of threshold voltage with gate width for uniform doping [17]

Including all of the above-mentioned effects the threshold voltage of a transistor can be modelled as:

$$V_{th} = V_{FB} + (\Phi_{s0} - \Delta\Phi_s) + \gamma\sqrt{\Phi_{s0} - V_{bs}}\left(1 - \lambda\frac{X_d}{L_{eff}}\right) + \Delta V_{NWE} + \Delta V_{QM}$$

(13.12)

The threshold voltage and the subthreshold current depend on the effective channel and source/drain doping [5,13–15]. In nano-scale MOSFET the effective channel doping strongly depends on the 'halo' doping concentration. Increasing the 'halo' doping concentration increases the effective channel doping. Figure 13.10 shows the variation of the subthreshold current with drain and substrate bias in a predictive 25 nm device. It can be observed that, the current increases with an increase in the drain bias (due to DIBL effect). Application of a reverse substrate bias reduces the current (due to 'body effect'). Moreover, it can be observed that, if the quantum effect is considered there is a significant reduction in the current. Figure 13.5 shows that an increase in the 'halo' strength results in a considerable reduction in the subthreshold leakage. This is due to the fact that, increasing 'halo' strength reduces the short-channel effect.

13.2.3   Modelling gate tunnelling current (I_{gate})

Reduction of gate oxide thickness results in an increase in the field across the oxide. The high electric field coupled with low oxide thickness results in tunnelling of electrons from inverted channel to gate (or vice versa) or from gate to source/drain overlap region (or vice versa) resulting in the gate oxide tunnelling current. In scaled devices (with oxide thickness <3 nm) this tunnelling occurs through the trapezoidal energy barrier (Figure 13.11) and is known as direct tunnelling. Direct tunnelling occurs when

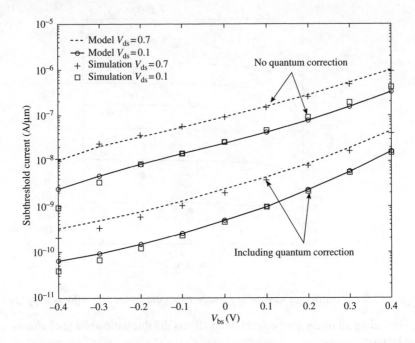

Figure 13.10   Variation of subthreshold leakage with substrate (V_{bs}) and drain bias (V_{ds}) for NMOS transistor N_{ref}. Analytical results obtained with the model from Reference 11 and simulation results are obtained using MEDICI device simulator

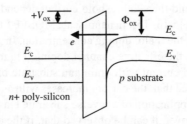

Figure 13.11   Direct tunnelling of electrons

the potential drop across the oxide (V_{ox}) is less than the SiO$_2$–Si conduction band energy difference (Φ_{ox}) (i.e. barrier height in Figure 13.11). The direct tunnelling current density is expressed as [5]:

$$J_{DT} = A_g (V_{ox}/T_{ox})^2 \exp\left(\frac{-B_g(1 - (1 - V_{ox}/\phi_{ox})^{3/2})}{V_{ox}/T_{ox}}\right)$$

$$\text{where,} \quad A_g = \frac{q^3}{16\pi^2\hbar\phi_{ox}} \quad \text{and} \quad B_g = \frac{4\sqrt{2m^*}\phi_{ox}^{3/2}}{3\hbar q} \tag{13.13}$$

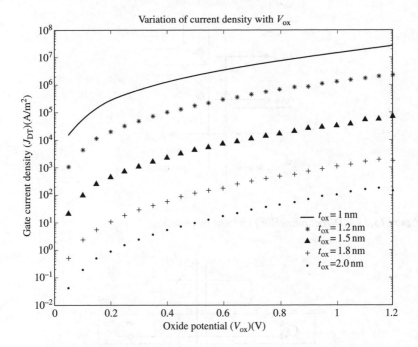

Figure 13.12 *Variation of tunnelling current density with potential drop across oxide*

where J_{DT} is the direct tunnelling current density, V_{ox} is the potential drop across the gate oxide, ϕ_{ox} is the barrier height of the tunnelling electron, m^* is the effective mass of an electron in the conduction band of silicon and T_{ox} is the oxide thickness. Figure 13.12 shows the variation of the direct tunnelling current density based on Equation (13.13). It can be observed that tunnelling current increases exponentially with oxide potential and oxide thickness. The potential drop across the oxide can be obtained from

$$V_{GS} = V_{FB} + V_{poly} + \Phi_S + V_{ox} \quad \text{where } V_{poly} = \frac{\varepsilon_{ox}^2 E_{ox}^2}{2q \varepsilon_{si} N_{poly}} \tag{13.14}$$

where N_{poly} is the doping density in the poly-silicon, ε_{si} is the permittivity of silicon and ε_{ox} is the permittivity of SiO_2.

There are three major mechanisms for direct tunnelling in MOS devices, namely, electron tunnelling from conduction band (ECB), electron tunnelling from valence band (EVB) and hole tunnelling from valance band (HVB) [18,19] (see Figure 13.13). In NMOS, ECB controls the gate-to-channel tunnelling current in inversion, whereas gate-to-body tunnelling is controlled by EVB in depletion-inversion and ECB in accumulation. In p-channel MOSs (PMOSs), HVB controls the gate-to-channel leakage in inversion, whereas gate-to-body leakage is controlled by EVB in depletion-inversion

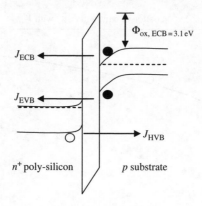

Figure 13.13   Direct tunnelling current mechanisms

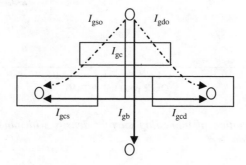

Figure 13.14   Components of tunnelling current [18,19]

and ECB in accumulation [18,19]. Since the barrier height for HVB (4.5 eV) is considerably higher than barrier height for ECB (3.1 eV), the tunnelling current associated with HVB is much less than the current associated with ECB. This results in lower gate leakage current in PMOS than in NMOS [20].

Major components of gate tunnelling in a scaled MOSFET device are [10,13]: (1) Gate to S/D overlap region current (edge direct tunnelling (EDT)) components (I_{gso} and I_{gdo}), (2) gate-to-channel current (I_{gc}), part of which goes to source (I_{gcs}) and rest goes to drain (I_{gcd}), (3) gate-to-substrate leakage current (I_{gb}) (Figure 13.14). Accurate modelling of each of the components is based on the following equation [18]:

$$J_{DT} = A_g \left(\frac{T_{oxref}}{t_{ox}} \right)^{ntox} \left(\frac{V_g V_{aux}}{t_{ox}^2} \right) \times \exp(-Bt_{ox}(\alpha - \beta|V_{ox}|)(1 + \gamma|V_{ox}|))$$

$$(13.15)$$

where T_{oxref} is the reference oxide thickness at which all parameters are extracted, ntox is a fitting parameter (default 1) and V_{aux} is an auxiliary function that approximates the density of tunnelling carriers and available states. The functional form of

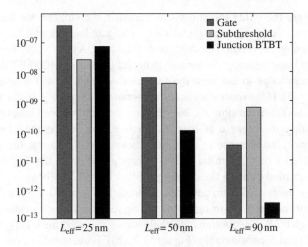

Figure 13.15   *Contribution of different leakage components in NMOS devices [21]*
at different technology generation (device simulation using MEDICI)

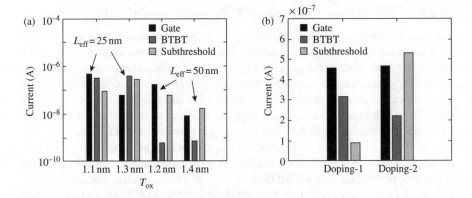

Figure 13.16   *Variation of different leakage components with (a) technology gener-*
ation and oxide thickness, and (b) doping profile. 'Doping-1' has a
stronger halo profile than 'Doping-2'

V_{aux} depends on the tunnelling mechanism and hence takes different values for the
different tunnelling components [18].

13.2.4   *Effect of technology scaling on different components of leakage*
currents

Figure 13.15 shows that with technology scaling each of the leakage components
increases drastically, resulting in increase in the total leakage current [3]. Figure 13.16
shows the different leakage component of NMOS devices of 25 nm and 50 nm physical
gate length [21] at different oxide thickness based on the results of device simulation

[12]. Also, only the oxide thickness was varied in the simulations for a particular technology node (keeping doping constant). The gate leakage and the subthreshold leakage are strongly correlated through oxide thickness. A high oxide thickness results in low gate leakage. Although according to the long-channel MOSFET theory, higher oxide thickness helps to increase the threshold voltage, it will worsen the short-channel effect [5]. If the short-channel effect is not very high (e.g. in the 50 nm device in Figure 13.16(a)) increasing T_{ox} may reduce the subthreshold leakage. However, in a nano-scale device where SCE is extremely severe (e.g. in the 25 nm device in the present case), an increase in the oxide thickness will increase the subthreshold leakage (Figure 13.16(a)). Similarly, the subthreshold leakage and the junction BTBT are strongly coupled through the doping profile. Figure 13.16(b) shows the different leakage components of a 25 nm device at different doping profile (oxide thickness and V_{DD} were kept constant). A strong 'halo' doping reduces the subthreshold current but results in a high BTBT. Reduction of the halo-strength lowers the BTBT, but increases subthreshold current considerably (Figure 13.16(b)). From the above discussion it can be concluded that magnitude of the leakage components and their relative dominance on each other depends strongly on device geometry and doping profile.

13.2.5   Effect of temperature on different components of leakage current

The basic physical mechanisms governing the different leakage current components have different temperature dependence. Subthreshold current is governed by the carrier diffusion that increases with an increase of temperature. Since tunnelling probability of an electron through a potential barrier does not depend directly on temperature, the gate and the junction band-to-band tunnelling is expected to be less sensitive to temperature variations. However, increase of temperature reduces the band-gap of silicon [22], which is the barrier height for tunnelling in BTBT. Hence, the junction BTBT is expected to increase with an increase in temperature. Figure 13.17 shows the effect of temperature variation on individual leakage component of the previously mentioned 25 nm NMOS device based on the device simulation. From Figure 13.17, it is observed that the subthreshold leakage increases exponentially with temperature, the junction BTBT increases slowly with temperature and the gate leakage is almost independent of temperature variation. Figure 13.17 shows that for that particular NMOS device, at $T = 300$ K (a possible temperature in the stand-by mode) the gate leakage is the dominant leakage component. However, the subthreshold and the BTBT become dominant at $T = 400$ K (a possible temperature in the active mode). Hence, it can be concluded that the individual leakage components and the total leakage depend strongly on temperature (or mode of operation).

13.3   Circuit techniques to reduce leakage in logic

Since circuits are mostly designed for the highest performance – say to satisfy overall system cycle time requirements – they are composed of large gates, highly parallel architectures with logic duplication. As such, the leakage power consumption is

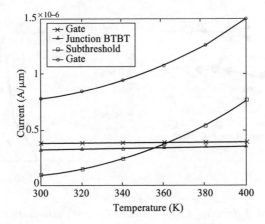

Figure 13.17   Simulation result for variation of different leakage components with temperature for NMOS device of $L_{eff} = 25$ nm [21]

Table 13.1   Circuit techniques to reduce leakage

| | Runtime techniques | |
Design time techniques	Standby leakage reduction	Active leakage reduction
Dual-V_{th}	Natural stacking Sleep transistor FBB/RBB	DVTS

substantial for such circuits. However, not every application requires a fast circuit to operate at the highest performance level all the time. Modules, in which computation is bursty in nature, e.g. functional units in a microprocessor or sections of a cache, are often idle. It is of interest to conceive of methods that can reduce the leakage power consumed by these circuits. Different circuit techniques have been proposed to reduce leakage energy utilising this slack without impacting performance. These techniques can be categorised based on when and how they utilise the available timing slack (Table 13.1), e.g. dual V_{th} statically assigns high V_{th} to some transistors in the non-critical paths at the 'design time' so as to reduce leakage current. The techniques, which utilise the slack in 'runtime', can be divided into two groups depending on whether they reduce standby leakage or active leakage. Standby leakage reduction techniques put the entire system in a low leakage mode when computation is not required. Active leakage reduction techniques slow down the system by dynamically changing the V_{th} to reduce leakage when maximum performance is not needed. In active mode, the operating temperature increases due to the switching activities of

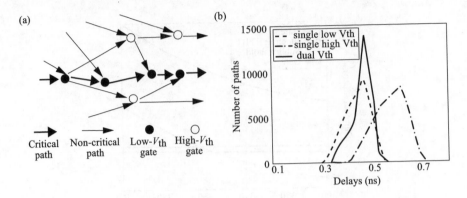

Figure 13.18 (a) *A dual V*th *CMOS circuit,* (b) *Path distribution of dual V*th *and single V*th *CMOS*

transistors. This has an exponential effect on subthreshold leakage making this the dominant leakage component during active mode and amplifying the leakage problem.

13.3.1 Design time techniques

Design time techniques exploit the delay slack in non-critical paths to reduce leakage. These techniques are static; once it is fixed, it cannot be changed dynamically while the circuit is operating.

13.3.1.1 Dual threshold CMOS

In logic, a high V_{th} can be assigned to some transistors in the non-critical paths so as to reduce subthreshold leakage current, while the performance is not sacrificed by using low V_{th} transistors in the critical path(s) [23,24]. No additional circuitry is required, and both high performance and low leakage can be achieved simultaneously. Figure 13.18(a) illustrates the basic idea of a dual V_{th} circuit. The path distribution of dual V_{th} and single V_{th} standard CMOS for a 32-bit adder is shown in Figure 13.18(b). Dual V_{th} CMOS has the same critical delay as the single low V_{th} CMOS circuit, but the transistors in the non-critical paths can be assigned high V_{th} to reduce leakage power. Dual threshold CMOS is effective in reducing leakage power during both standby and active modes. Many design techniques have been proposed, which consider upsizing of high V_{th} transistor [25] in dual V_{th} design to improve performance, or upsizing additional low V_{th} transistor to create more delay slack and then converting it to high V_{th} to reduce leakage power. Upsizing the transistor affects switching power and die area that can be traded off against using a low V_{th} transistor, which increases leakage power.

Domino logic can be susceptible to leakage – especially wide OR domino gates. Low threshold evaluation logic reduces noise immunity. Hence, for scaled technologies, domino may require larger keeper transistors which in turn can affect speed. Figure 13.19 shows a typical dual V_{th} domino logic [26] for low leakage noise immune

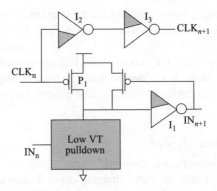

Figure 13.19 Dual V_{th} domino gate with low V_{th} devices shaded [26]

operations. Because of the fixed transition directions in domino logic, one can easily assign low V_{th} to all transistors that switch during the evaluate mode and high V_{th} to all transistors that switch during precharge modes. When a dual V_{th} domino logic stage is placed in standby mode, the domino clock needs to be high (evaluate) in order to shut off the high V_{th} devices (e.g. P_1, I_2 PMOS and I_3 NMOS). Furthermore, to ensure that the internal node remains at solid logic ZERO, which turns off the high V_{th} keeper and I_1 NMOS, the initial inputs into the domino gate must be set high.

A high V_{th} device can be achieved by varying different parameters, e.g. changing doping profile, using higher oxide thickness and increasing the channel length. Each parameter has its own trade-off in terms of process cost, effect on different leakage components and SCE.

13.3.1.1.1 Changing doping profile

Higher threshold voltages can be achieved by increasing the channel doping densities [27]. For this approach, two additional masks are required resulting in high process cost. This technique is commonly used to modify the threshold voltages. However, the threshold voltage can vary due to the non-uniform distribution of the doping density, making it difficult to achieve dual threshold voltages when the threshold voltages are very close to each other. High V_{th} can also be achieved by increasing the strength of halo by (1) increasing the peak doping A_p, (2) moving the position of the lateral peak of the halo close to the centre of the channel, i.e. by decreasing β_a and (3) moving the position of the vertical peak of the halo away from the bottom junction and towards the surface (Section 13.1). However, increasing the strength of the 'halo' increases the junction tunnelling (Figure 13.5), which might become severe in nano-scaled devices where junction tunnelling is a significant portion of total leakage.

13.3.1.1.2 Higher oxide thickness

A higher T_{ox} can be used to obtain a high V_{th} device for dual threshold CMOS circuits. Higher oxide thickness not only reduces the subthreshold leakage but also reduces gate oxide tunnelling, since the oxide tunnelling current decreases exponentially with increase in oxide thickness. Since higher oxide thickness reduces the gate capacitance, it is also beneficial for dynamic power reduction [27]. However, in a nano-scale device where SCE is extremely severe (e.g. in the 25 nm device), an increase in the oxide thickness will increase the subthreshold leakage (Figure 13.16(a)). In order to suppress the SCE, the high T_{ox} device needs to have a longer channel length as compared to the low T_{ox} device [27]. Advanced process technology is required for fabricating multiple T_{ox} CMOS.

13.3.1.1.3   *Larger channel length*

For short-channel transistors, the threshold voltage increases with the increase in channel length (V_{th} roll-off). Multiple channel length design uses the conventional CMOS technology. However, for the transistors with feature size close to 0.1 μm, halo implants [5] are used to suppress the short-channel effect. This causes the V_{th} roll-off to be very sharp; and hence, it is non-trivial to control the threshold voltages near the minimum feature size for such technologies. The longer transistor lengths for the high-threshold transistors will increase the gate capacitance, which has negative effect on the performance and power.

With the increase in V_{th} variation and supply voltage scaling, it is becoming difficult to maintain sufficient gap among low V_{th}, high V_{th} and supply voltage required for dual V_{th} design. Furthermore, dual V_{th} design increases the number of critical paths in a die. It has been shown in Reference 28 that as the number of critical paths on a die increases, within-die delay variation causes both mean and standard deviation of the die frequency distribution to become smaller, resulting in reduced performance.

13.3.2   *Runtime techniques*

13.3.2.1   **Standby leakage reduction techniques**

A common architectural technique to keep the power of fast, hot circuits within bounds has been to freeze the circuits – place them in a standby state – any time when they are not needed. Standby leakage reduction techniques exploit this idea to place certain sections of the circuitry in standby mode (low leakage mode) when they are not required.

13.3.2.1.1   *Natural transistor stacks*

Leakage currents in NMOS or PMOS transistors depend exponentially on the voltage at the four terminals of the transistor. Increasing the source voltage of an NMOS transistor reduces subthreshold leakage current exponentially due to negative V_{gs}, lowered signal rail (V_{DD}-V_s), reduced DIBL and body effect. This effect is also called self-reverse biasing for of transistor. The self-reverse bias effect can be achieved by turning off a stack of transistors [29]. Turning off more than one transistor in a stack raises the internal voltage (source voltage) of the stack, which acts as reverse biasing the source. Figure 13.20(a) depicts a simple pull down network of a four input NAND gate. This pull-down network forms a stack of four transistors. If some of the transistors are turned off for a long time, the circuit reaches a steady-state where

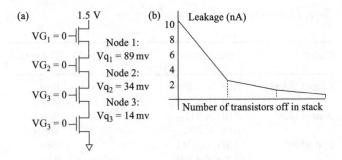

Figure 13.20   *(a) Effect of transistor stacking on source voltage, (b) leakage current versus number of transistors off in stack*

leakage through each transistor is equal and the voltage across each transistor settles to a steady-state value. In cases where only one NMOS device is off, the voltage at the source node of the off transistor would be virtually zero because all other on transistors will act as short circuits. Hence, there is no self-reverse biasing effect and the leakage across the off transistor is large. If more than one transistor is off, the source voltages of the off transistor, except the one connected to ground by on transistors, will be greater than zero. The leakage will be determined mainly by the most negatively self-reverse biased transistor (since subthreshold leakage is an exponential function of gate-source voltage). The voltages at the internal nodes depend on the input applied to the stack. Figure 13.20(a) shows the internal voltages when all four transistors are turned off. These internal voltages make the off transistors self-reverse biased. The reverse bias makes the leakage across the off transistor very small. Figure 13.20(b) shows the subthreshold leakage current versus number of off transistors in a stack. There is a large difference in leakage current between one off transistor and two off transistors. Turning off three transistors does improve subthreshold leakage, however, there is a diminishing return.

The voltages at the internal nodes depend on the input applied to the stack. Functional blocks such as NAND, NOR or other complex gates readily have a stack of transistors. Maximising the number of off transistors in a natural stack by applying proper input vectors can reduce the standby leakage of a functional block. A model and algorthim is proposed in Reference 10 to estimate leakage and to select the proper input vectors to minimise the leakage in logic blocks. Table 13.2 shows the quiescent current flowing into different functional blocks for the best- and worst-case input vectors. All the results are based on HSPICE simulation using 0.18 μm technology with $V_{DD} = 1.5$ V. Results show that application of a proper input vector can be efficient in reducing the total subthreshold leakage in the standby mode of operation [30].

Since gate and junction leakage are also important in scaled technologies, the input vector control technique using a stack of transistors needs to be reinvestigated to effectively reduce the total leakage. It has been shown that with high gate leakage, the traditional way of using stacking fails to reduce leakage and in the worst case might

Table 13.2   Input vector control

Circuit	Input vector	Iddq (nA)	Comments
4 input NAND	ABCD = 0000	0.60	Best
	ABCD = 1111	24.1	Worst
3 input NOR	ABC = 111	0.13	Best
	ABC = 000	29.5	Worst
Full adder	A,B,Ci = 111	7.8	Best
	A,B,Ci = 001	62.3	Worst
4 bit ripple adder	A = B = 0000, Ci = 0	91.3	Best
	A = B = 1111, Ci = 1	94.0	Best
	A = B = 0101, Ci = 1	282.9	Worst

increase the overall leakage [9]. The gate leakage depends on the voltage drop across different regions of the transistor. Applying '00' as the input to a two-transistor stack only reduces subthreshold leakage and does not change the gate leakage component. It has been shown that using '10' reduces the voltage drop across the terminals, where the gate leakage dominates, thereby lowering the gate leakage while offering marginal improvement in subthreshold leakage [9]. In scaled technologies where gate leakage dominates the total leakage, using '10' might produce more savings in leakage as compared to '00'. The source–substrate and drain–substrate junction BTBT leakage is a weak function of input voltage and hence, it can be neglected from the analysis.

13.3.2.1.2   Sleep transistor (forced stacking)

This technique inserts an extra series connected transistor (sleep transistor) in the pull-down/pull-up path of a gate and turns it 'off' in the standby mode of operation [31]. During regular mode of operation, the extra transistor is turned on. This provides substantial savings in leakage current during the standby mode of operation. However, due to the extra stacked transistor (sleep transistor), the drive current of forced-stack gate is lower resulting in increased delay. Hence, this technique can only be used for paths that are non-critical. If the V_{th} of the sleep transistor is high, extra leakage saving is possible. The circuit topology is known as MTCMOS (Figure 13.21) [32].

In fact, only one type (i.e. either PMOS or NMOS) of high V_{th} transistor is sufficient for leakage reduction. The NMOS insertion scheme is preferable, since the NMOS on-resistance is smaller at the same width and hence it can be sized smaller than a corresponding PMOS [33]. However, MTCMOS can only reduce leakage power in standby mode and the large inserted sleep transistors can increase the area and delay. Moreover, if data retention is required in standby mode, an extra high V_{th} memory circuit is needed to maintain the data. Instead of using high V_{th} sleep transistors, super cut-off CMOS (SCCMOS) circuit uses low V_{th} transistors with an inserted gate bias generator [34]. In standby mode, the gate is applied to $V_{DD} + 0.4$ V for PMOS ($V_{SS} - 0.4$ V for NMOS) by using the internal gate bias generator to fully

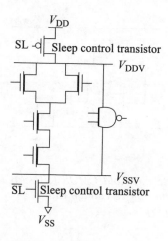

Figure 13.21   Schematic of MTCMOS circuit with low V_{th} device shaded [32]

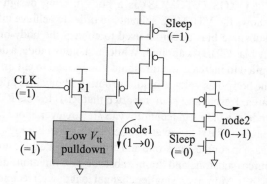

Figure 13.22   Domino gate with sleep transistor [35]

cut off the leakage current. Compared to MTCMOS where it becomes difficult to turn on the high V_{th} sleep transistor at very low supply voltages, SCCMOS circuits can operate at very low supply voltages.

A sleep transistor technique is proposed in Reference 35 to save leakage in domino gates. In Figure 13.22 two small sleep transistors are added to a conventional CMOS domino gate [35]. In standby mode the clock is left high and sleep signal is asserted. If the data input was high, node 1 would have been discharged. If the data input was low, node 1 would be high but leakage through NMOS dynamic pull-down stack would slowly discharge the node to ground. The NMOS sleep transistor is added to prevent any short circuit current in the static output logic while the dynamic node discharges to ground. Node 2 would rise as static pull up turns on which would cause the NMOS transistors in the pull-down stacks of the following domino gates to turn on, accelerating the discharge of their internal dynamic nodes. Since sleep transistors are not in the critical path (evaluation path), minimal performance loss is incurred.

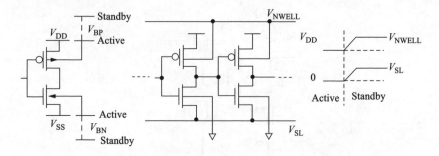

Figure 13.23   (a) Variable threshold CMOS [32], (b) Realising body biasing by changing the source voltage with respect to body voltage, which is grounded [39]

13.3.2.1.3   Forward/reverse body biasing

Variable threshold CMOS (VTCMOS) is a body biasing design technique [36]. Figure 13.23(a) shows the VTCMOS scheme. In order to achieve different threshold voltages, a self-substrate bias circuit is used to control the body bias. In the active mode, a zero body bias (ZBB) is applied. While in standby mode, a deep reverse body bias (RBB) is applied to increase the threshold voltage and to cut off the leakage current. Providing the body bias voltage requires routing a body bias grid and this adds to the overall chip area. Keshavarzi *et al.* reported that RBB lowers IC leakage by three orders of magnitude in a 0.35 μm technology [37]. However, more recent data shows that the effectiveness of RBB to lower I_{off} decreases as technology scales due to the exponential increase in source–substrate and drain–substrate band-to-band tunnelling leakage at the source–substrate and drain–substrate p–n junctions due to halo doping in scaled devices [37]. Moreover, smaller channel length with technology scaling and lower channel doping to reduce V_{th} worsens the short-channel effect and diminishes the body effect. This in turns weakens the V_{th} modulation capability of RBB.

For scaled technologies, recent design [38] has been proposed using forward body biasing (FBB) to achieve better current drive with less short-channel effect. Circuit is designed using high V_{th} transistor (high channel doping) reducing leakage in standby mode, while FBB is applied in active mode to achieve high performance. Both high-channel doping and FBB reduce short-channel effect relaxing the scalability limit of channel length due to V_{th} roll-off and DIBL. This results in higher I_{on} compared to low V_{th} design for same worst case I_{off}, improving performance. RBB can also be applied in standby mode together with FBB to further reduce the leakage current. It has been shown that FBB/high-V_{th} along with RBB reduces leakage by 20×, as opposed to 3× for the RBB/low-V_{th}. FBB devices, however, have larger junction capacitance and body effect, which reduces the delay improvement especially in stacked circuits. FBB can also be combined with lowering the V_{DD} to achieve the same performance as high V_{DD}, while reducing the switching and standby leakage power.

Raising the NMOS source voltage while tying the NMOS body to ground can produce the same effect as RBB. Forward body biasing can also be realised by applying a negative source voltage with respect to the body, which is tied to ground. Figure 13.23(b) illustrates the circuit diagram of this technique [39]. The main advantage is that it eliminates the need for a deep N-well or triple-well process since substrate of the target system and the control circuitry can be shared.

13.3.2.2 Active leakage reduction techniques

During active mode of operation the circuit works at higher temperature. It can be observed from Figure 13.17 that the subthreshold leakage increases exponentially with temperature, the junction BTBT increases slowly with temperature and the gate leakage is almost independent of temperature variation. Due to exponential increase in leakage, the active leakage power in sub-100 nm generations accounts for a large fraction of the total power consumption even during runtime. However, not every application requires a fast circuit to operate at the highest performance level all the time. Active leakage reduction techniques exploit this idea to intermittently slow down the fast circuitry and reduce the leakage power consumption as well as the dynamic power consumption when maximum performance is not required.

13.3.2.2.1 Dynamic V_{th} scaling (DVTS)

A DVTS scheme uses body biasing to adaptively change the V_{th} based on the performance demand. The lowest V_{th} is delivered via ZBB, if the highest performance is required. When performance demand is low, clock frequency is lowered and V_{th} is raised via RBB to reduce the run-time leakage power dissipation. In cases when there is no workload at all, the V_{th} can be increased to its upper limit to significantly reduce the standby leakage power. 'Just enough' throughput is delivered for the current workload by tracking the optimal V_{th} while leakage power is considerably reduced by intermittently slowing down the circuit.

Several different DVTS system implementations have been proposed in literature [40,41]. Figure 13.24 shows a DVTS hardware that uses continuous body bias control to track the optimal V_{th} for a given workload. A clock speed scheduler, which is embedded in the operating system, determines the (reference) clock frequency at runtime. The DVTS controller adjusts the PMOS and NMOS body bias so that

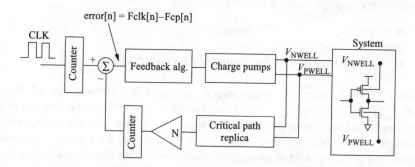

Figure 13.24   Dynamic V_{th} scaling system proposed in Reference 40

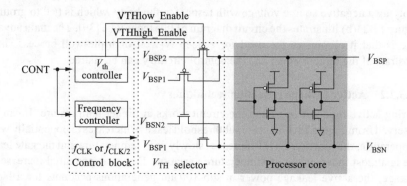

Figure 13.25 V_{th} hopping scheme proposed in Reference 41

the oscillator frequency of the critical path replica tracks the given reference clock frequency. The error signal, which is the difference between the reference clock frequency and the oscillator frequency, is fed into the feedback controller. The continuous feedback loop can also compensate for process, supply voltage and temperature variations. A simpler method called 'V_{th} hopping scheme', which dynamically switches between low V_{th} and high V_{th} depending on the performance demand, is proposed in Reference 41. The schematic diagram of the V_{th} hopping scheme is shown in Figure 13.25. Compared to the continuous body bias control in Figure 13.24, the discrete control has two levels of V_{th}. If control signal VTHlow_Enable is asserted, the transistors in the target system are forward body biased and the V_{th} is low. When performance can be traded off for lower power consumption, VTHhigh_Enable is asserted and a high V_{th} is applied. The operating frequency of the target system is set to f_{CLK} when V_{th} is low and to $f_{CLK/2}$ when the V_{th} is high. An algorithm that adaptively changes the V_{th} depending on the workload is also verified and applied to an MPEG4 video encoding system. As mentioned in the previous section, the effectiveness of RBB is expected to be low due to the worsening short-channel effect and increasing band-to-band tunnelling leakage at the source–substrate and drain–substrate junctions. FBB can be applied together with RBB to achieve a better performance-leakage tradeoff for DVTS systems.

13.4   Circuit techniques to reduce leakage in cache memories

Figure 13.26(a) shows the seven available terminals in a conventional 6T SRAM cell; V_{SL}, V_{PWELL}, V_{NWELL}, V_{DL}, V_{WL}, V_{BL} and V_{BLB}. Various SRAM cell architectures have been proposed in the past where one or more of the seven terminal voltages are controlled during standby mode for reducing the leakage components shown in Figure 13.26(b). Each technique exploits the fact that the active portion of a cache is very small, which gives the opportunity to put the large idle portion in a low-leakage sleep mode. Effectiveness and overhead of each technique are evaluated

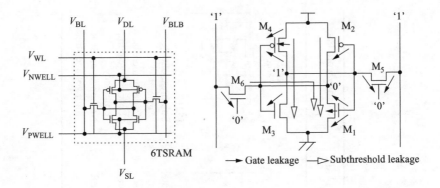

Figure 13.26   (a) *Severn terminals of the 6T SRAM cell, (b) dominant leakage components in a 6T SRAM*

based on the following discussions. First, the impact of the technique on various leakage components should be considered. Although subthreshold leakage still continues to dominate the I_{OFF} at high temperatures, ultra-thin oxides and high doping concentrations have led to a rapid increase in direct tunnelling gate leakage and BTBT leakage at the source and drain junctions in the nano-metre regime. Each leakage reduction technique needs re-evaluation in scaled technologies where subthreshold conduction is not the only leakage mechanism. Second, the impact of the leakage reduction technique on SRAM read/write delay should be considered. Third, the transition latency/energy overhead should be taken into account, because of the limited time and energy budget for the mode transition. Last, the leakage reduction technique should not have a noticeable impact on SRAM cell stability or soft error rate (SER). Based on these discussions, the different low-leakage SRAM cells are summarised in Table 13.3.

The source biasing scheme raises the source line voltage (V_{SL}) in sleep mode [42–46], which reduces subthreshold leakage due to the three effects described in Section 13.3.2.1.1 The gate leakage in the cell is also reduced due to the relaxed signal rail, V_{DD}- V_{SL} [46]. An extra NMOS has to be series connected in the pull-down path in order to cutoff the source line from ground during sleep mode, and this in turn imposes an extra access delay. The reduced signal charge in sleep mode also causes the SER to rise, requiring additional error correction coding circuits [30].

RBB the NMOS (or PMOS) can reduce subthreshold leakage via body effect, while not affecting the access time by switching to zero body-biasing (ZBB) in active mode [44,47,48]. A large latency/energy overhead is imposed for the body-bias transition due to the large V_{BB} swing and substrate capacitance. This scheme becomes less attractive in scaled technologies since the body coefficient decreases with smaller dimensions, and the source and drain junction BTBT leakage becomes enhanced by RBB. For scaled technologies, recent design [49] has been proposed using forward body biasing (FBB) to reduce subthreshold leakage and to achieve better current drive while maintaining reasonable junction BTBT. A new high V_{th} device optimised

Table 13.3 Low-leakage SRAM cell techniques

Scheme	Source biasing (V_{SL})	RBB/FBB (V_{PWELL}, V_{NWELL})	Dynamic V_{DD} (V_{DL})	Leakage biased (V_{BL}, V_{BLB})	Negative word line (V_{WL})
References	[42–46]	[44,47–49]	[44,50]	[51]	[52]
Leakage reduction	Subthreshold, gate: ⇓	Subthreshold: ⇓⇓	Subthreshold, gate: →	Subthreshold, gate: →	Subthreshold: ↓,
		*BTBT: ↑	*bitline leakage: ↑		*gate: ↑
Performance	*Delay increase	No delay increase	No delay increase	No delay increase	No delay increase
Overhead	Medium transition overhead	Large transition overhead	Large transition overhead	*Precharge latency overhead	*Low charge pump efficiency
Stability	Impact on SER	No impact on SER	*Worst SER	No impact on SER	No impact on SER, high voltage stress

for FBB is proposed which changes the doping profile by adjusting the peak halo doping (channel engineering) or uses gate material with a higher work function (work function engineering) [49]. This scheme resolves the drawback associated with RBB SRAM and suggest a viable solution for reducing leakage in nano-scale memories.

Supply voltage is lowered in a dynamic V_{DD} SRAM (DVSRAM) [44,50], which in turn reduces the subthreshold, gate and BTBT leakage. This scheme requires a smaller signal rail (V_{DL}-V_{GND}) compared to the SBSRAM for equivalent leakage savings. Although there is no impact on delay in the active mode, the large V_{DD} swing between sleep and active mode imposes a larger latency/energy transition overhead than the SBSRAM. Moreover, the greatest drawback of the DVSRAM is that it increases the bitline leakage in the sleep mode since the voltage level in the stored node also drops as the V_{DD} is lowered. Therefore, this scheme is not suitable for dual V_{th} designs where the speed-critical access transistors may already be using low V_{th} devices with high leakage levels.

A technique that biases the bitlines to an intermediate level has been proposed to reduce the access transistor leakage via the DIBL effect [51]. Since only the access transistors benefit from the leakage reduction, the overall leakage saving is moderate. Unlike the three previously mentioned techniques, this scheme has to be applied to the entire subarray because the bitline is shared across different cache lines. The main limitation comes from the fact that there is a precharge latency whenever a new subarray is accessed. This would mean that an architectural modification is required in order to resolve the multiple hit times in case the precharge instant is not known ahead of time.

The negative word line scheme [52] pulls down the V_{WL} to a negative voltage during standby in order to avoid the subthreshold leakage through the access transistors. However, it has issues such as increase in gate leakage and higher voltage stress in the access transistors. Although this technique has no impact on performance or SER, there is a power loss due to generating the negative bias using charge pumps. This becomes more serious as the supply voltage is scaled.

13.5 Conclusions

Semiconductor devices are aggressively scaled each technology generation to achieve high integration density while the supply voltage is scaled to achieve lower switching energy per device. However, to achieve high performance there is need for commensurate scaling of the transistor threshold voltage (V_{th}). Scaling of transistor threshold voltage is associated with exponential increase in subthreshold leakage current. Aggressive scaling of the devices in the nano-metre regime not only increases the subthreshold leakage but also has other negative impacts such as increased DIBL, V_{th} roll-off, reduced on-current to off-current ratio and increased source–drain resistance. To avoid these short-channel effects, oxide thickness scaling and higher and non-uniform doping needs to be incorporated as the devices are scaled in nano-metre regime, which results in exponential increase in gate and junction band-to-band tunnelling leakage. This increase in total leakage causes the leakage current to become a major component of total power consumption. Hence, leakage reduction techniques are becoming indispensable in future designs. This chapter explained the various leakage mechanisms and discussed different circuit level techniques to reduce leakage energy and design tradeoffs.

References

1   BORKAR, S.: 'Design challenges of technology scaling', *IEEE Micro*, 1999, **19**(4), 23

2   BREWS, J., and SZE, S.M.: 'High speed semiconductor devices' (John Wiley & Sons, New York, 1990)

3   *International Technology Roadmap for Semiconductors 2001 Edition*: Semiconductor Industry Association, Available: http://public.itrs.net/Files/2001ITRS/Home.htm

4   ROY, K., MUKHOPADHYAY, S., and MEIMAND, H.: 'Leakage current mechanisms andleakage reduction techniques in deep-submicron CMOS circuit', *Proceedings of IEEE*, February 2003, **91**(2), pp. 305–327

5   TAUR, Y., and NING, T.H.: 'Fundamentals of modern VLSI devices' (Cambridge University Press, New York, 1998)

6   KESHAVARZI, A., ROY, K., and HAWKINS, C.F.: 'Intrinsic leakage in deep submicron CMOS ICs-measurement-based test solutions', *IEEE Transaction on VLSI Sysytems*, 2000, **8**(6), pp. 717–723

7   ROY, K., and PRASAD, S.C.: 'Low-power CMOS VLSI circuit design' (Wiley Interscience Publications, New York, 2000)

8   AGARWAL, A., KIM, C.H., MUKHOPADHYAYA, S., and ROY, K.: 'Leakage in nano-scale technologies: mechanisms, impact and design considerations'. *Design automation* conference (DAC), 2004

9   MUKHOPADHYAY, S., NEAU, C., CAKICI, T., AGARWAL, A., KIM, C.H., and ROY, K.: 'Gate leakage reduction for scaled devices using transistor stacking', *IEEE Transactions on Very Large Scale Integration Systems*, 2003

10 LEE, Z., MCILRATH, M.B., and ANTONIADIS, D.A., 'Two-dimensional doping profile characterization of MOSFET's by inverse modeling using characteristics in the subthreshold Region', *IEEE Transaction on Electron Device*, August 1999, pp. 1640–1649

11 MUKHOPADHYAY, S., RAYCHOWDHURY, A., and ROY, K.: 'Accurate estimation of total leakage in nanometer scale bulk CMOS circuits based on device geometry and doping profile', *IEEE Transaction on CAD*, March 2005

12 MEDICI: Two-dimensional semiconductor device simulation program. AVANT! Corp., Fremont, CA, 2000

13 LIU, Z., HU, C., HUANG, J.-H., CHAN, T.-Y., JENG, M.-C., KO, P.K., and CHENG, Y.C.: 'Threshold voltage model for deep-submicrometer MOSFET's', *IEEE Transactions on Elecrons Devices*, 1993, **40**, pp. 86–95

14 ZHOU, X., LIM, K.Y., and LIM, D.: 'A general approach to compact threshold voltage formulation based on 2-D numerical simulation and experimental correlation for deep-submicron ULSI technology development', *IEEE Transactions on Elecrons Devices*, 2000, **47**, pp. 214–221

15 FOTTY, D.: 'MOSFET modelling with SPICE,' (Prentice Hall PTR, NJ, 1997)

16 BSIM3v3.2.2 MOSFET Model BSIM Group, University of California, Berkeley, CA http://www-device.eecs.berkeley.edu/~bsim3/

17 CHUNG, S., and LI, C.-T.: 'An analytical threshold-voltage model of trench-isolated MOS devices with nonuniformly doped substrates', *IEEE Transactions on Electron Devices*, 1992, **39**, pp. 614–622

18 CAO, K., LEE, W.-C., LIU, W. *et al.*: 'BSIM4 gate leakage model including source drain partition', *Tech. Digest IEDM*, 2000, pp. 815–818

19 BSIM4.2.1 MOSFET Model, BSIM Group, University of California, Berkeley, CA, http://www-device.eecs.berkeley.edu/~bsim3/

20 HAMZAOGLU, F., and STAN, M.: 'Circuit-level techniques to control gate leakage for sub-100 nm CMOS'. Proceedings of *ISLPED*, August 2002, pp. 60–63

21 *'Well-Tempered' Bulk-Si NMOSFET Device Home Page*, Microsystems Technology Laboratory, MIT, Available: http://www-mtl.mit.edu/Well/

22 PIERRET, R.: 'Advanced semiconductor fundamentals', in NEUDECK, G.W., and PIERRET, R.F. (Eds): 'Modular series on solid states devices, vol VI' (Addison-Wesley Publishing Company, MA, 1989)

23 KETKAR, M. *et al.*, 'Standby power optimization via transistor sizing and dual threshold voltage assignment'. Proceedings of the international conference on *Computer aided design*, November 2002, pp. 375–378

24 WEI, L., CHEN, Z., JOHNSON, M., ROY, K., YE, Y., and DE, V.: 'Design and optimization of dual threshold circuits for low voltage low power applications', *IEEE Transactions on VLSI Systems*, 1999, **7**, pp. 16–24

25 KARNIK, T. *et al.*: 'Total power optimization by simultaneous dual-Vt allocation and device sizing in high performance microprocessors'. ACM/IEEE *Design automation* conference, vol. 486, 2002

26 KAO, J.T., and CHANDRAKASAN, A.P.: 'Dual-threshold voltage techniques for low-power digital circuits,' *IEEE Journal of Solid-State Circuits*, **35**, 2000, p. 1009

27 SIRISANTANA, N., WEI, L., and ROY, K.: 'High-performance low-power CMOS circuits using multiple channel length and multiple oxide thickness'. Proceedings of the 2000 international conference on *Computer design*, vol. 227, 2000

28 BOWMAN, K.A. *et al.*: 'Impact of die-to-die and within die parameter fluctions on the maximum clock frequency distribution for gigascale integration', *IEEE Journal of Solid State Circuits*, 2002, **37**, pp. 183–190

29 YE, Y., BORKAR, S., and DE, V.: 'A new technique for standby leakage reduction in high performance circuits'. IEEE symposium on *VLSI circuits*, vol. 40, 1998

30   CHEN, Z., WEI, L., KESHAVARZI, A., and ROY, K.: 'IDDQ testing for deep submicron ICs: challenges and solutions', *IEEE Design and Test of Computers*, 2002, **19**, pp. 24–33

31   JOHNSON, M.C., SOMASEKHAR, D., and ROY, K.: 'Leakage control with efficient use of transistor stacks in single threshold CMOS'. Proceedings of ACM/IEEE *Design automation* conference, vol. 442, 1999

32   MUTOH, S. *et al.*: '1-V Power supply high-speed digital circuit technology with multi-threshold voltage CMOS', *IEEE Journal of Solid-State Circuits*, **30**, 1995, p. 847

33   KAO, J., CHANDRAKASAN, A., and ANTONIADIS, D.: 'Transistor sizing issues and tool for multi-threshold CMOS technology'. Proceedings of ACM/IEEE *Design automation* conference, vol. 495, 1997

34   KAWAGUCHI, H., NOSE, K., and SAKURAI, T.: 'A CMOS scheme for 0.5V supply voltage with pico-ampere standby current'. Digest of technical papers of IEEE international *Solid-state circuits conference*, vol. 192, 1998

35   HEO, S., and ASANOVIC, K.: 'Leakage-biased domino circuits for dynamic fine-grain leakage reduction'. Symposium on *VLSI circuits*, vol. 316, 2002

36   KURODA, T. *et al.*: 'A 0.9V 150MHz 10mW 4mm^2 2-D discrete cosine transform core processor with variable-threshold-voltage scheme'. Digest of technical papers of IEEE international *Solid-state circuits* conference, vol. 166, 1996

37   KESHAVARZI, A., HAWKINS, C.F., ROY, K., and DE, V.: 'Effectiveness of reverse body bias for low power CMOS circuits'. Proceedings of the 8th NASA symposium on *VLSI design*, vol. 231, 1999, pp. 2.3.1–2.3.9

38   NARENDRA, S. *et al.* 'Forward body bias for microprocessors in 130-nm technology generation and beyond', *IEEE Journal of Solid State Circuits*, May 2003, **38**, pp. 696–701

39   MIZUNO, H. *et al.*: 'An 18-μA standby current 1.8-V, 200-MHz microprocessor with self-substrate-biased data-retention mode', *IEEE Journal of Solid-State Circuits*, **34**, 1999, pp. 1492–1500

40   KIM, C.H., and ROY, K.: 'Dynamic V_{th} scaling scheme for active leakage power reduction'. *Design, automation and test in Europe*, vol. 163, 2002

41   NOSE, K. *et al.*: 'V_{th}-hopping scheme for 82% power saving in low-voltage processors'. Proceedings of IEEE *Custom integrated circuits* conference, vol. 93, 2001

42   AGARWAL, A., LI, H., and ROY, K.: 'A single-Vt low-leakage gated-ground cache for deep submicron', *IEEE Journal of Solid-State Circuits*, 2003, **38**, pp. 319–328

43   YAMAUCHI, H. *et al.*: 'A 0.8V/100MHz/sub-5mW-operated mega-bit SRAM cell architecture with charge-recycle offset-source driving (OSD) scheme'. Symposium on *VLSI circuits*, vol. 126, 1996

44   BHAVNAGARWALA, A.J., KAPOOR, A., and MEINDL, J.D.: 'Dynamic threshold CMOS SRAMs for fast, portable applications'. ASIC/SOC conference, vol. 359, 2000

45 OSADA, K. *et al.*: '16.7fA/cell tunnel-leakage-suppressed 16Mb SRAM for handling cosmic-ray-induced multi-errors'. International *Solid-state circuits* conference, vol. 302, 2003

46 AGARWAL, A., and ROY, K.: 'Noise tolerant cache design to reduce gate and subthreshold leakage in nanometer regime'. Accepted in international symposium of *Low power electronics and design (ISLPED2003)*, 2003

47 KAWAGUCHI, H., ITAKA, Y., and SAKURAI, T.: 'Dynamic leakage cut-off scheme for low-voltage SRAM's'. Symposium on *VLSI circuits*, vol. 140, 1998

48 KIM, C.H., and ROY, K.: 'Dynamic Vt SRAM: a leakage tolerant cache memory for low voltage microprocessors'. International symposium on *Low power electron and design*, vol. 251, 2002

49 KIM C.H. *et al.*: *ISLPED*, 2003

50 FLAUTNER, K. *et al.*: 'Drowsy caches: simple techniques for reducing leakage power'. International symposium on *Computer architecture*, vol. 148, 2002

51 HEO, S. *et al.*: 'Dynamic fine-grain leakage reduction using leakage-biased bitlines'. International symposium on *Computer architecture*, vol. 137, 2002

52 ITOH, K., FRIDI, A.R., BELLAOUAR, A., and ELMASRY, M.I.: 'A deep sub-V, single power-supply SRAM cell with multi-Vt, boosted storage node and dynamic load'. Symposium on *VLSI circuits digest of technical papers*, vol. 132, 1996

Part IV

Reconfigurable computing

Chapter 14

Reconfigurable computing: architectures and design methods

T.J. Todman, G.A. Constantinides, S.J.E. Wilton, O. Mencer, W. Luk and P.Y.K. Cheung

14.1 Introduction

Reconfigurable computing is rapidly establishing itself as a major discipline that covers various subjects of learning, including both computing science and electronic engineering. Reconfigurable computing involves the use of reconfigurable devices, such as Field Programmable Gate Arrays (FPGAs), for computing purposes. Reconfigurable computing is also known as configurable computing or custom computing, since many of the design techniques can be seen as customising a computational fabric for specific applications [1].

Reconfigurable computing systems often have impressive performance. Consider, as an example, the point multiplication operation in Elliptic Curve cryptography. For a key size of 270 bits, it has been reported [2] that a point multiplication can be computed in 0.36 ms with a reconfigurable computing design implemented in an XC2V6000 FPGA at 66 MHz. In contrast, an optimised software implementation requires 196.71 ms on a dual-Xeon computer at 2.6 GHz; so the reconfigurable computing design is more than 540 times faster, while its clock speed is almost 40 times slower than the Xeon processors. This example illustrates a hardware design implemented on a reconfigurable computing platform. We regard such implementations as a subset of reconfigurable computing, which in general can involve the use of runtime reconfiguration and soft processors.

Reconfigurable computing involves devices that can be reconfigured: their circuits can be changed after they are manufactured. This means that rather than using a single circuit for many applications, such as a microprocessor, specific circuits can be generated for specific applications. How can circuits be changed after manufacture? Typically, reconfigurable devices use memory whose state switches logical elements (e.g. flip-flops and function generators), and the wiring between them. The state of all these memory bits is known as the configuration of the device, and determines its function (e.g. image processor, network firewall). In Section 14.3, we survey different styles of reconfigurable logical elements and wiring.

New circuits for new applications can be uploaded to the reconfigurable device by writing to the configuration memory. An example of a reconfigurable device is the Xilinx Virtex 4 [3]. In this device, the configuration memory controls logical elements (which include flip-flops, function generators, multiplexors and memories), and wiring, arranged in a hierarchical scheme.

Designing circuits for reconfigurable devices is akin to designing application-specific integrated circuits, with the additional possibility that the design can change, perhaps in response to data received. Design methods can be general-purpose (e.g. using the C programming language), or special-purpose (e.g. using domain-specific tools such as MATLAB). We review such design methods in Section 14.4.

There are many commercial tools which support reconfigurable computing, including:

- Xilinx's ISE [4], an example of a reconfigurable device vendor tool, which generates configurations for Xilinx's families of reconfigurable hardware from inputs such as hardware description languages.
- Celoxica's DK [5] design suite, which allows descriptions based on the C programming language to be translated to configurations for reconfigurable hardware in the form of hardware description languages.
- Synplicity Synplify Pro [6], which allows reconfigurable designs in hardware description languages (perhaps the output of Celoxica's DK) to be optimised and converted into a netlist, which reconfigurable device vendor tools (such as Xilinx's ISE) can then convert into a configuration.

Is this speed advantage of reconfigurable computing over traditional microprocessors a one-off or a sustainable trend? Recent research suggests that it is a trend rather than a one-off for a wide variety of applications: from image processing [7] to floating-point operations [8].

Sheer speed, while important, is not the only strength of reconfigurable computing. Another compelling advantage is reduced energy and power consumption. In a reconfigurable system, the circuitry is optimised for the application, such that the power consumption will tend to be much lower than that for a general-purpose processor. A recent study [9] reports that moving critical software loops to reconfigurable hardware results in average energy savings of 35–70 per cent with an average speedup of 3–7 times, depending on the particular device used.

Other advantages of reconfigurable computing include a reduction in size and component count (and hence cost), improved time-to-market, and improved flexibility and upgradability. These advantages are especially important for embedded applications. Indeed, there is evidence [10] that embedded systems developers show a growing interest in reconfigurable computing systems, especially with the introduction of soft cores which can contain one or more instruction processors [11–16].

In this chapter, we present a survey of modern reconfigurable system architectures and design methods. Although we also provide background information on notable aspects of older technologies, our focus is on the most recent architectures and design methods, as well as the trends that will drive each of these areas in the near future. In other words, we intend to complement other surveys [17–21] by:

(1) providing an up-to-date survey of material that appears after the publication of the papers mentioned above;
(2) identifying explicitly the main trends in architectures and design methods for reconfigurable computing;
(3) examining reconfigurable computing from a perspective different from existing surveys, for instance classifying design methods as special-purpose and general-purpose;
(4) offering various direct comparisons of technology options according to a selected set of metrics from different perspectives.

The rest of the chapter is organised as follows. Section 14.2 contains background material that motivates the reconfigurable computing approach. Section 14.3 describes the structure of reconfigurable fabrics, showing how various researchers and vendors have developed fabrics that can efficiently accelerate time-critical portions of applications. Section 14.4 covers recent advances in the development of design methods that map applications to these fabrics, and distinguishes between those which employ special-purpose and general-purpose optimisation methods. Finally, Section 14.5 concludes and summarises the main trends in architectures, design methods and applications of reconfigurable computing.

14.2 Background

Many of today's compute-intensive applications require more processing power than ever before. Applications such as streaming video, image recognition and processing, and highly interactive services are placing new demands on the computation units that implement these applications. At the same time, the power consumption targets, the acceptable packaging and manufacturing costs, and the time-to-market requirements of these computation units are all decreasing rapidly, especially in the embedded hand-held devices market. Meeting these performance requirements under the power, cost and time-to-market constraints is becoming increasingly challenging.

In the following, we describe three ways of supporting such processing requirements: high-performance microprocessors, application-specific integrated circuits and reconfigurable computing systems.

High-performance microprocessors provide an off-the-shelf means of addressing processing requirements described earlier. Unfortunately for many applications, a single processor, even an expensive state-of-the-art processor, is not fast enough. In addition, the power consumption (100 W or more) and cost (possibly thousands of dollars) of state-of-the-art processors place them out-of-reach for many embedded applications. Even if microprocessors continue to follow Moore's Law so that their density doubles every 18 months, they may still be unable to keep up with the requirements of some of the most aggressive embedded applications.

Application-specific integrated circuits (ASICs) provide another means of addressing these processing requirements. Unlike a software implementation, an ASIC implementation provides a natural mechanism for implementing the large amount of parallelism found in many of these applications. In addition, an ASIC circuit does not need to suffer from the serial (and often slow and power-hungry) instruction fetch, decode and execute cycle that is at the heart of all microprocessors. Furthermore, ASICs consume less power than reconfigurable devices. Finally, an ASIC can contain just the right mix of functional units for a particular application; in contrast, an off-the-shelf microprocessor contains a fixed set of functional units which must be selected to satisfy a wide variety of applications.

Despite the advantages of ASICs, they are often infeasible or uneconomical for many embedded systems. This is primarily due to two factors: the cost of producing an ASIC often due to the mask's cost (up to $1 million [22]), and the time to develop a custom integrated circuit, can both be unacceptable. Only for the very highest-volume applications would the improved performance and lower per-unit price warrant the high non-recurring engineering (NRE) cost of designing an ASIC.

A third means of providing this processing power is a reconfigurable computing system. A reconfigurable computing system typically contains one or more processors and a reconfigurable fabric upon which custom functional units can be built. The processor(s) executes sequential and non-critical code, while code that can be efficiently mapped to hardware can be 'executed' by processing units that have been mapped to the reconfigurable fabric. Like a custom integrated circuit, the functions that have been mapped to the reconfigurable fabric can take advantage of the parallelism achievable in a hardware implementation. Also like an ASIC, the embedded system designer can produce the right mix of functional and storage units in the reconfigurable fabric, providing a computing structure that matches the application.

Unlike an ASIC, however, a new fabric need not be designed for each application. A given fabric can implement a wide variety of functional units. This means that a reconfigurable computing system can be built out of off-the-shelf components, significantly reducing the long design-time inherent in an ASIC implementation. Also unlike an ASIC, the functional units implemented in the reconfigurable fabric can change over time. This means that as the environment or usage of the embedded system changes, the mix of functional units can adapt to better match the new environment. The reconfigurable fabric in a handheld device, for instance, might implement large matrix multiply operations when the device is used in one mode, and large signal processing functions when the device is used in another mode.

Typically, not all of the embedded system functionality needs to be implemented by the reconfigurable fabric. Only those parts of the computation that are time-critical and contain a high degree of parallelism need to be mapped to the reconfigurable fabric, while the remainder of the computation can be implemented by a standard instruction processor. The interface between the processor and the fabric, as well as the interface between the memory and the fabric, are therefore of the utmost importance. Modern reconfigurable devices are large enough to implement instruction processors within the programmable fabric itself: soft processors. These can be general purpose, or customised to a particular application; application specific instruction processors and flexible instruction processors are two such approaches. Section 14.4.3 deals with soft processors in more detail.

Other devices show some of the flexibility of reconfigurable computers. Examples include graphics processor units and application specific array processors. These devices perform well on their intended application, but cannot run more general computations, unlike reconfigurable computers and microprocessors.

Despite the compelling promise of reconfigurable computing, it has limitations of which designers should be aware. For instance, the flexible routing on the bit level tends to produce large silicon area and performance overhead when compared with ASIC technology. Hence for large volume production of designs in applications without the need for field upgrade, ASIC technology or gate array technology can still deliver higher performance design at lower unit cost than reconfigurable computing technology. However, since FPGA technology tracks advances in memory technology and has demonstrated impressive advances in the last few years, many are confident that the current rapid progress in FPGA speed, capacity and capability will continue, together with the reduction in price.

It should be noted that the development of reconfigurable systems is still a maturing field. There are a number of challenges in developing a reconfigurable system. We describe two such challenges below.

First, the structure of the reconfigurable fabric and the interfaces between the fabric, processor(s) and memory must be very efficient. Some reconfigurable computing systems use a standard field-programmable gate array [3,23–27] as a reconfigurable fabric, while others adopt custom-designed fabrics [28–39].

Another challenge is the development of computer-aided design and compilation tools that map an application to a reconfigurable computing system. This involves determining which parts of the application should be mapped to the fabric and which should be mapped to the processor, determining when and how often the reconfigurable fabric should be reconfigured, which changes the functional units implemented in the fabric, as well as the specification of algorithms for efficient mappings to the reconfigurable system.

In this chapter, we provide a survey of reconfigurable computing, focusing our discussion on both of the issues described above. In the next section, we provide a survey of various architectures that are found useful for reconfigurable computing; material on design methods will follow.

14.3   Architectures

We shall first describe system-level architectures for reconfigurable computing. We then present various flavours of reconfigurable fabric. Finally, we identify and summarise the main trends.

14.3.1   *System-level architectures*

A reconfigurable system typically consists of one or more processors, one or more reconfigurable fabrics, and one or more memories. Reconfigurable systems are often classified according to the degree of coupling between the reconfigurable fabric and the CPU. Compton and Hauck [18] present the four classifications shown in Figure 14.1(a–d). In Figure 14.1(a), the reconfigurable fabric is in the form of one or more stand-alone devices. The existing input and output mechanisms of the processor are used to communicate with the reconfigurable fabric. In this configuration, the data

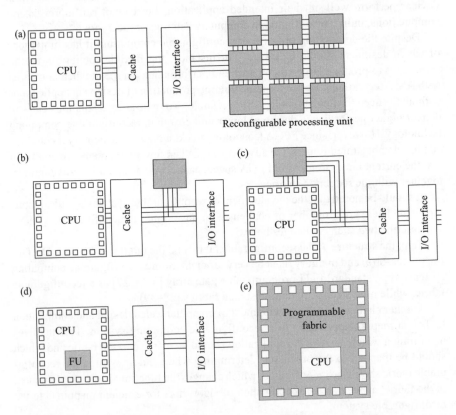

Figure 14.1   Five classes of reconfigurable systems: (a) external stand-alone processing unit, (b) attached processing unit, (c) co-processor, (d) reconfigurable fnctional unit and (e) processor embedded in a reconfigurable fabric. The first four are adapted from Reference 18

transfer between the fabric and the processor is relatively slow, so this architecture only makes sense for applications in which a significant amount of processing can be done by the fabric without processor intervention. Emulation systems often take on this sort of architecture [40,41].

Figure 14.1(b) and 14.1(c) shows two intermediate structures. In both cases, the cost of communication is lower than that of the architecture in Figure 14.1(a). Architectures of these types are described in References 31, 32, 36, 38, 42–45. Next, Figure 14.1(d) shows an architecture in which the processor and the fabric are very tightly coupled; in this case, the reconfigurable fabric is part of the processor itself; perhaps forming a reconfigurable sub-unit that allows for the creation of custom instructions. Examples of this sort of architecture have been described in References 33, 35, 39 and 46.

Figure 14.1(e) shows a fifth organisation. In this case, the processor is embedded in the programmable fabric. The processor can either be a 'hard' core [47,48], or can be a 'soft' core which is implemented using the resources of the programmable fabric itself [11–16].

A summary of the above organisations can be found in Table 14.1. Note the bandwidth is the theoretical maximum available to the CPU: for example, in Chess [33],

Table 14.1   Summary of system architectures

Class	CPU to memory bandwidth (MB/s)	Shared memory size	Fine grained or coarse grained	Example application
(a) External stand-alone processing unit				
RC2000 [49]	528	152MB	Fine grained	Video processing
(b)/(c) Attached processing unit/co-processsor				
Pilchard [50]	1064	20 kbytes	Fine grained	DES encryption
Morphosys [38]	800	2048 bytes	Coarse grained	Video compression
(d) Reconfigurable functional unit				
Chess [33]	6400	12288 bytes	Coarse grained	Video processing
(e) Processor embedded in a reconfigurable fabric				
Xilinx Virtex II Pro [27]	1600	1172 kB	Fine grained	Video compression

we assume that each block RAM is being accessed at its maximum rate. Organisation (a) is by far the most common, and accounts for all commercial reconfigurable platforms.

14.3.2   Reconfigurable fabric

The heart of any reconfigurable system is the reconfigurable fabric. The reconfigurable fabric consists of a set of reconfigurable functional units, a reconfigurable interconnect and a flexible interface to connect the fabric to the rest of the system. In this section, we review each of these components, and show how they have been used in both commercial and academic reconfigurable systems.

A common theme runs through this entire section: in each component of the fabric, there is a trade-off between flexibility and efficiency. A highly flexible fabric is typically much larger and much slower than a less flexible fabric. On the other hand, a more flexible fabric is better able to adapt to the application requirements.

In the following discussions, we will see how this trade-off has influenced the design of every part of every reconfigurable system. A summary of the main features of various architectures can be found in Table 14.2.

1. Reconfigurable functional units: Reconfigurable functional units can be classified as either coarse-grained or fine-grained. A fine-grained functional unit can typically implement a single function on a single (or small number) of bits. The most common kinds of fine-grained functional units are the small lookup tables that are used to implement the bulk of the logic in a commercial field-programmable gate array. A coarse-grained functional unit, on the other hand, is typically much larger, and may consist of arithmetic and logic units (ALUs) and possibly even a significant amount of storage. In this section, we describe the two types of functional units in more detail.

Many reconfigurable systems use commercial FPGAs as a reconfigurable fabric. These commercial FPGAs contain many three to six input lookup tables, each of which can be thought of as a very fine-grained functional unit. Figure 14.2(a) illustrates a lookup table; by shifting in the correct pattern of bits, this functional unit can implement any single function of up to three inputs – the extension to lookup tables with larger numbers of inputs is clear. Typically, lookup tables are combined into clusters, as shown in Figure 14.2(b). Figure 14.3 shows clusters in two popular FPGA families. Figure 14.3(a) shows a cluster in the Altera Stratix device; Altera calls these clusters 'logic array blocks' [24]. Figure 14.3(b) shows a cluster in the Xilinx architecture [27]; Xilinx calls these clusters 'configurable logic blocks' (CLBs). In the Altera diagram, each block labelled 'LE' is a lookup table, while in the Xilinx diagram, each 'slice' contains two lookup tables. Other commercial FPGAs are described in References 3, 23, 25 and 52.

Reconfigurable fabrics containing lookup tables are very flexible, and can be used to implement any digital circuit. However, compared to the coarse-grained structures in the next subsection, these fine-grained structures have significantly more area, delay and power overhead. Recognising that these fabrics are often used for arithmetic purposes, FPGA companies have added additional features such as carry-chains and

Table 14.2 Comparison of reconfigurable fabrics and devices

Fabric or device	Fine grained or coarse grained	Base logic component	Routing architecture	Embedded memory	Special features
Actel ProASIC+ [23]	Fine	3-input block	Horizontal and vertical tracks	256 × 9 bit blocks	Flash-based
Altera Excalibur [47]	Fine	4-input lookup tables	Horizontal and vertical tracks	2 kbit memory blocks	ARMv4T embedded processor
Altera Stratix II [24]	Fine/coarse	8-input adaptive logic module	Horizontal and vertical tracks	512 bits, 4 kbits, and 512 kbit blocks	DSP blocks
Garp [32]	Fine	Logic or arithmetic functions on four 2-bit input words	2-bit Buses in horizontal and vertical columns	External to fabric	
Xilinx Virtex II Pro [48]	Fine	4-input lookup tables	Horizontal and vertical tracks	18 kbit blocks	Embedded multipliers, PowerPC 405 processor
Xilinx Virtex II [27]	Fine	4-input lookup tables	Horizontal and vertical tracks	18 kbit blocks	Embedded multipliers
DReAM [51]	Coarse	8-bit ALUs	16-bit local and global buses	Two 16 × 8 Dual Port memory	Targets mobile applications
Elixent D-fabrix [20]	Coarse	4-bit ALUs	4-bit buses	256 × 8 memory blocks	
HP Chess [33]	Coarse	4-bit ALUs	4-bit buses	256 × 8 bit memories	
IMEC ADRES [34]	Coarse	32-bit ALUs	32-bit buses	Small register files in each logic component	
Matrix [35]	Coarse	8-bit ALUs	Hierarchical 8-bit buses	256 × 8 bit memories	
MorphoSys [38]	Coarse	ALU and multiplier, and shift units	Buses	External to fabric	
Piperench [31]	Coarse	8-bit ALUs	8-bit Buses	External to fabric	Functional units arranged in 'stripes'
RaPiD [29]	Coarse	ALUs	Buses	Embedded memory blocks	
Silicon Hive Avispa [37]	Coarse	ALUs, shifters, accumulators, and multipliers	Buses	Five embedded memories	

(a)   Inputs

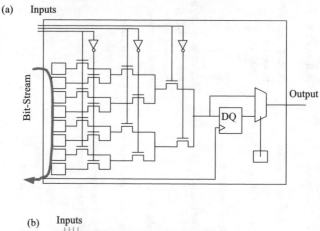

(b)   Inputs

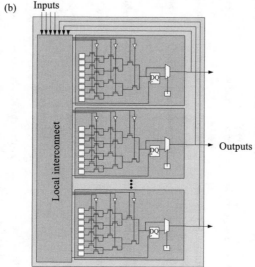

Figure 14.2   Fine-grained reconfigurable functional units: (a) three-input lookup table, (b) cluster of lookup tables

cascade-chains to reduce the overhead when implementing common arithmetic and logic functions. Figure 14.4 shows how the carry and cascade chains, as well as the ability to break a four-input lookup table into four two-input lookup tables can be exploited to efficiently implement carry-select adders [24]. The multiplexers and the exclusive-OR gate in Figure 14.4 are included as part of each logic array block, and need not be implemented using other lookup tables.

The example in Figure 14.4 shows how the efficiency of commercial FPGAs can be improved by adding architectural support for common functions. We can go much further than this though, and embed significantly larger, but far less flexible, reconfigurable functional units. There are two kinds of devices that contain coarse-grained

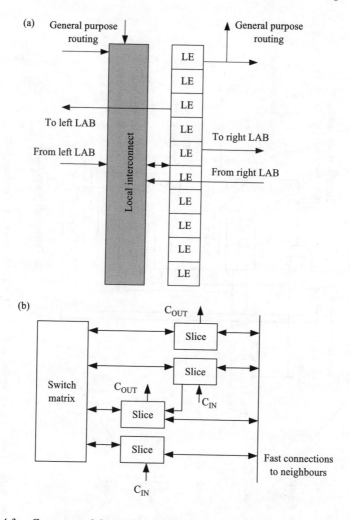

Figure 14.3 Commercial logic block architectures: (a) Altera logic array block [24], (b) Xilinx configurable logic block [27]

functional units; modern FPGAs, which are primarily composed of fine-grained functional units, are increasingly being enhanced by the inclusion of larger blocks. As an example, the Xilinx Virtex device contains embedded 18-bit by 18-bit multiplier units [27]. When implementing algorithms requiring a large amount of multiplication, these embedded blocks can significantly improve the density, speed and power of the device. On the other hand, for algorithms which do not perform multiplication, these blocks are rarely useful. The Altera Stratix devices contain a larger, but more flexible embedded block, called a DSP block, shown in Figure 14.5 [24]. Each of these blocks can perform accumulate functions as well as multiply operations. The comparison between the two devices clearly illustrates the flexibility and overhead

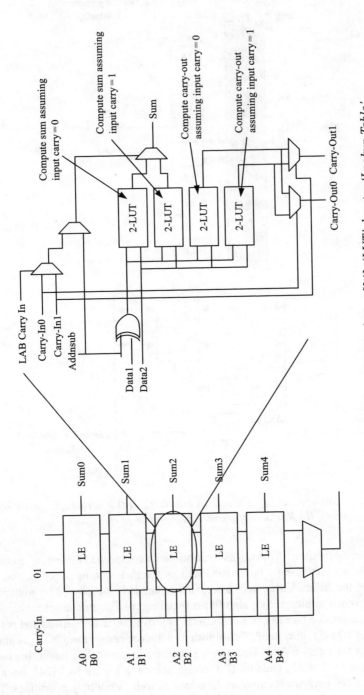

Figure 14.4 Implementing a carry-select adder in an Altera Stratix device [24]. 'LUT' denotes 'Lookup Table'

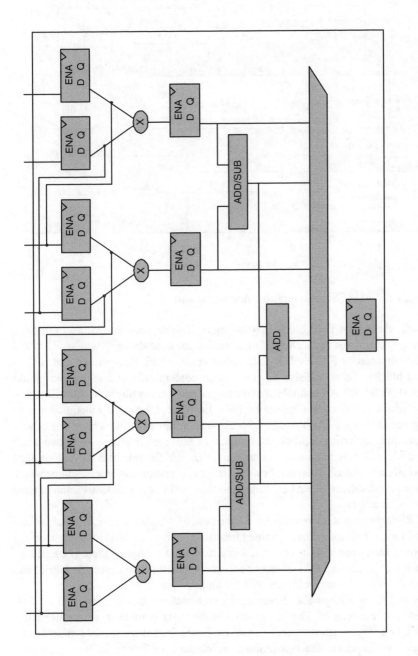

Figure 14.5   Altera DSP Block [24]

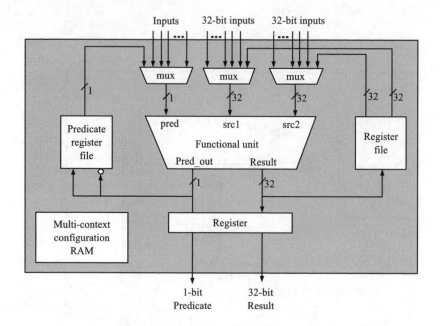

Figure 14.6   ADRES reconfigurable functional unit [34]

trade-off; the Altera DSP block may be more flexible than the Xilinx multiplier, however, it consumes more chip area and runs somewhat slower.

The commercial FPGAs described above contain both fine-grained and coarse-grained blocks. There are also devices which contain only coarse-grained blocks [28,29,31,33,34,38]. An example of a coarse-grained architecture is the ADRES architecture which is shown in Figure 14.6 [34]. Each reconfigurable functional unit in this device contains a 32-bit ALU which can be configured to implement one of several functions including addition, multiplication and logic functions, with two small register files. Clearly, such a functional unit is far less flexible than the fine-grained functional units described earlier; however if the application requires functions which match the capabilities of the ALU, these functions can be very efficiently implemented in this architecture.

2. Reconfigurable interconnects: Regardless of whether a device contains fine-grained functional units, coarse-grained functional units, or a mixture of the two, the functional units needed to be connected in a flexible way. Again, there is a trade-off between the flexibility of the interconnect (and hence the reconfigurable fabric) and the speed, area and power-efficiency of the architecture.

As before, reconfigurable interconnect architectures can be classified as fine-grained or coarse-grained. The distinction is based on the granularity with which wires are switched. This is illustrated in Figure 14.7, which shows a flexible interconnect between two buses. In the fine-grained architecture in Figure 14.7(a), each wire can be switched independently, while in Figure 14.7(b), the entire bus is switched as a unit. The fine-grained routing architecture in Figure 14.7(a) is more flexible,

⊠ Configuration bit

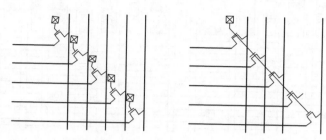

Figure 14.7   Routing architectures: (a) fine-grained and (b) coarse-grained

since not every bit needs to be routed in the same way; however, the coarse-grained architecture in Figure 14.7(b) contains far fewer programming bits, and hence suffers much less overhead.

Fine-grained routing architectures are usually found in commercial FPGAs. In these devices, the functional units are typically arranged in a grid pattern, and they are connected using horizontal and vertical channels. Significant research has been performed in the optimisation of the topology of this interconnect [53,54]. Coarse-grained routing architectures are commonly used in devices containing coarse-grained functional units. Figure 14.8 shows two examples of coarse-grained routing architectures: (a) the Totem reconfigurable system [28]; (b) the Silicon Hive reconfigurable system [37], which is less flexible but faster and smaller.

3. Emerging directions: Several emerging directions will be covered in the following. These directions include low power techniques, asynchronous architectures, and molecular microelectronics.

- *Low power techniques.* Early work explores the use of low-swing circuit techniques to reduce the power consumption in a hierarchical interconnect for a low-energy FPGA [55]. Recent work involves: (a) activity reduction in power-aware design tools, with energy saving of 23 per cent [56]; (b) leakage current reduction methods such as gate biasing and multiple supply-voltage integration, with up to two times leakage power reduction [57]; and (c) dual supply-voltage methods with the lower voltage assigned to non-critical paths, resulting in an average power reduction of 60 per cent [58].
- *Asynchronous architectures.* There is an emerging interest in asynchronous FPGA architectures. An asynchronous version of Piperench [31] is estimated to improve performance by 80 per cent, at the expense of a significant increase in configurable storage and wire count [59]. Other efforts in this direction include fine-grained asynchronous pipelines [60], quasi delay-insensitive architectures [61] and globally asynchronous locally synchronous techniques [62].
- *Molecular microelectronics.* In the long term, molecular techniques offer a promising opportunity for increasing the capacity and performance of reconfigurable computing architectures [63]. Current work is focused on developing programmable logic arrays based on molecular-scale nano-wires [64,65].

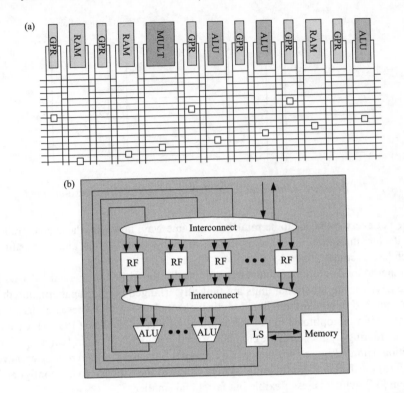

Figure 14.8   Example coarse-grained routing architectures: (a) Totem coarse-grained routing architecture [28] and (b) Silicon Hive coarse-grained routing architecture [37]

14.3.3   Architectures: main trends

The following summarises the main trends in architectures for reconfigurable computing.

1. Coarse-grained fabrics: As reconfigurable fabrics are migrated to more advanced technologies, the cost (in terms of both speed and power) of the interconnect part of a reconfigurable fabric is growing. Designers are responding to this by increasing the granularity of their logic units, thereby reducing the amount of interconnect needed. In the Stratix II device, Altera moved away from simple four-input lookup tables, and used a more complex logic block which can implement functions of up to seven inputs. We should expect to see a slow migration to more complex logic blocks, even in stand-alone FPGAs.

2. Heterogeneous functions: As devices are migrated to more advanced technologies, the number of transistors that can be devoted to the reconfigurable logic fabric increases. This provides new opportunities to embed complex non-programmable (or semi-programmable) functions, creating heterogeneous architectures with both general-purpose logic resources and fixed-function embedded blocks. Modern Xilinx parts have embedded 18 by 18-bit multipliers, while modern Altera parts have embedded DSP units which can perform a variety of multiply/accumulate functions. Again, we should expect to see a migration to more heterogeneous architectures in the near future.

3. Soft cores: The use of 'soft' cores, particularly for instruction processors, is increasing. A 'soft' core is one in which the vendor provides a synthesisable version of the function, and the user implements the function using the reconfigurable fabric. Although this is less area- and speed-efficient than a hard embedded core, the flexibility and the ease of integrating these soft cores makes them attractive. The extra overhead becomes less of a hindrance as the number of transistors devoted to the reconfigurable fabric increases. Altera and Xilinx both provide numerous soft cores, including soft instruction processors such as NIOS [11] and Microblaze [16]. Soft instruction processors have also been developed by a number of researchers, ranging from customisable JVM and MIPS processors [14] to ones specialised for machine learning [12] and data encryption [13].

14.4 Design methods

Hardware compilers for high-level descriptions are increasingly recognised to be the key to reducing the productivity gap for advanced circuit development in general, and for reconfigurable designs in particular. This section looks at high-level design methods from two perspectives: special-purpose design and general-purpose design. Low-level design methods and tools, covering topics such as technology mapping, floorplanning and place and route, are beyond the scope of this chapter – interested readers are referred to Reference 18.

14.4.1 General-purpose design

This section describes design methods and tools based on a general-purpose programming language such as C, possibly adapted to facilitate hardware development. Of course, traditional hardware description languages such as VHDL and Verilog are widely available, especially on commercial reconfigurable platforms.

A number of compilers from C to hardware have been developed. Some of the significant ones are reviewed here. These range from compilers which only target hardware, to those which target complete hardware/software systems; some also partition into hardware and software components.

We can classify different design methods into two approaches: the annotation and constraint-driven approach, and the source-directed compilation approach. The first approach preserves the source programs in C or C++ as much as possible and makes use of annotation and constraint files to drive the compilation process. The second approach modifies the source language to let the designer to specify, for instance, the amount of parallelism or the size of variables.

1. Annotation and constraint-driven approach: The systems mentioned below employ annotations in the source-code and constraint files to control the optimisation process. Their strength is that usually only minor changes are needed to produce a compilable program from a software description – no extensive re-structuring is required. Five representative methods are SPC [66], Streams-C [67], Sea Cucumber [68], SPARK [69] and Catapult-C [70].

SPC [66] combines vectorisation, loop transformations and retiming with automatic memory allocation to improve performance. SPC accelerates C loop nests with data dependency restrictions, compiling them into pipelines. Based on the SUIF framework [71], this approach uses loop transformations, and can take advantage of run-time reconfiguration and memory access optimisation. Similar methods have been advocated by other researchers [72,73].

Streams-C [67] compiles a C program to synthesisable VHDL. Streams-C exploits coarse-grained parallelism in stream-based computations; low-level optimisations such as pipelining are performed automatically by the compiler.

Sea Cucumber [68] compiles Java programs to hardware using a similar scheme to Handel-C, which we detail in the next section. Unlike Handel-C, no language extensions are needed; like Streams-C, users must call a library, in this case based on Communicating Sequential Processes (CSP [74]). Multiple circuit implementations of the library primitives enable trade-offs.

SPARK [69] is a high-level synthesis framework targeting multi-media and image processing. It compiles C code with the following steps: (a) list scheduling based on speculative code motions and loop transformations, (b) resource binding pass with minimisation of interconnect, (c) finite state machine controller generation for the scheduled datapath, (d) code generation producing synthesisable register-transfer level VHDL. Logic synthesis tools then synthesise the output.

Catapult C synthesises Register Transfer Level (RTL) descriptions from unannotated C++, using characterisations of the target technology from RTL synthesis tools [70]. Users can set constraints to explore the design space, controlling loop pipelining and resource sharing.

2. Source-directed compilation approach: A different approach adapts the source language to enable explicit description of parallelism, communication and other customisable hardware resources such as variable size. Examples of design methods following this approach include ASC [75], Handel-C [76], Haydn-C [77] and Bach-C [78].

ASC [75] adopts C++ custom types and operators to provide a C++ programming interface on the algorithm, architecture, arithmetic and gate levels. This enables the user to program on the desired level for each part of the application. Semi-automated design space exploration further increases design productivity, while supporting the optimisation process on all available levels of abstraction. The object-oriented design enables efficient code-reuse, and includes an integrated arithmetic unit generation library [79]. A floating-point library [80] provides over 200 different floating point units, each with custom bitwidths for mantissa and exponent.

Handel-C [76] extends a subset of C to support flexible width variables, signals, parallel blocks, bit-manipulation operations and channel communication. A distinctive feature is that timing of the compiled circuit is fixed at one cycle per

C statement. This allows Handel-C programmers to schedule hardware resources manually. Handel-C compiles to a 'one-hot' state machine using a token-passing scheme developed by Page and Luk [81]; each assignment of the program maps to exactly one control flip-flop in the state machine. These control flip-flops capture the flow of control (represented by the token) in the program: if the control flip-flop corresponding to a particular statement is active, then control has passed to that statement, and the circuitry compiled for that statement is activated. When the statement has finished execution, it passes the token to the next statement in the program.

Haydn-C [77] extends Handel-C for component-based design. Like Handel-C, it supports description of parallel blocks, bit-manipulation operations and channel communication. The principal innovation of Haydn-C is a framework of optional annotations to enable users to describe design constraints, and to direct source-level transformations such as scheduling and resource allocation. There are automated transformations so that a single high-level design can be used to produce many implementations with different trade-offs. This approach has been evaluated using various case studies, including FIR filters, fractal generators and morphological operators. The fastest morphological erosion design is 129 times faster and 3.4 times larger than the smallest design.

Bach-C [78] is similar to Handel-C but has an untimed semantics, only synchronising between parallel threads on synchronous communications between them, possibly giving greater scope for optimisation. It also allows asynchronous communications but otherwise resembles Handel-C, using the same basic one-hot compilation scheme.

Table 14.3 summarises the various compilers discussed in this section, showing their approach, source and target languages, target architecture and some example applications. Note that the compilers discussed are not necessarily restricted to the architectures reported; some can usually be ported to a different architecture by using a different library of hardware primitives.

14.4.2 Special-purpose design

Within the wide variety of problems to which reconfigurable computing can be applied, there are many specific problem domains which deserve special consideration. The motivation is to exploit domain-specific properties: (a) to describe the computation, such as using MATLAB for digital signal processing and (b) to optimise the implementation, such as using word-length optimisation techniques described later.

We shall begin with an overview of digital signal processing and relevant tools which target reconfigurable implementations. We then describe the word-length optimisation problem, the solution to which promises rich rewards; an example of such a solution will be covered. Finally we summarise other domain-specific design methods which have been proposed for video and image processing and networking.

Table 14.3 Summary of general-purpose hardware compilers

System	Approach	Source language	Target language	Target architecture	Example applications
Streams-C [67]	Annotation/ constraint-driven	C + library	RTL VHDL	Xilinx FPGA	Image contrast enhancement, pulsar detection [82]
Sea Cucumber [68]	Annotation/ constraint-driven	Java + library	EDIF	Xilinx FPGA	None given
SPARK [69]	Annotation/ constraint-driven	C	RTL VHDL	LSI, Altera FPGAs	MPEG-1 predictor, image tiling
SPC [66]	Annotation/ constraint-driven	C	EDIF	Xilinx FPGAs	String pattern matching, image skeletonisation
ASC [75]	Source-directed compilation	C++ using class library	EDIF	Xilinx FPGAs	Wavelet compression, encryption
Handel-C [76]	Source-directed compilation	Extended C	Structural VHDL, Verilog, EDIF	Actel, Altera Xilinx FPGAs	Image processing, polygon rendering [83]
Haydn-C [77]	Source-directed compilation	Extended C	Extended C (Handel-C)	Xilinx FPGAs	FIR filter, image erosion
Bach-C [78]	Source-directed compilation	Extended C	Behavioural and RTL VHDL	LSI FPGAs	Viterbi decoders, image processing

14.4.2.1 Digital signal processing

One of the most successful applications for reconfigurable computing is real-time Digital Signal Processing (DSP). This is illustrated by the inclusion of hardware support for DSP in the latest FPGA devices, such as the embedded DSP blocks in Altera Stratix II chips [24].

DSP problems tend to share the following properties: design latency is usually less of an issue than design throughput, algorithms tend to be numerically intensive but have very simple control structures, controlled numerical error is acceptable, and standard metrics, such as signal-to-noise ratio, exist for measuring numerical precision quality.

DSP algorithm design is often initially performed directly in a graphical programming environment such as Mathworks' MATLAB Simulink [84]. Simulink is widely used within the DSP community, and has been recently incorporated into the Xilinx System Generator [85] and Altera DSP builder [86] design flows. Design approaches such as this are based on the idea of data-flow graphs (DFGs) [87].

Tools working with this form of description vary in the level of user intervention required to specify the numerical properties of the implementation. For example, in the Xilinx System Generator flow [85], it is necessary to specify the number of bits used to represent each signal, the scaling of each signal (*namely* the binary point location) and whether to use saturating or wrap-around arithmetic [88].

Ideally, these implementation details could be automated. Beyond a standard DFG-based algorithm description, only one piece of information should be required: a lower-bound on the *output* signal to quantisation noise acceptable to the user. Such a design tool would thus represent a truly 'behavioural' synthesis route, exposing to the DSP engineer only those aspects of design naturally expressed in the DSP application domain.

14.4.2.2 The word-length optimisation problem

Unlike microprocessor-based implementations where the word-length is defined *a priori* by the hard-wired architecture of the processor, reconfigurable computing based on FPGAs allows the size of each variable to be customised to produce the best trade-offs in numerical accuracy, design size, speed and power consumption.

Given this flexibility, it is desirable to automate the process of finding a good custom data representation. The most important implementation decision to automate is the selection of an appropriate word-length and scaling for each signal [89] in a DSP system.

It has been argued that, often, the most efficient hardware implementation of an algorithm is one in which a wide variety of finite precision representations of different sizes are used for different internal variables [90]. The accuracy observable at the outputs of a DSP system is a function of the word-lengths used to represent all intermediate variables in the algorithm. However, accuracy is less sensitive to some variables than to others, as is implementation area. It is demonstrated in Reference 89 that by considering error and area information in a structured way using analytical and semi-analytical noise models, it is possible to achieve highly efficient DSP implementations.

In Reference 91 it has been demonstrated that the problem of word-length optimisation is NP-hard, even for systems with special mathematical properties that simplify the problem from a practical perspective [92]. There are, however, several published approaches to word-length optimisation. These can be classified as heuristics offering an area/signal quality trade-off [90,93,94], approaches that make some simplifying assumptions on error properties [93,95] or optimal approaches that can be applied to algorithms with particular mathematical properties [96].

Some published approaches to the word-length optimisation problem use an analytic approach to scaling and/or error estimation [94,97,98], some use simulation [93,95] and some use a hybrid of the two [99]. The advantage of analytic techniques is that they do not require representative simulation stimulus, and can be faster, however they tend to be more pessimistic. There is little analytical work on supporting data-flow graphs containing cycles, although in Reference 98 finite loop bounds are supported, while Reference 92 supports cyclic data-flow when the nodes are of a restricted set of types, extended to semi-analytic technique with fewer restrictions in Reference 100.

Some published approaches use worst-case instantaneous error as a measure of signal quality [94,95,97], whereas some use signal-to-noise ratio [90,93].

The remainder of this section reviews in some detail particular research approaches in the field.

The Bitwise Project [98] proposes propagation of integer variable ranges backwards and forwards through data-flow graphs. The focus is on removing unwanted most-significant bits (MSBs). Results from integration in a synthesis flow indicate that area savings of between 15 per cent and 86 per cent combined with speed increases of up to 65 per cent can be achieved compared to using 32-bit integers for all variables.

The MATCH Project [97] also uses range propagation through data-flow graphs, except variables with a fractional component are allowed. All signals in the model of Reference 97 must have equal fractional precision; the authors propose an analytic worst-case error model in order to estimate the required number of fractional bits. Area reductions of 80 per cent combined with speed increases of 20 per cent are reported when compared to a uniform 32-bit representation.

Wadekar and Parker [94] have also proposed a methodology for word-length optimisation. Like Reference 97, this technique also allows controlled worst-case error at system outputs, however each intermediate variable is allowed to take a word-length appropriate to the sensitivity of the output errors to quantisation errors on that particular variable. Results indicate area reductions of between 15 per cent and 40 per cent over the optimum uniform word-length implementation.

Kum and Sung [93] and Cantin et al. [95] have proposed several word-length optimisation techniques to trade-off system area against system error. These techniques are heuristics based on bit-true simulation of the design under various internal word-lengths.

In Bitsize [101,102], Abdul Gaffar *et al.* propose a hybrid method based on the mathematical technique know as automatic differentiation to perform bitwidth optimisation. In this technique, the gradients of outputs with respect to the internal variables are calculated and then used to determine the sensitivities of the outputs to the precision of the internal variables. The results show that it is possible to achieve an area reduction of 20 per cent for floating-point designs, and 30 per cent for fixed-point designs, when given an output error specification of 0.75 per cent against a reference design.

A useful survey of algorithmic procedures for word-length determination has been provided by Cantin *et al.* [103]. In this work, existing heuristics are classified under various categories. However the 'exhaustive' and 'branch-and-bound' procedures described in Reference 103 do not necessarily capture the optimum solution to the word-length determination problem, due to non-convexity in the constraint space: it is actually possible to have a *lower* error at a system output by reducing the word-length at an internal node [104]. Such an effect is modelled in the MILP approach proposed in Reference 96.

A comparative summary of existing optimisation systems is provided in Table 14.4. Each system is classified according to the several defining features described below.

- Is the word-length and scaling selection performed through analytic or simulation-based means?
- Can the system support algorithms exhibiting cyclic data flow? (Such as infinite impulse response filters.)
- What mechanisms are supported for Most Significant Bit (MSB) optimisations? (Such as ignoring MSBs that are known to contain no useful information, a technique determined by the scaling approach used.)
- What mechanisms are supported for Least Significant Bit (LSB) optimisations? These involve the monitoring of word-length growth. In addition, for those systems which support error-trade-offs, further optimisations include the quantisation (truncation or rounding) of unwanted LSBs.
- Does the system allow the user to trade-off numerical accuracy for a more efficient implementation?

14.4.2.3 An example optimisation flow

One possible design flow for word-length optmisation, used in the Right-Size system [100], is illustrated in Figure 14.9 for Xilinx FPGAs. The inputs to this system are a specification of the system behaviour (e.g. using Simulink), a specification of the acceptable signal-to-noise ratio at each output and a set of representative input signals. From these inputs, the tool automatically generates a synthesisable structural description of the architecture and a bit-true behavioural VHDL testbench, together with a set of expected outputs for the provided set of representative inputs. Also

Table 14.4 A comparison of wordlength and scaling optimisation systems and methods

System	Analytic/ simulation	Cyclic data flow?	MSB-optimisation	LSB-optimisations	Error trade-off	Comments
Benedetti [105]	Analytic	None	Through interval arithmetic	Through 'multi-interval' approach	No error	Can be pessimistic
Stephenson [98,106]	Analytic	For finite loop bounds	Through forward and backward range propagation	None	No error	Less pessimistic than [105] due to backwards propagation
Nayak [97]	Analytic	Not supported for error analysis	Through forward and backward range propagation	Through fixing number of fractional bits for all variables	User-specified or inferred absolute bounds on error	Fractional parts have equal wordlength
Wadekar [94]	Analytic	None	Through forward range propagation	Through genetic algorithm search for suitable wordlengths	User-specified absolute bounds	Uses Taylor series at limiting values to determine error propagation
Keding [107,108]	Hybrid	With user intervention	Through user-annotations and forward range propagation	Through user-annotations and forward wordlength propagation	Not automated	Possible truncation error
Cmar [99]	Hybrid for scaling simulation for error	With user intervention only	Through combined simulation and forward range propagation	Wordlength bounded through hybrid fixed or floating simulation	Not automated	Less pessimistic than Reference 105 due to input error propagation

Kum [93,109–111]	Simulation (hybrid for multiply-accumulate signals in References 93 and 109)	Yes	Through measurement of variable mean and standard deviation	Through heuristics based on simulation results	User-specified bounds and metric	Long simulation time possible
Constantinides [89]	Analytic	Yes	Through tight analytic bounds on signal range and automatic design of saturation arithmetic	Through heuristics based on an analytic noise model	User-specified bounds on noise power and spectrum	Only applicable to linear time-invariant systems
Constantinides [100]	Hybrid	Yes	Through simulation	Through heuristics based on a hybrid noise model	User-specified bounds on noise power and spectrum	Only applicable to differentiable non-linear systems
Abdul Gaffar [101,102]	Hybrid	With user intervention	Through simulation based range propagation	Through automatic differentiation based dynamic analysis	User-specified bounds and metric	Covers both fixed-point and floating-point

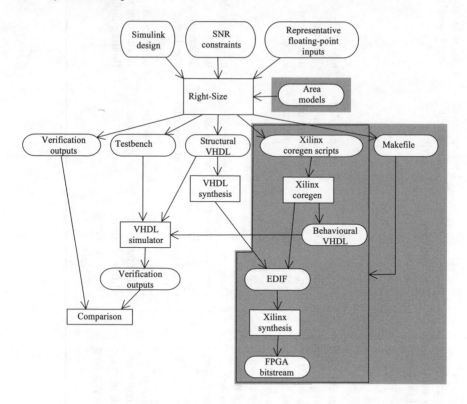

Figure 14.9   Design flow for the Right-Size tool [100]. The shaded portions are FPGA vendor-specific

generated is a makefile which can be used to automate the post-Right-Size synthesis process.

Application of Right-Size to various adaptive filters implemented in a Xilinx Virtex FPGA has resulted in area reduction of up to 80 per cent, power reduction of up to 98 per cent, and speed-up of up to 36 per cent over common alternative design methods without word-length optimisation.

14.4.2.4   Other design methods

Besides signal processing, video and image processing is another area that can benefit from special-purpose design methods. Three examples will be given to provide a flavour of this approach. First, the CHAMPION system [112] maps designs captured in the Cantata graphical programming environment to multiple reconfigurable computing platforms. Second, the IGOL framework [113] provides a layered architecture for facilitating hardware plug-ins to be incorporated in various applications in the Microsoft Windows operating system, such as Premiere, Winamp, VirtualDub and DirectShow. Third, the SA-C compiler [114] maps a high-level single-assignment language specialised for image processing description into hardware, using various optimisation methods including loop unrolling, array value propagation, loop-carried array elimination, and multi-dimensional stripmining.

Recent work indicates that another application area that can benefit from special-purpose techniques is networking. Two examples will be given. First, a framework has been developed to enable description of designs in the network policy language Ponder [115], into reconfigurable hardware implementations [116]. Second, it is shown [117] how descriptions in the Click networking language can produce efficient reconfigurable designs.

14.4.3 Other design methods

In the following, we describe various design methods in brief.

1. Runtime customisation: Many aspects of runtime reconfiguration have been explored [18], including the use of directives in high-level descriptions [118]. Effective runtime customisation hinges on appropriate design-time preparation for such customisation. To illustrate this point, consider a runtime customisable system that supports partial reconfiguration: one part of the system continues to be operational, while another part is being reconfigured. As FPGAs get larger, partial reconfiguration is becoming increasingly important as a means of reducing reconfiguration time. To support partial reconfiguration, appropriate circuits must be built at fabrication time as part of the FPGA fabric. Then at compile time, an initial configuration bitstream and incremental bitstreams have to be produced, together with runtime customisation facilities which can be executed, for instance, on a microprocessor serving as part of the runtime system [119]. Runtime customisation facilities can include support for condition monitoring, design optimisation and reconfiguration control.

Opportunities for runtime design optimisation include: (a) runtime constant propagation [120], which produces a smaller circuit with higher performance by treating runtime data as constant, and optimising them principally by Boolean algebra; (b) library-based compilation – the DISC compiler [121] makes use of a library of precompiled logic modules which can be loaded into reconfigurable resources by the procedure call mechanism; (c) exploiting information about program branch probabilities [122]; the idea is to promote utilisation by dedicating more resources to branches which execute more frequently. A hardware compiler has been developed to produce a collection of designs, each optimised for a particular branch probability; the best can be selected at runtime by incorporating observed branch probability information from a queueing network performance model.

2. Soft instruction processors: FPGA technology can now support one or more soft instruction processors implemented using reconfigurable resources on a single chip; proprietary instruction processors, such as MicroBlaze and Nios, are now available from FPGA vendors. Often such processors support customisation of resources and custom instructions. Custom instructions have two main benefits. First, they reduce the time for instruction fetch and decode, provided that each custom instruction replaces several regular instructions. Second, additional resources can be assigned to a custom instruction to improve performance. Bit-width optimisation, described in Section 14.4.2, can also be applied to customise instruction processors at compile time. A challenge of customising instruction processors is that the tools for producing and analysing instructions also need to be customised. For instance, the *flexible instruction processor* framework [14] has been developed to automate the steps in customising an instruction processor and the corresponding tools. Other researchers have proposed similar approaches [123].

Instruction processors can also run declarative langauges. For instance, a scalable architecture [12], consisting of multiple processors based on the Warren Abstract Machine, has been developed to support the execution of the Progol system [124], based on the declarative language Prolog. Its effectiveness has been demonstrated using the mutagenesis data set containing 12000 facts about chemical compounds.

3. Multi-FPGA compilation: Peterson *et al.* [125] have developed a C compiler which compiles to multi-FPGA systems. The available FPGAs and other units are specified in a library file, allowing portability. The compiler can generate designs using speculative and lazy execution to improve performance and ultimately they aim to partition a single program between host and reconfigurable resource (hardware/ software codesign). Duncan *et al.* [126] have developed a system with similar capabilities. This is also retargetable, using hierarchical architecture descriptions. It synthesises a VLIW architecture that can be partitioned across multiple FPGAs. Both methods can split designs across several FPGAs, and are retargetable via hardware description libraries. Other C-like languages that have been developed include MoPL-3, a C extension supporting data procedural compilation for the Xputer architecture which comprises an array of reconfigurable ALUs [127], and spC, a systolic parallel C variant for the Enable++ board [128].

4. Hardware/software codesign: Several research groups have studied the problem of compiling C code to both hardware and software. The Garp compiler [129] is intended to accelerate plain C, with no annotations to help the compiler, making it more widely applicable. The work targets one architecture only: the Garp chip, which integrates a RISC core and reconfigurable logic. This compiler also uses the SUIF framework. The compiler uses a technique first developed for VLIW architectures called hyperblock scheduling, which optimises for instruction-level parallelism across several common paths, at the expense of rarer paths. Infeasible or rare paths are implemented on the processor with the more common, easily parallelisable paths synthesised into logic for the reconfigurable resource. Similarly, the NAPA C compiler targets the NAPA architecture [130], which also integrates a RISC processor reconfigurable logic. This compiler can also work on plain C code but the programmer can add C pragmas to indicate large-scale parallelism and the bit-widths of variables to the code. The compiler can synthesise pipelines from loops.

5. Annotation-free compilation: Some researchers aim to compile a sequential program, without any annotations, into efficient hardware design. This requires analysis of the source program to extract parallelism for an efficient result, which is necessary if compilation from languages such as C is to compete with traditional methods for designing hardware. One example is the work of Babb *et al.* [131], targeting custom, fixed-logic implementation while also applicable to reconfigurable hardware. The compiler uses the SUIF infrastructure to do several analyses to find what computations affect exactly what data, as far as possible. A tiled architecture is synthesised, where all computation is kept as local as possible to one tile. More recently, Ziegler *et al.* [132] have used loop transformations in mapping loop nests onto a pipeline spanning several FPGAs. A further effort is given by the Garp project [129].

14.4.4 Emerging directions

1. Verification: As designs are becoming more complex, techniques for verifying their correctness are becoming increasingly important. Four approaches are described: (1) the InterSim framework [133] provides a means of combining software simulation and hardware prototyping. (2) The Lava system [134] can convert designs into a form suitable for input to a model checker; a number of FPGA design libraries have been verified in this way [135]. (3) The Ruby language [136] supports correctness-preserving transformations, and a wide variety of hardware designs have been produced. Fourth, the Pebble [137] hardware design language has been formally specified [138], so that provably correct design tools can be developed.

2. Customisable hardware compilation: Recent work [139] explains how customisable frameworks for hardware compilation can enable rapid design exploration, and reusable and extensible hardware optimisation. The framework compiles a parallel imperative language like Handel-C, and supports multiple levels of design abstraction, transformational development, optimisation by compiler passes and metalanguage facilities. The approach has been used in producing designs for applications such as signal and image processing, with different trade-offs in performance and resource usage.

14.4.5 Design methods: main trends

We summarise the main trends in design methods for reconfigurable computing below.

1. Special-purpose design: As explained earlier, special-purpose design methods and tools enable both high-level design as well as domain-specific optimisation. Existing methods, such as those compiling MATLAB Simulink descriptions into reconfigurable computing implementations [85,86,97,100,101,140], allow application developers without electronic design experience to produce efficient hardware implementations quickly and effectively. This is an area that would assume further importance in future.

2. Low power design: Several hardware compilers aim to minimise the power consumption of their generated designs. Examples include special-purpose design methods such as Right-Size [100] and PyGen [140], and general-purpose methods that target loops for configurable hardware implementation [9]. These design methods, when combined with low power architectures [58] and power-aware low-level tools [56], can provide significant reduction in power consumption.

3. *High-level transformations* Many hardware design methods [66,69,114] involve high-level transformations: loop unrolling, loop restructuring and static single assignment are three examples. The development of powerful transformations for design optimisation will continue for both special-purpose and general-purpose designs.

14.5 Summary

This chapter surveys two aspects of reconfigurable computing: architectures and design methods. The main trends in architectures are coarse-grained fabrics, heterogeneous functions and soft cores. The main trends in design methods are special-purpose design methods, low power techniques and high-level transformations. We wonder what a survey of reconfigurable computing, written in 2015, will cover?

Acknowledgements

Our thanks to Ray Cheung and Sherif Yusuf for their support in preparing this chapter. The support of Celoxica, Xilinx and UK EPSRC (Grant number GR/R 31409, GR/R 55931, GR/N 66599) is gratefully acknowledged.

References

1 LUK, W.: 'Customising processors: design-time and run-time opportunities', in PIMENTEL, A.D., and VASSILIADIS, S.: 'Computer systems: architectures, modeling, and simulation', LNCS 3133 (Springer, Berlin, 2004)

2 TELLE, N., CHEUNG, C.C., and LUK, W.: 'Customising hardware designs for elliptic curve cryptography', in PIMENTEL, A.D., and VASSILIADIS, S.: 'Computer systems: architectures, modeling, and simulation' LNCS 3133 (Springer, Berlin, 2004)

3 MORRIS, K.: 'Virtex 4: Xilinx details its next generation', *FPGA and Programmable Logic Journal*, June 2004

4 XILINX, Inc., *Logic Design Products*, http://www.xilinx.com/products/design_resources/design_tool/grouping/logic_design_prod.htm

5 CELOXICA, *DK Design Suite* http://www.celoxica.com/products/ tools/dk.asp

6 Synplicity, Synplify Pro, http://www.synplicity.com/products/synplifypro/index.html

7 GUO, Z., NAJJAR, W., VAHID, F., and VISSERS, K.: 'A quantitative analysis of the speedup factors of FPGAs over processors'. Proceedings of the international symposium on *FPGAs* (ACM Press, New York, 2004)

8 UNDERWOOD, K.: 'FPGAs vs. CPUs: trends in peak floating-point performance'. Proceedings of the international symposium on *FPGAs* (ACM Press, New York, 2004)

9 STITT, G., VAHID, F., and NEMATBAKHSH, S.: 'Energy savings and speedups from partitioning critical software loops to hardware in embedded systems', *ACM Transactions on Embedded Computing Systems*, 2004, 3(1), pp. 218–232

10 VEREEN, L.: 'Soft FPGA cores attract embedded developers' *Embedded Systems Programming*, 23 April 2004, http://www.embedded.com/ showArticle.jhtml?articleID=19200183

11 Altera Corp.: Nios II processor reference handbook. May 2004

12 FIDJELAND, A., LUK, W., and MUGGLETON, S.: 'Scalable acceleration of inductive logic programs'. Proceedings of the international conference on *Field-programmable technology* (IEEE, New York, 2002)

13 LEONG, P.H.W., and LEUNG, K.H.: 'A microcoded elliptic curve processor using FPGA technology', *IEEE Transactions on Very Large Scale Integration Systems*, 2002, 10(5), pp. 550–559

14 SENG, S.P., LUK, W., and CHEUNG, P.Y.K.: 'Flexible instruction processors'. Proceedings of the international conference on *Compilers, arch. and syn. for embedded systens* (ACM Press, New York, 2000)

15 SENG, S.P., LUK, W., and CHEUNG, P.Y.K.: 'Run-time adaptive flexible instruction processors', in GLESNER, M., ZIPF, P., and RENOVELL, M.: 'Field-programmable logic and applications', LNCS 2438 (Springer, Berlin, 2002)

16 XILINX, Inc., *Microblaze Processor Reference Guide*, June 2004

17 BONDALAPATI, K., and PRASANNA, V.K.: 'Reconfigurable computing systems', *Proceedings of the IEEE*, 2002, 90(7), pp. 1201–1217

18 COMPTON, K., and HAUCK, S.: 'Reconfigurable computing: a survey of systems and software', *ACM Computing Surveys*, 2002, 34(2), pp. 171–210

19 LUK, W., CHEUNG, P.Y.K., and SHIRAZI, N.: 'Configurable computing', in CHEN, W.K.: (Ed.): 'Electrical engineer's handbook' (Academic Press, New York, 2004)

20 SCHAUMONT, P., VERBAUWHEDE, I., KEUTZER, K., and SARRAFZADEH, M.: 'A quick safari through the reconfiguration jungle'. Proceedings of the *Design automation* conference (ACM Press, 2001)

21 TESSIER, R., and BURLESON, W.: 'Reconfigurable computing and digital signal processing: a survey', *Journal of VLSI Signal Processing*, May/June 2001, 28, pp. 7–27

22 SAXE, T., and FAITH, B.: ' Less is more with FPGAs', *EE Times*, September 13, 2004. http://www.eetimes.com/showArticle.jhtml?articleID=47203801

23 Actel Corp.: ProASIC Plus Family Flash FPGAs. v3.5, April 2004

24 Altera Corp.: Stratix II device handbook, February 2004

25 Lattice Semiconductor Corp, *ispXPGA Family*, January 2004

26 CANTLE, A.J., DEVLIN, M., LORD, E., and CHAMBERLAIN, R.: 'High frame rate, low latency hardware-in-the-loop image generation? An illustration of the Particle Method and DIME'. Proceedings of the *Aerosense* conference (SPIE, Bellingham, WA, 2004)

27 XILINX, Inc., *Virtex II Datasheet*, June 2004

28 COMPTON, K., and HAUCK, S., 'Totem: Custom reconfigurable array generation'. Proceedings of the symposium on *Field-programmable custom computing machines* (IEEE Computer Society Press, Washington, 2001)

29 EBELING, C., CONQUIST, D., and FRANKLIN, P.: 'RaPiD – reconfigurable pipelined datapath', in HARTENSTEIN, R.W., and GLESNER, M.: 'Field-programmable logic and applications', LNCS 1142 (Springer, Berlin, 1996)

30 Elixent Corporation, *DFA 1000 Accelerator Datasheet*, 2003

31 GOLDSTEIN, S.C., SCHMIT, H., BUDIU, M., CADAMBI, S., MOE, M., and TAYLOR, R.: 'PipeRench: a reconfigurable architecture and compiler', *IEEE Computer*, 2000, **33**(4), pp. 70–77

32 HAUSER, J.R., and WAWRZYNEK, J.: 'Garp: a MIPS processor with a reconfigurable processor'. IEEE symposium on *Field-programmable custom computing machines* (IEEE Computer Society Press, Washington, 1997)

33 MARSHALL, A., STANSFIELD, T., KOSTARNOV, I., VUILLEMIN, J., and HUTCHINGS, B.: 'A reconfigurable arithmetic array for multimedia applications'. ACM/SIGDA international symposium on *FPGAs*, 1999, pp. 135–143

34 MEI, B., VERNALDE, S., VERKEST, D., De MAN, H., and LAUWEREINS, R.: 'ADRES: An architecture with tightly coupled VLIW processor and coarse-grained reconfigurable matrix', in CHEUNG, P.Y.K., CONSTANTINIDES, G.A., and DE SOUSA, J. T.: 'Field-programmable logic and applications', LNCS 2778 (Springer, Berlin, 2003)

35 MIRSKY, E., and DEHON, A.: 'MATRIX: a reconfigurable computing architecture with configurable instruction distribution and deployable resources'. Proceedings of the symposium on *Field-programmable custom computing machines* (IEEE Computer Society Press, Washington, 1996)

36 RUPP, C.R., LANDGUTH, M., GARVERICK, T. *et al.*: 'The NAPA adaptive processing architecture'. IEEE symposium on *Field-programmable custom computing machines*, IEEE Computer Society, CA, May 1998, pp. 28–37

37 Silicon Hive: 'Avispa Block Accelerator', Product Brief, 2003

38 SINGH, H., LEE, M-H., LU, G., KURDAHI, F., BAGHERZADEH, N., and CHAVES, E.: 'MorphoSys: An integrated reconfigurable system for data-parallel and compute intensive applications', *IEEE Transactions on Computers*, 2000, **49**(5), pp. 465–481

39 TAYLOR, M. et al.: 'The RAW microprocessor: a computational fabric for software circuits and general purpose programs', *IEEE Micro*, 2002, **22**(2), pp. 25–35

40 Cadence Design Systems Inc: 'Palladium datasheet', 2004

41 Mentor Graphics, *Vstation Pro*: 'High performance system verification', 2003

42 Annapolis Microsystems, Inc.: 'Wildfire reference manual', 1998

43 LAUFER, R., TAYLOR, R., and SCHMIT, H.: 'PCI-PipeRench and the SwordAPI: a system for stream-based reconfigurable computing'. Proceedings of the symposium on *Field-programmable custom computing machines* (IEEE Computer Society Press, Washington, 1999)

44 VUILLEMIN, J., BERTIN, P., RONCIN, D., SHAND, M., TOUATI, H., and BOUCARD P.: 'Programmable active memories: reconfigurable systems come of age', *IEEE Transactions on VLSI Systems*, 1996, **4**(1), pp. 56–69

45 WITTIG, R.D., and CHOW, P.: 'OneChip: an FPGA processor with reconfigurable logic'. IEEE symposium on *FPGAs for custom computing machines*, 1996

46 RAZDAN, R., and SMITH, M.D.: 'A high performance microarchitecture with hardware programmable functional units'. International symposium on *Microarchitecture* (IEEE, New York, 1994) pp. 172–180

47 Altera Corp.: Excalibur device overview. May 2002

48 Xilinx, Inc., *PowerPC 405 Processor Block Reference Guide*, October 2003

49 Celoxica, *RC2000 Development and evaluation board data sheet*

50 LEONG, P., LEONG, M., CHEUNG, O. *et al.*: 'Pilchard – A reconfigurable computing platform with memory slot interface'. Proceedings of the symposium on *Field-programmable custom computing machines* (IEEE Computer Society Press, Washington, 2001)

51 BECKER, J., and GLESNER, M.: 'A parallel dynamically reconfigurable architecture designed for flexible application-tailored hardware/software systems in future mobile communication', *The Journal of Supercomputing*, 2001, **19**(1), pp. 105–127

52 Quicklogic Corp., *Eclipse-II Family Datasheet*, January 2004

53 BETZ, V., ROSE, J., and MARQUARDT, A.: 'Architecture and CAD for deep-submicron FPGAs' (Kluwer Academic Publishers, New York, 1999)

54 LEMIEUX, G., and LEWIS, D.: 'Design of interconnect networks for programmable logic' (Kluwer Academic Publishers, New York, 2004)

55 GEORGE, V., ZHANG, H., and RABAEY, J.: 'The design of a low energy FPGA'. Proceedings of the international symposium on *Low power electronics and design*, 1999.

56 LAMOUREUX, J., and WILTON, S.J.E.: 'On the interaction between power-aware FPGA CAD algorithms'. International conference on *Computer-aided design* (IEEE, 2003)

57 RAHMAN, A., and POLAVARAPUV, V.: 'Evaluation of low-leakage design techniques for field programmable gate arrays'. Proceedings of the international symposium on *Field-programmable gate arrays* (ACM Press, New York, 2004)

58 GAYASEN, A., LEE, K., VIJAYKRISHNAN, N., KANDEMIR, M., IRWIN, M.J., and TUAN, T.: 'A dual-V_{DD} low power FPGA architecture', in BECKER, J., PLATZNER, M., and VERNALDE, S.: 'Field programmable logic and applications', LNCS Series (Springer, Berlin, 2004)

59 KAGOTANI, H., and SCHMIT, H.: 'Asynchronous PipeRench: architecture and performance evaluations'. Proceedings of the symposium on *Field-programmable custom computing machines* (IEEE Computer Society Press, Washington, 2003)

60 TEIFE, J., and MANOHAR, R.: 'Programmable asynchronous pipeline arrays', in CHEUNG, P.Y.K., CONSTANTINIDES, G.A., and DE SOUSA, J.T.:

'Field programmable logic and applications', LNCS 2778 (Springer, Berlin, 2003)

61 WONG, C.G., MARTIN, A.J., and THOMAS, P.: 'An architecture for asynchronous FPGAs'. Proceedings of the international conference on *Field-programmable technology* (IEEE, New York, 2003)

62 ROYAL, A., and CHEUNG, P.Y.K.: 'Globally asynchronous locally synchronous FPGA architectures', in CHEUNG, P.Y.K., CONSTANTINIDES, G. A., and DE SOUSA, J.T.: 'Field programmable logic and applications', LNCS 2778 (Springer, Berlin, 2003)

63 BUTTS, M., DEHON, A., and GOLDSTEIN, S.: 'Molecular electronics: devices, systems and tools for gigagate, gigabit chips'. Proceedings of the international conference on *Computer-aided design* (IEEE, New York, 2002)

64 DeHon, A., and WILSON, M.J.: 'Nanowire-based sublithographic programmable logic arrays'. Proceedings of the international symposium on *FPGAs* (ACM Press, New York, 2004)

65 WILLIAMS, R.S., and KUEKES, P.J.: 'Molecular nanoelectronics'. Proceedings of the international symposium on *Circuits and systems* (IEEE, New York, 2000)

66 WEINHARDT, M., and LUK, W.: 'Pipeline vectorization', *IEEE Transactions on Computer-Aided Design*, 2001, **20**(2), pp. 234–248

67 GOKHALE, M., STONE, J.M., ARNOLD, J., and KALINOWSKI, M.: 'Stream-oriented FPGA computing in the Streams-C high level language'. Proceedings of the symposium on *Field-programmable custom computing machines* (IEEE Computer Society Press, Washington, 2000)

68 JACKSON, P.A., HUTCHINGS, B.L., and TRIPP, J.L.: 'Simulation and synthesis of CSP-based interprocess communication'. Proceedings of the symposium on *Field-programmable custom computing machines* (IEEE Computer Society Press, Washington, 2003)

69 GUPTA, S., DUTT, N.D., GUPTA, R.K., and NICOLAU A.: 'SPARK: a high-level synthesis framework for applying parallelizing compiler transformations'. Proceedings of the international conference on *VLSI design*, January 2003

70 McCLOUD S.: 'Catapult C synthesis-based design flow: speeding implementation and increasing flexibility'. White paper, Mentor Graphics, 2004

71 WILSON, R.P., FRENCH, R.S., WILSON, C.S. *et al.*: 'SUIF: an infrastructure for research on parallelizing and optimizing compilers', *ACM SIGPLAN Noticies*, 1994, **29**(12), pp. 31–37

72 HARRISS, T., WALKE, R., KIENHUIS, B., and DEPRETTERE, E.: 'Compilation from Matlab to process networks realized in FPGA', *Design Automation of Embedded Systems*, 2002, **7**(4), pp. 385–403

73 SCHREIBER, R. *et al.*: 'PICO-NPA: High-level synthesis of nonprogrammable hardware accelerators', *Journal of VLSI Signal Processing Systems*, 2002, **31**(2)

74 HOARE, C.A.R.: 'Communicating sequential processes' (Prentice Hall, New York, 1985)

75 MENCER, O., PEARCE, D.J., HOWES, L.W., and LUK, W.: 'Design space exploration with a stream compiler'. Proceedings of the international conference on *Field programmable technology* (IEEE, New York, 2003)

76 Celoxica: 'Handel-C language reference manual for DK2.0', Document RM-1003-4.0, 2003

77 DE FIGUEIREDO COUTINHO J.G., and LUK, W.: 'Source-directed transformations for hardware compilation'. Proceedings of the international conference on *Field-programmable technology* (IEEE, New York, 2003)

78 YAMADA, A., NISHIDA, K., SAKURAI, R., KAY, A., NOMURA, T., and KAMBE, T.: 'Hardware synthesis with the Bach system'. Proceedings of ISCAS (IEEE, New York, 1999)

79 MENCER, O.: 'PAM-Blox II: design and evaluation of C++ module generation for computing with FPGAs'. Proceedings of the symposium on *Field-programmable custom computing machines* (IEEE Computer Society Press, Washington, 2002)

80 LIANG, J., TESSIER, R., and MENCER, O.: 'Floating point unit generation and evaluation for FPGAs'. Proceedings of the symposium on *Field-programmable custom computing machines* (IEEE Computer Society Press, Washington, 2003)

81 PAGE, I., and LUK, W.: 'Compiling occam into FPGAs', *FPGAs* (Abingdon EE&CS Books, UK, 1991)

82 FRIGO, J., PALMER, D., GOKHALE, M., and POPKIN, M.-Paine, 'Gamma-ray pulsar detection using reconfigurable computing hardware'. Proceedings of the symposium on *Field programmable custom computing machines* (IEEE Computer Society Press, Washington, 2003)

83 STYLES, H., and LUK, W.: 'Customising graphics applications: techniques and programming interface'. Proceedings of the symposium on *Field-programmable custom computing machines* (IEEE Computer Society Press, Washington, 2000)

84 SIMULINK, http://www.mathworks.com

85 HWANG, J., MILNE, B., SHIRAZI, N., and STROOMER, J.D.: 'System level tools for DSP in FPGAs', in BREBNER, G., and WOODS, R.: 'Field-programmable logic and applications', LNCS 2147 (Springer, 2001)

86 Altera Corp.: 'DSP builder user guide'. Version 2.1.3 rev.1, July 2003

87 LEE, E.A., and MESSERSCHMITT, D.G.: 'Static scheduling of synchronous data flow programs for digital signal processing', *IEEE Transactions on Computers*, 1987

88 CONSTANTINIDES, G.A., CHEUNG, P.Y.K., and LUK, W.: 'Optimum and heuristic synthesis of multiple wordlength architectures', IEEE Transactions on *Computer Aided Design of Integrated Circuits and Systems*, 2003, **22**(10), pp. 1432–1442

89 CONSTANTINIDES, G.A., CHEUNG, P.Y.K., and LUK, W.: 'Synthesis and optimization of DSP algorithms' (Kluwer Academic, Dordrecht, 2004)

90 CONSTANTINIDES, G.A., CHEUNG, P.Y.K., and LUK, W.: 'The multiple wordlength paradigm'. Proceedings of the symposium on *Field-programmable*

custom computing machines (IEEE Computer Society Press, Washington, 2001)

91 CONSTANTINIDES, G.A., and WOEGINGER, G.J.: 'The complexity of multiple wordlength assignment', *Applied Mathematics Letters*, 2002, **15**(2), pp. 137–140

92 CONSTANTINIDES, G.A., CHEUNG, P.Y.K., and LUK, W.: 'Synthesis of saturation arithmetic architectures', *ACM Transactions on Design Automation of Electronic Systems*, 2003, **8**(3), pp. 334–354

93 KUM, K.-I., and SUNG, W.: 'Combined word-length optimization and high-level synthesis of digital signal processing systems', *IEEE Transactions on Computer Aided Design*, 2001, **20**(8), pp. 921–930

94 WADEKAR, S.A., and PARKER, A.C.: 'Accuracy sensitive word-length selection for algorithm optimization'. Proceedings of the international conference on *Computer design*, 1998

95 CANTIN, M.-A., SAVARIA, Y., and LAVOIE, P.: 'An automatic word length determination method'. Proceedings of the IEEE international symposium on *Circuits and systems*, 2001, pp. V-53–V-56

96 CONSTANTINIDES, G.A., CHEUNG, P.Y.K., and LUK, W.: 'Optimum wordlength allocation'. Proceedings of the symposium on *Field-programmable custom computing machines* (IEEE Computer Society Press, Washington, 2002)

97 NAYAK, A., HALDAR, M., CHOUDHARY, A., and BANERJEE, P.: 'Precision and error analysis of MATLAB applications during automated hardware synthesis for FPGAs'. Proceedings of the *Design automation and test in Europe*, 2001

98 STEPHENSON, M., BABB, J., and AMARASINGHE, S.: 'Bitwidth analysis with application to silicon compilation'. Proceedings of the *SIGPLAN programming language design and implementation* (ACM Press, New York, June 2000)

99 CMAR, R., RIJNDERS, L., SCHAUMONT, P., VERNALDE, S., and BOLSENS, I., 'A methodology and design environment for DSP ASIC fixed point refinement'. Proceedings of the *Design automation and test in Europe* (IEEE, Computer Society, CA, 1999)

100 CONSTANTINIDES, G.A.: 'Perturbation analysis for word-length optimization'. Proceedings of the symposium on *Field-programmable custom computing machines* (IEEE Computer Society Press, Washington, 2003)

101 ABDUL GAFFAR, A., MENCER, O., LUK, W., CHEUNG, P.Y.K., and SHIRAZI, N.: 'Floating-point bitwidth analysis via automatic differentiation'. Proceedings of the international conference on *Field-programmable technology*, IEEE, 2002

102 ABDUL GAFFAR, A., MENCER, O., LUK, W., and CHEUNG, P.Y.K., 'Unifying bit-width optimisation for fixed-point and floating-point designs'. Proceedings of the symposium on *Field-programmable custom computing machines* (IEEE Computer Society Press, Washington, 2004)

103 CANTIN, M.-A., SAVARIA, Y., and LAVOIE, P.: 'A comparison of automatic word length optimization procedures'. Proceedings of the IEEE international symposium on *Circuits and systems* (IEEE, New York, 2002)

104 CONSTANTINIDES, G.A.: 'High level synthesis and word length optimization of digital signal processing systems'. PhD thesis, Imperial College London, 2001

105 BENEDETTI, A., and PERONA, B.: 'Bit-width optimization for configurable DSP's by multi-interval analysis'. Proceedings of the 34th Asilomar conference on *Signals, systems and computers*, 2000

106 STEPHENSON, M.W.: 'Bitwise: Optimizing bitwidths using data-range propagation', Master's thesis, Massachussets Institute of Technology, Department of Electrical Engineering and Computer Science, May 2000

107 KEDING, H., WILLEMS, M., COORS, M., and MEYR, H.: 'FRIDGE: A fixed-point design and simulation environment'. Proceedings of the *Design automation and test in Europe*, 1998

108 WILLEMS, M., V. BÜRSGENS, KEDING, H., GRÖTKER, T., and MEYER, M.: 'System-level fixed-point design based on an interpolative approach'. Proceedings of 34th *Design automation conference* (ACM Press, New York, June 1997)

109 KUM, K., and SUNG, W.: 'Word-length optimization for high-level synthesis of digital signal processing systems'. Proceedings of the IEEE international workshop on *Signal processing systems*, 1998

110 SUNG, W., and KUM, K.: 'Word-length determination and scaling software for a signal flow block diagram'. Proceedings of the IEEE international conference on *Acoustics speech and signal processing*, 1994

111 SUNG, W., and KUM, K., 'Simulation-based word-length optimization method for fixed-point digital signal processing systems', *IEEE Transactions on Signal Processing*, 1995, **43**(12), pp. 3087–3090

112 ONG, S., KERKIZ, N., SRIJANTO, B., *et al.*: 'Automatic mapping of multiple applications to multiple adaptive computing systems'. Proceedings of the international symposium on *Field-programmable custom computing machines* (IEEE Computer Society Press, Washington, 2001)

113 THOMAS, D., and LUK, W.: 'A framework for development and distribution of hardware acceleration'. *Reconfigurable technology: FPGAs and reconfigurable processors for computing and communications*, Proceedings of the *SPIE*, 2002, **4867**

114 BOHM, W., HAMMES, J., DRAPER, B. *et al.*: 'Mapping a single assignment programming language to reconfigurable systems', *The Journal of Supercomputing*, 2002, **21**, pp. 117–130

115 DAMIANOU, N., DULAY, N., LUPU, E., and SLOMAN, M.: 'The Ponder policy specification language', in SLOMAN, M., LOBO, J., and LUPU, E.C.: Proceedings of the workshop on *Policies for distributed systems and networks*, LNCS 1995, Springer, 2001

116 LEE, T.K., YUSUF, S., LUK, W., SLOMAN, M., LUPU, E., and DULAY, N.: 'Compiling policy descriptions into reconfigurable firewall

processors'. Proceedings of the symposium on *Field-programmable custom computing machines* (IEEE Computer Society Press, Washington, 2003)

117 KULKARNI, C., BREBNER, G., and SCHELLE, G.: 'Mapping a domain specific language to a platform FPGA'. Proceedings of the *Design automation conference*, 2004

118 LEE, T.K., DERBYSHIRE, A., LUK, W., and Cheung, P.Y.K.: 'High-level language extensions for run-time reconfigurable systems'. Proceedings of the international conference on *Field-programmable technology* (IEEE, New York, 2003)

119 SHIRAZI, N., LUK, W., and CHEUNG, P.Y.K.: 'Framework and tools for run-time reconfigurable designs', *IEE Proc.-Comput. Digit. Tech.*, May 2000

120 DERBYSHIRE, A., and LUK, W.: 'Compiling run-time parametrisable designs'. Proceedings of the international conference on *Field-programmable technology* (IEEE, New York, 2002)

121 CLARK, D., and HUTCHINGS, B.: 'The DISC programming environment'. Proceedings of the symposium on *FPGAs for custom computing machines* (IEEE Computer Society Press, Washington, 1996)

122 STYLES, H., and LUK, W.: 'Branch optimisation techniques for hardware compilation', in CHEUNG, P.Y.K., CONSTANTINIDES, G.A., and DE SOUSA, J.T.: 'Field-programmable logic and applications', LNCS 2778 (Springer, Berlin, 2003)

123 KATHAIL, V. *et al.*: 'PICO: automatically designing custom computers', *Computer*, **35**(9), 2002

124 MUGGLETON, S.H.: 'Inverse entailment and Progol', *New Generation Computing*, 1995, **13**, pp. 245–286

125 PETERSON, J., O'CONNOR, B., and ATHANAS, P.: 'Scheduling and partitioning ANSI-C programs onto multi-FPGA CCM architectures'. International symposium on *FPGAs for custom computing machines* (IEEE Computer Society Press, Washington, 1996)

126 DUNCAN, A., HENDRY, D., and GRAY, P.: 'An overview of the COBRA-ABS high-level synthesis system for multi-FPGA systems'. Proceedings of the IEEE symposium on *FPGAs for custom computing machines* (IEEE Computer Society Press, Washington, 1998)

127 AST, A., BECKER, J., HARTENSTEIN, R., KRESS, R., REINIG, H., and SCHMIDT, K.: 'Data-procedural languages for FPL-based machines', in 'Field-programmable logic and applications', LNCS 849 (Springer, Berlin, 1994)

128 HÖGL, H., KUGEL, A., LUDVIG, J., MÄNNER, R. NOFFZ, K., and ZOZ, R.: 'Enable++: a second-generation FPGA processor'. IEEE symposium on 'FPGAs for custom computing machines' (IEEE Computer Society Press, Washington, 1995)

129 CALLAHAN, T., and WAWRZYNEK, J.: 'Instruction-level parallelism for reconfigurable computing', in HARTENSTEIN, R. W., and KEEVALIK, A.: 'Field-programmable logic and applications', LNCS 1482 (Springer, Berlin, 1998)

130 GOKHALE, M., and STONE, J.: 'NAPA C: compiling for a hybrid RISC/FPGA architecture'. Proceedings of the symposium on 'Field-programmable custom computing machines' (IEEE Computer Society Press, Washington, 1998)

131 BABB, J., REINARD, M., ANDRAS MORITZ, C. *et al*: 'Parallelizing applications into silicon'. Proceedings of the symposium on *FPGAs for custom computing machines* (IEEE Computer Society Press, Washington, 1999)

132 ZIEGLER, H., SO, B., HALL, M., and DINIZ, P.: 'Coarse-grain pipelining on multiple-FPGA architectures'. IEEE symposium on *Field-programmable custom computing machines* (IEEE, New York, 2002) pp. 77–88

133 RISSA, T., LUK, W., and CHEUNG, P.Y.K.: 'Automated combination of sim- ulation and hardware prototyping'. Proceedings of the international conference on *Engineering of reconfigurable systems and algorithms* (CSREA Press, San Diego, CA, 2004)

134 BJESSE, P., CLAESSEN, K., SHEERAN, M., and SINGH, S.: 'Lava: hard- ware design in Haskell'. Proceedings of the ACM international conference on *Functional programming* (ACM Press, New York, 1998)

135 SINGH, S., and LILLIEROTH, C.J.: 'Formal verification of reconfigurable cores'. Proceedings of the symposium on *Field-programmable custom comput- ing machines* (IEEE Computer Society Press, Washington, 1999)

136 GUO, S., and LUK, W.: 'An integrated system for developing regular array design', *Journal of Systems Architecture*, 2001, **47**, pp. 315–337

137 LUK, W., and MCKEEVER, S.W.: 'Pebble: a language for parametrised and reconfigurable hardware design', in 'Field-programmable logic and applica- tions', LNCS 1482 (Springer, Berlin, 1998)

138 MCKEEVER, S.W., LUK, W., and DERBYSHIRE, A.: 'Compiling hardware descriptions with relative placement information for parametrised libraries', in AAGAARD, M.D., and O'LEARY, J.W.: 'Formal methods in computer-aided design', LNCS 2517 (Springer, Berlin, 2002)

139 TODMAN, T., COUTINHO, J.G.F., and LUK, W.: 'Customisable hard- ware compilation'. Proceedings of the international conference on *Engineering of reconfigurable systems and algorithms* (CSREA Press, San Diego, CA, 2004)

140 OU, J., and PRASANNA, V.: 'PyGen: a MATLAB/Simulink based tool for synthesizing parameterized and energy efficient designs using FPGAs'. Proceed- ings of the international symposium on *Field-programmable custom computing machines* (IEEE Computer Society Press, Washington, 2004)

141 ATHANAS, P.M., and ABBOTT, A.L.: 'Real-time image processing on a custom computing platform', *IEEE Computer*, 1995, **28**, pp. 16–24

142 BERGMANN, N.W., and CHUNG, Y.Y.: 'Video compression with cus- tom computers', *IEEE Transactions on Consumer Electronics*, 1997, **43**, pp. 925–933.

143 BERRY, G., and KISHINEVSKY, M.: 'Hardware esterel language extension proposal'. http://www.esterel-technologies.com/v2/solutions/ iso_album/hardest.pdf

144 BLUM, T., and PAAR, C.: 'High-radix Montgomery modular exponentiation on reconfigurable hardware', *IEEE Transactions on Computers*, 2001, **50**(7), pp. 759–764

145 BONDALAPATI, K., and PRASANNA, V.K.: 'Dynamic precision management for loop computations on reconfigurable architectures'. Proceedings of the symposium on *Field-programmable custom computing machines* (IEEE Computer Society Press, Washington, 1999)

146 BOVE, V.M., WATLINGTON, J., and CHEOPS, J.: 'A reconfigurable data-flow system for video processing', *IEEE Transactions on Circuits and Systems for Video Technology*, 1995, **5**, pp. 140–149

147 CLARK, C.R., and SCHIMMEL, D.E.: 'A pattern-matching co-processor for network intrusion detection systems'. Proceedings of the international conference on *Field programmable technology* (IEEE, 2003)

148 DAO, M., COOK, T., SILVER, D., and P. D'Urbano: 'Acceleration of template-based ray casting for volume visualization using FPGAs'. Proceedings IEEE symposium on *Field-programmable custom computing machines*, 1995

149 DITMAR, J., TORKELSSON, K., and JANTSCH, A.: 'A dynamically reconfigurable FPGA-based content addressable memory for internet protocol characterization', in 'Field programmable logic and applications', LNCS 1896 (Springer, Berlin, 2000)

150 FENDER, J., and ROSE, J.: 'A high-speed ray tracing engine built on a field-programmable system'. Proceedings of the international conference on *Field-programmable technology* (IEEE, New York, 2003)

151 FRANKLIN, R., Carver D., and HUTCHINGS, B.L.: 'Assisting network intrusion detection with reconfigurable hardware'. Proceedings of the symposium on *Field-programmable custom computing machines* (IEEE Computer Society Press, Washington, 2002)

152 FRASER, C., and HANSON, D.: 'A retargetable C compiler: design and implementation' (Benjamin/Cummings Pub. Co, Menlo Park, CA, 1995)

153 GOKHALE, M., DUBOIS, D., DUBOIS, A., BOORMAN, M., POOLE, S., and HOGSETT, V.: 'Granidt: towards gigabit rate network intrusion detection technology', in GLESNER, M., ZIPF, P., and BENOVELL, M.: 'Field programmable logic and applications', LNCS 2438 (Springer, Berlin, 2002)

154 GURA, N., SHANTZ, S.C., EBERLE, H *et al.*: 'An end-to-end systems approach to elliptic curve cryptography', in KALISKI JR., B.S., KOÇ, Ç.K., and PARR, C.: Proceedings of the *Cryptographic hardware and embedded systems*, LNCS 2523, 2002, pp. 349–365

155 HAGLUND, P., MENCER, O., LUK, W., and TAI, B.: 'PyHDL: hardware scripting with Python'. Proceedings of the international conference on *Engineering of reconfigurable systems and algorithms*, 2003

156 HANKERSON, D., MENEZES, A., and VANSTONE, S.: 'Guide to elliptic curve cryptography' (Springer, 2004)

157 HANRAHAN, P.: 'Using caching and breadth-first search to speed up ray-tracing'. Proceedings of the *Graphics interface '86*, May 1986

158 HAYNES, S.D., STONE, J., CHEUNG, P.Y.K., and LUK, W.: 'Video image processing with the SONIC architecture', *IEEE Computer*, 2000, **33**, pp. 50–57

159 HAYNES, S.D., EPSOM, H.G., COOPER, R.J., and MCALPINE, P.L.: 'Ultra-SONIC: a reconfigurable architecture for video image processing', in 'Field programmable logic and applications', LNCS 2438 (Springer, Berlin, 2002)

160 HODJAT, A., and VERBAUWHEDE, I.: 'A 21.54 Gbit/s fully pipelined AES processor on FPGA'. Proceedings of the symposium on *Field-programmable custom computing machines* (IEEE Computer Society Press, Washington, 2004)

161 HOLZMANN, G.: 'The model checker SPIN', *IEEE Transactions on Software Engineering*, 1997, **23**(5), pp. 279–295

162 JAMES-ROXBY P.B., and DOWNS, D.J.: 'An efficient content addressable memory implementation using dynamic routing'. Proceedings of the symposium on *Field-programmable custom computing machines* (IEEE Computer Society Press, Washington, 2001)

163 JIANG, J., LUK, W., and RUECKERT, D.: 'FPGA-based computation of free-form deformations in medical image registration'. Proceedings of the international conference on *Field-programmable technology* (IEEE, 2003)

164 LEE, D.U., LUK, W., WANG, C., JONES, C., SMITH, M., and VILLASENOR, J.: 'A flexible hardware encoder for low-density parity-check codes'. Proceedings of the symposium on *Field-programmable custom computing machines* (IEEE Computer Society Press, Washington, 2004)

165 LEONARD, J., and MANGIONE-Smith, W.H.: 'A case study of partially evaluated hardware circuits: key-specific DES', 'Field programmable logic and applications, LNCS 1304 (Springer, Berlin, 1997)

166 MAHLKE, S. *et al.*: 'Bitwidth cognizant architecture synthesis of custom hardware accelerators', *IEEE Transactions on Computer-Aided Design*, 2001, **20**(11)

167 MAZZEO, A., ROMANO, L., and SAGGESE, G.P.: 'FPGA-based implementation of a serial RSA processor'. Proceedings of the *Design, automation and test in Europe conference* (IEEE, New York, 2003)

168 MONTGOMERY, P.: 'Modular multiplication without trial division', *Mathematics of Computation*, 1985, **44**, pp. 519–521.

169 MULLER, H., and WINCKLER, J.: 'Distributed image synthesis with breadth-first ray tracing and the Ray-Z-buffer', in *Data structures and efficient algorithms – final report on the DFG special initiative*, LNCS 594 (Springer, Berlin, 1992)

170 NAKAMARU, K., and OHNO, Y.: 'Breadth-first ray tracing using uniform spatial subdivision', *IEEE Transactions on Visualization and Computer Graphics*, 1997, **3**(4), pp. 316–328

171 NALLATECH: 'Ballynuey 2: Full-length PCI card DIME motherboard with four DIME slots, Nallatech, NT190-0045, 2002

172 National Bureau of Standards: 'Announcing the data encryption standard, Technical report FIPS Publication 46, US Commerce Department, National Bureau of Standards. January 1977

173   National Insititute of Standards and Technology: 'Advanced encryption standard, http://csrc.nist.gov/publication/drafts/dfips-AES.pdf

174   ORLANDO, G., and PAAR, C.: 'A high performance reconfigurable elliptic curve for $GF(2^m)$'. Proceedings of the workshop on *Cryptographic hardware and embedded systems*, LNCS 1965, 2000

175   ORLANDO, G., and PAAR, C.: 'A scalable $GF(p)$ Elliptic Curve processor architecture for programmable hardware', in KOÇ, Ç.K., NACCACHE, D. and PAAR, C.: Proceedings of the workshop on *Cryptographic hardware and embedded systems*, LNCS 2162, 2001

176   PARKER, S., MARTIN, W., SLOAN, P., SHIRLEY, P., SMITS, B., and HANSEN, C.: 'Interactive ray tracing'. Proceedings of the 1999 symposium on *Interactive 3D graphics* (ACM Press, New York, April 1999)

177   PASHAM, V., and TRIMBERGER, S.: 'High-speed DES and triple DES encryptor/decryptor'. Xilinx Application note: XAPP270, 2001. http://www.xilinx.com/bvdocs/appnotes/xapp270.pdf

178   PATTERSON, C.: 'High performance DES encryption in Virtex FPGAs using JBits'. Proceedings of the international conference on *Field-programmable custom computing machines*, IEEE, 2000, pp. 113–121

179   PLUNKETT, D., and BAILEY, M.: 'The vectorization of a ray-racing algorithm for increased speed', *IEEE Computer Graphics and Applications*, 1985, **5**(8), pp. 52–60

180   QUENOT, G.M., KRALJIC, I.C., SEROT, J., and ZAVIDOVIQUE, B.: 'A reconfigurable compute engine for real-time vision automata prototyping'. Proceedings of the IEEE workshop on *FPGAs for custom computing machines* (IEEE Computer Society Press, Washington, 1994)

181   RATHMAN, S., and SLAVENBURG, G.: 'Processing the new world of interactive media', *IEEE Signal Processing Magazine*, 1998, **15**(2), pp. 108–117

182   RIVEST, R.L., SHAMIR, A., and ADLEMAN, L.: 'A method for obtaining digital signatures and public-key cryptosystems', *Communications of ACM*, 1978, **21**(2), pp. 120–126

183   SAGGESE, G.P., MAZZEO, A., MAZZOCCA, N., and STROLLO, A.G.M.: 'An FPGA-based performance analysis of the unrolling, tiling, and pipelining of the AES Algorithm', in CHEUNG, P.Y.K., CONSTANTINIDES, G.A., and DE SOUSA, J.T.: 'Field-programmable logic and applications, LNCS 2778, 2003

184   SATOH, A., and TAKANO, K.: 'A scalable dual-field elliptic curve cryptographic processor', *IEEE Transactions on Computers*, 2003, **52**(4), pp. 449–460

185   SEDCOLE, P., CHEUNG, P.Y.K., CONSTANTINIDES, G.A., and LUK, W.: 'A reconfigurable platform for real-time embedded video image processing', in 'Field programmable logic and applications, LNCS 2778 (Springer, Berlin, 2003)

186   SIMA, M., COTOFANA, S.D., VASSILIADIS, S., VAN EIJNDHOVEN, J.T.J., and VISSERS, K.A.: 'Pel reconstruction on FPGA-augmented TriMedia', *IEEE Transactions on VLSI*, 2004, **12**(6), pp. 622–635

187 SOURDIS, I., and PNEVMATIKATOS, D.: 'Fast, large-scale string matching for a 10Gbps FPGA-based network intrusion system', in CHEUNG, P.Y.K., CONSTANTINIDES, G.A., and DE SOUSA, J.T.: 'Field programmable logic and applications', LNCS 2778 (Springer, Berlin, 2003)

188 Spin, http://spinroot.com/spin/whatispin.html.

189 STANDAERT, F.X., ROUVROY, G., QUISQUATER, J.J., and LEGAT, J.D.: 'Efficient implementation of Rijndael encryption in reconfigurable hardware: improvements and design tradeoffs', in WALTER, C.D., KOÇ, Ç.K., and PAAR, C.: Proceedings of the *Cryptographic hardware and embedded systems*, LNCS 2779, 2003

190 STYLES, H., and LUK, W.: 'Accelerating radiosity calculations using reconfigurable platforms'. Proceedings of the symposium on *Field-programmable custom computing machines* (IEEE Computer Society Press, Washington, 2002)

191 TANG, S.H., TSUI, K.S., and LEONG, P.H.W.: 'Modular exponentiation using parallel multipliers'. Proceedings of the international conference on *Field programmable technology* (IEEE, New York, 2003)

192 TODMAN, T., and LUK, W.: 'Reconfigurable designs for ray tracing'. Proceedings of the symposium on *Field-programmable custom computing machines* (IEEE Computer Society Press, Washington, 2001)

193 TODMAN, T., and LUK, W.: 'Combining imperative and declarative hardware descriptions'. Proceedings of the 36th Hawaii international conference on *System sciences* (IEEE, New York, 2003)

194 UDDIN, M.M., CAO, Y., and YASUURA, H.: 'An accelerated datapath width optimization scheme for area reduction of embedded systems'. Proceedings of the international symposium on *Systems synthesis* (ACM Press, New York, 2002)

195 WENBAN, A., O'LEARY, J.W., and BROWN, G.M.: 'Codesign of communication protocols', *IEEE Computer*, 1993, **26**(12), pp. 46–52

196 WENBAN, A., and BROWN, G.: 'A software development system for FPGA-based data acquisition systems'. IEEE symposium on *FPGAs for custom computing machines* (IEEE Computer Society Press, Washington, 1996)

197 WILCOX, D.C., PIERSON, L.G., ROBERTSON, P.J., WITZKE, E.L., and GASS, K.: 'A DES ASIC suitable for network encryption at 10 Gbps and beyond', in KOÇ, Ç.K., and PAAR, C.: Proceedings of the *Cryptographic hardware and embedded systems*, LNCS 1717 (Springer-Verlag, Berlin, 1999)

198 WILLIAMS, J.A., DAWOOD, A.S., and VISSER, S.J.: 'FPGA-based cloud detection for real-time onboard remote sensing'. Proceedings of the international conference on *Field-programmable technology* (IEEE, New York, 2002)

199 WOODS, R., TRAINOR, D., and HERON, J.P.: 'Applying an XC6200 to real-time image processing', *IEEE Design and Test of Computers*, 1998, **15**, pp. 30–38

Part V

Architectural synthesis

Chapter 15

CAD tools for embedded analogue circuits in mixed-signal integrated Systems-on-Chip

Georges G.E. Gielen

15.1 Introduction

With the evolution towards ultra-deep-submicron and nano-meter CMOS technologies [1], the design of complex Systems on a Chip (SoC) is emerging in consumer-market applications such as telecom and multimedia. These integrated systems are increasingly mixed-signal designs, embedding high-performance analogue or mixed-signal blocks and possibly sensitive RF frontends together with complex digital circuitry (multiple processors, some logic blocks and several large memory blocks) on the same chip. In addition, the growth of wireless services and other telecom applications increases the need for low-cost highly integrated solutions with very demanding performance specifications. This requires the development of intelligent front-end architectures that get around the physical limitations posed by the semiconductor technology. But also more traditional application domains like automotive or instrumentation show an increasing trend in integrating analogue sensor/actuator interfaces with digital electronics. And the emerging fields of miniaturised and possibly networked biomedical devices as well as sensor networks promises to be an even larger market for integrated mixed-signal systems.

This chapter addresses the problems and solutions that are posed by the design of such mixed-signal integrated systems. These include problems in:

- design methodologies and flows;
- simulation and modelling;
- design productivity (synthesis, yield);

- mixed-signal design verification, including analysis of signal integrity and crosstalk, for instance problems due to the embedding of analogue/RF blocks such as supply or substrate noise coupling.

The chapter explains the problems that are posed by these mixed-signal/RF SoC designs, describes the CAD solutions and their underlying methods already existing today, and outlines the challenges that still remain to be solved at present. Depending on how we can solve these problems, single-chip SoC integration will be a stairway to heaven or a highway to hell. In the latter case, two-chip solutions – possibly fabricated in different technologies and, maybe with a limited number of extra passives, stacked together in a System in a Package (SiP) – might economically be a more viable solution.

The chapter is organised as follows. Section 2 addresses the problems in mixed-signal design methodologies and describes the need for architectural exploration, analogue behavioural modelling and power/area estimation. Section 3 focuses on the issue of analogue design and layout productivity by discussing recent progress in analogue circuit and layout synthesis techniques. Also the issue of yield optimisation and design for manufacturability is addressed. Section 4 outlines bottlenecks in mixed-signal design verification, especially the problem of analysing crosstalk between digital and analogue circuits such as supply or substrate noise. Finally, Section 5 provides conclusions, followed by an extensive list of references.

15.2   Top-down mixed-signal design methodology

The growing complexity of the systems that can be integrated on a single chip today, in combination with the tightening time-to-market constraints, results in a growing design productivity gap for SoCs. That is why new design methodologies are being developed that allow designers to shift to a higher level of design abstraction, such as the use of platform-based design, object-oriented system-level hierarchical design refinement flows, hardware–software co-design and IP reuse, on top of the already established use of CAD tools for logic synthesis and digital place and route. However, these flows have to be extended to also incorporate the embedded analogue/RF blocks.

A typical top-down design flow for mixed-signal integrated systems may look as shown in Figure 15.1, where the following distinct phases can be identified: system specification, architectural design, cell design, cell layout and system layout assembly [2]. The advantages of adopting a top-down design methodology are:

- the possibility to perform system architectural exploration and a better overall system optimisation (e.g. finding an architecture that consumes less power) at a high level before starting detailed circuit implementations;
- the elimination of problems that often cause overall design iterations, like the anticipation of problems related to interfacing different blocks;
- the possibility to do early test development in parallel to the actual block design; etc.

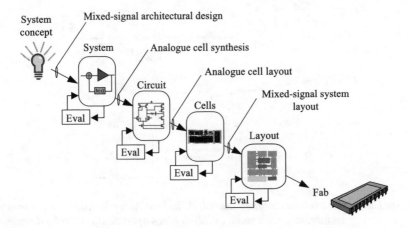

Figure 15.1   Top-down view of the mixed-signal IC design process

The ultimate advantage of top-down design therefore is to catch problems early in the design flow and as a result have a higher chance of first-time success with fewer or no overall design iterations, hence shortening the design time, while at the same time obtaining a better overall system design. The top-down design methodology, however, does not come for free and requires some investment from the design team, especially in terms of high-level modelling and setting up a sufficient model library for the targeted application domain. Even then there remains the risk that also at higher levels in the design hierarchy low-level details (e.g. matching limitations, circuit non-idealities, layout effects...) may be important to determine the feasibility or optimality of a solution. The high-level models used therefore must include such effects to the extent possible, but it remains difficult in practice to anticipate or model everything accurately at higher levels. Besides the models, also efficient simulation methods are needed at the architectural level in order to allow efficient interactive explorations. The issues of system exploration and simulation, as well as behavioural modelling and model generation will now be discussed in more detail.

15.2.1   System-level architectural exploration

The general objective of analogue architectural system exploration is twofold [3,4]. First of all, a proper (and preferrably optimal) architecture for the system has to be decided upon. Second, the required specifications for each of the blocks in the chosen architecture must be determined, so that the overall system meets its require-ments at minimum implementation cost (power, chip area, etc.). The aim of a system exploration environment is to provide the system designer with the platform and the supporting tool set to explore in a short time different architectural alternatives and to take the above decisions based on quantified rather than heuristic information.

Consider for instance the digital telecommunication link of Figure 15.2. It is clear that digital bits are going into the link to be transmitted over the channel, and that the

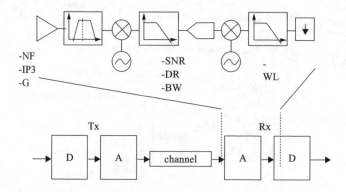

Figure 15.2   Digital telecommunication link, indicating a possible receiver frontend architecture with some building block specifications to be determined during frontend architectural exploration

received signals are being converted again in digital bits. One of the major considerations in digital telecom system design is the bit error rate, which characterises the reliability of the link. This bit error rate is impacted by the characteristics of the transmission channel itself, but also by the architecture chosen for the transmitter and receiver frontend and by the performances achieved and the non-idealities exhibited by the analogue/RF blocks in this frontend. For example, the noise figure and nonlinear distortion of the input low-noise amplifier (LNA) are key parameters. Similarly, the resolution and sampling speed of the used analogue-to-digital converter (ADC) may have a large influence on the bit error rate, but it also determines the requirements for the other analogue subblocks: a higher ADC resolution may relax the filtering requirements in the transceiver, resulting in simpler filter structures, though it will also consume more power and chip area than a lower-resolution converter. At the same time, the best trade-off solution, i.e. the minimum required ADC resolution and therefore also the minimum power and area, depends on the architecture chosen for the transceiver frontend.

Clearly, there is a large interaction between system-level architectural decisions and the performance requirements for the different subblocks, which on their turn are bounded by technological limits that shift with every new technology process being employed. Hence it is important to offer designers an exploration environment where they can define different frontend architectures and analyse and compare their performance quantitatively and derive the necessary building block specifications. Today the alternative architectures that are explored are still to be provided by the system designer, but future tools might also derive or synthesise these architectures automatically from a high-level language description [5].

The important ingredients that are needed to set up such an architectural exploration environment are [3,4]:

- a fast high-level simulation method that allows us to evaluate the performance (e.g. SNR or BER) of the frontend;

- a library of high-level (behavioural) models for the building blocks used in the targeted application domain, including a correct modelling of the important building block non-idealities (offset, noise, distortion, mirror signals, phase noise, etc.);
- power and area estimation models that, starting from the block specifications, allow us to estimate the power consumption and chip area that would be consumed by a real implementation of the block, without really designing the block.

The above ingredients allow a system designer to interactively explore frontend architectures. Combining this with an optimisation engine would additionally allow us to optimise the selected frontend architecture in determining the optimal building block requirements as to meet the system requirements at minimum implementation cost (power/area). Repeating this optimisation for different architectures then makes a quantitative comparison between these architectures possible before they are implemented down to the transistor level. In addition, the high-level exploration environment would also help in deciding on other important system-level decisions, such as determining the optimal partitioning between analogue and digital implementations in a mixed-signal system [6], or deciding on the frequency planning of the system, all based on quantitative data rather than ad hoc heuristics or past experiences.

As the above aspects are not sufficiently available in present commercial system-level simulators, more effective and more efficient solutions are being developed. To make system-level exploration really fast and interactive, dedicated algorithms can be developed that speed up the calculations by maximally exploiting the properties of the system under investigation and using proper approximations where possible. ORCA for instance is targeted towards telecom applications and uses dedicated signal spectral manipulations to gain efficiency [7]. A more recent development is the FAST tool which performs a time-domain dataflow type of simulation without iterations [8] and which easily allows dataflow co-simulation with digital blocks. Compared to commercial simulators like SPW, COSSAP, ADS or Matlab/SIMULINK, this simulator is more efficient by using block processing instead of point-by-point calculations for the different time points in circuits without feedback. In addition, the signals are represented as complex equivalent baseband signals with multiple carriers. The signal representation is local and fully optimised as the signal at each node in the circuit can have a set of multiple carriers and each corresponding equivalent baseband component can be sampled with a different time step depending on its bandwidth. Large feedback loops, especially when they contain non-linearities, are however more difficult to handle with this approach. A method to efficiently simulate bit error rates with this simulator has been presented in Reference 9.

15.2.1.1 Example

As an example [3,4], consider a frontend for a cable TV modem receiver, based on the MCNS standard. The MCNS frequency band for upstream communication on the CATV network is from 5 to 42 MHz (extended subsplit band). Two architectures are shown in Figure 15.3: (a) an all-digital architecture where both the channel

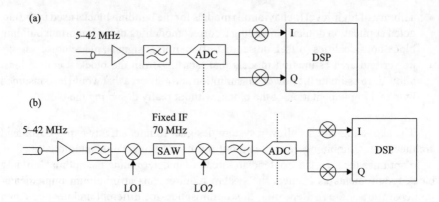

Figure 15.3   Two possible architectures for a cable TV application: (a) all-digital architecture, (b) classical architecture

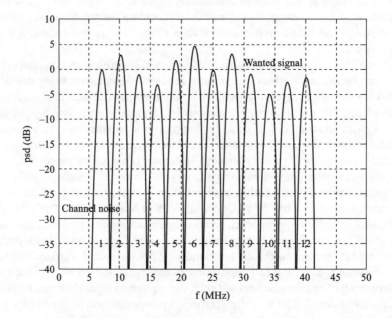

Figure 15.4   Typical input spectrum for a CATV frontend architecture using 12 QAM-16 channels

selection and the downconversion are done in the digital domain, and (b) the classical architecture where the channel selection is performed in the analogue domain.

A typical input spectrum is shown in Figure 15.4. For this example we have used 12 QAM-16 channels with a 3 MHz bandwidth. We assume a signal variation of the different channels of maximally ±5 dB around the average level. The average channel noise is 30 dB below this level. Figures 15.5 and 15.6 show the spectrum simulated

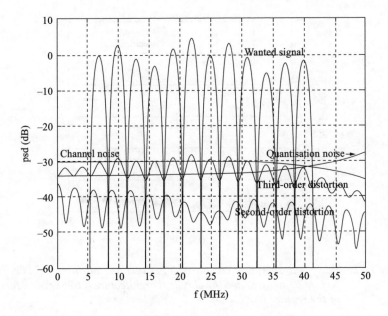

Figure 15.5   Simulated spectrum of the all-digital CATV architecture after the ADC

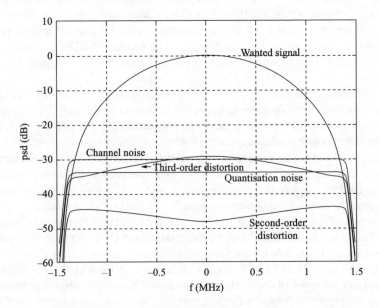

Figure 15.6   Simulated spectrum of the all-digital CATV architecture at the digital receiver output

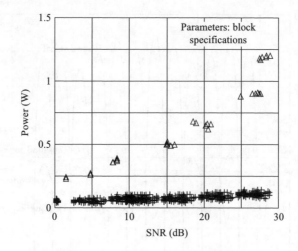

Figure 15.7   Power consumption comparison between the all-digital CATV archi-tecture (triangles) and the classical architecture (crosses) as a function of the required SNR [10]

by ORCA [7] for the all-digital architecture of Figure 15.3(a). Figure 15.5 shows the spectrum after the analogue-to-digital converter, whereas Figure 15.6 shows the spectrum at the output after digital channel selection and quadrature downconversion. The wanted channel signal and the effects of the channel noise, the ADC quantisation noise and the second- and third-order distortion are generated separately, provid-ing useful feedback to the system designer. The resulting SNDR is equal to 22.7 dB in this case, which corresponds to a symbol error rate of less than 10^{-10} for QAM-16.

By performing the same analysis for different architectures and by linking the required subblock specifications to the estimated power and/or chip area required to implement the subblocks, a quantitative comparison of different alternative architec-tures becomes possible with respect to (1) their suitability to implement the system specifications, and (2) the corresponding implementation cost in power consumption and/or silicon real estate. To assess the latter, high-level power and/or area estimators must be used to quantify the implementation cost. In this way the system designer can choose the most promising architecture for the application at hand.

Figure 15.7 shows a comparison between the estimated total power consump-tion required by the all-digital and by the classical CATV receiver architectures of Figure 15.3 as a function of the required SNR [10]. These results were obtained with the simulator FAST [8]. Clearly, for the technology used in the experiment, the classical architecture still required much less power than the all-digital solution.

Finally, Figure 15.8 shows the result of a BER simulation with the FAST tool for a 5 GHz 802.11 WLAN architecture [9]. The straight curve shows the result without taking into account non-linear distortion caused by the building blocks; the dashed curve takes this distortion into account. Clearly, the BER worsens a lot in the

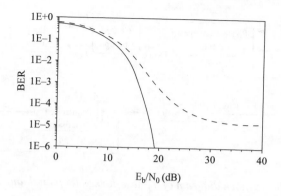

Figure 15.8 *Simulated BER analysis result for a 5-GHz 802.11 WLAN architecture with (dashed) and without (straight) non-linear distortion of the building blocks included [9]*

presence of non-linear distortion. Note that the whole BER analysis was performed in a simulation time which is two orders of magnitude faster than traditional Monte-Carlo analysis performed on a large number of OFDM symbols.

15.2.2 Top-down analogue block design

Top-down design is already heavily used in industry today for the design of complex analogue blocks like Delta-Sigma converters or PLL. In these cases first a high-level design of the block is done with the block represented as an architecture of subblocks, each modelled with a behavioural model that includes the major non-idealities as parameters, rather than a transistor schematic. This step is often done using Matlab/Simulink and it allows us to determine the optimal architecture of the block at this level, together with the minimum requirements for the subblocks (e.g. integrators, quantisers, VCO, etc.), so that the entire block meets its requirements in some optimal sense. This is then followed by a detailed device-level (SPICE) design step for each of the chosen architecture's subblocks, targeted to the derived subblock specifications. This is now illustrated for a PLL.

15.2.2.1 Example

The basic block diagram of a PLL is shown in Figure 15.9. If all subblocks like the phase-frequency detector or the voltage-controlled oscillator (VCO) are represented by behavioural models instead of device-level circuits, then enormous time savings in simulation time can be obtained during the design and verification phase of the PLL. For example, for requirements arising from a GSM-1800 design example (frequency range around 1.8 GHz, phase noise -121 dBc/Hz at 600 kHz frequency offset, settling time of the loop for channel frequency changes below 1 ms within $1e - 6$ accuracy), the following characteristics can be derived for the PLL subblocks using behavioural simulations with generic behavioural models for the subblocks [12]: $A_{LPF} = 1$,

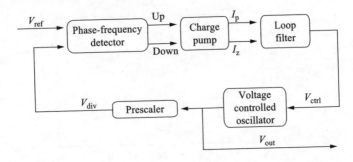

Figure 15.9 Basic block diagram of a phase-locked loop analogue block

$K_{\text{VCO}} = 1e6 \, \text{Hz/V}$, $N_{\text{div}} = 64$, $f_{\text{LPF}} = 100 \, \text{kHz}$. These specifications are then the starting point for the device-level design of each of the subblocks.

For the bottom-up system verification phase of a system, more detailed behavioural models have to be generated that are tuned towards the actual circuit design. For example, an accurate behavioural model for a designed VCO is given by the following equation set:

$$v_{\text{out}}(t) = A_0(v_{\text{in}}(t)) + \sum_{k=1}^{k=N} A_k(v_{\text{in}}(t)) \sin(\Phi_k(t))$$

$$\Phi_k(t) = \varphi_k(v_{\text{in}}(t)) + 2\pi \int_{t_0}^{t} k[h_{\text{stat 2dyn}}(\tau) \otimes f_{\text{stat}}(v_{\text{in}}(\tau))] \mathrm{d}\tau$$

(15.1)

where Φ_k is the phase of each harmonic k in the VCO output, A_k and φ_k characterise the (non-linear) static characteristic of a VCO, and h_{stat2dyn} characterises the dynamic voltage-phase behaviour of a VCO, both as extracted from circuit-level simulations of the real circuit. For example, Figure 15.10 shows the frequency response of both the original device-level circuit (light line) and the extracted behavioural model (dark line) for a low-frequency sinusoidal input signal. You can see that this input signal creates a side lobe near the carrier that is represented by the model within 0.25 dB accuracy compared to the original transistor-level circuit, while the gain in simulation time is more than 30× [12].

15.2.3 Analogue behavioural and performance modelling

The major workhorse for every analogue designer is the SPICE circuit simulator, which numerically solves the system of non-linear differential-algebraic equations that characterise the circuit by using traditional techniques of numerical analysis. Many variants of the SPICE simulator are now marketed by a number of CAD vendors and many IC manufacturers have in-house versions of the SPICE simulator that have been adapted to their own proprietary processes and designs. SPICE or its many derivatives have evolved into an established designer utility that is being used both

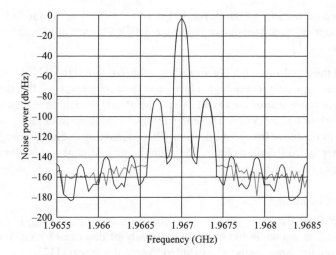

Figure 15.10 Frequency response of an extracted behavioural VCO model (dark line) compared to the underlying device-level circuit response (light line) [12]

during the design phase (often in a designer-guided trial-and-error fashion) and for extensive post-layout design verification.

The main problem with the standard SPICE simulator is that it is essentially a structural circuit simulator, and that its CPU time increases fast with the size of the circuit, making the simulation of really large designs infeasible. This is why in the past years the need has arisen for higher levels of abstraction to describe and simulate analogue circuits and mixed-signal systems.

There are three reasons for using higher-level analogue modelling (functional, behavioural or macro modelling) for systems-on-chip [2]:

- In a top-down design methodology based on hierarchical design refinement (like Figure 15.1) at higher levels of the design hierarchy, there is a need for higher-level models describing the pin-to-pin behaviour of the circuits in a mathematical format rather than representing it as a internal structural netlist of components. This is unavoidable during top-down design since at higher levels in the design hierarchy the details of the underlying circuit implementation are simply not yet known and hence only generic mathematical models can be used.
- A second use of behavioural models is during bottom-up system verification when these models are needed to reduce the CPU time required to simulate the block as part of a larger system. The difference is that in this case the underlying implementation is known in detail, and that peculiarities of the block's actual implementation can be incorporated as much as possible in the extracted model without slowing down the simulation too much.
- Third, when providing or using analogue IP macrocells in a SoC context, the virtual component (ViC) has to be accompanied by an executable model that efficiently models the pin-to-pin behaviour of the virtual component. This model

can then be used in system-level design and verification, by the SoC integrating company, even without knowing the detailed circuit implementation of the macrocell [11].

For all these reasons analogue/mixed-signal behavioural models are needed that describe analogue circuits at a higher level than the circuit level, i.e. that describe the input–output behaviour of the circuit in a mathematical model rather than as a structural network of basic devices. These higher-level models must describe the desired behaviour of the block (like amplification, filtering, mixing or quantisation) and simulate efficiently, while still including the major nonidealities of real implementations with sufficient accuracy.

15.2.3.1   Example

For example, the dynamic behaviour (settling time and glitch energy) of a current-steering DAC as shown in Figure 15.11 can easily be described by superposition of an exponentially damped sine and a shifted hyperbolic tangent [12]:

$$
i_{\text{out}} = A_{gl} \sin\left(\frac{2\pi}{t_{gl}}(t - t_0)\right) \exp\left(-\text{sign}(t - t_0)\frac{2\pi}{t_{gl}}(t - t_0)\right)
$$
$$
+ \frac{\text{level}_{i+1} - \text{level}_i}{2} \tanh\left(\frac{2\pi}{t_{gl}}(t - t_0)\right) + \frac{\text{level}_{i+1} + \text{level}_i}{2} \tag{15.2}
$$

where level_i and level_{i+1} are the DAC output levels before and after the considered transition, and where A_{gl}, t_0 and t_{gl} are parameters that need to be determined, e.g. by regression fitting to simulation results of a real circuit. Figure 15.12 compares the response of the behavioural model (with parameter values extracted from

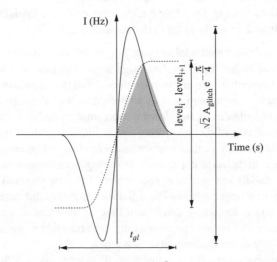

Figure 15.11   Typical dynamic behaviour of a current-steering digital-to-analogue converter output when switching the digital input code

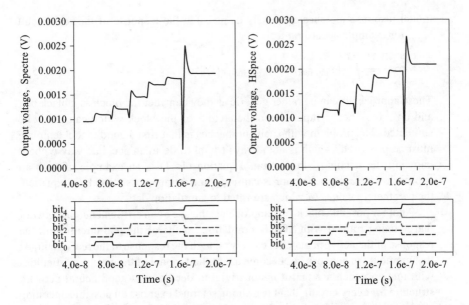

Figure 15.12   *Comparison between the device-level simulation results (on the right)*
and the response of the extracted behavioural model (on the left) [12]

SPICE simulations of the original circuit) with SPICE simulation results of the original circuit. The speed-up in CPU time is a factor 874 (!!) while the error is below 1 per cent [12].

The industrial use of analogue behavioural modelling is today leveraged by the availability of standardised mixed-signal hardware description languages such as VHDL–AMS [13,14] and VERILOG–AMS [15,16], both of which are extensions of the corresponding digital hardware description languages, and both of which are supported by commercial simulators today.

15.2.3.2 Behavioural model generation techniques

One of the largest problems today is the lack of systematic methods to create good analogue behavioural or performance models – a skill not yet mastered by the majority of analogue designers – as well as the lack of any tools to automate this process. Fortunately, in recent years research has started to develop methods that can automatically create models for analogue circuits, both behavioural models for behavioural simulation and performance models for circuit sizing. Techniques used here can roughly be divided into fitting or regression approaches, constructive approaches and model-order reduction methods.

- In the 'fitting or regression approaches' a parameterised model (e.g. a rational transfer function, a more general set of equations or even an artificial neural network model) is first proposed by the model developer and the values of the parameters are then fitted by some least-square error optimisation so that the model response matches as closely as possible the response of the real circuit [17], for example according to:

$$\text{error} = \int_0^T \|v_{\text{out,real}}(t) - v_{\text{out,model}}(t)\|^2 dt \tag{15.3}$$

These approaches can be rather generic as they consider the block as a black box and only look at the input–output behaviour of the block which can easily be simulated. The problem with such approaches is that first a good model template must be proposed, which is not always trivial to do in an accurate way without knowing the details of the circuit. Equation (15.1) is an example of such an approach where the behaviour is captured in a parameterised analytic equation, the parameters values of which are fitted to simulation data.

Another possible black-box approach is the use of an artificial neural network that is being trained with SPICE simulation results of the real circuit until the response of the network matches closely enough the response of the real circuit. At that moment the network has become an implicit model of the circuit. Difficulties with this approach are that it is not trivial to decide on a good neural network structure for every circuit, that the training set must exercise all possible operating modes of the circuit and that the resulting model is specific for one particular implementation of the circuit. To improve these methods, all progress made in other research areas such as in time series prediction could be applied here as well and is being explored.

- The second class of methods, the 'constructive approaches', try to generate or build a model from the underlying circuit description. This is then inherently a white-box model that is specific for the particular circuit at hand, but on the other hand it offers a higher guarantee of tracking the real circuit behaviour well in a wider range than the fitting methods. One approach for instance uses symbolic analysis techniques to first generate the exact set of describing algebraic/differential equations of the circuit, which are then simplified within a given error bound of the exact response using both global and local simplifications [18]. The resulting simplified set of equations then constitutes the behavioural model of the circuit and tracks nicely the behaviour of the circuit. The biggest drawback, however, is that the error estimation is difficult and for non-linear circuits heavily depends on the targeted response. Up till now, the gains in CPU time obtained in this way are not high enough for practical circuits. More research in this area is definitely needed.

- The third group of methods, the 'model-order reduction methods', are mathematical techniques that take as input a linear, time-invariant set of differential equations describing a state-space model of the circuit, for example :

$$\frac{dx}{dt} = Ax + Bu; \quad y = Cx + Du \tag{15.4}$$

where x represents the circuit state, u the circuit inputs, y the circuit outputs and the matrices A, B, C and D determine the circuit properties. As output model-order reduction methods produce a similar state-space model $\tilde{A}, \tilde{B}, \tilde{C}, \tilde{D}$, but with a state vector $\tilde{x}$ (thus matrix description) of lower dimensionality, i.e. of lower order:

$$\frac{d\tilde{x}}{dt} = \tilde{A}\tilde{x} + \tilde{B}u; \quad \tilde{y} = \tilde{C}\tilde{x} + \tilde{D}u \qquad (15.5)$$

These reduced-order models simulate much more efficiently, while approximating the exact response, for example matching the original model closely up to some specified frequency. Originally developed to reduce the complexity of linear interconnect networks for timing analysis [19], techniques such as asymptotic waveform evaluation (AWE) or related variants such as Padé via Lanczos (PVL), use moment matching and Padé approximation to generate a lower-order model for the response of the linear interconnect network. The early AWE efforts used explicit moment matching techniques which could generate unstable reduced-order models. Subsequent developments using Krylov-subspace-based iterative methods resulted in methods like PVL that overcame many of the deficiencies of the earlier AWE efforts, and passivity is now guaranteed using techniques like Arnoldi transformations [20], resulting in tools like PRIMA [21]. In recent years, similar techniques have also been applied to create reduced-order macro-models for analogue/RF circuits. Originally restricted to linear(ised) circuits, techniques were later developed or extended to cover also periodically time-varying circuits (e.g. time-varying Padé [22]), weakly non-linear circuits (e.g. Volterra-series-based polynomial reduction [23] and the NORM approach [24]) and strongly non-linear circuits (e.g. using trajectory piecewise-linear [25] or piecewise-polynomial approximations [26]).

Despite the progress made so far, still more research in the area of automatic or systematic behavioural model generation or model-order reduction is certainly needed, and the field is a hot research area at the moment.

15.2.3.3 Performance model generation techniques

Note that besides behavioural models that simplify the input–output behaviour of analogue circuits for purposes of faster simulation or verification, also performance models are needed. Performance models relate the achieveable performances of a circuit (e.g. gain, bandwidth, slew rate or phase margin) to the design variables (e.g. device sizes and biasing). Such performance models are used to speed up circuit sizing as will be discussed later on: in the circuit optimisation procedure, calls to the transistor-level simulation are replaced by performance model evaluations, resulting in substantial speedups (once the performance models have been created and calibrated).

Most approaches for performance model generation are based on fitting or regression methods where the parameters of a template model are fitted to have the model match as closely as possible a sample set of simulated data points. A recent example of such a fitting approach is the automatic generation of posynomial performance models for analogue circuits, that are created by fitting a pre-assumed posynomial equation template to simulation data created according to some design of experiments scheme [27]. Such a posynomial model could then for instance be used in the very efficient sizing of analogue circuits through convex circuit optimisation. To improve these methods, all progress made in other research areas such as in time series prediction (e.g. support vector machines [28]) or data mining techniques [29] could be applied here as well.

Despite the progress made so far, still more research in the area of automatic performance model generation is needed to reduce analogue synthesis times, especially for hierarchical synthesis of complex analogue blocks. This field is a hot research area at the moment.

15.2.4 Power/area estimation models

Besides behavioural models, the other crucial element to compare different architectural alternatives and to explore trade-offs during system-level exploration and optimisation are accurate and efficient power and area estimators [30]. They allow one to assess and compare the optimality of different design alternatives. Such estimators are functions that predict the power or area that is going to be consumed by a circuit implementation of an analogue block (e.g. an analogue-to-digital converter) with given specification values (e.g. resolution and speed). Since the implementation of the block is not yet known during high-level system design and considering the large number of different possible implementations for a block, it is very difficult to generate these estimators with high absolute accuracy. However, for the purpose of comparing different design alternatives during architectural exploration, the tracking accuracy of estimators with varying block specifications is of much more importance.

Such functions can be obtained in two ways:

- A first possibility is the derivation of analytic functions or procedures that return the power or area estimate given the block's specifications. An example of a general yet relatively accurate power estimator that was derived based on the underlying operating principles for the whole class of CMOS high-speed Nyquist-rate analogue-to-digital converters (such as flash, two-step, pipelined…architectures) is given by [30]:

$$\text{power} = \frac{V_{dd} \cdot L_{\min} \cdot (F_{\text{sample}} + F_{\text{signal}})}{10^{(-0.15 \cdot \text{ENOB} + 4.24)}} \tag{15.6}$$

where F_{sample} and F_{signal} are the clock and signal frequency, respectively, and where ENOB is the effective number of bits at the signal frequency. The estimator is technology scalable (V_{dd} and $L_{\min}$ are parameters of the model), and has been fitted with published data of real converters, and for more than 85 per cent of the designs checked, the estimator has an accuracy better than $2.2\times$. Similar functions are developed for other blocks, but of course often a more elaborate procedure is needed than a simple formula. For example, for the case of high-speed continuous-time filters [30], a crude filter synthesis procedure in combination with operational transconductor amplifier behavioural models had to be developed to generate accurate results, because the implementation details and hence the power and chip area vary quite largely with the specifications.

- A second possibility to develop power/area estimators is to extract them from a whole set of data samples from available or generated designs through interpolation or fitting of a predefined function or an implicit function like e.g. a neural network. As these methods do not rely on underlying operating principles, extrapolations of the models have no guaranteed accuracy.

In addition to power and area estimators also feasibility functions are needed that limit the high-level optimisation to realisable values of the building block specifications. These can be implemented under the form of functions (e.g. a trained neural network or a support vector machine [31]) that return whether a block is feasible or not, or of the geometrically calculated feasible performance space of a circuit (e.g. using polytopes [32] or using radial base functions [33]). These methods are also useful during automatic topology selection during circuit synthesis.

15.3 Analogue circuit and layout synthesis

Due to the knowledge-intensive nature of analogue design, most analogue designs today are still handcrafted manually by analogue expert designers, with only a SPICE-like simulation shell and an interactive layout environment (with parameterised procedural device generators) as supporting facilities. This makes the design cycle for analogue circuits long and error-prone. Therefore, although analogue circuits typically occupy only a small fraction of the total area of mixed-signal ICs, their design is often the bottleneck in mixed-signal systems, both in design time and effort as well as test cost, and they are often responsible for design errors and expensive reruns. This handcrafting is also increasingly at odds with the shortening time-to-market constraints of current consumer market products. This explains the growing need observed in industry today for analogue CAD tools that increase analogue design productivity by assisting designers with fast and first-time-correct design of analogue circuits, or even by automating certain tasks or the entire circuit design process where possible. Moreover, the performance of an analogue circuit is very much dependent on the characteristics of the technology used, making the use of fixed analogue cell libraries uneconomical. Therefore, for an analogue or RF design business to be economically viable, some form of 'soft' IP must be used, where the design knowledge is embedded in some sort of synthesis or generator tool, that can then spawn optimised designs in any specified target technology.

While the basic level of design abstraction for analogue circuits is mainly still the transistor level, commercial CAD tool support for analogue cell-level circuit and layout synthesis is currently emerging. There has been remarkable progress at research level over the past decade, and in recent years several commercial offerings have appeared on the market. Gielen and Rutenbar [2] offer a fairly complete survey of the area. Analogue synthesis consists of two major steps: (1) circuit synthesis followed by (2) layout synthesis. Most of the basic techniques in both circuit and layout synthesis rely on powerful numerical optimisation engines coupled to 'evaluation engines' that qualify the merit of some evolving analogue circuit or layout

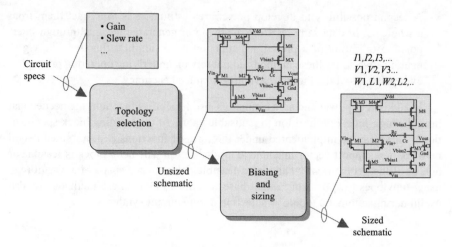

Figure 15.13   Flow of analogue circuit synthesis for a basic cell: topology selection and circuit sizing. For more complex cells these steps are repeated in an hierarchical refinement scheme down to the transistor level

candidate. State-of-the-art techniques in analogue circuit and layout synthesis will now be discussed in more detail.

15.3.1   Analogue circuit synthesis

The goal of analogue circuit synthesis is to create a sized circuit schematic from given circuit specifications. Therefore, as shown in Figure 15.13, circuit synthesis consists of two tasks: topology selection and specification translation/sizing [2]. Circuit synthesis is a critical step since most analogue designs require a custom optimised design and the design problem is typically underconstrained with many degrees of freedom and with many (often conflicting) performance requirements to be taken into account. Given a specified block performance, first an appropriate topology or circuit schematic has to be chosen to implement this block (this is the topology selection step). Subsequently, values for the subblock parameters have to be determined, so that the final block meets the specified performance constraints, preferrably in some optimised way according to the design criteria used. At the device level this step is called circuit sizing, in which case the sizes and biasing of all devices have to be determined. However, for more complex cells, the flow of Figure 15.13 is repeated in an hierarchical way with subsequent refinements down to the transistor level. At higher levels in the design hierarchy this sizing step is then called specification translation where performance specifications of the subblocks within the selected block topology have to be determined based on the block's overall specifications. The complete design flow is then an alternation of topology selection and specification translation down the design hierarchy [34]. In many cases the initial sizing produces a near-optimal design that is further fine-tuned with a circuit optimisation tool, e.g. to improve yield and design robustness. The performance of the resulting design is then verified using detailed circuit simulations with a simulator such as SPICE, and when needed the synthesis process is iterated to arrive at a close-fit design. We will now discuss the two basic tasks in more detail.

15.3.1.1 Topology selection

Given a set of performance specifications and a technology process, a designer or a synthesis tool must first select a circuit schematic that is most suitable to meet the specifications at minimal implementation cost (power, chip area). This problem can be solved by selecting a schematic from among a known set of alternative topologies such as stored in a library (topology selection), or by generating a new schematic, e.g. by modifying an existing schematic. Although the earliest synthesis approaches considered topology selection and sizing together, the task of topology selection has received less attention in recent years, where the focus was primarily on the circuit sizing. As finding the optimal circuit topology for a given set of performance specifications brings to bear the real expert knowledge of a designer, it was only natural that the first topology selection approaches like in OASYS [34], BLADES [35] or OPASYN [36] were rather heuristic in nature in that they used rules in one format or another to select a proper topology (possibly hierarchically) out of a predefined set of alternatives stored in the tool's library.

Later approaches worked in a more quantitative way in that they calculated the feasible performance space of each topology which fits the structural requirements, and then compared that feasible space to the actual input specifications during synthesis to decide on the appropriateness and the ordering of each topology. This can for instance be done using interval analysis techniques [32] or using interpolation techniques in combination with adaptive sampling [33]. In all these programs, however, topology selection is a separate step. There are also a number of optimisation-based approaches that integrate topology selection with circuit sizing as part of one overall optimisation loop, but typically only a limited number of predefined topological choices were allowed in the optimisation [37,38]. An interesting approach that uses a genetic algorithm to find the best topology choice was presented in DARWIN [39].

Of these methods, the quantitative and optimisation-based approaches are the more promising developments that address the topology selection task in a deterministic fashion as compared to the rather ad hoc heuristic methods, and they also open up the possibility for developing computer methods for structural or topological synthesis of analogue circuits, possibly leading to novel, yet unknown circuit structures.

15.3.1.2 Analogue circuit sizing

Once an appropriate topology has been selected, the next step is specification translation, where the performance parameters of the subblocks in the selected topology are determined based on the specifications of the overall block. At the lowest level in the design hierarchy this reduces to circuit sizing where the sizes and biasing of all devices have to be determined such that the final circuit meets the specified performance constraints. This mapping from performance specifications into proper, preferrably optimal, device sizes and biasing for a selected analogue circuit topology generally involves solving the set of physical equations that relate the device sizes to the electrical performance parameters. However, solving these equations explicitly is in general not possible, and 'analogue circuit sizing typically results in an underconstrained problem with many degrees of freedom'.

The two basic ways to solve for these degrees of freedom in the analogue sizing process are [2]:

- either by exploiting analogue design knowledge and heuristics;
- or by using today's powerful and robust optimisation techniques.

The first generation of analogue circuit synthesis systems presented in the mid to late 1980s like IDAC [40] and OASYS [34] were 'knowledge-based': specific heuristic design knowledge about the circuit topology under design (including the design equations but also the design strategy) was solicited from designers and encoded explicitly in some computer-executable form (e.g. a design plan), which was then executed during the synthesis run for a given set of input specifications to directly and fast obtain the design solution. However, the coverage range of these tools was found to be too small and the setup effort for introducing new schematics into the system was too large for real-life industrial practice, hence these tools failed on the commercial marketplace.

Therefore, starting from the late 1980s and until today, analogue circuit sizing methods are using robust numerical optimisation techniques to implicitly solve for the degrees of freedom in analogue design while optimising the performance of the circuit under the given specification constraints. The basic flow of such an optimisation-based sizing approach is schematically illustrated in Figure 15.14. At each iteration of the optimisation routine, i.e. for each set of proposed design variables (typically

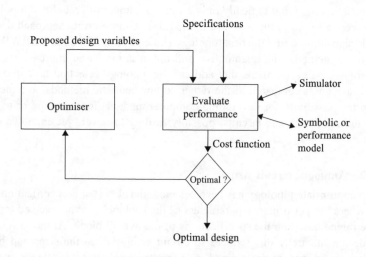

Figure 15.14   Basic flow of optimisation-based analogue circuit sizing

device sizes such as widths and lengths but also biasing currents and voltages of transistors, values of resistors and capacitors, etc.), the performance of the circuit has to be evaluated. Depending on which method (simulations, equations or performance models) is used for this performance evaluation, two different subcategories of methods, and recently also a compromise approach, can be distinguished:

- equation-based methods that use analytic equations to evaluate circuit performance;
- simulation-based methods that use numerical simulations to evaluate circuit performance;
- model-based methods that use performance models to evaluate circuit performance.

These different approaches are now described in more detail.

In the subcategory of 'equation-based optimisation approaches', (simplified) analytic design equations are used to describe the circuit performance. In approaches like OPASYN [36] and STAIC [41] the design equations still had to be derived and ordered by hand, but the degrees of freedom were resolved implicitly by optimisation. The OPTIMAN tool [42] added the use of a global simulated annealing algorithm, but also tried to solve the circuit design knowledge derivation problem: symbolic analysis techniques [43] were developed to automate the derivation of the (simplified) analytic design equations and constraint satisfaction techniques were used to automatically generate the design plans needed to evaluate the circuit performance at every iteration of the optimisation [44]. Together with a separate topology-selection tool based on boundary checking and interval analysis [32] and the performance-driven layout generation tool LAYLA [45], all these tools were integrated into the AMGIE analogue circuit synthesis system [46] that covers the complete design flow from specifications over topology selection and circuit sizing down to layout generation and automatic verification.

15.3.1.3   Example

An example of a circuit that has been synthesised with this AMGIE system is the particle/radiation detector frontend of Figure 15.15, which consists of a charge-sensitive amplifier (CSA) followed by an n-stage pulse-shaping amplifier (PSA) [46]. All opamps are complete circuit-level schematics in the actual design as indicated in the figure. A comparison between the specifications and the performances obtained by an earlier manual design of an expert designer and by the fully computer-synthesised circuit is given in Table 15.1. In the experiment, a reduction of the power consumption with a factor of 6 (from 40 to 7 mW) was obtained by the synthesis system compared to the manual solution. Also the final area is slightly smaller. Clearly, the computer-generated synthesised approach outperforms the manual design in power consumption!! The layout generated for this example is shown in Figure 15.16 and is very comparable to manual layout!

The technique of equation-based optimisation has also been applied to the high-level synthesis of $\Delta\Sigma$ modulators in the SD-OPT tool [47]. Recently a first attempt

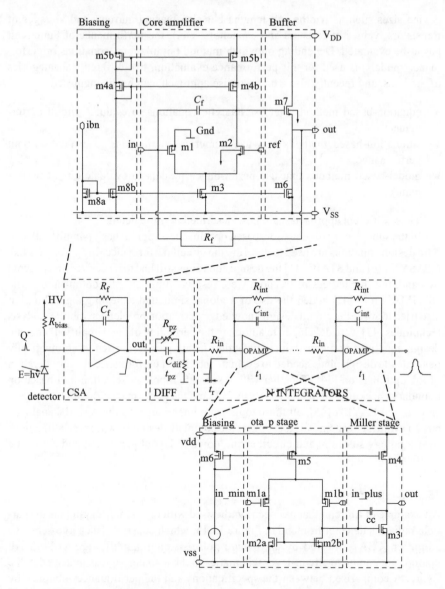

Figure 15.15   Particle/radiation detector frontend as example for analogue circuit synthesis (The opamp and filter stage symbols represent full circuit schematics as indicated.)

was presented towards the full behavioural synthesis of analogue systems from an (annotated) VHDL-AMS behavioural description. The VASE tool follows a hierarchical two-layered optimisation-based design-space exploration approach to produce sized subblocks from behavioural specifications [48].

Table 15.1   Results of analogue circuit synthesis experiment with the AMGIE system [46]

Performance	Specification	Manual design	Automated synthesis
Peaking time	<1.5 ms	1.1 ms	1.1 ms
Counting rate	>200 kHz	200 kHz	294 kHz
Noise	<1000 RMS e-	750 RMS e-	905 RMS e-
Gain	20 V/fC	20 V/fC	21 V/fC
Output range	>−1..1 V	−1..1 V	−1.5..1.5 V
Power	Minimal	40 mW	7 mW
Area	Minimal	0.7 mm^2	0.6 mm^2

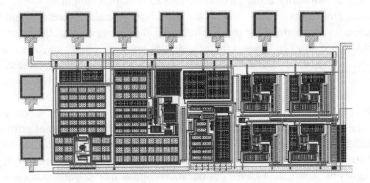

Figure 15.16   Layout of the particle/radiation detector frontend generated with the LAYLA analogue layout synthesis tool [45]

Recently, it has been shown that the design of CMOS opamps can be formulated (more precisely, it can be fairly well approximated) as a posynomial convex optimisation problem that can then be solved using geometric programming techniques, producing a close-by first-cut design in an extremely efficient way [49,50]. The initial optimisation time of minutes literally reduces to seconds. The same approach has been applied to some other circuits as well [51]. Unfortunately, not all circuit characteristics are posynomial, and approximating them accurately with posynomial functions is not always a simple task. Mixed solutions can be used as well, but computational efficiency goes down [52].

In general, the big advantages of the above analytic approaches are their fast evaluation time and their flexibility in manipulation possibilities. The latter is reflected in the freedom to choose the independent input variables, which has a large impact on the overall evaluation efficiency, as well as the possibility to perform more symbolic manipulations. The big drawback of the analytic methods, including the geometric programming ones, however, is that the design equations still have to be derived and, despite the progress in symbolic circuit analysis, not all design characteristics (such as transient or large-signal responses) are easy to capture in analytic equations with sufficient accuracy. For such characteristics either rough approximations have to be used, which undermines the sense of the whole approach, or one has to fall back on numerical simulations. Recent approaches use regression techniques to fit simulation data to an equation template for any kind of simulatable characteristic. For example, Daems *et al.* [27] uses a fitting approach for the automatic generation of symbolic posynomial performance models for analogue circuits, that are created by fitting a pre-assumed posynomial equation template to simulation data created according to some design-of-experiments scheme [27]. The accuracy of the results of course depends on the equation template that is used. The template not necessarily has to be posynomial, and in this way symbolic expressions can be generated for large-signal and transient characteristics. Very recently even a template-free approach has been presented where no predefined fitting template is used, but where the 'template' is evolved dynamically using genetic optimisation with a canonical-form grammar that adds extra terms or functions to the evolving symbolic expression until sufficient accuracy is obtained for the symbolic results with respect to the reference set of simulation data [97]. These kind of methods are very promising, since they are no longer limited to simple device models nor to small-signal characteristics only – they basically work for whatever characteristic can be simulated – but they still need further research.

All the above problems with equation-based methods have since the mid 1990s sparked research efforts to try to develop equation-free circuit optimisation approaches that use plain numerical simulators instead of symbolic models within the circuit sizing optimisation loop. Facilitated by improving computer power, a second subcategory of 'simulation-based optimisation approaches' towards analogue circuit synthesis has therefore emerged in recent years. These methods couple robust numerical optimisation with full SPICE simulation, making it possible to synthesise designs using the same modelling and verification tool infrastructure and accuracy levels that human experts use for manual design, be it at the expense of large CPU times (hours or days of optimisation time). These methods perform some form of full numerical simulation to evaluate the circuit's performance in the inner loop of the optimisation (see Figure 15.14). Although the idea of optimisation-based design for analogue circuits dates back at least 30 years [53], where tools like DELIGHT.SPICE [54] were used in fine-tuning an already designed circuit to better meet the specifications, the challenge in automated synthesis is to solve for all degrees of freedom when no good initial starting point can be provided. It is only recently that the computer power and numerical algorithms have advanced far enough to make this really practical.

The FRIDGE tool [55] calls a plain-vanilla SPICE simulation at every iteration of a simulated-annealing-like global optimisation algorithm. To cut down on the large synthesis time, more efficient optimisation algorithms are used and/or the simulations are executed as much as possible in parallel on a pool of workstations. The ANACONDA tool [56] for instance uses a global optimisation algorithm based on stochastic pattern search that inherently contains parallelism and therefore can easily be distributed over a pool of workstations, to try out and simulate 50 000 to 100 000 circuit candidates in a few hours. These brute-force approaches require very little

Table 15.2   Comparison between manual design result [59] and DAISY synthesis result [58] for a time-discrete $\Delta\Sigma$ modulator for ADSL specifications

Building block specifications	Published manual [59]	Synthesised DAISY [58]
Topology	Cascaded 2-1-1-	Cascaded-2-1-1
Oversampling ratio	24	24
OTA gain	>60 dB	>52 dB
OTA GBW	>160 MHz	>222 MHz
OTA output swing	≥1.8 V	>1.86 V
Switch on-resistance	<215 Ω	<217 Ω
Comparator offset	<100 mV	<130 mV
Comparator hysteresis	<40 mV	<19 mV

advance modelling work to prepare for any new circuit topology and have the same accuracy as SPICE. In Reference 57 ANACONDA/MAELSTROM, in combination with macromodelling techniques to bridge the hierarchical levels, was applied to an industrial-scale analogue system (the equaliser/filter frontend for an ADSL CODEC). Again, the experiments demonstrated that the synthesis results are comparable to or sometimes better than manual design!

The DAISY tool provides efficient high-level synthesis of discrete-time $\Delta\Sigma$ modulators [58] based on a simulation-based optimisation strategy. Simulations are now not performed at SPICE circuit level but at behavioural level. The high-level optimisation approach determines both the optimum modulator topology and the required building block specifications, such that the system specifications – mainly accuracy (dynamic range) and signal bandwidth – are satisfied at the lowest possible power consumption. A genetic-based differential evolution algorithm is used in combination with a fast dedicated $\Delta\Sigma$ behavioural simulator to realistically analyse and optimise the modulator performance. Table 15.2 shows the comparison between the results of a manual design [59] and the DAISY synthesis result [58] for ADSL specifications. Note that exactly the same modulator topology was decided upon by the tool (a cascaded 2-1-1 topology – see Figure 15.17) with the same oversampling ratio, while the synthesised building block specifications are also very similar to the manual design. Recently the DAISY tool was also extended to continuous-time $\Delta\Sigma$ modulators [60].

Although appealing, these simulation-based circuit optimisation methods still have to be used with care by designers because the runtimes (and therefore also the initial debug time) remain long, and because the optimiser may easily produce improper designs if the right design constraints are not added to the optimisation problem. Reducing the CPU time remains a challenging area for further research, and the use of performance models is one possible avenue being explored today to that end as these so-called 'model-based optimisation approaches' offer a good compromise between equation-based (with its speed) and simulation-based (with accuracy and no restriction to small-signal) optimisation. As discussed in Section 15.2.3, performance

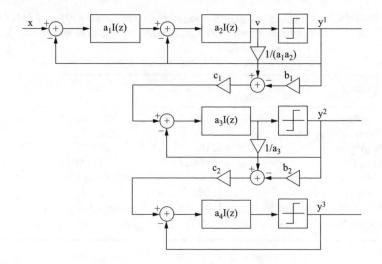

Figure 15.17   Cascaded 2-1-1 $\Delta\Sigma$ modulator topology

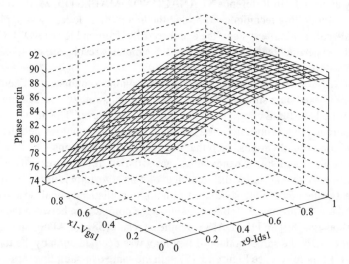

Figure 15.18   Performance model of the phase margin as a function of two design variables for an opamp (subset of the actual multidimensional performance model)

models relate the achievable performances of a circuit (e.g. gain, bandwidth, slew rate or phase margin) to the design variables (e.g. device sizes and biasing). Figure 15.18 for example shows part of such a performance model, displaying the phase margin as a function of two design variables for an operational amplifier [28]. Such performance models are used to speed up the circuit optimisation procedure, since calls to the transistor-level simulation are replaced by evaluations of the performance model, resulting in substantial speedups (once the performance models have been created and calibrated, which is a one-time up-front effort). Techniques to generate such models have been discussed in Section 15.2.3.

In addition, because of the large CPU times due to the circuit complexity, for more complex cells a hierarchical approach is needed, which requires behavioural or macromodels to bridge the different levels. This research is on-going at the moment. Another trend is the move towards multi-objective optimisation [61], which generates a set of design solutions, spread over the Pareto-optimal trade-off front, so that designers can a posteriori decide on the final design point taken instead of entering a priori weighting coefficients to the lumped cost function.

Although additional research is still needed, especially to reduce the CPU times for more complex cells, it can be concluded that a lot of research progress has been achieved over the last 10 years in the field of analogue circuit synthesis using optimisation techniques. Based on these results in recent years several commercial tools have been developed, that are now available or that will be introduced on the marketplace in the very near future, offering to industry the possibility to integrate analogue circuit optimisation in their design flows.

15.3.1.4 Yield and design for manufacturability

It has to be added that industrial design practice not only calls for fully optimised *nominal* design solutions, but also expects high robustness and yield in the light of varying operating conditions (supply voltage or temperature variations) and statistical manufacturing tolerances and mismatches [62,63]. Due to these fluctuations, the device parameters and consequently also the circuit performance characteristics will show fluctuations. The corresponding parametric yield is the ratio of the number of acceptable (i.e. functional and meeting all specifications) to all fabricated IC samples. The yield of course depends on the nominal design point chosen for the circuit, but unfortunately the relation between the (fluctuating) device parameters and the circuit performances is in general a non-linear transformation that is not known explicitly but has to be simulated. All this makes yield estimation a time-consuming task, which in practice is often obtained by Monte-Carlo simulations. An overview of more efficient techniques that trade-off accuracy versus CPU time can be found in Reference 2. Note that in practice not only the yield, but in general the robustness of the design against variations of both technological and environmental parameters has to be maximised. This implies techniques for variability minimisation and design centring. Both aspects can be captured in a characteristic like the capability index Cpk.

Here we briefly describe the efforts to integrate yield and Cpk optimisation in the analogue circuit synthesis process itself. Yield and robustness precautions were already hardcoded in the design plans of IDAC [40], but are more difficult to incorporate in optimisation-based approaches. Nevertheless, first attempts in this direction have already been presented. The ASTRX/OBLX tool has been extended with manufacturability considerations and uses a non-linear infinite programming formulation to search for the worst-case 'corners' at which the evolving circuit should be evaluated for correct performance [64]. The approach has been successful in several test cases but does increase the required CPU time even further (roughly by 4–10×). Also the OPTIMAN program has been extended by fully exploiting the availability of the analytic design equations to generate closed-form expressions for the sensitivities of the performances to the process parameters [65]. The impact of tolerances and mismatches on yield or Cpk can then easily be calculated at each optimisation iteration, which then allows to synthesise the circuits simultaneously for performance and for manufacturability (yield or Cpk). The accuracy of the statistical predictions still has to be improved. The approach in Reference 66 uses parameter distances as robustness objectives to obtain a nominal design that satisfies all specifications with as much safety margin as possible for process variations. The resulting formulation is the same as for design centring and can be solved efficiently using the generalised boundary curve. Design centring, however, still remains a second step after the nominal design. Therefore, more research in this direction is still needed, in order to develop techniques that can directly synthesise optimal and robust analogue designs in an efficient way.

15.3.2   Analogue layout synthesis

The next important step in the top-down mixed-signal design flow of Figure 15.1 after the circuit synthesis is the generation of the layout. The field of analogue layout synthesis is more mature than circuit synthesis, in large part because it has been able to leverage ideas from the mature field of digital layout, and several real commercial solutions have appeared on the market in recent years that can automate analogue layout generation. Below we distinguish analogue circuit-level layout synthesis, which has to transform a sized transistor-level schematic into a mask layout, and system-level layout assembly, in which the basic functional blocks are already laid out and the goal is to floorplan, place and route them, as well as to distribute the power and ground connections.

15.3.2.1   Analogue circuit-level layout synthesis

The earliest approaches to analogue cell layout synthesis relied on 'procedural module generation' [67], but these methods are mainly interesting at the device level only (transistors, spiral inductors, capacitor banks, etc.). Every layout engineer today uses such parameterised procedural device generators to create his or her layouts manually. To synthesise compact layouts of entire circuits, alternative methods have to be used. A first group of methods are the 'template-driven' approaches. For each circuit a geometric template (e.g. a sample layout [68] or a slicing tree [36]) is stored that fixes the relative position and interconnection of the devices. The layout is then completed by correctly generating the devices and the wires for the actual values of the design according to this fixed geometric template, thereby trying to use the area as efficiently as possible. These approaches are relatively fast but work best when the changes in circuit parameters result in little need for global alterations in the general circuit layout structure. This is the case for instance during technology migration or porting of existing layouts, but this is not the case in general. Because of their speed these methods are also typically used in a combined circuit and layout optimisation loop, as needed for RF circuits for instance (see below).

In practice changes in the circuit's device sizes often require large changes in the layout structure in order to get the best performance and the best area occupation. As the performance of an analogue circuit is negatively impacted by the parasitics introduced by the layout, such as the parasitic wire capacitance and resistance or the crosstalk capacitance between two neighbouring or crossing wires, it is of utmost importance to generate analogue circuit layouts such that (1) the resulting circuit still satisfies all performance specifications, and (2) the resulting layout is as compact as possible. This requires full-custom optimised layout synthesis, which today is typically implemented using an 'optimisation-based microcell-place-and-route layout generation approach' [2] where the layout solution is not predefined by some template, but where both placement and routing of basic devices or groups of devices (the 'microcells', e.g. current mirrors) are formulated as optimisation problems driven by some cost function. This cost function typically contains minimum area and net length and adherence to a given aspect ratio, but also other terms could be added (e.g. quantification of important performance degradations such as crosstalk). The advantage of the optimisation-based approaches is that they always look for the most optimum layout solution at runtime. The penalty to pay is their larger CPU times, and the dependence of the layout quality on the set-up of the cost function.

Examples of such tools are ILAC [69] and the different versions of KOAN/ANAGRAM [70,71]. The device placer KOAN relied on a very small library of device generators and migrated important layout optimisations into the placer itself. KOAN, which was based on an efficient simulated annealing algorithm, could dynamically fold, merge and abut MOS devices and thus discover desirable optimisations to minimise parasitic capacitance on the fly during optimisation. Its companion, ANAGRAM II, was a maze-style detailed area router capable of supporting several forms of symmetric differential routing, mechanisms for tagging compatible and incompatible classes of wires (e.g. noisy and sensitive wires), parasitic crosstalk avoidance and over-the-device routing. Also other device placers and routers operating in the macrocell-style have appeared (e.g. LADIES [72] and ALSYN [73]). Results from these tools can be quite impressive. For example, Figure 15.19 shows two versions of the layout of an industrial 0.25 μm CMOS comparator [2]. On the left is a manually created layout, on the right is a layout generated automically with a commercial tool operating in the microcell style. The automatic layout compares well to the manual one.

An important improvement in the next generation of optimisation-based layout tools was the shift from a rather qualitative consideration of analogue constraints to an explicit quantitative optimisation of the performance goals, resulting in the 'performance-driven' or 'constraint-driven' approaches. The degradation of the performance due to layout parasitics is quantified explicitly and the layout tools are driven such that this extra layout-induced performance degradation is within the margins allowed by the designer's performance specifications [74]. In this way, more optimum solutions can be found as the importance of each layout parasitic is weighed according to its impact on the circuit performance, and the tools can much better guarantee by construction that the circuit will meet the performance specifications also after the layout phase (if possible). Tools that adopt this approach include the

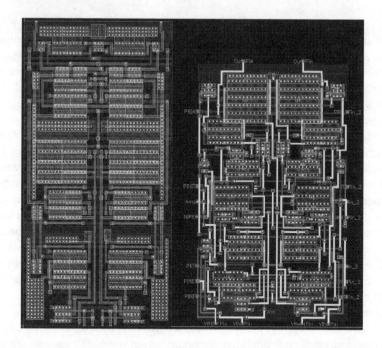

*Figure 15.19   Manual (left) versus automatic (right) layout for an industrial 0.25 μm
CMOS analogue comparator [2]*

area router ROAD [75], the placement tool PUPPY-A [76] and the compaction tool
SPARCS-A [77]. The routers ROAD [75] and ANAGRAM III [78] have a cost
function which drives them such that they minimise the deviation from acceptable
bounds on wire parasitics. These bounds are provided by designers or derived from
the margins on the performance specifications via sensitivities. The LAYLA system
[45,79] consists of performance-driven analogue placement and routing tools that
minimise the layout area while enforcing typical constraints such as symmetry and
that keep the performance degradation introduced by the layout parasitics within
the margins allowed by the user by penalising excess layout-induced performance
degradation. Effects considered include for instance the impact of device merging,
device mismatches, parasitic capacitance and resistance of each wire, parasitic coupling due to specific proximities, thermal gradients, etc. The router can manage not
just parasitic wire sensititivies, but also yield testability concerns [80]. A layout of a
particle-detector frontend circuit generated by means of the LAYLA tool was shown
in Figure 15.16. In all the above tools, sensitivity analysis is used to quantify the
impact on the final circuit performance of low-level layout decisions [74].

The above constraint-driven and performance-driven optimisation-based layout
synthesis methods for analogue circuits have matured significantly in recent years,
and are currently being offered commercially on the marketplace and can be integrated
in today's industrial design flows to increase analogue layout productivity.

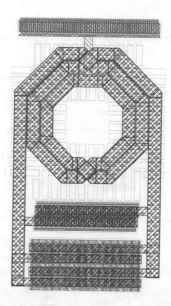

Figure 15.20   Full VCO layout generated with CYCLONE [81]

A problem for truly parasitic-sensitive circuits, such as RF circuits, is decoupling between circuit synthesis and layout synthesis which can no longer be treated as separate steps. Circuit sizing of RF circuits needs really accurate estimates of circuit wiring loads and other parasitics to obtain good sizing results. Therefore, the only possibility to achieve this is to merge layout synthesis into the circuit synthesis. To make this computationally tractable, typically template-driven (or procedural) layout generation techniques are used to generate the layout and extract the actual layout parasitics of the entire RF circuit at each iteration of the circuit sizing optimisation loop. For example the CYCLONE tool [81] generates optimal CMOS RF LC-tank VCOs. Both the circuit's device sizes and the inductor coil geometry parameters are globally optimised for the specified technology process as to meet the specifications (centre frequency, tuning range, phase noise) at minimum power consumption. The tool automatically performs electromagnetic simulations for the on-chip inductor to accurately calculate its losses during the circuit optimisation. It uses a template-based layout generation approach to obtain accurate predictions of the actual layout parasitics. Figure 15.20 shows an automatically generated VCO layout. The results depend on the characteristics of the target technology, as shown by the optimised coil parameters in Table 15.3. This tool is perfect for generating customised VCOs as IP macrocells.

On the other hand, for circuits with a more regular structure, other layout techniques are needed. This is true for the generation of ROMs or RAMs, but also for array-type analogue circuits like current-steering digital-to-analogue converters, folding/interpolating analogue-to-digital converters, etc. The MONDRIAAN tool [82] was developed for this purpose and it translates global layout specifications into a

*Table 15.3   VCO parameters resulting from two CYCLONE syn-
thesis runs for the same set of specifications but in two
different technologies [81]*

	Technology	
Parameter	Low-resistive substrate	High-resistive substrate
Ls	1.81 nH	2.85 nH
Rs	0.95 Ω	0.74 Ω
Inner rad, W, #Turns	134 μm, 22 μm, 2	178 μm, 18 μm, 2
Used metal layers	3 top layers	All 4 layers
Power	12.8 mW	8.8 mW

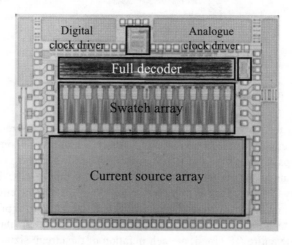

*Figure 15.21   Layout of a 14-bit digital-to-analogue converter generated using
MONDRIAAN [82] for the two analogue arrays and commercial digital place&route tools for the decoder [83]. The entire layout measures
3.2 × 4.1 mm^2*

detailed placement and interconnections of all basic cells in the array. The tool was
used to generate the layouts of the switch/latch array (middle) and the current cell
array (bottom) of the 14-bit current-steering DAC of Figure 15.21 [83]. The block at
the top is the thermodecoder which was synthesised from VDHL code with a logic
synthesis tool and the layout of which was generated with a commercial digital standard cell place&route tool. Use of tools such as MONDRIAAN resulted in a 3.5×
increase in design productivity for a truly high-performance analogue design [12].

15.3.2.2   Mixed-signal block place and route
After generating the layout of the individual blocks, the next step in the system design
flow (see Figure 15.1) is block place and route to assemble the system's overall layout.

For this, commercial tools exist but they still do not include all the constraints typically needed in mixed-signal designs, such as the handling of arbitrarily shaped blocks and complex symmetries, as well as the avoidance of signal interactions (crosstalk) and noise couplings. The academic WREN tool [84] comprises both a mixed-signal global router and channel router. The tool uses the notion of SNR-style (signal-to-noise ratio) constraints for incompatible signals, and strives to comply with designer-specified noise rejection limits on critical signals.

Critical in mixed-signal system layout is also the power grid design. In the mixed-signal case not only connectivity, ohmic drops and electromigration effects have to be considered, but also noise constraints (including transient effects like current spikes) and arbitrary (non-tree) grid topologies. The RAIL tool [85] addresses such concerns by casting mixed-signal power grid synthesis as a routing problem that uses fast AWE-based linear system evaluation to electrically model the entire power grid, package and substrate during layout while trying to satisfy dc, ac and transient performance constraints.

15.4 Mixed-signal verification and crosstalk analysis

The final step in the design flow is the detailed verification of the entire system layout. For mixed-signal systems this is today still a very big problem, both at the layout level and at the electrical level. At the layout level, DRC, ERC and LVS can easily be done for the different blocks. DRC and ERC can also easily be done for the entire chip, but due to the different tools typically used for analogue and digital blocks LVS of complete analogue–digital systems is not at all trivial.

The situation is even worse at the electrical level. After extraction of the parasitics from the layout the performance of the individual blocks can be verified using detailed simulations, but due to the complexity no complete device-level simulation of the entire system is feasible. Therefore, the performance of individual blocks has to be abstracted into behavioural models, which are then used to simulate and verify the system performance. The automatic extraction of analogue behavioural models that simulate fast, yet include the important non-idealities, is however a hot research area in full progress at this moment. See Section 15.2.3 for an overview. In industrial practice today, system verification is still merely a check of correct connectivity than a true proof of functionality and performance, being even far from any formal verification proof.

A difficult problem in mixed-signal designs, where sensitive analogue and RF circuits are integrated on the same die with large digital circuitry, is signal integrity analysis, i.e. the verification of all unwanted signal interactions through crosstalk or couplings at the system level that can cause parametric malfunctioning of the chip. Parasitic signals are generated (e.g. digital switching noise) and couple into the signal of interest, degrading or even destroying the performance of the analogue/RF circuitry. These interactions can come from capacitive or (at higher frequencies) inductive crosstalk, from supply line or substrate couplings, from thermal interactions, from coupling through the package, from electromagnetic interference (such

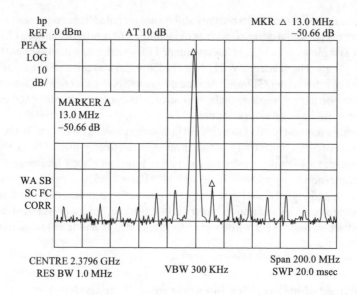

hp
REF .0 dBm   AT 10 dB

MKR △ 13.0 MHz
−50.66 dB

PEAK
LOG
10
dB/

MARKER △
13.0 MHz
−50.66 dB

WA SB
SC FC
CORR

CENTRE 2.3796 GHz
RES BW 1.0 MHz

VBW 300 KHz

Span 200.0 MHz
SWP 20.0 msec

Figure 15.22   Measured FM modulation due to substrate switching noise coupling [87]

as EMC/EMI), etc. Especially the analysis of digital switching noise that propagates through the substrate shared by the analogue and digital circuits has received much attention in recent years [86]. At the instants of switching, digital circuitry can inject spiky signals into the substrate, which then will propagate to and be picked up by the sensitive analogue/RF circuits. As an example, consider a VCO at 2.3 GHz and a digital circuit block (250 k gates) running at 13 MHz. As shown on the measurement plot of Figure 15.22, the digital clock is visible as FM modulation around the VCO frequency and may cause conflicts with out-of-band emission requirements [87].

In recent years research has been going on to find efficient yet accurate techniques to analyse these problems, which depend on the geometrical configuration and therefore are in essence three-dimensional field solving problems. Typically, finite difference methods or boundary element methods are used to solve for the substrate potential distribution due to injected noise sources [88–91], allowing to simulate the propagation of digital switching noise injected in the substrate to sensitive analogue nodes elsewhere in the same substrate. Recently these methods have been speeded up with similar acceleration techniques as in RF or interconnect simulation, e.g., using an eigendecomposition technique [92]. This propagation analysis, however, has to be combined with an analysis of the (signal-dependent) digital switching activity to know the actual (time-varying) injected signals, and with an analysis of the impact of the local substrate voltage variations on the analogue/RF circuit performance (e.g. the reduction of the effective number of bits of an embedded analogue-to-digital converter) in order to cover the entire problem.

The problem on the generation side is that the noise generating sources (i.e. the switching noise injected by the digital circuitry) are not known accurately but vary with time depending on the input signals or the embedded software being executed, and therefore have to be estimated statistically. Some attempts to solve this problem characterise every cell in a digital standard cell library by the current they inject in the substrate due to an input transition, and then calculate the total injection of a complex system by combining the contributions of all switching cells over time [93,94]. In the SWAN methodology [94] an equivalent macromodel of every standard cell is extracted which consists of capacitances, resistances and two time-varying current sources that model the current drawn between the two supplies and the current injected into the substrate when an input of the cell switches. These current waveforms are stored in a database. Once the library has been characterised, SWAN [94] extracts the actual switching data of a large complex system from VHDL simulations, and calculates the actual time-varying substrate-bounce voltage by combining the macromodels of all cells used in the design with a model for the package and external supply, and by efficiently simulating this network over time while applying the time-varying noise current source waveforms out of the database depending on the actual switchings of the cells as identified during the VHDL simulation. Figure 15.23 e.g., shows the comparison between time-domain SWAN simulations and measurements on a large experimental WLAN SoC with 220 k gates, that contains a scalable OFDM-WLAN baseband modem, a low-IF digital IQ (de)modulator, and an 8-bit embedded analogue-to-digital converter [95] fabricated in a 3.3 V 0.35 μm CMOS 2P5M process on an EPI-type substrate. Compared to the measurements, the simulated substrate-noise voltage from zero to 100 ns is within an error of 20 per cent in its RMS value and within an error of 4 per cent in its peak-to-peak value, which is a very good result for a difficult crosstalk effect like substrate noise couplings. Techniques to analyse the impact of this time-varying substrate and supply noise voltage on the

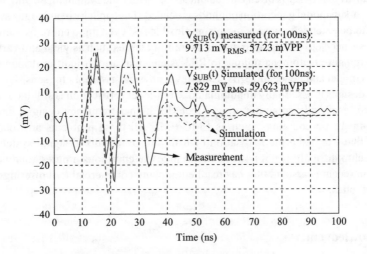

Figure 15.23   Measured and SWAN [94] simulated substrate noise in an experimental 220 k-gates WLAN SoC [95]

performance of the embedded analogue blocks are presently also being developed, but still require further work [96].

15.5   Conclusions

The last few years have seen significant advances in both design methodology and CAD tool support for analogue, mixed-signal and RF designs, enabling mixed-signal integrated systems on chip. The emergence of commercial AMS simulators supporting analogue behavioural modelling enables top-down design flows in many industrial scenarios. In addition, there is increasing research going on in system-level modelling and analysis, allowing architectural exploration of entire systems. Analogue cell synthesis tools, both for circuit sizing and for physical layout generation, all based on powerful optimisation methods, have appeared commercially on the market in several competing formulations. The use of synthesis or generation tools together with behavioural modelling also enables the soft reuse of analogue and RF blocks, and the fast migration of analogue blocks from one process to another. In addition, there is an increasing emphasis in the research community on mixed-signal verification and in particular on signal integrity analysis, to analyse problems related to embedding analogue blocks in a digital environment. Especially on the analysis of substrate and supply noise couplings in mixed-signal ICs a lot of progress has been made, with techniques developed that can predict substrate noise fluctuations in large digital systems within acceptable accuracy.

Despite this enormous progress in research and commercial offerings that today enables the efficient design of analogue blocks for embedding in mixed-signal SoCs, still several problems remain to be solved. Behavioural model generation remains a difficult art that today is often carried out ad hoc with little systematism, and therefore more work in model generation methods is needed. The capacity of analogue synthesis needs to be extended towards more complex blocks, even up to entire frontends, and synthesis needs to incorporate manufacturability issues such as yield and variability. There remain phenomena unique to RF systems that are difficult to design for and hard to model and hence to verify. Chip-level physical assembly for sensitive mixed-signal designs is essentially unautomated. Chip-level verification is still incompletely handled, especially some coupling effects for higher-frequency designs, including electromagnetic couplings such as EMC/EMI. All these problems are roadblocks ahead that can make the realisation of truly integrated mixed-signal systems on a single chip difficult, if not impossible, despite the enormous progress that has been made in recent years and that has resulted in many commercial tool offerings on the market today.

Acknowledgements

The author acknowledges all PhD researchers who have contributed to the reported results.

References

1 'International Technology Roadmap for Semiconductors 2003', http://public. itrs.net

2 GIELEN, G., and RUTENBAR, R.: 'Computer-aided design of analog and mixed-signal integrated circuits', *Proceedings of the IEEE*, 2000, **88**(12), pp. 1825–1854

3 GIELEN, G.: 'Modeling and analysis techniques for system-level architectural design of telecom frontends', *IEEE Transactions on Microwave Theory and Techniques*, 2002, **50**(1), pp. 360–368

4 GIELEN, G.: 'Top-down design of mixed-mode systems: challenges and solutions', in HUIJSING, J., SANSEN, W., and VAN DER PLASSCHE, R. (Eds): 'Advances in analog circuit design' (Kluwer Academic Publishers, 1999), ch. 18

5 'The MEDEA+ Design Automation Roadmap 2003', http://www.medea.org

6 DONNAY, S., GIELEN, G., and SANSEN, W.: 'High-level analog/digital partitioning in low-power signal processing applications'. Proceedings of the 7th international workshop on *Power and timing modeling, optimization and simulation (PATMOS)*, 1997, pp. 47–56

7 CROLS, J., DONNAY, S., STEYAERT, M., and GIELEN, G.: A high-level design and optimization tool for analog RF receiver front-ends'. Proceedings of the international conference on *Computer-aided design (ICCAD)*, 1995, pp. 550–553

8 WAMBACQ, P., VANDERSTEEN, G., ROLAIN, Y., DOBROVOLNY, P., GOFFIOUL, M., and DONNAY, S.: 'Dataflow simulation of mixed-signal communication circuits using a local multirate, multicarrier signal representation', *IEEE Transactions on Circuits and Systems*, Part I, 2002, **49**(11), pp. 1554–1562

9 VANDERSTEEN, G. *et al.*: 'Efficient bit-error-rate estimation of multicarrier transceivers'. Proceedings of the *Design, automation and test in Europe* (DATE) conference, 2001, pp. 164–168

10 WAMBACQ, P., VANDERSTEEN, G., DONNAY, S. *et al.*: 'High-level simulation and power modeling of mixed-signal front-ends for digital telecommunications'. Proceedings of the international conference on *Electronics, circuits and systems (ICECS)*, September 1999, pp. 525–528

11 Virtual Socket Interface Alliance, several documents including 'VSIA architecture document' and 'Analog/mixed-signal extension document', http://www.vsia.org

12 VANDENBUSSCHE, J. *et al.*: 'Systematic design of high-accuracy current-steering D/A converter macrocells for integrated VLSI systems', *IEEE Transactions on Circuits and Systems*, Part II, 2001, **48**(3), pp. 300–309

13 CHRISTEN, E., and BAKALAR, K.: 'VHDL-AMS – a hardware description language for analog and mixed-signal applications', *IEEE Transactions on Circuits and Systems, part II: Analog and Digital Signal Processing*, 1999, **46**(10)

14 IEEE 1076.1 Working Group: 'Analog and mixed-signal extensions to VHDL', http://www.eda.org/vhdl-ams/

15 KUNDERT, K., and ZINKE, O.: 'The designer's guide to Verilog-AMS' (Kluwer Academic Publishers, 2004)

16 'Verilog-AMS Language Reference Manual' version 2.1, http://www.eda.org/verilog-ams/

17 ANTAO, B., and EL-TURKY, F.: 'Automatic analog model generation for behavioral simulation'. Proceedings of the IEEE *Custom integrated circuits* conference (CICC), May 1992, pp. 12.2.1–12.2.4

18 BORCHERS, C., HEDRICH, L., and BARKE, E.: 'Equation-based behavioral model generation for nonlinear analog circuits'. Proceedings of the IEEE/ACM *Design automation* conference (DAC), 1996, pp. 236–239

19 PILLAGE, L., and ROHRER, R.: 'Asymptotic waveform evaluation for timing analysis', *IEEE Transactions on Computer-Aided Design*, 1990, **9**(4), pp. 352–366

20 SILVEIRA, L. *et al.*: 'A coordinate-transformed Arnoldi algorithm for generating guaranteed stable reduced-order models of RLC circuits'. Proceedings of the IEEE/ACM international conference on *Computer-aided design (ICCAD)*, 1996, pp. 288–294

21 ODABASIOGLU, A., CELIK, M., and PILEGGI, L.: 'PRIMA: passive reduced-order interconnect macromodeling algorithm', *IEEE Transactions on Computer-Aided Design*, 1998, **17**(8), pp. 645–654

22 ROYCHOWDHURY, J.: 'Reduced-order modeling of time-varying systems', *IEEE Transactions on Circuits and Systems*, Part II, 1999, **46**(10), pp. 1273–1288

23 WAMBACQ, P., GIELEN, G., KINGET, P., and SANSEN, W.: 'High-frequency distortion analysis of analog integrated circuits', *IEEE Transactions on Circuits and Systems*, Part II, 1999, **46**(3), pp. 335–345

24 PENG, L., and PILEGGI, L.: 'NORM: compact model order reduction of weakly nonlinear systems'. Proceedings of the *Design automation* conference, June 2003, pp. 472–477

25 REWIENSKI, M., and WHITE, J.: 'A trajectory piecewise-linear approach to model order reduction and fast simulation of nonlinear circuits and micromachined devices', *IEEE Transactions on Computer-Aided Design*, 2003, **22**(2), pp. 155–170

26 NING DONG, and ROYCHOWDHURY, J.: 'Piecewise polynomial nonlinear model reduction'. Proceedings of the *Design automation* conference, June 2003, pp. 484–489

27 DAEMS, W., GIELEN, G., and SANSEN, W.: 'Simulation-based generation of posynomial performance models for the sizing of analog integrated circuits', *IEEE Transactions on Computer-Aided Design*, 2003, **22**(5), pp. 517–534

28 KIELY TH., and GIELEN, G.: 'Performance modeling of analog integrated circuits using least-squares support vector machines'. Proceedings of the *Design, automation and test in Europe* (DATE) conference, February 2004, pp. 448–453

29 LIU, H., SINGHEE, A., RUTENBAR, R., and CARLEY, L.R.: 'Remembrance of circuits past: macromodeling by data mining in large analog design spaces'. Proceedings of the *Design automation* conference, June 2002, pp. 437–442

30 LAUWERS, E., and GIELEN, G.: 'Power estimation methods for analog circuits for architectural exploration of integrated systems', *IEEE Transactions on Very Large Scale Integration (VLSI) Systems*, 2002, **10**(2), pp. 155–162

31 DE BERNARDINIS F., JORDAN, M., and SANGIOVANNI-VINCENTELLI, A.: 'Support vector machines for analog circuit performance representation'. Proceedings of the *Design automation* conference (DAC), June 2003, pp. 964–969

32 VESELINOVIC, P. *et al.*: 'A flexible topology selection program as part of an analog synthesis system'. Proceedings of the IEEE *European design & test* conference (ED&TC), 1995, pp. 119–123

33 HARJANI, R., and SHAO, J.: 'Feasibility and performance region modeling of analog and digital circuits'. *Kluwer International Journal on Analog Integrated Circuits and Signal Processing*, 1996, **10**(1), pp. 23–43

34 HARJANI, R., RUTENBAR, R., and CARLEY, L.R.: 'OASYS: a framework for analog circuit synthesis'. *IEEE Transactions on Computer-Aided Design*, 1989, **8**(12), pp. 1247–1265

35 EL-TURKY F., and PERRY, E.: 'BLADES: an artificial intelligence approach to analog circuit design', *IEEE Transactions on Computer-Aided Design*, 1989, **8**(6), pp. 680–691

36 KOH, H., SÉQUIN C., and GRAY, P.: 'OPASYN: a compiler for CMOS operational amplifiers', *IEEE Transactions on Computer-Aided Design* 1990, **9**(2), pp. 113–125

37 MAULIK, P., CARLEY, L.R., and RUTENBAR, R.: 'Simultaneous topology selection and sizing of cell-level analog circuits', *IEEE Transactions on Computer-Aided Design*, 1995, **14**(4), pp. 401–412

38 NING, Z. *et al.*: 'SEAS: a simulated evolution approach for analog circuit synthesis'. Proceedings of the IEEE *Custom integrated circuits* conference (CICC), 1991, pp. 5.2.1–5.2.4

39 KRUISKAMP, W., and LEENAERTS, D.: 'DARWIN: CMOS opamp synthesis by means of a genetic algorithm'. Proceedings of the ACM/IEEE *Design automation* conference (DAC), 1995, pp. 550–553

40 DEGRAUWE, M. *et al.*: 'IDAC: an interactive design tool for analog CMOS circuits', *IEEE Journal of Solid-State Circuits*, 1987, **22**(6), pp. 1106–1115

41 HARVEY, J., ELMASRY, M., and LEUNG, B.: 'STAIC: an interactive framework for synthesizing CMOS and BiCMOS analog circuits', *IEEE Transactions on Computer-Aided Design*, 1992, **11**(11), pp. 1402–1416

42 GIELEN, G., WALSCHARTS, H., and SANSEN, W.: 'Analog circuit design optimization based on symbolic simulation and simulated annealing', *IEEE Journal of Solid-State Circuits*, 1990, **25**(3), pp. 707–713

43 GIELEN, G., WAMBACQ, P., and SANSEN, W.: 'Symbolic analysis methods and applications for analog circuits: a tutorial overview', *The Proceedings of the IEEE*, 1994, **82**(2), pp. 287–304

44 SWINGS, K., and SANSEN, W.: 'DONALD: a workbench for interactive design space exploration and sizing of analog circuits'. Proceedings of the IEEE *European design automation* conference (EDAC), 1991, pp. 475–479

45   LAMPAERT, K., GIELEN, G., and SANSEN, W.: 'Analog layout generation for performance and manufacturability' (Kluwer Academic Publishers, 1999)

46   VAN DER PLAS G., *et al.*: 'AMGIE – A synthesis environment for CMOS analog integrated circuits', *IEEE Transactions on Computer-Aided Design of Integrated Circuits and Systems*, 2001, **20**(9), pp. 1037–1058

47   MEDEIRO, F., PÉREZ-VERDÚ B., RODRÍGUEZ-VÁZQUEZ A., and HUERTAS, J.: 'A vertically-integrated tool for automated design of $\Sigma\Delta$ modulators', *IEEE Journal of Solid-State Circuits*, 1995, **30**(7), pp. 762–772

48   DOBOLI, A., and VEMURI, R.: 'Exploration-based high-level synthesis of linear analog systems operating at low/medium frequencies', *IEEE Transactions on Computer-Aided Design*, 2003, **22**(11), pp. 1556–1568

49   HERSHENSON, M., BOYD, S., and LEE, T.: 'Optimal design of a CMOS op-amp via geometric programming', *IEEE Transactions on Computer-Aided Design*, 2001, **20**(1), pp. 1–21

50   MANDAL, P., and VISVANATHAN, V.: 'CMOS op-amp sizing using a geometric programming formulation', *IEEE Transactions on Computer-Aided Design*, 2001, **20**(1), pp. 22–38

51   DEL MAR HERSHENSON, M.: 'Design of pipeline analog-to-digital converters via geometric programming'. Proceedings of the international conference on *Computer-aided design (ICCAD)*, pp. 317–324, November 2002.

52   VANDERHAEGEN, J., and BRODERSEN, R.: 'Automated design of operational transconductance amplifiers using reversed geometric programming'. Proceedings of the *Design automation* conference (DAC), pp. 133–138, June 2004

53   DIRECTOR, S., and ROHRER, R.: 'Automated network design – The frequency domain case', *IEEE Transactions on Circuit Theory*, 1969, **16**(5), pp. 330–337

54   NYE, W., RILEY, D., SANGIOVANNI-VINCENTELLI, A., and TITS, A.: 'DELIGHT.SPICE: an optimization-based system for the design of integrated circuits', *IEEE Transactions on Computer-Aided Design*, 1988, **7**(4), pp. 501–518

55   MEDEIRO, F. *et al.*: 'A statistical optimization-based approach for automated sizing of analog cells'. Proceedings of the ACM/IEEE international conference on *Computer-aided design (ICCAD)*, pp. 594–597, 1994.

56   PHELPS, R., KRASNICKI, M., RUTENBAR, R., CARLEY, L.R., and HELLUMS, J.: 'Anaconda: simulation-based synthesis of analog circuits via stochastic pattern search', *IEEE Transactions on Computer-Aided Design of Integrated Circuits and Systems*, 2000, **19**(6), pp. 703–717

57   PHELPS, R., KRASNICKI, M., RUTENBAR, R., CARLEY, L.R., and HELLUMS, J.: 'A case study of synthesis for industrial-scale analog IP: redesign of the equalizer/filter frontend for an ADSL CODEC'. Proceedings of the ACM/IEEE *Design automation* conference (DAC), pp. 1–6, 2000

58   FRANCKEN, K., and GIELEN, G.: 'A high-level simulation and synthesis environment for Delta-Sigma modulators', *IEEE transactions on computer-aided design*, 2003, **22**(8), pp. 1049–1061

59   GEERTS, Y., MARQUES, A., STEYAERT, M., and SANSEN, W.: 'A 3.3 V 15-bit delta-sigma ADC with a signal bandwith of 1.1 MHz for ADSL-applications', *IEEE J. Solid-State Circuits*, 1999, **34**, pp. 927–936

60   GIELEN, G., FRANCKEN, K., MARTENS, E., and VOGELS, M.: 'An analytical integration method for the simulation of continuous-time $\Delta\Sigma$ modulators', *IEEE Transactions on Computer-Aided Design*, 2004, **23**(3), pp. 389–399

61   DE SMEDT B., and GIELEN, G.: 'WATSON: design space boundary exploration and model generation for analog and RF IC design', *IEEE Transactions on Computer-Aided Design*, 2003, **22**(2), pp. 213–224

62   DIRECTOR, S., MALY, W., and STROJWAS, A.: 'VLSI design for manufacturing: yield enhancement' (Kluwer Academic Publishers, Dortchet, 1990)

63   ZHANG, J., and STYBLINSKI, M.: 'Yield and variability optimization of integrated circuits' (Kluwer Academic Publishers, Dortchet, 1995)

64   MUKHERJEE, T., CARLEY, L.R., and RUTENBAR, R.: 'Synthesis of manufacturable analog circuits'. Proceedings of the ACM/IEEE international conference on *Computer-aided design (ICCAD)*, November 1995, pp. 586–593

65   DEBYSER, G., and GIELEN, G.: 'Efficient analog circuit synthesis with simultaneous yield and robustness optimization'. Proceedings of the IEEE/ACM internation conference on *Computer-aided design (ICCAD)*, November 1998, pp. 308–311

66   ANTREICH, K., GRAEB, H., and WIESER, C.: 'Circuit analysis and optimization driven by worst-case distances', *IEEE Transactions on Computer-Aided Design*, 1994, **13**(1), pp. 57–71

67   KUHN, J.: Analog module generators for silicon compilation', *VLSI System Design*, May 1987

68   BEENKER, G., CONWAY, J., SCHROOTEN, G., and SLENTER, A.: 'Analog CAD for consumer ICs', in HUIJSING, J., VAN DER PLASSCHE, R., and SANSEN, W. (Eds): 'Analog circuit design', (Kluwer Academic Publishers, 1993), ch. 15, pp. 347–367

69   RIJMENANTS, J. *et al.*: 'ILAC: an automated layout tool for analog CMOS circuits', *IEEE Journal of Solid-State Circuits*, 1989, **24**(4), pp. 417–425

70   COHN, J., GARROD, D., RUTENBAR, R., and CARLEY, L.R.: 'Analog device-level layout generation' (Kluwer Academic Publishers, 1994)

71   COHN, J., GARROD, D., RUTENBAR, R., and CARLEY, L.R.: 'KOAN/ANAGRAM II: new tools for device-level analog placement and routing', *IEEE Journal of Solid-State Circuits*, 1991, **26**(3), pp. 330–342

72   MOGAKI, M., *et al.*: 'LADIES: an automated layout system for analog LSIs'. Proceedings of the ACM/IEEE international conference on *Computer-aided design (ICCAD)*, November 1989, pp. 450–453

73   MEYER ZU BEXTEN, V., MORAGA, C., KLINKE, R., BROCKHERDE, W., and HESS, K., 'ALSYN: flexible rule-based layout synthesis for analog ICs', *IEEE Journal of Solid-State Circuits*, 1993, **28**(3), pp. 261–268

74   MALAVASI, E., CHARBON, E., FELT, E., and SANGIOVANNI-VINCENTELLI, A.: 'Automation of IC layout with analog constraints', *IEEE Transactions on Computer-Aided Design*, 1996, **15**(8), pp. 923–942

75   MALAVASI, E., and SANGIOVANNI-VINCENTELLI, A.: 'Area routing for analog layout', *IEEE Transactions on Computer-Aided Design*, 1993, **12**(8), pp. 1186–1197

76   CHARBON, E., MALAVASI, E., CHOUDHURY, U., CASOTTO, A., and SANGIOVANNI-VINCENTELLI, A.: 'A constraint-driven placement methodology for analog integrated circuits'. Proceedings of the IEEE *Custom integrated circuits* conference (CICC), pp. 28.2.1–28.2.4, May 1992.

77   MALAVASI, E., FELT, E., CHARBON, E., and SANGIOVANNI-VINCENTELLI, A.: 'Symbolic compaction with analog constraints', *Wiley International Journal on Circuit Theory and Applications*, 1995, **23**(4), pp. 433–452

78   BASARAN, B., RUTENBAR, R., and CARLEY, L.R.: 'Latchup-aware placement and parasitic-bounded routing of custom analog cells'. Proceedings of the ACM/IEEE international conference on *Computer-aided design (ICCAD)*, November 1993

79   LAMPAERT, K., GIELEN, G., and SANSEN, W.: 'A performance-driven placement tool for analog integrated circuits', *IEEE Journal of Solid-State Circuits*, 1995, **30**(7), pp. 773–780

80   LAMPAERT, K., GIELEN, G., and SANSEN, W.: 'Analog routing for performance and manufacturability'. Proceedings of the IEEE custom integrated circuits conference (CICC), May 1996, pp. 175–178

81   DE RANTER C., VAN DER PLAS G., STEYAERT, M., GIELEN, G., and SANSEN, W.: 'CYCLONE: Automated design and layout of RF LC-oscillators', *IEEE Transactions on Computer-Aided Design*, October 2002, pp. 1161–1170

82   VAN DER PLAS G., VANDENBUSSCHE, J., GIELEN, G., and SANSEN, W.: 'A layout synthesis methodology for array-type analog blocks', *IEEE Transactions on Computer-Aided Design*, 2002, **21**(6), pp. 645–661

83   VAN DER PLAS, G. *et al.*: 'A 14-bit intrinsic accuracy Q^2 random walk high-speed CMOS DAC', *IEEE Journal of Solid-State Circuits*, 1999, **34**(12), pp. 1708–1718

84   MITRA, S., NAG, S., RUTENBAR, R., and CARLEY, L.R.: 'System-level routing of mixed-signal ASICs in WREN'. Proceedings of the ACM/IEEE international conference on *Computer-aided design (ICCAD)*, November 1992

85   STANISIC, B., VERGHESE, N., RUTENBAR, R., CARLEY, L.R., and ALLSTOT, D.: 'Addressing substrate coupling in mixed-mode ICs: simulation and power distribution synthesis', *IEEE Journal of Solid-State Circuits*, 1994, **29**(3)

86   DONNAY, S., and GIELEN, G. (Eds): 'Analysis and reduction techniques for substrate noise coupling in mixed-signal integrated circuits', *European Mixed-Signal Initiative for Electronic System Design* (Kluwer Academic Publishers, Dortchet, 2003)

87   LEENAERTS, D., GIELEN, G., and RUTENBAR, R.: 'CAD solutions and outstanding challenges for mixed-signal and RF IC design'. Proceedings of the international conference on *Computer-aided design (ICCAD)*, pp. 270–277, November 2001

88   GHARPUREY, R., and MEYER, R.: 'Modeling and analysis of substrate coupling in integrated circuits', *IEEE Journal of Solid-State Circuits*, 1996 **31**(3), pp. 344–353

89   VERGHESE, N., SCHMERBECK, T., and ALLSTOT, D.: 'Simulation techniques and solutions for mixed-signal coupling in integrated circuits' (Kluwer Academic Publishers, Dortchet, 1995)

90   VERGHESE, N., ALLSTOT, D., and WOLFE, M.: Verification techniques for substrate coupling and their application to mixed-signal IC design', *IEEE Journal of Solid-State Circuits*, 1996, **31**(3), pp. 354–365

91   BLALACK, T.: 'Design techniques to reduce substrate noise', in HUIJSING, J., VAN DE PLASSCHE, R., and SANSEN, W. (Eds) 'Advances in analog circuit design' (Kluwer Academic Publishers, Dortchet, 1999), ch. 9, pp. 193–217

92   COSTA, J., MIKE CHOU, and SILVEIRA, L.M.: 'Efficient techniques for accurate modeling and simulation of substrate coupling in mixed-signal IC's', *IEEE Transactions on Computer-Aided Design*, 1999, **18**(5), pp. 597–607

93   CHARBON, E., MILIOZZI, P., CARLONI, L., FERRARI, A., and SANGIOVANNI-VINCENTELLI, A.: 'Modeling digital substrate noise injection in mixed-signal ICs', *IEEE Transactions on Computer-Aided Design*, 1999, **18**(3), pp. 301–310

94   VAN HEIJNINGEN M., BADAROGLU, M., DONNAY, S., GIELEN, G., and DE MAN H.: 'Substrate noise generation in complex digital systems: efficient modeling and simulaton methodology and experimental verification', *IEEE Journal of Solid-State Circuits*, 2002, **37**(8), pp. 1065–1072

95   BADAROGLU, M. *et al.*: 'Modeling and experimental verification of substrate noise generation in a 220-kgates WLAN system-on-chip with multiple supplies', *IEEE Journal of Solid-State Circuits*, 2003, **38**(7), pp. 1250–1260

96   ZINZIUS, Y., GIELEN, G., and SANSEN, W.: 'Modeling the impact of digital substrate noise on analog integrated circuits', in DONNAY, and GIELEN (Ed.): 'Analysis and reduction techniques for substrate noise coupling in mixed-signal integrated circuits' (Kluwer Academic Publishers, 2003) ch. 7

97   MCCONAGHY, T., EECKELAERT, T., and GIELEN, G.: 'CAFFEINE: template-free symbolic model generation of analog circuits via canonical form functions and genetic programming'. Proceedings of the IEEE *Design automation and test in Europe* conference (DATE), 2005, pp. 1082–1087

Chapter 16

Clock-less circuits and system synthesis

Danil Sokolov and Alex Yakovlev

16.1 Introduction

In 1965 a co-founder of Intel Gordon Moore noticed that the number of transistors doubled every year since the invention of the integrated circuit. He predicted that this trend would continue for the foreseeable future [1]. In subsequent years the pace slowed down and now the functionality of the chip doubles every two years. However, the growth of circuit integration level is still faster than the increase in the designers' productivity. This creates a design gap between semiconductor manufacturing capability and the ability of electronic design automation (EDA) tools to deal with the increasing complexity, Figure 16.1.

The only way to deal with the increasing complexity of logic circuits is to improve the efficiency of the design process. In particular, design automation and component reuse help to solve the problem.

System-on-Chip (SoC) synthesis has proved to be a particularly effective way in which design automation and component reuse can be facilitated. An important role in the synthesis of SoCs is given to the aspects of modelling concurrency and timing [2]. These aspects have traditionally been dividing systems into synchronous (or clocked) and asynchronous (or self-timed). This division has recently become fuzzier because systems are built in a mixed style: partly clocked and partly clock-less. In fact, the argument about the way in which the system should be constructed, synchronously or asynchronously, is moving to another round of evolution. It is accepted that the timing issue should only be addressed in the context of the particular design criteria, such as speed, power, security, modularity etc. Given the complexity of the relationship between these criteria in every single practical case, the design of an SoC is increasingly going to be a mix of timing styles. While industrial designers have a clear and established notion of how to synthesise circuits with a global clock using EDA tools, there is still a lot of uncertainty and doubt about synthesis of clock-less

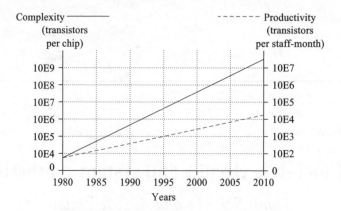

Figure 16.1   Design complexity and designer productivity

circuits. The latter remains a hot research field captivating many academics and graduate students. In the last two decades there have been dozens of research publications on asynchronous circuit synthesis, and it would be impossible to embrace them all in a single review. Readers without prior experience are invited to study them at an introductory level (e.g. [3] and http://www.cs.man.ac.uk/async/background) while the more experienced audience can delve into such methods in more detail by addressing monographs and papers (e.g. [4,5] and http://www.cs.man.ac.uk/async/pubwork).

The main goal of this review is to consider a coherent subset of synthesis methods for clock-less circuits based primarily on a common underlying model of computation and using a relatively simple example in which these methods can be compared. Such a model is Petri nets, used with various interpretations. The Petri nets can play a pivotal role in future synthesis tools for clock-less systems, exhibiting advanced concurrency and timing paradigms. This role can be as important as that of a finite state machine (FSM) in designing clocked systems. To make this review more practically attractive the use of Petri nets is considered in the context of a design flow with a front-end based on a hardware description language. Our running example will be a computation of the greatest common divisor (GCD) of two integers, which is a popular benchmark in the literature about digital circuit design.

16.1.1   Synchronous systems

The traditional design flow for synchronous systems is supported by design and verification tools, e.g. Cadence, Mentor Graphics, Synopsys, etc. However, a globally clocked SoC assembled from existing intellectual property (IP) cores (see Figure 16.2) suffers from the timing closure problems. Each IP core is designed for a certain clock period, assuming that the clock signal is delivered at the same time to all parts of the system. Finding a common clocking mode for the whole system is a very difficult obstacle on the way to component reuse.

The other problem of synchronous circuits is the clock skew caused by interconnect delays. In the past the transistors were the limiting factor of the circuit speed.

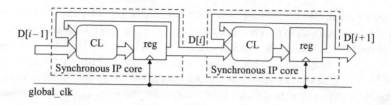

Figure 16.2   Synchronous system architecture

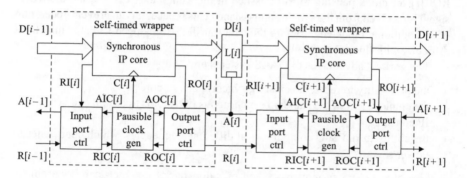

Figure 16.3   GALS system architecture

The increase of the circuit integration level resulted in the improvement of the transistor size and speed. However, the interconnect speed is not keeping the pace. Narrower wires have higher resistance for the same length, leading to slower signal edges and longer interconnect delays. Being proportional to interconnect delay, the clock skew becomes an increasing portion of the clock period. This means that eventually large circuits will need to get rid of global clocking in order to provide high speed.

16.1.2   Globally asynchronous locally synchronous systems

A promising method of composing systems from predesigned components is a globally asynchronous locally synchronous (GALS) approach [6]. In such systems the components are synchronous IP cores operating at their own clock speed, which allows the proven synchronous design methodologies to be employed. The interface between the components is converted to asynchronous style by putting them inside self-timed wrappers, as shown in Figure 16.3. This eliminates the need for a global clock with all of its associated problems.

The GALS self-timed wrapper whose basic structure is captured in Figure 16.3 is proposed in Reference 7. It contains a pausible clock generator and an asynchronous controller for each port. The data lines between two GALS modules are bundled with a pair of request-acknowledge signals. Any data transfer is initiated by the locally synchronous island on the transmitting side by activating the RO[i] request to the output port controller. The output port controller in turn instructs the clock generator to delay the next clock edge by using the ROC[i] request. After the clock of the module has been frozen, the local clock generator acknowledges it by the AOC[i] signal. Then the communication partner is notified by the request signal R[i]. Once the other GALS module has halted its clock using RIC[$i + 1$] and AIC[$i + 1$] handshake, it sets the acknowledge port signal A[i] and enables the input buffer latch L[i]. At this point, both modules have halted their clocks and can exchange data without any risk of timing violations. Once the data transfer is complete, the local clocks are released and the locally synchronous islands continue to operate in a normal synchronous mode.

Note an important timing assumption on the output port controller. The request ROC[i] for clock pausing must be issued in the same clock cycle when the RO[i] signal is received from the synchronous island. This is necessary to prevent generation of an additional clock edge before the data transfer. High-speed IP cores may have difficulty with this assumption.

The greatest advantages of GALS systems are:

- the possibility to reuse the existing synchronous IP cores;
- the employment of the standard synchronous EDA tools to design and verify new IP cores;
- the ability to run SoC components at different frequencies, which contributes to power savings.

However, GALS systems have their own drawbacks, e.g. metastability problem, when an asynchronous signal is sampled by a clock. In order to avoid metastability several methods are used.

One of the ways to minimise the probability of metastability is to path each asynchronous signal through a synchroniser, which is typically a pair of back-to-back connected flip-flops. Still, in a GALS system the number of connections between its synchronous blocks is large, which creates a non-negligible probability of system failure. The synchronisers also add extra latency to the signals which significantly impacts the system performance.

The other strategy to avoid metastability is the dynamic alteration of the local clock rate [8,9]. For this, a pausible clock generator is employed in each synchronous island. The clock generator is a ring oscillator with a control input for its stopping and starting. If some asynchronous channel of the synchronous island is not ready, then the inactive phase of the local clock is stretched until all channels are ready.

Several methods to ensure that metastability never occurs in a GALS system with pausible clocking are proposed in Reference 10. However, the alteration of the local clock may cause a deadlock when all components are waiting for the output of some other component. It is not trivial to guarantee that the system is deadlock free. It should be also noted that pausing the local clock slows down the entire synchronous island, and the slowdown may be exacerbated with multi-port GALS modules, where the probability of pausing the clock is higher. Furthermore, the local clock alteration may cause problems if the dynamic logic is used, as the length of the clock period becomes important there (as opposed to static logic). Finally, the ring oscillators which are used to form the pausible clocks (as opposed to crystal oscillators) suffer from significant jitter and frequency variation, which may result in a performance degradation.

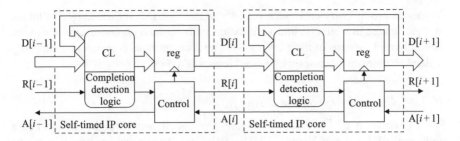

Figure 16.4   Asynchronous system architecture

There also other approaches to the design of GALS systems, such as module stalling via clock gating [11,12] and fine-grain synchronous handshaking [13–15].

16.1.3   Self-timed systems

The GALS approach minimises the designer work in the asynchronous domain, but does not completely escape it. Some of the components, particularly self-timed wrappers, are still asynchronous. At the same time, purely asynchronous circuits whose architecture is shown in Figure 16.4 offer a set of qualitative advantages which GALS systems do not have. In addition to better modularity and avoidance of clock distribution problem, self-timed systems can exhibit higher robustness, greater performance, power saving, lower electromagnetic noise, etc. [3,16–18].

The major obstacle on the mainstream use of asynchronous design techniques is the lack of a coherent design flow, compatible with conventional EDA tools and libraries. The compatibility issue is essential because a large part of the design flow is the mapping of the circuit netlist into silicon. For this task the traditional place and route tools can be reused. It is also possible to inherit the timing analysis and simulation tools. However, synthesis and verification tools intended for synchronous systems omit important features of asynchronous components. These parts of the traditional design flow have to be replaced.

The other impediment is that industry adheres to existing specification languages. The majority of industry designers think in terms of high-level hardware description languages (HDLs), such as Verilog and VHDL, which were created for synchronous designs. These languages require much more code to be written in order to specify an asynchronous component, compared to synchronous logic. Several new languages were developed for efficient asynchronous design [19–21]. However, adoption of a unique language in industry involves significant changes in the design flow and retraining the designers. These procedures are extremely costly and take valuable time, which makes the new languages difficult to accept for commercial companies.

Finally, all existing synchronous IP cores have to be abandoned in an asynchronous world. It will take years before all those components are replaced by asynchronous counterparts.

Even though the asynchronous techniques involve significant changes to the conventional design flow, the companies realise that this is the promising route to cover the design productivity gap. Such industry giants as IBM, Infineon, Intel, Philips, Sun etc. invest in synthesis and verification tools for asynchronous circuit design. They also replace parts of their new systems by asynchronous components, gradually replenishing design libraries with asynchronous IP cores.

16.1.4   General view on the design flow

For a designer it is convenient to specify the circuit behaviour in the form of a high-level HDL, such as Verilog, VHDL, Balsa, etc. This initial specification can be processed in two different ways:

- directly translated into the circuit structure analysing the syntax of the specification;
- transformed into an intermediate behavioural format convenient for subsequent verification and synthesis.

The former approach is the 'syntax-driven translation'. It is adopted by Tangram [19] and Balsa [20] design flows. The initial circuit specification for these tools is given in the languages based on the concept of processes, variables and channels, similar to communicating sequential processes (CSP) [21].

The latter approach is 'logic synthesis'. This approach is used in PipeFitter [22], TAST [23], PN2DCs [24,25], where Petri nets are used for intermediate design representation. Other examples are MOODs [26] and CASH [27]. The former starts from VHDL and uses a hardware assembly language ICODE for intermediate code. The latter starts from ANSI-C and uses a Pegasus dataflow graph for intermediate representation, which is further synthesised into control logic for micropipelines [28]. Some tools do not cover the whole design, but can be combined with the other tools to support the coherent design flow. For example, gate transfer level (GTL) [24], VeriMap [30], Theseus Logic NCL-D and NCL-X [31] are developed for synthesis of asynchronous datapath from register transfer level (RTL) specification. Other tools, such as Minimalist [32], 3D [33], Petrify [34] and OptiMist [35] are aimed at asynchronous controller synthesis from intermediate behavioural specifications. In turn, controller synthesis tools, can be combined with decomposition techniques to reduce the complexity of the specification. The survey focuses on the aspects of the logic synthesis approach.

The rest of the review is organised as follows. First, an overview of syntax-driven design flows is given. Then, the logic synthesis methods are discussed, the tools for synthesis of control and datapath are presented. Finally, the state of the asynchronous design automation is summarised and the ways of improvement are pointed out.

16.2   Syntax-driven translation

The basic design flow diagram for the syntax-driven approach is shown in Figure 16.5. The initial system specification is compiled into a parsing tree, which is subsequently mapped into a network of handshake components. The network can be used for

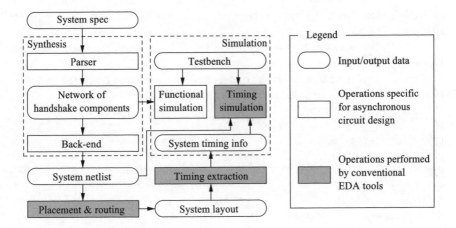

Figure 16.5   Syntax-driven design flow

behavioural simulation of the asynchronous system. The mapping of the network of handshake components into a gate netlist is performed by a back-end tool, which may vary for different technologies. The obtained gate netlist is mapped into silicon by conventional place and route tools. The timing information extracted from the layout can be used together with the gate netlist for timing simulation.

The syntax-driven approach was initially used in the Tangram group at Philips Research [19]. The Tangram design flow depends on a proprietary CSP-based language and private tool set. While being successfully used in the Philips research environment, the proprietary nature of the tools made practical widespread adoption of this methodology problematic.

The syntax-driven design flow became available for public use after the Manchester Amulet Group developed the Balsa design kit [36,37]. Similar to Tangram, it relies on the paradigm of handshake components as an intermediate representation of an asynchronous system. The Balsa language is created to provide a source for compiling handshake components and is also very similar to Tangram. In Balsa the circuits are described by procedures which contain the specification of processes. Procedures communicate by means of handshake ports. Most procedures consist of a body command whose behaviour is perpetually repeated using a loop.

Consider the Balsa description for the GCD problem:

```
01   import [balsa.types.basic]
02   procedure GCD (
03     input x : byte;
04     input y : byte;
05     output z : byte) is
06   local variable a, b : byte
07   begin
08     loop
```

```
09          x -> a || y -> b;
10          loop
11            while a /= b then
12            begin
13              if a > b then a := (a - b as byte)
14              else b := (b - a as byte)
15              end
16            end
17          end;
18          z <- b
19        end
20    end
```

The line numbers in the left column are shown for reference only and do not belong to the Balsa language. The first line of the code contains an inclusion of a pre-compiled module [balsa.types.basic], which only defines some common types, e.g. byte. The second line starts the procedure declaration with 8-bit input ports x, y and an 8-bit output port z, which are declared in lines 03–05. The local 8-bit variables a and b are declared in line 06. The procedure body is enclosed in an infinitely repeating loop. Inside the loop the concurrent communication on input channels x and y is expected, line 09. The concurrency is expressed by means of the || operator. The values of the input channels are saved into local variables using channel -> variable statements. After that the while-loop with a x /= y condition is started, where /= means 'not equal'. Note that sequential operations are separated by the ; operator. Inside the while-loop the if ... then ... else ... end statement is exploited, lines 13–15. In both its branches the assignment of an expression to a variable with type casting to byte is executed. In line 18, sequentially to the while-loop, the output communication is synchronised using channel <- variable statement. The handshake circuit obtained by compilation of this source code is shown in Figure 16.6.

A handshake circuit consists of handshake components (circles with the operation name inside) linked by channels (solid arcs). Each handshake component has one or more ports with which it can be connected point-to-point to a port of another handshake circuit by means of a channel. Each channel carries request and acknowledgement signalling as well as an optional data payload. The requests flow from the active component ports (filled circles) towards passive component ports (open circles). Acknowledgements flow in the opposite direction to requests. Where a channel carries data, the direction of the data is indicated by an arrow on that channel's arc. The direction of data may be different from the direction of signalling to support push and pull ports and channels.

A handshake component can be activated by sending a request to its passive port. When activated, it sends requests to a subset of its active ports and waits for acknowledgements. The subset of the ports activated by the component is determined by its function and may be data-dependent. The order in which the component activates its

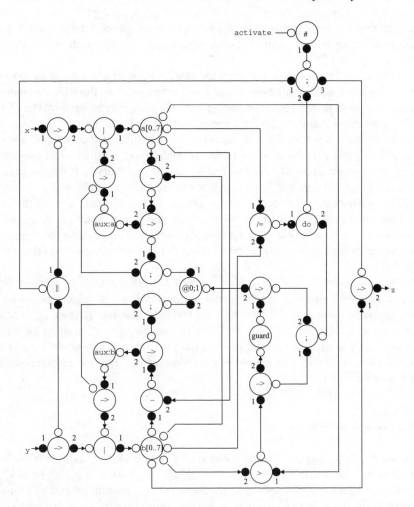

Figure 16.6   Handshake circuit for GCD

ports is shown by small numbers next to the ports. The ports of a handshake component which are marked with the same number are activated concurrently. When all
activated ports are acknowledged, the handshake component sends an acknowledgement to the passive port from which it was activated and finishes its operation until
the next activation [20].

One can notice correspondence between the syntax of the Balsa program and the
structure of the GCD handshake circuit in Figure 16.6. The operation of the GCD
circuit starts with the request on the channel marked as `activate`. It activates the
loop-component (#), which in turn sends a request to sequence-component (;).

First, the sequence-component activates the concur-component (| |). The concur-
component controls the fetching operation (->) for input channels x and y. The
data from input channels x and y is pushed through the multiplexers (|) to

the variables a and b respectively. When data is stored, the variables send acknowledgements back to the sequence-component (;), which then activates the while-component (do).

The while-component (do) requests the guard, which is the not-equal comparison (/=) between a and b variables. If the guard returns true, the while-component sends a request to the sequence-component, which controls the fetching of the a>b comparison result to the case-component (@ 0;1). If the result is true, the case-component activates the a-b function. The fetching of the subtraction result into a variable is performed using an intermediate aux:a variable and two fetch-components to avoid parallel reading and writing of a. Similarly, if the comparison returns false, the result of the b-a is fetched into the b variable.

The while-component continues to request the guard and activate the sub-traction procedure (described in the previous paragraph) until the guard value becomes false. After that, an acknowledgement is sent back to the sequence-component, which then activates the fetching (->) of the b variable to the output channel z.

The syntax-driven translation is attractive from the productivity point of view, as it avoids computationally hard global optimisation of the logic. Instead some local peephole optimisation is applied at the level of handshake circuits. Burst Mode synthesis tools (Minimalist [32], 3D [33]) can be used for the optimisation. However, the direct translation of the parsing tree into a circuit structure may produce very slow control circuits. The lack of global optimisation may not meet the requirements for high-speed circuits.

16.3   Logic synthesis

The design flow diagram for the logic synthesis approach to asynchronous sys-tem design is shown in Figure 16.7. The initial specification in a high-level HDL (Verilog or VHDL) is first split into two parts: the specification of control unit and specification of the datapath. Both parts are synthesised separately and subsequently merged into the system implementation netlist. An industrial EDA place and route tool can be used to map the system netlist into silicon. The existing simulation EDA tools can be reused for the behavioural simulation of the initial system specification. These tools can also be adopted for timing simulation of the system netlist back-annotated with timing information from the layout.

The variations in the design flow appear in the way of:

- extracting the specifications of control unit and datapath from the system specification;
- synthesis of the datapath;
- synthesis of the control unit either by direct mapping or by logic synthesis.

The following sections consider each of these issues separately. They all require some basic knowledge of the underlying formal model, Petri nets and their interpretations.

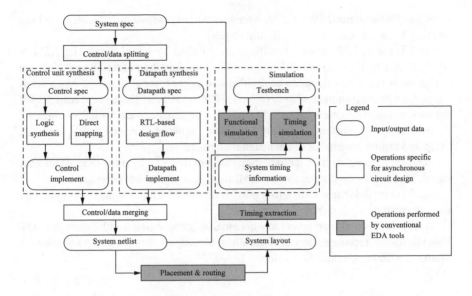

Figure 16.7   Logic-driven design flow

16.3.1   Petri nets and signal transition graphs

The convenient behavioural models for logic synthesis of asynchronous control circuits are 1-safe petri net (PN) and signal transition graph (STG). The datapath operations can be modelled by coloured Petri net (CPN).

A PN is formally defined as a tuple $\Sigma = \langle P, T, F, M_0 \rangle$ comprising finite disjoint sets of 'places' P and 'transitions' T, flow relation $F \subseteq (P \times T) \cup (T \times P)$ and initial marking M_0. There is an arc between x and y if $(x, y) \in F$. The 'preset' of a node x is defined as $\bullet x = \{y \mid (y, x) \in F\}$, and the 'postset' as $x\bullet = \{y \mid (x, y) \in F\}$. A 'marking' is a mapping $M \colon P \to N$ denoting the number of 'tokens' in each place ($N = \{0, 1\}$ for 1-safe PNs). It is assumed that $\bullet t \neq \emptyset \neq t\bullet$, $\forall t \in T$. A transition t is 'enabled' if $M(p) \neq 0$, $\forall p \in \bullet t$. The evolution of a PN is possible by 'firing' the enabled transitions. Firing of a transition t results in a new marking. M':

$$M'(p) = \begin{cases} M(p) - 1, & \forall\, p \in \bullet t, \\ M(p) + 1, & \forall\, p \in t\bullet, \\ M(p), & \forall\, p \notin \bullet t \cup t\bullet \end{cases}$$

Graphically places of a PN are represented as circles, transitions as boxes, arcs are shown by arrows, and tokens of the PN marking are depicted by dots in the corresponding places.

An extension of a PN model is a contextual net [38]. It uses additional elements such as non-consuming arcs, which only control the enabling of a transition and do not consume tokens. The reviewed methods use only one type of non-consuming arc, namely 'read-arcs'. A set of read-arcs R can be defined as follows: $R \subseteq (P \times T)$. There is an arc between p and t if $(p, t) \in R$.

A labelled petri net (LPN) is a PN whose transitions are associated with a labelling function λ, i.e. LPN $= \langle P, T, F, R, M_0, \lambda \rangle$ [39].

An STG is an LPN whose transitions are labelled by signal events, i.e. STG $= \langle P, T, F, R, M_0, \lambda \rangle$, where $\lambda: T \rightarrow A \times \{+, -\}$ is a labelling function and A is a set of signals. A set of signals A can be divided into a set of 'input signals' I and a set of 'output and internal signals' O, $I \cup O = A$, $I \cap O = \emptyset$. Note that a set of read-arcs R has been included into the model of STG, which is an enhancement w.r.t. Reference 40.

An STG is 'consistent' if in any transition sequence from the initial marking, rising and falling transitions of each signal alternate.

A signal is 'persistent' if its transitions are not disabled by transitions of another signal. An STG is 'output persistent' if all its output signals are persistent.

An STG is 'delay insensitive to inputs' if no event of input signal is switched by another event of input signal.

A CPN is a formal high-level net where tokens are associated with data types [41]. This allows the representation of datapath in a compact form, where each token is equipped with an attached data value.

16.4   Splitting of control and datapath

The first step in the logic synthesis of a circuit is the extraction of datapath and control specifications from the high-level description of the system. Often the partitioning of the system is performed manually by the designers. However, this might be impracticable for a large system or under a pressure of design time constraints. At the same time, the tools automating the extraction process are still immature and require a lot of investment to be used outside a research lab.

For example, the PipeFitter tool [22], which is based on Verilog HDL and PNs as an intermediate format, supports only a very limited subset of Verilog constructs (`module, function, initial, always, wait, if, case, fork, join`). Any high-level specification which contains a loop or a conditional jump cannot be processed by this tool. A simple GCD benchmark could not even be parsed because of the while-loop it contains. An attempt to modify the system specification so that it uses always-statement as a loop and then to synthesise it has also been unsuccessful.

The other tool, which works with VHDL high-level system specifications, is PN2DCs [24,25]. It supports the following language statements: `process, procedure, wait, if, case, loop, call, block`.

The VHDL specification is generated during the system-level synthesis. At this step the PN2DCs tool partitions the system into several subsystems based on their functionality and schedules them according to their interaction. Each subsystem is described by a VHDL process. The communication between processes is implemented using wait-statements over control variables. The relationship between processes is expressed in the form of a 'global net', which is a PN whose transitions are associated with system processes and the places divide the control into separate stages. Each global net transition is then refined according to the associated process description.

An LPN for control unit and a CPN for datapath are derived from the global net. The interface between control and datapath units is modelled by a 'local control net', which connects the generated LPN and CPN.

The extraction of control and datapath nets using PN2DCs tool is considered on the GCD benchmark which is described using the VHDL process notation:

```
01    entry gcd is port (
02       x: in STD_LOGIC_VECTOR (7 downto 0);
03       y: in STD_LOGIC_VECTOR (7 downto 0);
04       z: out STD_LOGIC_VECTOR (7 downto 0))
05    end gcd;
06    architecture gcd_func of gcd is
07    begin
08       process begin
09          wait on x;
10          wait on y;
11          while x< >y loop
12             if x>y then x := x - y;
13             else y := y - x;
14             end if;
15          end;
16          z <= y;
17       end process;
18    end gcd_func;
```

As the benchmark contains one process only, the global net consists of one transition representing this process. This transition is refined by parsing the system specification and using an ASAP scheduling algorithm.

The refined model of the system is shown in Figure 16.8. Two wait-statements for synchronisation on the x and y channels are scheduled concurrently. It is possible because both x and y are defined as inputs and are independent. The input operation, the while-loop and the result output on the z channel are scheduled in sequence. The while-loop is subject to further refinement together with the nested if-statement, during which the conditions of while-loop and if-statement are merged. The labels gt, eq and lt correspond to the result of comparison between x and y values and stand for 'greater than', 'equal' and 'less than' respectively. The assignment of the subtraction result to a signal is split into the subtraction operation (sub_gt, sub_lt) and storage of the result (store_x, store_y).

The global net in Figure 16.8 is used to generate the LPN for the control unit. The transitions of the LPN correspond to the atomic operations and places denote different stages of the GCD algorithm. Those places are subsequently mapped into memory elements.

The LPN for the GCD control unit produced by the PN2DCs tool is shown in Figure 16.9(a) using the solid lines. The dashed arcs and places represent the local

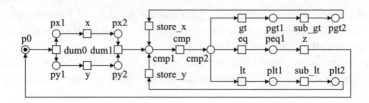

Figure 16.8   Refined global net for GCD

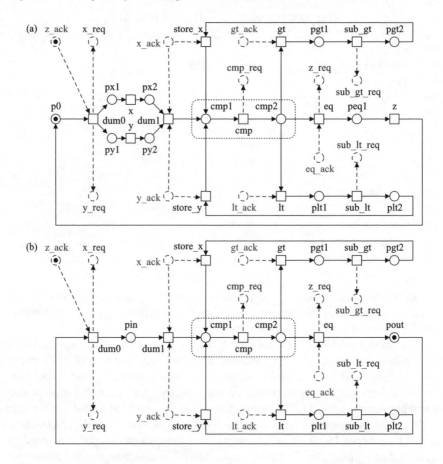

*Figure 16.9   LPN for GCD control unit: (a) LPN generated by the PN2DCs tool and
(b) Manually optimised LPN*

control net. One can notice that transitions x, y and z are not connected to the local
control net and therefore are redundant for the control unit. The detection and removal
of such redundant transitions can be automated. However, at the moment of the
experiments this optimisation was not implemented in the PN2DCs tool yet. The
result of manual optimisation applied to the GCD control unit LPN is shown in

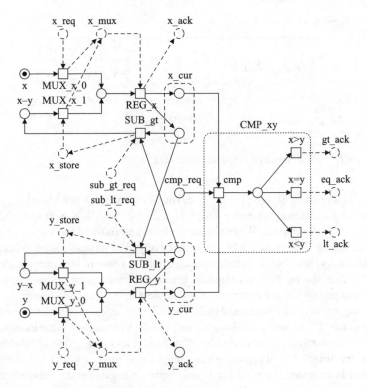

Figure 16.10   CPN for GCD datapath unit

Figure 16.9(b). In this LPN places px1, px2, py1 and py2 are merged into one place pin, which denotes the stage of data input. Places peq1 and p0 are also merged into place pout, which denotes the stage of data output. The optimised LPN is four places smaller than the original one, which results in area saving when LPN places are mapped into memory elements.

Signals z_ack and z_req compose the handshake interface to the environment. When set, the z_req signal means the computation is complete and output data is ready to be consumed. The z_ack signal is set when the output of the previous computation cycle is consumed and the new input data is ready to be processed.

The datapath CPN generated by PN2DCs is presented in Figure 16.10 using the solid lines. The dashed arcs and places represent the local control net. Transitions MUX_x_0 and MUX_x_1 are used for multiplexing input x or output from subtracter SUB_gt to REG_x. Similarly, MUX_y_0 and MUX_y_1 are multiplexing y or SUB_lt output to REG_y. The CMP_xy block of the net, framed in the dotted rectangle, is used for comparing the values of REG_x and REG_y. Depending on the comparison result one of the transitions x > y, x = y or x < y is fired.

In Figures 16.9, 16.10 the dashed arcs and places belong to the local control net. All the communication between the control unit and datapath unit is carried out by

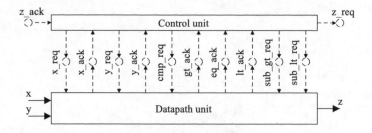

Figure 16.11   GCD control-datapath interface

this net, as shown in Figure 16.11. For example, when the z_ack signal is received, the control generates x_req and y_req, which enable the MUX_x_0 and MUX_y_0 transitions in the datapath. When the multiplexing is finished the values of x and y are stored using REG_x and REG_y, respectively. The datapath acknowledges this by x_ack and y_ack. These signals enable the *in* transition in the control unit.

After that, the control unit requests the comparison operation by means of the cmp_req signal. When the comparison is complete in the datapath, one of the signals gt_ack, eq_ack or lt_ack is returned to the control. If gt_ack is received, the control unit generates sub_gt_req request, which activates SUB_xy transition in the datapath. This results in subtracting the current value of y from x and storing the difference using REG_x transition. The datapath acknowledges this by x_ack and the comparison operation is activated again. If the lt_ack signal is issued by the datapath then the operation of the system is analogous to that of gt_ack. However, as soon as eq_ack is generated, the control unit issues the z_req signal to the environment, indicating that the calculation of GCD is complete.

Note that the local control net x_mux between MUX_x_0 and REG_x does not leave the datapath unit, thereby simplifying the control unit. Similarly, other signals, y_mux, x_store and y_store, in the local control net are kept inside the datapath.

16.5   Synthesis of datapath

The method of datapath synthesis employed in PN2DCs is based on the mapping of CPN fragments into predesigned hardware components. A part of the library of such components and corresponding CPN fragments is shown in Figure 16.12. The solid places and arcs in the CPN column correspond to data inputs and outputs; the dashed arcs and places denote the control signals (request and acknowledgement).

A block diagram for the GCD datapath is presented in Figure 16.13. It is mapped from the CPN specification shown in Figure 16.10. The CPN is divided into the following fragments, which have hardware implementations in the library shown in Figure 16.12: 2 multiplexers, 2 registers, 1 comparator and 2 subtracters. These hardware components are connected according to the arcs between the corresponding fragments of the CPN. To save the hardware, the output z is not latched in its own

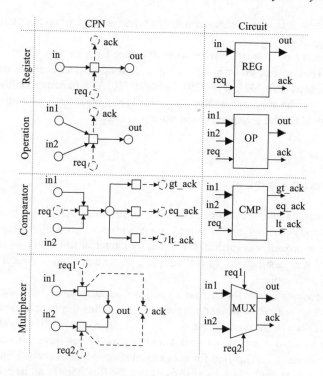

Figure 16.12   Mapping from CPN into circuit

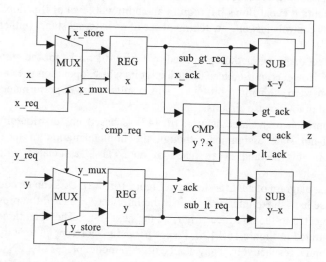

Figure 16.13   GCD datapath

register. Instead it is taken from the register y and is valid when the controller sets the z_req signal.

If the library of datapath components does not have an appropriate block, the latter should be constructed from RTL. One of the following tools can be used for this purpose: Theseus NCL [42], GTL [29], VeriMap [30], etc. In particular, the VeriMap tool is based on the approach of Kondratyev [31] and enriches it with an alternating spacer protocol for the enhancement of security features.

16.6 Direct mapping of control unit

The direct mapping approach originates from Huffman's work [43], where a method of 'the one-relay-per-row' realisation of an asynchronous sequential circuit was proposed. This approach was further developed by Unger in Reference 44 and had led to the '1-hot state assignment' of Hollaar [45], where a method of concurrent circuit synthesis was described.

The underlying model for Hollaar's circuits is an augmented finite state machine (AFSM), which is an FSM with added facilities, including timing mechanisms so that state changes can be delayed. These circuits have inputs that are logic values (signal levels as opposed to signal transitions), which is good for low-level interfacing. They use a separate set–reset flip-flop for every local state, which is set to 1 during a transition into the state, and which in turn resets to 0 the flip-flops of all its predecessor's local states. The main disadvantages of Hollaar's approach are the fundamental mode assumptions and the use of local state variables as outputs. The latter are convenient for implementing event flows but require an additional level of flip-flops if each of those events controls just one switching phase of an external signal (either from 0 to 1 or from 1 to 0).

The direct mapping method proposed by Patil [46] works for the whole class of 1-safe PNs. However, it produces control circuits whose operation uses a two-phase (no-return-to-zero) signalling protocol. This results in lower performance than can be achieved in four-phase circuits.

The approach of Kishinevisky *et al.* [47] is based on 'distributors' and also uses the 1-hot state assignment, though the implementation of local states is different. In this method every place of an STG is associated with a David cell (DC) [48].

A circuit diagram of a DC is shown in Figure 16.14(a). DCs can be coupled using a four-phase handshake interface, so that the interface $\langle r, a1 \rangle$ of the previous stage DC is connected to the interface $\langle r1, a \rangle$ of the next stage. The operation of a single DC is illustrated in Figure 16.14(b). Places p1 and p2 correspond to active levels of signals r1 and r, respectively. They can be used to model places of a PN as shown in Figure 16.14(c). The dotted rectangle depicts the transition between p1 and p2. This transition contains an internal place, where a token 'disappears' for the time $t_{r1-\rightarrow r+}$. In most cases this time can be considered as negligible, because it corresponds to a single gate delay.

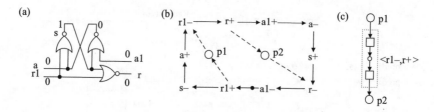

Figure 16.14 David cell: (a) circuit, (b) STG and (c) model

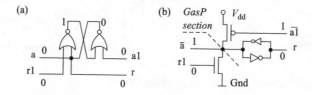

Figure 16.15 Fast David cell implementation: (a) gate-level, (b) transistor-level

The circuits built of DCs are speed independent (SI) [49] and do not need fundamental mode assumptions. On the other hand, these circuits are autonomous (no inputs/outputs). The only way of interfacing them to the environment is to represent each interface signal as a set of abstract processes, implemented as request-acknowledgement handshakes, and to insert these handshakes into the breaks in the wires connecting DCs. This restricts this method to high-level design.

An attempt to apply the direct-mapping method at a low-level, where inputs and outputs are signal events of positive or negative polarity, was done in Reference 50, where DC structures controlled output flip-flops. For this, a circuit converting a handshake event into the logic level was designed. Inputs were, however, still represented as abstract processes.

In Reference 51 the direct mapping from STGs and the problem of device-environment interface are addressed. This method converts the initially closed (with both input and output transitions) system specification into the open system specification. The open system specification consists of a 'tracker' and 'bouncer'. The tracker follows (or tracks) the state of the environment and is used as a reference point by the outputs. The bouncer interfaces the environment and generates output events in response to the input events according to the state of the tracker.

Faster and more compact solutions for a DC implementation were developed by introducing timing assumptions [51]. In DC implementations shown in Figure 16.15 the reset phase of state holding element happens concurrently with the token move into the next stage DC. An interesting feature of the transistor-level implementation shown in Figure 16.15(b) is that it internally contains a GasP interface [52], which uses a single wire to transmit a request in one direction and an acknowledgement in the other.

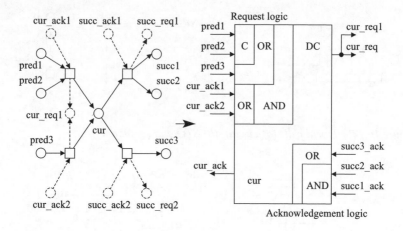

Figure 16.16   Mapping of LPN places into DCs

16.6.1   Direct mapping from LPNs

The PN2DCs tool maps the places of the control unit LPN into DCs. The request and acknowledgement functions of each DC are generated from the structure of the LPN in the vicinity of the corresponding place as shown in Figure 16.16. The request function of the DC is shown in its top-left corner and the acknowledgement function in the bottom-right conner.

The GCD control unit described by the LPN in Figure 16.9(b) is mapped into the netlist of DCs shown in Figure 16.17. Each DC in this netlist corresponds to the LPN place with the same name.

In this netlist the dotted wires can be actually removed thus simplifying the request functions of p1, pgt1, peq1, plt1 and p0. These wires are redundant because the trigger signals from the environment uniquely identify which DC among them should be activated even without a context signal from the preceding DCs. However, if the same set of input signals activates more than one DC, the context signal is required. For example, the request function for pgt2 must include both x_ack and sub_gt_req signals, because x_ack can be set in response to x_req and sub_gt_req and the controller should be able to distinguish between these situations.

16.6.2   Direct mapping from STGs

The method of direct mapping from STGs proposed in Reference 50 is implemented in the OptiMist software tool [35]. The tool first converts the system STG into a form convenient for direct mapping. Then it performs optimisation at the level of specification and finally maps the optimised STG into circuit netlist.

In order to transform the initially closed system specification into the open system specification, the concepts of 'environment tracking' and 'output exposure' should be applied. These concepts can be applied to an STG that is consistent, output persistent and delay insensitive to inputs.

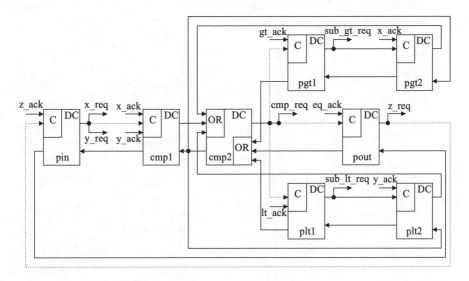

Figure 16.17   GCD control unit

The whole method is described on the basic example whose STG is depicted in Figures 16.18(a).

The first step in constructing the open system specification is splitting the system into 'device' and 'environment'. For this the original STG is duplicated as shown in Figure 16.18(b). Then, in the first copy, corresponding to the 'device', input events are replaced by dummies and in the second copy, corresponding to the 'environment', output events are replaced by dummies. The behaviour of the device and environment is synchronised by means of read-arcs between dummy transitions and successor places of their prototypes in the counterpart as shown in Figure 16.18(b).

At the second step the outputs of both device and environment are exposed by the following technique. Every interface signal is associated with a pair of complementary places representing the low and high levels of the signal. These places are inserted as transitive places between the positive and negative transitions of the signal, expressing the property of signal consistency. 'Trackers' of the device and environment use these 'exposed outputs' to follow (or track) the behaviour of the counterpart as shown in Figure 16.18(c).

After that, 'elementary cycles' are formed and read-arcs are introduced to represent the signals as shown in Figure 16.18(d). Read-arcs from the predecessor places of dummies to signal transitions and from the successor places of signal transitions to dummies preserve the behaviour of the system. The resultant system specification is 'weakly bisimular' [53] to the original. The elementary cycles are subsequently implemented as set–reset flip-flops (FF) and the places of the tracker as DCs, see Figure 16.18(e).

It is often possible to control outputs by the directly preceding interface signals without using intermediate states. Many places and preceding dummies can thus be

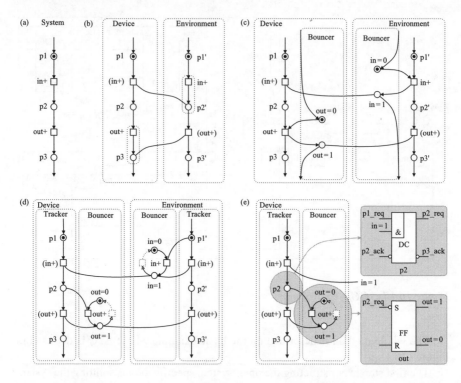

Figure 16.18   Method for the direct mapping from STGs: (a) system, (b) splitting the system, (c) output exposure, (d) elementary cycles and (e) mapping into circuit

removed, provided that the system behaviour is preserved w.r.t. the input–output interface (weak bisimulation). Such places are called 'redundant'. This way the place p2 is redundant in the considered example, Figure 16.19(a). It can be removed from the device tracker together with the preceding dummy (in+) as shown in Figure 16.19(b). Now the input $in1 = 1$ controls the output out+ transition directly, which results in latency reduction when the STG is mapped into the circuit, see Figure 16.19(c). Before the optimisation the output FF was set by the p2_req signal, which was generated in response to the input in1, see Figure 16.18(e). In the optimised circuit the output FF is triggered directly by the in1 input and the context signal p3_req is calculated in advance concurrently with the environment action.

The elimination of places is restricted, however, by potential 'coding conflicts'. Coding conflicts may cause tracking errors. Consider the system whose STG is depicted in Figure 16.20(a). The devices specification extracted from this STG by applying the above method is shown in Figure 16.20(b). The tracker part of the device can be further optimised. The removal of redundant places p2 and p4 does not cause any conflicts of the tracker, Figure 16.20(c). However, if the place $p3$ is eliminated as shown in Figure 16.20(d), then the tracker cannot distinguish between the output having not yet been set and the output already reset. Note the specifics of this direct

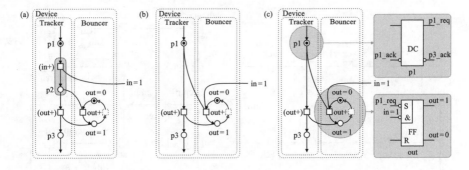

Figure 16.19   Optimisation of the device specification (a) redundant places, (b) optimisation and (c) mapping into circuit

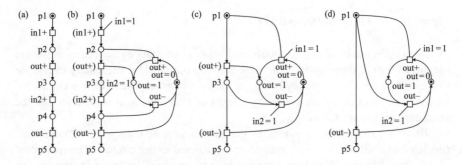

Figure 16.20   Preventing coding conflicts: (a) system, (b) output exposure, (c) correct optimisation and (d) coding conflict

mapping approach: only those signals whose switching directly precedes the given output or tracker transition are used in its support.

Consider the application of the OptiMist tool to the example of the GCD control unit. Its STG is obtained by refining the LPN generated from the HDL specification by PN2DCs tool, see Figure 16.8.

In order to produce the control unit STG shown in Figure 16.21 the events of the LPN are expanded to a four-phase handshake protocol. After that, the GCD datapath schematic shown in Figure 16.13 is taken into account to manually adjust the STG to the datapath interface. In the modified STG the request to the comparator cmp_req is acknowledged in 1-hot code by one of the signals: gt_ack, eq_ack or lt_ack. The request to the subtracter sub_gt_req is acknowledged by x_ack. This is possible because the procedure of storing the subtraction result into the register is controlled directly in the datapath and does not involve the control unit. Similarly sub_lt_req is acknowledged by ack_y.

When the OptiMist tool is applied to the original STG of GCD controller it produces the device specification which is divided into a tracker and a bouncer part,

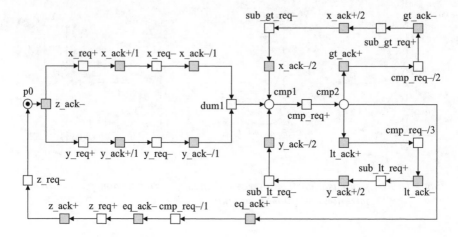

Figure 16.21   GCD control unit STG

as shown in Figure 16.22. The bouncer consists of elementary cycles representing the outputs of the GCD controller, one cycle for each output. The elementary cycles for the inputs are not shown as they belong to the environment. The tracker is connected to inputs and outputs of the system by means of read arcs, as it is described in the procedure of output exposure.

There are two types of places in the tracker part of the system: 'redundant' (depicted as small circles) and 'mandatory' (depicted as big circles). The redundant places can be removed without introducing a coding conflict while the mandatory places should be preserved. OptiMist tool determines the sets of redundant and mandatory places using the heuristics described in Reference 35.

The first heuristic, most important in terms of latency, states that each place whose all preceding transitions are controlled by inputs and all successor transitions are controlled by outputs can be removed from the tracker. Removal of such a place does not cause a coding conflict as the tracker can distinguish the state of the system before the preceding input-controlled transitions and after the succeeding output-controlled transitions. However, a place should be preserved if any of its preceding transitions is a direct successor of a choice place. Preserving such a place helps to avoid the situation when the conflicting transitions (direct successors of the choice place) are controlled by the same signal. The removal of the redundant places detected by the first heuristic reduces both the size and the latency of the circuit.

The redundant places detected by the above heuristic in the GCD example are px1, py1, px3, py3, px5, py5, pgt3, plt3, pgt5, plt5, peq3 and peq5. The places pgt1, plt1 and peq1 which follow the choice place cmp2 should be preserved. Their removal would cause an ambiguous situation when the first transitions to the three conflicting branches are controlled by the same signal cmp_req=0.

The next heuristic for redundant places detection traverses the chains of non-redundant places between input-controlled transitions. The traversing of a chain starts

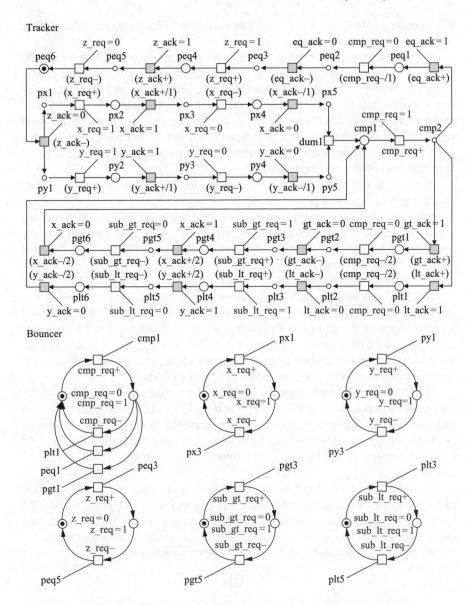

Figure 16.22 Exposure of the outputs and detection of the redundant places

from the place after an input-controlled transition and progresses in the direction of consuming–producing arcs. For each place in the chain it is checked if its removal causes a coding conflict. The potency of a coding conflict is checked assuming that all the places which are currently tagged as redundant are already removed from the tracker. If the coding conflict does not occur then the place is tagged as redundant.

The traversing of the chain stops when a non-redundant place is found. After that the next chain is processed.

The redundant places cmp2, pgt2, plt2 and peq2 are detected by this heuristic in the GCD example. Place cmp1 can also be tagged as redundant but it is kept by the OptiMist tool in order to preserve the simplicity of the cmp_req elementary cycle. Without this place the positive phase of the cmp_req would be controlled by two context signals from the tracker (read-arcs from px4 and py4) and two trigger signals from the environment (x_ack = 0 and y_ack = 0). The trade-off between the complexity of elementary cycles and the number of places in the tracker can be set as an OptiMist command line parameter.

Removal of any of the other places from the tracker causes the coding conflict and such places should be preserved.

After the outputs are exposed and the redundant places are detected, the OptiMist tool optimises the tracker by removing the redundant places and preceding dummy transitions. The removal of a place involves the change of the STG structure but preserves the behaviour of the system w.r.t. the input–output interface.

The result of GCD control unit optimisation is presented in Figure 16.23.

This STG can now be used for circuit synthesis. For this each tracker place is mapped into a DC and each elementary cycle is mapped into an FF. The request and acknowledgement functions of a DC are mapped from the structure of the tracker in the vicinity of the corresponding place as shown in Figure 16.24(a). The set and reset functions of an FF are mapped from the structure of the set and reset phases of the corresponding elementary cycle as shown in Figure 16.24(b).

The GCD controller circuit obtained by this technique is presented in Figure 16.25. This circuit will consist of 15 DCs and 6 FFs. If the FFs and DCs are implemented on transistor level as shown in Figure 16.15(b), then the maximum number of transistor levels in pull-up and pull-down stacks is four. This transistor stack appears in the request function of the DC for cmp1 and formed by the signals x_ack = 0, y_ack = 0, px4_req and py4_req.

The longest latency, which is the delay between an input change and reaction of the controller by changing some outputs, is exhibited by the cmp_req signal. The latency of its set and reset phases is equal to the delay of one DC and one FF. The other outputs are triggered directly by input signals which means that their latencies are equal to one FF delay plus the delay of one inverter when the trigger signal requires inversion.

16.7   Explicit logic synthesis

The explicit logic synthesis methods work with the low-level system specifications which capture the behaviour of the system at the level of signal transitions, such as STGs. These methods usually derive Boolean equations for the output signals of the controller using the notion of next state functions obtained from STGs [54].

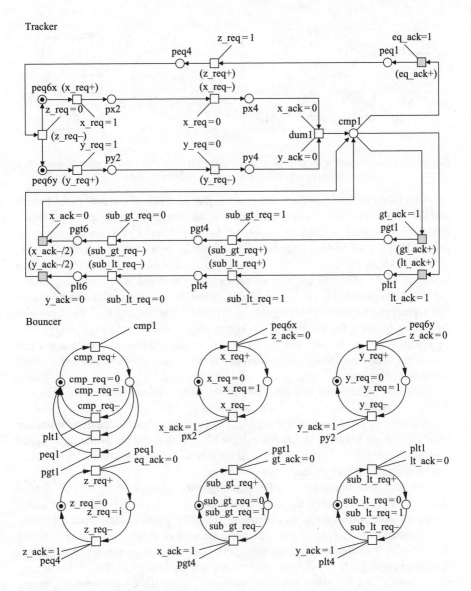

Figure 16.23   Optimisation

An STG is a succinct representation of the behaviour of an asynchronous control circuit that describes the causality relations among the events. In order to find the next state functions all possible firing orders of the events must be explored. Such an exploration may result in a state space which is much larger than the STG specification. Finding efficient representations of the state space is a crucial aspect in building logic synthesis tools.

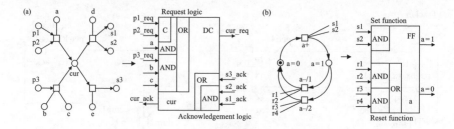

Figure 16.24   *Mapping STG into circuit: (a) mapping the tracke places into DCs and (b) mapping the elementary cycles into FFs*

A synthesis method based on state space exploration is implemented in the Petrify tool [34]. It represents the system state space in the form of a state graph (SG), which is a binary encoded reachability graph of the underlying PN. Then the theory of regions [56] is used to derive the Boolean equations for the output signals.

Figure 16.26 presents the SG for the GCD control unit, whose STG is shown in Figure 16.21. The SG consists of vertexes and directed arcs connecting them. Each vertex corresponds to a state of the system and is assigned a binary vector that represents the value of all signals in that state. The sequence of the signals in the binary vector is the following: <x_req, y_req, x_ack, y_ack, cmp_req, gt_ack, eq_ack, lt_ack, sub_gt_req, sub_lt_req, z_req, z_ack>. The initial state is marked with a box. The directed arcs are assigned with the signal events which are enabled in the preceding states.

Note that all possible combinations of the events in two concurrent branches x_req+ → x_ack+ → x_req− → x_ack− and y_req+ → y_ack+ → y_req− → y_ack− are expressed explicitly in the SG. The explicit representation of concurrency results in a huge SG for a highly concurrent STG. This is known as the 'state space explosion' problem, which puts practical bounds on the size of control circuits that can be synthesised using state-based techniques.

The other interesting issue is the unambiguous state encoding. The shadowed states in Figure 16.26 have the same binary code, but they enable different signal events. This means that the binary encoding of the SG signals alone cannot determine the future behaviour of the system. Hence, an ambiguity arises when trying to define the next-state function. Roughly speaking, this phenomenon appears when the system does not have enough memory to 'remember' in which state it is. When this occurs, the system is said to violate the complete state coding (CSC) property. Enforcing CSC is one of the most difficult problems in the synthesis of asynchronous circuits. The general idea of solving CSC conflicts is the insertion of new signals, which add more memory to the system. The signal events should be added in such a way that the values of inserted signals disambiguate the conflicting states.

16.7.1   Automatic CSC conflict resolution

One of the possibilities to resolve the CSC conflicts is to exploit the Petrify tool and the underlying theory of regions. In Petrify all calculations for finding the states

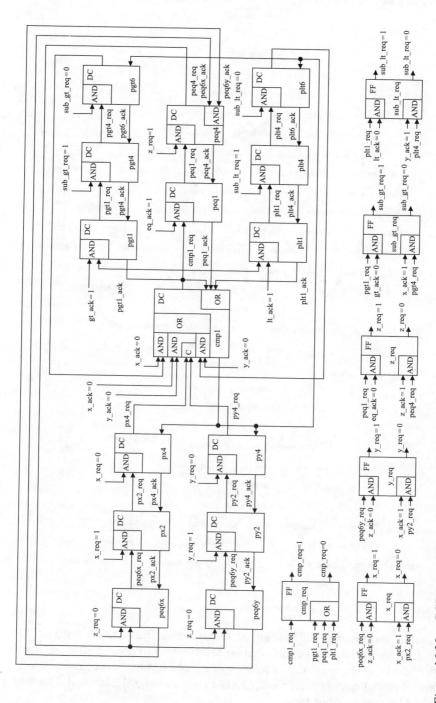

Figure 16.25   GCD controller circuit obtained by the OptiMist tool

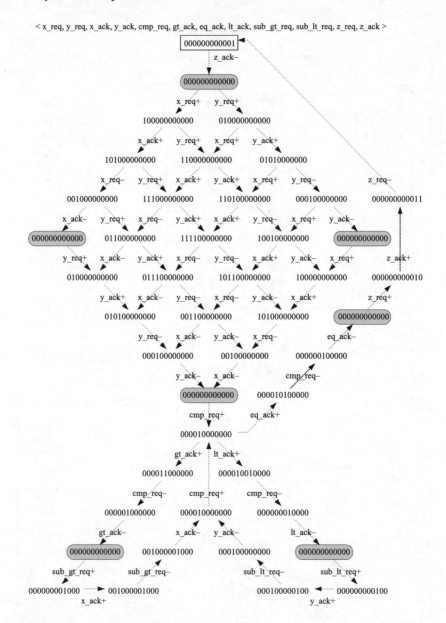

Figure 16.26   SG for GCD control unit

in conflict and inserting the new signal events are performed at the level of SG. The tool relies on the set of optimisation heuristics when deciding how to insert new transitions. However, the calculation of regions involves the computationally intensive procedures which are repeated when every new signal is inserted. This may result in long computation time.

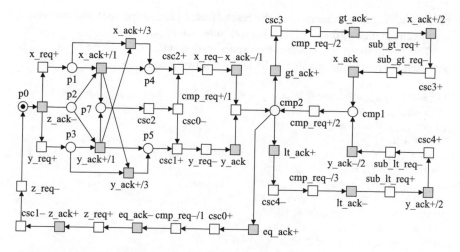

Figure 16.27   Resolution of CSC conflicts by Petrify

When the system becomes conflict-free, the SG is transformed back into STG. Often the structure of resultant STG differs significantly from the original STG, which might be inconvenient for its further manual modification. Actually, the STG look may change even after simple transformation to SG and back to STG, because the structural information is lost at the level of SG.

The conflict-free STG for the GCD control unit is shown in Figure 16.27. There are two changes to the structure of the STG which are not due to new signal insertion. First, the transition cmp_req+ is split into cmp_req+/1 and cmp_req+/2; second, the concurrent input of x and y is synchronised on cmp_req+/1 instead of dummy now.

Petrify resolves the CSC conflicts in GCD control unit specification adding five new signals, namely csc0, csc1, csc2, csc3, csc4. The insertion of signals csc0, csc3 and csc4 is quite predictable. They are inserted in three conflicting branches (one in each branch) in order to distinguish between the state just before cmp_req+/1 and just after eq_ack−, gt_ack−, lt_ack−, respectively.

For example, the state of the system before and after the following sequence of transitions is exactly the same: cmp_req + /1 $\rightarrow$ eq_ack+ $\rightarrow$ cmp_req − /1 $\rightarrow$ eq_ack−. In order to distinguish between these states csc0+ transition is inserted inside the above sequence. As the behaviour of the environment must be preserved, the new transition can only be inserted before the output transition cmp_req−/1. The are two possibilities for its insertion: sequentially or concurrently. The former type of insertion is usually (but not always) preferable for the smaller size of circuit, the latter for lower latency. Relying on its sophisticated heuristics Petrify decides to insert csc0+ sequentially. Signal csc0 is reset in the same branch outside the above sequence of transitions.

Similarly, signals csc1 and csc2 are inserted to distinguish the states before and after the sequence of transitions x_req+ $\rightarrow$ x_ack+ $\rightarrow$ x_req− $\rightarrow$ x_ack− and y_req+ $\rightarrow$ y_ack+ $\rightarrow$ y_req− $\rightarrow$ y_ack−, respectively. However, the resetting of the csc2 is not symmetrical to the reset of csc1 (as is expected) and involves a significant change of the original STG structure, see Figure 16.27.

The synthesis of the conflict-free specification with logic decomposition into gates with at most four literals results in the following equations:

```
[x_req] = z_ack' (csc0 csc1' + csc2');
[y_req] = z_ack' csc1';
[z_req] = csc0 eq_ack' csc1;
[3] = csc4' + csc3' + csc0 + csc2';
[cmp_req] = [3]' x_ack' y_ack' csc1;
[sub_gt_req] = csc3' gt_ack';
[sub_lt_req] = csc4' lt_ack';
[csc0] = csc2 csc0 + eq_ack;
[csc1] = csc0' y_ack + z_ack' csc1;
[9] = csc0' (csc2 + x_ack);
[csc2] = x_ack' csc2 y_ack' + [9];
[csc3] = gt_ack' (csc3 + x_ack);
[csc4] = lt_ack' (csc4 + y_ack);
```

The estimated area is 432 units and the maximum and average delay between the inputs is 4.00 and 1.75 events, respectively. The worst case latency is between the input x_ack+/1 and the output x_req−. The trace of the events is x_ack + /1 → csc_2− → csc_0− → csc_2+ → x_req−. Taking into account that CMOS logic is built out of negative gates these events correspond to the following sequence of gates switching: [x_ack ↑] → [x_ack' ↓] → [csc2' ↑] → [csc2 ↓] → [csc0' ↑] → [9' ↓] → [9 ↑] → [csc2' ↓] → [x_req' ↑] → [x_req ↓]. This gives the latency estimate equal to the delay of nine negative gates.

16.7.2 Semi-automatic CSC conflict resolution

A semi-automatic approach to CSC conflict resolution is adopted in the ConfRes tool [56]. The main advantage of the tool is its interactivity with the user during CSC conflict resolution. It visualises the cause of the conflicts and allows the designer to manipulate the model by choosing where in the specification to insert new signals.

The ConfRes tool uses STG unfolding prefixes [57] to visualise the coding conflicts. An unfolding prefix of an STG is a finite acyclic net which implicitly represents all the reachable states of the STG together with transitions enabled at those states. Intuitively, it can be obtained by successive firings of STG transitions under the following assumptions:

- for each new firing a fresh transition (called an event) is generated;
- for each newly produced token a fresh place (called a 'condition') is generated.

If the STG has a finite number of reachable states then the unfolding eventually starts to repeat itself and can be truncated (by identifying a set of 'cut-off events') without loss of information, yielding a finite and complete prefix.

In order to avoid the explicit enumeration of coding conflicts, they are visualised as 'cores', i.e. the sets of transitions causing one or more of conflicts. All such cores must eventually be eliminated by adding new signals.

The process of core resolution in the GCD control unit is illustrated in Figure 16.28. Actually, there are ten overlapping conflict cores in the STG. The ConfRes tool shows them in different colours similar to a geographical height-map. However, all ten cores would be hardly distinguishable on a grey-scale printout. That is the reason why only those cores whose resolution is currently discussed are shown. The cores are depicted as gray polygons covering the sets of sequential transitions. Each core has different brightness and is labelled with a name in rounded box to refer from the text.

The basic rules for new signal insertion are the following:

- In order to destroy a core one phase of a new signal should be inserted inside the core and the other phase outside the core.
- A new signal should be inserted into the intersection of several cores whenever possible, because this minimises the number of inserted signals, and thus the area and latency of the circuit.
- A new signal transition cannot be inserted before an input signal transition, because it would change the device–environment interface.
- Usually (but not always) the sequential insertion of a transition is preferable for smaller circuit size and concurrent insertion is advantageous for lower circuit latency.

Consider this procedure on the example of a GCD controller, Figure 16.28. Two experiments are conducted. In the first one the strategy of sequential signal insertion is exploited in order to compete automatic conflict resolution in circuit size. In the second experiment the new signals are inserted concurrently (where possible) in order to achieve lower latency.

In the experiment with sequential signals insertion, first, the *cores* C1 and C2 shown in Figure 16.28(a) are destroyed by inserting csc_x+ transition sequentially before x_req−. The reset phase of csc_x is inserted between eq_ack+ and cmp_req−/1 thereby destroying the core C3. Similarly, two other cores, symmetrical to C1 and C2 (not shown for readability of the diagram), are eliminated by inserting transition csc_y+ before y_req−. The reset phase of csc_y is inserted the same way as csc_x− (between eq_ack+ and cmp_req−/1) and destroys the core symmetrical to C3.

Second, the cores C4 and C5 are eliminated by inserting csc_lt+ sequentially before cmp_req−/3, see Figure 16.28(b). The reset phase of csc_lt is inserted outside the cores C4 and C5 sequentially before sub_lt_req−. Likewise, the core which is symmetrical to C4 (not shown for simplicity) is destroyed by inserting csc_gt+ before cmp_req+/2 and csc_gt− before sub_gt_req−.

Finally, only one core C6 is left, see Figure 16.28(c). It is destroyed by replacing the dummy dum1 by csc_eq− transition. The set phase of csc_eq is inserted outside the core before z_req−. The resultant conflict-free STG of the GCD controller is shown in Figure 16.29.

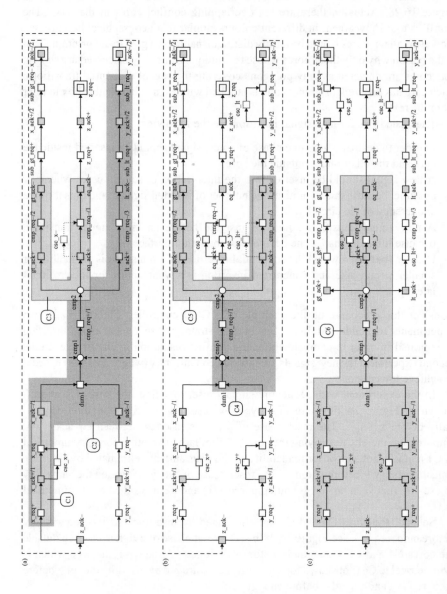

Figure 16.28 Visualisation of conflict cores in ConfRes

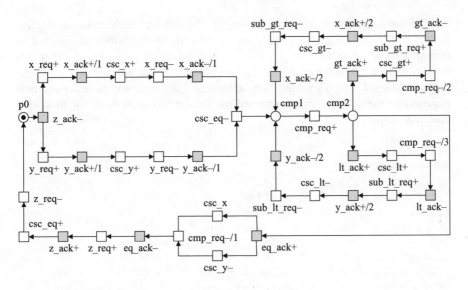

Figure 16.29   Resolution of CSC conflicts by ConfRes

Petrify synthesises this STG with logic decomposition into gates with at most four literals into the following equations:

```
[x_req] = csc_x' z_ack' csc_eq;
[y_req] = csc_y' z_ack' csc_eq;
[z_req] = csc_x' eq_ack' csc_eq';
[3] = csc_y' csc_x' + csc_gt + csc_lt;
[cmp_req] = [3]' y_ack' x_ack' csc_eq';
[sub_gt_req] = csc_gt gt_ack';
[sub_lt_req] = csc_lt lt_ack';
[csc_x] = eq_ack' (x_ack + csc_x);
[csc_y] = csc_y eq_ack' + y_ack;
[csc_gt] = x_ack' csc_gt + gt_ack;
[10] = csc_eq (csc_x' + csc_y') + z_ack;
[csc_eq] = csc_eq (x_ack + y_ack) + [10];
[csc_lt] = y_ack' (lt_ack + csc_lt);
```

The estimated area is 432 units, which is the same as when the coding conflicts are resolved automatically. However, the maximum and average delays between the inputs are significantly improved: 2.00 and 1.59 events, respectively.

The worst case latency of the circuit is between gt_ack+ and cmp_req−/2 (or between eq_ack+ and cmp_req−/1). If the circuit is implemented using CMOS negative gates then this latency corresponds to the following sequence of gates switching: [gt_ack ↑] → [csc_gt′ ↓] → [csc_gt ↑] → [3′ ↓] → [cmp_req′ ↑] → [cmp_req ↓]. This gives the latency estimate equal to the delay of five negative gates, which is significantly better than in the experiment with automatic coding conflict resolution.

The other experiment with semi-automatic CSC conflict resolution aims at lower latency of the GCD control circuit. Now the new signal transitions are inserted as concurrently as possible. Namely, csc_x+ is concurrent to x_ack+/1; csc_y+ is concurrent to y_ack+/1; csc_gt− is concurrent to x_ack+/2; and csc_lt− is concurrent to y_ack+/2. The other transitions are inserted the same way as in the previous experiment. Synthesis of the constructed conflict-free STG produces the following equations:

```
[0] = csc_x' z_ack' csc_eq;
[x_req] = x_req map0' + [0];
[2] = csc_y' z_ack' csc_eq;
[y_req] = y_ack' y_req + [2];
[z_req] = csc_y' eq_ack' csc_eq';
[5] = csc_y' csc_x' + map0 + csc_eq;
[cmp_req] = sub_lt_req' [5]' (map1 + eq_ack);
[sub_gt_req] = gt_ack' (sub_gt_req map1 + csc_gt);
[sub_lt_req] = sub_lt_req map1 + csc_lt lt_ack';
[csc_x] = eq_ack' (csc_x + x_req);
[csc_y] = eq_ack' (csc_y + y_req);
[csc_lt] = sub_lt_req' csc_lt + lt_ack;
[csc_gt] = sub_gt_req' (gt_ack + csc_gt);
[csc_eq] = map1' (csc_eq + z_ack);
map0 = sub_gt_req + csc_gt + csc_lt + x_ack;
[15] = csc_x' + x_req + csc_y';
map1 = [15]' y_ack' y_req' x_ack';
```

Two new signals, map0 and map1, are added by Petrify in order to decompose the logic into library gates with at most four literals. This results in larger estimated circuit size, 592 units. The average input-to-input delay of the circuit becomes 1.34 events, which is smaller than in the previous experiment. However, the maximum latency of the circuit is seven negative gates delay. It occurs, for example, between gt_ack+ and cmp_req− transitions. The gates switched between these transitions are: [gt_ack↑]→[csc_gt'↓]→[csc_gt↑]→[map0'↓]→[map0↑]→[5'↓]→[cmp_req'↑]→[cmp_req↓]. The worst case latency in this implementation is greater than the latency in the previous design due to the internal map0 and map1 signals, which are used for decomposition of non-implementable functions.

The complex gate implementation of the GCD controller, where CSC conflict is resolved manually by inserting new signals in series with the existing ones, is shown in Figure 16.30. This is the best solution (in terms of size and latency) synthesised by Petrify with the help of the ConfRes tool. It consists of 120 transistors and exhibits the latency of five negative gates delay.

Clearly, semi-automatic conflict resolution gives the designer a lot of flexibility in choosing between the circuit size and latency. The visual representation of conflict cores distribution helps the designer to plan how to insert each phase of a new signal optimally, thus possibly destroying several cores by one signal. The diagram of core distribution is updated after every new signal insertion. As all the modifications to

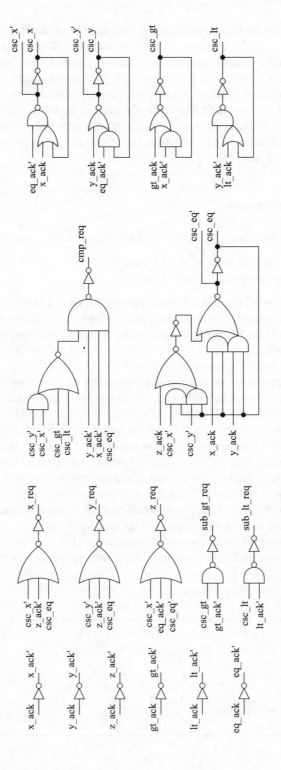

Figure 16.30 Complex gates implementation of GCD controller

the system are performed on its unfolding prefix, there is no need to recalculate the state space of the system, which makes the operation of ConfRes tool extremely fast.

Another approach to CSC conflicts resolution which avoids the expensive computation of the system state space is proposed in References 58 and 59. The approach is based on the structural methods, which makes it applicable for large STG specifications. Its main idea is to insert a new set of signals in the initial specification in a way that unique encoding is guaranteed in the transformed specification. The main drawback of this approach is that the structural methods are approximate and can only be exact for well-formed PNs.

16.8   Tools comparison

In this section the tools are compared using GCD benchmarks in two categories:

- system synthesis from high-level HDLs and
- synthesis of the control unit from STGs.

Table 16.1 presents characteristics of asynchronous GCD circuits synthesised by Balsa and PN2DCs tools from high-level HDLs. Both solutions are implemented using the AMS-0.35 μm technology and dual-rail datapath components. The size of each circuit is calculated using Cadence Ambit tool and the speed is obtained by circuit simulation in SPICE analogue simulator.

The benchmark shows that the circuit generated by PN2DCs tool is 16 per cent smaller and 33–42 per cent faster than the circuit synthesised by Balsa. The size and speed improvement in PN2DCs comparing to Balsa solution is due to different control strategies. Note that the intermediate controller specification for the PN2DCs tool is manually optimised by removing redundant places and transitions. This reduces the controller unit area by four DCs (732 μm^2). However, the optimisation algorithm is straightforward, the redundant places and transitions removal can be automated.

The time spent by Balsa and PN2DCs to generate the circuit netlists is negligible. This is because both tools use computationally simple mapping techniques, which allows us to process large system specifications in acceptable time.

The characteristics of the circuits synthesised from STGs are shown in Table 16.2. For the circuit generated by OptiMist tool, the number of transistors is counted for

Table 16.1   Comparison between Balsa and PN2DCs

Tool	Area (μm^2)	Speed (ns)		Computation time (s)
		$x = y$	$x = 12, y = 16$	
Balsa	119 647	21	188	<1
PN2DCs	100 489	14	109	<1
Improvement	16%	33%	42%	0

Table 16.2   Comparison between OptiMist and Petrify

Tool	Number of transistors	Latency (units)	Computation time (s)
OptiMist	174	4.5	<1
Petrify			
automatic	116	13.0	18
sequential	120	8.0	2
concurrent	142	11.0	4

the case of places being implemented as fast DCs shown in Figure 16.15(b). The request-acknowledgement logic of DCs and set–reset logic of FFs are implemented at transistor level. The number of transistors for the circuits generated by Petrify is counted for complex gate implementation. The technology mapping into the library gates with at most four literals is applied.

In all experiments, the latency is counted as the accumulative delay of negative gates switched between an input and the next output. The following dependency of a negative gate delay on its complexity is used. The latency of an inverter is associated with 1 unit delay. Gates which have maximum two transistors in their transistor stacks are associated with 1.5 units; 3 transistors – 2.0 units; 4 transistors – 2.5 units. This approximate dependency is derived from the analysis of the gates in AMS 0.35 μm library. The method of latency estimation does not claim to be very accurate. However, it takes into account not only the number of gates switched between an input and the next output, but also the complexity of these gates.

The Petrify tool was used to synthesise the circuits with three alternatives of CSC conflict resolution. In the first circuit the coding conflict is solved by inserting new signals automatically. In the second and the third circuits the semi-automatic method of conflict resolution is employed by using the ConfRes tool. In the second circuit the transitions of new signals are inserted sequentially, and in the third one concurrently.

The experiments show that the automatic coding conflict resolution may result in a circuit with high output latency which is due to non-optimal insertion of the new signals. The smallest circuit is synthesised when the coding conflicts are resolved manually by inserting the new signals sequentially. This solution also exhibits lower latency than in the case of automatic and concurrent signal insertion. The circuit with the new signals inserted concurrently lacks the expected law latency because of its excessive logic complexity.

The circuit with the lowest latency is generated by the direct mapping technique using the OptiMist tool. This tool also exhibits the smallest synthesis time which is due to low algorithmic complexity of the involved computations. This allows processing large specifications, which cannot be computed by Petrify in acceptable time because of the state space explosion problem.

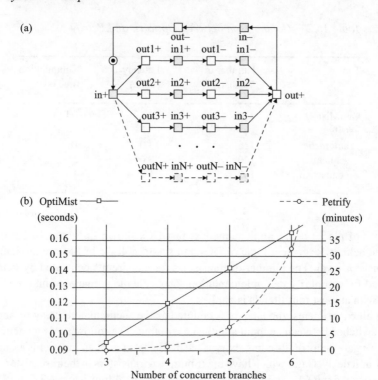

Figure 16.31   Dependency of computation time on STG complexity: (a) Scalable benchmark STG and (b) Computation time

This can be illustrated on the scalable benchmark whose STG is shown in Figure 16.31(a). Adding the concurrent branches as shown by dashed lines one can increase the complexity of the benchmark. When the concurrency increases the Petrify computation time grows exponentially, while the OptiMist computation time grows linearly on the same benchmark, see Figure 16.31(b).

However, the GCD controller synthesised by the OptiMist tool is about 45 per cent larger than Petrify's solutions.

16.9   Conclusions

The state of the art in the synthesis of asynchronous systems from high-level behavioural specifications has been reviewed. Two main approaches of circuit synthesis have been considered: syntax-driven translation and logic synthesis.

The syntax-driven approach is studied on the example of Balsa design flow. It uses a CSP-based language for the initial system specification. Its parsing tree is translated into a handshake circuit, which is subsequently mapped to the library of hardware components. This approach enables the construction of large-size asynchronous systems in a short time, due to its low computational complexity. However, the speed and area of the circuit implementations may not be the best possible. Therefore this approach benefits from peep-hole optimisations, which apply logic synthesis locally, to groups of components, as was demonstrated in Reference 32.

The logic synthesis approach is reviewed using the PN2DCs, OptiMist, Petrify and ConfRes tools. The PN2DCs tool partitions the VHDL system specification on control and datapath. Petri nets are used for their intermediate behavioural representation. The datapath PN is subsequently mapped into a netlist of datapath components. The controller PN can be either mapped into a David cell structure or further refined to an STG. The control unit can be synthesised by one of the above mentioned tools. Logic synthesis approach is computationally harder than the syntax-based one. The direct mapping of Petri nets and STGs in PN2DCs and OptiMist, helps to avoid state space explosion involved in the state encoding procedures used in Petrify. At the same time, this comes at the cost of more area.

It should be clear that tools like Petrify and ConfRes should be used for relatively small control logic, for instance in interfaces and pipeline stage controllers (see Reference 4), rather than complex data processing controllers, where PN2DCs is more appropriate. The latter is however not optimal for speed because it works at a relatively high-level of signalling. OptiMist, working at the STG level, combines the advantages of low computational complexity with high-speed due to its latency-aware implementation architecture with a bouncer and a tracker.

The greatest common divisor benchmark is used to evaluate all of the above mentioned tools. The size and speed of the resultant circuits are compared. They demonstrate the various possible enhancements in the design flow, such as the use of an interactive approach to state encoding in logic synthesis.

In the future a combination of techniques in a single tool flow might prove most advantageous. For example, at first, each output signal which has a complete state coding can be synthesised individually by Petrify, with or without use of ConfRes. Then, all the remaining outputs, whose CSC resolution is hard or impossible, can be mapped into logic at once by OptiMist. Here, the best trade-off between area and performance may be achieved.

Acknowledgements

The authors are grateful to Agnes Madalinski, Delong Shang, Julian Murphy and Alex Bystrov for their useful comments. EPSRC supports this work via GR/R16754 (BESST), GR/S12036 (STELLA) and GR/S81421 (SCREEN).

References

1   MOORE, G.: 'Cramming more components onto integrated circuits'. *Electronics*, 1965, **38**(8), pp. 68–70

2   JANTSCH, A. 'Modelling embedded systems and SoCs: concurrency and time in models of computation' (Morgan Kaufmann, San Francisco, CA, 2004)

3   HAUCK, S.: 'Asynchronous design methodologies: an overview', *Proceedings of the IEEE*, 1995, **83**(1), pp. 69–93

4   CORTADELLA, J., KISHINEVSKY, M., KONDRATYEV, A., LAVAGNO, L., and YAKOVLEV, A.: 'Logic synthesis for asynchronous controllers and interfaces' (Springer-Verlag, Berlin, Germany, March 2002)

5   CHRIS J. MYERS: 'Asynchronous circuit design' (Wiley-Interscience, John Wiley & Sons, Inc., 2001)

6   DANIEL M. CHAPIRO: 'Globally-asynchronous locally-synchronous systems'. PhD thesis, October 1984.

7   MUTTERSBACH, J.: 'Globally-asynchronous locally-synchronous architectures for VLSI systems'. PhD thesis, Swiss Federal Institute of Technology (ETH), Zürich, 2001.

8   KENNETH Y. YUN, and RYAN P. DONOHUE: 'Pausible clocking: a first step toward heterogeneous systems'. Proceedings of the international conference *Computer design (ICCD)*, October 1996

9   MUTTERSBACH, J., VILLIGER, T., and FICHTNER, W.: 'Practical design of globally-asynchronous locally-synchronous systems'. Proceedings of the international symposium on *Advanced research in asynchronous circuits and systems (ASYNC)* (IEEE Computer Society Press, CA, April 2000) pp. 52–59

10 DOBKIN, R., GINOSAR, R., and SOTIRIOU, C.P.: 'Data synchronization issues in GALS SoCs'. Proceedings of the international symposium on *Advanced research in asynchronous circuits and systems (ASYNC)* (IEEE Computer Society Press, Los Alamitos, CA, April 2004) pp. 170–179

11 LUCA P. CARLONI, KENNETH L. MCMILLAN, SALDANHA, A., and ALBERTO L. SAGIOVANNI-VINCENTELLI: 'A methodology for correct-by-construction latency insensitive design'. Proceedings of the international conference *Computer-aided design* (ICCAD), November 1999, pp. 309–315

12 SINGH, M., and THEOBALD, M.: 'Generalized latency-insensitive systems for gals architectures'. Proceedings of the workshop on *Formal methods for globally asynchronous locally synchronous architecture (FMGALS)* (within the *International formal methods* Europe symposium), September 2003

13 J. O'LEARY, and BROWN, G.: 'Synchronous emulation of asynchronous circuits', *IEEE Transactions on Computer-Aided Design*, 1997, **16**(2), pp. 205–209

14 PEETERS, A., and VAN BERKEL, K.: 'Synchronous handshake circuits'. Proceedings of the international symposium on *Advanced research in asynchronous circuits and systems (ASYNC)* (IEEE Computer Society Press, Los Alamitos, CA, March 2001) pp. 86–95

15 HANS M. JACOBSON, PRABHAKAR N. KUDVA, BOSE, P. *et al.*: 'Synchronous interlocked pipelines'. Proceedings of the international symposium on *Advanced research in asynchronous circuits and systems (ASYNC)* (IEEE Computer Society Press, Los Alamitos, CA, April 2002) pp. 3–12

16 DAVIS, A., and STEVEN M. NOWICK: 'An introduction to asynchronous circuit design'. Technical Report UUCS-97-013, University of Utah, September 1997

17 KESSELS, J., KRAMER, T., PEETERS, A., and TIMM, V.: 'DESCALE: a design experiment for a smart card application consuming low energy' (Kluwer Academic Publishers, Dortchet, 2001)

18 SPARSØ, J., and FURBER, S.: 'Principles of asynchronous circuit design – a system perspective' (Kluwer Academic, Dortchet, 2001)

19 VAN BERKEL, K., KESSELS, J., RONCKEN, M., SAEIJS, R., and F. SCHALIJ: 'The VLSI-programming language Tangram and its translation into handshake circuits'. Proceedings of the European conference on *Design automation (EDAC)*, 1991, pp. 384–389

20 BARDSLEY, A.: 'Implementing Balsa handshake circuits'. PhD thesis, Department of Computer Science, University of Manchester, 2000

21 HOARE, C.A.R.: 'Communicating sequential processes' (Prentice-Hall, NJ, 1985)

22 BLUNNO, I., and LAVAGNO, L.: 'Automated synthesis of micro-pipelines from behavioral Verilog HDL'. Proceedings of the international symposium on *Advanced research in asynchronous circuits and systems (ASYNC)* (IEEE Computer Society Press, Los Alamitos, CA, April 2000) pp. 84–92

23 DINDHUC, A., RIGAUD, J.-B., REZZAG, A. *et al.*: 'Tima asynchronous synthesis tools'. Communication to ACID workshop, January 2002

24 BURNS, F., SHANG, D., KOELMANS, A., and YAKOVLEV, A.: 'An asynchronous synthesis toolset using Verilog', in Proceedings of the *Design, automation and test in Europe (DATE)*, February 2004, pp. 724–725

25 SHANG, D., BURNS, F., KOELMANS, A., YAKOVLEV, A., and XIA, F.: 'Asynchronous system synthesis based on direct mapping using VHDL and Petri nets', *IEE Proceedings, Computers and Digital Techniques*, 2004, **151**(3), pp. 209–220

26 SACKER, M., BROWN, A., and RUSHTON, A.: 'A general purpose behavioural asynchronous synthesis system'. Proceedings of the international symposium on *Advanced research in asynchronous circuits and systems (ASYNC)* (IEEE Computer Society Press, Los Alamitos, April 2004) pp. 125–134

27 VENKATARAMANI, G., BUDIU, M., CHELCEA, T., and COPEN GOLDSTEIN, S.: 'C to asynchronous dataflow circuits: an end-to-end toolflow'. Proceedings of the international workshop on *Logic synthesis*, June 2004

28 SUTHERLAND, I.: 'Micropipelines'. *Communications of the ACM*, 1989, **32**(6), pp. 720–738

29 SMIRNOV, A., TAUBIN, A., and ROSENBLUM, L.: 'Gate transfer level synthesis as an automated approach to fine-grain pipelining'. Proceedings of the international conference on *Application and theory of Petri nets*, June 2004

30 SOKOLOV, D., MURPHY, J., BYSTROV, A., and YAKOVLEV, A.: 'Improving the security of dual-rail circuits'. Proceedings of the workshop on *Cryptographic hardware and embedded systems (CHES)*, August 2004

31   KONDRATYEV, A., and LWIN, K.: 'Design of asynchronous circuits using synchronous CAD tools', *IEEE Design & Test of Computers*, 2002, **19**(4), pp. 107–117

32   CHELCEA, T., BARDSLEY, A., EDWARDS, D., and STEVEN M. NOWICK: 'A burst-mode oriented back-end for the Balsa synthesis system'. Proceedings of the *Design, automation and test in Europe (DATE)*, March 2002, pp. 330–337

33   CHOU, W.-C., PETER A. BEEREL, and KENNETH Y. YUN: 'Average-case technology mapping of asynchronous burst-mode circuits', *IEEE Transactions on Computer-Aided Design*, 1999, **18**(10), pp. 1418–1434

34   CORTADELLA, J., KISHINEVSKY, M., KONDRATYEV, A., LAVAGNO, L., and YAKOVLEV, A.: 'Petrify: a tool for manipulating concurrent specifications and synthesis of asynchronous controllers'. Proceedings of the XI conference on *Design of integrated circuits and systems*, Barcelona, Spain, November 1996

35   SOKOLOV, D., BYSTROV, A., and YAKOVLEV, A.: 'STG optimisation in the direct mapping of asynchronous circuits'. Proceedings of the *Design, automation and test in Europe (DATE)*, Munich, Germany (IEEE Computer Society Press, Los Alamitos, CA, March 2003)

36   BARDSLEY, A., and EDWARDS, D.: 'Compiling the language Balsa to delay-insensitive hardware', in KLOOS, C.D., and CERNY, E. (Eds): 'Hardware Description Languages and their Applications (CHDL)', April 1997, pp. 89–91

37   EDWARDS, D., and BARDSLEY, A.: 'Balsa: an asynchronous hardware synthesis language', *The Computer Journal*, 2002, **45**(1), pp. 12–18

38   MONTANARI, U., and ROSSI, F.: 'Acta informacia'. Technical report, 1995

39   YAKOVLEV, A., KOELMANS, A., and LAVAGNO, L.: 'High-level modeling and design of asynchronous interface logic', 1995, **12**(1), pp. 32–40

40   ROSENBLUM, L., and YAKOVLEV, A.: 'Signal graphs: from self-timed to timed ones'. Proceedings of international workshop on *Timed Petri nets* (IEEE Computer Society Press, Torino, Italy, July 1985) pp. 199–207

41   JENSEN, K.: 'Coloured Petri nets: basic concepts, analysis methods and practical use', Vol. 1 (Springer-Verlag, 1997)

42   LIGTHART, M., FANT, K., SMITH, R., TAUBIN, A., and KONDRATYEV, A.: 'Asynchronous design using commercial HDL synthesis tools'. Proceedings of the international symposium on *Advanced research in asynchronous circuits and systems (ASYNC)* (IEEE Computer Society Press, Los Alamitos, CA, April 2000) pp. 114–125

43   HUFFMAN, D.A.: 'The synthesis of sequential switching circuits', in MOORE, E.F. (Ed.): 'Sequential machines: selected papers' (Addison-Wesley, 1964). Reprinted from J. Franklin Institute, 1954, **257**(3), pp. 161–190 and (4), pp. 275–303, April 1954

44   UNGER, S.H.: 'Asynchronous sequential switching circuits' (Wiley-Interscience, John Wiley & Sons, Inc., New York, USA, 1969)

45   LEE A. HOLLAAR: 'Direct implementation of asynchronous control units', *IEEE Transactions on Computers*, 1982, **C-31**(12), pp. 1133–1141

46   SUHAS S. PATIL, and DENNIS, J.B.: 'The description and realization of digital systems'. Proceedings of the *IEEE COMPCON*, 1972, pp. 223–226

47 KISHINEVSKY, M., KONDRATYEV, A., TAUBIN, A., and VARSHAVSKY, V.: 'Concurrent hardware: the theory and practice of self-timed design'. Series in Parallel Computing (Wiley-Interscience, John Wiley & Sons, Inc., 1994)

48 DAVID, R.: 'Modular design of asynchronous circuits defined by graphs', *IEEE Transactions on Computers*, 1997, **26**(8), pp. 727–737

49 DAVID, E. MULLER, and BARTKY, W.S.: 'A theory of asynchronous circuits'. Proceedings of an international symposium on the *Theory of switching* (Harvard University Press, April 1959) pp. 204–243

50 VARSHAVSKY, V., and MARAKHOVSKY, V.: 'Asynchronous control device design by net model behavior simulation', in BILLINGTON, J. (Ed.): 'Application and theory of Petri nets', volume 1091 of 'Lecture notes in computer science' (Springer-Verlag, June 1996) pp. 497–515

51 BYSTROV, A., and YAKOVLEV, A.: 'Asynchronous circuit synthesis by direct mapping: interfacing to environment'. Proceedings of the international symposium on *Advanced research in asynchronous circuits and systems (ASYNC)* (IEEE Computer Society Press, Manchester, UK, April 2002) pp. 127–136

52 SUTHERLAND, I., and FAIRBANKS, S.: 'GasP: a minimal FIFO control'. Proceedings of the international symposium on *Advanced research in asynchronous circuits and systems (ASYNC)* (IEEE Computer Society Press, Los Alamitos, CA, March 2001) pp. 46–53

53 MILNER, R.: 'Communication and concurrency' (Prentice-Hall, NJ, 1989)

54 CHU, T.-A.: 'Synthesis of self-timed VLSI circuits from graph-theoretic specifications'. PhD thesis, June 1987

55 CORTADELLA, J., KISHINEVSKY, M., KONDRATYEV, A., LAVAGNO, L., and YAKOVLEV, A.: 'A region-based theory for state assignment in speed-independent circuits', *IEEE Transactions on Computer-Aided Design*, 1997, **16**(8), pp. 793–812

56 MADALINSKI, A., BYSTROV, A., KHOMENKO, V., and YAKOVLEV, A.: 'Visualization and resolution of coding conflicts in asynchronous circuit design'. Proceedings of the design, automation and test in Europe (DATE) (IEEE Computer Society Press, Los Alamitos, CA, March 2003)

57 KHOMENKO, V.: 'Model checking based on Petri net unfolding prefixes'. PhD thesis, School of Computer Science, University of Newcastle upon Tyne, 2002

58 CARMONA, J., and CORTADELLA, J.: 'A structural encoding technique for the synthesis of asynchronous circuits'. International conference on *Application of concurrency to system design*, June 2001, pp. 157–166

59 CARMONA, J.: 'Structural methods for the synthesis of well-formed concurrent specifications'. PhD thesis, Software Department, Universitat Politècnica de Catalunya, March 2004

Part VI

Network-on-chip

Chapter 17

Network-on-chip architectures and design methods

Luca Benini and Davide Bertozzi

17.1 Introduction

Increasing integration densities made possible by shrinking device geometries will have to be fully exploited to meet the computational requirements of applications in domains such as multimedia processing, automotive, ambient intelligence. For instance, the computational load of typical ambient intelligence tasks will be ranging from 10 MOPS for lightweight audio processing, 3 GOPS for video processing, 20 GOPS for multilingual conversation interfaces and up to 1 TOPS for synthetic video generation. These workloads will have to be delivered with tightly constrained power levels (from a few watts, for wall-plugged appliances, to a few milliwatts, for portable and wearable devices), affordable cost and high reliability [1]. System architecture and design technology must adapt to the critical challenges posed by both the large-scale integration and the small features of elementary devices.

To tackle the application and integration complexity challenges, and the ensuing design productivity gap, SoCs are and will increasingly be designed by re-using large-scale programmable components, such as microprocessors, micro-controllers and media-processors, as well as large embedded memory macros and numerous standard peripherals and specialised co-processors. Design methodologies have to support component re-use in a plug-and-play fashion in order to be effective. In this reuse-dominated context, there is little doubt on the fact that the most critical factor in system integration will be the scalability of the communication fabric among components. This conclusion is further strengthened if we focus on the 'challenges of the small' posed by the unrelenting pace of scaling. Whereas computation and storage power-delay products (i.e. energy) benefit from device scaling (smaller gates, smaller memory cells), the energy for global communication does not scale down, hence, propagation delays on global wires will greatly exceed the clock period, and the power consumed to drive wires will dominate the power breakdown. Moreover, estimating delays accurately will become increasingly harder, as wire geometries may be determined late in the design flow. Electrical noise due to cross-talk, delay variations and synchronisation failures will be likely to produce bit upsets. Thus, the transmission of digital values on wires will be slow, power-hungry and inherently unreliable.

Network-on-chips (NoCs) are viewed by many researchers and designers as a response to the 'interconnect showstopper'. The basic premise of the 'network-on-chip revolution' is fundamentally simple: the on-chip interconnect fabric should be designed using the same guiding principles that drive the development of macroscopic communication networks, which have demonstrated sustainable scalability, exponentially improving performance, remarkable robustness and reliability over many years of rapid evolution. The NoC literature has flourished in between 2002 and 2005, with many strong contributions on development, analysis and implementation. This chapter does not attempt a complete overview, but it aims at providing a survey on the evolution of the field, moving from state-of-the art communication fabrics (SoC buses), to forward-looking NoC research prototypes. We will underline the many elements of continuity as well as the key differences between SoC buses and NoCs, in an effort to extract some general guiding principles in a very dynamic landscape.

The chapter is organised as follows. In Section 17.2, on-chip buses and their evolutions are presented. Section 17.3 provides a quantitative assessment of the performance of basic on-chip interconnects and their state-of-the-art evolutions. Section 17.4 and 17.5 focus on network on chip architectures, components and prototypes. Section 17.6 is dedicated to design technology for NoCs. Conclusions and perspectives for future evolutions are depicted in Section 17.7.

17.2   On-chip buses

On-chip buses have originally been architected to mimic their off-chip counterparts, relying on the analogy between building a board with commodity components and building a system-on-chip with IP cores. Ultimately, buses rely on shared communication channels and on an arbitration mechanism which is in charge of serialising bus access requests (time division multiplexing). This widely adopted solution obviously suffers from power and performance scalability limitations, but it has the advantage of low complexity and reduced area for the interface, communication and control logic.

Besides scalability, another limitation of early on-chip buses is poor decoupling between core interfaces and bus protocols, which greatly weakens modularity and composability of complex designs. To better understand this issue, we can use as an example LAN/WAN interfaces in traditional computer networks, where the access protocol to the network is completely standardised (TCP/IP), and independent from the physical implementation (e.g. a shared medium, as in wireless networks, or complex multistage networks on cable or fibre). In order to test the level of decoupling between interconnect access protocol and core interfaces, a simple conceptual test can be performed, using a communication 'initiator' (also called 'master') and a 'target' (also called 'slave'). If master and slave can be connected directly to each other (i.e. with a point-to-point connection), then obviously the topology and internal protocol of the interconnect are completely decoupled from the core interface. We call PP (point-to-point) these network interfaces. PP is a very desirable property from the design integration viewpoint, because it completely decouples communication fabric design from core design.

On-chip buses have evolved in an effort to address the above-mentioned limitations. We can distinguish three main directions of evolution: (1) enhancements in the parallelism and efficiency of the bus access protocol, which help in fully exploiting the bandwidth of the available interconnect resources; (2) enhancements in topology, to increase the available interconnect bandwidth; (3) re-definition and standardisation of PP target and initiator interfaces. We will follow these trends with the help of a case study, namely advanced Micro-Controller Bus Architecture (AMBA), which is probably the most widely deployed on-chip communication protocol.

17.2.1 Tracking the evolutionary path: the AMBA bus

Advanced micro-controller bus architecture is a bus standard which was originally conceived by ARM to support communication among ARM processor cores [2]. The AMBA specification provides standard bus protocols for connecting on-chip components, custom logic and specialised functions.

AMBA defines a segmented bus architecture, where bus segments are connected with each other via a bridge that buffers data and drives the control signals across segments. A 'system bus' is defined, which provides a high-speed, high-bandwidth communication channel between embedded processors and high-performance peripherals. Two system buses are actually specified: the 'AMBA High-Speed Bus' (AHB) and the 'Advanced System Bus' (ASB).

Moreover, a low-performance and low power 'peripheral bus' (called 'Advanced Peripheral Bus', APB) is specified, which accommodates communication with general purpose peripherals and is connected to the system bus via a bridge, acting as the only APB master. The overall AMBA architecture is illustrated in Figure 17.1.

Even though AMBA defines three bus protocols, we will focus only on the most advanced one, namely, AMBA AHB. The main features of AMBA AHB can be summarised as follows:

- *Non-tristate implementation:* AMBA-AHB implements a separate read and write data bus in order to avoid the use of tristate drivers. In particular, master and slave signals are multiplexed onto the shared communication resources (read and write data buses, address bus, control signals).
- *Support for multiple initiators:* A control logic block, called *arbiter* ensures that only one bus master is active on the bus and also that when no masters request the bus a default master is granted. A simple request–grant mechanism is implemented between the arbiter and each bus master.
- *Pipelined and burst transfers:* Address and data phases of a transfer occur during different clock periods. In fact, the address phase of any transfer occurs during

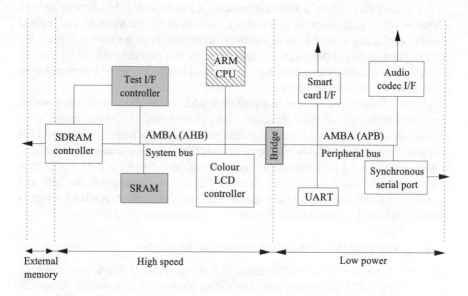

Figure 17.1   AMBA bus architecture

the data phase of the previous transfer. Overlapping of address and data enables full exploitation of bus bandwidth if slaves are fast enough to respond in a single clock cycle.

- *Wide data bus configurations:* Support for high-bandwidth data-intensive applications is provided using wide on-chip memories. System buses support 32, 64 and 128-bit data-bus implementations with a 32-bit address bus, as well as smaller byte and half-word designs.

In a normal bus transaction, the arbiter grants the bus to the master until the transfer completes and the bus can then be handed over to another master. However, in order to avoid excessive arbitration latencies, the arbiter can break up a burst. In that case, the master must re-arbitrate for the bus in order to complete the remaining data transfers.

A basic AHB transfer consists of four clock cycles. During the first one, the request signal is asserted, and in the best case at the end of the second cycle a grant signal from the arbiter can be sampled by the master. Then, address and control signals are asserted for slave sampling on the next rising edge, and during the last cycle the data phase is carried out (read data bus driven or information on the write data bus sampled). A slave may insert wait states into any transfer, thus extending the data phase, and a ready signal is available for this purpose.

Four, eight and sixteen-beat bursts are defined in the AMBA AHB protocol, as well as undefined-length bursts. During a burst transfer, the arbiter re-arbitrates the bus when the penultimate address has been sampled, so that the asserted grant signal can be sampled by the relative master at the same point where the last address of the burst is sampled. This makes bus master handover at the end of a burst transfer very efficient.

For long transactions, the slave can decide to split the operation warning the arbiter that the master should not be granted access to the bus until the slave indicates it is ready to complete the transfer. This transfer 'splitting' mechanism is supported by all advanced on-chip interconnects, since it prevents high latency slaves from keeping the bus busy without performing any actual transfer of data.

As a result, split transfers can significantly improve bus efficiency, i.e. reduce the number of bus busy cycles used just for control (e.g. protocol handshake) and not for actual data transfers. Advanced arbitration features are required in order to support split transfers, as well as more complex master and slave interfaces.

The main limitations of the AHB protocol are (1) no complete support for multiple outstanding transactions and out-of-order completion, which greatly limit bandwidth in the case of slow slaves (if a slave is not ready to respond, no other transaction can bypass the blocked one, and the bus is unused during the wait cycles); (2) no PP interface definition (for instance, initiators have to drive directly the arbitration request signals, hence, they are directly exposed to the interconnect-specific time division multiplexing protocol); (3) limited intrinsic scalability caused by the presence of shared single master-to-slave and slave-to-master channels.

17.2.2 AMBA evolutions

To address the limitations outlined in the previous section, advanced specifications of the AMBA bus have been proposed, featuring increased performance and better link utilisation. In particular, the 'Multi-Layer AHB' and the 'AMBA AXI' interconnect schemes will be briefly reviewed in the following sub-sections. 'Multi-Layer AHB' can be seen as an evolution of bus topology while keeping the AHB protocol unchanged. In contrast, 'AMBA AXI' represents a significant advancement of the protocol. It should be observed that all these interconnect performance improvements can be achieved at the expense of silicon area and complexity.

17.2.2.1 Multi-layer AHB

The multi-layer AHB specification aims at increasing the overall bus bandwidth and providing a more flexible interconnect architecture with respect to AMBA AHB. This is achieved by using a more complex interconnection matrix (also called a crossbar) which enables parallel access paths between multiple masters and slaves in a system [3].

Therefore, the multi-layer bus architecture allows the interconnection of unmodified standard AHB master and slave modules with an increased available bus bandwidth. The resulting architecture becomes very simple and flexible: each AHB layer only has one master and no arbitration and master-to-slave muxing is needed. Moreover, the interconnect protocol implemented in these layers can be very simple: it does not have to support request and grant, nor retry or split transactions.

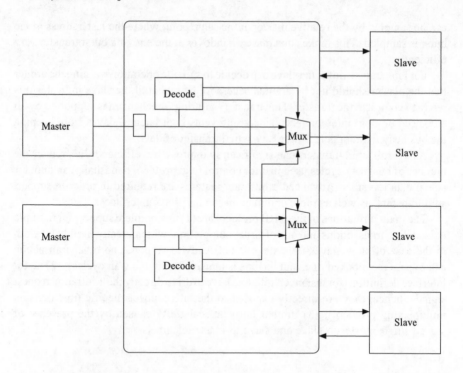

Figure 17.2   Schematic view of the multi-layer AHB interconnect

The additional hardware needed for this architecture with respect to the AHB is a multiplexer to connect the multiple masters to the peripherals and some arbitration is also required when more than one master tries to access the same slave simultaneously.

Figure 17.2 shows a schematic view of the multi-layer concept. The interconnect matrix contains a decode stage for every layer in order to determine which slave is required during the transfer. The multiplexer is used to route the request from the specific layer to the desired slave.

The arbitration protocol decides the sequence of accesses of layers to slaves based on a priority assignment. The layer with lowest priority has to wait for the slave to be freed by higher priority layers. Different arbitration schemes can be used, and every slave port has its own arbitration. Input layers can be served in a round-robin fashion, changing every transfer or every burst transaction, or based on a fixed priority scheme. It is also interesting to outline the capability of this topology to support multi-port slaves. Some devices, such as SDRAM controllers, work much more efficiently when processing transfers from different layers in parallel.

The number of input/output ports on the interconnect matrix is completely flexible and can be adapted to suit to system requirements. However, as the number of masters and slaves in the system increases, the complexity of the crossbar interconnect rapidly becomes unmanageable. In essence, while a shared bus has limited scalability in terms of available bandwidth, crossbars do not scale well in hardware complexity (which impacts silicon area, cycle time and power). To limit crossbar complexity blowup, some optimisation techniques have to be used, such as defining multiple masters on a single layer, multiple slaves appearing as a single slave to the interconnect matrix or defining local slaves to a particular layer.

17.2.2.2 AMBA AXI

AXI is the most recent evolution of the AMBA interface. It significantly enhances protocol performance and it also includes optional extensions for low power operation [4]. This high-performance protocol provides flexibility in the implementation of interconnect architectures while still keeping backward-compatibility with existing AHB and APB interfaces.

AMBA AXI is a fully PP connection. It decouples masters and slaves from the underlying interconnect, by defining only 'master interfaces' and symmetric 'slave interfaces'. This approach, besides allowing backward compatibility and interconnect topology independence, has the advantage of simplifying the handshake logic of attached devices, which only need to manage a PP link.

To provide higher parallelism, four different logical monodirectional channels are provided in AXI interfaces: an address channel, a read channel, a write channel and a write response channel. Activity on different channels is mostly asynchronous (e.g. data for a write can be pushed to the write channel before or after the write address is issued to the address channel), and can be parallelised, allowing multiple outstanding read and write requests, with out-of-order completion.

Figure 17.3(a) shows how a read transaction uses the read address and read data channels. The write operation over the write address and write data channels is presented in Figure 17.3(b). Data is transferred from the master to the slave using a write data channel, and it is transferred from the slave to the master using a read data channel. In write transactions, where all the data flows from the master to the slave, the AXI protocol has an additional write response channel to allow the slave to signal to the master regarding the completion of the write transaction. The rationale of this split-channel implementation is based upon the observation that, usually, the required bandwidth for addresses is much lower than that for data (e.g. a burst requires a single address but maybe four or eight data transfers). Thus, it might be possible to allocate more interconnect bandwidth to the data bus than the address bus.

The mapping of channels, as visible by the interfaces, to interconnect resources is decided by the interconnect designer; single resources might be shared by all channels of a certain type in the system, or a variable amount of dedicated wires may be available, up to a full crossbar.

To conclude, we observe that on-chip buses have come a long way. On one hand, PP protocols act now fully as network interfaces, on the other hand, multi-layer topologies can provide much higher bandwidth than a single shared channel. We used AMBA as a case study for these trends, but the landscape of evolutionary interconnects is very diverse, and many alternatives do exist [5,6].

Still, these evolutionary approaches do not address in full the fundamental scalability limitation of any single-hop interconnect. Networks-on-chip, as described in

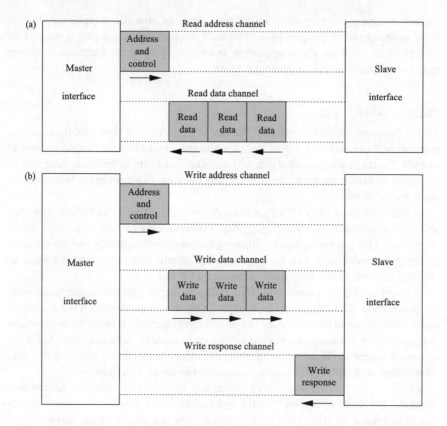

Figure 17.3   Architecture of transfers: (a) Read operation, (b) Write operation

Section 17.4, aim precisely at providing sustainable scalability by making it possible to define multi-hop topologies and providing efficient support to switching, routing and flow-control.

17.3   Quantitative analysis

This section focuses on providing some quantitative evidence of the performance benefits provided by enhanced protocols and high-bandwidth topologies. At first, scalability of evolving interconnect fabric protocols is assessed. Then, we will focus on speed enhancements due to multi-channel topologies.

17.3.1   Protocol efficiency

SystemC models of AMBA AHB and AMBA AXI (provided within the Synopsys CoCentric/Designware® [7] suites) are used within the framework of the MPARM simulation platform [8–10].

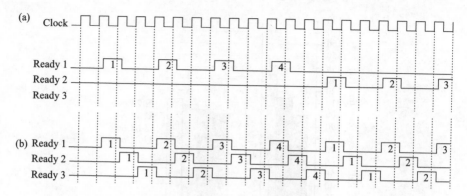

Figure 17.4   Concept waveforms showing burst interleaving for AMBA AHB and AXI interconnects. (a) AMBA AHB, (b) AMBA AXI

The simulated on-chip multiprocessor consists of a configurable number of ARM cores attached to the system interconnect. Traffic workload and pattern can easily be tuned by running different benchmark code on the cores, by scaling the number of system processors, or by changing the amount of processor cache, which leads to different amounts of cache refills.

An AMBA AHB link and a more advanced, but also more expensive, AMBA AXI interconnect with shared bus topology are tested under heavy load. Figure 17.4 shows an example of the efficiency improvements made possible by advanced interconnects in the test case of slave devices having two wait states, with three system processors and four-beat burst transfers. AMBA AHB has to pay two cycles of penalty per transferred datum. AMBA AXI is capable of interleaving transfers, by sharing data channel ownership in time. Under conditions of peak load, when transactions always overlap, AMBA AHB is limited to a 33 per cent efficiency (transferred words over elapsed clock cycles), while AMBA AXI can theoretically reach a 100 per cent throughput.

In order to assess interconnect scalability, a benchmark is independently but concurrently run on every system processor performing accesses to its private memory (involving bus transactions). This means that, while producing real functional traffic patterns, the test setup was not constrained by bottlenecks due to shared slave devices. Private memories are assumed to introduce one wait state before responses.

Scalability properties of the system interconnects can be observed in Figure 17.5, reporting the execution time variation with an increasing number of cores attached to a single shared interconnect under heavy traffic load. Core caches are kept very small (256 bytes) in order to cause many cache misses and therefore significant levels of interconnect congestion. Execution times are normalised against those for a two-processor system, trying to isolate the scalability factor alone. The heavy bus congestion case is considered here because the same analysis performed under light traffic conditions (e.g. with 1 kB caches) shows that both interconnects perform very well, with AHB showing a moderate performance decrease of 6 per cent when moving from two to eight running processors.

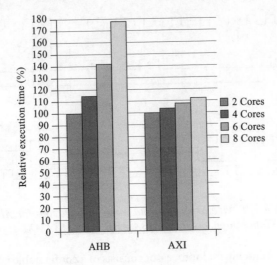

Figure 17.5   Execution times with 256B caches

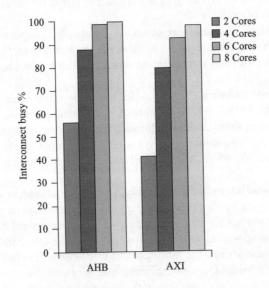

Figure 17.6   Bus busy time with 256B caches

With 256 B caches, the resulting execution times, as Figure 17.5 shows, get 77 per cent worse for AMBA AHB when moving from two to eight cores, while AXI manages to stay within 12 per cent and 15 per cent. The reason behind the behaviour pointed out in Figure 17.5 is that under heavy traffic load and with many processors, interconnect saturation takes place. This is clearly indicated in Figure 17.6, which

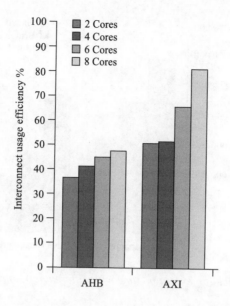

Figure 17.7   Bus usage efficiency with 256B caches

reports the fraction of cycles during which some transaction was pending on the bus with respect to total execution time.

In such a congested environment, as Figure 17.7 shows, AMBA AXI can achieve transfer efficiencies (defined as data actually moved over bus contention time) of up to 81 per cent, while AMBA AHB reaches 47 per cent only – near to its maximum theoretical efficiency of 50 per cent (one wait state per data word). These plots stress the impact that comparatively low-area-overhead optimisations can sometimes have in complex systems.

It must be pointed out, however, that protocol improvements alone cannot overcome the intrinsic performance limitations due to the shared nature of the interconnect resources. While protocol features can push the saturation boundary further, and get near to a 100 per cent efficiency, traffic loads taking advantage of more parallel topologies will always exist. The charts reported here already show some traces of saturation even for the most advanced protocols.

17.3.2   Multi-channel topologies

Topology enhancement can provide additional steam to bandwidth-saturated buses. To illustrate this point with some experimental evidence, a test based on functional simulation of a complete multi-processor architecture with an AMBA AHB compliant interconnect and 8 ARM7 cores will be presented. Two applications were run on the platform, namely independent matrix multiply and independent matrix multiply with semaphore synchronisation upon completion. In the first application, each processor performs matrix multiplication and it is completely independent from the other processors. Matrices are stored in different memories, which are connected as slaves to

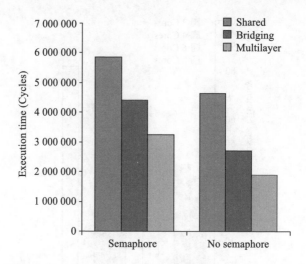

Figure 17.8   Application execution time with different topologies

the system interconnect, hence there is no contention among processors for the same memory slave. However, execution time is heavily impacted by contention for the shared communication resource. In the second application, data processing is exactly the same as for the first application, but processors synchronise after computing every element of the product matrix using a counting semaphore which is contained in a dedicated slave device.

Two applications were run with different interconnects, namely: a shared bus, a multi-layer implementation based on a full crossbar, and a bridged solution that splits the bus in two segments. Results are summarised in Figure 17.8. Data transfers have no destination conflicts, hence, very significant speedups can be achieved by advanced topologies. Note that for the independent matrix multiply benchmark, a speedup of more than a factor of 2 is achieved. The bridged bus gives lower speedup. This is expected because splitting the bus in two segments gives a maximum theoretical throughput enhancement of 2, in the case of no traffic between segments, while theoretical crossbar speedup is N (where N is the number of independent channels).

Speedups are inferior for the synchronised application. This demonstrates that the application-level speedup provided by a given topology strongly depends on the nature of the traffic generated by applications. In this case, there is a traffic bottleneck created by the single counting semaphore in shared memory: all processors will contend for semaphore access, and this application-level contention significantly impacts both execution time and speedups. The bottleneck is even more evident for the bridged solution, because many transactions have to cross the bridge (namely, all semaphore accesses to processors on the opposite side of the bridge).

As a conclusive note, it is important to stress that these application-level speedups are achieved at the price of significantly increased cost in crossbar logic. In fact, crossbar complexity scales quadratically with N, and a full crossbar would not be

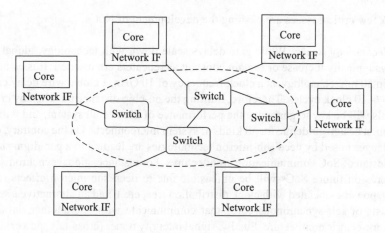

Figure 17.9   NoC architecture

usable in practice for *N* much larger than 10. The bridged solution, which is the most commonly used in today's designs, has even more limited scalability, and it is highly sensitive to traffic patterns: in fact, it can perform even worse than a single shared bus if many transactions have to traverse the bridge. Hence, this solution should be evaluated very carefully for possible traffic mismatches.

17.4   Packet-switched interconnection networks

The previous sections have described evolutionary communication architectures. We now focus on a more revolutionary approach to on-chip communication, known as 'Network-on-Chip (NoC)' [11,12]. NoCs are packet-switched, multi-hop intercon-nection networks integrated onto a single chip. Cores access the network by means of PP interfaces, and have their packets forwarded to destinations through a number of hops (see Figure 17.9). NoCs differ from wide area networks in their local proximity and because they exhibit less non-determinism. Local, high-performance networks – such as those developed for large-scale multiprocessors – have similar requirements and constraints. However, some distinctive features, such as energy constraints and design-time specialisation, are unique to SoC networks.

The scalable and modular nature of NoCs and their support for efficient on-chip communication potentially lead to NoC-based multi-core systems characterised by high structural complexity and functional diversity. On one hand, these features need to be properly addressed by means of new design methodologies, while on the other hand more efforts have to be devoted to modelling on-chip communication archi-tectures and integrating them into a single modelling and simulation environment combining both processing elements and communication architectures. The develop-ment of NoC architectures and their integration into a complete MPSoC design flow is the main focus of an ongoing worldwide research effort [13–15].

A few critical issues are pushing the development of NoCs.

Technology issues: While gate delays scale down with technology, global wire delays typically increase or remain constant as repeaters are inserted. It is estimated that in 50 nm technology, at a clock frequency of 10 GHz, a global wire delay can be up to 6–10 clock cycles. Therefore, limiting the on-chip distance travelled by critical signals will be key to guarantee the performance of the overall system, and will be a common design guideline for all kinds of system interconnects. On the contrary, other challenges posed by deep sub-micron technologies are leading to a paradigm shift in the design of SoC communication architectures. For instance, global synchronisation of cores on future SoCs will be unfeasible due to deep sub-micron effects (clock skew, power associated with clock distribution tree, etc.), and an alternative scenario consists of self-synchronous cores that communicate with one another through a network-centric architecture. Finally, signal integrity issues (cross-talk, power supply noise, soft errors, etc.) will lead to more transient and permanent failures of signals, logic values, devices and interconnects, thus raising the reliability concern for on-chip communication. In many cases, on-chip networks can be designed as regular structures, allowing electrical parameters of wires to be optimised and well controlled. This leads to lower communication failure probabilities, thus enabling the use of low-swing signalling techniques, and to the capability of exploiting performance optimisation techniques such as wavefront pipelining.

Design productivity issues: It is well known that synthesis and compiler technology development do not keep up with IC manufacturing technology development. Moreover, times-to-market need to be kept as low as possible. Reuse of complex pre-verified design blocks is an efficient means to increase productivity, and regards both computation resources and the communication infrastructure. It would be highly desirable to have processing elements that could be employed in different platforms by means of a plug-and-play design style. To this purpose, a scalable and modular on-chip network represents a more efficient communication infrastructure compared with shared bus-based architectures. However, the reuse of processing elements is facilitated by the definition of standard network interfaces, which also make the modularity property of the NoC effective. The Virtual Socket Interface Alliance (VSIA) has attempted to set the characteristics of this interface industry-wide. OCP is another example of standard interface sockets for cores. It is worth remarking that such network interfaces also decouple the development of new cores from the evolution of new communication architectures. The core developer will not have to make assumptions about the system, when the core will be plugged into. Similarly, designers of new on-chip interconnects will not be constrained by the knowledge of detailed interfacing requirements for particular legacy SoC components. Finally, let us observe that NoC components (e.g. switches or interfaces) can be instantiated multiple times in the same design (as opposed to the arbiter of traditional shared buses, which is instance-specific) and reused in a large number of products targeting a specific application domain. The developments of NoC architectures and protocols is fuelled by the aforementioned arguments, in spite of the challenges represented by the need for new design methodologies and an increased complexity in system design.

17.4.1   NoC basic principles

Next, an overview of NoC basic principles and degrees of freedom for designers will be provided.

17.4.1.1   Topology

An interconnection network can be viewed as a collection of router nodes connected by shared channels. The connection pattern of these nodes defines the 'network topology'. This latter should be designed so as to meet the bandwidth and latency requirements of applications at minimum cost. The bandwidth can be maximised by saturating bisection bandwidth, the bandwidth across the midpoint of the system.

Latency minimisation can be another network design requirement. To achieve low latency, a topology must balance the desire for a small average distance between nodes against a low serialisation latency. The hop count between a source and a destination node is critical with respect to latency: it represents the number of channels and nodes a packet must traverse on average to reach its destination.

Mesh topology is frequently used for networks-on-chip. It exhibits a regular structure and is most suitable for tile-based architectures. Regular meshes can be designed with better control on electrical parameters and therefore on communication noise sources (e.g. crosstalk), although they might result in link under-utilisation or localised congestion. In fact, not all computation units have the same communication requirements, thus leading to mapping inefficiencies on regular topologies. When considering tile-based architectures [16], another issue is given by the different physical size of computation units that leads to an inherently irregular floorplanning. Nostrum NoC architecture [17] makes use of a mesh topology (see Figure 17.10 (top)), wherein each switch is connected to four switches and to one core. Motivation for this choice was regularity of layout, predictable electrical properties and the expected locality of traffic. Switch-to-switch connections consist of 256 shielded and differential data signals.

The SPIN micronetwork [18] is another example of network architecture, making use of a fat tree topology. Every node has four sons and the father is replicated four times at any level of the fat tree, as reported in Figure 17.10 (bottom). This topology is intrinsically redundant, since the four fathers offer four equivalent paths in order to route a message between two sons of the same father. The shortest path between two subscribers is the one that goes through by the nearest common ancestor. A fat tree topology has the following advantages: its diameter (maximum number of links between two subscribers) remains reasonable ($2 \times \log_4 n$, where n is the number of layer of network), the topology is scalable and uses a small number of routers for a given number of subscribers. It has a natural hierarchical structure which can be useful in many embedded systems.

Irregular topologies have to deal with more complex physical design issues but are more suitable to implement customised, domain-specific communication architectures. A higher performance and a lower energy dissipation for a specific application domain are likely to be paid with a higher design complexity (e.g. selection of an optimised custom topology, deadlock-free efficient routing algorithms, etc.). NoC

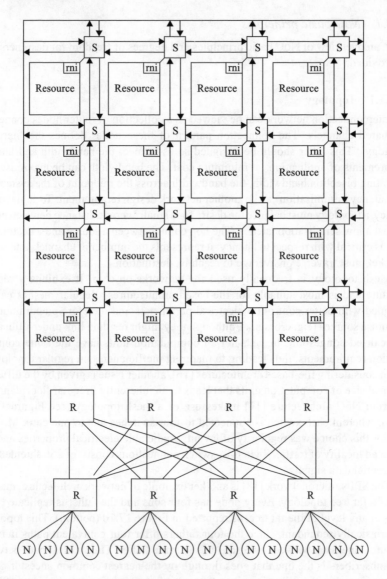

Figure 17.10   Mesh architecture (top) and Fat Tree topology (bottom)

design methodologies supporting irregular topologies require soft macros as basic
network building blocks, which can be customised at instantiation time for a specific
application. For instance, the Proteo NoC [19] consists of a small library of predefined, parameterised components that allow the implementation of a large range of
different topologies, protocols and configurations. xpipes interconnect [20] and its
synthesiser xpipesCompiler [21] push this approach to the limit, by instantiating
an application-specific NoC from a library of soft macros (network interface, link and

switch). The components are highly parameterisable and provide reliable and latency insensitive operation.

17.4.1.2 Flow control

Flow control determines how network resources (such as channel bandwidth, buffer capacity and control state) are allocated to packets traversing the network. Inefficient implementation of flow control has side-effects in terms of wastes of bandwidth and unproductive resource occupancy, thus leading to the utilisation of a small fraction of the ideal bandwidth and to a high and variable latency. Flow control can be viewed both as a resource allocation mechanism (to allow the packet to reach its destination) or as a contention resolution mechanism.

Circuit switching is the basic solution for flow control, wherein only packet headers are buffered and traverse the network ahead of any packet payload, reserving the appropriate resources along the path. The established 'circuit' is then torn down by deallocating resources. This approach requires a lower storage space at the cost of a lower link utilisation. It is, however, widely adopted for contention-free packet propagation across the network, resulting in quality-of-service guarantees (latency and throughput).

On the contrary, when providing support for best effort traffic, buffering plays a critical role in determining performance. In fact, data can be stored while waiting to acquire network resources. This buffering can be done either in units of packets, as with store-and-forward and cut-through flow control, or at the finer granularity of *flits*, as in the case of wormhole flow control. By breaking a packet into smaller, fixed-sized flits, the amount of storage needed at any particular node can be greatly reduced.

In 'store-and-forward' flow control, a packet is forwarded from a switch to the next one in the network only when this latter has enough storage space for the entire packet. The packet has to be entirely stored at a switch before being transmitted forward. Of course, this approach poses the highest buffering requirements to the network switches and incurs a very high communication latency. This approach is rarely used for MPSoC communication architectures, but there are exceptions. In fact, Æthereal routers support both best effort and guaranteed throughput traffic. For this latter case, output channel utilisation is split into time slots, each of which is selectively reserved to input ports based on an initial programming phase. During each slot time, store-and-forward of the input packet is the actual flow control mechanism.

'Virtual cut-through' flow control overcomes the latency penalty of store-and-forward flow control by forwarding a packet as soon as the header is received and resources (buffer and channel) are acquired, without waiting for the entire packet to be received. As with store-and-forward, virtual cut-through allocates resources (especially buffers) at the coarse granularity of packets. This per-packet buffer allocation accounts for the inefficient utilisation of storage space as well as for an increase of contention latency. For instance, a high-priority packet colliding with a low-priority packet must wait for the entire low-priority packet to be transmitted before it can acquire the channel.

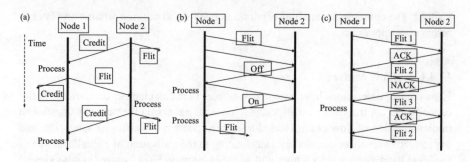

Figure 17.11   Flow control techniques: (a) credit-based, (b) on-off and (c) ACK-NACK

A workaround for the two above shortcomings consists of a per-flit (rather than per-packet) resource allocation, and is the strategy followed by 'wormhole' flow control. When the head flit of a packet arrives at a node, it must acquire an output channel, one flit buffer and one flit of channel bandwidth. Body flits take the same channel acquired by the head flit and hence need only acquire a flit buffer and a flit of channel bandwidth to advance. The tail flit of a packet is handled like a body flit, but also releases the output channel as it passes.

Compared with the previous approaches, wormhole flow control makes far more efficient use of buffer space, as only a small number of flit buffers are required per channel. This comes at the cost of some throughput, since wormhole flow control may block a channel mid-packet. In fact, the channel may be owned by a packet, but buffers are allocated on a per-flit basis. When a flit cannot acquire a buffer, the channel goes idle even though there is another packet that could potentially use the channel bandwidth. Although channel bandwidth is allocated flit-by-flit, it can only be used by flits of one packet. Wormhole switching is the most widely adopted mechanism for flow control.

All control flow techniques using buffering need a means to communicate the availability of buffers to the downstream nodes. This buffer management informs the upstream nodes when they must stop transmitting flits because all of the downstream flit buffers are full. There are three such backpressure mechanisms (see Figure 17.11).

In 'credit-based flow control' the upstream router keeps a count of the number of free flit buffers in each channel downstream. Then, each time the upstream router forwards a flit, thus consuming a downstream buffer, it decrements the appropriate count. If the count reaches zero, all of the downstream buffers are full and no further flits can be forwarded until a buffer becomes available. Once the downstream router forwards a flit and frees the associated buffer, it sends a credit to the upstream router, causing a buffer count to be incremented. For each flit sent downstream, a corresponding credit is eventually sent upstream. This requires a significant amount of upstream signalling and, especially for small flits, can represent a large overhead.

'On-off flow control' greatly reduces the amount of upstream signalling, since the upstream state is a single control bit that represents whether the upstream node is permitted (on) to send or not (off). A signal is sent upstream only when it is necessary to change this state (e.g. an off can be sent to indicate that the number of free buffers falls below a certain threshold). Like credit-based flow control, the on-off mechanism requires a round-trip delay between the time a buffer becomes empty, triggering a credit or an on signal, and when a flit arrives to occupy that buffer. However, with an adequate number of buffers, on/off flow control can operate with very little upstream signaling.

Finally, 'ACK/NACK flow control' reduces the minimum and average buffer vacancy time. In fact, the upstream node optimistically sends flits whenever they become available. If the downstream node has a buffer available, it accepts the flit and acknowledges to the upstream node. If no buffers are available when the flit arrives, it is dropped and a NACK is notified to the upstream node. This latter holds each flit until it receives an ACK, therefore it is able to retransmit a NACKed flit. Although effective in systems with large buffering resources, this mechanism holds such resources for a longer time with respect to the previous mechanisms, and incurs useless link utilisation in case of blocking of downstream resources for a long time.

17.4.1.3 Routing

The objective of routing is to find a path from a source node to a destination node on a given topology. Routing is one of the key components that determine the performance of the network, and aims at reducing the number of hops and overall latency and at balancing the load of network channels. In general, performance of routing algorithms strongly depends on topology.

Let us classify routing algorithms as in Reference 22:

- Deterministic algorithms always choose the same path between two nodes. They are easy to implement and deadlock issues can be easily solved. On the other hand, they do not use path diversity and therefore show poor performance with respect to load balancing.
- Oblivious algorithms always choose a route without considering any information about the network's present state. They include deterministic algorithms as a subset. Random splitting of traffic among different paths is always an oblivious algorithm.
- Adaptive algorithms adapt to the state of the network, using this state information to make routing decisions. In theory, adaptive routing should be better than an oblivious algorithm, however, in practice they are not, since only local information can be easily accessed. As a result, network load can turn out to be locally balanced but globally imbalanced. As a workaround, adaptive routing can be coupled with flow control mechanisms to have access to a broader range of network state information and to take a course of action with larger impact on the network (e.g. backpressure). Please note that fully adaptive routing may result in live-lock, which should be carefully addressed.

The mechanism used to implement any routing algorithm is referred to as 'routing mechanics'. Many routers use 'routing tables' either at the source or at each hop along the route to implement the routing algorithm. With a single entry per destination, a table is restricted to deterministic routing, but oblivious and adaptive routing can be implemented by providing multiple table entries for each destination. An alternative to tables is 'algorithmic routing', in which specialised hardware computes the route or next hop of a packet at runtime. However, algorithmic routing is usually restricted to simple routing algorithms and regular topologies.

With 'source routing', all routing decisions are entirely taken in the source node by table lookup of a precomputed route. There is at least one table entry per destination, and the predefined route is embedded into the packet. This allows rapid propagation of the packet through the network across the selected path since no further computation has to be carried out at the switches. This routing mechanism is frequently used for its simplicity and scalability, although it cannot be used to implement adaptive routing since it cannot take advantage of network state information. Routing tables can also be stored at the intermediate network nodes (node table routing) in order to make up for this inconvenience. However, this approach significantly increases the latency for a packet to pass through a router.

Finally, instead of storing the routing path in a table, it can be computed using an algorithm, and this is usually done by means of a computational logic circuit. The algorithmic approach comes at the cost of a loss of generality with respect to the table-based approach. In fact, the algorithm is only specific to one topology and to one routing strategy on that topology. On the other side, this mechanism turns out to be more efficient (area- and performance-wise) than table-based routing.

17.4.2   NoC architecture

Messages that have to be transmitted across the network are partitioned into fixed-length packets. Packets in turn are often broken into message flow control units called flits. In the presence of channel width constraints, multiple physical channel cycles can be used to transfer a single flit. A phit is the unit of information that can be transferred across a physical channel in a single step. Flits represent logical units of information, as opposed to phits that correspond to physical quantities. In many implementations, a flit is set to be equal to a phit. The basic building blocks for packet-switched communication across NoCs are network link, switch and network interface.

17.4.2.1   The link

The performance of interconnects is a major concern in scaled technologies. As geometries shrink, gate delay improves much faster than the delay in long wires. It has been estimated that only a fraction of the chip area (between 0.4 and 1.4 per cent) will be reachable in one clock cycle [23]. Therefore, the long wires increasingly determine the maximum clock rate, and hence performance, of the entire design. The problem becomes particularly serious for domain-specific heterogeneous SoCs, where the wire structure is highly irregular and may include both short and extremely long switch-to-switch links.

A solution to overcome the interconnect-delay problem consists of pipelining interconnects [24]. Wires can be partitioned into segments bounded by relay stations, which have a function similar to the one of latches on a pipelined data path.

Segment length satisfies pre-defined timing requirements (e.g. desired clock speed of the design). In this way, link delay is changed into latency, but data introduction rate becomes decoupled from the link delay. This requires the system to be made of modules whose behaviour does not depend on the latency of the communication channels (latency insensitive operation). As a consequence, the use of interconnect pipelining can be seen as a part of a new and more general methodology for deep sub-micron (DSM) designs, which can be envisioned as synchronous distributed systems composed by functional modules that exchange data on communication channels according to a latency-insensitive protocol. This protocol ensures that functionally correct modules behave correctly independently of the channel latencies [24]. The effectiveness of the latency-insensitive design methodology is strongly related to the ability of maintaining a sufficient communication throughput in the presence of increased channel latencies.

The International Technology Roadmap for Semiconductors (ITRS) 2001 [25] assumes that interconnect pipelining is the strategy of choice in its estimates of achievable clock speeds. Some industrial designs already make use of interconnect pipelining. For instance, the NETBURST micro-architecture of Pentium 4 contains instances of a stage dedicated exclusively to handle wire delays: in fact, a so-called drive stage is used only to move signals across the chip without performing any computation and, therefore, can be seen as a physical implementation of a relay station [26].

As an example, the xpipes NoC supports pipelined links and latency-insensitive operation in the implementation of its building blocks, and related design issues are now briefly reported. Switch-to-switch links are subdivided into basic segments whose length guarantees that the desired clock frequency (i.e. the maximum speed provided by a certain technology) can be used. According to the link length, a certain number of clock cycles is needed by a flit to cross the interconnect. These design choices are at the basis of latency insensitive operation of the NoC and allow the construction of an arbitrary network topology and hence support for heterogeneous architectures, without creating clock cycle bottlenecks on long links.

The link model is equivalent to a pipelined shift register. Pipelining has been used both for data and control lines, hence also for ACK lines used by ACK flits to propagate from the destination switch back to the source one. This architecture impacts the way link-level error control is performed in order to ensure robustness against communication errors. In fact, multiple outstanding flits propagate across the link during the same clock cycle. When flits are correctly received at the destination switch, an ACK is propagated back to the source, and after N clock cycles (where N is the length of the link expressed in number of repeater stages) the flit will be discarded from the buffer of the source switch. On the contrary, a corrupted flit is NACKed and will be retransmitted in due time. The implemented retransmission policy is GO-BACK-N, to keep the switch complexity as low as possible.

Communication-related power minimisation is another critical NoC design objective. 'Dynamic voltage scaling' has been recently applied to buses [27,28]. In Reference 28 the voltage swing on communication buses is reduced, even though signal integrity is partially compromised. Encoding techniques can be used to detect corrupted data which is then retransmitted. The retransmission rate is an input to a closed-loop DVS control scheme, which sets the voltage swing at a trade-off point between energy saving and latency penalty (due to data retransmission).

The 'On-Chip Network' (OCN) for low power heterogeneous SoC platforms illustrated in Reference 14 employs some advanced techniques for low power physical interconnect design. OCN consists of global links connecting clusters of tightly connected IPs which are several millimeters long. By using overdrivers, clocked sense amplifiers and twisted differential signalling, packets are transmitted reliably with less than 600 mV swing. The size of a transceiver and the overdrive voltage are chosen to obtain a 200 mV separation at the receiver end. A 5 mm global link of 1.6 μm wire-pitch can carry a packet at 1.6 GHz with 320 ps wire-delay and consumes 0.35 pJ/bit. On the contrary, a full-swing link consumes up to 3$\times$ more power and additional area of repeaters.

An on-chip serialisation technique [29] is also used in OCN, thus significantly reducing area. However, the number of signal transitions on a link is increased since the temporal locality between adjacent packets is removed. An ad hoc serialised low-energy transmission coding scheme was therefore designed as an attempt to exploit temporal locality between packets. The encoder generates a '1' only when there is difference between a current packet and a previous packet before it is serialised. The decoder then uses this encoded packet to reconstruct the original input, using its previously stored packet. A 13.4 per cent power saving is obtained for a multi-media application. The power overhead associated with the encoder/decoder is only 0.4 mW.

17.4.2.2 Switch architecture

The task of the switch is to carry packets injected into the network to their final destination, following a statically defined or dynamically determined routing path. The switch transfers packets from one of its input ports to one or more of its output ports. Switch design is usually characterised by a power–performance trade-off: power-hungry switch memory resources can be required by the need to support high-performance on-chip communication. A specific design of a switch may include both input and output buffers or only one type of buffer. Input queuing uses fewer buffers, but suffers from head-of-line blocking. Virtual output queuing has a higher-performance, but at the cost of more buffers. Network flow control specifically addresses the limited amount of buffering resources in switches.

Guaranteeing quality of service in switch operation is another important design issue, which needs to be addressed when time-constrained (hard or soft real time) traffic is to be supported. Throughput guarantees or latency bounds are examples of time-related requirements. Contention-related delays are responsible for large fluctuations of performance metrics, and a fully predictable system can be obtained only by means of contention-free routing schemes. With circuit switching, a connection is set up over which all subsequent data is transported. Therefore, contention resolution takes place during connection setup, and time-related guarantees during data transport can be given. In time division circuit switching, bandwidth is shared by time division multiplexing connections over circuits. In packet switching, contention is unavoidable since packet arrival cannot be predicted. Therefore arbitration mechanisms and buffering resources must be implemented at each switch, thus delaying data in an unpredictable manner and making it difficult to provide guarantees. Best effort NoC architectures can mainly rely on network oversizing to bound fluctuations of performance metrics.

17.4.2.3 Network interface

The network interface (NI) is entrusted with several critical tasks: (1) providing a standardised set of PP transactions to cores, (2) efficient mapping of PP transactions into a (possibly large) set of network transactions, (3) interfacing with the packet-based network fabric (packet assembly, delivery and disassembly).

The first objective requires the definition of a standardised PP interface. AMBA AXI is an example of such an interface, but its definition and evolution is controlled by a single company. To avoid the captivity risks associated with proprietary standards, several core interface standardisation initiatives have been promoted. For instance, the VSIA vision [30] is to specify open standards and specifications that facilitate the integration of software and hardware virtual components from multiple sources. Different complexity interfaces are described in the standard, from Peripheral Virtual Component Interfaces (VCI) to Basic VCI and Advanced VCI. Another example of standard socket to interface cores to networks is represented by Open Core Protocol (OCP) [31]. Its main characteristics are a high degree of configurability to adapt to the core's functionality and the independence of request and response phases, thus supporting multiple outstanding requests and pipelining of transfers (VCI and OCP have recently announced a merger).

Data packetisation is a critical task for the NI, and has an impact on the communication latency, besides the latency of the communication channel. The packet preparation process consists of building packet header, payload and packet tail. The header contains the necessary routing and network control information (e.g. source and destination address). When source routing is used, the destination address is ignored and replaced with a route field that specifies the route to the destination. This overhead in terms of packet header is counterbalanced by the simpler routing logic at the network switches: they simply have to look at the route field and route the packet over the specified switch output port. The packet tail indicates the end of a packet and may contain redundant checksum bits for error-detecting or error-correcting codes.

A very critical issue in the design of advanced network interfaces is the degree of support for multiple outstanding transactions. Given the multi-hop nature of NoCs, packet delivery latency (even without considering congestion-related latency) can be high. For instance, even if xpipes is highly tuned for low-latency operation, each switch inserts two clock cycles' latency, and pipelined links can significantly increase latency. If every PP transaction blocks the core interface until completion, network latency may seriously impact the bandwidth available to the cores, especially if cores issue posted transactions (i.e. transactions which do not require a response phase) or can initiate multiple outstanding transactions. It is important to note, however, that supporting multiple outstanding transactions significantly increases control complexity and buffering in the NI, and the hardware cost is justified only when interfacing with advanced cores.

Before closing this section on NoC architecture and components, we need to mention the important issue of synchronisation. Even though it may be possible to take the simplifying assumption that an entire SoC is synchronised by a single clock, in reality there is little doubt that all large-scale SoCs and their communication fabric will need to support much more flexible synchronisation schemes. In fact, the cost (in terms of area, power, design effort) of distributing a single clock at the frequency needed to provide adequate performance is already unmanageable in current technology. Even though some authors are investigating fully asynchronous communication schemes, the most likely solution for NoC synchronisation is a 'globally asynchronous, locally synchronous' (GALS) paradigm [32,33]. Flexible GALS synchronisation provides an additional degree of freedom for NoC optimisation. For instance, if the NoC is clocked faster than the core, very wide data transfers from cores can be serialised in time over narrow network links [29]. This approach can greatly help in reducing wiring congestion, especially for crossbars and for NoCs with reduced number of hops and switches with a large number of input and output ports.

17.5   Overview of NoC prototypes

17.5.1   x*pipes*

x*pipes* is a SystemC library of highly parameterisable, synthesiseable network building blocks, optimised for low-latency and high-frequency operation. Communication is packet switched, with source routing (based upon street-sign encoding) and wormhole flow control. x*pipes* utilises OCP as a means to interface with the SoC cores.

The x*pipes* switch models the basic building block of the NoC switching fabric. It implements a two-cycle-latency, output-queued router that supports fixed and round robin priority arbitration on the input lines, and a flow control protocol with ACK/NACK, Go-Back-N semantics.

Allocation of inputs towards specific output lines is handled at the 'Allocator' module. Multiple allocators exist in a switch, each driving one of its output ports (O_i).

Assuming that one input is currently owning access to O_i, it is maintained in its state until a tail flit arrives. Arbitration is subsequently performed upon receipt of a header flit with routing information dictating that the incoming packet should exit through O_i.

An input flit can be rejected, and therefore NACKed, due to one or more of the following reasons:

- The output line is occupied by a previous transmitting packet.
- The buffering space for output O_i is already filled.
- Another header flit requesting the same output is concurrently appearing on another input port, and arbitration is won by the latter.

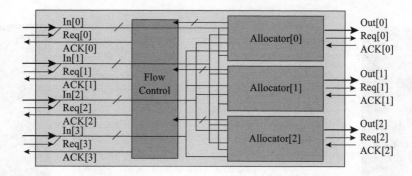

Figure 17.12 ×pipes 4 × 3 *switch*

If a flit A is dropped, then all subsequent incoming flits are dropped as well, until flit A reappears at the input, adhering to a Go-Back-N flow control mechanism.

After a packet has won the arbitration, its header flit is properly adjusted in order to prepare for the next switch along the routed path to the slave. More specifically, routing information pertaining to the current switch is rotated away; this allows positioning of the per-hop routing bits at a fixed offset within the flits.

The switch is parameterisable in the number of its inputs and outputs, as well as in the size of the buffering at the outputs. A 4 × 3 instantiation and the 4 × 3 allocator are depicted in Figure 17.12.

The ×pipes NI is designed as a bridge between an OCP interface and the NoC switching fabric. Its purposes are the synchronisation between OCP and ×pipes timings, the packeting of OCP transactions into ×pipes flits and vice versa, the computation of routing information and the buffering of flits to improve performance.

The ×pipes NI is designed to comply with version 2.0 of the OCP specifications. In addition to the core OCP signals, support includes, e.g. the ability to perform both non-posted or posted writes (i.e. writes with or without response) and various types of burst transactions, including reads with single request and multiple responses. This allows for thorough exploration of bandwidth/latency trade-offs in the design of a system.

To provide complete deployment flexibility, the NI is parameterisable in both the width of OCP fields and of ×pipes flits. Depending on the ratio between these parameters, a variable amount of flits is needed to encode an OCP transaction.

For any given transaction, some fields (such as the OCP MAddr wires, specific control signals, routing information) can be transmitted just once; in contrast, other fields (such as the OCP MData or SData wires) need to be transmitted repeatedly, for example during a burst transaction. Thus, the NI is built around two registers; one holds the transaction header, while the second one holds the transaction payload. The first register samples OCP signals once per transaction, while the second is refreshed on every burst beat.

A set of flits encodes the header register; subsequently, multiple sets of flits are pushed out, each set encoding one update of the payload register. Sets of payload

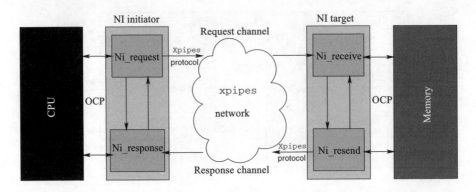

Figure 17.13   x p i p e s *network interfaces*

flits are pushed out until transaction completion. Header and payload content are never allowed to mix in the same flit, thus simplifying the required logic. Routing information is attached to the header flit of a packet by checking the transaction address against a Look-Up Table (LUT).

As shown in Figure 17.13, two NIs are implemented in xpipes, named 'initiator' (attached to system masters) and 'target' (attached to system slaves). A master–slave device will need two NIs, an initiator and a target, for operation. Each NI is additionally split in two submodules, one for the request and one for the response channel. These submodules are loosely coupled: whenever a transaction requiring a response is processed by the request channel, the response channel is notified; whenever the response is received, the request channel is unblocked. The mechanism is currently supporting only one outstanding non-posted transaction.

The xpipes interface of the NI is bidirectional; for example, the initiator NI has an output port for the request channel and one input port for the response channel (the target NI is dual). The output stage of the NI is identical to that of the xpipes switches, for increased performance. The input stage is implemented as a simple dual-flit buffer with minimal area occupation, but still makes use of the same flow control used by the switches.

17.5.2 Æthereal

The 'Philips Æthereal' NoC addresses the communication needs of consumer electronics SoCs with real-time requirements, such as those used in digital video set-top boxes. Conceptually, the Æthereal router module consists of two independent routers. The best effort router offers uncorrupted lossless (flow controlled) ordered data transport. The guaranteed throughput router adds hard throughput and latency guarantees over a finite time interval. Combining GT and BE routers ensures efficient resource utilisation.

The GT router sub-system is based on a time-division multiplexed circuit switching approach. A router uses a slot table to (1) avoid contention on a link, (2) divide up bandwidth per link between connections and (3) switch data to the correct output.

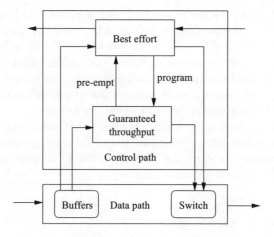

Figure 17.14   Hardware view of the combined BE-GT router

Every slot table T has S time slots (rows), and N router outputs (columns). There is a logical notion of synchronicity: all routers in the network are in the same fixed-duration slot. In a slot *s* at most one block of data can be read/written per input/output port. In the next slot, the read blocks are written to their appropriate output ports. Blocks thus propagate in a store-and-forward fashion. The latency a block incurs per router is equal to the duration of a slot and bandwidth is guaranteed in multiples of block size per *S* slots.

The BE router uses packet switching, and it has been shown that both input queuing with wormhole flow-control or virtual cut-through routing and virtual output queuing with wormhole flow-control are feasible in terms of buffering cost. The BE and GT router sub-systems are combined in the Æthereal router architecture. The GT router offers a fixed end-to-end latency for its traffic, which is given the highest priority by the arbiter. The BE router uses all the bandwidth (slots) that has not been reserved or used by GT traffic. GT router slot tables are programmed by means of BE packets. Negotiations, resulting in slot allocation, can be done at compile time, and be configured deterministically at run time. The hardware view of the final router is reported in Figure 17.14. The control paths of the BE and GT routers are separate, yet interrelated. Moreover, the arbitration unit (including link level flow control for the BE router) has been merged with the BE router itself. The data path, mainly consisting of the switch matrix, is shared. In computer network router architectures, the buffers of BE and GT traffic would be stored in a shared RAM. For the small amount of buffering in on-chip routers (3 words/GT queue and 24 words/BE queue) using either RAMs or register file memories would be very area inefficient. By using dedicated GT and BE hardware first-in first-out buffers, the area of the router is reduced by two-thirds.

A comparison between the hardware implementation of the xpipes switch and the Æthereal switch is quite instructive. Both switches have been targeted to similar 130 nm technologies. A xpipes switch with 4 ports and 64b flits uses 0.19 mm^2 of silicon area and can be clocked at 800 MHz. An Æthereal switch with 5 ports (32b phits) uses 0.26 mm^2 of silicon area and is clocked at 500 MHz. The internal buffering in the xpipes switch is 6 flits per output port while Æthereal is 24 phits per input port. We note that the Æthereal switch achieves better buffer density, mainly because it uses custom-designed FIFO macros, while the xpipes switch is fully synthesised. However, xpipes is faster. This is probably due to the QoS support in Æthereal, which impacts control complexity and ultimately cycle time.

The network interface is the bridge between a core and a router. It implements end-to-end flow control, admission control and traffic shaping, connection setup and teardown, and transaction reordering. Like the router, it contains a slot table, but has dedicated hardware FIFOs per connection.

17.6   NoC design technology

NoC architectures are pushing the evolution of traditional IC design methodologies in order to more effectively deal with functional diversity and complexity. At the application level, the key design challenge is to expose task-level parallelism and to formally capture concurrent communication in models of computation. Then, high-level concurrent tasks have to be mapped to the underlying communication and computation resources. At this level, an abstract model of the hardware architecture is usually exposed to the mapping tool, so that area and power estimates can be given in the early design stage, and different objective functions (e.g. minimisation of communication energy) can be considered to evaluate the feasibility of alternative mappings.

For NoC-based MPSoCs, a critical step in communication mapping is the network topology selection for its significant impact on overall system performance, which is increasingly communication-dominated. In this area, we can distinguish two different approaches: namely, mapping onto pre-defined, regular topologies with homogeneous nodes, and mapping onto ad hoc, application-specific topologies with heterogeneous nodes. The first approach can leverage a large body of research from traditional parallel computing, where the key problem is how to effectively map complex parallel applications on given regular topologies (which are typically used in highly parallel large-scale multiprocessors), and it is conceptually more tractable, because it decouples topology definition and instantiation from mapping. A few early approaches to this problem in a NoC setting have recently been proposed [34].

It is important to note, however, that the individual components of SoCs are inherently heterogeneous with widely varying functionality and communication requirements. The communication infrastructure should optimally match communication patterns among these components accounting for the individual component needs. As an example, consider the implementation of an MPEG4 decoder [35], depicted in Figure 17.15(b), where blocks are drawn roughly to scale and links represent inter-block communication. First, the embedded memory (SDRAM) is much larger than all other cores and it is a critical communication bottleneck. Block sizes are highly non-uniform and the floorplan does not match the regular, tile-based floorplan

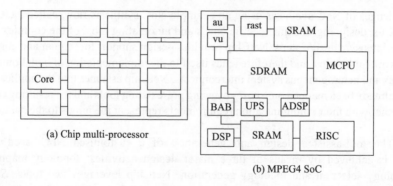

(a) Chip multi-processor   (b) MPEG4 SoC

Figure 17.15 (a) Tile based architecture; (b) MPEG 4 decoder implementation

as shown in Figure 17.15(a). Second, the total communication bandwidth to/from the embedded SDRAM is much larger than that required for communication among the other cores. Third, many neighbouring blocks do not need to communicate. Even though it may be possible to implement MPEG4 onto a homogeneous fabric, there is a significant risk of either under-utilising many tiles and links, or, at the opposite extreme, of achieving poor performance because of localised congestion. These factors motivate the use of an application-specific on-chip network.

With an application-specific network, the designer is faced with the additional task of designing network components (e.g. switches) with different configurations (e.g. different I/Os, virtual channels, buffers) and interconnecting them with links of uneven length. These steps require significant design time and the need to verify network components and their communications for every design. The library-based nature of network building blocks seems to be the more appropriate solution to support domain-specific custom NoCs. The xpipes NoC takes this approach. As described in the previous section, its network building blocks have been designed as highly configurable and design-time composable soft macros described in SystemC at the cycle-accurate level. An optimal system solution will also require an efficient mapping of high-level abstractions on to the underlying platform. This mapping procedure involves optimisations and trade-offs between many complex constraints, including quality of service, real-time response, power consumption, area, etc. Tools are urgently needed to explore this mapping process, and assist and automate optimisation where possible. The first challenge for these tools is to bridge the gap in building custom NoCs that optimally match the communication requirements of the system. The network components they build should be highly optimised for that particular NoC design, providing large savings in area, power and latency with respect to standard NoCs based on regular structures.

17.6.1 NoC synthesis case study: xpipes

The design methodology has to partition the design problem into manageable tasks and to define the tools and practices for those tasks. In this section, we illustrate the challenges of NoC synthesis using an example NoC synthesis flow, called NetChip [36], for designing domain-specific NoCs and automating most of the complex and time-intensive design steps. NetChip provides design support for regular and custom network topologies, and therefore lends itself to the implementation of both homogeneous and heterogeneous system interconnects. NetChip assumes that the application has already been mapped onto cores by using pre-existing tools and the resulting cores together with their communication requirements represent the inputs to the synthesis flow.

The tool-assisted design and generation of a customised NoC-based system is achieved by means of three major design activities: topology mapping, topology selection and topology generation. NetChip leverages two tools: *SUNMAP*, which performs the network topology mapping and selection functions, and `xpipesCompiler`, which performs the topology generation function. SUNMAP produces a mapping of cores onto various NoC topologies that are defined in a topology library. The mappings are optimised for the chosen design objective (such as minimising area, power or latency) and satisfy the design constraints (such as area or bandwidth constraints). SUNMAP uses floorplanning information early in the mapping process to determine the area-power estimates of a mapping and to produce feasible mappings (satisfying the design constraints). The tool supports various routing functions (dimension ordered, minimum-path, traffic splitting across minimum-paths, traffic splitting across all paths) and chooses the mapping onto the best topology from the library of available ones.

A design file describing the chosen topology is input to the `xpipesCompiler`, which automatically generates the SystemC description of the network components (switches, links and network interfaces) and their interconnection with the cores. A custom hand-mapped topology specification can also be accepted by the NoC synthesiser, and the network components with the selected configuration can be generated accordingly. The resulting SystemC code for the whole design can be simulated at the cycle-accurate and signal-accurate level.

The complete `xpipesCompiler` flow is summarised as follows. From the specification of an application, the designer (or a high-level analysis and exploration tool, such as SUNMAP) creates a high-level view of the SoC floorplan, including nodes (with their network interfaces), links and switches. Based on clock speed target and link routing, the number of pipeline stages for each link is also specified. The information on the network architecture is specified in an input file for the `xpipesCompiler`. Routing tables for the network interfaces are also specified. The tool takes as additional input the SystemC library of soft network components, based on the architectural templates described in Section 17.4.2. The output is a SystemC hierarchical description, which includes all switches, links, network nodes and interfaces and specifies their topological connectivity. The final description can then be compiled and simulated at the cycle-accurate and signal-accurate level. At this point, the description can be fed to back-end RTL synthesis tools for silicon implementation. In a nutshell, the `xpipesCompiler` generates a set of network component instances that are custom-tailored to the specification contained in its input network description file. This tool allows comparison of the effects (in terms of area, power

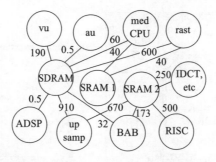

Figure 17.16   Core graph representation of an example MPEG4 with annotated average communication requirements

and performance) of mapping applications on customised domain-specific NoCs and regular (e.g, mesh) NoCs.

As an example, let us focus on the MPEG4 decoder. Its core graph representation together with its communication requirements are reported in Figure 17.16. The edges are annotated with the average bandwidth requirements of the cores in MB/s. Customised application-specific NoCs that closely match the application's communication characteristics have been manually developed and compared with a regular mesh topology. The different NoC configurations are reported in Figure 17.17. In the MPEG4 design considered, many of the cores communicate with each other through the shared SDRAM. Therefore, a large switch is used for connecting the SDRAM with other cores (Figure 17.17(b)), while smaller switches are employed for other cores. An alternate custom NoC is also considered (Figure 17.17(c)): it is an optimised mesh network, with superfluous switches and switch I/Os removed. Area (in 0.1 μm technology) and power estimates for the different NoC configurations are reported in Table 17.1. The area calculations are based on analytical models of xpipes switch area, including crossbar area, buffer and logic area. Although all cores communicate with many other cores and therefore many switches are needed, area savings for custom NoCs are significant.

The power dissipation for the NoC designs has been estimated using the analytical models proposed in Reference 37. These models account for the hardware complexity of the switches as well as the traffic passing through them. Power savings for the custom NoC1 are not relevant, as most of the traffic traverses the larger switches connected to the memories. As power dissipation on a switch increases non-linearly with increase in switch size, there is more power dissipation in the switches of custom NoC1 (that has an 8 × 8 switch) than the mesh NoC. However, most of the traffic traverses short links in this custom NoC, thereby giving marginal power savings for the whole design. In contrast, the NoC2 solution is much more power-efficient.

Figure 17.18 reports the variation of average packet latency (for 64B packets, 32 bit flits) with link bandwidth. Custom NoCs, as synthesised by xpipesCompiler, have lower packet latencies as the average number of switches and link traversals is lower. At the minimum plotted bandwidth value, almost 10 per cent savings in latency

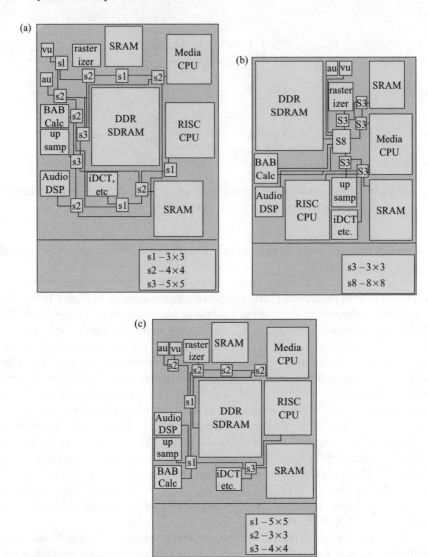

Figure 17.17   NoC configurations for MPEG4 decoder

are achieved. Area, power and performance optimisations by means of custom NoCs turn out to be more difficult for MPEG4 than for other applications such as Video Object Plane Decoders and Multi-Window Displayer, where more significant savings have been obtained [21].

To conclude, we observe that the custom NoC synthesis approach is viable and competitive only if supported by a complete and robust design flow and toolset. Even though fully automated NoC synthesis enables re-use of pre-designed components

*Table 17.1   Area and power estimates for NoC config-
urations*

Instance	Area (mm^2)	A Ratio	Power (mW)	P Ratio
Mesh	1.31	1.00	114.36	1.00
Custom 1	0.86	1.52	110.66	1.03
Custom 2	0.71	1.85	93.66	1.22

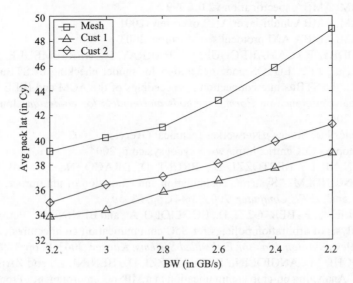

Figure 17.18   Average packet latency as a function of the link bandwidth

(i.e., the soft macros) without compromising flexibility, the quality of components synthesised starting from soft macros can be lower than that of highly optimised custom-designed hard macros. Hence, much work has to be done, especially in the synthesis backend (RTL and logic optimisation, placement and routing), to fully exploit the advantages of this approach with respect to regular and homogeneous NoC architectures.

17.7   Conclusions

This chapter reviewed the guiding principles that are driving the evolution of SoC communication architectures from state-of-the-art shared buses to forward-looking NoC architectures. It shows how the large gap (in terms of design technology) between these two solutions is currently being bridged by means of bus protocols, aiming at a better exploitation of the available bandwidth for on-chip communication, as well as bandwidth-enhancing bus topology evolutions. Finally, NoC design issues are discussed and some early research prototypes are described as case studies, pointing out the need for new design skills and methodologies in order to fully exploit the benefits of these architectures.

References

1   BOEKHORST, F.: 'Ambient intelligence, the next paradigm for consumer electronics: How will it affect silicon?', *ISSCC 2002*, **1**, pp. 28–31

2   ARM: AMBA specification v2.0, 1999

3   ARM: AMBA multi-layer AHB overview, 2001

4   ARM: AMBA AXI protocol specification, 2003

5   WODEY, P., CAMARROQUE, G., BARRAY, F., HERSEMEULE, R., and Cousin, J.P.: 'LOTOS code generation for model checking of STBus based SoC: the STBus interconnection'. Proceedings of the ACM and IEEE international conference on *Formal methods and models for co-design*, June 2003, pp. 204–213

6   Sonics Inc: Sonics μNetworks. Technical Overview, 2002

7   Synopsys CoCentric: http://www.synopsys.com, 2004

8   BENINI, L., BERTOZZI, D., BRUNI, D, DRAGO, N., FUMMI, F., and PONCINO, M.: 'SystemC cosimulation and emulation of multiprocessor SoC designs', *IEEE Computer*, 2003, **36**(4), pp. 53–59

9   POLETTI, F., BERTOZZI, D., BOGLIOLO, A., and BENINI, L.: 'Performance analysis of arbitration policies for SoC communication architectures', *Journal of Design Automation for Embedded Systems, Kluwer*, 2003, **8**, pp. 189–210

10   LOGHI, M., ANGIOLINI, F., BERTOZZI, D., BENINI, L., and ZAFALON, R.: 'Analyzing on-chip communication in a MPSoC environment'. Proceedings of IEEE *Design automation and test in Europe* conference (DATE04), 2004, pp. 752–757

11   HENKEL, J., WOLF, W., and CHAKRADHAR, S.: 'On-chip networks: A scalable, communication-centric embedded system design paradigm'. Proceedings of the international conference on *VLSI design*, 2004, pp. 845–851

12   BENINI, L., and DE MICHELI, G.: 'Networks on chips: a new SoC paradigm', *IEEE Computer*, **35**(1), 2002, pp. 70–78

13   RIJPKEMA, E., GOOSSENS, K., and RADULESCU, A.: 'Trade-offs in the design of a router with both guaranteed and best-effort services for networks on chip'. Proceedings of *Design automation and test in Europe*, March 2003, pp. 350–355

14   LEE, K., *et al.*: 'A 51mw 1.6 ghz on-chip network for low power heterogeneous SoC platform', *ISSCC Digest of Tech. Papers*, 2004, pp. 152–154

15   BOLOTIN, E., CIDON, I., GINOSAR, R., and KOLODNY, A.: 'QNoC: QoS architecture and design process for network on chip', *The Journal of Systems Architecture, Special Issue on Networks on Chip*, December 2003, **50**(2–3), pp. 105–128

16 DALLY, W.J., and LACY, S.: 'VLSI architecture: past, present and future' Conference on *Advanced research in VLSI*, 1999, pp. 232–241

17 MILLBERG, M., NILSSON, E., THID, R., KUMAR, S., and JANTSCH, A.: 'The Nostrum backbone – A communication protocol stack for networks on chip'. International conference on *VLSI design*, 2004, pp. 693–696

18 GUERRIER, P., and GREINER, A.: 'A generic architecture for on-chip packet-switched interconnections'. Proceedings of the *Design automation and test in Europe* conference, March 2000, pp. 250–256

19 SAASTAMOINEN, I., SIGUENZA-TORTOSA, D., and NURMI, J.: 'Interconnect IP node for future systems-on-chip designs'. IEEE Workshop on *Electronic design, test and applications*, January 2002, pp. 116–120

20 DALL'OSSO, M., BICCARI, G., GIOVANNINI, L., BERTOZZI, D., and BENINI, L.: 'Xpipes: a latency insensitive parameterized network-on-chip architecture for multi-processor SoCs'. *ICCD 2003*, October 2003, pp. 536–539

21 JALABERT, A., MURALI, S., BENINI, L., and DE MICHELI, G.: 'XpipesCompiler: a tool for instantiating application specific networks on chip'. *DATE 2004*, 2004, pp. 884–889

22 DALLY, W.J., and TOWLES, B.: 'Principles and practices of interconnection networks' (Morgan Kaufmann Publishers, San Mateo, CA, 2004)

23 AGARWAL, V., HRISHIKESH, M.S., KECKLER, S.W., and BURGER, D.: 'Clock rate versus IPC: the end of the road for conventional microarchitectures'. International symposium on *Computer architecture*, June 2000, pp. 248–250

24 CARLONI, L.P., MCMILLAN, K.L., and SANGIOVANNI VINCENTELLI, A.L.: 'Theory of latency-insensitive design'. *IEEE Transactions on CAD of ICs and Systems*, 2001, **20**(9), pp. 1059–1076

25 ITRS 2001: http://public.itrs.net/Files/2001ITRS/Home.htm

26 GLASKOWSKY, P.: 'Pentium4 (partially) previewed'. *Microprocessor Report*, 2000, **14**(8), pp. 10–13

27 Li SHANG, LI-SHIUAN PEH, and NIRAJ K. JHA.: 'Dynamic Voltage Scaling with Links for Power Optimisation of Interconnection Networks'. HPCA – proceedings of the international symposium on *High performance computer architecture*, Anaheim, February 2003, pp. 91–102

28 WORM, F., IENNE, P., THIRAN, P., and DE MICHELI, G.: 'An adaptive low-power transmission scheme for on-chip networks'. ISSS, Proceedings of the International symposium on *System synthesis*, Kyoto, October 2002, pp. 92–100

29 LEE, S.J., *et al.*: 'An 800 MHz star-connected on-chip network for application to systems on a chip', *ISSCC Digest of Technical Papers*, 2003, pp. 468–469

30 VSI ALLIANCE: 'Virtual component interface standard', 2000

31 OCP INTERNATIONAL PARTNERSHIP: 'Open core protocol specification', 2001

32 LINES, A.: 'Asynchronous interconnect for synchronous SoC design', *IEEE Micro*, 2004, **24**(1), pp. 32–41

33 MUTTERSBACH, J., VILLIGER, T., KAESLIN, H., FELBER, N., and FICHTNER, W.: 'Globally-asynchronous locally-synchronous architectures to

simplify the design of on-chip systems'. IEEE *ASIC/SOC* conference, September 1999, pp. 317–321

34   MURALI, S., and DE MICHELI, G.: 'SUNMAP: a tool for automatic topology selection and generation for NoCs'. Proceedings of the *Design automation* conference, 2004, pp. 914–919

35   BERTOZZI, D., JALABERT, A., MURALI, S., *et al.*: 'NoC synthesis flow for customized domain specific multi-processor Systems-on-Chip', *IEEE Transactions on Parallel and Distributed Systems*, Special issue on on-chip networks, to be published

36   VAN DER TOL, E.B., and JASPERS, E.G.T.: 'Mapping of MPEG4 decoding on a flexible architecture platform', *SPIE 2002*, January 2002, pp. 1–13

37   WANG, H.S. *et al.*: 'Orion: A power-performance simulator for interconnection networks', *IEEE Micro* November 2002

Chapter 18

Asynchronous on-chip networks

Manish Amde, Tomaz Felicijan, Aristides Efthymiou,
Douglas Edwards and Luciano Lavagno

18.1 Introduction

The main idea of an SoC design methodology is to 'divide' complex chips into several independent functional blocks and 'conquer' each of them using standard synchronous methodologies and existing CAD tools. These functional blocks are then connected by the means of an on-chip communication infrastructure to form a functional system.

Dividing a chip into smaller blocks keeps the technology scaling problems, such as clock-skew, manageable, however this is only true for each individual block, while the problems aggravate drastically for the interconnect itself. This is because the network elements may be scattered all over the chip connected by relatively long wires which do not scale well in deep sub-micron technologies [1]. Synchronising such a network with a single clock source is problematic at best.

There are major problems in having various synchronous on-chip communication, namely:

- *Modularity and design reuse*: In the synchronous world, a complete redesign on the chip is needed if a component of the chip is modified or if the frequency of operation is changed, thus making the design non-modular. Normally, all the components have to be redesigned at the same new clock frequency or at rationally related ones. This leads to waste of design effort. Globally asynchronous locally synchronous (GALS) IP cores with asynchronous interfaces would make them amenable for design reuse.
- *Electromagnetic interference (EMI)*: All the switching activity in a synchronous chip takes place at a given clock tick making the circuit prone to EMI effects. In comparison, switching activity is distributed over time in a clock-less chip.

- *Worst-case performance*: The circuit is always designed for the worst-case performance, since the critical path in the circuit determines the clock period.
- *Clock power consumption*: Large clock buffer trees present in current design lead to a high power consumption. Studies show that high-speed processors have power consumption dominated by clock and the average clock power consumption by the clock is 45 per cent of the total power consumed [2]. Similar statistics are reported for high- and medium-speed application-specific integrated circuits (ASICs) as well.
- *Clock skew*: The problem of distributing the global clock in a chip with minimal clock skew is getting difficult to solve due to the increase in clock frequencies, smaller feature sizes and growing design complexities. Few ASIC designers can afford the sophisticated calibration techniques used in leading edge microprocessors [3], and would like to enjoy the intrinsic robustness with respect to manufacturing and runtime variability that asynchronous circuits exhibit.

Due to the above-mentioned problems in using a synchronous design style, efforts are being made to design chips asynchronously. A significant advantage of asynchronous design is smoother handling of both fabrication-time inter-chip and runtime intra-chip variability (the latter requires completion detection, the former only delay matching). Also, all the aforementioned problems associated with the distribution of global clock over the entire chips like – clock power consumption, clock skew and EMI – are eliminated. Moreover, the designs become modular since timing assumptions are explicit in the hand-shaking protocols. Hence no redesign is needed if an asynchronous component is modified. Furthermore, the circuit would work faster, exploiting average case rather than worst-case performance.

There has been a remarkable resurgence of interest in asynchronous design since the mid-1980s. Since the early to mid-1990s, a number of asynchronous chips and designs have been successfully fabricated for substantial designs in both industry and academia. These include an infrared communication chip at HP Laboratories [4], an instruction-length decoder at Intel [5], a configurable self-timed digital signal processor (DSP) [6,7], high-speed first-in first-outs (FIFOs) and routing fabric at Sun Laboratories, and microprocessors at University of Manchester, Caltech and University of Tokyo (see Reference 8, and the accompanying issue, for more details). In the late 1990s, Philips Semiconductors introduced asynchronous microcontroller chips into commercial pagers and cell phones. All of this activity suggests two trends: (1) a growing acknowledgement by industry and academia of some potential advantages of asynchronous design and (2) the beginnings of maturity in establishing practical design styles and sound synthesis techniques for asynchronous controllers and datapaths.

But asynchronous design strategies also come with their own set of problems. Asynchronous design is a more difficult task compared to synchronous design. Glitch-free circuits have to be generated as compared to synchronous domain where the data only has to be stable before the arrival of the clock. Also, asynchronous design suffers from the absence of industrial tool support. Lack of a mature tool flow has prevented

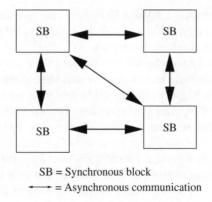

SB = Synchronous block
←——→ = Asynchronous communication

Figure 18.1   System level view of GALS

this methodology from being widely adopted by designers in industry. (For more information on asynchronous synthesis, see Part V, Chapter 16.)

Moreover, several asynchronous circuit implementation techniques have a very high overhead in terms of area, delay and possibly even power consumption. This is due to the fact that truly asynchronous datapaths require implementing each signal in dual rail, and collecting acknowledgements from every gate output in the circuit. In this chapter we will survey techniques that avoid such large overhead, at the expense of fewer gains in terms of, e.g. EMI and average case performance.

The GALS and desynchronisation design styles that are described below, are aimed at filling the gap between the purely synchronous and asynchronous domains. They consist of synchronous modules communicating asynchronously as shown in the system-level view in Figure 18.1. These methodologies are promising because they allow synchronous design of components at their own optimum clock frequency, but facilitate asynchronous communication between modules. This leads to a design flow fairly similar to the synchronous flow but with a few additional components which enable asynchronous communication. It eliminates the global clock leading to a reduction of power consumption and alleviating the clock skew problem. It facilitates modular system design which is scalable. Close resemblance to synchronous design also makes it amenable to attract the attention of synchronous designers who are not willing to experiment with asynchronous design.

GALS refers to a communication framework in which local clocks are either unsynchronised or paused. This means that there is a risk of metastability at the interfaces which is not present in 'traditional' speed-independent or delay-insensitive asynchronous circuits [19]. Metastability is a condition where the voltage level of signal is at an intermediate level – neither 0 nor 1 – and which may persist for an indeterminate amount of time.

Desynchronisation bears some similarity to GALS techniques, in that the datapath remains essentially synchronous and its clocks are locally generated, but it prevents metastability completely by using handshakes. As such, a desynchronised circuit can be obtained automatically from a synchronous one. It has approximately the same area, power and performance, but lower EMI, due to the spreading over time of clock edges, and better modularity, due to the explicit handshakes between components that automatically satisfy local timing constraints.

However, GALS and desynchronisation are not able to provide the designer with the full power of asynchronous techniques. We thus also review some logic synthesis methods that have been proposed in the past to deal with asynchronicity at the gate level. The full domain of asynchronous synthesis techniques is too broad to be covered in a survey chapter. The interested reader is referred to Reference 8 for a recent collection of papers.

In this chapter we first present formal frameworks for the analysis of transformations from synchronous to asynchronous systems, and their implementation in the desynchronisation flow. Next we discuss speed-independent circuits and their logic synthesis techniques. We then proceed to explain various schemes for implementing GALS-based systems. We finally conclude with a discussion on asynchronous NoCs, and with a case study.

The chapter is organised as follows. Section 18.2 describes formal frameworks for analysing the relationship between synchronous and asynchronous circuits, as implemented by desynchronisation strategies. Section 18.3 describes speed-independent asynchronous circuit specification and synthesis methods, which are often used to synthesise the controllers for the other techniques discussed in this chapter. Section 18.4 discusses some mixed-mode synchronous and asynchronous strategies, mostly based on ad hoc approaches. Section 18.5 investigates interfacing schemes based on pausable clocks. Section 18.6 describes GALS control blocks, their circuit implementations and key figures such as latency and throughput. Section 18.7 gives an insight into the state-of-the-art in asynchronous NoCs. Section 18.8 describes quality of service (QoS) for NoCs. Section 18.9 presents a case study of the NoC circuitry implemented in the Asynchronous Open Source Processor IP of the DLX Architecture (ASPIDA) project. Section 18.10 compares and contrasts the various approaches.

18.2   Formal models

18.2.1   *Signal transition graphs and state graphs*

Speed-independent circuits are designed at the gate level to be hazard free using the unbounded gate delay model. The most popular specification style for speed-independent circuits are signal transition graphs (STGs) [9]. These graphs are a class of interpreted Petri Nets [10] that allow the designer to comfortably capture the behaviour of an asynchronous circuit in a manner that is quite similar to timing diagrams.

As an example consider Figure 18.2(a), which depicts the interface of a device with a VME bus. The behaviour of the controller is as follows: a request to read from or write into the device is received by one of the signals DSr or DSw, respectively. In a read cycle, a request to read is done through signal LDS. When the device has the data ready (LDTACK), the controller must open the transceiver to transfer data

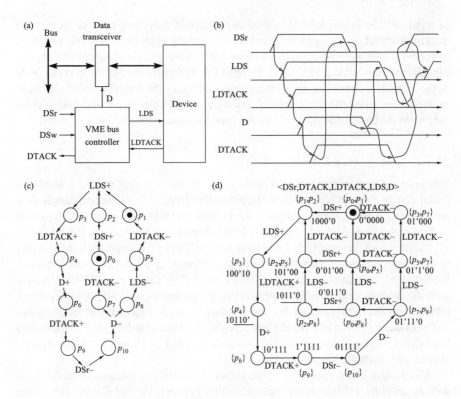

Figure 18.2   (a) VME bus controller, (b) timing diagram, (c) STG and (d) SG for the read cycle

to the bus (signal D). In the write cycle, data is first transferred to the device (D). Next, a request to write is done (LDS). Once the device acknowledges the reception of the data (LDTACK), the transceiver must be closed to isolate the device from the bus. Each transaction must be completed by a return-to-zero of all interface signals, seeking for a maximum parallelism between the bus and the device operations.

Figure 18.2(b) shows a timing diagram of the read cycle and Figure 18.2(c) the corresponding STG. All events in this STG are interpreted as signal transitions: rising and falling edges are labelled with '+' and '−', respectively (we also use the notation a^* if we are not specific about the direction of the signal transition).

An STG has two types of vertices: transitions and places (circles). Places can be marked with tokens (black dots). The set of all places currently marked is called a marking. A transition is enabled if all its input places contain a token. In the initial marking of the STG in Figure 18.2(c) only one transition, DSr+, is enabled; LDS+ is not enabled because its input place p_2 does not have a token. Every enabled transition can fire. Firing removes one token from every input place of the transition and adds one token to each of its output places. After the firing of transition DSr+ the net moves to a new marking $\{p_1, p_2\}$ and then LDS+ becomes enabled, while other transitions (none in this case) sharing the same input place(s) may be disabled due to the lack of input tokens. Transitions are called concurrent if they both can fire from some marking without disabling each other. By exhaustively exploring reachable markings of the STG and associating each of them with a binary code of signal values, one can generate the 'state graph' (SG). Figure 18.2(d) depicts the SG for the read cycle of the VME bus controller, with binary codes labelling the states (enabled signals in each state are marked with a prime). State graphs are of primary importance since they form the basis of logic synthesis for speed-independent circuits.

18.2.2 Multi-clock Esterel

Synchronous design tools have a wide range of tools giving rise to a tried and tested design flow. Asynchronous circuits suffer from lack of mature design flow and efforts are being made to capture the asynchronous behaviour of GALS system in the synchronous domain. One effort in this direction is Multi-clock Esterel [11].

Synchronous languages [12,13] have a significant advantage with their ability to prove correctness of the hardware circuits before they are actually implemented. Esterel is a synchronous language used for modelling reactive systems interacting with the environment. It is an imperative language and hence uses variables which retain their value until updated. It is used mainly for modelling controller applications and provides synchronous parallelism. Hence, it could also be used for modelling hardware systems. Esterel inherently assumes a global clock and it cannot handle a system with multiple clocks.

Multi-clock Esterel provides a framework for modelling multiple local clocks as well as enabling asynchronous communication between various components in the design. It also provides a clean model for integrating Verilog/VHDL features in a design. It aims to retain the existing features of reactive languages like pre-emption and more importantly verifiability. It can be considered to satisfy the 'synchrony hypothesis' as its reactions can be associated with local clock ticks.

The asynchronous communication between concurrently running locally clocked reactive components is based on latches with limited memory. In Reference 11 the authors show an example design of a Micropipeline in a modular fashion and show Multi-clock Esterel modules could be composed in a hierarchical fashion.

Multi-clock Esterel could also be used to model a subset of Verilog Hardware Description Language (VHDL) code enabling the possibility of hardware software codesign using VHDL while verifying the entire design in the synchronous paradigm of Multi-clock Esterel.

18.2.3 Signal/Polychrony framework

The goal of this research is to model GALS in a multi-clock synchronous environment and map it to an asynchronous system preserving all the properties proven in the synchronous domain.

Signal [12] is a programming framework which provides a formal way of modelling various synchronous components running on different clocks and validating that the asynchronous composition of the various components would lead to a functionally correct behaviour. It achieves this by transforming the asynchronous composition of the various synchronous components having different clocks to a fully synchronous multi-clock model preserving behavioural equivalence. The synchronous model takes advantage of verification tools available for the synchronous languages. The correct behaviour can thus be checked by extensive simulation and model-checking in the synchronous domain.

This methodology could be used in integrating various intellectual property (IP) cores designed at different clock frequencies using a desynchronisation protocol and formally verifying the functional correctness of this GALS network.

In Reference 14, the authors provided a formal way of capturing asynchrony in the synchronous framework of Signal. They prove that an ideal asynchronous model can be completely mapped in Signal with unbounded first-in first-out (FIFOs) for inter-component communication. They also show that the class of synchronous models that can be implemented asynchronously without any loss in semantics, i.e. while preserving the deterministic behaviour which is a key characteristics of synchronous models, must satisfy the properties of endochrony and isochrony. Roughly speaking, endochrony means that a component, whose interface is going to be made asynchronous, must be able to tell from the values of its inputs which inputs must be read next. This approximately corresponds to the sufficient property stated by Kahn [15] to ensure determinate behaviour for the data flow networks, namely that processes cannot probe input FIFOs for presence of data. Isochrony, on the other hand, means that if two components share a variable, they must agree on the values which are assigned to it at each step. Note that, unfortunately, the identification of bounds to the size of FIFO channels in Kahn Process Networks is undecidable [16], and hence the problem of correctly deploying an arbitrary synchronous system onto an asynchronous architecture must be solved by a human, using a lot of simulation, iteration and guesswork.

The high-level system specification is transformed into a low-level circuit representation through a series of steps. At each step, the transformation from a higher to lower level of abstraction should preserve the correctness across the transformation. Polychrony [17] is a platform which along with the synchronous programming framework of Signal provides formal refinement of multi-clocked models from high-level behavioural specification to the low-level synthesis and implementation of these models using formal verification techniques. Polychrony takes a high-level SystemC/SpecC specification and refines it in a semantic-preserving manner towards a GALS implementation. This allows one to leverage the implementation of various synchronous components with multiple clocks with assurance of a functionally correct asynchronous communication between different clocked synchronous components.

The advantage of using Polychrony in a high-level design flow is that it automates the complex task of formal design verification at each stage of refinement and renders the low-level implementation formally correct. The polychronous model of Signal formally captures the behavioural abstractions from SystemC/SpecC programs as well as behavioural specifications from IP cores. The Polychrony platform aids in automating the synthesis of behavioural specifications while formally verifying the correctness of the transformation at each design flow step. Hence, one can rapidly

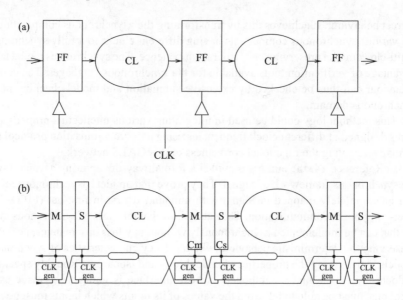

Figure 18.3   Synchronous and desynchronised pipelined circuit

codesign hardware/software GALS architectures while being assured of formally conforming to the original behavioural specifications.

18.2.4   Desynchronisation

Desynchronisation [18] builds upon these theoretical foundations in order to provide the designer with the option to derive a medium-grained asynchronous implementation from a traditional synchronous specification. Assuming an initial design implemented with edge-triggered flip-flops, it requires the following steps:

1   Conversion of the flip-flop-based synchronous circuit into a latch-based one (M and S latches in Figure 18.3(b)),
2   Generation of matched delays for combinational logic (rounded rectangles in Figure 18.3(b)) and
3   Interconnection of controllers for local clocks.

The method for desynchronising an arbitrary netlist relies on composition of the controllers. It requires to identify direct connections, via combinational logic, between adjacent groups of latches, and then the overall clock generation circuit is obtained through composition of timing diagrams corresponding to these partial descriptions.

The specification of a pairwise interaction between even–odd and odd–even latches for overlapping desynchronisation is shown in Figure 18.4. It models the communication of data from latch A to latch B. The latches are transparent when the control signal is high. Initially, only half of the latches contain data ($\mathcal{D}$). Data items

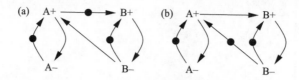

Figure 18.4 Synchronisation between latches: (a) even → odd, (b) odd → even

Table 18.1 Synchronous vs. desynchronised DLX

	Synchronous DLX	Desynchronised DLX
Cycle time	4.4 ns	4.45 ns
Dyn. power cons.	70.9 mW	71.2 mW
Area	372 656 μm^2	378 058 μm^2

flow in such a way that a latch never captures a new item before its successor latches have captured the previous one.

Data overwriting can never occur, even though the pulses for the latch control can overlap. This model is based on the observation that a data item can ripple through more than one latch, as long as the previous values stored in those rippling latches have already been captured by the successor latches. As an example, event B+ can fire as soon as data is available in A (arc A+ → B+) and the previous data in B has been captured by C (arc C− → B+).

Reference 18 suggests that desynchronisation results in circuits with almost identical area, performance and power consumption as the original synchronous ones. Desynchronised circuits, however have smaller EMI, due to the out-of-phase clocks, and better modularity, due to the explicit handshakes encapsulating timing constraints. A comparison between a synchronous and a desynchronised version of the same processor is shown in Table 18.1

The electromagnetic emission advantages can be seen by looking at the spectrum of the current absorbed by the circuit from the power rails, shown in Figure 18.5.

18.3 Speed-independent circuit implementation

This section describes implementation techniques for asynchronous controllers, which are used by the other methods described in this chapter. Speed-independent design techniques use fully asynchronous models for both specification and implementation, and hence can be used, due to the state explosion problem, only for small components.

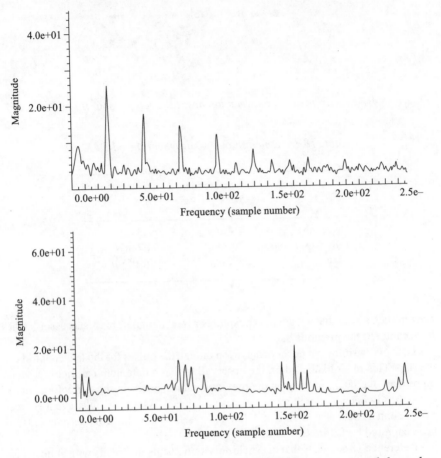

Figure 18.5   FFT of current consumption in synchronous (above) and desynchronised (below) DLX

The following properties must hold in an SG, e.g. derived from an STG, to be implementable as a speed-independent circuit [19]:

- Consistency holds when rising and falling transitions alternate for each signal.
- Complete state coding ensures that any two states with the same binary code have the same set of enabled output signals, as discussed below.
- Output-persistency holds when no output signal transition can be disabled by another signal transition, thus ensuring that no hazard can appear at the gate outputs.

The latter property is often associated with the notion of 'acknowledgement'. Informally, we say that transition b^* acknowledges transition a^* if the fact that b^* fires after a^* has been enabled indicates that a^* has already fired. We say that a^* is acknowledged if any firing sequence starting from $a*$ enabled is acknowledged by some transition.

The main steps in the logic synthesis assume that the SG is consistent and output-persistent, and are as follows.

1 Encode the SG in such a way that the complete state coding property holds (this may require the addition of internal signals).
2 Derive the 'next-state' functions for each output and internal signal of the circuit.
3 Map the functions onto a netlist of gates.

The next-state function for a signal a maps the binary code of each SG state s into:

- 1 if the signal has value $0'$ or 1 in the binary code of s (it is either excited to go to 1, or stable at 1)
- 0 if the signal has value $1'$ or 0 in the binary code of s
- – (don't care) for all binary codes that do not correspond to any reachable SG state.

For example, the next-state function of signal LDS in Figure 18.6(a) has value 1 in states labelled $1000'01$ and $10'1111$, value 0 in states labelled $0'011'00$ and $0'00000$, and value – in states labelled 010100. The set of states in which a signal has value $0'$ (resp. $1'$) is called the 'excitation region' (ER) for its rising (resp. falling) transition. For example, the excitation region of LDS+ in Figure 18.6(a) is the state labelled $1000'01$. The set of states in which a signal has value 1 (resp. 0) is called the 'quiescent region' (QR) for its rising (resp. falling) transition. Thus the next state function for signal LDS has value 1 in ER (LDS+) $\cup$ QR (LDS+) and value 0 in ER (LDS−) $\cup$ QR(LDS−).

This definition, however, has a problem, as shown by the two underlined states in the SG in Figure 18.2(d). These states correspond to different markings, $\{p4\}$ and $\{p2, p8\}$, but their binary codes are equal, 10110. Moreover, enabling conditions in these two states for output signal LDS are different. Therefore, the value of the next state Boolean function for signal LDS for vector 10110 should be 1 for the first state and 0 for the second state. A similar problem holds for signal D. The result is a conflict in the definition of the function. A possible method to solve this problem is to insert new state signals that disambiguate the encoding conflicts and ensure the satisfaction of complete state coding. This is the equivalent of the state assignment step in the case of synchronous circuits. Figure 18.6(a) depicts a new SG in which a new signal, csc0, has been inserted. Now the next-state functions for signals LDS and D can be uniquely defined. The insertion of new signals must be done in such a way that the resulting SG satisfies consistency and output-persistency, as discussed in References 20 and 21.

Once the next-state function has been derived, Boolean minimisation can be performed to obtain a logic equation that implements the behaviour of the signal. In this step, it is crucial to make an efficient use of the don't care conditions. For the example

(a)   (b)

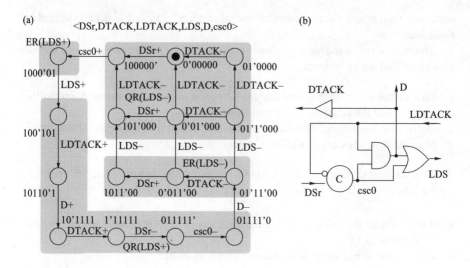

Figure 18.6   SG for the read cycle with complete state coding (a), and its implementation (b)

of Figure 18.6, the following equations can be obtained:

$$D = LDTACK \cdot csc0$$

$$LDS = D + csc0$$

$$DTACK = D$$

$$csc0 = DSr \cdot LDTACK + csc0 \cdot (DSr + LDTACK).$$

The implementability properties of the SG ensure that any circuit implementing the next-state function of each signal with only one complex gate does not have any hazard under the unbounded gate delay model [19]. A possible hazard-free gate implementation for the next-state function of the read cycle example is shown in Figure 18.6(b), where the sequential gate shown as a circle with 'C' is a so-called C element [19] with next state function $c = ab + c(a + b)$.

The design flow discussed above has an essential problem, because logic functions for signals might be too complex to be mapped into single gates available in the library, and hence must be decomposed in order to make them implementable. Unfortunately, this step introduces new internal signals that may cause hazards. On the other hand, 'merging gates' does not cause new hazards in speed-independent (SI) circuits. Hence, classical methods for combinational logic technology mapping can be used to combine gates, after an appropriate decomposition has been found [22].

The approach discussed in this section splits the problem of 'hazard-free logic decomposition' of a gate into two subproblems: (1) combinational decomposition (aiming at decompositions of the following type: $C = F \cdot G + R$), and (2) insertion of a new hazard-free signal. This process is iterated until all gates of the circuit can be mapped onto library gates or no more progress can be achieved, e.g. because no

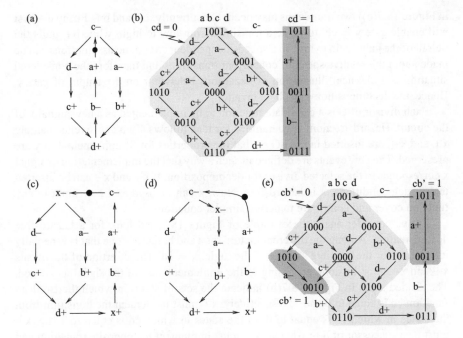

Figure 18.7   Example of signal insertion

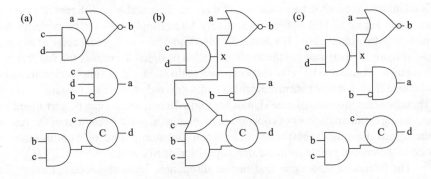

Figure 18.8   Implementation of the STGs of Figure 18.7

hazard-free decomposition can be found for any of the complex gates. Figure 18.7(a) and (b) depicts the STG and the SG of the specification of a circuit. A complex gate implementation of the circuit is shown in Figure 18.8(a).

Let us assume that only two-input gates are available in the library. Thus, signals a and b are not directly mappable and must be decomposed without violating output-persistency. To illustrate this point, let us decompose the gate a in Figure 18.8(a) by extracting the factor $y = cb'$. The ON- and OFF-sets of the function for y are shown in Figure 18.7(e) by the shaded areas. When the circuit enters state 0000 (underlined in Figure 18.7(e)) two transitions may occur concurrently: c+ and b+. Firing c+ first will enable gate y = cb′ to make a transition from low to high, while b+ pulls the output of the gate again to low. In a speed-independent circuit, no assumptions can be made about the relative speed of concurrent transitions and therefore the considered situation is a classical illustration of hazardous behaviour on the output of gate y. Hence, the decomposition y = cb′ is invalid.

Each divisor of C is a candidate function to be implemented as a new signal x of the circuit. Hazard-freedom is guaranteed for it as follows. Two new events, namely x+ and x−, are inserted in the SG so that the properties for SI implementability are preserved. The new events are defined in such a way that the implementation of signal x corresponds to the selected divisor for decomposition. If x+ and x− can be inserted under such conditions, x is hazard-free. Now, x can be used as a new input to any function cover and contribute to derive simpler equations.

Let us consider again the example of Figure 18.7 and look for a hazard-free decomposition. Among the different factors for a and b, there is one that is especially interesting for the possible sharing of the logic: x = cd. The insertion of the events x+ and x− must be done according to the implementation of the signal as x = cd. The shaded areas in Figure 18.7(b) indicate the sets of states in which the Boolean function cd is equal to 0 and 1, respectively. x+ must implement the transition from the states in which cd is equal to 0, to the states in which cd is equal to 1, i.e. x+ must be a successor of d+; whereas x− must implement the opposite transition and therefore is inserted after c−.

Figure 18.7(c) and (d) depict two possible insertions of signal x at the STG level. Both insertions result in specifications that are implementable as different SI circuits, shown in Figure 18.8(b) and (c), respectively. Interestingly, both can be implemented with only two-input gates. However, the insertion of x− as a predecessor of a− and d− (Figure 18.7(c)) changes the implementation of signal d, because the fact that x− triggers d− forces x to be in the support of any realisation of d. A simpler circuit can be obtained if x− is made concurrent with d− and thus only trigger a− (Figure 18.7(d)). Therefore, the insertion of new signals for logic decomposition can be performed by exploring different degrees of concurrency with regard to the behaviour of the rest of the signals. Finding the best trade-off between concurrency and logic optimisation is one of the crucial problems in the decomposition of SI circuits.

The interested reader can find further discussion, including decomposition and matching techniques for 'sequential' gates, in Reference 23.

18.4   Mixed synchronous/asynchronous solutions

The Pentium 4™ processor [3] uses 47 different clock domains, whose skew relative to a global reference clock is programmable. Domain clocks were intentionally skewed to improve operating frequency and up to one speed bin improvement is reported. The design uses two phase-locked loops (PLLs) – one for the core and one for the input/output (I/O) logic. From these, six different clock frequencies are derived. The Pentium 4™ also has critical portions (e.g. the arithmetic and logic unit (ALU)) working at twice the clock frequency of the rest of the chip [3]. Non-critical ones work at half the clock frequency, in order to save area, power and design effort.

The Alpha processor [24] illustrates the need for flexible clocking schemes in order to enable core reuse in system-on-chip (SoC) designs. The entire chip is partitioned into 11 clock domains, where one domain is a migration of a processor core from an older design. The existing clock distribution in this embedded core is used as a reference clock. Four major clocks (one reference and three derived) are used to clock separate chip sections. Delay-locked loops (DLLs) are used to maintain small phase alignment errors among major clocks.

An example of mixed synchronous and asynchronous implementation is given in Reference 25, which presents the design of a digital finite impulse response (FIR) filter used in read channels of modern disk drives. The degree of pipelining in the filter is dynamically variable and depends on the input data rate. The performance of this filter was found to be better than existing read channel filters.

The high-speed asynchronous portion of the chip is sandwiched between two synchronous portions. The asynchronous datapath in the chip uses dual-rail dynamic logic and the synchronous datapath in the chip uses single-rail static logic. The asynchronous section relies upon handshakes for communication whereas the synchronous section is dependent on global clocking. Thus, the interface circuitry between asynchronous and synchronous datapaths is responsible for data conversion. It also needs to adapt to different control signals on either side of the interface. The first interface requires conversion from synchronous to asynchronous domain and the second interface requires asynchronous to synchronous conversion. The interface circuitry achieves this by having special latches for performing data conversion and pulse generators for implementing the handshaking protocol for the asynchronous section. In order to resynchronise and avoid metastability at the second interface, a delayed version of the Req handshake signal generated at the first interface is passed directly to the second interface using a programmable delay element. The programmable delay should be greater than the delay for the correct data computation by the asynchronous section.

18.5 Pausable clock interfacing schemes

Pausable clocking schemes are proposed as mechanisms for data transmission between synchronous modules running at different clock frequencies. In this scheme, the receiver clock is paused whenever the sampling of data lines by the receiver could lead to potential metastability. The sender clock is paused till the data is correctly sampled by the receiving module. This avoids synchronisation failure at the receiving end and flow control at the sender end.

A similar approach is also followed by recent work on the Razor processor [26], in which a comparator (including a metastability detector) identifies when a register incorrectly latches a value, due to a critical timing problem. In the next clock cycle the pipeline is simply restarted with the correct data copied back in every register from a shadow latch, and processing continues synchronously, with 'skipped' clock cycles.

Pausable clock schemes listed below, on the other hand, generally stretch clock cycles and do not ensure phase alignment with an external reference clock. Razor is a very promising approach to tackling variability, in that it allows one to clock a processor very close to its true speed. However, reliable operation over extended periods of time, due to inherent risk of metastability, still needs to be demonstrated.

In Reference 27, the authors comment that previously proposed schemes [28] do not scale well for high clock frequencies of locally synchronous (LS) components and multiple cycle delay in clock distribution due to large clock buffer trees. Due to the presence of large clock buffer trees in the LS components, the assumption of previous schemes of data transfer being stalled within one clock cycle of pausing the sender clock does not hold and leads to extra transmissions in what the authors call the 'clock overrun window', which denotes the skew between pausing the clock and actual stopping of data transmission by the sender module.

They propose a circuit for interfacing two high-frequency LS modules using a partial handshake protocol which achieves high data rates and has small probability of failure. A partial handshake is used as it provides faster data transfer than a complete handshake protocol. They propose a 'direct path' FIFO to account for long interconnect delay and an additional 'buffered' FIFO to capture data transferred in the clock overrun window. This scheme does not pause the receiver clock for synchronisation but pauses the sender clock to achieve flow control.

In Reference 27, transistor level sender and receiver interface circuits are given and the models are verified by SPICE simulation. The timing analysis of the interface circuits proves that under certain circumstances of bad signal timings of the signal with respect to sender and receiver clock, the synchronisation circuit would fail with a small probability of failure thus improving on previous schemes.

Chakraborty *et al.* in Reference 29 discuss using abstract timing diagrams to reason about correctness of interfacing techniques between synchronous modules. They point out that there are various different interfacing techniques available but it is difficult to compare them due to difference of analysis carried out for each of them. Reference 29 uses abstract timing diagrams for analysing specific interfacing schemes and understand why certain schemes work under restricted conditions and fail otherwise. They further point out that robust asynchronous interfaces could be built if certain new circuits could be implemented.

18.6   GALS implementations

According to the GALS methodology used in Reference 30, the asynchronous circuits required to convert LS modules to conform to GALS standard are restricted to implementing 'self-timed' wrappers around each module. Each LS module is driven by a pausable clock in its self-timed wrapper avoiding metastability and data corruption. The self-timed wrappers consist of a pausable local clock generator, port controllers and test structures as shown in Figure 18.9. They have implemented five wrapper elements in technology-independent VHDL. The port controllers are implemented as asynchronous finite state machines using the extended burst mode paradigm of

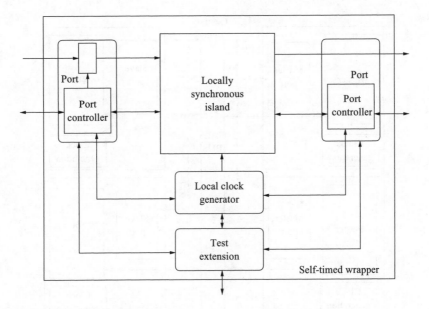

Figure 18.9   Self-timed wrapper

Reference 31. These are synthesised using the 3D tool which results in a synthesisable AND–OR implementation [32].

In Reference 30 the authors describe two different types of port controllers:

1 *Poll-type, or non-blocking, port*: This port is used whenever a data item is needed but computation could proceed without it arriving immediately. The LS modules keep functioning while the data transfer is handled by the port.
2 *Demand-type, or blocking, port*: This port is used when the LS module cannot continue computation till the arrival of data on the port. While waiting for data, the demand-type port suspends the local clock reducing power consumption of the module.

Various tunable local-clock generators are compared in Reference 33. The current research is directed towards high frequency tunable local oscillators for better performance of individual GALS modules.

Point to point interconnects have been implemented to allow asynchronous communication between GALS modules [34]. In Reference 34, interface wrapper circuits are presented for communication between LS modules. The wrapper interface consists of an arbiter and a calibrated delay line. This ensures that a stable local clock signal is generated. Metastability is avoided as clocking is done only after the data is ready. Circuits with sleep mode, where the local clock is stopped due to unavailability of data, are also presented. This could lead to reduced power consumption. FIFOs with various depths have been used as asynchronous channels between modules. The simulations show that designing for FIFO depths greater than 2 does not improve bandwidth of communication between modules.

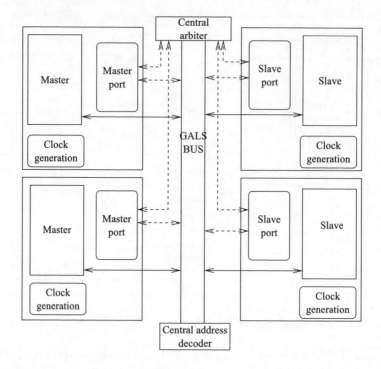

Figure 18.10   GALS bus architecture

Multi-point interconnects are also required for efficient SoC GALS systems. Two different interconnection topologies have been proposed in Reference 35:

1 *GALS bus*: In this architecture all port controllers of LS modules(master/slave) in a design are connected to the same bus as shown in Figure 18.10. The arbitration and address decoding is central for power efficiency.
2 *Ring structure*: A transceiver is associated with the port controller of every LS module as shown in Figure 18.11. The arbitration is done by the transceivers and it decides which request to grant, either from the previous transceiver or from its own port controller, when it wants to insert some packets into the ring.

The ring structure leads to higher latency as each packet has to encounter one or more transceivers. But it leads to lower interconnect length between modules and possibly reduces power consumption. One aspect that is not discussed by the authors of Reference 35 is fault tolerance. Failure of one module in a ring architecture would lead to the failure of communication between all modules, whereas the GALS bus would be more tolerant to such faults.

A GALS test chip with three million transistors was implemented in 0.25 μm technology [36]. The chip contains 25 GALS modules and occupies a total area of 25 mm^2. A design flow has been presented for automating the design of GALS chip. This is facilitated by using a library of self-timed elements which can be used

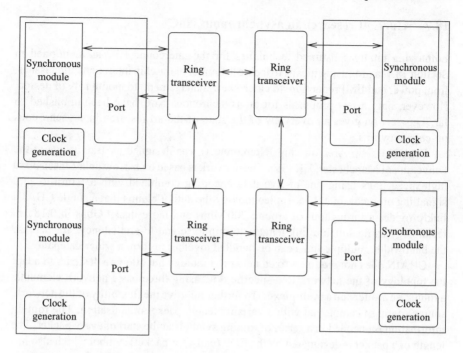

Figure 18.11   Ring architecture

to convert synchronous modules to GALS modules. The requirement of a pausable local clock led to the addition of a programmable delay element and Mutex element to the standard cell library. Timing verification was carried out hierarchically at three levels:

1   Inside the self-timed wrapper library;
2   Within each GALS module;
3   For the handshake signals between GALS modules.

The authors claim that since their GALS methodology requires a limited number of self-timed sub-circuits, most of the design process can be handled by using customised design automation scripts.

In Reference 37, a study has been carried out to measure the performance and power consumption of the GALS methodology for a hypothetical super-scalar processor architecture. The results show that GALS design does not lower power consumption appreciably and the overheads of using multi-clocked synchronous blocks leads to a performance drop in the range of 5–15 per cent. It further says that voltage scaling techniques for each synchronous block would help bridge the performance gap. The authors of Reference 38, propose a strategy for optimally partitioning the synchronous logic into synchronous blocks for maximising power reduction. They report average power reduction of 30 per cent.

18.7   Current research in asynchronous NoC

Employing totally self-timed techniques for the interconnect is, as mentioned in Section 18.1, a promising means to tackle a number of on-chip interconnection issues, from power and EMI reduction, to clock skew management, to modularity of design. However, only a few proposals for an asynchronous NoC have been published so far. This section gives an overview of the state-of-the-art research in asynchronous on-chip networks.

CHAIN (CHip Area on-chip INterconnect) was designed by Bainbridge at the University of Manchester, UK [39]. The network is based on narrow delay-insensitive high-speed links using one-of-five data encoding combined with a return-to-zero signalling protocol. In a 0.35 μm technology the author claims that a single CHAIN link provides a throughput of around 700 Mbps and more than 1 Gbps in 0.18 μm technology using suitable link lengths to minimise end-to-end latency. To increase the bandwidth, multiple links can be bundled together to form a wider datapath.

CHAIN does not require a fixed network topology but allows a designer to adapt the topology of the network to a specific SoC using three basic network elements: a router, an arbiter and a multiplexer. To further improve the flexibility of the network, source routing is employed with a variable length packet organisation. The routing information is encoded in a series of routing symbols at the start of every packet. The length of a packet is designated by the EOP (end-of-a-packet) symbol which also has a function to tear down the route set by the header of the packet.

CHAIN implements a split transaction protocol typically employing two separate networks for the command and response in order to improve the performance of the interconnect. Also, the network supports atomic sequences of multiple commands.

NEXUS is another asynchronous on-chip network developed at Fulcrum Microsystems, USA [40]. Their approach is based on a 16-port, 36-bit asynchronous crossbar that connects synchronous modules through asynchronous channels and clock-domain converters. Nexus is a quasi delay-insensitive (QDI) on-chip interconnect infrastructure using one-of-four encoding and pre-charge domino logic. It also supports a split transaction protocol with a request burst going out and a completion burst returning. Implemented in a 0.13 μm low-voltage CMOS process Nexus runs at 1.35 GHz and exhibits latency of 2 ns.

Liljeberg *et al.* from the University of Turku, Finland proposed a self-timed ring architecture as a replacement for on-chip buses [41]. They implemented a 12-stage bi-directional ring network with 36 pipeline sections. The network employs a two-phase signalling protocol between stages to accommodate relatively long wire segments with less transitions within a transaction cycle, and a four-phase signalling protocol for internal control within a stage to enable design of fast and relatively simple control logic circuitry. The datapath is encoded using a standard single-encoding scheme.

The authors compared three closely related structures: a bi-directional ring, a bi-directional folded ring and a bi-directional open ring against different types of traffic. The simulation results show that the peak throughput of a single segment in one direction is between 0.8 and 1.0 Gwords/s in 0.18 μm technology with the segment lengths of 1 and 4 mm, respectively. The maximum measured throughput of the whole ring is 6.61 Gwords/s.

An asynchronous ring-based network was also proposed in Reference 42.

A related research direction is that of asynchronous communication mechanisms (ACMs [43]), which implement various degrees of synchronisation between communicating parties, from fully independent to fully interlocked. ACMs are an important mechanism, especially in the form which does not block the reader nor the writer of a communication channel, in order to implement hard real-time with asynchronous techniques. In other words they go beyond QoS-based soft real-time, in order to provide full timing guarantees to safety-critical systems. The cost is slightly higher than traditional FIFO-based mechanisms, which block the reader when empty and the writer when full, and the performance is comparable to that of traditional FIFOs [44].

18.8 QoS for NoCs

A modern SoC may consist of many different components and IP blocks interconnected by an NoC. These components can exhibit disparate traffic characteristics and constraints, such as requirements for guaranteed throughput and bounded communication latency.

As an example consider a connection between a video camera and a Motion Picture Experts Group (MPEG) encoder. Such a connection has to maintain a constant throughput with bounded jitter (variation in end-to-end latency) in order to support the required quality of the system. If the camera and the encoder are a part of a complex SoC interconnected by an on-chip network, the connection has to share the network bandwidth with the rest of the traffic. In order to maintain the quality of the system, the network has to provide the required bandwidth for the connection at any given time.

It is therefore essential for a modern NoC to support QoS in order to accommodate such components sharing the same communication medium. Furthermore, the ability of an NoC to provide guaranteed services enables a designer to make critical timing decisions early in the design process thus avoiding unnecessary design iterations [45].

In synchronous networks QoS is often provided by time division multiplexing (TDM). TDM partitions the time axis into time-slots where each time-slot presents a unit of time in which a single flow can transmit data over a physical channel. Guaranteed throughput is provided by reserving a proportion of time-slots for a particular flow. For example, if a connection requires 50 per cent of the available bandwidth, a network has to ensure that every other time-slot is available for that particular connection. Reserved slots traverse the network in a well-synchronised manner without having to arbitrate for the output link with the rest of the traffic. The Aethereal NoC developed at Philips and the Sonics on-chip bus employ a TDM technique to support guaranteed throughput [46,47].

Although TDM provides a high level of QoS it is unsuitable for asynchronous implementation because it requires global synchronisation between network elements.

Another way to provide QoS is to employ a packet scheduling algorithm that will prioritise input traffic in terms of the level of QoS required. In Reference 48 Felicijan and Furber proposed a QoS architecture suitable for asynchronous on-chip networks using virtual channels [49] where a connection with QoS requirements uses a virtual channel in order to reserve buffer space. The bandwidth of the network is distributed using a priority-based asynchronous arbiter according to the priority level of each individual virtual channel. The same authors also proposed a low latency asynchronous arbiter suitable for QoS applications [50] which overcomes the problem of allowing a contender to obtain over 50 per cent of the resource allocation in a self-timed system by using downstream knowledge to trigger the arbitration.

18.9   Case study: the ASPIDA NoC

ASPIDA is a project which aims to demonstrate the feasibility of designing and delivering an asynchronous IP in a portable, reusable manner.

With regard to asynchronous networks on chip, one of the main contributions of this project is the creation of an asynchronous interface specification aiming to become the asynchronous equivalent of WISHBONE [51], a synchronous SoC interconnection architecture for reusable IP cores. This interface specification is heavily influenced from both CHAIN and WISHBONE. Most of the interface signals are named following the WISHBONE convention. The major difference from WISHBONE is that the model for inter-core communication is based on split transactions. Thus there are two separate interconnect fabrics: one for commands and another for responses. The asynchronous request–acknowledge handshake signals make this interface specification robust and easy to reuse without a need to verify complex timing assumptions.

ASPIDA will produce a demonstrator chip containing an asynchronous system on a chip. Figure 18.12 shows its main components, which include an asynchronous open-source DLX processor core (obtained using the desynchronisation techniques described above), two memories dedicated for instructions (IMEM) and data (DMEM), a test interface controller (TIC) for initialisation/debugging and three interfaces with the external world: a synchronous WISHBONE interface for connection to synchronous peripherals (WB), a novel asynchronous, general-purpose interface (GP) and a 'bare' CHAIN interface (BC) for adding more CHAIN networks and/or debugging the interconnection.

In the ASPIDA system there are three initiators (masters) and five targets (slaves). The DLX core has a Harvard architecture, so it has two initiators in the interconnection, one for the instruction port and another for the data port. The remaining initiator is attached to the TIC so that it can access the memories and the external interfaces, as well as being able to test the interconnection.

Two of the five targets are the system's memory: 1K words of SRAM each. Their main purpose in the system is to provide fast, on-chip memory space for the DLX so that it can run simple programs at a high speed without the need to access a slow external memory and without taking-up too much area. The remaining targets

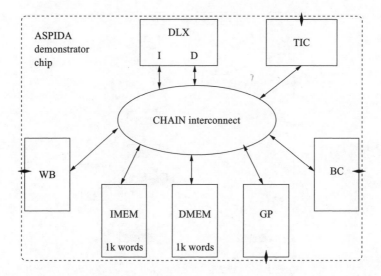

Figure 18.12   Top-level block diagram of the ASPIDA demonstrator chip

essentially extend the interconnection off-chip, so that synchronous (WB port) or asynchronous (GP port) 'peripherals' can be accessed by the processor.

In the above system the most common situation is that the processor instruction port will communicate with the instruction memory and the data port with the data memory. All other initiator–target communications should be made possible by the interconnect, but their performance is not as crucial. The above observation led to the interconnection architecture shown in Figure 18.13 which optimises the two commonly used paths.

18.9.1   Design for testability features of the interconnect

One of the major challenges of the project was to include circuits that guarantee full test coverage, as a typical synchronous system would. Although well-known techniques can be applied to the circuits implementing the processor datapaths and the SRAMs, for the delay insensitive circuits of CHAIN, these techniques would dramatically increase the area and reduce the circuit speed.

The approach followed for the interconnect fabric is to insert scan-latches in the acknowledge paths of the CHAIN pipeline stages, so that the common input of the parallel C elements can be controllable. In comparison, the standard approach would require a scan-latch for each C element. Thus our approach resulted in considerable area savings and performance improvement.

The test patterns are manually generated for each of the four basic building blocks of CHAIN interconnects and a computer program has been developed which, given the network topology and the patterns for each of the component types, produces a full test sequence that gives over 99.5 per cent stuck-at fault coverage for the interconnect. The test strategy is explained in References 52 and 53.

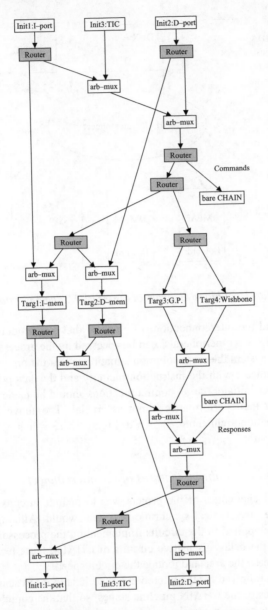

Figure 18.13   The ASPIDA interconnection network

18.9.2   Implementation

Since one of the aims of ASPIDA is to produce portable, reusable asynchronous IP, the implementation is standard cell based, using a 0.18-μm technology. It should be noted that area and performance improvements can be gained by using even a small number of special asynchronous cells.

The processor core is built using the desynchronisation techniques described earlier. The interconnection is drawn as a schematic and later passed on to a standard synthesis tool to optimise the gate mapping.

For the results presented here, only the network was placed and routed. The area of the network, including the interfaces, is $0.63 \, mm^2$ with the cell density set at approximately 70 per cent. Actually, most of the above area is taken up by the interfaces; the actual interconnection fabric occupies less than 15 per cent of the total core area.

18.9.3 Evaluation

In order to evaluate the performance of the ASPIDA interconnect, two sets of simulations were conducted creating different traffic scenarios in the network.

Each master interface is connected to a traffic generator programmable to inject different types of traffic into the network. The masters generate two types of commands: read and write in a proportion of 70 and 30 per cent, respectively. The length of a write command packet is ten bytes and the length of a read command packet is six bytes. Similarly, each slave interface is connected to a dummy client to emulate the behaviour of a slave client. The length of a response packet depends on the type of the command received from a master and represents six bytes in the case of a read command and two bytes in the case of a write command. Therefore, the total number of bytes transferred in a single command–response cycle is 12.

The performance of the network was assessed by measuring throughput and latency of each individual master. The throughput corresponds to the number of bytes a single master is able to transmit and receive per unit of time, and the latency represents the time between when a master sends a command through the network and when it receives the response back from the client. Note that the dummy clients exhibit zero service time thus the latency represents only the time packets spend traversing the network.

As mentioned above, two sets of simulations were conducted in order to evaluate the performance of the network. The first set was designed to mimic the traffic characteristics of the environment the network was designed for. In this case masters IP and DP were set to generate commands for slaves IM and DM, respectively, as fast as possible. Furthermore, master TIC was set to generate commands to IM and DM in order to disrupt the throughput of the masters IP and DP. The TIC generates commands at random intervals – according to the exponential distribution function – with an average packet rate between 0 and 100 per cent of the physical bandwidth. Figure 18.14 shows how the traffic generated by the TIC affects the throughput of the IP.

When there is no TIC traffic present the network dedicates the whole physical bandwidth to the IP master, however when the TIC traffic is introduced, the IP throughput decreases almost linearly until it reaches around 50 per cent of the maximum bandwidth. Figure 18.14 shows that the network guarantees approximately half of the physical bandwidth to the IP master. In terms of latency Figure 18.15 shows a similar situation. Note that the masters do not have any buffering capabilities, thus the results shown in Figure 18.15 do not include any queuing time.

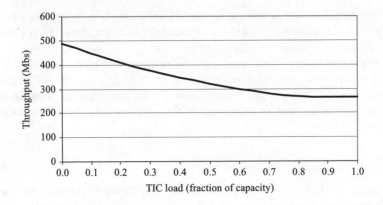

Figure 18.14   The impact of the TIC load on the throughput of the IP. Note that the TIC has the same impact on the throughput of the DP master

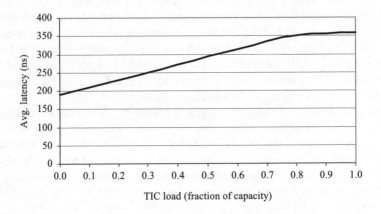

Figure 18.15   Latency of the IP master versus the TIC load

The second set of simulations generates a more generic traffic scenario in the network. In this case all three masters (IP, DP and TIC) were programmed to issue command packets to randomly chosen slave targets (WB, IM, DM, BC and GP) for every transaction cycle. Furthermore, each master generates commands exponentially distributed across the time axis. The throughput and latency of each master was measured against different traffic loads. Figures 18.16 and 18.17 show the throughput and the latency of the IP, DP and TIC masters, respectively.

It is interesting to note that the relative ranking of the three masters in this set of simulations reflects the topology of the network. As the targets are selected randomly, most of the traffic will follow the main trunk of the fabric, as shown in Figure 18.13. So IP will be 'fighting' with TIC and the combined flow with DP. Thus the performance of DP is significantly better when the network is congested. Among IP and TIC, the former is connected to IM through a relatively short route, while the latter does not

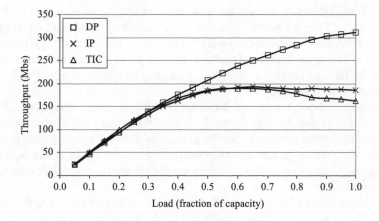

Figure 18.16   Throughput of the IP, DP and TIC masters

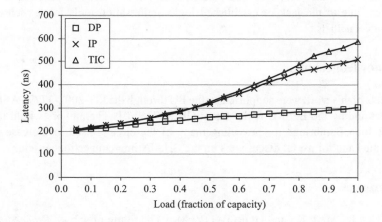

Figure 18.17   Latency of the IP, DP and TIC masters

have such special connections. As in one out of five times the IP sends packets to IM, it manifests, on aggregate, a somewhat better performance than TIC.

18.10   Conclusions

With the Semiconductor Industry Association (SIA) roadmap pointing to increasing clock frequencies and smaller feature sizes, distributing a global clock across an entire chip is becoming less and less feasible. While progress in clocking structures continues, several research groups convincingly argue that a complete paradigm shift would ensure significant advantages.

We first surveyed techniques which minimally depart from the synchronous scheme, or in some sense are loosely coupled synchronous schemes, as in the case of desynchronisation. These are the most likely candidates to be picked up first by design teams. They are very easy to use, fully automated or at least tool-supported, but also provide little incentive beyond EMI reduction. The latter is a significant issue only for mixed signal digital and analogue circuits, for very cheap integrated circuits, due to the reduced packaging cost, and in security applications, due to the reduced data-correlated emissions.

We then considered stoppable clocks and GALS schemes, which retain a fully synchronous design methodology for the LS blocks, while using standardised wrappers, produced by module generators, for the interfacing. They provide more independence between the modules, in that the overall performance need not be determined by the slowest stage, but may exhibit metastability, thus resulting in potentially unpredictable performance.

Finally, we looked at truly asynchronous NoCs, which again, due to the need for standardised design flows, use pre-defined modules and module generators, whose output is then assembled to determine the overall network logic. These asynchronous structures have the best power and performance, but are often less efficient in terms of area, due to the lack of established logic optimisation tools for asynchronous gate-level netlists.

Acknowledgements

Aspects of this work were supported by the EU through the IST-2002-37796 ASPIDA project and by the EPSRC under Grant GR/R47363/01. The authors would like to thank John Bainbridge for providing many of the circuits used in the case study presented and for useful discussions on the CHAIN interconnection system.

References

1   HO, R., MAI, K., and HOROWITZ, M.: 'The future of wires', *Proceedings of the IEEE*, 2001, **89**(4), pp. 490–504

2   TIWARI, V., SINGH, D., RAJGOPAL, S., MEHTA, G., PATEL, R., and BAEZ, F.: 'Reducing power in high-performance microprocessors'. Proceedings of the 35th annual conference on *Design automation*, 1998, pp. 732–737

3   KURD, N., BARKATULLAH, J., DIZON, R., FLETCHER, T., and MADLAND, P.: 'Multi-GHZ clocking scheme for Intel Pentium 4^{TM} microprocessor'. IEEE international solid-state circuits conference, 2001, pp. 404–405

4   STEVENS, K.S., ROBISON, S.V., and DAVIS, A.L.: 'The post office – communication support for distributed ensemble architectures'. Sixth international conference on *Distributed computing systems*, 1986

5   STEVENS, K., GINOSAR, R., and ROTEM, S.: 'Relative timing'. Proceedings of the IEEE international symposium on *Advanced research in asynchronous circuits and systems* (IEEE Computer Society Press, April 1999)

6   PAVER, N.C., DAY, P., FARNSWORTH, C., JACKSON, D.L., LIEN, W.A., and LIU, J.: 'A low-power, low-noise configurable self-timed DSP'. Proceedings

of the IEEE international symposium on *Advanced research in asynchronous circuits and systems* (IEEE Computer Society Press, Los Alamitos, CA, March 1998)

7 FURBER, S.B. and DAY, P.: 'Four-phase micropipeline latch control circuits', *IEEE Transactions on VLSI Systems*, 1996, **4**(2), pp. 247–253

8 VAN BERKEL, K., JOSEPHS, M., and NOWICK, S.M.: 'Scanning the technology: applications of asynchronous circuits', *Proceedings of the IEEE*, 1999, **87**(2), pp. 223–233

9 CHU, T.-A.: 'Synthesis of self-timed VLSI circuits from graph-theoretic specifications'. PhD thesis, MIT, June 1987

10 MURATA, T.: 'Petri nets: properties, analysis and applications', *Proceedings of the IEEE*, 1989, pp. 541–580

11 RAJAN, B., and SHYAMASUNDAR, R.K.: 'Multiclock ESTEREL: a reactive framework for asynchronous design'. International conference on *Parallel and distributed processing symposium*, 2000, pp. 201–210

12 HALBWACHS, N.: 'Synchronous programming of reactive systems' (Kluwer Academic Publishers, Boston, MA, 1993)

13 EDWARDS, S., LAVAGNO, L., LEE, E.A., and SANGIOVANNI-VINCENTELLI, A.: 'Design of embedded systems: formal models, validation, and synthesis', *Proceedings of the IEEE*, 1997, **85** (3), pp. 366–390

14 MOUSAVI, M.R., LE GUERNIC, P., TALPIN, J.-P., SHUKLA, S.K., and BASTEN, T.: 'Modeling and validating globally asynchronous design in synchronous frameworks'. Proceedings of the conference on *Design automation and test in Europe*, 2004, pp. 384–389

15 KAHN, G.: 'The semantics of a simple language for parallel programming'. Proceedings of IFIP congress, August 1974

16 BUCK, J.T.: 'Scheduling dynamic dataflow graphs with bounded memory using the token flow model'. PhD thesis, U.C. Berkeley, 1993. UCB/ERL Memo M93/69

17 TALPIN, J.-P., LE GUERNIC, P., SHUKLA, S.K., GUPTA, R., and DOUCET, F.: 'Polychrony for formal refinement-checking in a system-level design methodology'. Third international conference on *Application of concurrency to system design*, 2003, pp. 9–19

18 BLUNNO, I., CORTADELLA, J., KONDRATYEV, A., LAVAGNO, L., LWIN, K., and SOTIRIOU, C.: 'Handshake protocols for desynchronization'. Proceedings of the international symposium on *Advanced research in asynchronous circuits and systems* (IEEE Computer Society Press, Los Alamitos, CA, April 2004)

19 MULLER, D.E., and BARTKY, W.C.: 'A theory of asynchronous circuits', in 'Annals of computing laboratory of Harvard University' (Harvard University Press, 1959), pp. 204–243

20 VANBEKBERGEN, P., LIN, B., GOOSSENS, G., and DE MAN, H.: 'A generalized state assignment theory for transformations on signal transition graphs'. Proceedings of the IEEE international conference on *Computer-aided design*, 1992, pp. 112–117

21   CORTADELLA, J., KISHINEVSKY, M., KONDRATYEV, A., LAVAGNO, L., and YAKOVLEV, A.: 'Complete state encoding based on the theory of regions'. Proceedings of the IEEE international symposium on *Advanced research in asynchronous circuits and systems*, March 1996

22   SIEGEL, P., DE MICHELI, G., and DILL, D.: 'Automatic technology mapping for generalized fundamental mode asynchronous designs'. Proceedings of the IEEE *Design automation* conference, June 1993

23   CORTADELLA, J., KISHINEVSKY, M., KONDRATYEV, A., LAVAGNO, L., PASTOR, E., and YAKOVLEV, A.: 'Decomposition and technology mapping of speed-independent circuits using Boolean relations', *IEEE Transactions on Computer-Aided Design*, 1999, **18**(9), pp. 1221–1236

24   XANTHOPOULOS, T., BAILEY, D., GANGWAR, A., GOWAN, M., JAIN, A., and PREWITT, B.: 'The design and analysis of the clock distribution network for a 1.2 GHZ Alpha microprocessor'. IEEE international *Solid-state circuits* conference, 2001, pp. 402–403

25   MONTEK SINGH, JOSE A. TIERNO, ALEXANDER RYLYAKOV, SERGEY RYLOV, and STEVEN M. NOWICK: 'An adaptively-pipelined mixed synchronous–asynchronous digital FIR filter chip operating at 1.3 Gigahertz'. Proceedings of the international symposium on *Advanced research in asynchronous circuits and systems*, April 2002, pp. 84–95

26   AUSTIN, T., BLAAUW, D., MUDGE, T., and FLAUTNER, K.: 'Making typical silicon matter with razor', *IEEE Computer*, 2004, **37**(3), pp. 41–49

27   MEKIE, J., SUPRATIK CHAKRABORTY, and SHARMA, D.K.: 'Evaluation of pausible clocking for interfacing high speed IP cores in GALS framework'. Proceedings of international conference on *VLSI Design*, 2004, pp. 559–564

28   KENNETH Y. YUN and AYOOB E. DOOPLY: 'Pausible clocking-based heterogeneous systems', *IEEE Trans. Very Large Scale Integr. Syst.*, 1999, **7**(4), pp. 482–488

29   CHAKRABORTY, S., MEKIE, J., and SHARMA, D.K.: 'Reasoning about synchronization issues in GALS systems: a unified approach'. Invited paper in Proceedings of the workshop on *Formal methods in GALS architectures (FMGALS)*, Formal Methods Europe symposium, September 2003

30   GURKAYNAK, F.K., VILLIGER, T., and OETIKER, S.: 'An introduction to the GALS methodology at ETH Zurich'. Proceedings of the *Formal methods for globally asynchronous locally synchronous (GALS) architecture (FMGALS)*, September 2003, pp. 32–41

31   YUN, K.Y., and DILL, D.L.: 'Automatic synthesis of extended burst-mode circuits: part I (specification and hazard-free implementations)', *IEEE Transactions on Computer-Aided Design*, 1999, **18**(2), pp. 101–117

32   YUN, K.Y., and DILL, D.L.: 'Automatic synthesis of 3D asynchronous state machines'. Proceedings of the 1992 IEEE/ACM international conference on *Computer-aided design* (IEEE Computer Society Press, Washington, 1992), pp. 576–580

33   OETIKER, S., VILLIGER, T., GURKAYNAK, F.K., KAESLIN, H., FELBER, N., and FICHTNER, W.: 'High resolution clock generators for

globally-asynchronous locally-synchronous designs'. Handouts of the second *Asynchronous circuit design* workshop, ACiD 2002, January 2002

34 MOORE, S., TAYLOR, G., MULLINS, R., and ROBINSON, P.: 'Point to point GALS interconnect'. Eighth international symposium on *Advanced research in asynchronous circuits and systems*, 2002

35 VILLIGER, T., GURKAYNAK, F.K., OETIKER, S., KAESLIN, H., FELBER, N., and FICHTNER, W.: 'Multi-point interconnect for globally-asynchronous locally synchronous systems'. Handouts of the second *Asynchronous circuit design* workshop, ACiD, January 2002

36 OETIKER, S., GURKAYNAK, F.K., VILLIGER, T., KAESLIN, H., FELBER, N., and FICHTNER, W.: 'Design flow for a 3-million transistor GALS test chip'. Handouts of the third *Asynchronous circuit design* workshop, ACiD, January 2003

37 IYER, A., and MARCULESCU, D.: 'Power and performance evaluation of globally asynchronous locally synchronous processors'. Proceedings of the 29th annual international symposium on *Computer architecture* (IEEE Computer Society, Washington, 2002), pp. 158–168

38 HEMANI, A., MEINCKE, T., KUMAR, S., *et al.*: 'Lowering power consumption in clock by using globally asynchronous locally synchronous design style'. Proceedings of the 36th ACM/IEEE conference on *Design automation* (ACM Press, New York, 1999), pp. 873–878

39 BAINBRIDGE, J., and FURBER, S.: 'CHAIN: a delay-insensitive chip area interconnect', *IEEE Micro*, 2002, **22**(5), pp. 16–23

40 LINES, A.: 'Asynchronous interconnect for synchronous SoC design', *IEEE Micro*, 2004, **24**(1), pp. 32–41

41 LILJEBERG, P., PLOSILA, J., and ISOAHO, J.: 'Self-timed architecture for SoC applications'. Proceedings of the 16th IEEE international *SoC* conference, September 2003, pp. 359–361

42 CARRION, C., and YAKOVLEV, A.: 'Design and evaluation of two asynchronous token ring adapters'. Technical report, CS-TR: 562, Department of Computing Science, University of Newcastle, 1997

43 SIMPSON, H.: 'Four-slot fully asynchronous communication mechanism', *Proceedings of the IEE*, 1990, **137**(E)(1), pp. 17–30

44 XIA, F., YAKOVLEV, A., CLARK, I., and SHANG, D.: 'Data communication in systems with heterogeneous timing', *IEEE Micro*, 2002 **22**(6), pp. 58–69

45 GOOSSENS, K., VAN MEERBERGEN, J., PETEERS, A., and WIELAGE, P.: 'Networks on silicon: combining best effort and guaranteed services'. *Design automation and test in Europe* conference (DATE'02), March 2002, pp. 423–425

46 RIJPKEMA, E., GOOSSENS, K., DIELISSEN, J., *et al.*: 'Trade offs in the design of a router with both guaranteed and best-effort services for networks on chip', *IEE Proceedings: Computers and Digital Technique*, 2003, **50**(5), pp. 294–302

47 WINGARD, D.: 'Micronetwork-based integration for SoCs'. Proceedings of the *Design automation* conference, DAC 2001, Las Vegas, USA, June 2001

48   FELICIJAN, T., and FURBER, S.: 'Quality of service (QoS) for asynchronous on-chip networks'. *Formal methods for globally asynchronous locally synchronous architecture* (FMGALS 2003), September 2003

49   DALLY, W.J.: 'Virtual-channel flow control', *IEEE Transactions on Parallel Distribution Systems*, 1992, 3(2), pp. 194–205

50   FELICIJAN, T., BAINBRIDGE, J., and FURBER, S.: 'An asynchronous low latency arbiter for quality-of-service (QoS) applications'. Proceedings of the 15th IEEE international conference on *Microelectronics*, December 2003, pp. 123–126

51   OpenCores Organization. WISHBONE system-on-chip (SoC) interconnection architecture for portable IP cores. http://www.opencores.org/wishbone/doc/specs/wbspec_b3.pdf, 2002

52   EFTHYMIOU, A., SOTIRIOU, C., and EDWARDS, D.: 'Automatic scan insertion and pattern generation for asynchronous circuits'. *Design automation and test in Europe* conference (DATE'04), February 2004, p. 672

53   EFTHYMIOU, A., BAINBRIDGE, J., and EDWARDS, D.: 'Adding testability to an asynchronous interconnect for globally-asynchronous, locally-synchronous systems-on-chip'. IEEE Asian *Test* symposium (IEEE CS Press, Los Alamitos, CA, 2004), pp. 20–23

Part VII

Simulation and verification

Chapter 19

Covalidation of complex hardware/software systems

Ian G. Harris

19.1 Hardware/software systems

A hardware/software system can be defined as one in which hardware and software must be designed together, and must interact to properly implement system functionality. To be considered a hardware/software system, the design of hardware and software components must be dependent on each other. For example, a word processor application for a desktop computer is not a hardware/software system because the word processor software design is largely independent of the underlying hardware platform on which it is executed. The vast majority of practical electronic devices involve closely interacting hardware and software components and can therefore be classified as hardware/software. This includes virtually all consumer electronics (e.g. cellphones, MP3 players, automotive systems), medical electronics (e.g. CT scanners, heath monitoring devices) and military electronics (e.g. guidance systems, vehicle control systems).

The main benefit of using both hardware and software components is that using a variety of components enables tight design constraints to be satisfied by using components whose properties most exactly match the given design requirements. Hardware/software systems are built from a wide range of hardware and software components which are associated with different trade-offs in design characteristics, such as performance, cost and reliability. Typical computational components used in hardware/software codesign include the following.

- *Interpreted/scripting software*: This class includes code in scripting language such as Perl and CSH, and also interpreted languages such as Java and Matlab. Components of this type have the advantage of low development costs because many implementation details are hidden from the programmer and addressed by the interpreter (e.g. dynamic memory management). These components are also portable to many different hardware platforms and they often allow higher security because the interpreter can evaluate security concerns at runtime. The main disadvantages of interpreted components is that their execution is slow due to the overhead of the interpreter. Interpreted components also give the programmer very little control of resource usage such as memory, power and performance. In the context of tightly constrained systems, this is a serious weakness.

- *Compiled software*: This class includes software in languages like C and C++. Development costs are a bit higher for these languages than for interpreted languages because details such as dynamic memory management are left to the programmer. However this also means that the programmer has more direct control over the resource usage. For this reason, compiled software is very useful in the development of systems with tight design constraints.

- *Assembly code*: Assembly coding requires the programmer to write assembly code directly without the help of a compiler. Assembly coding incurrs a high development cost since low-level details of the software execution must be determined by the programmer. Assembly coding has the great advantage that it enables cycle-accurate control of processor resourses which is often necessary when interfacing with hardware components.

- *Microprocessor*: If any software component is used in the system then at least one microprocessor is needed to execute the software. Microprocessors come in several different varieties. General purpose central processing units (CPUs) have significant resources in terms of performance, memory and instruction set completeness, but they are also expensive and have large power requirements. Smaller embedded microprocessors and digital signal processors have sufficient resources to satisfy most embedded applications and are typically designed to consume less power. Microprocessor design is a difficult task so hardware/software system designers purchase pre-designed microprocessors to include in their systems. The microprocessor may also be pre-fabricated so that it only needs to be installed into a printed circuit board.

- *Field-programmable gate array*: Field-programmable gate arrays (FPGAs) are pre-fabricated integrated circuits which contains interconnect and generic functional units which can be 'programmed' by blowing fuses or setting memory bits. FPGAs offer the programming flexibility of software together with much of the performance advantage of a hardware implementation. Design costs are significantly higher for FPGAs than software because hardware design is complex, but no fabrication cost is required since the FPGA is pre-fabricated.

- *Application-specific integrated circuit*: An application-specific integrated circuit (ASIC) is an integrated circuit which is designed and fabricated to perform a particular task. ASIC development requires not only a complex design process but also an expensive fabrication process. Developing an ASIC for a hardware/software system is the most expensive implementation option but it enables the most effective optimisation of performance, power, reliability and cost. The high cost of ASIC development can be amortised if the system being developed is a high-volume product. If the task to be implemented is a common one then a

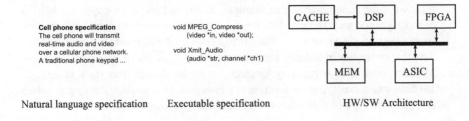

Figure 19.1 shows three boxes labelled: "Cell phone specification — The cell phone will transmit real-time audio and video over a cellular phone network. A traditional phone keypad ..." under "Natural language specification"; "void MPEG_Compress (video *in, video *out); void Xmit_Audio (audio *str, channel *ch1)" under "Executable specification"; and a block diagram with CACHE, DSP, FPGA, MEM, ASIC under "HW/SW Architecture".

Figure 19.1   Hardware/software codesign overview

pre-designed core or a pre-fabricated ASIC may be available from an appropriate vendor.

19.1.1   Codesign stages

In order to understand the errors which appear in hardware/software systems it is useful to understand the basic steps in the codesign process. An overview of the codesign process is depicted in Figure 19.1. The goal is to generate a hardware/software architecture from a specification. Figure 19.1 shows a subset of the process for a cell phone design with video processing features. The figure shows a natural language specification of the design, an executable specification and a final hardware software architecture which is the result of the codesign process.

Here we give an outline of the principle stages of the codesign process to provide context for our discussion of design errors. The ordering of the application of design stages is not fixed. In fact, design stages are often intertwined in practice. A more complete survey of the codesign process can be found in Reference 1.

- *Natural language specification*: This is a description of the system behaviour which is intended to be read by a human designer. All systems should have a natural language specification to be used as a reference during the design process.
- *Executable system description*: An executable system description is a refinement of the specification which is complete enough to enable simulation of the system being designed. The key feature of such a description is that it is written using a language with precisely defined execution semantics.
- *System partitioning*: The tasks in the system description must be partitioned across a set of hardware and software components which will be used to implement each task. For example, suppose we wish to design a cell phone which captures video and encodes it in moving pictures expects group (MPEG) format in real-time. MPEG encoding is a compute-intensive task so it might be mapped to an ASIC designed explicitly for that purpose. Other cell phone tasks like the processing of the audio data might be implemented as a C program executing on a digital signal processor.
- *Communication synthesis*: The different tasks in a system description are never completely independent of each other, so they must exchange data in order to

satisfy system functionality. For example, in our cell phone example, the MPEG encoding block must receive data from a camera and receive control information from a keypad via a processor. The MPEG encoder block must also send encoded data to a trasmitter in order to send the data to the cell phone network.

- *Memory synthesis*: A memory hierarchy must be designed to store intermediate data while satisfying performance, power and cost goals. This step involves selecting the type of memories to be used and their configuration.

From the description of the stages of codesign it should be clear that there are many places where mistakes can be made and design errors can be introduced. Design errors incur design cost and increase time-to-market because they require effort to detect and redesign effort to correct. To reduce the impact of design errors covalidation must be efficient and it must be applied frequently so that errors are detected early in the process.

19.1.2   Covalidation issues

Several features of the hardware/software problem make it unique and difficult. Each covalidation technique addresses these issues to different degrees.

19.1.2.1   Component reuse

Component reuse is an established design paradigm in the hardware domain and in the software domain. Hardware reuse may involve the use of pre-fabricated integrated circuits which can be general-purpose processors or application-specific components. The use of pre-designed components has the great advantage that the design and validation of the components is not the responsibility of the system designer. This has the potential to greatly reduce the system design and test effort required. The main disadvantage of reuse is that the system designer must completely understand the behaviour of components that he/she did not design in order to integrate them into the system.

19.1.2.2   Varied design styles

There are many different ways to design any component in both the hardware and software domains. Each different approach to design leads to a different set of likely errors which must be identified during testing. Each design style also limits test access in different ways, requiring the use of different testing techniques. In software for instance code might be written in an object-oriented style or not. Objects often contain data and functions/methods which are 'private' and cannot be accessed externally. Data privacy must often be eliminated for the purposes of testing in order to observe incorrect results during testing. A hardware behaviour might be described procedurally as is common in software, or as a finite state machine. The choice of coverage metric should be guided by the hardware design style, so a transition coverage might be applied to a state machine while branch coverage might be applied to a procedural description.

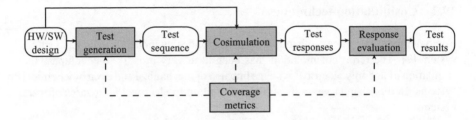

Figure 19.2   Hardware/software covalidation steps

19.1.2.3   Varied design abstraction levels

Hardware/software components are described at a variety of abstraction levels to match the need for design quality and flexibility. On the software side, the bulk of the code might be written in a high-level language such as Java while performance-critical sections are written in assembly. Behavioural hardware descriptions in Verilog might be used to describe most hardware blocks while power-critical components are described at the transistor-level to implement dynamic voltage scaling.

19.1.3   Stages of covalidation

An outline of the steps involved in the covalidation process is shown in Figure 19.2. The covalidation process starts with a hardware/software design which is to be tested. Covalidation may be performed on an incomplete design, so the design may be only an executable specification, or the final architecture may be available. Covalidation involves three major steps, **test generation**, **cosimulation** and **response evaluation**. The test generation process typically involves a loop in which the test sequence is progressively evaluated and refined until coverage goals are met. Cosimulation (or emulation) is then performed using the resulting test sequence, and the cosimulation test responses are evaluated for correctness. A key component of test generation is the set of **coverage metrics** used which abstractly describe the testing goals. The coverage metric is needed to provide detection goals for the automatic test generation process, and the coverage metric enables the error detection qualities of a test sequence to be evaluated. Response evaluation is also a bottleneck because it typically requires manual computation of correct responses for all test stimuli.

In this chapter we describe the stages of covalidation involved with cosimulation, test generation, response evaluation and the coverage metrics which support the covalidation process. The techniques for cosimulation are summarised in Section 19.2. We describe coverage metrics used to describe design defects in Section 19.3, as well as the automatic test generation techniques which are based on those coverage metrics in Section 19.4. Section 19.5 describes automatic approaches to evaluate test responses, including assertions and self-checkers. Conclusions and future directions of the field are discussed in Section 19.6.

19.2   Cosimulation techniques

Cosimulation is the process of simulating disparate design models together as a single system [2]. The term 'cosimulation' has been used very broadly to encompass the simulation of not only electrical systems but also of mechanical and even biochemical systems. In this chapter we will limit our definition to electrical hardware/software systems.

The goal of cosimulation is to model the behaviour of a hardware/software system when it is driven by a set of input stimuli. The characteristics of the behaviour which are modelled determine the cosimulator complexity and accuracy. Cosimulation can be used to estimate 'dynamic' characteristics of the system, those charcteristics which vary depending on the input sequence applied during execution. Minimally cosimulation should be **functionally correct**; it should provide the correct output sequence for a given input sequence. In addition to functional correctness, cosimulation is used to estimate timing, power, communication bandwidth/latency and memory use.

19.2.1   Timing correctness

The **timing correctness** of a simulation describes the accuracy with which each simulation event is known. A functionally correct simulation of an adder will correctly perform addition and a timing correct simulation will additionally indicate the delay required to perform addition. The required accuracy of the timing information has a strong impact on simulator complexity and performance. The timing accuracy of cosimulation tools can be organised in the following categories.

1   *Pico-second accurate simulation*: This type of model has the highest accuracy and the lowest performance.
2   *Cycle-accurate simulation*: This model provides accurate register contents at each clock cycle boundary.
3   *Transaction-level simulation*: In a transaction-level model (TLM), the details of communication among computation components are separated from the details of computation components. Details of communication and computation are hidden in a TLM and may be added at later design stages. TLMs speed up simulation and allow exploring and validating design alternatives at a higher level of abstraction.

19.2.2   Abstraction level

The modelling ability of cosimulation is strongly influenced by the **abstraction level** at which cosimulation is performed. Simulation at a high level of abstraction means that few internal details about each operation are used during simulation, while a low level of abstraction means that detailed structural information is available. Low level simulation has access to more detail which can be used to generate more accurate estimates of design characteristics such as timing and power. Internal structural detail is also useful during debugging in order to locate the source of a design error. This is often the case while debugging embedded software running on a processor. A software bug may manifest itself as incorrect values in hidden registers inside the processor,

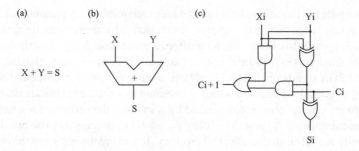

Figure 19.3   Abstractions of an adder, (a) add instruction, (b) adder block, (c) adder, gate-level

so that the availability of accurate internal register values is needed. However, the use of information detail for accurate estimation incurs a performance penalty on the cosimulation process.

Figure 19.3 shows three abstraction levels at which an adder can be simulated. Figure 19.3(a) shows a single add instruction. The behaviour of the instruction is clear but the delay is largely unknown without further information. The documentation for the processor on which the statement is executed may indicate the number of clock cycles required to perform an addition, so cycle-accurate simulation is possible at this level. Figure 19.3(b) shows a register-transfer level (RTL) adder block. Documentation on this component will provide the maximum delay from an input to an output in terms of picoseconds. In a ripple-carry adder the carry chain is the longest path, so the carry chain delay would indicate the worst-case adder delay. The most accurate simulation is possible using the gate-level description in Figure 19.3(c) which shows a single full adder stage in a ripple-carry adder. With this level of detail it is possible to determine the path which is critical for the specific input sequence. Although the carry chain is the longest path in a ripple-carry adder, an addition which does not impact the carry chain will have a shorter critical path. If two additions are performed sequentially, both of which result in no carry chain activity ($0 + 1$ and $1 + 2$ for instance), then the carry chain does not impact the adder's performance.

19.2.3   Component simulation methods

A challenge of cosimulation is the efficient management of the interaction between components which are described with very different computational models. Simulation techniques exist for each type of component found in a hardware/software system in isolation but fundamental differences in abstraction level make the simulation techniques difficult to use together. For instance, digital hardware may be simulated at a relatively low level of abstraction using an event-driven simulator with pico-second accuracy. On the other hand, software may be written at a high level of abstraction (possibly in an interpreted language like Java) and a real-time operating system (RTOS) might be used to abstract performance issues from the programmer. Software 'simulation' could be performed by running the code on any processor which supports the RTOS and the interpreter for the language.

Many pure hardware simulators exist which can simulate a generic circuit description. The simulators vary in the level of abstraction used, from analogue circuit-level simulation up to behavioural simulation from a procedural design description. There also exist simulators of specific classes of hardware designs including memory simulators and FPGA simulators. These simulators achieve good performance and accuracy by using structural information which is common to a class of hardware designs.

Software simulation is accomplished by compiling the software for a target processor and simulating the processor using a model. The processor is also pre-designed and usually pre-fabricated intellectual property (IP). In order to preserve the confidentiality of IP design, detailed information required for simulation may not be provided. The following techniques are often used to model processors with varying levels of timing accuracy.

- *Instruction set simulation*: The contents of memory elements are correctly modelled at machine at instruction boundaries. Cycle-to-cycle timing effects such as pipeline stalls are ignored.
- *Host processor*: Rather than model the target processor, software can be compiled to a host processor and simulation is performed by executing the software as processes which communicate with hardware simulator processes. No processor model is needed but timing accuracy suffers because the software timing is not related to the timing of the actual target processor.
- *Bus functional model*: A bus functional model does not model the complete behaviour of the processor, but only the different bus cycles the processor can execute. For this reason it cannot be used to simulate and debug software components. Bus functional models are used to debug hardware and its interactions with the processor by replicating the processor's bus interactions.
- *Hardware modeller*: This describes the use of a real processor part as the hardware model. This technique can be applied to model any pre-fabricated hardware including processors as well as ASICs.

Subsets of the cosimulation problem are well studied [3] and a number of industrial tools exist which enable the cosimulation of a variety of system types. Managing the difficult trade-off between performance and timing accuracy is still a problem for large systems.

19.3   Coverage metrics

A coverage metric provides a fast approximation of the error detection ability of a given test sequence. When generating a test sequence an empirical evaluation of an existing test sequence is required to direct the process and to provide a goal for completion. In order to describe a coverage metric we must first define the concept of a 'design error'. A design error is the difference between the designer's intent and an executable specification of the design. The designer's intent is most commonly expressed as a natural language specification. An executable specification is a precise description of the design which can be simulated. Executable specifications are often expressed using high-level hardware/software languages. Design errors may range from simple syntax errors confined to a single line of a design description, to a fundamental misunderstanding of the design specification which may impact a large segment of the description. The nature of design errors is not well understood but several studies exist in the hardware domain [4,5] and in the software domain [6,7] which attempt to classify them.

The purpose of testing is to detect all possible design errors, so the correct way to evaluate a test sequence is to determine the fraction of all possible design errors which are detected by the sequence. This fraction, which we will refer to as 'error coverage', can be determined by injecting each possible set of design errors into a design and simulating the erroneous designs with the test sequence. If the output of an erroneous design differs from the output of the correct design then the corresponding set of design errors is detected by the test sequence. The error coverage is not computable in practice for two reasons. The first reason is the time complexity; the set of all possible design errors is far too large to consider for any reasonably sized system. The second reason is the unpredictability of design errors. The source of design errors is often cognitive, relating to the thought process of individual designers. The cognitive processes of designers, and humans in general, is not well understood and cannot be completely predicted. Human factors such as training, experience, intelligence and even emotional mood all have a bearing on the type and frequency of the design errors which will be present.

A coverage metric defines a set of 'coverage goals' which must be satisfied during simulation. Ideally, satisfaction of all coverage goals should indicate that all possible design errors are detected. A coverage metric can be used to evaluate a test sequence by determining the fraction of coverage goals that are satisfied when the design is simulated with the test sequence.

The coverage goals defined by a coverage metric are meant to approximate the detection requirements of potential design errors. The ideal property which a coverage metric should guarantee is that the satisfaction of all coverage goals during testing should ensure the detection of all design errors. The degree to which a coverage metric guarantees this property for is the measure of the effectiveness of the metric. The relationship between a coverage metric and the set of design errors is depicted in Figure 19.4. In Figure 19.4 the set of coverage goals is shown to be much smaller than the set of design errors. This is an essential property for a coverage metric since the main reason for using a coverage metric is to manage complexity. The mapping from coverage goals to design errors indicates the design errors which are detected when the coverage goal is satisfied. The 'uncovered errors' are those whose detection is not guaranteed by any of the coverage goals. A good coverage metric will produce very few uncovered errors.

19.3.1 Coverage metric classification

The majority of hardware/software codesign systems are based on a top-down design methodology which begins with a behavioural system description. As a result, the majority of covalidation fault models are behavioural-level coverage metrics. Existing

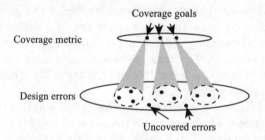

Figure 19.4   Relationship between coverage metrics and design errors

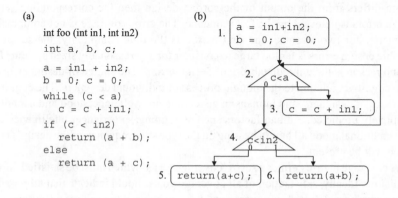

Figure 19.5   Behavioural descriptions: (a) textual description, (b) CDFG

coverage metrics can be classified by the style of behavioural description upon which the models are based.

Many of the coverage metrics currently applied to hardware/software designs have their origins in either the hardware [8] or the software [7] domains. As a tool to describe covalidation coverage metrics we will use the simple system example shown in Figure 19.5. Figure 19.5(a) shows a simple behaviour, and Figure 19.5(b) shows the corresponding control-dataflow graph (CDFG). The example in Figure 19.5 is limited because it is composed of only a single process and it contains no signals which are used to model real time in most hardware description languages. In spite of these limitations, the example is sufficient to describe the relevant features of many coverage metrics.

Table 19.1 presents a taxonomy of covalidation coverage metrics classified according to the abstraction level of the behaviour on which they operate. Each class of coverage metrics is described in the following subsections.

19.3.1.1   Textual coverage metrics

A textual coverage metric is one which is applied directly to the original textual behavioural description. The simplest textual fault model is the statement coverage metric introduced in software testing [7] which associates a potential error with each

Table 19.1   Taxonomy of covalidation coverage metrics

Model class	Coverage metric
Textual	Mutation analysis
	Statement coverage
Control-dataflow	Branch coverage
	Path coverage
	Domain coverage
	OCCOM
State machine	State coverage
	Transition coverage
Gate-level	Stuck-at coverage
	Toggle (bit flip) coverage
Application-specific	Microprocessor metrics
	User-defined
Interface	Communication faults
	Timing-induced faults

line of code, and requires that each statement in the description be executed during testing. This metric is very efficient since the number of potential faults is equal to the number of lines of code. Coverage evaluation for this metric is very low complexity, requiring only that an array be updated after each statement is executed. However, this coverage metric is accepted as having limited accuracy in part because fault effect observation is ignored. In spite of its limitations, statement coverage is well used in practice as a minimal testing goal.

Mutation analysis is a textual coverage metric which was originally developed in the field of software testing [6], but has also been applied to hardware validation [9]. In mutation analysis terminology, a 'mutant' is a version of a behavioural description which differs from the original by a single potential design error. A 'mutation operator' is a function which is applied to the original program to generate a mutant. A set of mutation operators describes all expected design errors. Coverage is computed by simulating all mutants and determining how many have output sequences which are not correct.

Since behavioural hardware descriptions share many features in common with procedural software programs, previous researchers [9] have used a subset of the software mutation operations presented in Reference 6. A typical mutation operation is 'arithmetic operator replacement' (AOR), which replaces each arithmetic operator with another operator. For example, if we assume the existence of four arithmetic operators, $+, -, *, /$, then applying AOR to the first line of the design in Figure 19.5(a) would produce three mutants. Each mutant would correspond to the replacement of the first line of code with a = in1 + in2, with a = in1 − in2, a = in1 * in2 and a = in1/in2, respectively. To compute coverage each mutant would be simulated with the test data and its results would be compared to the correct results. If the test data for the design in Figure 19.5(a) assigned the two inputs in1 and in2 to 2 and 0, respectively then the correct final value of 'out' would be 2. The three mutants produce the following final values for 'out': 2, 0, and 'no value' (a divide-by-zero error would occur). Of these three results, two are different from the correct result, so mutation coverage is 67 per cent.

19.3.1.2   Control-dataflow metrics

A number of coverage metrics are based on the traversal of paths through the CDFG representing the system behaviour. In order to apply these metrics to a hardware/software design, both hardware and software components must be converted into a CDFG description. The earliest control-dataflow coverage metrics include branch coverage and path coverage [7] models used in software testing.

The branch coverage metric associates potential faults with each direction of each conditional in the CDFG. Branch coverage requires that the set of all CDFG paths covered during covalidation includes both directions of all binary-valued conditionals. Branch coverage is commonly used in hardware validation and software testing, but it is also accepted to be insufficient to guarantee correctness alone. The time complexity of computing branch coverage is linear in the number of branch instructions in the program, and is therefore linear in the size of the description. Branch coverage evaluation is performed by recording the direction of each branch as it is taken during simulation.

The path coverage metric is more demanding than the branch coverage metric because path coverage requires the simulation of paths through the control flow which involve a number of individual branches. The assumption is that an error is associated with some path through the control-flow graph and therefore all control paths must be executed to guarantee error detection. The number of control paths can be infinite when the CDFG contains a loop as in Figure 19.5(b), so the path coverage metric may be used with a limit on path length [10]. Since the total number of control-flow paths grows exponentially with the number of conditional statements, several researchers have attempted to select a subset of all control-flow paths which are sufficient for testing. In dataflow testing, each variable occurrence is classified as either a definition occurrence or a use occurrence. Paths are selected which connect a definition occurrence to a use occurrence of the same variable. For example in Figure 19.5(b), node 1 contains a definition of signal a and nodes 2, 5, and 6 contain uses of signal a. In this example, paths 1, 2, 4, 5 and 1, 2, 4, 6 must be executed in order to cover both of these definition-use pairs. The dataflow testing criteria have also been applied to behavioural hardware descriptions [11].

The majority of control-dataflow coverage metrics consider the control-flow paths traversed with minimal constraints on the values of variables and signals. For example in Figure 19.5(b), in order to traverse path 1, 2, 3, the value of c must be minimally constrained to be less than a, but no additional constraints are required. This can be contrasted with variable/signal-oriented coverage metrics which place more stringent constraints on signal values to ensure fault detection. The domain analysis technique in software testing [7,12] considers not only the control-flow path traversed, but also the variable and signal values during execution. A domain is a subset of the input space of a program in which every element causes the program to follow a common control path. A domain fault causes program execution to switch to an incorrect domain. Domain faults may be stimulated by test points anywhere in the input space, but they are most likely to be stimulated by inputs which cause the program to be in a state which is 'near' a domain boundary. An example of this property can be seen in Figure 19.5(b) in the traversal of path 1, 2, 3. The only constraint required is that $c < a$, but if the difference between c and a is small, then there is a greater likelihood that a small change in the value of c will cause the incorrect path to be traversed. Researchers have applied this idea to develop a domain coverage metric which can be applied to hardware and software descriptions [13].

Many control-dataflow coverage metrics consider the requirements for fault activation without explicitly considering fault effect observability. Researchers have developed observability-based behavioural coverage metrics [14] to alleviate this weakness. The observability-based code coverage metric (OCCOM) has been applied for hardware validation [14] and for software validation [15]. The OCCOM approach inserts faults called 'tags' at each variable assignment which represent a positive or negative offset from the correct signal value. The sign of the error is known but the magnitude is not. Observability analysis along a control-flow path is done probabilistically by using the algebraic properties of the operations along the path and simulation data. As an example, in Figure 19.5 we will assume that a positive tag is inserted on the value of variable c and we must determine if the tag is propagated through the condition $c < in2$ in node 4 of Figure 19.5(b). Since the tag is positive, it is possible that the conditional statement will execute incorrectly in the presence of the tag, so the OCCOM approach optimistically assumes tag propagation in this case. Notice that a negative tag could not affect the execution of the conditional statement. A metric proposed in Reference 16 more accurately determines observability by exploring tag propagation along all possible control-flow paths which could be executed as a result of a design error.

19.3.1.3   State machine coverage metrics

Finite state machines (FSMs) are the classic method of describing the behaviour of a sequential system and fault models have been defined to be applied to state machines. State machine coverage metrics assume that a design error impacts the structure of the state machine, the states and the transitions between them. A design error in a state machine might add or remove states, add or remove edges, or alter the inputs or outputs associated with existing edges. Any of these changes could only be detected if the affected portion of the state machine is exercised during test.

The commonly used fault models are the 'state coverage' models which require that all states be reached, and 'transition coverage' which requires that all transitions be traversed. Calculation of coverage using these models requires the update of a table containing all states and transitions in the behaviour. The act of updating these tables is not time consuming but the size of these tables will be large for realistic state machines. The problems associated with state machine testing are understood from classical switching theory and are summarised in a thorough survey of state machine testing [17].

19.3.1.4   Gate-level coverage metrics

A gate-level coverage metric is one which was originally developed for and applied to gate-level circuits. Manufacturing testing research has defined several gate-level fault models which are now applied at the behavioural level. For example, the stuck-at fault model assumes that each wire may be held to a constant value of 0 or 1 due to an error. The stuck-at fault model has also been applied at the behavioural level for manufacturing test [18] and for hardware–software covalidation [19]. Behavioural designs often use variables which are represented with many bits and gate-level fault models are typically applied to each bit, individually. For example, if we assume that an integer as declared in Figure 19.5(a) is 32 bits long, then applying the single stuck-at fault model to a variable would produce 32 stuck-at-1 faults and 32 stuck-at-0 faults. Gate-level fault models have the potential weakness that they are structural in nature rather than behavioural. Gate-level fault models were intended to describe physical defects which could occur during the very large scale integration (VLSI) manufacturing process.

19.3.1.5   Application-specific coverage metrics

A coverage metric which is designed to be generally applicable to arbitrary design types may not be as effective as a metric which targets the behavioural features of a specific application. To justify the cost of developing and evaluating an application-specific coverage metric, the market for the application must be very large and the error modes of the application must be well understood. For this reason, application-specific coverage metrics are seen in microprocessor test and validation [20–22]. Early microprocessor metrics target relatively generic microprocessor features. For example, researchers define a metric for instruction-sequencing functions [20] by describing the fault effects (i.e. activation of erroneous microorders), and describing the fault detection requirements. More metrics target the modern processor features such as pipelining [21,22].

Another alternative to the use of a traditional coverage metric is to allow the designer to define the coverage metric. This option relies on the designer's expertise at expressing the characteristics of the metric in order to be effective. The manual definition of a fault model, also known as functional verification, is best applied in well-understood domains such as microprocessor validation [23]. Several tools have been developed which automatically evaluate user-specified properties during simulation to identify the existence of faults.

19.3.1.6   Interface errors

To manage the high complexity of hardware/software design and covalidation, efforts have been made to separate the behaviour of each component from the communication architecture [24]. Interface covalidation becomes more significant with the onset

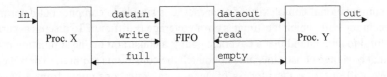

Figure 19.6   Interprocess communication via a FIFO

of core-based design methodologies which utilise pre-designed, pre-verified cores. Since each core component is pre-verified, the system covalidation problem focuses on the interface between the components.

A case study of the interface-based covalidation of an image compression system has been presented [25]. Researchers classify the interface faults which occurred during the design process into three groups: (1) COMP2COMP faults involving communication between pairs of components, (2) COMP2COMM faults involving the interaction between each component and the communication architecture and (3) COMM faults involving the coordinated interactions between the communication architecture and all components. In Reference 25, test benches are developed manually to target each of these interface fault classes.

Additional interface complexity is introduced by the use of multiple clock domains in large systems. The interfaces between different clock domains must be essentially asynchronous. Unless a high-overhead timing-independent circuit implementation is used (such as differential cascode voltage switch logic), asynchronous interfaces are particularly vulnerable to timing-induced faults. Timing-induced faults are described in Reference 26 as faults which cause the definition of a signal value to occur earlier or later than expected. An example of the occurrence of this type of fault would be an increased delay on the 'empty' status signal of a first-in-first out (FIFO) buffer.

Interprocess communication using a FIFO is shown in Figure 19.6. Process X sends data to Process Y through an intervening FIFO which is used to manage intermittent mismatch in the sending and receiving rates of X and Y, respectively. The FIFO interface contains datain and dataout signals which carry data. The interface contains two control signals write and read which indicate when new data is placed into the FIFO and removed from the FIFO. The status signals full and empty are used to indicate when the FIFO cannot be written to or read from without causing an error. When the FIFO becomes empty the empty signal should be asserted within a known amount of time. If the empty signal is issued later than expected, the FIFO may be read from while it is empty. In Reference 13 a timing coverage metric is presented and a technique for coverage evaluation is introduced.

19.3.2   Metric accuracy

An ideal coverage metric will accurately reflect the coverage of potential design errors. In addition, the coverage computation process should be much faster than the process of computing error coverage through brute-force simulation of erroneous designs.

The computational complexity of each metric is simple to compute but determining the accuracy with which it estimates error coverage is difficult. To discuss accuracy it is helpful to understand that a coverage goal is a constraint on the simulation process and that coverage constraints may not match the constraints required to detect errors.

We will describe the process of generating state constraints from a coverage goal. The process of generating coverage constraints from a design and a metric can be arbitrarily complex depending on the metric used. It is not tractable to generate coverage constraints in practice but examining coverage constraints is useful in the context of this chapter to help evaluate metric accuracy.

19.3.2.1   Coverage goals as constraints

In order to understand the strengths and weaknesses of different coverage metrics, it is useful to see each coverage goal as a constraint on the simulation of the system under test. Once coverage goals are expressed as constraints, a coverage metric can be evaluated by determining how closely its coverage goals match the detection requirements of various design errors. We will describe how each coverage goal can be viewed as a test constraint and we will evaluate the effectiveness of each constraint in detecting design errors.

We will describe coverage constraints by considering the application of statement coverage when testing the design in Figure 19.5. A statement coverage goal for the design in Figure 19.5 is to 'execute statement 6, return $(a + c)$'. This goal can be stated as a constraint on the variables in the design by examining the control-flow paths which must be traversed in order to execute the statement. Examining the control-flow graph in Figure 19.5(b) reveals that all paths which execute statement 6 share the following two properties.

1   At some point the condition at statement 2 must become FALSE.
2   The condition at statement 4 must be FALSE.

Before we can express these constraints formally it is necessary to introduce some terminology. We cannot discuss the value of variable c statically because c may take several different values during the course of simulation. To refer to the value of c at a particular point in simulation we need to describe the statement which most recently assigned a value to c and the number of times that the assignment statement has been executed. In Figure 19.5 the variable c is assigned by two statements, $c = 0$ and $c = c + in1$. We will refer to the value of c after these two statements are executed as c1 and c2, respectively. In general a statement can be executed many times in the presence of a loop, as is the case with statement $c = c + in2$. To describe multiple executions of the same statement we will add an index to the c variable which indicates the iteration of the statement which last assigned a value to c. So the name c1[i] will refer to the value of variable c immediately after the statement $c = 0$ has been executed $i + 1$ times. The name c2[i] is defined similarly for statement $c = c + in2$. Notice that statement $c = 0$ can only be executed once, so c1[0] = 0 and c1[i] is undefined for $i > 0$.

Using this terminology we can express the execution of statement 6 with the constraint on the design variables shown in Equation (19.1),

$$((c1[0] \geq in1 + in2) \cap (c1[0] \geq in2)) || ((c2[n] \geq in1 + in2)$$

$$\cap (c2[n] \geq in2)) \tag{19.1}$$

where n is a positive integer. Equation (19.1) contains four clauses, each of which corresponds to a FALSE evaluation of one of the two conditional predicates in the description. The first two clauses describe the control-flow path which avoids the while loop completely. Along this path the value of c is only assigned once, by the first assignment, so the value of c is c1[0]. The second two clauses describe the paths that enter the while loop and then exit after the nth iteration.

Constraints can be used to describe not only the coverage goals but also the requirements for detecting individual design errors. Consider the design error for the design in Figure 19.5 where the statement 'return(a + c)' was incorrectly substituted with the statement 'return(a − c)'. In order to detect this error it is first necessary that the statement be executed during simulation; the constraints for that are expressed in Equation (19.1). Additionally, the incorrect output a − c must not equal the correct output $a + b$, so it must be true that a − c $\neq$ a + c.

19.3.2.2 Coverage goals and error detection

For an ideal coverage metric, the coverage value achieved by simulating a design with a test sequence would always equal the fraction of potential design errors detected by the test sequence. The extent to which a coverage metric satisfies this goal can be judged by comparing the coverage constraints to the error detection constraints. A coverage metric satisfies the detection constraints of a design error e if the satisfaction of some coverage goal g must imply the satisfaction of the detection constraints of the error e. So the coverage goal constraints g must be 'more strict' than the detection constraints d. In practice, however, coverage goals are often 'less strict' than detection constraints, causing the coverage metric to overestimate the fraction of errors detected. This is the case for the statement coverage constraint in Equation (19.1) and the design error where the statement 'return(a + c)' was switched with 'return (a − c)'. The statement coverage goal ensures that the erroneous statement is executed but detection additionally requires that a + c $\neq$ a − c. We can infer from the relationship between detection and coverage constraints that the statement coverage metric will be optimistic, consistently overestimating the coverage of design errors.

The coverage goals may be insufficiently strict to ensure error detection, but coverage goals may also be so strict that they cannot be satisfied. In this case the coverage metric can be pessimistic because coverage can never reach 100 per cent, even after all design errors are detected. An example of an unsatisfiable coverage goal can be seen in the satisfaction of a path coverage goal for the design in Figure 19.5. It is not possible to execute a control path which enters the 'while' loop and then follows the TRUE branch of the following 'if' statement. This is because the 'while' loop cannot be exited unless c $\geq$ in1 + in2 and in1 > 0. It is not possible for c < in2 under these conditions. If 100 per cent path coverage is used as a goal for test generation then test generation will never complete in this case.

19.4   Automatic test generation techniques

Several automatic test generation (ATG) approaches have been developed which vary in the class of search algorithm used, the coverage metric used, the search space technique used and the design abstraction level used. In order to perform test generation for the entire system, both hardware and software component behaviours must be described in a uniform manner. Although many behavioural formats are possible, previous ATG approaches have focused on CDFG and FSM behavioural models.

Table 19.2 presents a taxonomy of covalidation test generation techniques classified according to the coverage goal of the search algorithm. Each class of test generation techniques is described in the following subsections.

Two classes of search algorithms have been explored, 'goal directed' and 'coverage directed'. Figure 19.7 shows an outline of both of these classes of algorithms. Goal directed techniques successively target a specific coverage goal and construct a test sequence to detect that goal. Each new test sequence is merged with the current test sequence (typically through concatenation) and the resulting coverage is evaluated to determine if test generation is complete. This class of algorithms suffers in terms of time complexity because it directly solves the test generation problem for individual goals, requiring a complex search of the space of input sequences. However, goal directed algorithms have the advantage that they are 'complete' in the sense that a test sequence will be found for a goal if a test sequence exists. Another class of search algorithms are the coverage directed algorithms which seek to improve coverage without targeting any specific fault. These algorithms heuristically modify

Table 19.2   Taxonomy of covalidation test generation techniques

Test gen class	Solving technique
Goal directed	Linear programming + SAT
	Integer linear programming + SAT
	Constraint logic programming
	Model checking counterexample
	Switching theory
	Implicit state enumeration
Coverage directed	Genetic algorithms
	Random mutation hill climbing
	Directed-random tests

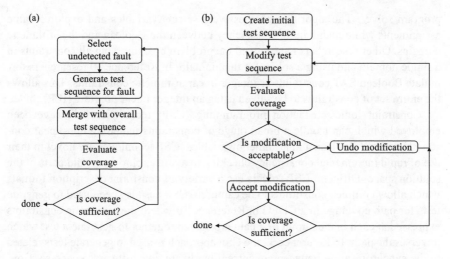

Figure 19.7   Classes of test generation algorithms: (a) fault directed, (b) coverage directed

an existing test set to improve total coverage, and then evaluate the coverage produced by the modified test set. If the modified test set corresponds to an improvement in coverage then the modification is accepted. Otherwise the modification is either rejected or another heuristic is used to determine the acceptability of the modification. Coverage directed techniques have the potential to be much less time consuming than goal directed techniques because they may use fast heuristics to modify the test set. The drawback of coverage directed techniques is that they are not guaranteed to detect any particular fault although the fault may be detectable.

19.4.1   Goal directed techniques

Section 19.3.2.1 discussed how the coverage goals defined by a coverage metric can be described as a set of constraints on the system behaviour. Once the constraints have been identified, the test generation problem is equivalent to the problem of solving the constraints simultaneously to produce a test sequence at the system inputs. Because the operations found in a hardware/software description can be either Boolean or arithmetic, the solution method chosen must be able to handle both types of operations. The Boolean version of the problem is traditionally referred to as the SATISFIABILITY (SAT) problem and has been well studied as the fundamental NP-complete problem. A great deal of work has been done on SAT solving techniques [27,28]. Handling both Boolean and arithmetic operations poses an efficiency problem because classical solutions to the two problems have been presented separately.

In Reference 16 researchers define the HSAT problem as a hybrid version of the SAT problem which considers linear arithmetic constraints together with Boolean SAT constraints. Researchers in Reference 29 present an algorithm to solve the HSAT problem which combines a SAT solving technique [28] with a traditional linear program solver. The algorithm progressively selects variables and explores value assignments while maintaining consistency between the Boolean and the arithmetic domains. Other researchers have solved the problem by expressing all constraints in a single domain and using a solver for that domain. In Reference 30 researchers formulate Boolean SAT constraints as integer linear arithmetic constraints. This allows the entire set of constraints to be solved using an integer linear program (ILP) solver.

Constraint logic/satisfaction programming (CSP) techniques [31] have been employed which can handle a broad range of constraints including non-linear constraints on both Boolean and arithmetic variables. CSP techniques are novel in their use of rapid incremental consistency checking to avoid exploring invalid parts of the solution space. Different CSP solvers use a variety of constraint description formats which allow complex constraints to be captured. CSP has also been used to generate tests for path coverage in a CDFG in Reference 10 where the arithmetic constraints expressed at each branch point of a path are solved together to generate a test which traverses the path. In Reference 32 the CSP approach is used to generate tests related to the synchronisation between concurrent hardware and software processes. Constraints are generated which describe the behaviour of the hardware–software system and which describe the conditions which would activate a potential synchronisation fault.

State machine testing has been accomplished by defining a 'transition tour' which is a path which traverses each state machine transition at least once [33]. Transition tours have been generated by iteratively improving an existing partial tour by concatenating on to it the shortest path to an uncovered transition [33]. In Reference 34, a test sequence is generated for each transition by asserting that a given transition does not exist in a state machine model, and then using a model checking tool to disprove the assertion. A byproduct of disproving the assertion is a counterexample which is a test sequence which includes the transition. Since this technique relies on model checking technology, it shares its performance and memory requirement characteristics with model checking approaches.

If a fault effect can be observed directly at the machine outputs, then covering each state and transition during test is sufficient to observe the fault. In general, a fault effect may cause the machine to be in an incorrect state which cannot be immediately observed at the outputs. In this case, a 'distinguishing sequence' must be applied to differentiate each state from all other states based on output values. The testing problems associated with state machines, including the identification of distinguishing, synchronising and homing sequences, are well understood [17].

19.4.2   *Coverage directed techniques*

Several techniques have been developed which generate test sequences without targeting any specific coverage goal. Coverage is improved by modifying an existing test sequence, and then evaluating the coverage of the new sequence. These techniques differ in the method used to modify the test sequence, the cost function used to evaluate a sequence and the criteria used to accept a new sequence. The modification method is typically either random or directed random.

An example of such a technique is presented in Reference 35 and 36 which uses a genetic algorithm to successively improve the population of test sequences. In the terminology of genetic algorithms, a 'chromosome' describes a test sequence. Many test sequences are initially generated randomly. Random matings can occur between the chromosomes which describe the test sequences, but the mating process defines and restricts the way in which two test sequences are merged. The cost function (or 'fitness' function) used to evaluate a test sequence is the total number of elementary operations (variable read/write) which are executed. In this technique, the total number of elementary operations is being used as an approximation of the likelihood of error detection.

Work presented in Reference 37 uses a random mutation hill climber (RMHC) algorithm which randomly modifies a test sequence to improve a testability cost function. The test sequence modification is completely random and the criteria for accepting a new sequence is that the cost function is improved. The coverage metric targeted using this approach is the single stuck-at metric applied to the individual bits of each variable in the behavioural description. The cost function used contains two parts: (1) the number of statements executed by the sequence, and (2) the number of outputs which contain a fault effect. Results show that the CPU time required using this approach is nearly an order of magnitude less than the time required using commercial gate-level test generation tools.

In Reference 38 researchers generate directed-random pattern sequences to be used for test. No particular coverage metric is assumed in this approach, so it is up to the user to provide the directives for pattern generation. Two types of directives are used: (1) 'constraints' which define the boundaries of the space of feasible test patterns, and (2) 'biases' which direct assignments of values to signals in a non-random way. For example, a constraint might indicate that the following relationship between variables must hold, in1 < in2. Because this is a constraint, no test can be generated which violates this condition. A bias expresses the desired probability distribution for the values of a signal throughout the set of all patterns. For example, a bias of the form (in2, 0.9) would indicate that the probability that input in2 is equal to 1 should be 0.9. It is the task of the test engineer to develop a set of constraints and biases which will reveal a particular class of errors.

19.5   Test response analysis

Detection of errors requires that the test responses gathered during cosimulation be compared to the correct test responses. Since any signal value can be observed at any point during simulation, the potential amount of test data is enormous. Manual evaluation of test responses requires that the correct signal values be predetermined. Manually computing correct test results is time consuming and efforts have been made to automate the process. The use of assertions and self-checkers to evaluate test responses have been investigated in both the hardware and software domains.

19.5.1   Assertions

An assertion is a logical statement which expresses a fact about the system state at some point during the system's execution. Assertions have been proposed and used in both software [39] and hardware domains [40]. The use of assertions is well accepted and has been integrated into the design proccess of many large-scale systems [41,42]. Assertions are primarily useful in evaluating the correctness of the system state during simulation but the use of assertions has also extended to supporting functional test generation [38] and performance analysis [43].

An assertion is typically written as a logical relationship between the values of different storage elements which would be variables in a software program, but would also include registers, flip-flops, latches and time-varying signals in a hardware description. Many languages for the description of assertions have been proposed but we will use a simple first-order predicate calculus in most of our examples. In this discussion we are interested in the concepts behind the use of assertions, rather than any specific implementation details. For this reason our examples will use a generic syntax which may not match any specific assertion language but is sufficient to describe the concepts related to assertions. In a system which describes the operation of traffic lights at an intersection, we might want to express the fact that both lights cannot be green at the same time as follows:

$$(\text{colorNS} == \text{`green'}) \cap (\text{colorEW} == \text{`green'}) \qquad (19.2)$$

In Equation (19.2) variables colourNS and colourEW represent the the colours of the north–south and east–west signal lights, respectively. The assertion in Equation (19.2) can be referred to as a 'positive' assertion because it expresses a relationship which must be satisfied. Notice that when Equation (19.2) is negated then it is a 'negative' assertion which expresses a relationship which must not be satisfied. Negative assertions are essentially the same as error handling code, commonly used in software, which throw an exception when the system enters an incorrect state. Notice that positive and negative assertions are equivalent in their information content so we will assume the use of positive assertions through the remainder of this section, without loss of generality.

Assertions can be evaluated during simulation to determine whether or not an error occurred which forces the system into a state which is known to be incorrect. Assertions may be defined 'globally' which must be satisfied at all times, or assertions may be defined 'locally' which are only satisfied at the point in the description where the assertion is placed. The traffic light controller assertion in Equation (19.2) is an example of a global assertion. A local assertion has the potential to specify more fine-grained properties because the assertion can be defined to use state information derived from its position in the description. Figure 19.8 shows an example of a local assertion in a traffic light controller code segment. The assertion in Figure 19.8 states that the colour of the north–south light must be yellow. This statement is clearly not always true in general but the assertion is added just before the statement which changes the light colour to red. If we assume that the light must be yellow before it is red, then the assertion must be true at the point in the description where it is asserted.

```
TrafficLight()
    .
    .
assertion(colorNS == 'yellow')
colorNS = "red";
    .
    .
```

Figure 19.8   Local assertion in a traffic light controller

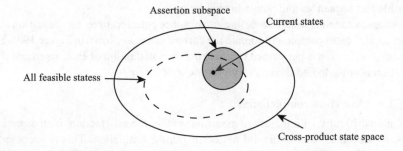

Figure 19.9   State space hierarchy

An assertion which describes the state of a system at a point in its execution can be referred to as an 'instantaneous' assertion. The assertions in Equation (19.2) and Figure 19.8 are both instantaneous because they both describe properties at one time step. Although the assertion in Equation (19.2) is globally true at all time steps, it is considered to be instantaneous because it expresses a statement about each time step individually, independent of all other time steps. An assertion is referred to as being 'temporal' if it expresses constraints on sequences of time steps. In order to express temporal constraints a logic must be used which expresses temporal relationships. To give an example of a temporal assertion we will introduce the 'next' operator used in property specification language (PSL) [44]. The statement $p \rightarrow next\ q$ states that if statement p is true at time step t then statement q must be true at time step $t + 1$. Using this temporal operator we can state the fact that the north–south traffic light must turn red one time step after it becomes yellow with the following expression.

$$(\text{colorNS} == \text{'yellow'}) \rightarrow next\ (\text{colorNS} == \text{'red'}) \tag{19.3}$$

An assertion defines boundaries on the correct execution of the system. An instantaneous assertion defines a subset of the state space and a temporal assertion defines a subset of the set of all execution sequences. The discussion here will be limited to instantaneous assertions but the same argument could be extended to temporal assertions as well. The state space subset defined by an instantaneous assertion must contain the actual system state at the point in execution where the assertion is evaluated. Figure 19.9 is used to show the state space hierarchy during system execution. The largest space in Figure 19.9, called the 'cross-product state space', is the space defined by the cross-product of the states of all individual state elements in the system. Only a subset of these states, referred to as 'all feasible states', may be entered during the operation of the system if it is free of design errors. At any given point during the operation of the system there is a subset of the all feasible states set called 'current states' which must contain the current system state if the system is error free. If the system is completely deterministic then the set of current states must have cardinality 1. The current states set must be a subset of the all feasible states set because the current states set is dependent on the input sequence and the point in simulation being evaluated, while the all feasible states set is the union of feasible states over all possible test sequences and points in simulation.

An instantaneous assertion defines a subspace referred to as the 'assertion subspace' which must completely contain the current states as shown in Figure 19.9. Test response evaluation is performed by checking the satisfaction of each assertion; if an assertion is not satisfied then a design error exists.

19.5.1.1   Assertion completeness

The main difficulty with the use of assertions is that the satisfaction of all assertions does not guarantee that errors did not occur during simulation. This is because the assertion subspace is a superset of the set of current states. In the traffic light controller, e.g. both north–south and east–west traffic lights may become yellow at the same time due to a design error without violating the assertion in Equation (19.2). To increase the chances that errors are detected, the set of assertions must be as 'complete' as possible. In terms of the state space, this means that the assertion subspace must be as small as possible while still containing the correct states set. This requires that the assertions be written as strictly as possible to reduce the number of incorrect states which can satisfy the assertion. For example, the assertion for the traffic light controller in Equation (19.2) can be replaced by the stronger assertion in Equation (19.4).

$$(\text{colorNS} == \text{`red}') \cup (\text{colorEW} == \text{`red}') \tag{19.4}$$

The assertion in Equation (19.4) is stronger than Equation (19.2) because the subspace that Equation (19.4) defines is a proper subset of the subspace defined by Equation (19.2). Notice that the assertion in Equation (19.4) catches the erroneous condition when both light directions are yellow at the same time.

Defining a set of assertions which is complete is difficult because the task of assertion definition is largely manual, so the completeness of a set of assertions depends on the abilities of individual designers. Definition of a complete set of assertions is typically very expensive due to the rigorous and manual nature of the process. The cost investment is worthwhile for some highly standardised and reused applications such as floating point division [45] and the peripheral component interface (PCI) bus protocol [46].

19.5.2   Self-checking

A self-checking component is one which automatically evaluates its correctness by comparing its results to the results of one or more other redundant components which implement the same function. Using only two redundant components enables the detection of errors but not correction, since it is impossible to know which redundant version is the one with the correct result. At least three or more components allow correction as well. If it is assumed that the likelihood of a majority of components producing incorrect answers is very small, then correction can be achieved by selecting the result produced by the majority of redundant components.

Self-checkers are distinguished from assertions in a number of ways including their description style. While assertions are described declaratively, as logical statements, self-checkers are described 'procedurally' as a sequence of operations. Also, a self-checker does not simply restrict the space of correct results as an assertion would. A self-checker actually computes the correct result(s). In terms of the state space hierarchy shown in Figure 19.9, a self-checker computes the set of correct states, just as an ordinary component would. Defining features of a self-checking technique include the implementation of the redundant components and the number of redundant components used.

An important distinction between self-checking techniques is the point in the system's lifecycle when they are applied. Self-checking can be applied prior to deployment of the system in the field for the purpose of 'validation'. Self-checking can also be applied post-deployment to enhance 'reliability'. Self-checking incurs some overhead in terms of cost, performance and power, which can be difficult to justify in tightly constrained systems. High overhead is one reason that self-checking is not well used in standard hardware and software projects today.

A key requirement of any self-checking technique is that the redundant components used for comparison must not operate in exactly the same way as the original component so that they all do not manifest the same errors. One way to accomplish this is by assigning completely different design teams to implement the same system. This approach is referred to as N-version programming in the software domain [47]. A significant limitation of this approach is the exorbitant cost of multiple design teams. The reliability provided using this approach relies on the independence of the design teams. Such independence is difficult to establish in practice because programmers are likely to be trained in the same industrial environment and using the design tools. Designer independence also contradicts the current trend toward design reuse to reduce design times.

A theoretical framework for self-checking has been developed by Blum and Kanna [48] and has been applied to several practical programming examples [49]. In Reference 49 a general technique is presented to create a self-checking program from a non-self-checking program for numerical programs including matrix multiplication and integer division. The self-checking technique exploits a property of many numerical functions referred to as **random self-reducibility**. A function f is random self-reducible if $f(x)$ can always be computed as $F(f(a_1), \ldots, f(a_c))$ where F is an easily computable function and the numbers $a_1, \ldots, a_c$ are randomly distributed and are also easily computable, given x. The key idea is that $f(x)$ can be computed as a function of $f(a_1), \ldots, f(a_c)$. If the numbers $a_1, \ldots, a_c$ are randomly distributed then it is very unlikely that the implementation of f would produce an incorrect result for the majority of values $a_1, \ldots, a_c$.

The advantage of the technique presented in Reference 49 is that the self-checking function is created in a straightforward way from the original program without the need for alternate design teams. This greatly reduced design cost as compared to N-version programming, and it simplifies reliability analysis since the level of independence between different design teams is not an issue.

19.6   Conclusions and future directions

It is clear that the field is maturing as researchers have begun to identify and agree on the essential problems to be solved. Our understanding of covalidation has developed to the point that industrial tools are available which provide practical solutions to test generation, particularly at the state machine level. Although automation tools are available, they are not fully trusted by designers and as a result, a significant amount of manual test generation is required for the vast majority of design projects. By examining the state of previous work we can identify areas which should be studied in future in order to increase the industrial acceptance of covalidation techniques.

A significant obstacle to the widespread acceptance of available techniques is the lack of faith in the correlation between covalidation coverage metrics and real design errors. Automatic test generation techniques have been presented which are applicable to large scale designs, but until the underlying coverage metrics are accepted, the techniques will not be applied in practice. Coverage metrics must be evaluated by identifying a correlation between fault coverage and detection of real design errors. Essential to this evaluation is the compilation of design errors produced by real designers. Research has begun in this direction [50,51] and should be used to evaluate existing covalidation coverage metrics. Once coverage metrics are empirically evaluated we can expect to see large increases in covalidation productivity through the automation of test generation.

Analysis of test responses is a bottleneck in the covalidation process because the definition of assertions and self-checkers requires design understanding that only a designer can have. Assertions, express properties of the design which must be satisfied, but developing these properties requires an understanding of the specification. It is possible to generate assertions which are generically applicable to a class of designs such as microprocessors (e.g. 'all RAW hazards are illegal in any pipeline') but properties unique to a design must be expressed manually.

A great deal of research in hardware/software covalidation is extended from previous research in the hardware and software domains, but communication between hardware and software components is a problem unique to hardware/software covalidation. The interfaces between hardware and software introduce many new design issues which can result in errors. For example, software may be executed on an embedded processor which is in a different clock domain than other hardware blocks with which it communicates. Such communication requires the use of some asynchronous communication protocol which must be implemented in hardware and software. Asynchronous communication is a difficult concept for both hardware and software designers, so it can be expected to result in numerous design errors.

Hardware/software communication complexity has also increased because inter-processor communication is handled very differently in hardware as compared to software. Hardware description languages typically provide only the most basic synchronisation mechanisms, such as the 'wait' expression in Verilog hardware description language (VHDL). More complicated protocols (e.g. two-way handshake) must be implemented manually and are therefore vulnerable to design errors. Inter-process communication in software tends to use high-level communication primitives such as monitors (e.g. the 'synchronised' statement in Java). Although the implementation of each primitive may be known to be correct, the primitive itself may be used incorrectly by the designer, resulting in design errors. Relatively little research has investigated testing the interfaces between hardware and software components, but this research area is essential.

References

1 DEMICHELI, G., ERNST, R., and WOLF, W.: 'Readings in hardware/software co-design' (Morgan Kaufmann Publishers, San Francisco, CA, 2002)

2 FUMMI, F., LOGHI, M., PONCINO, M., MARTINI, S., PERBELLINI, G., and MONGUZZI, M.: 'Virtual hardware prototyping through timed hardware-software co-simulation'. Proceedings of *Design automation and test in Europe*, 2005, pp. 798–803

3 JAMES A. ROWSON: 'Hardware/software co-simulation'. Proceedings of *Design automation* conference, 1994, pp. 439–440

4 VAN CAMPENHOUT, D., MUDGE, T., and HAYES, J.P.: 'Collection and analysis of microprocessor design errors', *IEEE Design and Test of Computers*, 2000, **17** pp. 51–60

5 BENTLEY, B.: 'Validating the intel pentium 4 microprocessor'. Proceedings of *Design automation* conference, 2001

6 KING, K.N., and OFFUTT, A.J.: 'A fortran language system for mutation-based software testing', *Software Practice and Engineering*, 1991, **21**(7), pp. 685–718

7 BEIZER, B.: 'Software testing techniques' (Van Nostrand Reinhold, New York, 1990, 2nd edn)

8 TASIRAN, S., and KEUTZER, K.: 'Coverage metrics for functional validation of hardware designs', *IEEE Design and Test of Computers*, 2001, **18**(4), pp. 36–45

9 AL G. HAYEK, and ROBACH, C.: 'From specification validation to hardware testing: a unified method'. International *Test* conference, 1996, pp. 885–893

10 VEMURI, R., and KALYANARAMAN, R.: 'Generation of design verification tests from behavioral vhdl programs using path enumeration and constraint programming', *IEEE Transactions on Very Large Scale Intergration Systems*, 1995, **3**(2), pp. 201–214

11 ZHANG, Q., and HARRIS, I.G.: 'A data flow fault coverage metric for validation of behavioral hdl descriptions'. International conference on *Computer-aided design*, 2000

12   WHITE, L., and COHEN, E.: 'A domain strategy for computer program testing', *IEEE Transactions on Software Engineering*, 1980, **SE-6**(3), pp. 247–247

13   ZHANG, Q., and HARRIS, I.G.: 'A domain coverage metric for the validation of behavioral vhdl descriptions'. International Test conference, 2000

14   DEVADAS, S., GHOSH, A., and KEUTZER, K.: 'An observability-based code coverage metric for functional simulation'. International conference on *Computer-aided design*, 1996, pp. 418–425

15   COSTA, J.C., DEVADAS, S., and MONTIERO, J.C.: 'Observability analysis of embedded software for coverage-directed validation'. International conference on *Computer-aided design*, 2000, pp. 27–32

16   VERMA, S., RAMINENI, K., and HARRIS, I.G.: 'An efficient control-oriented coverage metric'. Asian-pacific *Design automation* conference, 2005

17   LEE, D., and YANNAKAKIS, M.: 'Principles and methods of testing finite state machines – a survey', *IEEE Transactions on Computers*' 1996, **84**(8), pp. 1090–1123

18   THAKER, P.A., AGRAWAL, V.D., and ZAGHLOUL, M.E.: 'Register-transfer level fault modeling and test evaluation techniques for vlsi circuits'. International *Test* conference, 2000, pp. 940–949

19   FIN, A., FUMMI, F., and SIGNORETTO, M.: 'SystemC: a homogenous environment to test embedded systems'. International workshop on *Hardware/ software codesign* (CODES), 2001

20   BRAHME, D., and ABRAHAM, J.A.: 'Functional testing of microprocessors'. *IEEE Transactions on Computers*, 1984, **C-33**, pp. 475–485

21   MISHRA, P., DUTT, N., and NICOLAU, A.: 'Automatic verification of pipeline specifications'. *High-level design validation and test* workshop, 2001, pp. 9–13

22   UTAMAPHETAI, N., BLANTON, R.D., and SHEN, J.P.: 'Relating buffer-oriented microarchitecture validation to high-level pipeline functionality'. *High-level design validation and test* workshop, 2001, pp. 3–8

23   ZIV, A.: 'Cross-product functional coverage measurement with temporal properties-based assertions'. Proceedings of *Design automation and test in Europe*, 2004, pp. 834–839

24   ROWSON, J.A., and SANGIOVANNI-VINCENTELLI, A.: 'Interface-based design'. Proceedings of *Design automation* conference, 1997, pp. 178–183

25   PANIGRAHI, D., TAYLOR, C.N., and DEY, S.: 'Interface based hardware/ software validation of a system-on-chip'. *High level design validation and test* workshop, 2000, pp. 53–58

26   ZHANG, Q., and HARRIS, I.G.: 'A validation fault model for timing-induced functional errors'. International *Test* conference, 2001

27   ARORA, R., and HSIAO, M.S.: 'Cnf formula simplification using implication reasoning'. *High level design validation and test* workshop, 2004, pp. 129–134

28   LARRABEE, T.: 'Test pattern generation using boolean satisfiability', *IEEE Transactions on Computer-Aided Design*, 1992, **11**, pp. 4–15

29   FALLAH, F., DEVADAS, S., and KEUTZER, K.: 'Functional vector generation for hdl models using linear programming and 3-satisfiability'. Proceedings of *Design automation* conference, 1998, pp. 528–533

30 ZENG, Z., KALLA, P., and CIESIELSKI, M.: 'Lpsat: a unified approach to rtl satisfiability'. Proceedings of *Design, automation and test in Europe* conference, 2000

31 RINA DECHTER: 'Constant processing' (Kaufman, San Francisco, CA, 2003)

32 XIN, F., and HARRIS, I.G.: 'Test generation for hardware-software covalidation using non-linear programming'. *High-level design validation and test* workshop, 2002

33 HO, R.C., YANG, C.H., HOROWITZ, M.A., and DILL, D.L.: 'Architecture validation for processors'. International symposium on *Computer architecture*, 1995, pp. 404–413

34 GEIST, D., FARKAS, M., LANDVER, A., UR, S., and WOLFSTHAL, Y.: 'Coverage-directed test generation using symbolic techniques'. Proceedings of the first international conference on *Formal methods in computer-aided design*, 1996, pp. 143–158

35 CORNO, F., PRINETTO, P., and SONZA REORDA, M.: 'Testability analysis and ATPG on behavioral RT-level VHDL'. International *Test* conference, 1997, pp. 753–759

36 CORNO, F., SONZE REORDA, M., SQUILLERO, G., MANZONE, A., and PINCETTI, A.: 'Automatic test bench generation for validation of RT-level descriptions: an industrial experience'. Proceedings of *Design automation and test in Europe*, 2000, pp. 385–389

37 LAJOLO, M., LAVAGNO, L., REBAUDENGO, M., SONZA, M. REORDA, and VIOLANTE, M.: 'Behavioral-level test vector generation for system-on-chip designs'. *High level design validation and test* workshop, 2000, pp. 21–26

38 YUAN, J., SHULTZ, K., PIXLEY, C., MILLER, H., and AZIZ, A.: 'Modeling design constraints and biasing in simulation using BDDs'. International conference on *Computer-aided design*, 1999, pp. 584–589

39 HOARE, C.A.R.: 'Assertion: a personal perspective', *IEEE Annals of the History of Computing*, 2003, **25**(2), pp. 14–25

40 FOSTER, H.D., KROLNIK, A. C., and LACEY, D.J.: 'Assertion-based design' (Kluwer Academic Publishers, Boston, MA, 2003)

41 KANTROWITZ, M., and NOACK, L.M.: 'I'm done simulating: now what? verification coverage analysis and correctness checking of the decchip 21164 alpha microprocessor'. Proceedings of *Design automation* conference, 1996, pp. 325–330

42 TAYLOR, S., QUINN, M., BROWN, D. *et al.*: 'Functional verification of a multiple-issue, out-of-order, superscalar alpha processor-the dec alpha 21264 microprocessor'. Proceedings of *Design automation* conference, 1998, pp. 338–343

43 CHEN, X., LUO, Y., HSIEH, H., BHUYAN, L., and BALARIN, F.: 'Utilizing formal assertions for system design of network processors'. Proceedings of *Design automation and test in Europe*, 2004

44 'Accelera proposed standard property specification language (psl) 1.0', January 2003

45   CLARKE, E.M., KHAIRA, M., and XIAO, X.: 'Word level model checking – avoiding the pentium fdiv error'. Proceedings of *Design automation* conference, 1996, pp. 245–248

46   CHAUHAN, P., CLARKE, E.M., LU, Y., and WANG, D.: 'Verifying ip-core based system-on-chip designs'. ASIC/SOC conference, 1999, pp. 27–31

47   LEVESON, N.G., CHA, S.S., KNIGHT, J.C., and SHIMEALL, T.J.: 'The use of self checks and voting in software error detection: an empirical study', *IEEE Transactions on Software Engineering*, 1990, 16(4), pp. 432–443

48   BLUM, M., and KANNA, S.: 'Designing programs that check their work'. Annual ACM symposium on *Theory of computing*, 1989, pp. 86–97

49   BLUM, M., LUBY, M., and RUBINFELD, R.: 'Self-testing/correcting with applications to numerical problems'. Annual ACM symposium on *Theory of computing*, 1990, pp. 73–83

50   TASIRAN, S., and KEUTZER, K.: 'Coverage metrics for functional validation of hardware designs', *IEEE Design and Test of Computers*, 2000, 17(4), pp. 51–60

51   GAUDETTE, E., MOUSSA, M., and HARRIS, I.G.: 'A method for the evaluation of behavioral fault models'. *High level design validation and test* workshop, 2003, pp. 169–172

Chapter 20

Hardware/software cosimulation from interface perspective

Sungjoo Yoo and Ahmed A. Jerraya

20.1 Introduction

20.1.1 HW/SW cosimulation example

HW/SW cosimulation is to validate the functionality and timing of both software (SW) and hardware (HW) in a single simulation. Figure 20.1 shows an example of HW/SW cosimulation. Assume a system consisting of two processors (processor A and B) and an on-chip communication network as shown in Figure 20.1(a). To simulate the system, we may need two simulators (instruction set simulators, ISSs) for the two processors and a simulator of communication network as shown in Figure 20.1(b).

HW/SW cosimulation runs as the simulators communicate with each other by exchanging events (denoted with bold arrows in Figure 20.1(b)). Examples of events are memory accesses by processors, interrupts from network to processors. In order to obtain timing-accurate simulation, the simulators need to synchronise their simulation with each other. For that purpose, they exchange synchronisation events (denoted with dashed arrows in Figure 20.1(b)) at every clock cycle.

Figure 20.2(a) shows more details of the interface between processor and communication network. We assume ARM processor as the processor (processor A in Figure 20.1(a)). Figure 20.2(b) illustrates the signal waveforms of the interface. At time 1, the processor makes a write request to the address 0×1000 with the data of 0×0010 while setting control signals (nMREQ and nRW). The request can be terminated only when the HW grant (nWait = 1) arrives at time 4. Then, at time 6, HW interrupt (nIRQ = 0) arrives at the processor.

Figure 20.3 exemplifies an instruction set simulator (ISS A in Figure 20.1(b)) communicating with HW simulation. The function SimulateOneCycle( ) is called at every clock. When called, it exchanges events with HW simulation as shown at line 3. If an interrupt arrives, it sets the program counter (pc) to the start address of the interrupt

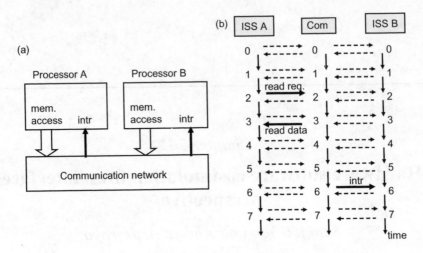

(a)

(b)

Figure 20.1   *An example of HW/SW communication (a) An example of interrupt-based inter-processor communication, (b) Lock-step synchronisation*

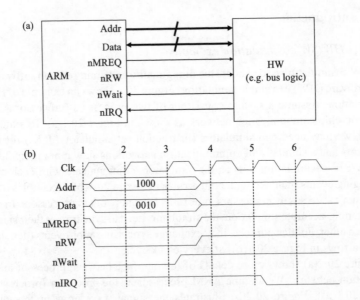

Figure 20.2   *An example of an interface between processor and HW and its waveform (a) Interface between processor and HW, (b) Waveform of interface signals*

service routine. Otherwise, it simulates the execution of the current assembly instruction. If the current (or previous) instruction is a load/store instruction, it may take more than one cycle depending on when the HW grant arrives. Line 14–20 shows the case that the current instruction is a store instruction to the address 0×1000 with

```
 1 SimulateOneCycle() {
 2 {
 3      ExchangeEvents(Addr, Data, nMREQ, nRW, nWait, nIRQ);
 4
 5      if (nIRQ == 0) pc = 0x18; // simulate the jump to interrupt service routine
 6
 7      switch(state) {
 8      case core: // simulate processor pipeline execution
 9          InstructionFetch();
10          Decode();
11          Execute(state); // may enter memory access states when load/store
12          break;
13      case wr_start:
14          Addr = Cur_Addr; // 0x1000
15          Data = Cur_Data; //0x0010
16          nMREQ = 0;
17          nRW = 1;
18          if(nWait == 0) state = wr_wait;
19          else state = core;
20          break;
21      case wr_wait:
22          if(nWait == 0) state = wr_wait;
23          else          nMREQ = 1;
24                        nRW = 0;
25                        state = core;
26          break;
27          ...
28      }
29 }
```

Figure 20.3   A code example of instruction set simulator communicating with HW simulation

the data, 0×0010. The simulator will set address/data buses and control signals (line 14–17), and wait for the grant (line 18). The simulation of a single cycle continues in this way.

Most of current commercial HW/SW simulators (e.g. Mentor Graphics Seamless CVE) have been implemented in this way i.e. exchanging events between SW and HW simulators at every clock cycle. In this chapter, based on this knowledge, we will address the current and future issues of HW/SW cosimulation.

20.1.2   Issues of HW/SW cosimulation

As system complexity grows, the validation becomes more and more time-consuming thereby becoming a bottleneck in shortening time-to-market. One of the effective ways to speed up HW/SW cosimulation is to raise the abstraction levels of simulation from the cycle-accurate level. However, while raising abstraction levels of simulation, designers face a new problem in handling high abstraction levels: mixed abstraction level (in short, mixed-level) cosimulation. The mixed-level cosimulation problem is how to manage different abstraction levels of SW and HW models (e.g. OS level SW model and transaction level HW model, etc.) in HW/SW cosimulation.

In this chapter, we explain mixed-level cosimulation in a unified manner using the concept of an 'HW/SW interface'. Then, we introduce a new challenge, i.e. cosimulation for MPSoC. For a survey on traditional issues and techniques of HW/SW cosimulation, readers are recommended to refer to References 1 and 2.

This chapter is organised as follows. In Section 20.2, we explain abstraction levels used in SoC design in terms of function, SW interface and HW interface. We present the concept, applications and techniques of mixed-level cosimulation in Section 20.3. We address the issue of cosimulation performance in raising abstraction levels in Section 20.4. We introduce a new challenge of cosimulation for MPSoC in Section 20.5. We summarise the chapter in Section 20.6.

20.2   Abstraction levels in HW/SW cosimulation

HW/SW cosimulation captures both application behaviour (SW application, HW design) and architectural component behaviour (bus, network-on-chip, memory, DMA, interrupt controller, etc.). HW/SW cosimulation is needed at every step of HW/SW refinement. The refinement is performed in two ways: function and interface refinement as shown in Figure 20.4.

Figure 20.4(a) illustrates function refinement. Function refinement is to transform functional specification into implementation in SW or in HW. For instance, SW functionality described in a sequential program (e.g. reference code of H.264 decoder) may be refined to multiple tasks (e.g. Entropy decoder, deblock filter, etc.) to exploit the parallelism of underlying hardware (e.g. multi-threaded processor). Functionality may also be written in a parallel fashion (e.g. multi-task) and can be implemented as a sequential code to run on a conventional single-threaded processor without a task scheduler. HW functionality may be refined (manually or by a behavioural synthesis tool) from an architecture-independent description (e.g. synthesisable C code)

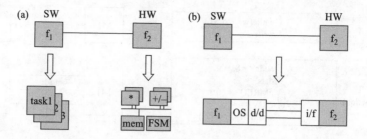

Figure 20.4   (a) Function and (b) interface refinement

to RTL (register transfer level) design which includes functional units (multipliers, adder/subtractors, etc.), memory elements and control units (as shown in the figure), and to a gate-level netlist.

Figure 20.4 (b) shows interface refinement. Interface refinement is to allow refined functionality to run and to communicate with each other. For instance, multiple tasks may need an operating system (OS) to run on the target processor and device driver (d/d) to communicate with external HW modules. HW design needs also interface logic for it to communicate with other HW or SW modules via on-chip communication network (e.g. on-chip bus). The interface logic can be a simple one to interpret a communication protocol (e.g. Advance eXtensible Interface (AXI) [3], Open Core Protocol (OCP) [4]) or a complex one such as a network interface [5].

Both function and interface refinement can take several steps, respectively. Each step may need an abstraction level. For instance, function refinement can go through algorithm, task/process, instruction accurate/RTL and cycle accurate levels. Interface refinement can be performed in two ways depending on the SW or HW side of interface. We call each side of interfaces 'SW' or 'HW interface'. In the following subsections, we explain the abstraction levels of the two types of interface.

20.2.1   Abstraction levels of HW interface

The abstraction levels of HW interface are known as transaction level models (TLMs). There are a few slightly different TLMs, e.g. message, transaction, transfer, and cycle accurate levels [6] and algorithm, programmer's view (PV), PV with timing (PVT), and cycle callable (CC) levels [7]. Figure 20.5 exemplifies the abstraction levels of HW interface according to the definition in Reference 6.[1]

Figure 20.5(a) shows an example of two modules (M1 and M2) which are connected with each other at cycle-accurate (C/A) level. Their interconnect has address, data, control signals and clock. The figure shows also a code section of module M1 to access the interconnect. It shows a cycle-by-cycle behaviour of HW interface. The address and data bus signals are handled in a bit-by-bit manner.

---

[1] We use transaction level models in Reference 6 only as an example. The arguments described in this chapter are not limited to the used examples, but include also other TLMs, e.g. Reference 7.

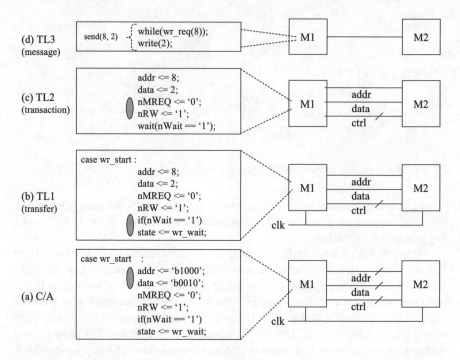

Figure 20.5   Abstraction levels of HW interface

Figure 20.5(b), denoted by TL1, gives a higher abstraction level than C/A level by abstracting away the bit-by-bit behaviour of address and data signals (a code section denoted by an oval in Figure 20.5(a)). In Reference 6, this level is called 'transfer level'. Avoiding the bit-by-bit manipulation of address and data signals, this level gives simulation speed up without losing simulation accuracy since it abstracts only the representations (i.e. data types in simulation models) of address and data signals.

The simulation model of Figure 20.5(c), denoted by TL2, abstracts away the clock signal (a code section denoted by an oval in Figure 20.5(b)). The model becomes an event-driven one. It may give faster simulation than that of Figure 20.5(b) since event-driven simulation is known to be superior to cycle-based simulation in terms of simulation performance especially when event activity is low.[2] Simulation accuracy can be still cycle-accurate at this level.

Figure 20.5(d) gives a very different abstraction than the other two ones in Figure 20.5(b) and (c). Compared with Figure 20.5(c), it gives an abstraction of control signals (Memory Request (nMREQ) and Read Write (nRW) in this case). Since the behaviour of control signals specifies a protocol of on-chip interconnect, e.g. AXI [3] and OCP [4], the abstraction makes the HW interface independent of

---

[2] The performance difference between cycle-based and event-driven simulation is also determined by the efficiency of the simulation kernel. In other words, an event-driven simulation kernel with high scheduling overhead may yield worse performance than a highly optimised cycle-based simulation kernel.

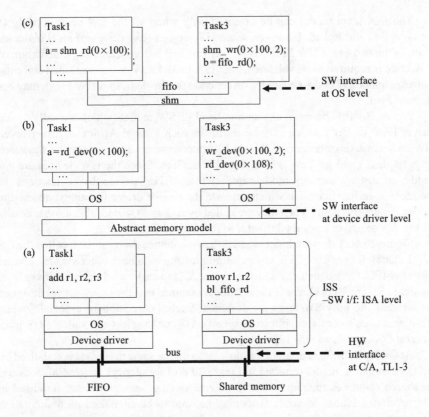

Figure 20.6   Abstraction levels of SW interface

on-chip communication protocol. Thus, the simulation accuracy of this model is not guaranteed to be cycle-accurate. This level is called the 'message level' (TL3). Since an HW module with its interface at message level is not limited to a specific on-chip communication protocol, it can be easily reused over different SoC designs with different on-chip communication protocols.

20.2.2   Abstraction levels of SW interface

Figure 20.6(a) shows an example of SoC architecture which consists of two processors, on-chip bus, shared memory and a dedicated FIFO (first-in, first-out). SW tasks, Task 1 and 2 (and associated OS and device drivers) run on one processor and Task 3 on the other processor. The SW code (task, OS, and device driver code) is assumed to be compiled using a target compiler (e.g. armcc in the case of the ARM processor) and simulated on an instruction set simulator (ISS). In terms of a simulation model in HW/SW cosimulation, we call the compiled SW code 'instruction set architecture (ISA) level model'. The ISA is a SW interface to the SW code. From the viewpoint of HW simulation model, the processor may be represented by a bus functional model (BFM) [8].

The ISA level model can be obtained only when all the SW code is ready to be compiled and linked. However, when HW design (e.g. design of an HW device such as the dedicated HW FIFO) is not yet finished but a current implementation of SW code is required to be validated,[3] the ISA model may not be exploited since the corresponding SW code, i.e. device driver code for the dedicated HW FIFO, may not be ready yet.

Figure 20.6(b) shows an abstraction level of SW model (which we call 'device driver level model') which may be useful in such a case. At device driver level, SW code consists of task and OS code.[4] Both codes can call device driver functions (e.g. rd_dev( ) and wr_dev( )) to access HW devices. Since the HW devices are not ready, an abstract memory model replaces them. Thus, device driver functions (to be more specific, functions which emulate the device driver functions) access the abstract memory model when they are called by task and OS code. For further details of the device driver level model, refer to References 9–14.

Figure 20.6(c) shows a higher model than the device driver model, called the OS level model. It consists of SW tasks communicating with each other via OS services such as FIFO, shared memory, semaphore, etc. Compared with the device driver level model in Figure 20.6(b), the OS level model assumes that the OS is not yet designed or selected and only SW tasks are designed. Such a case happens especially when designers want to design application-specific OSs or to select one suitable OS from several OS candidates [15–18].

At OS level, the execution of multiple tasks on the same processor is serialised by an OS simulation model (omitted in Figure 20.6(c) for clarity). In general, accesses to shared resources (including shared objects such as semaphore) are serialised in OS level simulation. Multiple tasks may be able to be simulated on a simulation environment (i.e. functional simulation) such as SystemC [19] without task serialisation which could be enabled by the OS simulation model. However, in this case, simulation accuracy is inferior to what OS simulation gives. Figure 20.7 exemplifies the case.

Figure 20.7 illustrates a typical case where OS level simulation helps to reveal design problems that cannot otherwise be 'detected' by a functional simulation of the system (e.g. simulating multiple tasks in a simulation environment such as SystemC without OS simulation model). Figure 20.7 (a) shows an example of system specification composed of three tasks T_1, T_2 and T_3. T_1 and T_2 access a shared resource protected by a semaphore. T_3 is activated by an external asynchronous event. Upon the reception of that event, a signal is emitted to T_2. Then, T_2 executes some computation (using the shared resource) and then sends a signal back to T_3 in order to notify the end of the computation.

Figure 20.7(b) shows an execution trace of the example obtained in functional simulation. At time t_1, task T_1 starts to run and acquires the semaphore. At time t_2,

---

[3] Recently, such cases are encountered more and more often since SW code is getting more and more complex and it needs to be validated even before the HW prototype is available.

[4] OS code is written on top of API, called the hardware abstraction layer (HAL) or board support package (BSP) API.

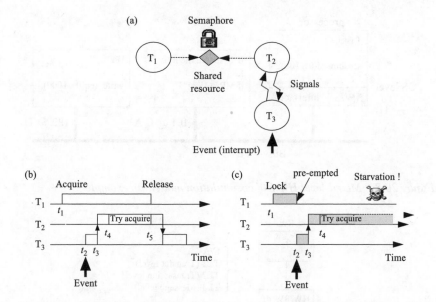

Figure 20.7   Comparison between functional and OS level simulation

an external event arrives and task T_3 starts to run. Since accesses to shared resources are not serialised in functional simulation, both tasks can run concurrently. At t_3, task T_2 receives an event from T_3 and starts to run. At t_4, T_2 tries to acquire the semaphore. Since the semaphore is already locked by T_1, T_2 keeps trying to acquire the semaphore until T_1 releases it at t_5. T_2 acquires the semaphore at t_5 and continues its execution.

Figure 20.7(c) shows the execution trace obtained in the simulation with an OS simulation model. In this case, we assume that the three tasks are mapped on a single processor, a static priority-based pre-emptive scheduling is used and task T_3 (T_1) has the highest (lowest) priority. The execution is the same with the case of Figure 20.7(b) by time t_2. At t_2, due to the external event, the execution of T_1 is pre-empted by the OS simulation model and T_3 starts to run. Then, after receiving the signal from T_3, T_2 starts to run. At t_4, T_2 tries to acquire the semaphore. Since the semaphore is already locked by T_1, T_2 fails to acquire it. However, since it has the highest priority, T_2 keeps trying to acquire the semaphore holding the processor (forever). As shown in this example, OS level simulation can reveal design errors (especially, related to multi-task synchronisation) that might not have been detected in functional simulation.

20.3   Mixed-level cosimulation

Figure 20.8 exemplifies an HW/SW cosimulation model which consists of SW and HW simulation models at different abstraction levels. In the figure, an SW task

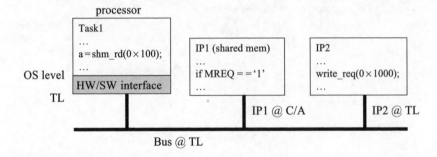

Figure 20.8   Mixed-level HW/SW cosimulation model: an example

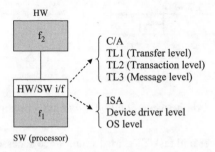

Figure 20.9   Abstraction levels of HW/SW interface

running on a processor is at OS level and the processor is modelled at transaction level (TL) on its HW interface. The on-chip bus is at TL, two HW modules are at TL and cycle-accurate level (C/A), respectively.

In the case of the processor, we have a special interface for mixed-level cosimulation which is located between SW code and the HW interface of the processor. It is called the 'HW/SW interface for mixed-level cosimulation', in short, the HW/SW interface in this chapter[5] (shaded rectangle in Figure 20.8). It has two interfaces: SW interface for SW code and HW interface for connection with the other HW parts. In HW/SW cosimulation, the HW/SW interface serves to enable SW code (e.g. multiple tasks) to run (e.g. by an OS simulation model) and to communicate with HW modules.

HW/SW interface can be specified by the abstraction levels at both sides of the HW/SW interface as shown in Figure 20.9. For instance, assuming that the HW interface has four abstraction levels and the SW interface three abstraction levels as shown in the figure, we may need up to 12 cases for the HW/SW interface. In this regard, the HW/SW interface is a generalised model of conventional BFM since it covers a wider range of abstraction levels than the conventional BFM. The conventional BFM

---

[5] In a more precise terminology, the HW/SW interface represents real SW code (OS and device driver) and HW interface logic which enables SW tasks running on the processor to communicate with HW modules. In this chapter, we use the terms only in simulation perspective.

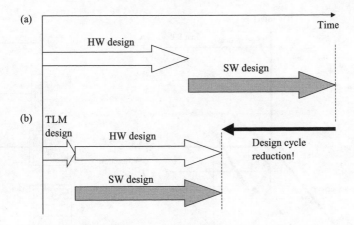

Figure 20.10   HW/SW   co-development   (a)   Sequential   HW/SW   development,
(b) HW/SW co-development based on TLM

covers only the cases with an SW interface at algorithm level or ISA level and an
HW interface at cycle-accurate level. Each of the existing solutions in mixed-level
HW/SW cosimulation covers only a subset of the total combinations of an HW/SW
interface. For instance, Reference 20 covers only the combinations of device driver
level (SW interface) and TLMs (HW interface) while Reference 17 covers only those
of OS level (SW interface) and TLMs (HW interface).

The mixed-level cosimulation model exemplified in Figure 20.8 is quite common
in current SoC design flows. In this section, we explain how mixed-level cosimulation
models are produced in the design flow and how to simulate those models.

20.3.1   Where do mixed-level HW/SW cosimulation models come from?

Designers encounter various mixed-level cosimulation models in the HW/SW co-
development flow as shown in Figure 20.10. In conventional sequential design flow
(shown in Figure 20.10 (a)) where SW design starts only after HW design is finished,
in HW/SW cosimulation, the abstraction level of HW design is fixed (mostly at C/A
level). Only the abstraction levels of SW (though there are usually only two levels
of SW abstraction, algorithm and ISA levels) can change. In such a case, mixed-
level cosimulation is enabled by a bus functional model (BFM) which transforms
memory accesses from SW code into cycle-accurate events on processor interface
ports (address/data buses and control signals) [8].

Figure 20.10(b) illustrates HW/SW co-development exploiting the TLM concept.
In this flow, first the transaction level model of HW design is created. Then, SW
design can start using the HW TLM as a virtual platform [21–24]. SW design can be
refined from algorithm level down to ISA level. SW code can be compiled, loaded
and executed on the virtual platform. The debugging and optimisation of SW design
can be performed exploiting the virtual platform. In the meantime, HW design can be
refined from TLM to C/A in a simultaneous way. Figure 20.10(b) exemplifies the

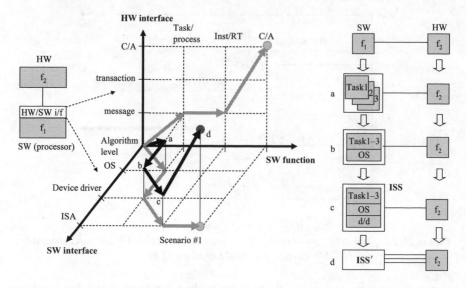

Figure 20.11   (a) HW/SW function and interface refinement space and (b) a scenario of #1 refinement

reduction in design cycle obtained by HW/SW co-development. According to recent industrial applications of TLM, the design cycle can be reduced up to 30 per cent by early SW development using TLMs [25,26].

In terms of the number of abstraction levels, designers encounter more cases of mixed-level HW/SW cosimulation model in the HW/SW co-development flow than in the conventional sequential design flow. Considering a general design flow where we can encounter all the possibilities of mixed-level cosimulation models, we can imagine a 'refinement space' of function and interface as shown in Figure 20.11(a).

The space has three dimensions: one of SW function refinement and two of HW and SW interface refinement. Given a processor or HW module, we can imagine one refinement space like Figure 20.11(a). Figure 20.11(a) represents a refinement space of a processor on which function f_1 runs as shown in the left-hand side of Figure 20.11(a).

The abstraction levels of the function are shown to be task/process level, instruction/RT (register transfer) level and cycle-accurate level. SW and HW interface dimensions in the space represent those of the HW/SW interface[6] as the two dashed arrows in the left-hand side of Figure 20.11(a).

In the space, a point is denoted by a tuple ⟨AS, AH, AF⟩, where AS, AH and AF represent the abstraction levels of the SW interface, HW interface and SW function. The origin of the space represents an algorithm level code (which is not yet partitioned into HW and SW).

---

[6] In the case of the HW module, the refinement space will be two dimensional since only function and HW interface dimensions are required.

A scenario of refinement corresponds to a walk in the refinement space. Figure 20.11(b) shows a scenario of refinement. Figure 20.11(a) shows the corresponding walk from the space origin to a point, ⟨ISA, C/A, C/A⟩ denoted with a solid circle in the figure.

Point a (in both figures) represents the case that the algorithm level code is mapped to SW and refined to multiple SW tasks. A solid arrow from the space origin to point a represents the refinement. The figure shows also the projections (shaded arrows) of the arrow on the sub-space of function and HW interface and that of function and SW interface. In terms of SW interface abstraction, the multiple tasks are at OS level. The HW interface (of the processor on which SW tasks will run) is at message level.

Refinement from point a to b represents that an OS is designed or selected for the multiple tasks. However, device drivers are not yet fixed since HW device design is not yet finished. Since only the SW interface is refined from OS level to device driver level, there is no projected arrow on the sub-space of function and HW interface in this case. The HW interface is still at message level.

Refinement from point b to c represents that the device drivers are designed (maybe since the corresponding HW devices design is finished). Since all the SW code is ready, it can be compiled and run on an ISS which provides for instruction accuracy, e.g. Reference 24. However, the HW interface of the processor is at message level in this case (maybe since the HW interface of HW module f_2 in Figure 20.11 is not yet refined but remains at message level). The projected arrow on the sub-space of function and SW interface corresponds to the refinement taken in this step.

Refinement from point c to d represents a new ISS (denoted by ISS' in the figure) with cycle accuracy. The HW interface is refined to cycle-accurate level (maybe since the HW interface of module f_2 is refined to cycle-accurate level).

As shown in Figure 20.11, the refinement of function and interface can give various combinations of mixed-level cosimulation models. In the next sub-section, we will explain how to simulate mixed-level cosimulation. Before going to the next sub-section, note that another important application where mixed-level cosimulation is required is to enhance simulation performance by raising the abstraction levels of some parts of the system which are already refined to a low level. Section 20.4 will address the performance issue.

20.3.2 How to simulate mixed-level cosimulation?

First, we will introduce mixed-level HW simulation methods. Then, we explain a key technology in mixed-level HW/SW cosimulation, simulation of handling interrupts in OS level simulation.

20.3.2.1 Mixed-level HW simulation

We will handle cases where HW modules with different abstraction levels of interface communicate with each other. There are two approaches to tackle this issue. One is to simulate all the abstraction levels of HW interface during simulation [27]. The other is to design/generate a simulation wrapper that adapts different abstraction levels [28].

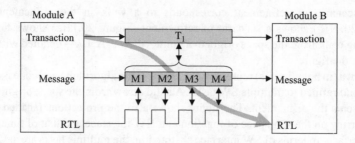

Figure 20.12   Simulation for mixed-level HW interface: SystemC^{SV} case

Figure 20.12 shows a case where all the abstraction levels are simulated [27]. In this model called 'SystemC^{SV}', three abstraction levels of the HW interface are assumed: RTL, message and transaction.[7] The figure exemplifies the meaning of each abstraction level. Assume that the two modules A and B have only a single interconnection between them and that, as shown in the figure, the HW interface of module A is at transaction level and that of module B at RTL. We need to make both sides of transaction level and RTL communicate with each other for the simulation to work. The method simulates all the three levels of the HW interface though only a part of the simulation (denoted by the shaded arrow in the figure) is really required.

The advantages of this method are two-fold. One is that the communication protocol being simulated at high level can be validated against the simulation results obtained at low levels. The other is that interconnecting modules with different abstraction levels of the HW interface do not need an additional adaptation since all the possible abstraction levels are already supported in the HW interface.

In Reference 28, the authors present a method of generating only the necessary simulation wrapper for an HW interface. Figure 20.13 shows the internal architecture of the simulation wrapper of an HW interface. It consists of two parts. One is called 'simulator interface' which adapts different simulation languages and simulators. The other is called 'communication interface' which adapts different abstraction levels of HW interface. Communication interface is divided into three parts: port adapter, channel adapter and internal communication media.

A port adapter is connected to the port(s) of the HW module (via the simulator interface, if different simulation languages/simulators are used). For each pair of port of HW module and its abstraction level, a port adapter is pre-designed in a simulation model library. For each communication protocol, a channel adapter is pre-designed for each abstraction level of the communication protocol. Internal communication media does not carry any information on abstraction level between port and channel adapters. Thus, a port/channel adapter depends only on the abstraction level of HW module port/communication channel.

---

[7] The names of abstraction levels are different from one model to another. The abstraction levels with the same name do not correspond to each other, e.g. transaction levels in References 6 and in 7.

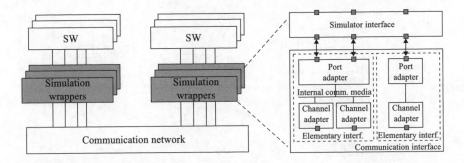

Figure 20.13   *Automatic generation of simulation model for mixed-level simulation*

Given an HW module and a set of communication channels, a simulation model is generated according to the abstraction levels of ports of the HW module and those of the communication channels. This approach aims at automatically generating simulation wrappers for mixed-level HW simulation while minimising the number of required library components.

20.3.2.2   Mixed-level HW/SW cosimulation

As the abstraction level of SW design is raised, integrating OS level simulation of SW code into HW/SW cosimulation gets more and more attention [11,12,17,18,20]. Even a commercial tool for this purpose has been released [15].

In terms of simulation model composition, in this case, SW simulation model consists of SW code, OS/device driver simulation model (e.g. VxSim [29], SoCOS [30], ITRON [31]) and a bus functional model (BFM) which supports TLMs as well as cycle-accurate level. Figure 20.14(a) exemplifies HW/SW cosimulation model where SW model consists of SW task code, OS model and BFM.

Compared with conventional simulation with OS models [29] where we use OS simulators on top of which application tasks run, HW/SW cosimulation with OS simulation differs in that (1) HW simulation is performed and (2) SW simulation is timed [18]. Especially, timed SW simulation enables us to validate performance as well as functionality, which is not allowed in conventional simulation with OS models.

Timed SW simulation is enabled by annotating SW execution delay on the target processor into the SW task code. Figure 20.14(a) exemplifies the annotation with function delay(). Delay values can be estimated using existing methods of estimating SW execution delay on the target processor [32].

The key technology in HW/SW cosimulation with OS (device driver) models is the simulation of interrupt handling. Function delay() needs to simulate interrupt handling. Figure 20.14(b) shows how function delay() works in a simulation environment such as SystemC. When it is called by task code, function wait()[8] is called. When function wait() returns, we can have one of two cases. One is that there was no

---

[8] In this case, we use SystemC wait(**event, time**) which returns when **time** elapses without **event** or when **event** occurs before **time** elapses.

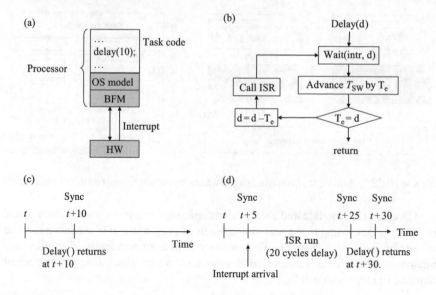

Figure 20.14   Simulation of interrupt handling

interrupt event which arrived at the processor on which task code runs during the time period of delay d. Figure 20.14(c) exemplifies this case. At time t, function delay(10) is called by task code. There is no interrupt event from time t to $t + 10$. In this case, function delay(10) returns after advancing the simulation time of SW, T_{SW}[9] by the amount of elapsed time, T_e ($= 10$) as Figure 20.14(b) shows.

The other case that function wait( ) returns is when an interrupt event arrived at the processor before the delay period elapses. Figure 20.14(d) exemplifies this case. In this case, an interrupt event (e.g. nIRQ changes from '1' to '0' in the case of ARM7 processor) arrives at time $t + 5$. Upon the interrupt, an interrupt service routine (ISR) is executed in reality. In HW/SW cosimulation, a simulation model of ISR is executed as shown in Figure 20.14(b).

After the ISR simulation is finished, function delay( ) needs to continue to complete the remaining delay. To do that, before calling the ISR simulation, the remaining delay value is calculated ($d = d - T_e$). As Figure 20.14(d) shows, after finishing the ISR simulation, the remaining time period needs to be elapsed as if a new function delay( ) is called with the remaining period.

During ISR simulation, the ISR simulation model can invoke OS scheduling (in the OS model) thereby yielding a task context switch. The ISR simulation can be even pre-empted by the arrival of another higher priority interrupt (e.g. FIQ in the case of ARM7 processor). That is, nested interrupts need to be supported. For further details

---

[9] We assume that SW simulation time and HW simulation time are managed separately. They are synchronised when function delay( ), to be more specific, function wait( ) is called.

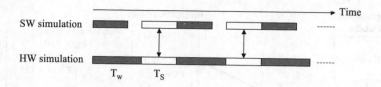

Figure 20.15 synchronisation overhead in HW/SW cosimulation

of OS/device driver level simulation in HW/SW cosimulation, refer to References 13 and 14.

20.4 Performance of mixed-level cosimulation

In this section, we address performance issues that designers face while raising the abstraction levels of HW/SW cosimulation. Simulation performance is determined by simulation workload (of SW and HW simulation) and simulation kernel overhead. It is known that simulation workload decreases as the abstraction level is raised. Simulation kernel overhead is inversely proportional to the number of synchronisations between SW and HW simulation.

Figure 20.15 illustrates the synchronisation overhead in HW/SW cosimulation. In the figure, dark rectangles represent simulation workload (T_w) and blank rectangles synchronisation overhead (T_s) in terms of simulation runtime. Synchronisation overhead may include IPC (inter-process communication) overhead in conventional multi-process HW/SW cosimulation [33] and scheduler overhead, e.g. SystemC scheduler overhead in uni-process HW/SW cosimulation [22,23].

20.4.1 Amdahl's law in HW/SW cosimulation performance

Raising abstraction levels of simulation does not always yield expected speedup. It is because if one item (e.g. cycle-accurate HW simulation workload) is improved (by raising the abstraction level to that of algorithm or TLM), speedup is not as high as expected. It is because the dominance in simulation runtime changes as we raise the abstraction level of SW and HW simulation. For instance, in the case that SW simulation is performed by instruction set simulators and HW simulation by cycle-accurate RTL simulation, HW simulation may dominate total simulation runtime (while yielding simulation performance, $\sim$1k cycles/s). If we raise the abstraction level of HW simulation from cycle-accurate level to cycle-approximate task level (function) and TLM level (HW interface), then, dominance may move from HW simulation to SW simulation thereby giving simulation performance ($\sim$100k cycles/s) which may be inferior to what the simulation speedup in HW simulation alone can give. In this case, if we raise again the abstraction level of SW simulation from ISA level to device driver or OS level, then, dominance may move again from SW simulation to synchronisation overhead while giving higher simulation performance, e.g. $\sim$10M cycles/s. The simulation speedup is inferior to what device driver/OS level

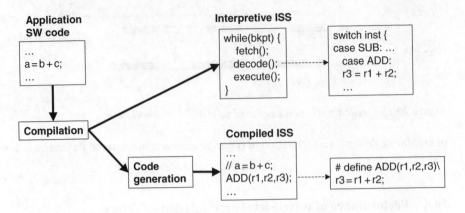

Figure 20.16   Two types of instruction set simulation

simulation alone gives (e.g. ~1000 times speedup). The reason why synchronisation overhead dominates is that HW and SW simulation must synchronise with each other frequently to check interrupt events going from HW to SW simulation.

20.4.2   Techniques for improving simulation performance

The most powerful way to improve simulation performance is raising abstraction levels of simulation. However, in each case of possible combinations of abstraction levels in HW/SW cosimulation, we need techniques to further improve simulation performance. In the following, we explain existing techniques of speeding up SW simulation and those of reducing synchronisation overhead thereby improving HW/SW cosimulation performance. In terms of HW simulation speedup, except in the case of transaction level modelling, there is little work at abstraction levels higher than RTL, i.e. algorithm or task/process level since existing simulation techniques [34] which are independent from HW/SW partitioning can be applied to high-level HW simulation.

Most techniques for SW simulation speed up focus on instruction set simulation [35–37]. Figure 20.16 illustrates two types of instruction set simulation: interpretive and compiled simulation.

Interpretive instruction set simulator (ISS) takes as input compiled assembly code. It interprets each of the assembly instructions. The figure exemplifies the interpretive ISS of a processor with three pipeline stages, e.g. ARM7 processor. The ISS (e.g. ARMulator) simulates each pipeline stage. In the case of interpretive ISS, the simulation of the decoding stage takes most of the simulation runtime since it consists of a very long switch-case statement which interprets each instruction as the figure shows.

Compiled ISS overcomes the problem of instruction decoding in the interpretive ISS by pre-interpreting assembly instructions (in Code Generation step in Figure 20.16) and by generating the simulation model of each assembly instruction in the source code of ISS as shown in the figure [35]. By removing the overhead of instruction decoding during simulation run, the compiled ISS can improve the simulation performance significantly. However, it lacks in supporting the simulation of self-modifying code and interrupts. It suffers also from the overhead of code size increase.

There have been several approaches to preserve the speed of compiled ISS while keeping the capability of interpretive ISS. In Reference 36, SW simulation is based on an interpretive ISS. The key idea of this method is to reuse the information of instruction decoding. First, an assembly instruction is simulated by the interpretive ISS. The information of instruction decoding is stored in a table together with the address of this instruction. Later, when the same instruction (identified by its address) needs to be decoded, the saved information of instruction decoding is used instead of executing the interpretive ISS. Since the instruction decoding is performed only once (if the table is large enough) during the simulation run, the method is called JIT (just-in-time) cache compiled simulation.

In Reference 38, a concept called 'cached simulation' is presented. The main idea of this method is to replace the ISS execution in HW/SW cosimulation as often as possible by that of the source code of the application function. To do that, the delay of the application function is obtained by executing the ISS when the application function is called for the first time. Then, the delay value is stored in a table called the 'delay cache' together with the information of execution path in the code of application function. When the same execution path of the application function is simulated, the delay value is reused instead of executing the ISS. Thus we can minimise executing the ISS.

Synchronisation overhead reduction can be achieved by optimistic approaches [39,40] or by a concept of message grouping [41]. Figure 20.17 shows two types of synchronisation between SW and HW simulation: lock step and optimistic synchronisation. In lock step synchronisation, at every system clock cycle, both SW and HW simulation synchronises with each other to exchange events (e.g. memory read/write request signals from SW to HW simulation and interrupts from HW to SW

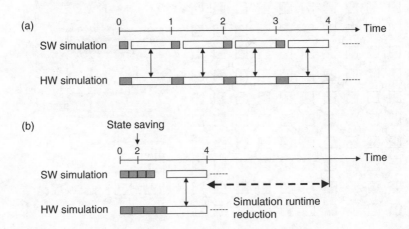

Figure 20.17 Reducing synchronisation overhead by optimistic simulation: (a) lock step synchronisation, (b) optimistic simulation

simulation) as shown in Figure 20.17(a). The overhead of such a synchronisation is inhibitively high especially when multiple processes are involved in simulation or when the abstraction levels of both SW and HW simulation are high (e.g. OS level for SW simulation and task/process level for HW simulation, respectively).

Optimistic simulation reduces the synchronisation overhead by skipping synchronisation for a certain amount of simulation cycles. In Figure 20.17(b), SW simulation is assumed to advance its execution by four clock cycles optimistically, i.e. without synchronisation with HW simulation. During that period, it saves its simulation state (at time 2). At time 4, both simulations synchronise with each other. If there is an event that should have been exchanged between SW and HW simulation before time 4, SW simulation rolls back to one of saved simulation states whose timestamp is earlier than that of the missed event. If not, the simulation continues in the same way. The figure illustrates simulation speedup by optimistic simulation. In this method, a key consideration is to decide when to save simulation states since too frequent state saving may offset the speedup by state saving overhead.

20.5   A challenge: cosimulation for MPSoC

MPSoC is introducing new concepts and problems into SoC design methodology including network-on-chip [42], parallel processors [43], distributed memory [44], parallel programming model [43], distributed OS [45], etc. HW/SW cosimulation faces a new challenge in the MPSoC era. Figure 20.18 illustrates HW/SW cosimulation of MPSoC. A simplified architecture of MPSoC is depicted in the left-hand side of the figure. It consists of sub-systems (big rectangles in the figure) and a network connecting sub-systems via routers (small rectangles in the figure).

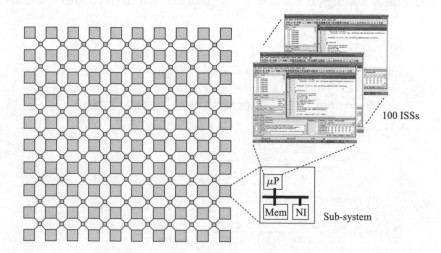

Figure 20.18   HW/SW cosimulation of MPSoC

The interconnection among routers is omitted in the figure for clarity. A sub-system may consist of processor(s), local interconnect, local memory and network interface (NI) as exemplified in the figure.

Assume that we perform HW/SW cosimulation of the MPSoC with a cycle-accurate ISS for each processor. We may need up to 100 ISSs for the simulation as exemplified in the figure. Such a simulation may suffer from very low speed (less than ∼1kcycle/s) due to high simulation workload of SW simulation (e.g. 100 ISSs run) though HW simulation may be performed at a high abstraction level. Such a low performance may not be acceptable in application software development and in architecture exploration (e.g. buffer size optimisation for DMA or network interface, etc.).

To overcome the problem of poor simulation performance, we may need to apply higher abstraction levels to HW/SW cosimulation (e.g. OS level simulation [15–18,46,47] instead of cycle-accurate ISS execution) or apply other methods such as parallel simulation.

In the near future, a slightly higher abstraction level model, e.g. instruction accurate (IA) ISS may be practically applied together with functional/timed models of peripherals. The performance of IA ISS ranges between 10 and 100 MIPS on high-performance PCs. Such a high performance even allows for simulating an entire board consisting of a complex processor, bus and peripherals almost in real time [47]. Embedded software developers will benefit from the simulation model since they can validate application SW code in real time before the hardware prototype is available. In the cases that tens of processors need to be simulated, IA ISSs may not give enough performance to allow for application SW validation. A set of higher abstraction levels for both hardware and software simulation [13–16] may need to be applied in this case.

In order to improve simulation performance, high-level simulation may not be always the best solution since it lacks in simulation accuracy (which might be needed in performance estimation) and visibility of simulated system (which might be needed in functional validation or in debugging). One desirable solution may be to be able to change abstraction levels of simulation dynamically during the simulation run. Figure 20.19 exemplifies this idea.

Assume that we simulate a MPSoC with N processors. The figure shows that the abstraction levels of simulation of processors change dynamically during simulation between a high abstraction level (HL), e.g. OS level and a low abstraction level (LL), e.g. ISA level. The abstraction level may be lowered when a detailed simulation is needed for debugging purpose (e.g. ISS execution to check stack overflow) or for simulation accuracy (e.g. ISS execution to simulate interrupt handling at cycle accuracy). The figure illustrates that the abstraction levels of processors #1 and #N are lowered between times t_2 and t_3 to accurately simulate inter-processor communication (e.g. DMA launch, waiting on interrupt, etc.). After the low-level simulation, the abstraction levels may be raised to speed up simulation.

Changing the abstraction levels of simulation is not a new idea. However, the existing technique [48] is limited to changing the abstraction levels of the HW interface. In the case of MPSoC cosimulation, we need a method which enables us to

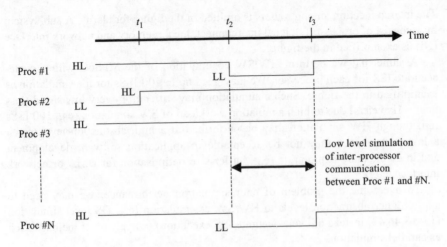

Figure 20.19   Dynamic change of abstraction levels in HW/SW cosimulation of MPSoC

change all the abstraction levels of HW/SW interfaces and function. To the best of authors' knowledge, there is little work for this issue.

Considering Amdahl's law in HW/SW cosimulation explained in Section 20.4, if both the abstraction levels of SW and HW simulation are raised, synchronisation overhead may dominate again total HW/SW simulation runtime. For instance, recalling OS level simulation in Figure 20.14, if the granularity of delay annotation is very small, e.g. a few clock cycles, then the synchronisation overhead by function delay( ) will dominate the entire simulation runtime, which may give a poor simulation performance masking off the benefit of high-level simulation.

Possible solutions to overcome this problem will be (1) to increase the granularity of delay annotation (hundreds or thousands of cycles delay) in order to reduce the number of synchronisation, (2) to predict timing points only necessary for synchronisation [40], or (3) to apply optimistic simulation approaches [39,41]. More research is required for novel techniques to reduce synchronisation overhead in high-level HW/SW cosimulation of MPSoC.

20.6   Conclusions

In this chapter, we explained mixed-level HW/SW cosimulation as the current issue of HW/SW cosimulation and addressed the performance problem of MPSoC cosimulation. First, we introduced the abstraction levels of function, HW interface and SW interface. Mixed-level cosimulation is required in HW/SW co-development and for the purpose of enhancing cosimulation performance. To better understand how mixed-level cosimulation models are produced in HW/SW co-development, we introduced the concept of refinement space. Techniques of mixed-level cosimulation were explained for both cases of HW interface and SW interface.

In terms of cosimulation performance, raising abstraction levels of simulation may not always give as much performance improvement as expected due to Amdahl's law. Thus, in addition to raising abstraction levels, we need techniques to improve each of SW and HW simulation and to reduce synchronisation overhead in HW/SW cosimulation. In this chapter, we explained techniques to improve SW simulation based on instruction set simulation and to reduce synchronisation overhead exploiting optimistic simulation.

MPSoC prevents a new problem to HW/SW cosimulation. Due to a high number of processors in MPSoC, current solutions of HW/SW cosimulation based on ISS execution may not give a sufficient simulation performance. In this chapter, we addressed the direction of some potential research to devise new techniques to improve HW/SW cosimulation performance for MPSoC with a trade-off between performance and accuracy.

References

1 RAWSON, J.A.: 'Hardware/software co-simulation'. Proceedings of the *Design automation conference*, 1994, pp. 439–440

2 VALDERRAMA, C., *et al*.: 'Automatic generation of interfaces for distributed c-vhdl cosimulation of embedded systems: an industrial experience'. Proceedings of the international workshop on *Rapid system prototyping*, June 1996

3 AXI PROTOCOL: http://www.arm.com

4 OCP PROTOCOL: http://www.ocpip.org

5 RADULESCU, A., *et al*.: 'An efficient on-chip network interface offering guaranteed services, shared-memory abstraction, and flexible network programming', *IEEE Transactions on CAD of Integrated Circuits and Systems*, 2005, 24(1), pp. 4–17

6 HAVERINEN, A., *et al*.: 'SystemC based SoC communication modeling for the OCP protocol', http://www.ocpip.org/, October 2002

7 VANTHOURNOUT, B., GOOSSENS, S., and KOGEL, T.: 'Developing transaction-level models in systemC', available at http://www.us.design-reuse.com/articles/article8523.html

8 SÉMÉRIA, L., and GHOSH, A.: 'Methodology for hardware/software co-verification in C/C++'. Proceedings of the *Asia and South Pacific design automation* conference, January 2000, pp. 405–408

9 TAN, S.M., *et al*.: 'Virtual hardware for operating system development'. Technical report, UIUC, http://choices.cs.uiuc.edu/uChoices/Papers/uChoices/vchoices/vchoices.pdf, September 1995

10 CARBON KERNEL: http://www.carbonkernel.org/

11 BRADLEY, M., and XIE, K.: 'Hardware/software co-verification with RTOS application code', http://www.techonline.com/community/tech_topic/21082

12 HONDA, S., *et al*.: 'RTOS-centric hardware/software cosimulator for embedded system design'. Proceedings of the international conference on *Hardware/software codesign and system synthesis (CODES-ISSS)*, 2004

13   YOO, S., BOUCHHIMA, A., BACIVAROV, I., and JERRAYA, A.A.: 'Building fast and accurate SW simulation models based on SoC hardware abstraction layer and simulation environment abstraction layer'. Proceedings of the *Design automation and test in Europe*, 2003

14   BACIVAROV, I., *et al.*: 'ChronoSym – a new approach for fast and accurate SoC cosimulation', *International Journal of Embedded Systems*, 2005, **1/2**

15   COFLUENT STUDIO: http://www.cofluentdesign.com/

16   MADSEN, J.: 'RTOS modelling for multi-processor SoC using SystemC'. MPSoC School, 2003

17   YU, H., GERSTLAUER, A., and GAJSKI, D.: 'RTOS scheduling in transaction level models'. International conference on *Hardware/software codesign and system synthesis*, October 2003, pp. 31–36

18   BOUCHHIMA, A., YOO, S., and JERRAYA, A.: 'Fast and accurate timed execution of high level embedded software using HW/SW interface simulation model'. Proceedings of the *Asia and South Pacific design automation conference*, 2004

19   SystemC: http://www.systemc.org

20   YOO, S., *et al.*: 'Automatic generation of fast timed simulation models for OS in SoC design'. Proceedings of the conference on *Design automation and test in Europe*, 2002

21   System Studio:
http://www.synopsys.com/products/cocentric_studio/cocentric_studio.html

22   ConvergenSC: http://www.coware.com

23   MaxSim: http://www.axysdesign.com/products.html

24   Platform development kits: www.virtio.com

25   KONG, J.: 'SoC IN NANOERA: Challenges and endless possibility'. Keynote Speech. Proceedings of the DATE, March 2005

26   'System-Level IC Design Accelerates SoC Delivery'. Nikkei Electronics ASIA, February 2005

27   SIEGMUND, R., and MÜLLER, D.: 'SystemC*SV*: An extension of SystemC for mixed multi-level communication modeling and interface-based system design'. Proceedings of the DATE, 2001

28   NICOLESCU, G., *et al.*: 'Validation in a component-based design flow for multicore SoCs'. Proceedings of the international symposium on *System synthesis*, 2002

29   VxSim, Windriver Systems Inc.:
http://www.windriver.com/products/html/vxsim.html

30   DESMET, D., *et al.*: 'Operating system based software generation for systems-on-chip'. Proceedings of the *Design automation* conference, 2000

31   HASSAN, M., *et al.*: 'RTK-Spec TRON: a simulation model of an ITRON Based RTOS Kernel in SystemC'. Proceedings of the DATE, March 2005

32   LAJOLO, M., *et al.*: 'A compilation-based software estimation scheme for hardware/software co-simulation'. Proceedings of the international conference on *Hardware/software codesign*, 2000

33 KLEIN, R.: 'Miami: a hardware software co-simulation environment'. Proceedings of the international workshop on *Rapid system prototyping*, 1996

34 Ptolemy project: http://ptolemy.eecs.berkeley.edu/

35 VOJIN ZIVOJNOVIC, and HEINRICH MEYR: 'Compiled HW/SW co-simulation'. Proceedings of the *Design automation* conference, 1996

36 NOHL, A., *et al.*: 'A universal technique for fast and flexible instruction-set architecture simulation'. Proceedings of the *Design automation* conference, 2002

37 RESHADI, M., *et al.*: 'Instruction set compiled simulation: a technique for fast and flexible instruction set simulation'. Proceedings of the DAC, 2003

38 LIU, J., *et al.*: 'Software timing analysis using HW/SW cosimulation and instruction set simulator'. Proceedings of the international conference on *Hardware/software codesign*, 1999

39 YOO, S., and CHOI, K.: 'Optimizing timed cosimulation by hybrid synchronisation', *Design Automation for Embedded Systems*, 2000, **5**(2), pp. 129–152

40 JUNG, J., YOO, S., and CHOI, K.: 'Performance improvement of multiprocessor systems cosimulation based on SW analysis'. Proceedings of the *Design automation and test in Europe*, 2001

41 YOO, S., CHOI, K., and DONG S. HA,: 'Performance improvement of geographically distributed cosimulation by hierarchically grouped messages', *IEEE Transactions on VLSI Systems*, 2000, **8**(5), pp. 492–502

42 BENINI, L., and DE MICHELI, G,: 'Networks on chips: a new SoC paradigm', *IEEE Computer*, 2002, **35**(1), pp. 70–78

43 MAGARSHACK, P., and PAULIN, P.: 'System on chip beyond the nanometer wall'. Proceedings of the *Design automation* conference, 2003

44 MACII, E., and PATEL, K.: 'Synthesis of partitioned shared memory architectures for energy-efficient multiprocessor SoC'. Proceedings of the *Design automation and test in Europe*, 2004

45 YOUSSEF, M., *et al.*: 'Debugging HW/SW interface for MPSoC: video encoder system design case study'. Proceedings of the *Design automation* conference, 2004

46 HONDA, S., *et al.*: 'RTOS-centric hardware/software cosimulator for embedded system design'. Proceedings of the CODES+ISSS, September 2004

47 Virtio Co. Press Release: 'Virtio jumpstarts software development for Texas instruments OMAP2420 platform', October 2004

48 HINES, K., and BORRIELLO, G.: 'Selective focus as a means of improving geographically distributed embedded system co-simulation'. Proceedings of the international workshop on *Rapid system prototyping*, 1997

Chapter 21

System-level validation using formal techniques

Rolf Drechsler and Daniel Große

21.1 Introduction

With increasing design complexity, verification becomes a more and more important aspect of the design flow. Modern circuits contain up to several hundred million transistors. In the meantime it has been observed that verification becomes the major bottleneck, i.e. up to 80 per cent of the overall design costs are due to verification. This is one of the reasons why recently several methods have been proposed as alternatives to classical simulation, since it cannot guarantee sufficient coverage of the design. For example, in Reference 1 it has been reported that for the verification of the Pentium IV more than 200 billion cycles have been simulated, but this only corresponds to 2 CPU minutes, if the chip is run at 1 GHz.

Formal verification techniques have gained large attention, since they allow us to prove the correctness of a circuit, i.e. they ensure 100 per cent functional correctness. Beside being more reliable, formal verification approaches have also been shown to be more cost effective in many cases, since test bench creation – usually a very time consuming and error prone task – becomes superfluous [2].

In this chapter, first in Section 21.2 some of the application domains, where formal techniques have successfully been used, are briefly described. Links to further literature are given where the interested reader can get more information. Then, as an example, equivalence checking based on formal techniques is described from an industrial perspective in Section 21.3. This gives a better understanding of the problems that should have to be considered while applying formal verification. Aspects of system level verification are discussed in Section 21.4, where SystemC is used as the modelling platform. After some preliminaries on SystemC and the property language used, two approaches for verification at the system level are discussed. An approach which allows property checking for SystemC descriptions at the block level is presented in Section 21.5. The design and a property is transformed into a Boolean satisfiability problem and then SAT techniques are applied for the proof.

An example is discussed in detail to demonstrate the overall flow. But for very large systems a formal proof may fail. For the block integration on the system level the communication needs to be verified. Section 21.6 introduces an approach for SystemC descriptions which allows the generation of checkers from temporal properties. By this, it becomes possible to check properties where formal proof methods fail due to complexity. Furthermore the properties can not only be checked during simulation but also after fabrication. Together, the two approaches enable a concise circuit and system verification methodology. In Section 21.7 a list of challenging problems is given, including an outline of some topics that need further investigation in the context of formal hardware verification and Section 21.8 concludes the chapter.

21.2   Formal verification

The main idea of formal hardware verification is to prove the functional correctness of a design instead of simulating some vectors. For the proof process different techniques have been proposed. Most of them work in the Boolean domain, like binary decision diagrams (BDDs) or SAT solvers.

The typical hardware verification scenarios, where formal proof techniques are applied, are

- equivalence checking (EC)
- property checking (PC), also called model checking (MC)

The goal of EC is to ensure the equivalence of two given circuit descriptions. (This is discussed in more detail in the next section.)

In contrast to EC, where two circuits are considered, for PC a single circuit is given and properties are formulated in a dedicated 'verification language'. It is then formally proven whether these properties hold under all circumstances. While 'classical' computational tree logic (CTL)-based model checking [3] can only be applied to medium sized designs, approaches based on 'bounded model checking' (BMC) as discussed in Reference 4 give very good results when used for complete blocks with up to 100k gates.

Nevertheless, all these approaches can run into problems caused by complexity, e.g. if the circuit becomes too large or if the function being represented turns out to be 'difficult' for formal methods. The second problem often arises in cases of complex arithmetics, such as multipliers.

Motivated by this, hybrid methods have been proposed, e.g. symbolic simulation and assertion checking. These methods try to bridge the gap between simulation and correctness proofs. But these approaches also make use of formal proof techniques.

21.3   Equivalence checking in an industrial environment

This section describes the equivalence checker 'gatecomp' that is part of the tool set CVE developed in the formal verification group of Infineon Technologies as

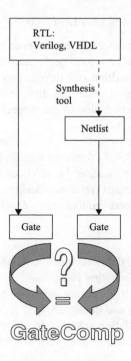

Figure 21.1   Synthesis verification flow

presented in [5]. The information on the tool gives the status of the year 2000. In the meantime more advanced techniques are included, while we restrict ourselves in the following to the basic concepts.

Most designers will first experience formal verification when using an equivalence checking tool for sign-off, e.g. to check that the final netlist has the same behaviour as previous netlists and even the original registered transfer level (RTL). As an example, the general flow for the synthesis verification, i.e. checking the equivalence of a RTL description and a netlist, is shown in Figure 21.1. Starting from the RTL description a netlist is generated by a synthesis tool. Then both descriptions are translated into an internal gate format that is used by gatecomp to prove functional equivalence. The translation is done by the frontends. This independence from the synthesis tool guarantees a further improvement of quality of the overall design. In a similar way equivalence of RTL versus RTL and netlist versus netlist descriptions is proven. The steps in general are as follows:

1   Translate both designs to an internal format.
2   Establish the correspondence between the two designs in a matching phase.
3   Prove equivalence or inequivalence.
4   In the case of an inequivalence a counter-example is generated and the debugging phase starts.

Several powerful features of gatecomp support the user during these steps.

21.3.1 Advanced equivalence checking

The CVE toolset contains the advanced equivalence checker gatecomp. The algorithms used are highly efficient; a typical performance figure for gatecomp is 100k gates per minute; many multi-million gate designs have been verified so far. Before describing the examples in more detail, we first review the main features of the tool and also show the differences due to more precise modelling compared with alternative implementations.

Gatecomp is used to compare netlist versus netlist, RTL versus netlist or RTL versus RTL. While other tools only focus on bug finding, in addition gatecomp is targeted towards simulation verification, i.e. to check that what is simulated on the RTL is also simulated on the netlist (reference design).

Gatecomp is an advanced equivalence checker due to the following differentiators:

- Speed
  - An efficient hash based data structure allows it to handle complete designs by very low memory consumption [6,7].
  - Multi-engines with multi-threading guarantee beside the very fast execution, and also the robustness and quality [8,9].
  - The denotational translation schemes on word level in language frontends support the use of RTL information for the equivalence proof.
- Capacity
  - The intelligent control of multi-engines ensures a tight integration of the different proof engines.
  - The frontends make use of compositional translation.
- Language coverage
  - The long-term experience with various description formats, such as VHDL (including VITAL), Verilog (including UDPs) and EDIF, results in robust frontends and very user-friendly linting tools.
- Debugging
  - A graphical user interface (GUI) allows for easy handling of the results (see Figure 21.2).
  - A link to ModelSim and Debussy for source code/netlist debugging is provided.
  - A fast reachability analysis for eliminating spurious sequences is integrated.
- Flexibility
  - The multi-engine concept can easily be extended.
  - Gatecomp supports full Boolean constraints.
  - A third party transistor extraction tool can be accessed.
  - Automatical generation of controllability and observability don't cares.
- Rich set of features
  - Multipliers of large bit-length can be handled.
  - Matching techniques: name based, simulation and prover based, structural, user defined, change name file from Synopsys.
  - Automatic removal of redundant states.

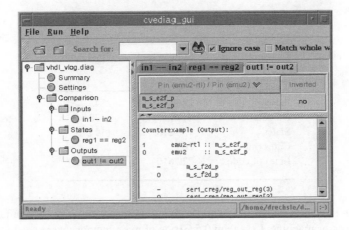

Figure 21.2   Graphical user interface

- Support for clock-gating, tri-states, black boxing, compression of counter-examples, assertions and constraints, scan path insertion, etc.

21.3.2   Precision

The idea is that what you simulate on RTL is also simulated on the netlist. Gatecomp uses a different approach for the proof algorithms. Instead of a two- a four-valued logic is used, that allows it to use synthesis and simulation semantics. While other tools reduce the simulation to two values only, gatecomp can model the precise language semantics, e.g. 9-valued in VHDL. The main features are

- A formal library qualification tool guarantees conformance of synthesis and simulation view and issues warnings otherwise. The libraries are compiled into simple functional replacements. This – in addition to being very robust and reliable – results in very fast runtimes during the equivalence checking phase.
- The simulation/synthesis mismatches are proven/highlighted.
- Gatecomp is independent of the internal workings of the synthesis tools.
- The tool allows a formal handling of internal don't cares for RTL/RTL and RTL/netlist comparison.

21.3.3   Experiments

To demonstrate the usage of the tool in equivalence checking of ASICs we report on three verification scenarios: a netlist versus netlist comparison, a RTL versus netlist comparison, and a RTL versus RTL comparison.

21.3.3.1   Netlist versus netlist comparison

First, we describe the verification of a synthesised netlist against its description after test logic insertion. The verified designs contained approximately 2.6 million gates.

Table 21.1   Information on ASIC complexity

Characteristics	Synthesised netlist	Final netlist
Inputs	2843	2843
Outputs	4178	4178
States	150 218	150 215
Gates (million)	2635	2634
Lines of code	222 610	3 861 939

Details are given in Table 21.1. The verification times were <20 CPU minutes on a four processor machine. Less than 0.5 GByte of main memory were used.

21.3.3.2   RTL versus netlist comparison

This is the typical scenario for synthesis verification as shown in Figure 21.1. In our example the RTL description had more than 50 000 lines of code and the resulting netlist consists of over 2 million gates. Gatecomp took <23 CPU minutes and <420 MByte of main memory to prove functional correctness.

21.3.3.3   RTL versus RTL comparison

A Verilog design was automatically translated into a VHDL description. After translation each module was checked by equivalence checking for functional correctness. For almost all blocks the verification was done in no time and fully automatic. In only a few cases – where the Verilog-VHDL-translation was erroneous – the tool took a few CPU minutes. In the case of a block with more than 600 outputs and over 1000 state variables the verification took 7 CPU minutes and <80 MByte were used. Based on the counter-example generated by gatecomp the design bug, that was due to a wrong assignment of don't care values, could easily be fixed. In all three cases, this high performance is to be seen as a result of tight interaction between different tool components, i.e. the frontends, the proof engine and the debugging environment. The multi-engine concept used in gatecomp and its intelligent control guarantees high flexibility and robustness also on large designs with several million gates.

21.3.4   Comparison

Finally, we report about a comparison of various equivalence checking tools. All experiments were carried out on a SUN Sparc 2 with 256 MByte running SunOS 5.7. An initial netlist is compared with a post layout netlist including routing that has been obtained by application of Magma *Blast Fusion*™, one of the leading physical design systems that also applies logic synthesis techniques. By this, the comparison often becomes more difficult. The initial netlist consists of 370 k gate equivalents. The netlist has more than 4500 outputs and more than 21 000 states bits. The runtimes

Table 21.2   Information on runtime

Characteristics	Tool 1	Tool 2	Tool 3	Gatecomp
Runtime	>1 week*	≈21 h	≈18 h	≈3.5 h

* After a finetuning of the parameters for this tool a reduction of runtime to 4 h has been achieved. But similar results can be expected for the other tools by variation of the parameters, e.g. gatecomp can do the comparison in <1 h, if specialised parameters are chosen.

of gatecomp in comparison to three other commercially available tools[1] are given in Table 21.2. All tools are started with their default settings, i.e. no tuning of the parameters is done. As can be seen, significant reductions in runtime can be obtained.

The examples described above show the application to real-world examples. Using the powerful tool gatecomp, equivalence checking of multi-million gate designs can be performed within minutes, and by this is superior to classical simulation – not only with respect to quality – but also regarding runtime. This has a direct impact on the costs of the verification process that can be reduced significantly based on formal techniques.

21.4   System-level verification

While classical approaches to circuit design make use of 'Hardware Description Languages' (HDLs), such as VHDL or Verilog, there is a strong interest in C-like description languages for system level modelling [10]. These languages allow for higher abstraction and fast simulation in an early stage of the design process. Furthermore, hardware/software co-design can be performed in the same system environment. One of the most popular languages of this type is SystemC [11].[2] But so far, most verification approaches for SystemC are based on simulation only [12,13]. Of course, due to the reasons discussed in the introduction, it would be desirable to have formal verification techniques also at the system level.

Before the verification approaches are described in detail, the main features of SystemC and the property language used are reviewed.

21.4.1   Preliminaries

In the following circuits and systems are modelled in SystemC. Therefore, first a short overview on SystemC is given. Then the formalism for specification of temporal properties is described.

---

[1] Names not given to guarantee anonymity.
[2] All techniques discussed in the following can also be transferred to other system-level languages, e.g. SystemVerilog.

21.4.1.1   SystemC

The main features of SystemC for modelling a system are based on the following:

- Modules are the basic building blocks for partitioning a design. A module can contain processes, ports, channels and other modules. Thus, a hierarchical design description becomes possible.
- Communication is realised with the concept of interfaces, ports and channels. An interface defines a set of methods to access channels. Through ports a module can send or receive data and access channel interfaces. A channel serves as a container for communication functionality, e.g. to hide communication protocols from modules.
- Processes are used to describe the functionality of the system, and allow expressing concurrency in the system. They are declared as special functions of modules and can be sensitive to events, e.g. an event on an input signal.
- Hardware specific objects are supplied, e.g. signals, which represent physical wires, clocks and a set of data-types useful for hardware modelling.

Besides this, SystemC provides a simulation kernel. The functionality is similar to traditional event-based simulators. Note that a SystemC description can be compiled with a standard C++ compiler to produce an executable specification. The output of a system can be textual, using C++ routines such as `cout` for instance, or waveforms. As a C++ class library SystemC can easily be extended by using the facilities of C++.

21.4.1.2   Property language

Describing temporal properties for verification can be done in many different ways, since there exist several languages and temporal logics. We use the notation of the property checker from Infineon Technologies (e.g. see References 14 and 15 for more details). A property consists of two parts: a list of assumptions 'assume part' and a list of commitments 'proof part'. An assumption/commitment has the form

```
      at t+a: expression;
   or during[t+a,t+b]: expression;
   or within[t+a,t+b]: expression;
```

where t is a time point, and a, $b \in \mathbb{N}$ are offsets. If all assumptions hold, all commitments in the proof part have to hold as well. Since a and b are finite a property argues only over a finite interval, which is called the 'observation window'.

Example 21.1 The property `test` says that whenever signal x becomes 1, two clock cycles later signal y has to be 2.

```
theorem test is
assume:
  at t: x = 1;
prove:
  at t+2: y = 2;
end theorem;
```

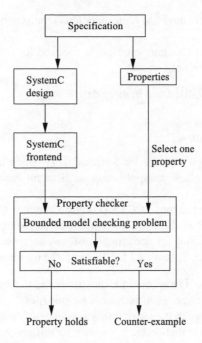

Figure 21.3   Property checking work flow

In general a property states that whenever some signals have a given value, some other (or the same) signals assume specified values. Of course, it is also possible to describe symbolic relations of signals. Furthermore the property language allows to argue over time intervals, e.g. that a signal has to hold in a specified interval. This is expressed by using the keywords `during` and `within`; whereas `during` states that the expression has to hold all the time in the interval, with `within` the expression has to hold at least once in the specified interval. Also a set of advanced operators and constructs is provided to allow for expressing complex constraints more easily.

21.5   Property checking

In this section, we present an efficient approach to property checking of SystemC designs.[3] Before a detailed description is given the overall flow is outlined. After design implementation and formalisation of the specification into temporal properties the proposed approach works as follows (see also Figure 21.3):

1   The SystemC design is transformed into an internal finite state machine (FSM) representation by the SystemC frontend.

---

[3] A first approach to property checking of SystemC designs based on BDDs has been presented in Reference 16. Here an extension of this technique is presented. A larger set of SystemC constructions is supported and a more powerful prover based on a SAT engine is used.

2 A single property and the FSM representation is translated into a bounded model checking problem.
3 The bounded model checking problem is checked for satisfiability to decide if the property holds or not.

These steps are now discussed in more detail.

21.5.1 SystemC frontend

The frontend is based on a parser for SystemC [17] which has been developed to be generic in order to allow an application in different areas, e.g. verification or visualisation.

To build the parser the tool PCCTS (Purdue Compiler Construction Tool Set) [18] was used. PCCTS enables the description of the syntax of a language by a grammar, provides facilities for construction of easy-to-process data structures and finally generates a parser. Specialised for SystemC the parser was built as follows:

- A preprocessor is used to account for directives and to filter out header-files that are not part of the design, such as system-header-files.
- A lexical analyser splits the input into a sequence of 'tokens'. These are given as regular expressions that define keywords, identifiers etc. of SystemC descriptions. Besides C++ keywords also essential keywords of SystemC are added, e.g. SC_MODULE or sc_int.
- A syntactical analyser checks if the sequence of tokens conforms to the 'grammar' that describes the syntax of SystemC. Terminals in this grammar are the tokens.

PCCTS creates the lexical and syntactical analyser from tokens and grammar, respectively. Together they are referred to as the parser. The result of parsing a SystemC description is an abstract syntax tree (AST). At this stage no semantic checks have been performed, e.g. for type conflicts. The AST is constructed using a single node type, that can have a pointer to the list of children and a pointer to one sibling. Additional tags at each node are used to store the type of a statement, the string for an identifier and other necessary information.

In the following the overall procedure for the transformation of a SystemC description into a FSM representation is given (see also Figure 21.4):

- After preprocessing the parser is used to build the AST from the SystemC description of a design.
- The AST is traversed to build an intermediate representation of the design. All nodes in an AST have the same type, all additional information is contained in attached tags. Therefore different cases have to be handled at each node while traversing the AST. By transforming the AST into the intermediate representation the information is made explicit in the new representation by the analyser for further processing. The intermediate representation is built using classes to represent building blocks of the design, e.g. modules, statements or blocks of statements. During this traversal semantic consistency checks are carried out. This

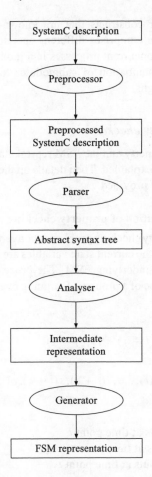

*Figure 21.4   Overall procedure for transformation of a SystemC description into a
FSM representation*

includes checking for correct typing of operands, consistency of declarations and
definitions, etc.

- The FSM representation is generated by traversing the intermediate representation
  recursively. Classes dedicated to certain constructs enable the construction of
  output functions and transition functions from the intermediate representation.

The SystemC frontend is restricted to SystemC register transfer level (RTL)
descriptions, i.e. to a subset of possible C++ and SystemC constructs [19]. To prevent
difficulties already known from high-level-synthesis C++ features such as dynamic
memory allocation, pointers, recursions or loops with variable bounds are not allowed.
In the same way some SystemC constructs have no direct correspondence on the RTL
and are excluded, e.g. SystemC channels. For channels that obey certain restrictions
the FSM transformation can be extended by providing a library of RTL realisations.

Supported are all other constructs that are known from traditional hardware description languages. This comprises different operators for SystemC-datatypes, hierarchical modelling or concurrent processes in a module. Additionally, the new-operator is allowed for instantiation of submodules to allow e.g. for a compact description of scalable designs.

21.5.2   SystemC property checker

First the translation of a property and the FSM representation of the SystemC design into a Boolean problem are explained. Then details on the SystemC property checker are given and an example is provided.

21.5.2.1   Boolean formulation of property checking

The initial sequential property checking problem is converted into a combinational one by unrolling the design, i.e. the current state variables are identified with the previous next state variables of the underlying FSM. The process of unrolling is shown in Figure 21.5. A BMC instance of a property P arguing over the finite interval $[t, t+c]$ for a design D is given by:

$$b = \bigwedge_{j=0}^{c-1} T_\delta(i(t+j), s(t+j), s(t+j+1))$$

$$\wedge \neg P(i(t), s(t), o(t), \dots, i(t+c), s(t+c), o(t+c))$$

with

- $i(t) = (i_1^t, \dots, i_m^t)$ inputs at time point t
- $s(t) = (s_1^t, \dots, s_n^t)$ states at time point t
- $o(t) = \lambda(i(t), s(t))$ outputs at time point t
- T_δ the transition relation

The BMC instance b depends only on the states $s(t)$ and the inputs $i(t), \dots, i(t+c)$. It is unsatisfiable if for all states $s(t)$ and all input sequences $i(t), \dots, i(t+c)$ the property P over the interval $[t, t+c]$ holds for the design D. If b is satisfiable a counter-example for the property P has been found.

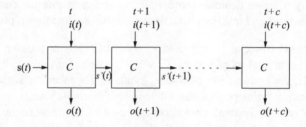

Figure 21.5   Unrolling

The SystemC property checker takes the FSM representation of the SystemC design and a property as input. Then the property is translated into an expression using only inputs, states and outputs of the SystemC design annotated with time points. The unrolled FSM representation and the property expression are converted into a bit-level representation. Here hashing and merging techniques for minimisation are used. The bit-level representation is given to the SAT solver zChaff [20] which has been integrated into the property checker. In the case of a counter-example a waveform in VCD format is generated to allow for an easy debugging.

21.5.2.2 Example

To illustrate the transformation of a property and the unrolled FSM representation into a BMC instance we provide a small example.

In Figure 21.6 a SystemC description of a two-bit counter is shown. Besides the clock input the counter has a reset input `reset` and the output `out`. The current value of the counter is stored in `count_val`. Figure 21.7 shows the underlying FSM of the counter. Since the output and the state of the counter are identical only the state is given. Basically, this FSM is the result of the SystemC frontend. Besides the FSM other information of the SystemC design is stored, e.g. data-types of variables. If the reset is represented by the Boolean variable r and `count_val` by the two state

```
SC_MODULE(counter) {
  sc_in_clk clock;
  sc_in<bool> reset;
  sc_out< sc_uint<2> > out;

  // counter value
  sc_uint<2> count_val;

  void do_count() {
    if (reset.read()) {
      count_val = 0;
    } else {
      count_val = count_val + 1;
    }
    out = count_val;
  }

  SC_CTOR(counter) {
    SC_METHOD(do_count);
    sensitive << clock.pos();
  }
};
```

Figure 21.6   A 2-bit counter

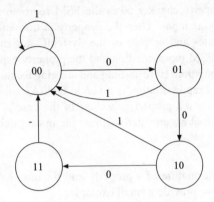

Figure 21.7   FSM of the 2-bit counter

```
theorem reset is
  assume:
  at t: reset = 1;
prove:
  at t+1: count_val = 0;
end theorem;
```

Figure 21.8   Property reset *for module* counter

```
theorem count is
  assume:
  at t: reset = 0;
  at t: count_val < 3;
prove:
  at t+1: count_val = prev(count_val) + 1;
end theorem;
```

Figure 21.9   Property count *for module* counter

variables *h* (high) and *l* (low), the transition function of the FSM is given by:

$$\delta_h(r^t, h^t, l^t) = \neg\, r^t \wedge (h^t \oplus l^t)$$
$$\delta_l(r^t, h^t, l^t) = \neg\, r^t \wedge \neg l^t$$

For the counter three properties have been formulated. The first property reset describes the reset behaviour of the counter (Figure 21.8). With the second property count the normal operation of the counter is characterised (Figure 21.9). With the prev operator values of previous time points can be accessed. The last property states that the counter counts from three to zero (Figure 21.10). Obviously these properties hold for the two-bit counter.

```
theorem modulo is
  assume:
  at t: reset = 0;
  at t: count_val = 3;
prove:
  at t+1: count_val = 0;
end theorem;
```

Figure 21.10   Property modulo *for module* counter

Now we consider the property modulo in more detail. Since this property argues over the interval $[t, t + 1]$ the underlying FSM is unrolled only once, i.e. the first part of the resulting BMC instance is

$$T_\delta(\, r^t, h^t, l^t, h^{t+1}, l^{t+1}\,) =$$
$$h^{t+1} \equiv \neg\, r^t \wedge (h^t \oplus l^t) \wedge l^{t+1} \equiv \neg\, r^t \wedge \neg l^t$$

Based on this unrolling step the values of variables at time point $t + 1$ are available (e.g. count_val at time point $t + 1$ via h^{t+1} and l^{t+1}). Then the property modulo corresponds to the Boolean formulation:

$$\neg(\; \underbrace{\neg r^t}_{reset@t=0} \;\wedge\; \underbrace{h^t \wedge l^t}_{count\_val@t=3} \;) \quad \text{(assume part)}$$
$$\vee\; (\underbrace{\neg h^{t+1} \wedge \neg l^{t+1}}_{count\_val@t+1=0}) \qquad \text{(prove part)}$$

Finally, as explained above the BMC instance is converted into a SAT instance and checked for satisfiability.

21.5.3   Experiments

The algorithms have been implemented in C++. All experiments have been carried out on an Intel Pentium IV 3 GHz with 1 GB RAM running Linux. A runtime limit of 2 CPU hours has been set.

In a first example we studied a scalable hardware realisation of the bubble sort algorithm. The SystemC description is shown in Figure 21.11. This module implements the sort algorithm for eight data words. The bit size of each data word is determined by a typedef. Note that the approach from Reference 16 did not support constructs, e.g. typedefs or for-loops. In total the correctness of sorting has been proven with nine properties. The first property sorted states that the resulting sequence is ordered correctly, i.e. that the value of an output is greater or equal compared to values at outputs with smaller indices (see Figure 21.12). In Table 21.3 the results are given for the property sorted and increasing bit sizes of data words (column 'Bit size'). The next two columns provide information about the SAT instance, i.e. the number of clauses and literals, respectively. In the last column the overall CPU time needed (CPU seconds) is reported. Due to the heuristic nature of the SAT solver the proof time might slightly vary as can be seen in case of bit size 8.

```
typedef sc_uint<4> T;

SC_MODULE( bubble )
{
  sc_in< T >   in[8];
  sc_out< T >  out[8];
  T  buf[8];

  void do_it() {
    for(int i = 0; i < 8; i++)
      buf[i] = in[i];
    for(int i = 0; i < 8-1; i++) {
      for(int j = 0;j < (8-i)-1; j++) {
        if( buf[j] > buf[j+1] ) {
          T tmp;
          tmp = buf[j];
          buf[j] = buf[j+1];
          buf[j+1] = tmp;
        }
      }
    }
    for(int i = 0; i < 8; i++)
      out[i] = buf[i];
  }

  SC_CTOR( bubble ) {
    SC_METHOD(do_it);
    sensitive << in[0] << in[1]
     << in[2] << in[3] << in[4]
     << in[5] << in[6] << in[7];
  }
};
```

Figure 21.11   Bubble sort

But in general the run time needed increases with the bit size and is moderate even for larger bit sizes.

Finally, another eight properties have been proven for the SystemC module bubble. These properties formalise that all input values of the module bubble can be found at the outputs. The summarised results for different bit sizes are shown in Table 21.4. Again the first column gives the bit size. In the next two columns details of a single SAT instance are shown. These numbers are identical for each of the eight properties, since the properties are symmetric, i.e. only the according input differs within the eight properties. The last column provides the sum of the run times

```
theorem sorted is
    prove:
    at t: out[0] <= out[1];
    at t: out[1] <= out[2];
    at t: out[2] <= out[3];
    at t: out[3] <= out[4];
    at t: out[4] <= out[5];
    at t: out[5] <= out[6];
    at t: out[6] <= out[7];
end theorem;
```

Figure 21.12   Property sorted *for module* bubble

Table 21.3   Results for different input sizes of module bubble *and property* sorted

Bit size	Clauses	Literals	Time
4	6390	14458	17.18
8	12754	28894	286.93
16	25482	57766	125.25
32	50938	115510	560.48

Table 21.4   Results for different bit sizes of module bubble *and input properties*

Bit size	Clauses	Literals	Time
4	6298	14262	58.49
8	12570	28502	681.52
16	25114	56982	845.76
32	50202	113942	3662.07

for all eight properties. As can be seen, the correctness of the implementation of the bubble sort algorithm can be proven for up to 32 bits in 1 CPU hour.

While SystemC 1.x is focused more on RTL descriptions, SystemC 2.0 supports several constructs for system-level modelling. In this context channels are of high relevance. An important example of a channel provided with the SystemC distribution are FIFOs. In the refinement step, these FIFOs are then translated to the RTL. In a second series of experiments synchronous FIFOs with variable depth have been

Table 21.5   Results for different FIFO depths

Depths	Property	Clauses	Literals	Time
8	reset	2708	6264	0.26
8	nochange	11 145	25 631	0.51
8	write	13 302	30 612	0.81
16	reset	5181	12 025	0.52
16	nochange	22 309	51 327	1.78
16	write	26 158	60 248	2.75
32	reset	9958	23 162	1.08
32	nochange	44 557	102 539	7.90
32	write	51 741	119 229	14.24
64	reset	19 343	45 051	2.35
64	nochange	88 865	204 531	40.64
64	write	102 680	236 670	58.63
128	reset	37 944	88 444	6.50
128	nochange	177 377	408 279	247.82
128	write	204 415	471 227	283.74

studied. The FIFO uses a register bank, a read pointer, a write pointer and a counter. It supports simultaneous read and write. Different properties have been developed which describe e.g. the behaviour after reset, no change of the FIFO content if no data is written to the FIFO and more details on the write access to the FIFO. For a bit size of 32 bits and increasing FIFO depths results are shown in Table 21.5.

In the first and second column the depth of the FIFO and the property are given, respectively. In the next two columns details on the SAT instance are provided, i.e. the number of clauses and the literals. Finally, the run time is given in the last column. The results clearly show that for high depth, i.e. FIFOs with more than 100 registers, the verification time needed is in the range of a few minutes. This demonstrates that even though it cannot be expected that complete systems can be checked, also complex system level constructs can be formally verified using this approach.

21.6   Generation of checkers

On the system level often property checkers may not be able to formally prove a behaviour due to the complexity of the underlying modules. For this, alternatives have been proposed that try to combine the techniques of simulation and formal verification.

There are several approaches to system-level verification which are based on assertions [21]. The key idea is to describe expected or unexpected behaviour directly in the device under test. These conditions are checked dynamically during simulation. An approach to check temporal assertions for SystemC has been presented in [13]. There, the specified properties are translated to a special kind of finite state

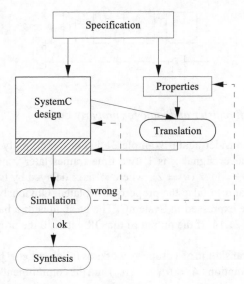

Figure 21.13   Work flow

machines (AR-automata). These automata are then checked during the simulation run by algorithms, which have been integrated into the SystemC simulation kernel. In contrast in Reference 22 a method has been proposed to synthesise properties for circuits into hardware checkers. Properties which have been specified for (formal) verification are directly mapped onto a very regular hardware layout.

Following the latter idea in this section a method is presented which allows checking of temporal properties for circuits and systems described in SystemC not only during simulation [24,23]. A property is translated into a synthesisable SystemC checker and embedded into the circuit description. This enables the evaluation of the properties during the simulation and after fabrication of the system. Of course, with this approach a property is not formally proven and only parts of the functionality are covered. But the proposed method is applicable to large circuits and systems and supports the checking of properties in the form of an on-line test. This on-line test is applicable, even if formal approaches fail due to limited resources.

Before the details are given, the work flow of the approach is illustrated in Figure 21.13. At first the design has to be built and the specification has to be formalised into properties. Then the properties are translated to checkers and embedded into the design description (hatched area in the figure). If all checkers hold during simulation the next step of the design flow can be entered.

21.6.1   SystemC checker

21.6.1.1   Basic idea of checker generation

The basic idea of the translation of a property into a checker is illustrated by the following example.

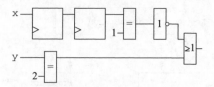

Figure 21.14   Shift register and logic for property test

Example 21.2 Consider again Example 21.1. For the property test it has to be checked that whenever signal x is 1, two time frames later y has to be 2. This is equivalent to $\neg(x'' = 1) \vee (y = 2)$, where x'' is x delayed by two clock cycles. If the equation evaluates to false the property is violated. Obviously the translation of the property can be expressed in SystemC. The basic idea of a hardware realisation is shown in Figure 21.14. If the output of the OR gate is 0 the property fails.

In general the translation of a property works as follows: Let P be a property which consists of the assumptions $A = (a_1, \ldots, a_m)$ and the commitments $C = (c_1, \ldots, c_n)$. Then the translation algorithm is based on four steps:

1   Parse P and determine the maximum offset o_{max} of the property by analysing the time points of all a_i and c_j.
2   For each signal used in P generate a shift register of length o_{max}. Then the values of a signal at time points $t, t+1, \ldots, t+o_{max}$ are determined by the outputs of the flip-flops in the corresponding shift register. The offset i of a time point can directly be identified with the ith flip-flop, if the flip-flops are enumerated in descending order. This is illustrated in Figure 21.15.
3   Combine the signals of each a_i (and c_j) as stated by the logic operations in its expression. Thereby the variables of the appropriate time points are used. In the case of the interval operators during and within an AND and an OR of the resulting expressions is computed. The results of this step are the equations $\hat{a}_1, \ldots, \hat{a}_m, \hat{c}_1, \ldots, \hat{c}_n$ corresponding to the assumptions and commitments of P.
4   The final equation is $check_P = \neg \bigwedge_{i=1}^{m} \hat{a}_i \vee \bigwedge_{j=1}^{n} \hat{c}_j$.

Of course all described transformations from the property description into the resulting equation $check_P$ have to be performed by using SystemC constructs, i.e. the use of different data types and operators has to be incorporated. Finally, the property P can be checked by evaluating $check_P$ in each clock cycle during simulation or operation. In the next section some details about the transformation into SystemC code are given.

21.6.1.2   Transformation into SystemC checkers

A property is assumed to use only port variables and signals of a fixed SystemC module or from its sub-modules. During the translation for the variables of the properties shift registers have to be created as described in the previous section (Step 2). For this purpose a generic register has been modelled as shown in Figure 21.16. The register delays an arbitrary data type for one clock cycle. If such a templated register is not

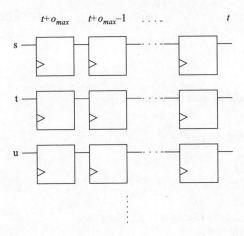

Figure 21.15   *Mapping of time points*

```
template<class T >
class regT : public sc_module {
public:
    sc_in_clk clock;
    sc_in<T > in;
    sc_out<T > out;

    SC_CTOR(regT) {
        SC_METHOD(doit);
        sensitive_pos << clock;
    }
    void doit() {
        out = in.read();
    }
};
```

Figure 21.16   *Generic register*

directly supported by the synthesis tool, it is possible to replace every templated register with a version where the concrete input and output types are explicitly specified. The generic register can be used as shown in the example in Figure 21.17. There a register with an `sc_int<8>` input and output is declared and instantiated.

During the generation of the shift registers of length o_{max} for a variable, o_{max} generic registers have to be declared and instantiated. This is done in the constructor of the considered module. The necessary `sc_signals` (output variables of the registers) for the different time points are declared as member variables of the considered module. Their names are produced by adding the number of delays to the variable

```
regT<sc_int<8> > *r = new regT<sc_int<8> >("reg");
r->clock(clock);
r->in(a);
r->out(a_d);
```

Figure 21.17   Usage of generic register

```
SC_MODULE(module) {
public:
     // ports
     sc_in_clk clock;
     ...
     // sc_signals for different
     // time points
     sc_signal<T> x_d1,x_d2;

     SC_CTOR(module) {
          // shift register
          regT<T> rx_d1 = ...
          rx_d1->clock(clock);
          rx_d1->in(x);
          rx_d1->out(x_d1);
          regT<T> rx_d2 = ...
          rx_d2->clock(clock);
          rx_d2->in(x_d1);
          rx_d2->out(x_d2);
          ...

     }
};
```

Figure 21.18   Insertion of a shift register for property test

name. The absolute time points cannot be used, because if a variable is employed in at least two properties the delay of the same time points may differ.

Example 21.3   Consider again Example 21.1. Let the data type of x be T. Let the property test be written for the SystemC module module. As has been explained above x has to be delayed two times. Then the resulting shift register is inserted into the module as shown in Figure 21.18.

As can be seen in Figure 21.18 the data type of a variable used in a property has to be known for declaration of the sc_signals and shift registers. Thus, with a simple parser the considered SystemC module is scanned for the data types of the property variables.

```
// theorem: test
bool check_test = !( ( x_d2.read() == 1 ) ) |
    ( y.read() == 2 ) ;
if (check_test == false) {
    cout<<"@"<<sc_simulation_time()<<
    ":
    THEOREM test FAILS!"<<endl;
}
```

Figure 21.19   Checker for property test

The resulting code to check a property (equivalent to the equation $check_P$) is embedded into an SC_METHOD process of the module, which is sensitive to the module clock, i.e. the process is triggered every clock cycle. In the final step of writing SystemC code for the translated property the following is taken into account:

- The shift register for each variable used in a property is shared between different checkers.
- In the case of an array access it is necessary to distinguish between an access to an array of ports and an access to a port which contains an array type. An array of ports is mapped onto different variables each representing an index of the array. Furthermore the access operator [ ] has to be replaced accordingly.
- The operators of the property language have to be mapped onto its counterparts in C++, e.g. = to ==.
- The resulting checker formula is assigned to a Boolean variable check_ <property name>. If this variable is false during simulation the property is violated and an output is given using the cout routine. For the synthesis part an output port for the considered module has to be generated, which assumes zero if the property fails.

Example 21.4 In Figure 21.19 the translated equation check _test for the property test is shown. If the property fails, this is prompted directly to the designer.

21.6.1.3   Optimisations

All shift registers for different properties of one concrete module which are driven by the same clock, can be integrated into one clocked process. Then in the constructor SC_CTOR of the module instead of the shift registers only one clocked process has to be declared. In this process the output variables are written, e.g. in the case of the property test the process statements are

```
x_d1.write(x); x_d2.write(x_d1);
```

So the number of SC_METHODs is reduced and the simulation speed increases (see also Section 21.6.2.2).

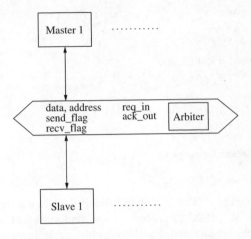

Figure 21.20   Bus architecture

As explained above if the checkers are synthesised one-to-one for each property an output port is generated, which assumes the value zero if the property fails. This leads to a trade-off between good diagnosis and number of output pins. Diagnosis is easy if each property directly corresponds to an output pin, while many outputs require more chip area.

21.6.2   Experimental results

The technique described above is experimentally studied by generating checkers that are included during simulation. For this task a bus architecture has been modelled. In Figure 21.20 a block diagram of the bus architecture is shown. The bus is described as a SystemC module, and masters and slaves can connect to the bus. The bus is divided into a data part, an address part and a flag part. These are all `sc_inout`-ports and they have the type `sc_uint` with a scalable size to allow for variable data width, number of slaves and number of masters. The address is used by the masters to address a slave. The flags `send_flag` and `recv_flag` are set during a bus transaction (see below). Furthermore the bus contains a scalable arbiter. Thus the bus also has a request input and an acknowledge output for each master. The arbiter consists of n cells (one for each master) and combines priority arbitration with a round robin scheme. This guarantees that every master will finally get access to the bus. In Figure 21.21 the arbiter is shown. Summarised, the features of the bus are

- Only masters can write to the bus and each master has a unique ID.
- A slave has a unique address. This address is given at instantiation of the slave.
- A bus transaction works as follows:

  1   A master requests the bus via its request output. If access is granted see Step 2, otherwise the master waits for an acknowledgement.

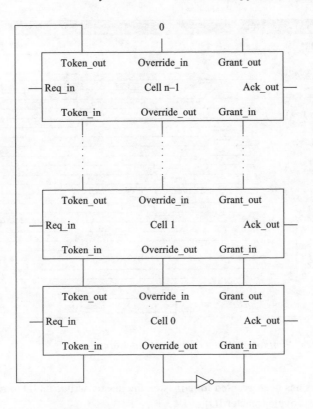

Figure 21.21   The integrated arbiter

2   The master writes the target address and the data to the bus. Furthermore, the master writes its ID to the send_flag. Then the master waits for an acknowledgement that the slave has received the data via the recv_flag (ID of the master at the recv_flag).

3   If a slave detects its address on the bus, the slave reads the data and writes the ID from the send_flag to the recv_flag of the bus.

4   If the master detects its ID on the bus, the data transmission was successful.

A waveform example of a bus with five masters and eight slaves is shown in Figure 21.22.

21.6.2.1   Checkers

An informal description of the properties which have been embedded as checkers into the bus module is given as:

1   Two output signals of the arbiter can never become 1 at the same time (mutual exclusion).

2   The acknowledge is only set if there has been a request (conservativeness).

3   Each request is confirmed by an acknowledge within $2n$ time frames (liveness).

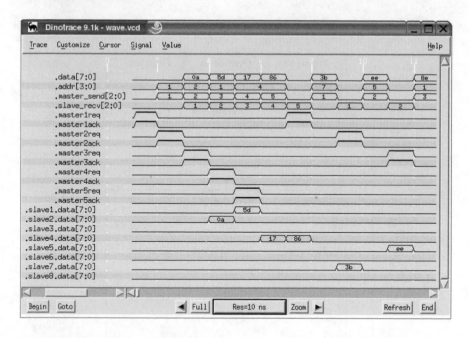

Figure 21.22   Simulation trace of a small bus example

4   If the bus has been granted for a master, the master writes its ID to `send_flag` in the next cycle (master ID).
5   If a slave has been addressed, the slave writes the master ID (available at the `send_flag`) to the `recv_flag` (acknowledge master).

21.6.2.2   Simulation results

Again all experiments have been carried out on an Intel Pentium IV 3 GHz with 1 GB RAM running Linux. Checkers have been generated for all described properties. For each property the simulation performance in case of no checkers, the simple approach, and the optimised approach are compared. For this task the bus model has been simulated for 100 000 clock cycles for various numbers of masters. Note that the number of masters connected to the bus is equal to the number of arbiter cells. For the checkers described above we obtained the following results:

1   In Figure 21.23 the performance comparison for the checker mutual exclusion is shown. As can be seen the simulation time for the simple and the optimised approach increases with the number of masters. Both approaches behave similar since the observation window of the mutual exclusion property is 1, so no registers have to be created. For this reason no optimisation is possible. The total runtime overhead is moderate, i.e. within a factor of two for 40 cells.
2   The simulation performance with and without the checkers for the conservativeness properties is nearly identical (see Figure 21.24). This is an expected

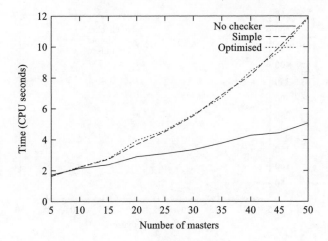

Figure 21.23   Comparison of simulation performance for checker mutual exclusion

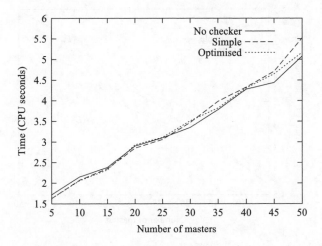

Figure 21.24   Comparison of simulation performance for checker conservativeness

behaviour, because each conservativeness property only argues over two signals of each arbiter cell.

3   In Figure 21.25 the results for the liveness checkers are shown. The figure shows that the optimised approach leads to significantly better results than the simple approach. Since the observation window of the liveness property is $2n$ (where n is the number of masters) the number of SC_METHODs has been reduced effectively by optimisation. However, the runtime overhead compared with pure simulation is due to the significantly increasing size of the observation windows of this properties.

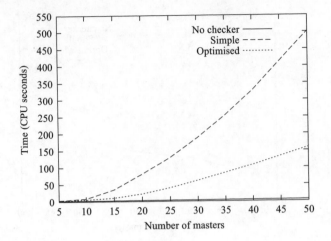

Figure 21.25   Comparison of simulation performance for checker liveness

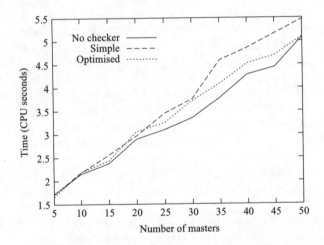

Figure 21.26   Comparison of simulation performance for checker master ID

4   The results for the checkers of master ID show that there is a small benefit of the optimised approach over the simple approach (see Figure 21.26). In total these properties can be checked very quickly during simulation.

5   As expected the acknowledge master property leads to the same performance as pure simulation, because this property could be described as being very compact. Figure 21.27 shows the diagram.

The experiments demonstrate that the overhead during simulation for properties with large observation windows is moderate, and negligible for properties with smaller observation windows.

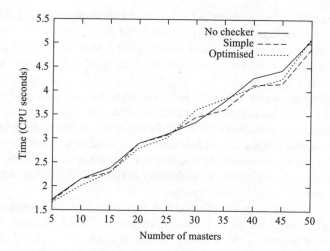

Figure 21.27   Comparison of simulation performance for checker acknowledge master

21.7   Challenges

Even though formal verification techniques are very successfully applied and have become the state-of-the-art in many design flows, still many problems exist. In this section a list of these problems is given. The list is not complete in the sense that all difficulties are covered, but many important ones are mentioned. This gives a better understanding of current problems in hardware verification and shows directions for future research.

Complexity: According to Moore's law the complexity of the circuits steadily increases. For this, the underlying data structures are very important. For EC and BMC often dedicated data structures are used. For representation of the state space BDDs have been shown to work well, but if the size of the circuit becomes too large the BDDs often suffer from 'memory explosion'.

Proof technology: While BDDs and SAT are the most popular techniques in hardware verification and have also been applied to many domains, there is still a lot of research going on. Besides the classical monolithic approaches modern EC tools make use of multi-engine approaches that combine different techniques, such as SAT, BDD, term rewriting, ATPG and random pattern simulation. How to successfully combine these – often orthogonal – approaches is not fully understood today.

Word-level approaches: Even though most proof techniques today work on the bit-level, many studies have shown that significant improvements can be achieved if the proof engine makes use of high-level information or even completely works on a higher level of abstraction. In this context ILP solvers also showed promise.

Matching in EC: As described in Section 21.2, before the proof process starts the correspondence between the circuits has to be established. Here, several techniques exist, such as name-based, structural or prover-based, but still for large industrial designs these methods often fail. This results in very time-consuming user-defined matching.

Reachability of counter-examples: In EC and BMC the generated counter-example might not be reachable in normal circuit operation. This results from the modelling of the circuit, i.e. instead of a FSM only the combinational part is considered. Thus, it has to be checked that the counter-example is 'valid' after it has been generated, or the prover has to ensure that it is reachable. Techniques have to be developed ensure this without a complete reachability analysis of the FSM, which is usually not feasible due to complexity reasons.

Arithmetic: Industrial practice has shown that today's proof techniques, such as BDD and SAT, have difficulties with arithmetic circuits, such as multipliers. Word-level approaches have been proposed as an alternative, but these methods often turned out to be difficult to integrate in fully automatic tools. For this, arithmetic circuits – often occurring in circuit design – are still difficult to handle.

System integration: PC works best on the module level, i.e. for blocks with up to 100k gates. But in multi-chip modules many of these blocks are integrated to build a system. Due to complexity the modules cannot be verified as one large block and for this models and approaches are needed.

Hybrid approaches: For complex blocks or on the system level PC might be a very complex task and for this simpler alternatives have been studied, i.e. techniques that are more powerful than classical simulation but need less resources than PC. Techniques, such as symbolic simulation or assertion-based verification, in this context also make use of formal verification techniques.

Analog/mixed signal: Most EC and PC models assume that the circuit is purely digital, while in modern system-on-chip designs many analogue components are integrated. For this, also models and proof mechanisms need to be developed for analogue and mixed signal devices.

Retiming: For EC retimed circuits are still difficult to handle, since in this case the state matching cannot be performed. Thus, the problem remains sequential and by this becomes far too complex.

Multiple clocks: Many circuits have different clocking domains, while verification tools can often only work with a single clock.

Coverage: To check the completeness of a verification process coverage metrics have to be defined. While typical methods, such as state coverage, are much too weak in the context of formal verification, there still does not exist a good measure that is comfortable to use for PC.

Diagnosis: After a fault has been identified by a formal verification tool a counter-example is generated. The next step is to identify the fault location or a reason for the failing proof process. Here, also formal proof techniques can be applied.

Most solutions to these problems are still in a very early stage of development, but these fields have to be addressed to make formal hardware verification successful in industrial applications. To orient the reader, some recent references are provided to give a starting point for further studies: see References 15, 21, 22 and 25–50.

21.8 Conclusions

In this chapter formal verification with a special focus on system-level verification has been discussed. While EC works very well on complete designs with several million transistors, PC approaches are so far mainly applicable at the block level.

For a solution for complete systems, still many problems have to be solved, where some of the most important solutions were given in the previous section.

In future design projects verification will become more and more important and the creation of a concise verification methodology decides about successful tape-outs.

Acknowledgement

The list of challenging problems has been developed in the context of the book project *Advanced Formal Verification* [32]. We like to thank all the contributors for the interesting discussions.

References

1   BENTLEY, B.: 'Validating the Intel Pentium 4 microprocessor'. Proceedings of the *Design automation* conference, 2001, pp. 244–248
2   WINKELMANN, K., TRYLUS, H.-J., STOFFEL, D., and FEY, G.: 'A cost-efficient block verification for a UMTS up-link chip-rate coprocessor'. Proceedings of the *Design, automation and test in Europe*, 2004, volume 1, pp. 162–167
3   BURCH, J.R., CLARKE, E.M., MCMILLAN, K.L., and DILL, D.L.: 'Sequential circuit verification using symbolic model checking'. Proceedings of the *Design automation* conference, 1990, pp. 46–51
4   BIERE, A., CIMATTI, A., CLARKE, E.M., FUJITA, M., and ZHU, Y.: 'Symbolic model checking using SAT procedures instead of BDDs'. Proceedings of the *Design automation* conference, 1999, pp. 317–320
5   DRECHSLER, R., and HÖRETH, S.: 'Gatecomp: Equivalence checking of digital circuits in an industrial environment'. Proceedings of the international workshop on *Boolean problems*, 2002, pp. 195–200
6   KUEHLMANN, A., and KROHM, F.: 'Equivalence checking using cuts and heaps'. Proceedings of the *Design automation* conference, 1997, pp. 263–268
7   VAN EIJK, C.: 'Formal methods for the verification of digital circuits'. PhD thesis, Eindhoven University of Technology, 1997
8   GOLDBERG, E., PRASAD, M., and BRAYTON, R.: 'Using SAT for combinational equivalence checking'. Proceedings of the international workshop on *Logic synth.*, 2000, pp. 185–191

9   PARUTHI, V., and KUEHLMANN, A.: 'Equivalence checking combining a structural SAT-solver, BDDs, and simulation'. Proceedings of the international conference on *Computer design*, 2000, pp. 459–464

10 GUPTA, R. (moderator): 'IEEE design and test roundtable on C++-based design'. *IEEE Design & Test of Computer*, 2001, **18**(3), pp. 115–123

11 Synopsys Inc., CoWare Inc., and Frontier Design Inc.: http://www.systemc.org. 'Functional specification for systemC 2.0', 2002

12 FERRANDI, F., RENDINE, M., and SCUITO, D.: 'Functional verification for SystemC descriptions using constraint solving'. Proceedings of the *Design, automation and test in Europe*, 2002, pp. 744–751

13 RUF, J., HOFFMANN, D.W., KROPF, T., and ROSENSTIEL, W.: 'Simulation-guided property checking based on multi-valued ar-automata'. Proceedings of the *Design, automation and test in Europe*, 2001, pp. 742–748

14 BORMANN, J., and SPALINGER, C.: 'Formale Verifikation für Nicht-Formalisten (Formal verification for non-formalists)'. *Informationstechnik und Technische Informatik*, 2001, **43**, pp. 22–28

15 JOHANNSEN, P., and DRECHSLER, R.: 'Formal verification on register transfer level – utilizing high-level information for hardware verification'. Proceedings of the IFIP international conference on *VLSI*, 2001, pp. 127–132

16 GROßE, D., and DRECHSLER, R.: 'Formal verification of LTL formulas for SystemC designs'. IEEE international symposium on *Circuits and systems*, 2003, pp. V:245–V:248

17 FEY, G., GROßE, D., CASSENS, T., GENZ, C., WARODE, T., and DRECHSLER, R.: 'ParSyC: An efficient SystemC parser'. Proceedings of the workshop on *Synthesis and system integration of mixed information technologies (SASIMI)*, 2004, pp. 148–154

18 PARR, T.: 'Language translation using PCCTS and C++: a reference guide' (Automata Publishing Co., 1997)

19 Synopsys. 'Describing synthesizable RTL in systemC™, Vers. 1.1'. Synopsys Inc., 2002. Available at http://www.synopsys.com

20 MOSKEWICZ, M.W., MADIGAN, C.F., ZHAO, Y., ZHANG, L., and MALIK, S.: 'Chaff: Engineering an efficient SAT solver'. Proceedings of the *Design automation* conference, 2001, pp. 530–535

21 FOSTER, H., KROLNIK, A., and LACEY, D.: 'Assertion-based design' (Kluwer Academic Publishers, New York, 2003)

22 DRECHSLER, R.: 'Synthesizing checkers for on-line verification of system-on-chip designs'. IEEE international symposium on *Circuits and systems*, 2003, pp. IV:748–IV:751

23 GROßE, D., and DRECHSLER, R.: 'Checkers for SystemC designs'. Proceedings of the second ACM & IEEE international conference on *Formal methods and models for codesign (MEMOCODE)*, 2004, pp. 171–178

24 KREBS, A., and RUF, J.: 'Optimized temporal logic compilation', *Journal of Universal Computer Science*, 2003, **9**(2), pp. 120–137

25 ALI, M., VENERIS, A., SAFARPOUR, S., ABADIR, M., DRECHSLER, R., and SMITH, A.: 'Debugging sequential circuits using Boolean satisfiability'. Proceedinngs of the international conference on *CAD*, 2004

26 BENING, L., and FOSTER, H.: 'Principles of verifiable RTL design' (Kluwer Academic Publishers, New York, 2001)

27 BERGERON, J.: 'Writing testbenches: functional verification of HDL models' (Kluwer Academic Publishers, New York, 2003)

28 BRINKMANN, R., and DRECHSLER, R.: 'RTL-datapath verification using integer linear programming'. Proceedings of the *ASP design automation* conference, 2002, pp. 741–746

29 CHOCKLER, H., KUPFERMAN, O., KURSHAN, R., and VARDI, M.: 'A practical approach to coverage in model checking', in 'Computer aided verification', vol. 2102 of *LNCS*. (Springer Verlag, Heidelberg, 2001), pp. 66–77

30 COPTY, F., IRRON, A., WEISSBERG, O., KROPP, N., and KAMHI, G.: 'Efficient debugging in a formal verification environment', *Software Tools for Technology Transfer*, 2003, **4**, pp. 335–348

31 DRECHSLER, R.: 'Formal verification of circuits' (Kluwer Academic Publishers, New York, 2000)

32 DRECHSLER, R.: 'Advanced formal verification' (Kluwer Academic Publishers, New York, 2004)

33 DRECHSLER, R., and DRECHSLER, N.: 'Evolutionary algorithms for embedded system design' (Kluwer Academic Publisher, New York, 2002)

34 DRECHSLER, R., and SIELING, D.: 'Binary decision diagrams in theory and practice', *Software Tools for Technology Transfer*, 2001, **3**, pp. 112–136

35 HASSOUN, S., and SASAO, T.: 'Logic synthesis and verification' (Kluwer Academic Publishers, New York, 2001)

36 HO, P.-H., SHIPLE, T., HARER, K., KUKULA, J., DAMIANO, R., BERTACCO, V., TAYLOR, J., and LONG, J.: 'Smart simulation using collaborative formal and simulation engines'. Proceedings of the international conference on *CAD*, 2000, pp. 120–126

37 HOSKOTE, Y., KAM, T., HO, P., and ZHAO, X.: 'Coverage estimation for symbolic model checking'. Proceedings of the *Design automation* conference, 1999, pp. 300–305

38 HSU Y.-C., TABBARA, B., CHEN Y.-A., and TSAI, F.: 'Advanced techniques for RTL debugging'. Proceedings of the *Design automation* conference, 2003, pp. 362–367

39 JONES, R.: 'Symbolic simulation methods for industrial formal verification' (Kluwer Academic Publishers, New York, 2002)

40 KÖLBL, A., KUKULA, J., and DAMIANO, R.: 'Symbolic RTL simulation'. Proceedings of the *Design automation* conference, 2001, pp. 47–52

41 KROPF, Th.: 'Introduction to formal hardware verification' (Springer, Heidelberg, 1999)

42 KUEHLMANN, A., GANAI, M., and PARUTHI, V.: 'Circuit-based Boolean reasoning'. Proceedings of the *Design automation* conference, 2001, pp. 232–237

43   MOHNKE, J., MOLITOR, P., and MALIK, S.: 'Limits of using signatures for permutation independent Boolean comparison', *Formal Methods in System Design: An International Journal*, 2002, **2**(21), pp. 167–191

44   MOUNDANOS, D., ABRAHAM, J., and HOSKOTE, Y.: 'Abstraction techniques for validation coverage analysis and test generation', *IEEE Transactions on Computers*, January 1998, pp. 2–14

45   RASHINKAR, P., PATERSON, P., and SINGH, L.: 'System-on-a-chip verification' (Kluwer Academic Publishers, New York, 2000)

46   REDA, S., DRECHSLER, R., and ORAILOGLU, A.: 'On the relation between SAT and BDDs for equivalence checking'. Proceedings of the international symposium on *Quality electronic design*, 2002, pp. 394–399

47   SAFARPOUR, S., VENERIS, A., DRECHSLER, R., and HANG, J.: 'Managing don't cares in Boolean satisfiability'. Proceedings of *Design, automation and test in Europe*, 2004, **1**, pp. 260–265

48   STOFFEL, D., and KUNZ, W.: 'Equivalence checking of arithmetic circuits on the arithmetic bit level', *IEEE Transactions on CAD*, 2004, **23**(5), pp. 586–597

49   STOFFEL, D., WEDLER, M., WARKENTIN, P., and KUNZ, W.: 'Structural FSM traversal', *IEEE Transactions on CAD*, 2004, **23**(5), pp. 598–619

50   VENERIS, A., SMITH, A., and ABADIR, M.S.: 'Logic verification based on diagnosis techniques'. Proceedings of the *Design automation* conference, 2003

Part VIII

Manufacturing test

Chapter 22

Efficient modular testing and test resource partitioning for core-based SoCs

Krishnendu Chakrabarty

22.1 Introduction

Shrinking process technologies and increasing design sizes have led to billion-transistor integrated circuits (ICs). To reduce IC design and manufacturing costs, test development and test application must be quick as well as effective. High transistor counts in ICs result in large test data sets, long test development and application times and the need for expensive test equipment. Effective test development techniques that enhance the utilisation of test data, testing time and test equipment are therefore necessary to increase production capacity and reduce test cost.

Test resource partitioning (TRP) deals with the partitioning and optimisation of test resources to enhance test effectiveness and reduce test cost. This chapter describes the use of TRP for complex ICs and presents recent advancements in test access mechanisms, test scheduling and test data compression.

This chapter begins with an introduction to the latest IC design philosophy – system-on-a-chip (SoC). Increasing SoC sizes lead to greater test resource requirements for manufacturing test. This growth in test resource requirements motivates the need for efficient TRP techniques during SoC test development.

22.1.1 SoC design

SoCs are crafted by system designers who purchase intellectual property (IP) circuits, known as embedded cores, from core vendors and integrate them into large designs. Embedded-cores are complex, pre-designed and pre-verified circuits that can be purchased off-the-shelf and reused in designs. While SoCs have become popular as a means to integrate complex functionality into designs in a relatively short amount of time, there remain several roadblocks to rapid and efficient system integration. Primary among these is the lack of core interface and testability standards upon which core design and system development can be based. Importing core designs from different IP sources and stitching them into designs often entails cumbersome format translation. Testing SoCs is equally challenging in the absence of standardised test structures. Hence a number of SoC and core development working groups have been formed, notable among these being the Virtual Socket Interface Alliance (VSIA) [1] and the IEEE P1500 working group on embedded-core test [2].

The VSIA was formed in September 1996 with the goal of establishing a unifying vision for the SoC industry, and the technical standards required to facilitate system integration. VSIA specifies interface standards, which will allow cores to fit quickly into 'virtual sockets' on the SoC, at both the architectural level and the physical level [1]. This will allow core vendors to produce cores with a uniform set of interface features, rather than having to support different sets of features for each customer. SoC integration is in turn simplified since cores may be imported and plugged into standardised 'sockets' on SoCs with relative ease.

The IEEE P1500 working group was established to draft a test standard for digital logic and memory cores. The activities of the P1500 working group include the development of (1) a standardised core test language, (2) a test wrapper interface from cores to on-chip test access mechanisms and (3) guidelines for the test of mergeable cores.

22.1.2   Testing a system-on-a-chip

An SoC test is essentially a composite test composed of the individual tests for each core, the user defined logic (UDL) tests and interconnect tests. Each individual core or UDL test may involve surrounding components and may imply operational constraints (e.g. safe mode, low power mode, bypass mode) which necessitate special isolation modes.

The SoC test development is especially challenging because of several reasons. Embedded-cores represent intellectual property and core vendors are reluctant to divulge structural information about their cores to users. Thus, users cannot access core netlists and insert design-for-testability (DFT) hardware that can ease test application from the surrounding logic. Instead, a set of test patterns is provided by the core vendor that guarantees a specific fault coverage. These test patterns must be applied to the cores in a given order, using a specific clocking strategy. Care must often be taken to ensure that undesirable patterns and clock skews are not introduced into these test streams. Furthermore, cores are often embedded in several layers of user-designed or other core-based logic, and are not always directly accessible from chip I/Os. Propagating test stimuli to core inputs may therefore require dedicated test transport mechanisms. Moreover, translation of test data is necessary at the inputs and outputs of the embedded-core into a format or sequence suitable for application to the core.

A conceptual architecture for testing embedded-core-based SoCs is shown in Figure 22.1 [3]. It consists of three structural elements:

1   *Test pattern source and sink.* The test pattern source generates the test stimuli for the embedded-cores, and the test pattern sink compares the response(s) to the expected response(s).

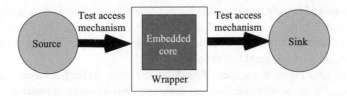

Figure 22.1   *Overview of the three elements in an embedded-core test approach: (1) test pattern source and sink, (2) test access mechanism and (3) core test wrapper [3]*

2   *Test access mechanism (TAM)*. The TAM transports test patterns. It is used for on-chip transport of test stimuli from test pattern source to the core under test, and for the transport of test responses from the core under test to a test pattern sink.

3   *Core test wrapper*. The core test wrapper forms the interface between the embedded-core and its environment. It connects the terminals of the embedded-core to the rest of the IC and to the TAM.

Once a suitable test data transport mechanism and test translation mechanism have been designed, the next major challenge confronting the system integrator is test scheduling. This refers to the order in which the various core tests and tests for user-designed interface logic are applied. A combination of BIST and external testing is often used to achieve high fault coverage [4, 5], and tests generated by different sources may therefore be applied in parallel, provided resource conflicts do not arise. Effective test scheduling for SoCs is challenging because it must address several conflicting goals: (1) SoC testing time minimisation, (2) resource conflicts between cores arising from the use of shared TAMs and on-chip BIST engines, (3) precedence constraints among tests and (4) power constraints.

The increasing complexity of core-based SoCs has led to an enormous rise in the test data volume necessary to attain the desired test coverage for the SoC. For example, the test data volume can be as high as several Gbits for an industrial ASIC [6]. Since automatic test equipment (ATE) costs range in the millions of dollars, most IC manufacturers do not replace ATEs with each new SoC design. However, the memory and channel bandwidth of older ATEs are limited; hence storing the entire test for a new-generation SoC in tester memory is often infeasible. Furthermore, the additional memory add-on capability for older ATEs is often limited. Hence a small increase in test data volume can result in either having to purchase a new tester that costs millions of dollars, or executing several test data 'load-apply-reload' sessions, thus adversely affecting testing time and test cost. Test set compression techniques that can reduce tester memory requirements are therefore highly desirable [7, 8]. However, decompressing compressed test data on-chip often requires additional hardware overhead, and SoC designers may be reluctant to provide large decompression circuits. Test data compression methods, whose corresponding decompression circuits are small, form an important area of research in SoC test.

Finally, analogue and mixed-signal cores are increasingly being integrated onto SoCs with digital cores. Testing mixed-signal cores is challenging because their failure mechanisms and testing requirements are not as well modelled as they are for digital cores. It is difficult to partition and test analogue cores, because they may be prone to crosstalk across partitions. Capacitance loading and complex timing issues further exacerbate the mixed-signal test problem.

This chapter presents a survey of modular test methods for SoCs that enhance the utilisation of test resources such as test data, testing time and test hardware.

22.2   Test resource partitioning

The new modular IC design paradigm partitions the SoC design process into a step-by-step procedure that can be spread out over several organisations, thus exploiting their individual core competencies. This design paradigm naturally extends to a modular test development approach for SoCs, in which test resources are partitioned and optimised to achieve robust, high quality and effective testing at low cost.

TRP refers to the process of partitioning monolithic test resources, such as the test data set or the top-level TAM into sub-components that can be optimised to achieve significant gains in test resource utilisation [9]. For example, large test data sets can be partitioned into subsets, some of which can be generated by on-chip hardware, thus reducing ATE complexity and cost. The top-level TAM can be partitioned into several sub-TAMs that fork out to test cores in parallel, thus increasing test concurrency and reducing testing time. Finally, a modular test schedule for the SoC can be crafted such that idle time on each test data delivery component is minimised, thus leading to improved test hardware utilisation and a lower vector memory requirement for the tester.

SoC test resources are of three main types: (1) test hardware, (2) testing time and (3) test data. 'Test hardware' refers to special-purpose hardware used for test generation and application. This can be either on- or off-chip. Off-chip test hardware refers to external test equipment such as ATE, wafer probes, analogue instrumentation, etc., while on-chip hardware refers to test wrappers for cores, TAMs and DFT structures such as LFSRs, boundary scan and test points. 'Testing time' refers to the time required for manufacturing test. This includes the time for wafer test, contact test, digital vectors and DC/AC parametrics. The time required to apply digital vectors includes test data download time from workstations to the ATE across a network. Finally, 'test data' refers to the sequences of test patterns, test responses and control signals that are applied to the SoC. These may be in the form of either digital signals or analogue waveforms.

TRP techniques that address each SoC test resource can be classified as shown in Figure 22.2. These are briefly described here.

22.2.1   TRP techniques to optimise test hardware

The external test equipment required to test a $100 SoC can cost of the order of millions of dollars; efforts to reduce dependence on expensive external test equipment have therefore gained considerable importance. On-chip test hardware overhead must

Test resource partitioning

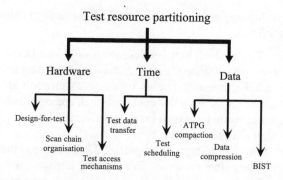

Figure 22.2   A classification of TRP techniques [9]

also remain below acceptable levels (generally of the order of a few per cent of total chip area) to keep SoC costs competitive. There is thus a compelling need for techniques that bolster the utilisation of test hardware, while providing a high level of test coverage. Test hardware partitioning techniques are described below.

Design for test: DFT techniques involve the addition of optimised test logic within the cores and at the chip level to enhance testability. Test logic includes scan chains, test access ports, test points, boundary scan, structures that partition large modules, I_{DDQ} support logic and test clock support circuitry. DFT logic facilitates test pattern generation and application, and also assists in the support test environments such as wafer probe, functional test, burn-in and life-cycle test [10]. The dependence on external test equipment is therefore significantly reduced. Test application is facilitated by the availability of scan chains. The organisation of these scan chains is an important TRP issue because it affects test data volume and testing time.

Scan chain organisation: Scan design can be used for TRP to support external test equipment and simplify test generation. Effective scan chain organisation is required to address issues such as contention-free shifting and sampling, clock skew in multiple time domains, scan chain balancing and power constraints [10].

Test access mechanisms: TAMs and test wrappers have been proposed as important components of an SoC test access architecture [3]. Test wrapper and TAM design is a critical TRP technique in SoC system integration since it directly impacts test cost. This is because a TAM architecture that minimises the idle time spent by TAMs and wrappers during test directly reduces the number of don't-care bits in vectors stored on the tester, thereby reducing ATE vector memory depth and testing time [11].

22.2.2   TRP techniques to minimise testing time

Testing time is a major contributor to test cost, which is now widely believed to be close to 50 per cent of the SoC manufacturing cost. Efficient means of downloading and scheduling tests that can identify faulty components quickly are therefore vital to reduce manufacturing cost and time-to-market. TRP methods to reduce testing time are described as follows.

Test data transfer: TRP methods based on data compression can be used to address the issue of reducing test data download time [12] between workstations and ATE. New generation ATE includes several test heads, memory modules and add-on devices on the ATE to address the test of SoCs that contain digital, analogue and memory cores. This reduces the time required to switch between different ATE or download different test programs during test application.

Test scheduling: Test scheduling is an important TRP technique that enhances the utilisation of the testing time resource. This directly impacts product quality and time-to-market, and is directly related to the manufacturer's economic performance. The testing time can be reduced by increasing test parallelism, aborting the test of SoCs as soon as the first failing pattern is detected, and ensuring that power constraints are not violated during test application.

22.2.3   TRP techniques to reduce test data volume

Test data sets for large SoCs now require of the order of several gigabytes of tester memory, thus contributing significantly to overall test cost. TRP techniques that address test data volume reduction are described as follows.

Test compaction: This technique reduces test data volume by compacting the partially specified test cubes generated by automatic test pattern generation (ATPG) algorithms. It requires no additional hardware investment. The test set is compacted through dynamic or static compaction procedures [13, 14].

Test data compression: Test data volume can be significantly reduced through statistical data compression techniques such as run-length, Golomb [7], VIHC [15], exponential Golomb and subexponential Golomb [16], and frequency-directed run-length (FDR) codes [17]. These techniques compress the precomputed test set T_D provided by the core vendor into a much smaller test set T_E, which is stored in ATE memory. An on-chip decoder performs pattern decompression to generate T_D from T_E during pattern application.

Built-in self test (BIST): BIST offers several advantages when used for TRP. It lets precomputed test sets be embedded in test sequences generated by on-chip hardware, supports test reuse and at-speed testing, and protects IP.

Figure 22.3 illustrates the use of TRP approaches for SoC test. The figure presents an ATE applying tests to an SoC composed of several cores. The test schedule for the SoC is also illustrated. The test data for the SoC has been transformed into (1) a compressed data set stored in ATE memory, and (2) a test pattern decoder implemented on-chip to decompress test patterns for application to the cores. In this way, test resources available in one form (data) can be partitioned into several subcomponents of differing types (data and time) to enhance test resource utilisation. The TAM in Figure 22.3 delivers test patterns to each core according to the test schedule. The TAM includes the on-chip test generation logic for BISTed cores. The test schedule shown presents the start and end times for the test for each core. Each

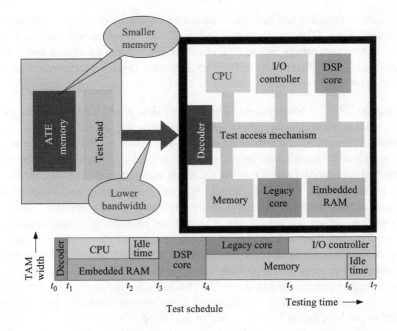

Figure 22.3   SoC test based on TRP [9]

rectangle in the test schedule represents a test for a core. The height of the rectangle corresponding to each test in the schedule represents the width of the TAM that is used for that test. The CPU and the embedded RAM each require only a part of the total TAM width; hence they can be tested in parallel. In this manner, the test schedule can be used to represent the different TAM partitions at various instants of time.

Finally, sophisticated test control mechanisms are often needed during the application of compressed patterns to the SoC and on-chip decompression. Test control is an often overlooked issue in the literature on test compression and on-chip pattern decompression. In order to concurrently test multiple embedded-cores with different test requirements (bandwidth, data rates, capture clock frequencies, etc.), a set of ATE channels must be used exclusively for control, or preferably, additional control logic must be implemented on the SoC.

22.3   Modular testing of SoCs

Modular testing of embedded-cores in an SoC is being increasingly advocated to simplify test access and test application [3]. To facilitate modular test, an embedded-core must be isolated from surrounding logic, and test access must be provided from the I/O pins of the SoC. Test wrappers are used to isolate the core, while TAMs transport test patterns and test responses between SoCs pins and core I/Os [3].

Effective modular test requires efficient management of the test resources for core-based SoCs. This involves the design of core test wrappers and TAMs, the assignment of test pattern bits to ATE channels, the scheduling of core tests and the assignment of ATE channels to SoCs. The challenges involved in the optimisation of SoC test resources for modular test can be divided into three broad categories.

1. *Wrapper/TAM co-optimisation.* Test wrapper design and TAM optimisation are of critical importance during system integration since they directly impact hardware overhead, testing time and tester data volume. The issues involved in wrapper/TAM design include wrapper optimisation, core assignment to TAM wires, sizing of the TAMs and routing of TAM wires. As shown in References 11, 18 and 19, most of these problems are $\mathcal{NP}$-hard. Figures 22.4(a) and (b) illustrate the position of TAM design and test scheduling in the SoC design for test and test generation flows.

2. *Constraint-driven test scheduling.* The primary objective of test scheduling is to minimise testing time, while addressing one or more of the following issues: (1) resource conflicts between cores arising from the use of shared TAMs and BIST resources, (2) precedence constraints among tests and (3) power dissipation constraints. Furthermore, testing time can often be decreased further through the selective use of test pre-emption [21]. As discussed in References 4 and 21, most problems related to test scheduling for SoCs are also $\mathcal{NP}$-hard.

3. *Minimising ATE re-load under memory depth constraints.* Given test data for the individual cores, the entire test suite for the SoC must be made to fit in a minimum number of ATE memory loads (preferably one memory load). This is important because, while the time required to apply digital vectors is relatively small, the time required to load several gigabytes of data to the ATE memory from workstations is significant [22, 23]. Therefore, to avoid splitting the test into multiple ATE load-apply sessions, the number of bits required to be stored on any ATE channel must not exceed the limit on the channel's memory depth.

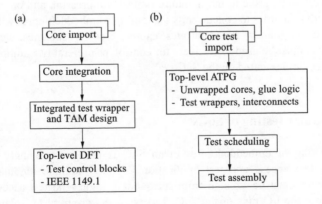

Figure 22.4   *The (a) DFT generation flow and (b) test generation flow for SoCs [20] (©IEEE, 2002)*

In addition, the rising cost of ATE for SoC devices is a major concern [24]. Due to the growing demand for pin counts, speed, accuracy and vector memory, the cost of high-end ATE for full-pin, at-speed functional test is predicted to rise to over $20M by 2010 [24]. As a result, the use of low-cost ATEs that perform structural rather than at-speed functional test is increasingly being advocated for reducing test costs. Multi-site testing, in which multiple SoCs are tested in parallel on the same ATE, can significantly increase the efficiency of ATE usage, as well as reduce testing time for an entire production batch of SoCs. The use of low-cost ATE and multi-site test involve test data volume reduction and test pin count (TAM width) reduction, such that multiple SoC test suites can fit in ATE memory in a single test session [23, 25].

As a result of the intractability of the problems involved in test planning, test engineers have adopted a series of simple ad hoc solutions in the past [23]. For example, the problem of TAM width optimisation is often simplified by stipulating that each core on the SoC has the same number of internal scan chains, say W; thus, a TAM of width W bits is laid out and cores are simply daisy-chained to the TAM. However, with the growing size of SoC test suites and rising cost of ATE, more aggressive test resource optimisation techniques that enable effective modular test of highly complex next-generation SoCs using current-generation ATE is critical.

22.3.1   Wrapper design and optimisation

A core test wrapper is a layer of logic that surrounds the core and forms the interface between the core and its SoC environment. Wrapper design is related to the well-known problems of circuit partitioning and module isolation, and is therefore a more general test problem than its current instance (related to SoC test using TAMs). For example, earlier proposed forms of circuit isolation (precursors of test wrappers) include boundary scan and BILBO [13].

The test wrapper and TAM model of SoC test architecture was presented in Reference 3. In this paper, three mandatory wrapper operation modes listed were (1) normal operation, (2) core-internal test and (3) core-external test. Apart from the three mandatory modes, two optional modes are 'core bypass' and 'detach'.

Two proposals for test wrappers have been the 'test collar' [26] and TestShell [27]. The test collar was designed to complement the Test Bus architecture [26] and the TestShell was proposed as the wrapper to be used with the TestRail architecture [27]. In Reference 26, three different test collar types were described: combinational, latched and registered. For example, a simple combinational test collar cell consisting of a 2-to-1 multiplexer can be used for high-speed signals at input ports during parallel, at-speed test. The TestShell described in Reference 27 is used to isolate the core and perform TAM width adaptation. It has four primary modes of operation: function mode, IP test mode, interconnect test mode and bypass mode. These modes are controlled using a test control mechanism that receives two types of control signals: pseudo-static signals (that retain their values for the duration of a test) and dynamic control signals (that can change values during a test pattern).

An important function of the wrapper is to adapt the TAM width to the core's I/O terminals and internal scan chains. This is done by partitioning the set of core-internal scan chains and concatenating them into longer wrapper scan chains, equal in number to the TAM wires. Each TAM wire can now directly scan test patterns into a single wrapper scan chain. TAM width adaptation directly affects core testing time and has been the main focus of research in wrapper optimisation. Note that to avoid problems related to clock skew, internal scan chains in different clock domains must either not be placed on the same wrapper scan chain, or anti-skew (lock-up) latches must be placed between scan FFs belonging to different clock domains.

The issue of designing balanced scan chains within the wrapper was addressed in Reference 28; see Figure 22.5. The first techniques to optimise wrappers for test time reduction were presented in Reference 19. To solve the problem, the authors proposed two polynomial-time algorithms that yield near-optimal results. The LPT (largest processing time) algorithm is taken from the multi-processor scheduling literature and solves the wrapper design problem in very short computation times. At the expense of a slight increase in computation time, the COMBINE algorithm yields even better results. It uses LPT as a start solution, followed by a linear search over the wrapper scan chain length with the first fit decreasing heuristic.

To perform wrapper optimisation, Iyengar *et al.* [11] proposed 'Design_wrapper', an algorithm based on the best fit decreasing heuristic for the bin packing problem. The algorithm has two priorities: (1) minimising core testing time and (2) minimising the TAM width required for the test wrapper. These priorities are achieved by balancing the lengths of the wrapper scan chains designed, and identifying the number of wrapper scan chains that actually need to be created to minimise testing time. Priority (2) is addressed by the algorithm since it has a built-in reluctance to create a new wrapper scan chain, while assigning core-internal scan chains to the existing wrapper scan chains [11].

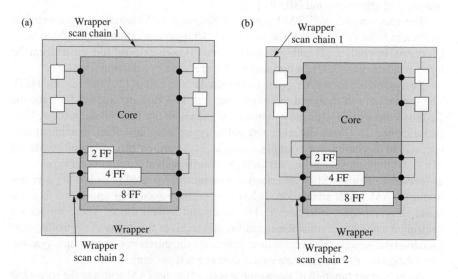

Figure 22.5   Wrapper chains: (a) unbalanced, (b) balanced

Wrapper design and optimisation continue to attract considerable attention. Recent work in this area has focused on 'light wrappers', i.e. the reduction of the number of register cells [29], and the design of wrappers for cores and SoCs with multiple clock domains [30].

22.3.2 TAM design and optimisation

Many different TAM designs have been proposed in the literature. TAMs have been designed based on direct access to cores multiplexed onto the existing SoC pins [31], reusing the on-chip system bus [32], searching transparent paths through and/or around neighbouring modules [33–35], and one-bit boundary scan rings around cores [36, 37].

Recently, the most popular appear to be the dedicated, scalable TAMs such as Test Bus [26] and TestRail [27]. Despite the fact that their dedicated wiring adds to the area costs of the SoC, their flexible nature and guaranteed test access have proven successful. Three basic types of such scalable TAMs have been described in Reference 38 (see Figure 22.6): (a) the 'Multiplexing' architecture, (b) the 'Daisy-chain' architecture and (c) the 'Distribution' architecture. In the Multiplexing and Daisychain architectures, all cores get access to the total available TAM width, while in the Distribution architecture, the total available TAM width is distributed over the cores.

In the Multiplexing architecture, only one core wrapper can be accessed at a time. Consequently, this architecture only supports serial schedules, in which the cores are tested one after the other. An even more serious drawback of this architecture is that testing the circuitry and wiring in between cores is difficult with this architecture; interconnect test requires simultaneous access to multiple wrappers. The other two basic architectures do not have these restrictions; they allow for both serial as well as parallel test schedules, and also support interconnect testing.

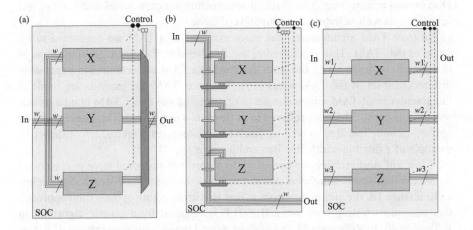

Figure 22.6   The (a) Multiplexing, (b) Daisychain and (c) Distribution architectures [20, 38] (©IEEE, 2002)

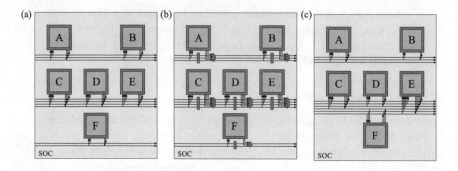

*Figure 22.7   The (a) fixed-width Test Bus architecture, (b) fixed-width TestRail
architecture and (c) flexible-width Test Bus architecture [20] (©IEEE,
2002)*

The 'Test Bus' architecture [26] (see Figure 22.7(a)) is a combination of the
Multiplexing and Distribution architectures. A single Test Bus is in essence the same
as what is described by the Multiplexing architecture; cores connected to the same Test
Bus can only be tested sequentially. The Test Bus architecture allows for multiple Test
Buses on one SoC that operate independently, as in the Distribution architecture. Cores
connected to the same Test Bus suffer from the same drawback as in the Multiplexing
architecture, i.e. their wrappers cannot be accessed simultaneously, hence making
core-external testing difficult or impossible.

The 'TestRail' architecture [27] (see Figure 22.7(b)) is a combination of the Daisychain and Distribution architectures. A single TestRail is in essence the same as what
is described by the Daisychain architecture: scan-testable cores connected to the same
TestRail can be tested simultaneously, as well as sequentially. A TestRail architecture
allows for multiple TestRails on one SoC, which operate independently, as in the
Distribution architecture. The TestRail architecture supports serial and parallel test
schedules, as well as hybrid combinations of those.

In most TAM architectures, the cores assigned to a TAM are connected to *all*
wires of that TAM. This is referred to as 'fixed-width' TAMs. A generalisation of
this design, is one in which the cores assigned to a TAM each connect to a (possibly
different) subset of the TAM wires [39]. The core–TAM assignments are made at
the granularity of TAM wires, instead of considering the entire TAM bundle as one
inseparable entity. These are referred to as flexible-width TAMs. This concept can be
applied to both Test Bus as well as TestRail architectures. Figure 22.7(c) shows an
example of a flexible-width Test Bus architecture.

Most SoC test architecture optimisation algorithms proposed have concentrated
on fixed-width Test Bus architectures and assume cores with fixed-length scan chains.
In Reference 18, the author describes a Test Bus architecture optimisation approach
that minimises testing time using ILP. ILP is replaced by a genetic algorithm in
Reference 40. In Reference 41, the authors extend the optimisation criteria of Reference 18 with place-and-route and power constraints, again using ILP. In References 42
and 43, Test Bus architecture optimisation is mapped to the well-known problem of

two-dimensional bin packing and a Best Fit algorithm is used to solve it. Wrapper design and TAM design both influence the SoC testing time, and hence their optimisation needs to be carried out in conjunction in order to achieve the best results. The authors in Reference 11 were the first to formulate the problem of integrated wrapper/TAM design; despite its $\mathcal{NP}$-hard character, it is addressed using ILP and exhaustive enumeration. In Reference 44, the authors presented efficient heuristics for the same problem.

Idle bits exist in test schedules when parts of the test wrapper and TAM are under-utilised leading to idle time in the test delivery architecture. In Reference 45, the authors first formulated the testing time minimisation problem both for cores having fixed-length as well as cores having flexible-length scan chains. Next, they presented lower bounds on the testing time for the Test Bus and TestRail architectures and then examined three main reasons for under-utilisation of TAM bandwidth, leading to idle bits in the test schedule and testing times higher than the lower bound [45]. The problem of reducing the amount of idle test data was also addressed in Reference 46.

The optimisation of a flexible-width Multiplexing architecture (i.e. for one TAM only) was proposed in Reference 39. This work again assumes cores with fixed-length scan chains. The paper describes a heuristic algorithm for co-optimisation of wrappers and Test Buses based on rectangle packing. In Reference 39, the same authors extended this work by including precedence, concurrency and power constraints, while allowing a user-defined subset of the core tests to be pre-empted.

Fixed-width TestRail architecture optimisation was investigated in Reference 47. Heuristic algorithms have been developed for the co-optimisation of wrappers and TestRails. The algorithms work both for cores with fixed-length and flexible-length scan chains. TR-ARCHITECT, the tool presented in Reference 47, is currently in actual industrial use.

22.3.3 Test scheduling

Test scheduling for SoCs involving multiple test resources and cores with multiple tests is especially challenging, and even simple test scheduling problems for SoCs have been shown to be $\mathcal{NP}$-hard [4]. In Reference 5, a method for selecting tests from a set of external and BIST tests (that run at different clock speeds) was presented. Test scheduling was formulated as a combinatorial optimisation problem. Re-ordering tests to maximise defect detection early in the schedule was explored in Reference 48. The entire test suite was first applied to a small sample population of ICs. The fault coverage obtained per test was then used to arrange tests that contribute to high fault coverage earlier in the schedule. The authors used a polynomial-time algorithm to re-order tests based on the defect data as well as execution time of the tests [48]. A test scheduling technique based on the defect probabilities of the cores has recently been reported [49].

Macro Test is a modular testing approach for SoC cores in which a test is broken down into a 'test protocol' and list of test patterns [50]. A test protocol is defined at the terminals of a macro and describes the necessary and sufficient conditions to test the macro [51]. The test protocols are expanded from the macro-level to the SoC pins and can either be applied sequentially to the SoC, or scheduled to increase parallelism. In Reference 51, a heuristic scheduling algorithm based on pairwise composition of test protocols was presented. The algorithm determines the start times for the expanded test protocols in the schedule, such that no resource conflicts occur and test time is minimised [51].

System-on-chips in test mode can dissipate up to twice the amount of power they do in normal mode, since cores that do not normally operate in parallel may be tested concurrently [52]. 'Power-constrained' test scheduling is therefore essential in order to limit the amount of concurrency during test application to ensure that the maximum power budget of the SoC is not exceeded. In Reference 53, a method based on approximate vertex cover of a resource-constrained test compatibility graph was presented. In Reference 54, the use of list scheduling and tree-growing algorithms for power-constrained scheduling was discussed. The authors presented a greedy algorithm to overlay tests such that the power constraint is not violated. A constant additive model is employed for power estimation during scheduling [54]. The issue of re-organising scan chains to trade-off testing time with power consumption was investigated in Reference 55. The authors presented an optimal algorithm to parallelise tests under power and resource constraints. The design of test wrappers to allow for multiple scan chain configurations within a core was also studied.

In Reference 21, an integrated approach to test scheduling was presented. Optimal test schedules with precedence constraints were obtained for reasonably sized SoCs. For precedence-based scheduling of large SoCs, a heuristic algorithm was developed. The proposed approach also includes an algorithm to obtain pre-emptive test schedules in $O(n^3)$ time, where n is the number of tests [21]. Parameters that allow only a certain number of pre-emptions per test can be used to prevent excessive BIST and sequential circuit test pre-emptions. Finally, a new power-constrained scheduling technique was presented, whereby power-constraints can be easily embedded in the scheduling framework in combination with precedence constraints, thus delivering an integrated approach to the SoC test scheduling problem.

22.3.4   Integrated TAM optimisation and test scheduling

Both TAM optimisation and test scheduling significantly influence the testing time, test data volume and test cost for SoCs. Furthermore, TAMs and test schedules are closely related. For example, an effective schedule developed for a particular TAM architecture may be inefficient or even infeasible for a different TAM architecture. Integrated methods that perform TAM design and test scheduling 'in conjunction' are therefore required to achieve low-cost, high-quality test.

In Reference 56, an integrated approach to test scheduling, TAM design, test set selection and TAM routing was presented. The SoC test architecture was represented by a set of functions involving test generators, response evaluators, cores, test sets, power and resource constraints, and start and end times in the test schedule modelled as Boolean and integral values [56]. A polynomial-time algorithm was used to solve these equations and determine the test resource placement, TAM design and routing, and test schedule, such that the specified constraints are met.

The mapping between core I/Os and SoC pins during the test schedule was investigated in Reference 42. TAM design and test scheduling was modelled as two-dimensional bin-packing, in which each core test is represented by a rectangle. The height of each rectangle corresponds to the testing time, the width corresponds to the core I/Os and the weight corresponds to the power consumption during test. The objective is to pack the rectangles into a bin of fixed width (SoC pins), such that the bin height (total testing time) is minimised, while power constraints are met. A heuristic method based on the Best Fit algorithm was presented to solve the problem [42]. The authors next formulated constraint-driven pin mapping and test scheduling as the chromatic number problem from graph theory and as a dependency matrix partitioning problem [43]. Both problem formulations are $\mathcal{NP}$-hard. A heuristic algorithm based on clique partitioning was proposed to solve the problem.

The problem of TAM design and test scheduling with the objective of minimising the 'average' testing time was formulated in Reference 57. The problem was reduced to one of minimum-weight perfect bipartite graph matching, and a polynomial-time optimal algorithm was presented. A test planning flow was also presented.

In Reference 39, a new approach for wrapper/TAM co-optimisation and constraint-driven test scheduling using rectangle packing was described. Flexible-width TAMs that are allowed to fork and merge were designed. Rectangle packing was used to develop test schedules that incorporate precedence and power constraints, while allowing the SoC integrator to designate a group of tests as pre-emptable. Finally, the relationship between TAM width and tester data volume was studied to identify an effective TAM width for the SoC.

The work reported in Reference 39 was extended in Reference 58 to address the minimisation of ATE buffer re-loads and include multi-site test. The ATE is assumed to contain a pool of memory distributed over several channels, such that the memory depth assigned to each channel does not exceed a maximum limit. Furthermore, the sum of the memory depth over all channels equals the total pool of ATE memory. Idle bits appear on ATE channels whenever there is idle time on a TAM wire. These bit positions are filled with don't-cares if they appear between useful test bits; however, if they appear only at the end of the useful bits, they are not required to be stored in the ATE.

The SoC test resource optimisation problem for multi-site test was stated as follows. Given the test set parameters for each core, and a limit on the maximum memory depth per ATE channel, determine the wrapper/TAM architecture and test schedule for the SoC, such that (1) the memory depth required on any channel is less than the maximum limit, (2) the number of TAM wires is minimised and (3) the idle bits appear only at the end of each channel. A rectangle packing algorithm was developed to solve this problem.

A new method for representing SoC test schedules using k-tuples was discussed in Reference 59. The authors presented a p-admissible model for test schedules that is amenable to several solution methods such as local search, two-exchange, simulated annealing and genetic algorithms that cannot be used in a rectangle-representation environment.

Finally, recent work on TAM optimisation has focused on the use of ATEs with port scalability features [60–62]. In order to address the test requirements of SoCs, ATE vendors have recently announced a new class of testers that can simultaneously drive different channels at different data rates. Examples of such ATEs include the Agilent 93000 series tester based on port scalability and the test processor-per-pin architecture [63] and the Tiger system from Teradyne [64] in which the data rate can be increased through software for selected pin groups to match SoC test requirements. However, the number of tester channels with high data rates may be constrained in practice due to ATE resource limitations, the power rating of the SoC and scan frequency limits for the embedded-cores. Optimisation techniques have been developed to ensure that the high data-rate tester channels are efficiently used during SoC testing [62].

The availability of dual-speed ATEs was also exploited in References 60 and 61, where a technique was presented to match ATE channels with high data rates to core scan chain frequencies using virtual TAMs. A 'virtual TAM' is an on-chip test data transport mechanism that does not directly correspond to a particular ATE channel. Virtual TAMs operate at scan-chain frequencies; however, they interface with the higher-frequency ATE channels using bandwidth matching. Moreover, since the virtual TAM width is not limited by the ATE pin-count, a larger number of TAM wires can be used on the SoC, thereby leading to lower testing times. A drawback of virtual TAMs, however, is the need for additional TAM wires on the SoC as well as frequency division hardware for bandwidth matching. In Reference 62, the hardware overhead is reduced through the use of a smaller number of on-chip TAM wires; ATE channels with high data rates directly drive SoC TAM wires, without requiring frequency division hardware.

22.3.5   Modular testing of mixed-signal SoCs

Prior research on modular testing of SoCs has focused almost exclusively on the digital cores in an SoC. However, most SoCs in use today are mixed-signal circuits containing both digital and analogue cores [65–67]. Increasing pressure on consumer products for small form factors and extended battery life is driving single-chip integration, and blurring the lines between analogue/digital design types. As indicated in the 2001 International Technology Roadmap for Semiconductors [24], the combination of these circuits on a single die compounds the test complexities and challenges for devices that fall in an increasing commodity market. Therefore, an effective modular test methodology should be capable of handling both digital and analogue cores, and it should reduce test cost by enabling test reuse for reusable embedded modules.

In traditional mixed-signal SoC testing, tests for analogue cores are applied either from chip pins through direct test access methods, e.g. via multiplexing, or through a dedicated analogue test bus [68, 69], which requires the use of expensive mixed-signal testers. For mid- to low-frequency analogue applications, the data is often digitised at the tester, where it is affordable to incorporate high quality data converters. In most mixed-signal ICs, analogue circuitry accounts for only a small part of the total silicon ('big-D/small-A'). However, the total production testing cost is dominated by analogue testing costs. This is because of the fact that expensive mixed-signal testers are employed for extended periods of time resulting in high overall test costs. A natural solution to this problem is to implement the data converters on-chip. Since most SoC applications do not push the operational frequency limits, the design of such data converters on-chip appears to be feasible. Until recently, such an approach has not been deemed desirable due to its high hardware overhead. However, as the cost of on-chip silicon is decreasing and the functionality and the number of cores in a typical SoC are increasing, the addition of data converters on-chip for testing analogue cores now promises to be cost-efficient. These data converters eliminate the need for expensive mixed-signal test equipment.

Recently, results have been reported on the optimisation of a unified test access architecture that is used for both digital and analogue cores [70]. Instead of treating the digital and analogue portions separately, a global test resource optimisation problem is formulated for the entire SoC. Each analogue core is wrapped by a DAC–ADC pair and a digital configuration circuit. Results show that for 'big D/small A' SoCs, the testing time and test cost can be reduced considerably if the analogue cores are wrapped, and the test access and test scheduling problems for the analogue and digital cores are tackled in a unified manner.

Each analogue core is provided with a test wrapper where the test information includes only digital test patterns, clock frequency, the test configuration and pass/fail criteria. This analogue test wrapper converts the analogue core to a virtual digital core with strictly sequential test patterns, which are the digitised analogue signals. In order to utilise test resources efficiently, the analogue wrapper needs to provide sufficient flexibility in terms of required resources with respect to all the test needs of the analogue core. One way to achieve this uniform test access scheme for analogue cores is to provide an on-chip ADC–DAC pair that can serve as an interface between each analogue core and the digital surroundings, as shown in Figure 22.8.

Analogue test signals are expressed in terms of a signal shape, such as sinusoidal or pulse, and signal attributes, such as frequency, amplitude and precision. These tests are provided by the core vendor to the system integrator. In the case of analogue testers, these signals are digitised at the high precision ADCs and DACs of the tester. In the case of on-chip digitisation, the analogue wrapper needs to include the lowest cost data converters that can still provide the required frequency and accuracy for applying the core tests. Thus, on-chip conversion of each analogue test to digital patterns imposes requirements on the frequency and resolution of the data converters

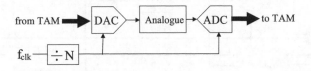

Figure 22.8   *On-chip digitisation of analogue test data for uniform test access [70] (©IEEE, 2003)*

of the analogue wrapper. These converters need to be designed to accommodate all the test requirements of the analogue core.

Analogue tests may also have a high variance in terms of their frequency and test time requirements. While tests involving low-frequency signals require low bandwidth and high test times, tests involving high-frequency signals require high bandwidth and low test time. Keeping the bandwidth assigned to the analogue core constant results in under-utilisation of the precious test resources. The variance of analogue test needs have to be fully exploited in order to achieve an efficient test plan. Thus, the analogue test wrapper has to be designed to accommodate multiple configurations with varying bandwidth and frequency requirements.

Figure 22.9 shows the block diagram of an analogue wrapper that can accommodate all the abovementioned requirements. The control and clock signals generated by the test control circuit are highlighted in this figure. The registers at each end of the data converters are written and read in a semi-serial fashion depending on the frequency requirement of each test. For example, for a digital TAM clock of 50 MHz, 12-bit DAC and ADC resolution and an analogue test requirement of 8 MHz sampling frequency, the input and output registers can be updated with a serial-to-parallel ratio of 6. Thus, the bandwidth requirement of this particular test is only 2 bits. The digital test control circuit selects the configuration for each test. This configuration includes the divide ratio of the digital TAM clock, the serial to parallel conversion rate of the input and output registers of the data converters and the test modes.

'Analogue test wrapper modes' – In the normal mode of operation, the analogue test wrapper is completely by-passed; the analogue circuit operates on its analogue input/output pins. During testing, the analogue wrapper has two modes, a 'self-test' mode and a 'core-test' mode. Before running any tests on the analogue core, the wrapper data converters have to be characterised for their conversion parameters, such as the non-linearity and the offset voltage. The self-test mode is enabled through the analogue multiplexer at the input of the wrapper ADC, as shown in Figure 22.9. The parameters of the DAC–ADC pair are determined in this mode and are used to calibrate the measurement results. Once the self-test of the test wrapper is complete, core test can be enabled by turning off the 'self-test' bits.

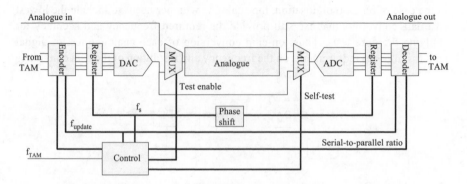

Figure 22.9   Block diagram of the analogue test wrapper [70] (©IEEE, 2003)

For each analogue test, the encoder has to be set to the corresponding serial-to-parallel conversion ratio (cr), where it shifts the data from the corresponding TAM inputs into the register of the ADC. Similarly, the decoder shifts data out of the DAC register. The update frequency of the input and output registers, $f_{update} = f_s \times cr$, is always less than the TAM clock rate, f_{TAM}. For example, if the test bandwidth requirement is two bits and the resolution of the data converters is 12 bits, the input and output registers of the data converters are clocked at a rate six times less than the clock of the encoder, and the input data is shifted into the encoder and out of the decoder at a two-bits/cycle rate. The complexity of the encoder and the decoder depends on the number of distinct bandwidth and TAM assignments (the number of possible test configurations). For example, for a 12-bit resolution, the bandwidth assignments may include 1, 2, 3, 4, 6 and 12 bits, where in each case the data may come from distinct TAMs. Clearly, in order to limit the complexity of the encoder–decoder pair, the number of such distinct assignments has to be limited. This requirement can be imposed in the test scheduling optimisation algorithm.

The analogue test wrapper transparently converts the analogue test data to the digital domain through efficient utilisation of the resources, thus this obviates the need for analogue testers. The processing of the collected data can be done in the tester by adding appropriate algorithms, such as the FFT algorithm. Further details and experimental results can be found in References 70 and 71.

22.3.6 Modular testing of hierarchical SoCs

A hierarchical SoC is designed by integrating heterogeneous technology cores at several layers of the hierarchy [24]. The ability to re-use embedded-cores in a hierarchical manner implies that 'today's SoC is tomorrow's embedded-core' [72]. Two broad design transfer models are emerging in hierarchical SoC design flows.

1 *Non-interactive.* The non-interactive design transfer and hand-off model is one in which there is limited communication between the core vendor and the SoC integrator. The hard cores are taken off-the-shelf and integrated into designs as optimised layouts.

2 *Interactive.* The interactive design transfer model is typical of larger companies where the business units producing IP cores may be part of the same organisation as the business unit responsible for system integration. Here, there is a certain amount of communication between the core vendor and core user during system integration. The communication of the core user's requirements to the core vendor can play a role in determining the core specifications.

Hierarchical SoCs offer reduced cost and rapid system implementation; however, they pose difficult test challenges. Most TAM design methods assume that the SoC hierarchy is flattened for the purpose of test. However, this assumption is often unrealistic in practice, especially when older-generation SoCs are used as hard cores in new SoC designs. In such cases, the core vendor may have already designed a TAM within the 'mega-core' that is provided as an optimised and technology-mapped layout to the SoC integrator.

A 'mega-core' is defined as a design that contains non-mergeable embedded-cores. In order to ensure effective testing of an SoC based on mega-cores, the top-level TAM must communicate with lower level TAMs within mega-cores. Moreover, the system-level test architecture must be able to reuse the existing test architecture within cores; redesign of core test structures must be kept to a minimum and it must be consistent with the design transfer model between the core designer and the core user [73].

A TAM design methodology that closely follows the design transfer model in use is necessary because if the core vendor has implemented 'hard' (i.e. non-alterable) TAMs within mega-cores, the SoC integrator must take into account these lower-level TAM widths while optimising the widths and core assignment for higher-level TAMs. On the other hand, if the core vendor designs TAMs within mega-cores in consultation with the SoC integrator, the system designer's TAM optimisation method must be flexible enough to include parameters for lower-level cores. Finally, multi-level TAM design for SoCs that include reused cores at multiple levels is needed to exploit 'TAM reuse' and 'wrapper reuse' in the test development process.

It is only recently that the problem of designing test wrappers and TAMs of multi-level TAMs for the 'cores within cores' design paradigm [74, 75] has been considered. Two design flows have been considered for the scenario in which mega-cores are wrapped by the core vendor prior to delivery. In an alternative scenario, the mega-cores can be delivered to the system integrator in an unwrapped fashion, and the system integrator appropriately designs the mega-core wrappers and the SoC-level TAM architecture to minimise the overall testing time.

Figure 22.10 illustrates a mega-core that contains four embedded-cores and additional logic external to the embedded-cores. The core vendor for this mega-core has wrapped the the four embedded-cores, and implemented a TAM architecture to acess the embedded-cores. The TAM architecture consists of two test buses of widths 3 bits and 2 bits, respectively, that are used to access the four embedded-cores. It is assumed here that the TAM inputs and outputs are not multiplexed with the functional pins. Next, Figure 22.11 shows how a two-part wrapper (Wrapper 1 and Wrapper 2) for the mega-core can be designed to drive not only the TAM wires within the mega-core, but also to test the logic that is external to the embedded-cores. In this design, the TAM inputs for Wrapper 1 and Wrapper 2 are multiplexed in time, such that the embedded-cores within the mega-core are tested before the logic external to them, or vice versa. Test generation for the top-level logic is done by the mega-core vendor with the wrappers for the embedded-cores in functional mode. During the testing of the top-level logic in the mega-core using Wrapper 1, the wrappers for the embedded-cores must therefore be placed in functional mode to ensure that the top-level logic can be tested completely through the mega-core I/Os and scan terminals.

Mega-cores may be supplied by core vendors in varying degrees of readiness for test integration. For example, the IEEE P1500 proposal on embedded-core test defines two compliance levels for core delivery: 1500-wrapped and 1500-unwrapped [51]. Here we describe three other scenarios, based in part on the P1500 compliance levels. These scenarios refer to the roles played by the system integrator and the core vendor in the design of the TAM and the wrapper for the mega-core. For each scenario, the design transfer model refers to the type of information about the mega-core that is

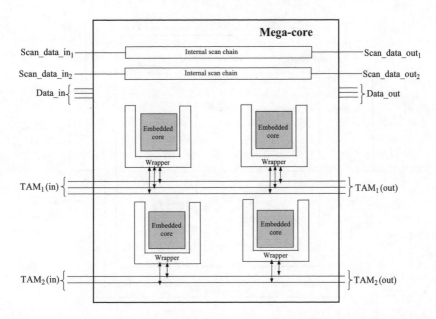

Figure 22.10   An illustration of a mega-core with a pre-designed TAM architecture

provided by the core vendor to the system integrator. The term 'wrapped' is used to denote a core for which a wrapper has been pre-designed, as in Reference 51. The term 'TAM-ed' is used to denote a mega-core that contains an internal TAM structure.

1 *Scenario 1*: Not TAM-ed and not wrapped: In this scenario, the system integrator must design a wrapper for the mega-core as well as TAMs within the mega-core. The mega-cores are therefore delivered either as soft cores or before final netlist and layout optimisation, such that TAMs can be inserted within the mega-cores.

2 *Scenario 2*: TAM-ed and wrapped: In this scenario, we consider TAM-ed mega-cores for which wrappers have been designed by the core vendor. This scenario is especially suitable for a mega-core that was an SoC in an earlier generation. It is assumed that such mega-cores are wrapped by the core vendors prior to design transfer and test data for the mega-core cannot be further serialised or parallelised by the SoC integrator. This implies that the system integrator has less flexibility in top-level TAM partitioning and core assignment. At the system level, only structures that facilitate normal/test operation, interconnect test and bypass are created. This scenario includes both the interactive and non-interactive design transfer models.

3 *Scenario 3*: TAM-ed but not wrapped: In this scenario, the mega-core contains lower-level TAMs, but it is not delivered in a wrapped form; therefore, a wrapper for it must be designed by the system integrator. In order to design a wrapper as sketched in Figure 22.11, the core vendor must provide information about the number of functional I/Os, the number and lengths of top-level scan chains in the mega-core, the number of TAM partitions and the size of each partition, and the testing time for each TAM partition. Compared to the non-interactive design transfer model in

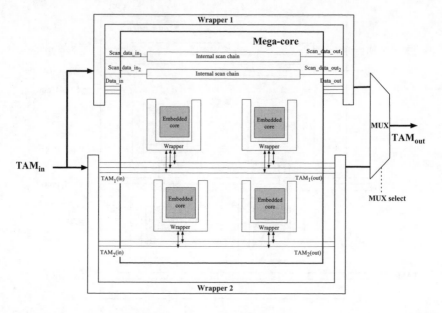

*Figure 22.11   An illustration of a two-part wrapper for the mega-core that is used
to drive the TAMs in the mega-core and to test the logic external to
the embedded-cores*

Scenario 2, the system integrator in this case has greater flexibility in top-level TAM
partitioning and core assignment. Compared to the interactive design transfer model
in Scenario 2, the system integrator here has less influence on the TAM design for a
mega-core; however, this loss of flexibility is somewhat offset by the added freedom
of being able to design the mega-core wrapper. Width adaptation can be carried out
in the wrapper for the mega-core such that a narrow TAM at the SoC-level can be
used to access a mega-core that has a wider internal TAM.

Optimisation techniques for these scenarios are described in detail in References
74–76. As hierarchical SoCs become more widespread, it is expected that more
research effort will be devoted to this topic.

22.4   Test data compression

The increased density of SoCs and the need to test for new types of defects in nanometer technologies have resulted in a tremendous increase in test data volume and test
application time. The test data volume for ICs in 2014 is projected to be as much as
150 times the test data volume in 1999 [77].

In addition to the increasing density of ICs, today's SoC designs also exacerbate
the test data volume problem. The increase in test data volume not only leads to the
increase of testing time, but the high test data volume may also exceed the limited
memory depth of ATE. Multiple ATE reloads are time-consuming since data transfer

from a workstation to the ATE hard disk or from the ATE hard disk to ATE channels are relatively slow; the upload time ranges from tens of minutes to hours [78]. While test application time for scan vectors can be reduced by using a large number of internal scan chains, the number of internal scan chains that can be driven by an ATE is limited in practice due to pin count constraints.

As discussed in Section 22.2, test data volume for IP cores can be reduced by compressing the precomputed test set T_D provided by the core vendor to a much smaller data set T_E, which is stored in ATE memory. An on-chip decoder is used for pattern decompression to generate T_D from T_E during pattern application [15,17,79–81]. Such techniques are typically based on run-length codes and their variants, e.g. FDR codes. However, most methods based on compression codes target single scan chains and they require complex synchronisation between the ATE and the circuit under test. Test data volume reduction techniques based on on-chip linear decompression hardware [82, 83], multiplexer-based switches [84], ATE/EDA synergies [85], as well as BIST [86] and hybrid BIST [87] have also been presented. Test data compression is now a mature research area and commercial tools for 'embedded test' such as TestKompress and OPMISR are now available [88]. However, to achieve high compression, these tools and techniques utilise structural information about the circuit under test, which limits their applicability for IP cores.

22.4.1   Use of data compression codes

In this subsection, we describe the use of exponential-Golomb codes and subexponential codes for compressing scan test data [16]. These codes often provide greater compression of test data than other codes proposed for test data compression. Moreover, only a small amount of hardware is required for on-chip decompression of test data encoded using exponential-Golomb and subexponential codes. (The decompression logic synthesised using Synopsys tools requires less than 50 gates and less than 100 gates, respectively.) The decompression logic is independent of the core under test and the test set. The proposed compression/decompression scheme requires no modifications to the core under test.

The underlying assumption here is that a single data channel is used to deliver the compressed test patterns from the tester to the core under test. This approach is therefore targeted towards a reduced pin-count test and low-cost DFT tester [89] environment, where a narrow interface between the tester and the SoC is desired.

In the following description, we assume that the precomputed SoC test data consists of n test patterns $t_1, t_2, \ldots, t_n$. The don't-care bits in the test set are set to 0s. The test patterns are suitably reordered and serialised before compression. The reordered and serialised test set is denoted as T_D, which is then compressed using these codes.

Exponential-Golomb code: The exponential-Golomb code provides a variable-to-variable encoding method, i.e. the runs of 0s are encoded as variable-length codewords. A run of 0s is divided into successive sub-runs of length 2^k, 2^{k+1}, $2^{k+2}, \ldots, 2^{k+i-1}$, until the number of remaining 0s is less than 2^{k+i}, where k is

Table 22.1   Codewords of the exponential-Golomb code

Run-length	$k = 0$		$k = 1$ (FDR)		$k = 2$	
	i	Codeword	i	Codeword	i	Codeword
0	0	0	0	00	0	000
1	1	100		01		001
2		101	1	1000		010
3	2	11000		1001		011
4		11001		1010	1	10000
5		11010		1011		10001
6		11011	2	110000		10010
7	3	1110000		110001		10011
8		1110001		110010		10100
9		1110010		110011		10101
10		1110011		110100		10110

the code parameter of the exponential-Golomb code. The rest of the run is encoded as a $(k + i)$-bit binary number. The power of the exponential-Golomb code lies in the fact that it can encode both short and long runs efficiently because the successive sub-runs grow exponentially. Let l denote the run-length to be encoded. The encoding steps are as follows.

1 Determine i such that $\sum_{j=0}^{i-1} 2^{j+k} \leq l < \sum_{j=0}^{i} 2^{j+k}, i \geq 0$;
2 Form the prefix of i 1s;
3 Insert the separator 0;
4 Form the tail: express the value of $(l - \sum_{j=0}^{i-1} 2^{j+k})$ as a $(k+i)$-bit binary number.

Table 22.1 shows the codewords of the exponential-Golomb code for run-lengths varying from 0 to 10 with code parameter $k = 0, 1$ and 2. Note that the FDR code described in Reference 17 corresponds to a special case of the exponential-Golomb code ($k = 1$). In the table, we have separated the codewords into groups. The group index i is also the number of 1s in the prefix. From the table, we note the following properties of the exponential-Golomb code:

1 The length of prefix in group A_i is i (excluding the separator 0);
2 In group A_i, the run-length represented by the prefix is $l_{\text{prefix}} = \sum_{j=0}^{i-1} 2^{k+j} = 2^k(2^i - 1)$, which is the value of the continuous 1s in the prefix shifted left by k bits;
3 The length of tail in group A_i is $(k + i)$;
4 In group A_i, the run-length represented by the tail is between 0 to $2^{k+i} - 1$;
5 The size of group A_i is 2^{k+i}.

Based on these properties, it is straightforward to derive the decoding algorithm as follows:

Table 22.2   Codewords of the subexponential code

Run-length	$k = 0$		$k = 1$		$k = 2$	
	i	Codeword	i	Codeword	i	Codeword
0	0	0	0	00	0	000
1	1	10		01		001
2	2	1100	1	100		010
3		1101		101		011
4	3	111000	2	11000	1	1000
5		111001		11001		1001
6		111010		11010		1010
7		111011		11011		1011
8	4	11110000	3	1110000	2	110000
9		11110001		1110001		110001
10		11110010		1110010		110010

1 Let i be the number of leading 1s (prefix) in the codeword;
2 Form a run of 0s of length $\sum_{j=0}^{i-1} 2^{j+k}$;
3 Skip the next 0 (separator);
4 The next $(k + i)$ bits make up the tail. Form a run of 0s of length represented by the tail;
5 Append 1 to the run of 0s;
6 Go to step 1 to process the next codeword.

Subexponential code: As in the case of the exponential-Golomb code, the codewords of the subexponential code can also be divided into groups and we use the number of 1s in the prefix as the group index. Table 22.2 shows the codewords of the subexponential code for run-lengths varying from 0 to 10 with the code parameter $k = 0, 1, 2$. From the table, we can find that the size of group A_0 is 2^k. For $i \geq 1$, the size of group A_i is $2^{(k+i-1)}$. This property is closely tied to the encoding procedure, which is described as follows. Let l denote the run-length to be encoded.

1 Determine the group index i using the following rules:
  - if $l < 2^k$, then $i = 0$.
  - if $l \geq 2^k$, then determine i such that $2^{i+k-1} \leq l < 2^{i+k}$.
2 Form the prefix of i 1s;
3 Insert the separator 0;
4 Form the tail: express the value of $(l - 2^{i+k-1})$ as a $(i + k - 1)$-bit binary number.

We can see that A_0 is a special group in the subexponential code. Its size is the same as that of group A_1. This property makes the encoding and decoding for the subexponential code a little more complex than that for the exponential-Golomb code. A few additional steps are needed during encoding and decoding. The decoding procedure for the subexponential code is as follows.

1   Let i be the number of leading 1s (prefix) in the codeword;

2   Form a run of 0s of length $\begin{cases} 0, & \text{if } i = 0 \\ 2^{i+k-1}, & \text{otherwise} \end{cases}$

3   Skip the next 0 (separator);

4   Compute the length of the tail, c_{tail}, as
$$\begin{cases} k, & \text{if } i = 0 \\ k + i - 1, & \text{if } i \geq 1 \end{cases}$$

5   The next c_{tail} bits are the tail. Form a run of 0s of length represented by the tail;

6   Append 1 to the run of 0s;

7   Go to step 1 to process the next codeword.

Information-theoretic analysis: We next use basic principles of information theory to explain why the exponential-Golomb and subexponential codes are more suitable than the conventional run-length code and Golomb codes for test data compression. The entropy of a data source is expressed as $H = \sum_{i=1}^{n} -p_i \log_2 p_i$, where p_i is the probability of occurrence of symbol i, and n is the number of different symbols that can appear in the data. During the compression of an SoC test set, we are encoding runs of 0s. Hence we can regard the runs of 0s in the test data as different symbols, for which we can compute probabilities of occurrence, and thereby obtain the entropy of this test set. Since entropy denotes the average amount of information carried by a single symbol and the unit of entropy is a binary bit, we can consider the product of the entropy and the total number of runs in the test set as a theoretical lower bound on the size of the compressed test set T_E.

While the entropy bound is a useful measure, we can get additional guidelines from information theory on how to select the best compression code for a given test set. Suppose the probability of occurrence of run-length i is p_i, and a c_i-bit codeword is used to encode this run. The amount of information carried by this run-length is $-\log_2 p_i$ bits. In order to obtain optimum compression, we should use a $(-\log_2 p_i)$-bit codeword to encode this run-length, i.e. the ideal value of c_i is $-\log_2 p_i$. In this way, we can compare the compression effectiveness of various codes (e.g. FDR, exponential-Golomb and subexponential) relative to an ideal code that assigns a codeword of size $-\log_2 p_i$ bits to a run of length i.

Figure 22.12 shows the codeword length for run-lengths 0 to 100 for a number of compression codes, as well as the codeword length for an ideal code. The test set used in Figure 22.12 is that of the s13207 ISCAS-89 benchmark circuit. The results show that the codeword sizes for the Golomb code (parameter $m = 4$) and the conventional run-length code (block size $b = 4$) are significantly different from the ideal codeword sizes; in particular, they diverge significantly for large run lengths. On the other hand, the codeword sizes for the exponential-Golomb code with $k = 1$ (FDR) and the subexponential code with $k = 1$ are quite close to the ideal codeword size; thus we can expect these codes to outperform Golomb and run-length codes. Similar results are observed for the test sets of other benchmark circuits (Figure 22.13 shows the comparison for s38584). Note that the ideal code is 'ideal' when the symbols are runs of 0s. If the symbols are formed in other ways, the ideal code should also be modified accordingly. Note also that for certain values of the run-length, the ideal code does

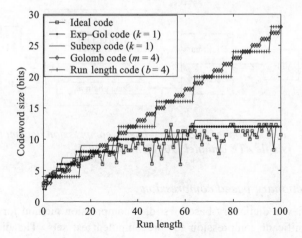

Figure 22.12   The comparison of codeword length between various codes and the ideal codeword length for circuit s13207

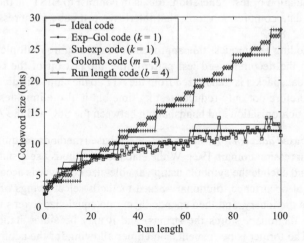

Figure 22.13   The comparison of codeword length between various codes and the ideal codeword length for circuit s38584

worse than some of the coding methods. This is not unexpected, since the ideal code is optimal over the set of all the run-lengths in the test data, and it provides greater compression for the given test data set.

Experimental results and details of the decompression architectures for these codes are described in Reference 16.

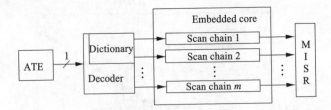

Figure 22.14   Illustration of dictionary-based compression/decompression for a single ATE channel [90] (©ACM, 2003)

22.4.2   Dictionary-based compression

Next we review a dictionary-based test data compression method for IP cores that provides significant compression for precomputed test sets. The dictionary uses fixed-length indices, and its entries are carefully selected such that the dictionary is efficiently utilised. The proposed method is based on the use of a small number of ATE channels to drive a large number of internal scan chains in the core under test; see Figure 22.14. This technique does not require a gate-level circuit model for fault simulation or test generation; this is in contrast to BIST methods and commercial test data compression tools that interleave test cube compression with test generation [82, 83].

Unlike coding techniques, this approach does not require multiple clock cycles to determine the decompressed test pattern after the last bit of the corresponding compressed data packet is transferred from the ATE to the chip. The dictionary-based approach therefore not only reduces testing time but it also eliminates the need for additional synchronisation and handshaking between the SoC and the ATE.

Dictionary-based data compression: Dictionary-based methods are quite common in the data compression domain [92]. While statistical methods use a statistical model of the data and encode the symbols using variable-size codewords according to their frequencies of occurrence, dictionary-based methods select strings of the symbols to establish a dictionary, and then encode them into equal-size tokens using the dictionary. The dictionary stores the strings, and it may be either static or dynamic (adaptive). The former is permanent, sometimes allowing for the addition of strings but no deletions, whereas the latter holds strings previously found in the input stream, allowing for additions and deletions of strings as new input is processed.

A simple example of a static dictionary is an English dictionary used to encode English text that consists of words. A word in the input text is encoded as an index to the dictionary if it appears in the dictionary. Otherwise it is encoded as the size of the word followed by the word itself. In order to distinguish between the index and the raw word, a flag bit needs to be added to each codeword. We present an example next to illustrate the encoding of a word. Suppose the dictionary contains 2^{20} words and thus needs a 20-bit index to specify an entry. A value of 0 for the flag bit indicates that this codeword is composed of the size of the word and the word itself following the flag bit. A value of 1 for the flag bit implies that the 20 bits of data following it

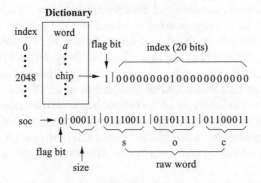

Figure 22.15   An example of dictionary-based data compression [90] (©ACM, 2003)

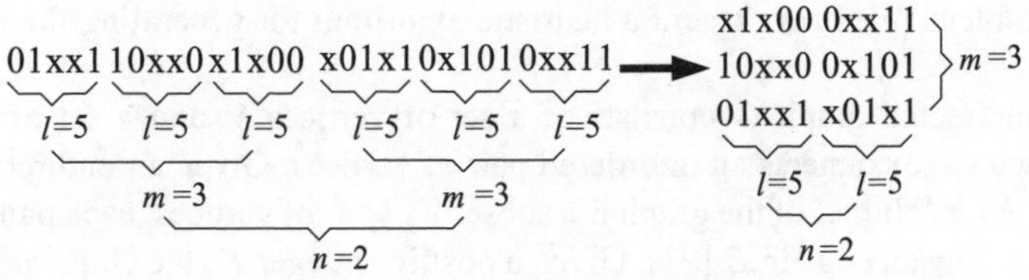

$$01xx1 \underbrace{10xx0}_{l=5} \underbrace{x1x00}_{l=5} \quad \underbrace{x01x1}_{l=5} \underbrace{0x101}_{l=5} \underbrace{0xx11}_{l=5} \rightarrow \begin{matrix} x1x00 \; 0xx11 \\ 10xx0 \; 0x101 \\ 01xx1 \; x01x1 \end{matrix} \Big\} m=3$$

Figure 22.16   An example of formatting the test data for multiple scan chains [90] (©ACM, 2003)

is a dictionary index. Suppose a 5-bit field is used to specify the size of the word. As shown in Figure 22.15, the word 'chip', which is present in the dictionary with index 2048, is encoded as 1|00000000100000000000. The word 'soc', which is not in the dictionary, is encoded as 0|00011|01110011|01101111|01100011, where the 5-bit field 00011 indicates that three more bytes follow it.

Dictionary-based compression of test data: In the following description, we assume that the precomputed SoC test data T_D consists of n test patterns $t_1, t_2, \ldots, t_n$. The scan elements of the core under test are divided into m scan chains in as balanced a manner as possible. Each test vector can therefore be viewed as m subvectors. If one or more subvectors are shorter than the others, don't-cares are padded to the end of these subvectors so that all the subvectors have the same length, which is denoted by l. The m-bit data at the same position of each subvector constitute an m-bit word. A total of nl m-bit words thus are formed and encoded during the compression procedure. Figure 22.16 illustrates the formatting of the given test data for multiple scan chains. During test application, after a codeword is shifted into the decoder, an m-bit word $u_1, u_2, \ldots, u_m$ is immediately generated by the decoder and fed into the scan chains (one bit for each scan chain).

In a representative dictionary-based test data compression method [90], each codeword is composed of a prefix and a stem. The prefix is a 1-bit identifier that indicates whether the stem is a dictionary index or a word of uncompressed test data.

If it equals 1, the stem is viewed as a dictionary index. On the other hand, if the prefix equals 0, the stem is an uncompressed word and it is m bits long. The length of the dictionary index depends on the size of the dictionary. If D is the set of the entries in the dictionary, the length of the index $l_{\text{index}} = \lceil \log_2 |D| \rceil$, where $|D|$ is the size of the dictionary. Since l_{index} is much smaller than m, the compression efficiency is greater if more test data words can be obtained from the dictionary. However, the dictionary must be reasonably small to keep the hardware overhead low. Fortunately, since there are many don't-care bits in scan test data for typical circuits, we can appropriately map these don't-care bits to binary values and carefully select the entries for the dictionary, so that as many words as possible are mapped to the entries in the dictionary.

An important step in the compression procedure is that of selecting the entries in the dictionary. This problem can be easily mapped to a variant of the clique partitioning problem from graph theory [91,92]. We next describe the clique partitioning problem and then show how the problem of determining dictionary entries can be mapped to this problem. We then present a heuristic algorithm for generating the dictionary entries.

An undirected graph G consists of a set of vertices V and a set of edges E, where each edge connects an unordered pair of vertices. Given an undirected graph $G = (V, E)$, a 'clique' of the graph is a subset $V' \subseteq V$ of vertices, each pair of which is connected by an edge in E [91]. Given a positive integer K, the clique partitioning problem refers to the partitioning of V into k cliques, where $k \leq K$. The clique partitioning problem is $\mathcal{NP}$-hard [93],[1] hence heuristic approaches must be used to solve it in reasonable time for large problem instances.

Recall that in dictionary-based data compression, we obtain nl m-bit words after placing the test set in a multiple scan chain format. Two words $u_1u_2 \cdots u_m$ and $v_1v_2 \cdots v_m$ are defined to be 'compatible' to each other if for any position i, u_i and v_i are either equal to each other or at least one of them is a don't-care bit. A undirected graph G is constructed to reflect the compatible relationships between the words as follows. First, a vertex is added to the graph for each word. Then we examine each pair of words. If they are mutually compatible, an edge is added between the corresponding pair of vertices. A clique in G refers to a group of test data words that can be mapped to the same dictionary entry. If the dictionary can have at most $|D|$ entries and the total number of words is nl, the goal of the compression procedure is to find the largest subset of G that can be partitioned into $|D|$ cliques; the remaining vertices in G denote test data words that are not compressed. This problem can easily be shown to be $\mathcal{NP}$-hard by contradiction. If the compression can be optimally solved in polynomial time then it provides a yes/no answer to the decision version of the clique partitioning problem in polynomial time. The following heuristic procedure has been used in Reference 90.

1   Copy the graph G to a temporary data structure G';
2   Find the vertex v with the maximum degree in G';

---

[1] The decision version of the clique partitioning problem is $\mathcal{NP}$-complete.

Table 22.3 *An example of test data for multiple scan chains [90] (©ACM, 2003)*

Scan chain index	Word index															
	16	15	14	13	12	11	10	9	8	7	6	5	4	3	2	1
1	1	1	0	1	1	1	1	0	0	0	X	X	0	X	0	1
2	X	0	1	1	0	0	1	X	X	X	1	X	X	0	1	0
3	X	X	X	X	0	X	0	1	0	0	1	1	0	X	X	X
4	X	0	X	0	X	X	0	X	0	0	0	0	0	X	0	1
5	0	0	0	0	X	0	X	0	X	X	X	0	X	1	0	X
6	0	X	1	0	1	0	X	X	1	X	0	0	X	0	X	X
7	1	0	1	X	X	X	X	1	1	0	X	1	0	0	1	0
8	1	X	0	X	0	1	X	1	0	X	X	X	X	X	X	1

3 Establish a subgraph that consists of all the vertices connected to v. Copy this subgraph to G' and add v to a set C (the subgraph thus formed does not include the vertex v);

4 If G' is not empty, go to Step 2. Otherwise, a clique C has been formed consisting of all the vertex found in Step 2;

5 Remove the vertices in the clique C from G and copy $G - C$ to G'. Go to Step 2 and repeat until $|D|$ cliques are found.

The complexity of this procedure is $O(N^3)$, where $N = nl$ is the number of vertices in the graph. Table 22.3 shows an example of test data formatted for multiple scan chains. The number of scan chains m is 8 in this example. There are a total of 16 words, each of which has 8 bits. Figure 22.17 shows the corresponding graph G for the test data. Let us assume that a dictionary of size four is to be formed, i.e. $|D| = 4$. Using the greedy algorithm described above, we obtain four cliques: $\{5, 6, 13, 16\}$, $\{2, 8, 14\}$, $\{3, 4, 7\}$ and $\{1, 11\}$. (Here we use the word indices of Table 22.3 to represent the vertices.) After finding the cliques, we obtain the corresponding dictionary entry for each clique by merging the words in this clique. In this example, the four dictionary entries are $\{11100011, 01000110, 0000100X, 10X10001\}$. Three bits are then needed to encode the words in the cliques; an additional 1 bit is needed for the prefix, and 2 bits are required for the dictionary index. For words that are not in any clique, a total of 9 bits each must be transferred from the ATE. Since there are 12 words that can be generated from the dictionary, the size of the compressed data is $3 \times 12 + 9 \times 4 = 72$ bits, which corresponds to a compression of 43.75%. Moreover, the dictionary entries still contain some don't-care bits, which can reduce the hardware for the decoder.

The clique partitioning procedure introduces a certain degree of randomness in the way the don't-care bits in T_D are filled; the resulting 'random fill' can be expected to increase the fortuitous detection of non-modelled faults. This is in contrast to

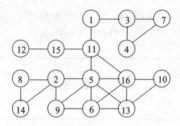

Figure 22.17   The graph G for the example of Table 22.3 [90] (©ACM, 2003)

coding methods in which the don't-cares are all mapped to 0s. Further details about the compression procedure, decompression architecture and results for benchmark circuits are presented in Reference 90.

22.5   Conclusions

Rapid advances in test development techniques are needed to reduce the test cost of million-gate SoC devices. This survey chapter has presented a number of state-of-the-art techniques for reducing test time and test data volume, thereby decreasing test cost. Modular test techniques for digital, mixed-signal and hierarchical SoCs must develop further to keep pace with design complexity and integration density. The test data bandwidth needs for analogue cores are significantly different than those for digital cores, therefore unified top-level testing of mixed-signal SoCs remains a major challenge. Most SoCs today include embedded-cores that operate in multiple clock domains. Since the forthcoming P1500 standard does not address wrapper design for at-speed testing of such cores, research is needed to develop wrapper design techniques for multi-frequency cores. There is also a pressing need for test planning methods that can efficiently schedule tests for these multi-frequency cores. The work reported in Reference 30 is a promising first step in this direction. In addition, compression techniques for embedded-cores also need to be developed and refined. Of particular interest are techniques that can combine TAM optimisation and test scheduling with test data compression. Some preliminary studies on this problem have been reported recently [94, 95].

Acknowledgements

This survey is based on joint work and papers published with several students and colleagues. In particular, the author acknowledges Anshuman Chandra, Vikram Iyengar, Lei Li, Erik Jan Marinissen, Sule Ozev and Anuja Sehgal. This research was supported in part by the National Science Foundation under grants CCR-9875324 and CCR-0204077, and in part by the Semiconductor Research Corporation under contract no. 2004-TJ-1174.

References

1   Virtual Socket Interface Alliance: http://www.vsi.org
2   IEEE P1500 Standard for Embedded Core Test: http://grouper.ieee.org/groups/1500
3   ZORIAN, Y., MARINISSEN, E.J., and DEY, S.: 'Testing embedded-core-based system chips', *IEEE Computer*, 1999, **32**, pp. 52–60
4   CHAKRABARTY, K.: 'Test scheduling for core-based systems using mixed-integer linear programming', *IEEE Transactions on Computer-Aided Design*, 2000, **19**, pp. 1163–1174
5   SUGIHARA, M., DATE, H., and YASUURA, H.: 'A novel test methodology for core-based system LSIs and a testing time minimization problem'. Proceedings of the international *Test* conference, 1998, pp. 465–472
6   HETHERINGTEN, G., FRYARS, T., TAMARAPALLI, N., KASSAB, M., HASSAN, A., and RAJSKI, J.: 'Logic BIST for large industrial designs'. Proceedings of the international *Test* conference, 1999, pp. 358–367
7   CHANDRA, A., and CHAKRABARTY, K.: 'System-on-a-chip test data compression and decompression architectures based on Golomb codes', *IEEE Transactions on Computer-Aided Design*, 2001, **20**, pp. 355–368
8   IYENGAR, V., CHAKRABARTY, K., and MURRAY, B.T.: 'Deterministic built-in pattern generation for sequential circuits', *Journal of Electronic Testing: Theory and Applications*, 1999, **15**, pp. 97–115
9   CHAKRABARTY, K., IYENGAR, V., and CHANDRA, A.: 'Test resource partitioning for system-on-a-chip' (Kluwer Academic Publishers, Norwell, MA, 2002)
10 CROUCH, A.L.: 'Design-for-test for digital IC's and embedded core systems' (Prentice Hall, Upper Saddle River, NJ, 1999)
11 IYENGAR, V., CHAKRABARTY, K., and MARINISSEN, E.J.: 'Test wrapper and test access mechanism co-optimization for system-on-chip', *Journal of Electronic Testing: Theory and Applications*, 2002, **18**, pp. 213–230
12 ISHIDA, M., HA, D.S., and YAMAGUCHI, T.: 'COMPACT: A hybrid method for compressing test data'. Proceedings of the IEEE *VLSI test* symposium, 1998, pp. 62–69
13 ABRAMOVICI, M., BREUER, M.A., and FRIEDMAN, A.D.: 'Digital systems testing and testable design' (Computer Science Press, New York, NY, 1990)
14 HAMZAOGLU, I., and PATEL, J.H.: 'Test set compaction algorithms for combinational circuits'. Proceedings of the international conference on *CAD*, 1998, pp. 283–289
15 GONCIARI, P.T., AL-HASHIMI, B., and NICOLICI, N.: 'Variable-length input Huffman coding for system-on-a-chip test', *IEEE Transactions on Computer-Aided Design of ICs and Systems*, 2003, **22**, pp. 783–783
16 LI, L., and CHAKRABARTY, K.: 'On using exponential-Golomb codes and subexponential codes for system-on-a-chip test data compression', *Journal of Electronic Testing: Theory and Applications*, 2004, 20

17   CHANDRA, A., and CHAKRABARTY, K.: 'Test data compression and test resource partitioning for system-on-a-chip using frequency-directed run-length (FDR) codes', *IEEE Transactions on Computers*, 2003, **52**, pp. 1076–1088

18   CHAKRABARTY, K.: 'Optimal test access architectures for system-on-a-chip', *ACM Transactions on Design Automation of Electronic Systems*, 2001, **6**, pp. 26–49

19   MARINISSEN, E.J., GOEL, S.K., and LOUSBERG, M.: 'Wrapper design for embedded core test'. Proceedings of the international *Test* conference, 2000, pp. 911–920

20   IYENGAR, V., CHAKRABARTY, K., and MARINISSEN, E.J.: 'Recent advances in TAM optimization, test scheduling, and test resource management for modular testing of core-based SOCs'. Proceedings of the IEEE Asian *Test* symposium, 2002, pp. 320–325

21   IYENGAR, V., and CHAKRABARTY, K.: 'System-on-a-chip test scheduling with precedence relationships, preemption, and power constraints', *IEEE Transactions on Computer-Aided Design of ICs and Systems*, 2002, **21**, pp. 1088–1094

22   BARNHART, C., *et al.*: 'OPMISR: The foundation for compressed ATPG vectors'. Proceedings of the international *Test* conference, 2001, pp. 748–757

23   MARINISSEN, E.J., and VRANKEN, H.: 'On the role of DfT in IC – ATE matching'. International workshop on *TRP*, 2001

24   Semiconductor Industry Association. 'International technology roadmap for semiconductors (ITRS)', 2001, http://public.itrs.net/

25   VOLKERINK, E., *et al.*: 'Test economics for multi-site test with modern cost reduction techniques'. Proceedings of the *VLSI test* symposium, 2002, pp. 411–416

26   VARMA, P., and BHATIA, S.: 'A structured test re-use methodology for core-based system chips'. Proceedings of the international *Test* conference, 1998, pp. 294–302

27   MARINISSEN, E.J., *et al.*: 'A structured and scalable mechanism for test access to embedded reusable cores'. Proceedings of the international *Test* conference, 1998, pp. 284–293

28   CHAKRABORTY, T.J., BHAWMIK, S., and CHIANG, C.-H.: 'Test access methodology for system-on-chip testing'. Proceedings of the international workshop on *Testing embedded core-based system-chips*, 2000, pp. 1.1-1–1.1-7

29   XU, Q., and NICOLICI, N.: 'On reducing wrapper boundary register cells in modular SOC testing'. Proceedings of the international *Test* conference, 2003, pp. 622–631

30   XU, Q., and NICOLICI, N.: 'Wrapper design for testing IP cores with multiple clock domains'. Proceedings of the *Design, automation and test in Europe* (DATE) conference, 2004, pp. 416–421

31   IMMANENI, V., and RAMAN, S.: 'Direct access test scheme – Design of block and core cells for embedded ASICs'. Proceedings of the international *Test* conference, 1990, pp. 488–492

32 HARROD, P.: 'Testing re-usable IP: A case study'. Proceedings of the international *Test* conference, 1999, pp. 493–498

33 GHOSH, I., DEY, S., and JHA, N.K.: 'A fast and low cost testing technique for core-based system-on-chip'.Proceedings of the *Design automation* conference, 1998, pp. 542–547

34 CHAKRABARTY, K.: 'A synthesis-for-transparency approach for hierarchical and system-on-a-chip test', *IEEE Transactions on VLSI Systems*, 2003, **11**, pp. 167–179

35 NOURANI, M., and PAPACHRISTOU, C.: 'An ILP formulation to optimize test access mechanism in system-on-chip testing'. Proceedings of the international *Test* conference, 2000, pp. 902–910

36 WHETSEL, L.: 'An IEEE 1149.1 based test access architecture for ICs with embedded cores'. Proceedings of the international *Test* conference, 1997, pp. 69–78

37 TOUBA, N.A., and POUYA, B.: 'Using partial isolation rings to test core-based designs', *IEEE Design and Test of Computers*, 1997, **14**, pp. 52–59

38 AERTS, J., and MARINISSEN, E.J.: 'Scan chain design for test time reduction in core-based ICs'. Proceedings of the international *Test* conference, 1998, pp. 448–457

39 IYENGAR, V., CHAKRABARTY, K., and MARINISSEN, E.J.: 'Test access mechanism optimization, test scheduling and tester data volume reduction for system-on-chip', *IEEE Transactions on Computers*, 2003, **52**, pp. 1619–1632

40 EBADI, Z.S., and IVANOV, A.: 'Design of an optimal test access architecture using a genetic algorithm'. Proceedings of the Asian *Test* symposium, 2001, pp. 205–210

41 IYENGAR, V., and CHAKRABARTY, K.: 'Test bus sizing for system-on-a-chip', *IEEE Transactions on Computers*, 2002, **51**, pp. 449–459

42 HUANG, Y., *et al.*: 'Resource allocation and test scheduling for concurrent test of core-based SOC design'. Proceedings of the Asian *Test* symposium, 2001, pp. 265–270

43 HUANG, Y., *et al.*: 'On concurrent test of core-based SOC design', *Electronic Journal of Testing: Theory and Applications*, 2002, **18**, pp. 401–414

44 IYENGAR, V., CHAKRABARTY, K., and MARINISSEN, E.J.: 'Efficient test access mechanism optimization for system-on-chip', *IEEE Transactions on Computer-Aided Design of Integrated Circuits & Systems*, 2003, **22**, pp. 635–643

45 MARINISSEN, E.J., and GOEL, S.K.: 'Analysis of test bandwidth utilization in test bus and TestRail architectures in SOCs', *Digest of papers of DDECS*, 2002, pp. 52–60

46 GONCIARI, P.T., AL-HASHIMI, B., and NICOLICI, N.: 'Addressing useless test data in core-based system-on-a-chip test', *IEEE Transactions on Computer-Aided Design of ICs and Systems*, 2003, **22**, pp. 1568–1590

47   GOEL, S.K., and MARINISSEN, E.J.: 'Effective and efficient test architecture design for SOCs'. Proceedings of the international *Test* conference, 2002, pp. 529–538

48   JIANG, W., and VINNAKOTA, B.: 'Defect-oriented test scheduling'. Proceedings of the *VLSI test* symposium, 1999, pp. 433–438

49   LARSSON, E., POUGET, J., and PENG, Z.: 'Defect-aware SOC test scheduling'. Proceedings of the *VLSI test* symposium, 2004, pp. 359–364

50   BEENKER, F., BENNETTS, B., and THIJSSEN, L.: 'Testability concepts for digital ICs – the macro test approach'. Frontiers in Electronic Testing, vol. 3 (Kluwer Academic Publishers, Boston, MA, 1995)

51   MARINISSEN, E.J., *et al.*: 'On IEEE P1500's standard for embedded core test', *Journal of Electronic Testing: Theory and Applications*, 2002, **18**, pp. 365–383

52   ZORIAN, Y.: 'A distributed BIST control scheme for complex VLSI devices'. Proceedings of the *VLSI test* symposium, 1993, pp. 6–11

53   CHOU, R.M., SALUJA, K.K., and AGRAWAL, V.D.: 'Scheduling tests for VLSI systems under power constraints', *IEEE Transactions on VLSI Systems*, 1997, **5**(2), pp. 175–184

54   MURESAN, V., WANG, X., and VLADUTIU, M.: 'A comparison of classical scheduling approaches in power-constrained block-test scheduling'. Proceedings of the international *Test* conference, 2000, pp. 882–891

55   LARSSON, E., and PENG, Z.: 'Test scheduling and scan-chain division under power constraint'. Proceedings of the Asian *Test* symposium, 2001, pp. 259–264

56   LARSSON, E., and PENG, Z.: 'An integrated system-on-chip test framework'. Proceedings of the *DATE* conference, 2001, pp. 138–144

57   KORANNE, S.: 'On test scheduling for core-based SOCs'. Proceedings of the international conference on *VLSI design*, 2002, pp. 505–510

58   IYENGAR, V., GOEL, S.K., MARINISSEN, E.J., and CHAKRABARTY, K.: 'Test resource optimization for multi-site testing of SOCs under ATE memory depth constraints'. Proceedings of the international *Test* conference, 2002, pp. 1159–1168

59   KORANNE, S., and IYENGAR, V.: 'A novel representation of embedded core test schedules'. Proceedings of the international *Test* conference, 2002, pp. 538–539

60   SEHGAL, A., IYENGAR, V., KRASNIEWSKI, M.D., and CHAKRABARTY, K.: 'Test cost reduction for SOCs using virtual TAMs and Lagrange multipliers'. Proceedings of the IEEE/ACM *Design automation* conference, 2003, pp. 738–743

61   SEHGAL, A., IYENGAR, V., and CHAKRABARTY, K.: 'SOC test planning using virtual test access architectures', *IEEE Transactions on VLSI Systems*, 2004, **12**, pp. 1263–1276

62   SEHGAL, A., and CHAKRABARTY, K.: 'Efficient modular testing of SOCs using dual-speed TAM architectures'. Proceedings of the IEEE/ACM *Design, automation and test in Europe* (DATE) conference, 2004, pp. 422–427

63 Agilent Technologies: 'Winning in the SOC market', available online at: http://cp.literature.agilent.com/litweb/pdf/5988-7344EN.pdf

64 Teradyne technologies: 'Tiger: advanced digital with ilicon germanium technology', available at http://www.teradyne.com/tiger/digital.html

65 YAMAMOTO, T., GOTOH, S.-I., TAKAHASHI, T., IRIE, K., OHSHIMA, K., and MIMURA, N.: 'A mixed-signal 0.18-65m CMOS SoC for DVD systems with 432-MSample/s PRML read channel and 16-Mb embedded DRAM', *IEEE Journal of Solid-State Circuits*, 2001, **36**, pp. 1785–1794

66 KUNDERT, H., CHANG, K., JEFFERIES, D., LAMANT, G., MALAVASI, E., and SENDIG, F.: 'Design of mixed-signal systems-on-a-chip', *IEEE Transactions on Computer-Aided Design of Integrated Circuits and Systems*, 2000, **19**, pp. 1561–1571

67 LIU, E., WONG, C., SHAMI, Q., *et al.*: 'Complete mixed-signal building blocks for single-chip GSM baseband processing'. Proceedings of the IEEE custom integrated circuits conference, 1998, pp. 11–14

68 CRON, A.: 'IEEE P1149.4 – almost a standard'. Proceedings of the international *Test* conference, 1997, pp. 174–182

69 SUNTER, S.K.: 'Cost/benefit analysis of the P1149.4 mixed-signal test bus', *IEE Proceedings-Circuits, Devices and Systems*, 1996, **143**, pp. 393–398

70 SEHGAL, A., OZEV, S., and CHAKRABARTY, K.: 'TAM optimization for mixed-signal SOCs using test wrappers for analog cores'. Proceedings of the IEEE international conference on *CAD*, 2003, pp. 95–99

71 SEHGAL, A., LIU, F., OZEV, S., and CHAKRABARTY, K.: 'Test planning for mixed-signal SoCs with wrapped analog cores'. Proceedings of the *Design, automation and test in Europe* (DATE) conference, 2005, pp. 50–55

72 CHICKERMANE, V., *et al.*: 'A building block BIST methodology for SOC designs: A case study'. Proceedings of the international *Test* conference, 2001, pp. 111–120

73 PARULKAR, I., *et al.*: 'A scalable, low cost design-for-test architecture for UltraSPARC™ chip multi-processors'. Proceedings of the international *Test* conference, 2002, pp. 726–735

74 IYENGAR, V., CHAKRABARTY, K., KRASNIEWSKI, M.D., and KUMAR, G.N.: 'Design and optimization of multi-level TAM architectures for hierarchical SOCs'. Proceedings of the IEEE *VLSI test* symposium, 2003, pp. 299–304

75 CHAKRABARTY, K., IYENGAR, V., and KRASNIEWSKI, M.: 'Test planning for modular testing of hierarchical SOCs', *IEEE Transactions on Computer-Aided Design of Integrated Circuits & Systems*, 2005, **24**, pp. 435–448

76 SEHGAL, A., GOEL, S.K., MARINISSEN, E.J., and CHAKRABARTY, K.: 'IEEE P1500-compliant test wrapper design for hierarchical cores'. Proceedings of the international *Test* conference, 2004, pp. 1203–1212

77 KHOCHE A., and RIVOIR J.: 'I/O bandwidth bottleneck for test: is it real?'. *Test resource partitioning* workshop, 2002

78   VRANKEN, H., HAPKE, F., ROGGE, S., CHINDAMO, D., and VOLKERINK, E.: 'ATPG padding and ATE vector repeat per port for reducing test data volume'. Proceedings of the international *Test* conference, 2003, pp. 1069–1076

79   JAS, A., and TOUBA, N.A.: 'Test vector decompression via cyclical scan chains and its application to testing core-based design'. Proceedings of the international *Test* conference, 1998, pp. 458–464

80   TEHRANIPOUR, M., NOURANI, M., and CHAKRABARTY, K.: 'Nine-coded compression technique with application to reduced pin-count testing and flexible on-chip decompression'. Proceedings of the *DATE* conference, 2004, pp. 1284–1289

81   WURTENBERGER, A., TAUTERMANN, C.S., and HELLEBR, S.: 'A hybrid coding strategy for optimized test data compression'. Proceedings of the international *Test* conference, 2003, pp. 451–459

82   KOENEMANN, B., *et al.*: 'A SmartBIST variant with guaranteed encoding'. Proceedings of the Asian *Test* symposium, 2001, pp. 325–330

83   RAJSKI, J., *et al.*: 'Embedded deterministic test for low-cost manufacturing test'. Proceedings of the international *Test* conference, 2002, pp. 301–310

84   TANG, H., REDDY, S.M., and POMERANZ, I.: 'On reducing test data volume and test application time for multiple scan chain designs'. Proceedings of the international *Test* conference, 2003, pp. 1079–1088

85   KHOCHE, A., VOLKERINK, E., RIVOIR, J., and MITRA, S.: 'Test vector compression using EDA-ATE synergies'. Proceedings of the *VLSI test* symposium, 2002, pp. 97–102

86   LIANG, H.-G., HELLEBRAND, S., and WUNDERLICH, H.-J.: 'Two-dimensional test data compression for scan-based deterministic BIST'. Proceedings of the international *Test* conference, 2001, pp. 894–902

87   JAS, A., KRISHNA, C.V., and TOUBA, N.A.: 'Hybrid BIST based on weighted pseudo-random testing: a new test resource partitioning scheme'. Proceedings of the *VLSI test* symposium, 2001, pp. 2–8

88   CHANDRA, A., and CHAKRABARTY, K.: 'Efficient test data compression and decompression for system-on-a-chip using internal scan chains and Golomb coding'. Proceedings of the *Design, automation and test in Europe* (DATE) conference, 2001, pp. 145–149

89   VRANKEN, H., WAAYERS, T., FLEURY, H., and LELOUVIER, D.: 'Enhanced reduced pin-count test for full-scan designs'. Proceedings of the international *Test* conference, 2001

90   LI, L., CHAKRABARTY, K., and TOUBA, N.A.: 'Test data compression using dictionaries with selective entries and fixed-length indices', *ACM Transactions on Design Automation of Electronic Systems*, 2003, **8**, pp. 470–490

91   CORMEN, T.H., LEISERSON, C.E., and RIVEST, D.L.: 'Introduction to algorithms' (McGraw-Hill, New York, NY, 2001)

92 SALOMON, D.: 'Data compression: the complete reference' (Springer-Verlag New York, Inc., New York, NY, 2000)

93 GAREY, M.R., and JOHNSON, D.S.: 'Computers and intractability: a guide to the theory of NP-completeness' (Freeman, New York, 1979)

94 IYENGAR, V., CHANDRA, A., SCHWEIZER, S., and CHAKRABARTY, K.: 'A unified approach for SOC testing using test data compression and TAM optimization'. Proceedings of the IEEE/ACM *Design, automation and test in Europe* (DATE) conference, 2003, pp. 1188–1189

95 GONCIARI, P.T., and AL-HASHIMI, B.: 'A compression-driven test access mechanism design approach'. Proceedings of the European *Test* synposium, 2004, pp. 100–105

Chapter 23

On-chip test infrastructure design for optimal multi-site testing

Sandeep Kumar Goel and Erik Jan Marinissen

23.1 Introduction

The manufacturing test costs of subsequent generations of systems-on-chip (SoCs) threaten to increase beyond what is acceptable, if no proper countermeasures are taken. Factors that drive the digital test costs up are the increases in pin count, test data volume, speed and corresponding required automatic test equipment (ATE) accuracy. Especially the test data volume has risen dramatically, due to a combination of growth in transistor count and new advanced test methods (such as delay-fault testing) which add significantly to the test set size. As a consequence, testing of 'monster chips' [1] requires expensive ATEs with a large channel count and deep test vector memory [2].

Several methods are applied to reduce the test costs. With built-in self test (BIST), SoCs test (parts of) themselves and hence eliminate the need for ATE altogether. BIST for embedded memories has become a mainstream approach [3]. For logic, however, BIST is expensive to implement on chip, and hence its usage is typically limited to applications which require in-field testing [4]. Test data compression (TDC) techniques still require the presence of an ATE, but reduce the demands on both vector memory and test application time by exploiting the many 'don't care' bits in the test set to compress that test set [5, 6].

Another effective approach to reduce test cost is 'multi-site testing', in which multiple instances of the same SoC are tested in parallel on a single ATE [7–11]. More 'sites' means more devices are tested in parallel. Multi-site testing amortises the ATE's fixed costs over multiple SoCs. It can be used in addition to BIST or TDC. Rivoir [12] showed that multi-site testing is more effective in reducing the overall test cost than simply using low-cost ATEs, because it reduces all test cost contributors and not just the capital cost of ATE.

Efficient multi-site testing requires effective management of test resources such as the number and depth of ATE channels and the on-chip DfT, while taking into account parameters such as test time, index time, contact yield, etc. One way to allow an increase in the number of sites is to increase the number of ATE channels. However, this solution not only brings substantial extra costs, but also is not scalable to SoCs with high pin counts. The other way to increase the number of sites is to narrow down the SoC–ATE test interface, i.e. the number of SoC terminals that need to be contacted during testing. Reduced-pin-count-test (RPCT) [8, 13, 14] is a well-known DfT technique that exactly does this.

In this chapter, we propose to optimise the throughput for both wafer test and final test by means of multi-site testing. For wafer test, we maximise the throughput by testing a relatively large number of sites through a narrow enhanced-RPCT (E-RPCT) [14] interface. For final test, we contact all SoC pins, and hence the number of multi-sites is limited. We present a generic throughput model for multi-site testing, valid for both wafer test and final test, which considers the effects of test time, index time, abort-on-fail and re-test after contact fails. Subsequently, we present an algorithm that for a given SoC with a fixed target ATE and a probe station, designs and optimises an on-chip test infrastructure, DfT, that allows for maximal-throughput wafer-level multi-site testing. In case the given SoC uses a flattened top-level test, our algorithm determines the design of an E-RPCT [14]. In case the given SoC uses a modular test approach [15], in addition to the E-RPCT wrapper, the algorithm determines the on-chip test architecture consisting of test access mechanisms (TAMs) and core wrappers [16, 17]. Next, we present a second algorithm that for a given fixed ATE and SoC handler, and for the SoC with DfT optimised for wafer testing, determines the multi-site number for maximal throughput at final test.

The number of sites at which the test throughput is maximum, is referred to as 'optimal multi-site'. These numbers can be (and typically are) different for wafer test and final test. Note that maximal throughput is a different optimisation criterion from simply maximising the number of multi-sites. For wafer test, a large number of sites means less ATE channels per SoC, which in turn increases the test application time per SoC. For final test, a large number of sites means that more SoCs need be handled in parallel, which can increase the handling time. Consequently, in order to minimise test costs through multi-site testing, the number of sites should be tuned such that the test throughput is maximised.

The outline of this chapter is as follows. Section 23.2 reviews the prior work in this domain. Section 23.3 describes our multi-site test flow. Section 23.4 details our model for multi-site test throughput. Section 23.5 formally defines the problems of optimal multi-site testing of SoCs for wafer and final test, while Section 23.6 describes two algorithms to solve them. Section 23.7 contains experimental results for the Philips SoC PNX8550 [1] and SoCs taken from the 'ITC'02 SoC Test Benchmarks' [18]. Finally, Section 23.8 concludes the chapter.

23.2   Prior work

Reduced-pin-count testing is a DfT technique used to reduce the number of integrated circuit (IC) pins that need to be contacted by the ATE. RPCT assumes the presence

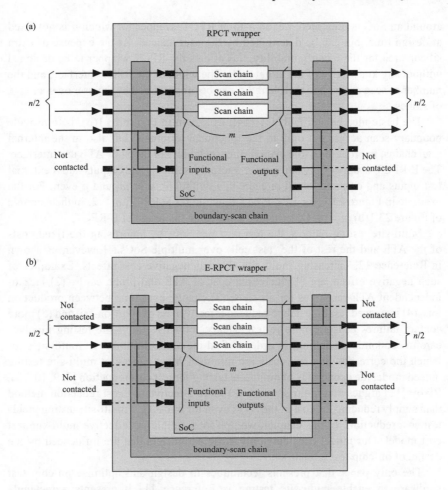

Figure 23.1   Example of (a) an RPCT wrapper and (b) an E-RPCT wrapper

of internal and boundary-scan. The basic principle of RPCT is that only the input and output terminals of the scan chains (including the boundary-scan chain), test control pins and clock pins need to be connected to the ATE channels. Access to all other functional pins is achieved via the boundary-scan chain. An RPCT wrapper around the SoC converts m internal test inputs and outputs into n external test inputs and outputs, for all integers n, m with $2s < n < m$ and n even, where s denotes the number of internal-scan chains. Figure 23.1(a) shows an example RPCT wrapper. In this example, $m = 13$ and $s = 3$, and consequently, $n \geq 8$.

First use of RPCT with level-sensitive scan design (LSSD) boundary-scan was reported by IBM to enable the use of low-cost ATE for application specific integrated circuits (ASICs) [13]. Since then, several extensions have been made to the basic RPCT technique. Two such extensions are reconfigurable RPCT [8] and E-RPCT [14]. In Reference 8, a technique to design a reconfigurable RPCT wrapper around an SoC is presented, i.e. an *n*-to-*m* RPCT wrapper for which *n* is not fixed at design time, but can be determined by the user instead. At the expense of extra silicon area for the reconfigurability, this allows the RPCT wrapper to be designed without any knowledge of the target ATE. The width of the RPCT interface, and the number of scan chains and their lengths, can be programmed by the user over a range of values.

The basic idea behind E-RPCT [14], as shown in Figure 23.1(b), is to provide boundary-scan access not only to the functional terminals, but also to the internal scan chains, in order to enable even further scalability of the SoC–ATE test interface. The E-RPCT wrapper truly converts *m* internal test inputs and outputs into *n* external test inputs and outputs, for all integers n, m with $0 < n < m$, and *n* even. For the example in Figure 23.1, for RPCT, $n \geq 8$, while for E-RPCT, $n \geq 2$; in the example of Figure 23.1(b), $n = 4$. Our chapter is based on the usage of E-RPCT.

Multi-site testing reduces the test cost per SoC, by amortising the fixed costs of the ATE and the rest of the 'test cell' over multiple SoCs. However, as shown in Reference 12, increasing multi-site also has negative cost effects. Examples of such negative effects are (1) increased cost of ATE and probe cards, (2) lack of independent ATE resources for all sites, (3) change-over time between production lots, (4) increased test time due to reduced effectiveness of abort-on-fail and (5) more contact failures causing more re-test. Hence, effective multi-site testing involves careful economic modelling, in order to find the optimal number of multi-sites for which the corresponding test costs are minimal. Most papers on multi-site testing indeed model the economics of multi-site testing for test cost reduction [7, 9, 10, 12]. Rivoir [12] argues that multi-site test is a more effective test cost reduction method than simply reducing the cost of the ATE; even for free ATEs, multi-site testing yields test cost reduction! In this chapter, we present a simple yet effective multi-site test cost model. Our model considers only those parameters that are influenced by the design of on-chip test infrastructure.

The only paper that presents techniques to design and optimise on-chip test hardware to enable multi-site testing is Reference 11. It presents a rectangle bin-packing-based technique to design the test architecture (consisting of TAMs and core wrappers) for a modularly tested SoC with a target ATE, such that the test architecture requires a minimum number of ATE channels and the SoC test data volume fits on the given ATE. A minimum number of ATE channels per device enables the maximum multi-site testing possible for the given SoC. While this paper was the first one in this domain, it has several limitations. The paper only discusses the design of core wrappers and TAMs for modularly tested, core-based SoCs, and ignores the design of a chip-level E-RPCT wrapper. It maximises the number of sites that can be tested in parallel, while we show in this chapter that this does not always yield maximum throughput. To maximise the number of sites, the paper assumes that a common set of input channels can be used to broadcast test stimuli to all sites, which is often not practical (see Section 23.4.1). And finally, Reference 11 considers test time only, and does not take into account the effects of index time, abort-on-fail and re-test rate due to contact fails.

23.3   Multi-site test flow

Multi-site testing can be done at wafer test as well as at final ('packaged IC') test. In this chapter, we assume the following two-step test flow.

1 During 'wafer test', the internal circuitry of the SoC die in question is tested. This is done through a narrow E-RPCT interface, in order to enable a large number of sites, as well as to reduce the chances for contact test fails. The non-E-RPCT pins of the SoC are not contacted.
2 During 'final test', the IOs of the packaged SoC are tested. For this purpose, 'all' pins of the SoC are contacted. The SoC-internal circuitry is tested again; we assume that this is done through the same E-RPCT interface designed for wafer test.

The pins of the SoC need to be contacted for test at least once during the test flow. We propose to do that during the final test, as then also the bonding wire and assembly result can be included in the test. Consequently, the number of sites during final test is limited by the number of available ATE channels divided by the number of pins of the SoC. In addition, this number is limited by the maximum number of multi-sites possible on the available device handler.

Multi-site testing during wafer test does not have the restrictions described above for final test. Today's high channel-count ATEs and corresponding high pin-count probe technologies enable massive multi-site testing for wafer test by using a narrow E-RPCT interface. The only limitation at wafer test is that full multi-site testing is not possible at the periphery of the wafer, due to its circular shape. However, in our calculations in this chapter we ignore this effect. We focus on the design of test infrastructure to maximise the multi-site test throughput during wafer testing.

We assume that for testing the internals of the SoC, during final test the same narrow E-RPCT interface that was designed for enabling optimal multi-site during wafer testing is reused again. An alternative could have been to allow the final test to use all available SoC pins for the SoC-internal tests. However, this would have required an area-expensive reconfigurable E-RPCT wrapper [8], as well as reconfigurable test wrappers [19] around all cores. Moreover, it would require regeneration of the tests over the new, wider SoC–ATE interface. Note that a negative consequence of our assumption is that the SoC-internal test during final test requires deeper ATE vector memory and takes more clock cycles than strictly necessary. In Section 23.7, we evaluate the impact of our choice on test throughput during final test.

23.4   Multi-site test throughput model

In this section, we present a test throughput model for multi-site testing. We take into account the effects of test time, index time, abort-on-fail and re-test rate due to contact fails.

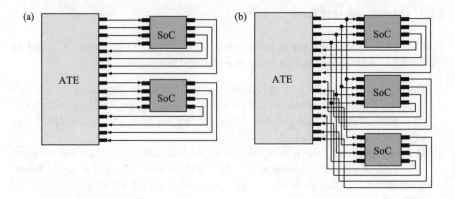

Figure 23.2   Multi-site test example (a) 'without' and (b) 'with' stimuli broadcast

23.4.1   Stimuli broadcast discussion

For an SoC that requires k channels for its test, an ATE with N channels can do at most n_{max} multi-sites, with

$$n_{max} = \left\lfloor \frac{N}{k} \right\rfloor \tag{23.1}$$

A technique to increase the maximum number of multi-sites on a given, fixed ATE is to apply 'stimuli broadcast' [11]. The probe card or load board are prepared such that a single set of ATE stimulus channels is used to provide stimuli to all instances of an SoC. Figure 23.2 shows an example ATE with $N = 16$ and an SoC with $k = 8$. Figure 23.2(a) shows that two instances of the SoC can be tested on this ATE 'without' stimuli broadcast, while Figure 23.2(b) depicts that 'with' stimuli broadcast the same ATE can test three instances of the SoC.

For an SoC with i test inputs and o test outputs (with $i + o = k$), with stimuli broadcast the maximum amount of multi-sites n_{max} becomes

$$n_{max} = \left\lfloor \frac{N - i}{o} \right\rfloor \tag{23.2}$$

In the case of (E-)RPCT, the test interface is formed by scan chains only, and hence $i = o = k/2$, and hence

$$n_{max} = \left\lfloor \frac{N - k/2}{k/2} \right\rfloor = \left\lfloor \frac{2N}{k} \right\rfloor - 1 \tag{23.3}$$

The benefit of stimuli broadcast for multi-site testing is obvious: it almost doubles the amount of multi-sites possible for a given SoC–ATE combination. However, next to the cost of probe card or load board adaptation, stimuli broadcast is not always practical. Some ATEs simply do not support broadcasting; they assign a channel to a site, and, if that site fails, no more stimuli are sent to the SoC under test. Furthermore, broadcasting can cause undesired side effects, such as a fault at the bonding pad of one site causing incorrect test results on other sites.

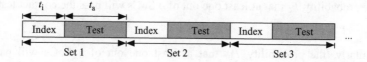

Figure 23.3   Wafer testing time consists of index time t_i and test time t_a

In this chapter, our solution approach accommodates explicitly both cases, i.e. 'without' and 'with' stimuli broadcast.

23.4.2   Wafer test

In the case of the wafer test, the complete wafer is located near the ATE, and the E-RPCT bonding pads of the SoCs under test are physically probed. The total time spent on a set of SoCs to be tested in parallel is the sum of the index time t_i and the test application time t_a, as depicted in Figure 23.3.

The 'index time' t_i is the time required to position the probe interface in order to make contact with the bonding pads of the SoC(s) under test. We assume the index time to be a constant, dependent on the type of probe station, but 'not' on the multi-site number. A typical value is $t_i = 0.7$ s.

The test consists of a contact test and a manufacturing test. In the 'contact test', it is checked whether all terminals required for the subsequent manufacturing test are properly connected to the ATE. If one or more of these terminals are not properly connected, the SoC fails the contact test. The probability p_c of a single terminal to pass the contact test, also referred to as the 'contact yield', needs to be high to be able to successfully test high pin-count SoCs. All terminals undergo their contact test simultaneously, and hence the contact test time t_c is a constant. A typical value is $t_c = 10$ ms.

During the 'manufacturing test', the SoC is checked for manufacturing defects. In this chapter, we only consider digital tests that can be applied through an E-RPCT scan chain; examples of such tests are logic and memory tests. The probability p_m of a single SoC to pass the manufacturing test is also referred to as the 'yield'. The manufacturing test time t_m depends on the width of the E-RPCT test interface, the test data volume and how well the various SoC tests can be scheduled.

The total test time can now be written as

$$t = t_i + t_a = t_i + t_c + t_m \tag{23.4}$$

In high-volume production testing, where faulty chips are often not analysed, but simply discarded, it is possible to abort the test as soon as the first failing test vector is observed. This 'abort-on-fail' strategy can significantly reduce the average test time per device, especially in the case of low yield. As shown in Section 23.7.6, multi-site testing reduces the effect of the abort-on-fail strategy on the average test time; in a multi-site testing environment, tests can only be aborted if all n sites have started failing, which is simply less likely to happen. For an SoC with k pins involved in its test, the probability P_c that at least one out of n SoCs will pass the contact test is

$$P_c = 1 - (1 - p_c^k)^n \tag{23.5}$$

Similarly, the probability P_m that at least one out of n SoCs will pass the manufacturing test is

$$P_m = 1 - (1 - p_m)^n \tag{23.6}$$

We assume that the failing SoCs do not take any test time ($t_m = 0$). Based on this, a theoretical lower bound on the total test application time for a set of n devices can be written as

$$t_a = t_c + P_c \times P_m \times t_m \tag{23.7}$$

In reality, a failing device does take some test time ($t_m > 0$) before a fault is found in it. The choice for the obviously unrealistic assumption that $t_m = 0$ for a failing SoC is motivated by the fact that it allows us to make a strong conclusion about the reduced effectiveness of abort-on-fail in multi-site testing in Section 23.7.6.

Assuming a full utilisation of the ATE, the total number of devices tested per hour D_{th} for n multi-site testing can be written as

$$D_{th} = \frac{3600 \times n}{t_i + t_a} \tag{23.8}$$

In this equation, both t_i and t_a are in seconds. Furthermore, t_a can be either the original test application time, or, when abort-on-fail is used, the reduced test application time of Equation (23.7).

In many companies, it is common practice to re-test those devices that failed on their contact test. The premise of re-testing is that the chances are high that the failure was caused by a wrong probe contact, rather than that the SoC itself was faulty. If that is indeed the case, it would be a waste to discard basically good devices. Excluding the unlikely event of multiple failing terminal contacts per SoC, D_{th} SoCs with k terminals each and a contact yield p_c per terminal, will require $D_r = (1 - p_c) \times k \times D_{th}$ SoCs/h to be re-tested. While the number of devices tested per hour D_{th} remains unaffected, re-testing has an impact on the number of unique devices tested per hour D_{th}^u. Assuming at most one failing terminal contact per SoC, and that devices are re-tested at most once, D_{th}^u can be written as

$$D_{th}^u = D_{th} - D_r = (1 - (1 - p_c) \times k) \times D_{th} \tag{23.9}$$

23.4.3   Final test

A device handler for final test has the following main components: an 'input rail', 'loader', 'unloader', 'sockets' and 'bins'. Figure 23.4(a) shows a handler for final test. Figures 23.4(b) and (c) show a conceptual view of the handling and contacting during multi-site testing for final test, in this case for four sites (also referred to as 'quad-site' testing). Basically, the handling consists of two alternating steps. In Step 1, shown in Figure 23.4(b), devices are moved serially from the input rail to the loading arm ('loader') of the device handler. Simultaneously, devices that are already placed in the

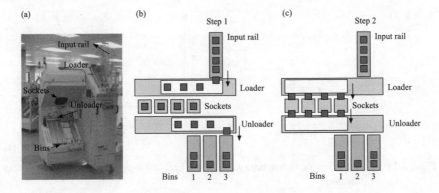

Figure 23.4   Example of (a) a handler and (b + c) a conceptual representation of its two-step quad-site operation

sockets are tested, and already tested devices (if any) are moved from the unloader to their respective bins. Examples of bins are 'good', 'bad contact' and 'functional failure'. Subsequently, in Step 2, shown in Figure 23.4(c), untested devices are moved in parallel from the loader to the sockets, while tested devices are moved from the sockets to the unloader. Steps 1 and 2 are repeated in a (virtually) endless iteration.

In our approach, during final test, all SoC pins need to be contacted and the full contact test is carried out. This limits the amount of multi-site to at most the number of available ATE channels divided by the number of pins K per SoC. Please note that variables K and k are different. For a given SoC, K represents the total number of functional pins at the SoC boundary, while k (with $k \leq K$) is the width of the E-RPCT interface for the SoC.

For multi-site testing at final test, the device handler for packaged SoCs might also be a limiting factor. Currently, for logic ICs, handlers up to eight sites are available in the market. For memory ICs, where massive multi-site testing is common due to their very large test times, handlers up to 128 sites are available. These handlers require multiple pick-ups, such that the index time increases in a step-wise fashion with the number of sites. Alternatively, large multi-site can also be achieved by so-called strip handlers [20], where ICs are tested while still in their lead frame strips, i.e. before being singulated into individual units. For strip handlers, the handling time does not vary with the number of sites. Figure 23.5 shows the variation in index time with multi-site for both types of handlers.

A generic expression for the index time t_i as a function of the number of sites n can be written as follows

$$t_i(n) = t_1 + \left\lfloor \frac{n-1}{\sigma} \right\rfloor \times \tau \tag{23.10}$$

where t_1 is the minimum index time (considering only single-site testing), σ is the 'pick-up size', i.e. the number of ICs that can be picked up at once and τ is the 'pick-up time', i.e. the time per individual pick-up. For memory handlers, typical values of these parameters are $t_1 = 0.4$ s, $\sigma = 4$, and $\tau = 0.2$ s. For strip handlers, $\tau = 0$.

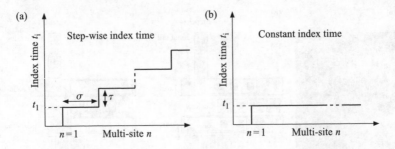

*Figure 23.5   Variation in index time for (a) multiple pick-up handlers and (b) strip
handlers*

With the assumption that both contact and manufacturing tests are executed for
the packaged devices, the total test time for final test can now be written as

$$t = t_i(n) + t_c + t_m \tag{23.11}$$

Assuming a full utilisation of the ATE, the total number of devices tested per hour
D_{th} for n multi-site testing for the final test can be written as

$$D_{th} = \frac{3600 \times n}{t_i(n) + t_c + t_m} \tag{23.12}$$

Similarly, with the assumptions that there is at most one failing terminal contact
per SoC, and the failing devices are re-tested at most once, the number of unique
devices tested per hour D_{th}^u can be written as

$$D_{th}^u = (1 - (1 - p_c) \times K) \times D_{th} \tag{23.13}$$

23.5   Problem definitions

For wafer test, we distinguish between the problems of test infrastructure design for
flat and modularly tested, core-based SoCs. For flat SoCs, we design an E-RPCT
wrapper around the SoC. For modularly tested SoCs, in addition to the E-RPCT
wrapper, we also design TAMs and core wrappers.

The problem of test infrastructure design for optimal multi-site testing (i.e. with
maximal throughput) of flat SoCs for wafer test can be formally defined as follows.

Problem 1 [Test infrastructure design for flat SoCs].   Given an SoC with a number
of test patterns p, a number of functional input terminals i, a number of functional
output terminals o, a number of functional bidirectional terminals b, a number of scan
chains s and for each scan chain j, the length of the scan chain in flip flops $l(j)$. Also
given a target ATE with N channels, each with vector memory depth V. Furthermore are given the test clock frequency f and a target SoC probe station with index
time t_i.

Determine the number of multi-sites n, the number of ATE channels per site k (with k even), and a k-to-x E-RPCT wrapper for the SoC (with $x = i + o + 2b + 2s$ and $k \leq x$), resulting in $T(k)$ test clock cycles per SoC and $t_\mathrm{m} = T(k)/f$, such that during n multi-site testing

1   the number of required ATE channels does not exceed the number of available ATE channels, i.e.

    (a)   without stimuli broadcast: $n \times k \leq N$

    (b)   with stimuli broadcast: $(n + 1) \times k/2 \leq N$,

2   the required ATE vector memory depth does not exceed the available depth, i.e. $T(k) \leq V$,

3   the test application time is calculated as follows

    (a)   without abort-on-fail: $t_\mathrm{a} = t_\mathrm{c} + t_\mathrm{m}$

    (b)   with abort-on-fail: $t_\mathrm{a} = t_\mathrm{c} + P_\mathrm{c} \times P_\mathrm{m} \times t_\mathrm{m}$,

4   the test throughput is maximum, i.e.

    (a)   without re-test: D_th 'is maximum'

    (b)   with re-test: D_th^u 'is maximum'. ∎

For the example SoC shown in Figure 23.6(a), an example E-RPCT wrapper that needs to be designed in Problem 1 is shown in Figure 23.6(b). The example SoC contains three scan chains, three functional input terminals and four functional output terminals. For clarity, control and clock terminals are not shown in the figure. Without E-RPCT wrapper, in total 13 ATE channels, i.e. six input and seven output ATE channels, are required to test this SoC. However, the E-RPCT wrapper shown in Figure 23.6(b) only requires four ATE channels ($k = 4$), i.e. two input and two output channels.

Next, we discuss the problem of test infrastructure design for modularly tested, core-based SoCs. Modular testing of an SoC requires a test infrastructure consisting of wrappers and TAMs [16, 17]. Therefore, for modularly tested SoCs, in addition to the E-RPCT wrapper, we also design core wrappers and TAMs. The problem of test infrastructure design for optimal multi-site testing of core-based SoCs for wafer test can be formally defined as follows.

Problem 2 [Test infrastructure design for core-based SoCs].   Given all parameters as specified in Problem 1. Furthermore is given a set of modules M in the SoC, and for each module $m \in M$ the number of test patterns $p(m)$, the number of functional input terminals $i(m)$, the number of functional output terminals $o(m)$, the number of functional bidirectional terminals $b(m)$, the number of scan chains $s(m)$ and for each scan chain j, the length of the scan chain in flip flops $l(m, j)$.

Determine the number of multi-sites n, the number of ATE channels per site k (with k even and $k = u + 2w$), an u-to-x E-RPCT wrapper for the SoC (with $x = i + o + 2b + 2s$ and $u \leq x$) such that u channels are used for testing the top-level SoC itself and a test architecture with w TAM wires for testing the embedded modules

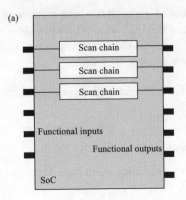

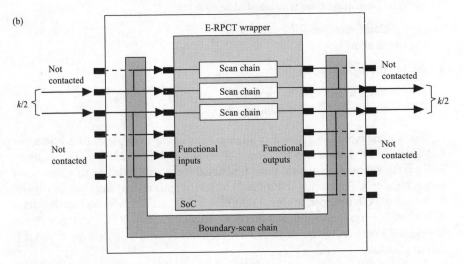

Figure 23.6   An example of (a) a flat SoC and (b) the E-RPCT wrapper around this SoC

(i.e. determine the number of TAMs, the width of these TAMs, the assignment of modules to TAMs and the core wrapper design per module such that the summed TAM width is w [16, 17]), resulting in $T(k)$ test clock cycles for the entire SoC including embedded modules and $t_m = T(k)/f$, such that during n multi-site testing

1   the number of required ATE channels does not exceed the number of available ATE channels, i.e.

(a)   without stimuli broadcast: $n \times k \leq N$
(b)   with stimuli broadcast: $(n + 1) \times k/2 \leq N$,

2   the required ATE vector memory depth does not exceed the available depth, i.e. $T(k) \leq V$,

3 the test application time is calculated as follows

    (a) without abort-on-fail: $t_a = t_c + t_m$

    (b) with abort-on-fail: $t_a = t_c + P_c \times P_m \times t_m$,

4 the test throughput is maximum, i.e.

    (a) without re-test: D_{th} 'is maximum'

    (b) with re-test: D_{th}^u 'is maximum'.   ■

For the example modularly tested SoC shown in Figure 23.7(a), an example DfT solution that needs to be designed in Problem 2 is shown in Figure 23.7(b). Like the flat SoC in the previous example, this example SoC contains three scan chains, three functional input terminals and four functional output terminals. This SoC also contains an embedded Core A, which has two scan chains and two input and output terminals. We need to design an E-RPCT wrapper, as well as a wrapper [21] for Core A, and assign TAMs to it. In the test infrastructure shown in Figure 23.7(b), four ATE channels are used to connect Core A, while another four ATE channels are used in the E-RPCT wrapper around the core. Therefore, for this case $k = 8$.

Problem 2 is actually a generalised version of Problem 1. For a flat SoC, we simply deal with zero embedded modules, i.e. $M = \emptyset$. Hence, if we address Problem 2, Problem 1 is implicitly solved.

As mentioned earlier we use the test infrastructure that is designed and optimised for wafer test also for final test. Therefore, Problems 1 and 2 do not require solutions tailored for final test. However, for final test also, we need to find the number of sites for which the test throughput is maximum. Therefore, for final test, we solve another optimisation problem as described in Problem 3.

Problem 3 [Optimal multi-site testing for final test].   Given an SoC with I functional input pins, O functional output pins and the DfT infrastructure (i.e. E-RPCT wrapper for the SoC, and the TAMs and wrappers for the internal modules) with the corresponding test length T (in test clock cycles) and manufacturing test time t_m. Also given a target ATE with N channels, each with vector memory depth V. Furthermore given a target SoC handler with minimum index time t_1, pick-up size σ and pick-up time τ.

Determine the number of multi-sites n, such that during n multi-site testing

1 the number of used ATE channels does not exceed the number of available ATE channels, i.e.

    (a) without stimuli broadcast: $n \times (I + O) \leq N$

    (b) with stimuli broadcast: $(n \times O + I) \leq N$,

2 the required ATE vector memory depth does not exceed the available depth, i.e. $T \leq V$,

3 the test application time is calculated as follows:

    (a) without abort-on-fail: $t_a = t_c + t_m$

    (b) with abort-on-fail: $t_a = t_c + P_c \times P_m \times t_m$,

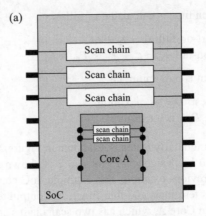

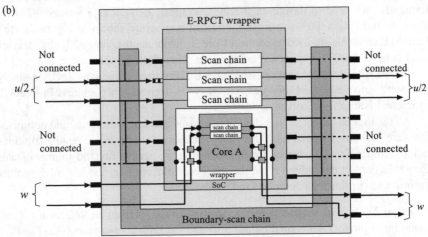

Figure 23.7   An example of (a) a modularly tested SoC, and (b) the example DfT for this SoC

4   the test throughput is maximum, i.e.

   (a)   without re-test: D_{th} 'is maximum'
   (b)   with re-test: D_{th}^{u} 'is maximum'.   ∎

23.6   Proposed solutions

23.6.1   Optimal multi-site testing for wafer test

In this section, we present a two-step algorithm for Problem 2. In Step 1 of the algorithm, we determine the maximum multi-site n_{max} and the corresponding test infrastructure for the given SoC and ATE. In Step 2, we use linear search to find the

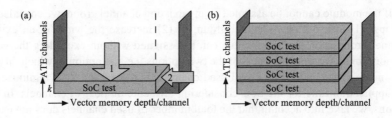

Figure 23.8   Fitting SoC test data on the target ATE with as few ATE channels as possible (a) in order to allow the maximum number of multi-sites (b)

number of sites n_{opt} ($n_{\max} \geq n_{\text{opt}} \geq 1$) for which the test throughput is maximum and we modify the test infrastructure accordingly.

In the algorithm, we use the notion of 'channel group'. A channel group is a group of ATE channels, that belong together because they exclusively serve one on-chip TAM. In our algorithm G denotes the set of channel groups. For all $g \in G$, $m(g)$ denotes the set of modules connected to channel group g, $k(g)$ denotes the number of channels assigned to group g, $V(g)$ denotes the maximum vector memory filling over all $k(g)$ channels of group g and $V_{\text{free}}(g)$ denotes the summed free vector memory over all $k(g)$ channels of group g.

Step 1: In this step, we first assign channels to all internal modules and then assign separate channels to the top-level SoC. While determining $n_{\max}$, we use two optimisation criteria, as illustrated in Figure 23.8(a). Criterion 1 is the minimisation of the number of ATE channels k utilised by one SoC, such that the test still fits into the vector memory depth V of the ATE. Criterion 2 is the minimisation of the actual filling of the vector memory. Criterion 1 has priority, as it maximises the number of sites, as shown in Figure 23.8(b). Criterion 2 is meant to reduce the test application time per SoC.

In this step, we first calculate the minimum number of ATE channels $k_{\min}(m)$ required for every module $m \in M$ such that the module's test time $t(m, k_{\min})$ does not exceed the ATE vector memory depth per channel V. To calculate the test time for a module, we need to design a wrapper around the module. To design the wrapper around a module m for a given number of ATE channels k, we use the COMBINE algorithm presented in Reference 21. If for any $m \in M$, $k_{\min}(m) > N$ then the SoC cannot be tested on the target ATE and the procedure is exited. Otherwise, modules are sorted in decreasing order of their $k_{\min}(m)$. Initially, there are no channel groups, i.e. $G = \emptyset$. Now we start with the assignment of modules to channel groups. For every module, we first try to find out whether the module can be assigned to an already existing channel group without exceeding its vector memory depth limit. For the very first module, there are no existing channel groups, therefore it is assigned $k_{\min}(m)$ number of channels and a channel group of width $k_{\min}(m)$ is formed. Iteratively, we move to the next module. If more than one channel group is found to which a module can be assigned, then the module is assigned to the group g^* that requires the smallest vector memory depth, i.e. $V(g^*)$ is minimum after assignment of the module.

If the module cannot be assigned to any existing channel group, we consider two options: (1) create a new channel group, or (2) increase the width of an existing channel group such that the module can be assigned without exceeding the vector memory limit. We select the best of the two options, i.e. the option in which the total free memory V_{free} available on all used channels is maximum. This minimises the test application time for the SoC considering the same number of channels. In both options, we take into account that the total number of used channels does not exceed N; if the assignment of a module leads to the violation of this constraint, then the SoC cannot be tested on the ATE and the procedure is exited. The procedure is repeated until all modules are assigned.

Next, we assign ATE channels to the top-level SoC. We first calculate the minimum number of ATE channels $k_{\text{min}}(\text{SoC})$ required for the top-level SoC test such that the test time for the top-level test does not exceed the ATE channel depth. If the sum of already assigned ATE channels $\left(\sum_{g \in G} k(g)\right)$ for the internal cores tests and these new channels $k_{\text{min}}(\text{SoC})$ exceed the total available number of channels, then the SoC cannot be tested on this tester and the procedure is exited. Otherwise, a new channel group is formed and the top-level SoC is assigned to the channel group.

The summed width of all channel groups determines the total number of channels k for the SoC. The test application time for the SoC corresponds to the maximum of the filled vector memory $(\max_{g \in G} V(g))$ over all channel groups. Based on the test clock frequency f, actual test application time t_{m} can be calculated as $t_{\text{m}} = \max_{g \in G} V(g)/f$ s. The maximum multi-site possible is $n_{\text{max}} = \lfloor N/k \rfloor$ in the case without stimuli broadcast, and $n_{\text{max}} = \lfloor 2N/k \rfloor - 1$ in the case with stimuli broadcast.

An example with one iteration of Step 1 is illustrated in Figure 23.9. Figure 23.9(a) shows a situation in which Cores A and B have already been assigned to a channel group that requires $k1$ ATE channels, while Core C is assigned to another channel group that requires $k2$ ATE channels. Figure 23.9(b) corresponds to the case, in which the algorithm tries to add Core D to either one of the two already existing channel groups. Unfortunately, both alternatives exceed the vector memory depth limit V. Hence, the algorithm is forced to start using more ATE channels in order to add Core D. In Figure 23.9(c), the three alternatives considered are depicted. Alternative (1) is to add a new channel group for Core D, in this case with $k3 = k_{\text{min}}(D)$ ATE channels. Alternative (2) extends channel group 1 iteratively from $k1$ to $k1 + k3$ channels, and is only valid if Core D can now be added without exceeding the vector memory depth V. Similarly, Alternative (3) extends channel group 2 iteratively from $k2$ to $k2 + k3$ channels. The alternative which yields the smallest vector memory filling is selected.

Step 2: In this step, we identify the number of sites n_{opt} for which the throughput D_{th} or D_{th}^{u} is maximum. Here, we only consider D_{th}; in case of maximising D_{th}^{u}, D_{th} can be replaced by D_{th}^{u}.

We use linear search from n_{max} down to 1 to calculate the corresponding D_{th} value. In every iteration, we try to redistribute the ATE channels k_{free} freed-up by giving up one site over the remaining sites. Only if $k_{\text{free}} \geq 2n$ (for the case without stimuli broadcast) or $k_{\text{free}} \geq n + 1$ (for the case with stimuli broadcast), redistribution

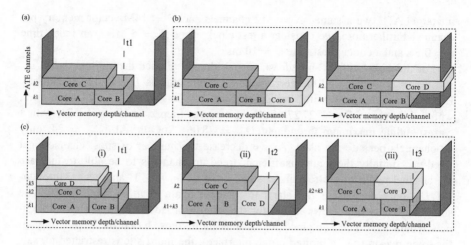

Figure 23.9 Illustrative example of one iteration of Step 1 for an SoC with four cores

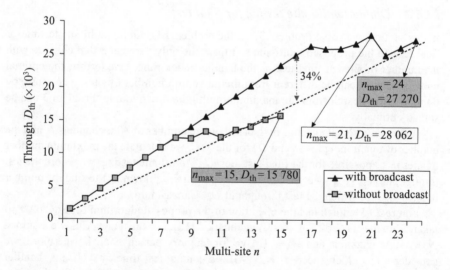

Figure 23.10 Example illustrating the operation of the proposed two-step algorithm for Philips SoC PNX8550

makes sense. If so, for each site, we assign iteratively free channels to the channel group that is maximally filled. This can reduce the test application time per site. We record the throughput for the value of n. Finally, after the linear search, we find n_{opt} as the number of sites for which the throughput D_{th} is maximum.

Figure 23.10 illustrates the operation of the proposed two-step algorithm for the Philips SoC PNX8550 [1], for both the cases with and without stimuli broadcast. For the target ATE, we assumed $N = 512$ channels and $V = 14M$ vector memory per channel. Furthermore, we consider a test clock speed $f = 5$ MHz, an index time $t_i = 0.5$ s and a contact test time $t_c = 10$ ms.

For the case 'without' broadcast, Step 1 already yielded the optimal result, i.e. $n_{max} = n_{opt} = 15$, and the corresponding throughput is $D_{th} = 15\,780$ SoCs/h. However, for the case 'with' broadcast, Step 1 results in $n_{max} = 24$ multi-sites, with a throughput $D_{th} = 27\,270$ SoCs/h, whereas Step 2 finds $n_{opt} = 21$, with a corresponding maximum throughput $D_{th} = 28\,062$ SoCs/h. At $n_{max} = 24$, only 40 channels per site are used. As we decrease the number of sites, channels get freed-up. Initially there are insufficient freed-up channels to be able to increase the channel width to the remaining sites, i.e. $k_{free} < n + 1$, and hence, throughput D_{th} only decreases. D_{th} starts to increase again at $n = n_{opt} = 21$. The straight, dashed line shows, again for the stimuli broadcast case, which throughputs would have been obtained for various multi-sites, based on Step 1 only. If for some reason (e.g. a limited probe interface), the multi-site is restricted to, say, $n \leq 16$, Steps $1 + 2$ together result in 34 per cent more throughput than Step 1 alone.

23.6.2   Optimal multi-site testing for final test

In this section, we solve Problem 3, i.e. the problem of optimal multi-site testing for final test. In the proposed multi-site test flow, the only parameter that changes with the number of sites at final test, is the handler index time. Due to this, the optimal number of sites can be different from the maximum number of sites. Therefore, by doing a linear search from the maximum multi-site down to one, Problem 3 can be solved optimally.

In the proposed algorithm, first, based on the number of ATE channels N and the number of functional pins (I and O) for the SoC, we calculate the maximum number of sites n_{max} possible for the final test. For n_{max} multi-site testing, we record the test throughput D_{th}^*. Next, we use a linear search from n_{max} down to 1 to find the number of sites n_{opt} for which the test throughput D_{th}^* is maximum.

Figure 23.11 illustrates the operation of the proposed algorithm for SoC p22810 taken from the 'ITC'02 SoC Test Benchmarks' [18]. For the target ATE, we assumed $N = 1024$ channels and $V = 1M$ vector memory per channel. Furthermore, we consider a test clock speed $f = 5$ MHz, a contact test time $t_c = 10$ ms, handler index time $t_1 = 0.4$ s, pick-up size $\sigma = 4$ and pick-up time $\tau = 0.2$ s. For the given ATE and without stimuli broadcast, the on-chip test infrastructure design algorithm presented in Section 23.6.1 results in a DfT infrastructure with a manufacturing test time $t_m = 0.204$ s. For SoC p22810, we take the number of terminals of the top-level SoC module (referred to as Module 0 in Reference 18) as the number of functional pins to be contacted during final test. Therefore, $K = 173$ for SoC p22810. Based on this, maximum five ($n_{max} = 5$) multi-site testing can be done on the given ATE during final test. For $n_{max} = 5$, the test throughput is $D_{th} = 22\,113$ SoCs/h. However, the algorithm presented in this section finds $n_{opt} = 4$, with a corresponding maximum throughput $D_{th} = 23\,453$ SoCs/h.

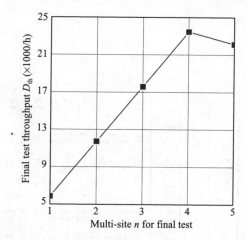

Figure 23.11   Example illustrating the operation of the proposed algorithm for SoC p22810 [18]

23.7   Experimental results

This section presents experimental results for the proposed approach and how it compares to the prior work in Reference 11 . Subsequently, we use our throughput model and algorithm to evaluate the effects of index time, ATE parameters, contact yield and abort-on-fail on multi-site test throughput.

For most of our experiments, we consider a 'small' and a 'large' SoC. The small SoC is the Philips SoC p22810, taken from the 'ITC'02 SoC Test Benchmarks' [18]. It embeds 22 logic and six memory modules. The large SoC is the Philips PNX8550 [1]. PNX8550 is based on the Philips Nexperia™ Home Platform and embeds 62 logic and 212 memory modules. Unless specified otherwise, we assume a test clock $f = 5$ MHz, contact test time $t_c = 10$ ms, wafer-level index time $t_i = 0.7$ s, and 'no' stimuli broadcast.

23.7.1   Algorithmic performance

First, we compare the performance of the proposed algorithm to the results published by Iyengar *et al.* [11] for three large 'ITC'02 SoC Test Benchmarks' [18]. Iyengar *et al.* [11] did not include the top-level SoC module in their calculations, assumed stimuli broadcast and calculated n_{max} instead of n_{opt}. In order to compare the two algorithms on a fair and equal basis, for this comparison we also ignored the top-level SoC module, assumed stimuli broadcast and have only applied Step 1 of Algorithm 1 in order to calculate n_{max}.

Table 23.1 lists for four SoCs the number of ATE channels k used for a single SoC, and the maximum multi-site n_{max}. For each SoC, an ATE with a fixed number of channels N and a range of vector memory depths V were considered, just as in Reference 11. For k, Table 23.1 presents a theoretical lower bound value LB for k

Table 23.1 Experimental results for maximum multi-site for the rectangle bin-packing algorithm in Reference 11 and our new algorithm

SoC p22810	N = 512				SoC p34392	N = 512				SoC p93791	N = 512			
	k		n_{max}			k		n_{max}			k		n_{max}	
V	LB [11]	Us	[11]	Us	V	LB [11]	Us	[11]	Us	V	LB [11]	Us	[11]	Us
384k	36	36	23	27	768k	40	40	21	24	1.000M	54	58	16	16
448k	30	32	27	31	896k	34	34	24	29	1.256M	44	46	21	21
512k	26	28	31	35	1.000M	30	30	31	33	1.512M	36	38	24	25
576k	24	24	35	41	1.128M	26	26	33	38	1.768M	32	36	29	27
640k	22	22	38	45	1.256M	24	24	35	41	2.000M	28	28	33	35
704k	20	20	45	50	1.384M	22	22	38	45	2.256M	24	26	38	38
768k	18	18	45	55	1.512M	20	20	41	50	2.512M	22	24	41	41
832k	16	18	50	55	1.640M	18	18	45	55	2.768M	20	20	45	50
896k	16	16	50	63	1.768M	18	18	50	55	3.000M	18	20	50	50
960k	14	14	55	72	1.896M	16	16	50	63	3.256M	18	18	55	55
1M	14	14	63	72	2.000M	16	16	55	63	3.512M	16	18	63	63

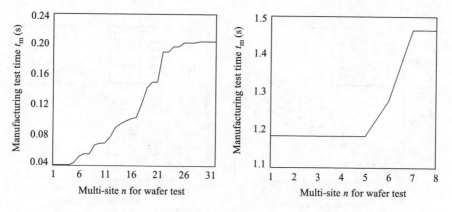

Figure 23.12 *For wafer test, variation in manufacturing test time t_m with the number of sites n for (a) SoC p22810 and (b) PNX8550*

from Reference 11 and the actual value of k obtained by our approach. For n_{max}, Table 23.1 presents the values obtained by Reference 11 and our new approach. In all cases, except for SoC p93791 with $V = 1.768M$ channel depth, our algorithm obtains a higher multi-site than that in Reference 11. Furthermore, in most cases, our algorithm matches the lower bound on the number of ATE channels for a single SoC. This shows that our algorithm is very effective and usually achieves the maximum multi-site possible for the given SoC.

23.7.2 Index time

Next, we analyse the impact of index time on multi-site test throughput for both wafer test and final test. For wafer testing, our test throughput model assumes the index time to be constant, independent from the number of multi-sites. The relative impact of the index time t_i on the test throughput D_{th} depends on the ratio between t_i and the manufacturing test time t_m. A small SoC will have a small t_m, such that t and hence D_{th} are dominated by index time t_i. A large SoC will have a large t_m, such that t and hence D_{th} are dominated by t_m. In the latter case, the impact of a varying index time t_i is reduced.

This point is illustrated by Figures 23.12 and 23.13. Figure 23.12(a) shows the manufacturing test time t_m for SoC p22810 for a varying number of multi-sites on a fixed ATE with $N = 512$ channels and $V = 1M$ vector memory depth per channel. Figure 23.12(b) shows a similar graph for PNX8550 and a fixed ATE with $N = 512$ channels and $V = 7M$ vector memory depth per channel. The figures show that manufacturing test times of SoC p22810 are relatively small, while the manufacturing test times of PNX8550 are relatively large, compared to the typical probe station index time $t_i = 0.7$ s. Figure 23.13(a) and (b) shows the variation in wafer-level test throughput D_{th} with the number of sites n for varying index times t_i for the same two cases.

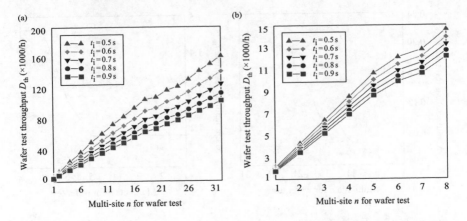

Figure 23.13   *For wafer test, variation in test throughput D_{th} with the number of sites n for a range of index times t_i for (a) SoC p22810 and (b) SoC PNX8550*

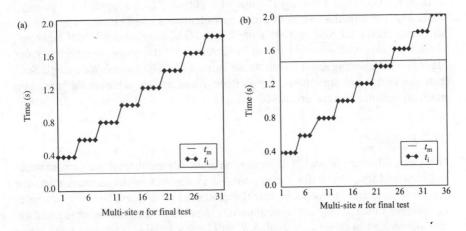

Figure 23.14   *For final test, variation in manufacturing test time and index time with number of sites for (a) SoC p22810 and (b) SoC PNX8550*

For final testing, our approach assumes that the manufacturing test is executed via the E-RPCT interface designed for wafer testing. Therefore, during final test, the manufacturing test time t_m is constant, and independent from the number of multi-sites. However, our test throughput model assumes the index time t_i to increase with the number of multi-sites in a step-wise fashion.

Figure 23.14 depicts both t_m and t_i for a varying number of multi-sites n for SoC p22810 (Figure 23.14(a)) and PNX8550 (Figure 23.14(b)). For SoC p22810, $t_m = 0.204$ s, corresponding to the optimal multi-site number in Figures 23.12(a) and 23.13(a). For PNX8550, $t_m = 1.464$ s, corresponding to the optimal multi-site

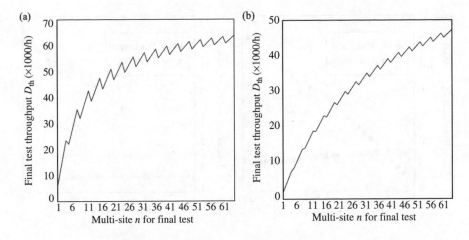

Figure 23.15   For final test, variation in test throughput D_{th} with the number of sites n for (a) SoC p22810 and (b) SoC PNX8550

number in Figures 23.12(b) and 23.13(b). In both graphs, t_i increases in a step-wise fashion with $t_1 = 0.4$ s, pick-up size $\sigma = 4$ and pick-up time $\tau = 0.2$ s.

For final test, the number of multi-sites n is limited to at most the ratio of ATE channels N and the number of functional pins K per SoC. For SoC p22810 and PNX8550, $K = 173$ and 584, respectively. Hence, even for an ATE with $N = 1024$ channels, the maximal number of multi-site is very limited. Consequently, it is difficult to show the impact of the increasing index time in final test on the test throughput of these two SoCs. In order to still be able to do so, we (unrealistically) assume that we have an ATE with sufficient channels to allow up to 64 multi-sites for both SoCs. Under that assumption, Figure 23.15 shows the variation in the test throughput D_{th} for a varying number of multi-sites n. Unlike the throughput graphs for wafer test in Figure 23.13, the throughput curve for the final test in Figure 23.15 starts to saturate at large multi-site numbers. This is especially true for a small SoC like p22810, for which, due to the multiple pick-ups of the handler, the index time becomes very large compared to the manufacturing test time. The conclusion is that multi-site testing in final test does pay off for smaller multi-site numbers and for final test with strip handlers. Massive multi-site testing with a multiple pick-up handler does not pay off in terms of test throughput.

23.7.3   Reusing the on-chip infrastructure for final test

Next, we consider the effect of designing and optimising the on-chip test infrastructure for maximum wafer test throughput on the test throughput for final test.

For SoC p22810, wafer testing reaches maximal throughput for $n_{opt} = 32$ multi-sites, with a manufacturing test time $t_m = 0.204$ s. Due to the fact that all $K = 173$ pins need to be contacted, we can only do quad-site testing in final test, whereas we reuse the same narrow SoC–ATE test interface that allows for 32 multi-sites in

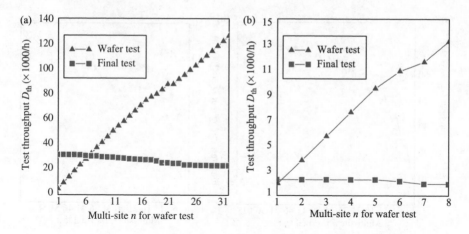

Figure 23.16   For both wafer and final test, variation in test throughput with number of sites at wafer level for (a) SoC p22810 and (b) SoC PNX8550

wafer test. Consequently, the test time in final test is also $t_m = 0.204$ s, which yields a final test throughput $D_{th} = 23\,512$ SoCs/h. If the test infrastructure would have been optimised for four sites, the manufacturing test time would have been $t_m = 0.040$ s and the corresponding final test throughput would have been $D_{th} = 31\,934$ SoCs/h. Hence, our (well-motivated) choice to reuse the on-chip wafer test infrastructure for final test has a negative impact of 26 per cent on the final test throughput.

A similar story is true for PNX8550. The optimal final test throughput $D_{th} = 2259$ SoCs/h, for an SoC–ATE interface optimised for single-site testing. The actual final test throughput $D_{th} = 1921$ SoCs/h, as the SoC–ATE interface is optimised for octal-site testing. This amounts to a negative impact of 15 per cent. Figure 23.16 shows the test throughput during both wafer test and final test as a function of the number of wafer test multi-sites for which the SoC–ATE interface is designed. Despite the negative impact of reusing the wafer test interface during final test, the figure shows that the summed benefit of multi-site testing is clearly a positive one.

In the remaining part of experimental results, we focus on wafer consider SoC PNX8550 only.

23.7.4   Economical ATE extensions

Next, we examine what happens to the wafer test throughput D_{th} of PNX8550 if we extend its 'basic' ATE (with $N = 512$ and $V = 7M$) with either more channels or deeper vector memory.

Figure 23.17(a) shows the increase in test throughput D_{th} for an increasing number of ATE channels N. The figure illustrates that the test throughput increases linearly with the number of ATE channels; by doubling the number of ATE channels, the test

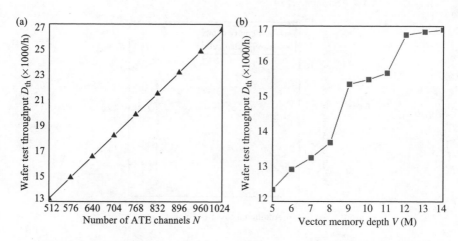

Figure 23.17   Variation in test throughput with (a) the number of ATE channels N and (b) the vector memory depth V

throughput can be doubled. This is due to the fact that the number of sites increases linearly with the number of channels, while the test time remains constant.

Figure 23.17(b) shows the increase in test throughput D_{th} for increasing deeper vector memory per ATE channel V. The figure shows that the test throughput does not increase linearly with the vector memory depth. This is due to the fact that an increase in test vector memory depth not only leads to an increase in multi-site, but also to an increase in test application time. Therefore, doubling the test vector memory does 'not' result in a double throughput.

Based on the results shown above, it seems that increasing the number of ATE channels is more attractive than increasing the vector memory depth, in order to get the test throughput up. However, the cost of increasing the vector memory depth is small compared to the cost of increasing the number of ATE channels. According to standard market prices, buying 16 additional ATE channels with 7M memory depth would cost roughly $8000; upgrading test vector memory for 16 channels from 7M to 14M would cost only $1500. Therefore, if we double the test vector memory for all 512 channels, it will cost around $(512/16) \times \$1500 = \$48\,000$. For this money, the increase in test throughput is 27 per cent, as shown in Figure 23.18. For the same amount of money, we can buy about 96 channels. This will result in an 18 per cent increase in test throughput, as shown in Figure 23.18. Therefore, for the same cost, increasing the test vector memory depth is more economical than increasing the number of ATE channels; if only and if we are allowed to change the DfT.

23.7.5   Contact yield and re-test rate

Next, we analyse the impact of the contact yield on the unique test throughput during wafer test. Figure 23.19 shows, for PNX8550, the unique test throughput D_{th}^{u} as a function of the vector memory depth V for varying contact yield p_c. As expected, a decreasing contact yield results in more re-testing, and hence a lower unique test

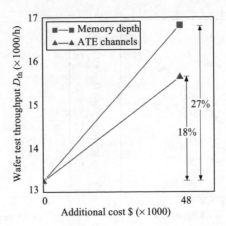

Figure 23.18   Increasing ATE vector memory depth can be more economical than increasing the number of ATE channels

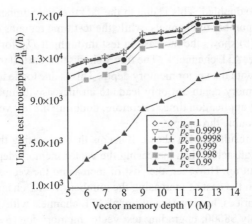

Figure 23.19   Impact of contact yield p_c and memory depth V on the unique test throughput D_{th}^u

throughput. However, the figure shows that the negative impact of the contact yield on D_{th}^u is worsened for small vector memory depths. This can be explained as follows. With small vector memory depths, the E-RPCT interface needs to be wider, in order to fit the test data volume into the ATE vector memories. Consequently, the number of pins k to be contacted grows. Contacting more pins increases the chances for contact failure. Hence the re-test rate increases and the unique test throughput decreases. Therefore, it can be concluded that deep vector memories are useful, not only to increase the regular test throughput, but also to increase the unique test throughput.

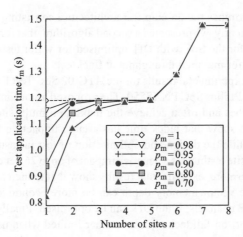

Figure 23.20   The average manufacturing test time t_m as a function of multi-site n for varying SoC yields p_m

23.7.6   Abort-on-fail

Finally, we demonstrate the influence of multi-site testing on the effectiveness of applying abort-on-fail. For single-site testing, abort-on-fail reduces the average manufacturing test time t_m; the actual reduction depends on the SoC yield p_m. In Figure 23.20, we show the average manufacturing test time t_m as a function of the number of multi-sites n, for varying SoC yields p_m. The figure illustrates that increasing multi-site testing quickly reduces the positive effect of applying abort-on-fail. Even at a low yield of $p_m = 70$ per cent (and under the overly optimistic assumption that for failing devices $t_m = 0$, see Equation (23.7)), the effectiveness of abort-on-fail becomes invisible for $n \geq 5$. This is due to the fact that if five or more SoCs are tested in parallel, the probability is quite low that all SoCs will fail their test.

23.8   Conclusions

To reduce test cost, multi-site testing is popularly used. In this chapter, we considered multi-site testing for both wafer and final test, and modelled the test throughput considering the effects of test time, index time, abort-on-fail and re-test after contact fails. We showed that, for a given, fixed ATE, multi-site testing requires optimisation of the on-chip test infrastructure and it should be test throughput and 'not' the amount of sites that should be maximised.

We presented an algorithm that for a given SoC with a fixed target ATE and probe station, designs and optimises an on-chip test infrastructure, DfT, that allows for maximal-throughput wafer-level multi-site testing. In case the given SoC uses a flattened top-level test, our algorithm determines the design of an E-RPCT wrapper. In case the given SoC uses a modular test approach, in addition to the E-RPCT wrapper, the algorithm determines the on-chip test architecture consisting of TAMs and core wrappers. Subsequently, we presented a second algorithm, that for a given fixed ATE, SoC handler, and for the SoC with DfT optimised for wafer testing, determines the multi-site number for maximal throughput at final test.

We presented experimental results for the 'ITC'02 SoC Test Benchmarks' as well as for the complex Philips SoC PNX8550. The proposed algorithms outperform other published approaches and often achieve the maximum multi-site possible. Experimental results also show that index time has a large impact on the test throughput for small SoCs, while the design and optimisation of test infrastructure for optimal wafer-level multi-site testing has a negative impact of up to 26 per cent on the final test throughput. Furthermore, experimental results show that to increase the test throughput, increasing the vector memory depth can be more economical than increasing the number of ATE channels, if the DfT can be optimised. Finally, we conclude that benefits of the abort-on-fail technique are rather limited when used in combination with multi-site testing.

Acknowledgements

We thank Peter Lagner of Advantest in Munich, Germany and Stefan Eichenberger and Robert van Rijsinge of Philips Semiconductors in Nijmegen, The Netherlands for fruitful discussions during the conception of this work.

References

1   GOEL, S.K., CHIU, K., MARINISSEN, E.J., NGUYEN, T., and OOSTDIJK, S.: 'Test infrastructure design for the Nexperia™ home platform PNX8550 system chip'. Proceedings of the *Design, Automation, and Test in Europe* (DATE) designers forum, Paris, France, February 2004, pp. 108–113

2   MARINISSEN, E.J., and VRANKEN, H.: 'On the role of DfT in IC-ATE matching'. Digest of papers of IEEE international workshop on *Test Resource Partitioning* (TRP), Baltimore, MD, November 2001

3   DEKKER, R., BEENKER, F., and THIJSSEN, L.: 'Realistic built-in self-test for static RAMs', *IEEE Design & Test of Computers*, 1989, **6**(1), pp. 26–34

4   VRANKEN, H., MEISTER, F., and WUNDERLICH, H.-J.,: 'Combining deterministic logic BIST with test point insertion'. Proceedings of the IEEE European *Test* Workshop (ETW), Corfu, Greece, May 2002, pp. 105–110

5   POEHL, F., BECK, M., ARNOLD, R. *et al.*: 'Industrial experience with adoption of EDT for low-cost test without concessions'. Proceedings of the IEEE International *Test* Conference (ITC), Charlotte, NC, September 2003, pp. 1211–1220

6   MITRA, S., and KIM, K.S.: 'X-Compact: an efficient response compaction technique for test cost reduction'. Proceedings of the IEEE

International *Test* Conference (ITC), Baltimore, MD, October 2002, pp. 311–320

7   EVANS, A.C.: 'Applications of semiconductor test economics, and multi-site testing to lower cost of test'. Proceedings of the IEEE International *Test* Conference (ITC), Atlantic City, NJ, September 1999, pp. 113–123

8   KHOCHE, A., KAPUR, R., ARMSTRONG, D., WILLIAMS, T.W., TEGETHOFF, M., and RIVOIR, J.: 'A new methodology for improved tester utilization'. Proceedings of the IEEE International *Test* Conference (ITC), Baltimore, MD, October 2001, pp. 916–923

9   VOLKERINK, E., KHOCHE, A., KAMAS, L.A., RIVOIR, J., and KERKHOFF, H.G.: 'Tackling test trade-offs from design, manufacturing to market using economic modeling'. Proceedings of the IEEE International *Test* Conference (ITC), Baltimore, MD, October 2001, pp. 1098–1107

10 VOLKERINK, E., KHOCHE, A., RIVOIR, J., and HILLIGES, K.D.: 'Test economics for multi-site test with modern cost reduction techniques'. Proceedings of the IEEE *VLSI test* symposium (VTS), Monterey, CA, April 2002, pp. 411–416.

11 IYENGAR, V., GOEL, S.K., CHAKRABARTY, K., and MARINISSEN, E.J.: 'Test resource optimization for multi-site testing of SoCs under ATE memory depth constraints'. Proceedings of the IEEE International *Test* Conference (ITC), Baltimore, MD, October 2002, pp. 1159–1168.

12 RIVOIR, J.: 'Parallel test reduces cost of test more effectively than just a cheap tester'. Proceedings of the IEEE International *Electronics Manufacturing Technology* symposium (SEMI-THERM), San Jose, CA, March 2004, pp. 263–272

13 BASSET, R.W., *et al.*: 'Low-cost testing of high-density logic components'. Proceedings of the IEEE International *Test* Conference (ITC), October 1989, pp. 550–557

14 VRANKEN, H., WAAYERS, T., FLEURY, H., and LELOUVIER, D.: 'Enhanced reduced pin-count test for full scan design'. Proceedings of the IEEE International *Test* Conference (ITC), Baltimore, MD, October 2001, pp. 738–747

15 ZORIAN, Y., MARINISSEN, E.J., and DEY, S.: 'Testing embedded-core based system chips'. Proceedings of the IEEE International *Test* Conference (ITC), Washington, DC, October 1998, pp. 130–143

16 IYENGAR, V., CHAKRABARTY, K., and MARINISSEN, E.J.: 'Co-optimization of test wrapper and test access architecture for embedded cores', *Journal of Electronic Testing: Theory and Applications (JETTA)*, 2002, **18**(2), pp. 213–230

17 GOEL, S.K., and MARINISSEN, E.J.: 'SoC test architecture design for efficient utilization of test bandwidth', *ACM Transactions on Design Automation of Electronic Systems (TODAES)*, 2003 **8**(4), pp. 399–429

18 MARINISSEN, E.J., IYENGAR, V., and CHAKRABARTY, K.: 'A set of benchmarks for modular testing of SoCs'. Proceedings of the IEEE International *Test* Conference (ITC), Baltimore, MD, October 2002, pp. 519–528. See http://www.hitech-projects.com/itc02socbenchm/

19   KORANNE, S.: 'Design of reconfigurable core wrappers for embedded core based SoC test'. Proceedings of the International Symposium on *Quality of Electronic Design* (ISQED), San Jose, CA, March 2002

20   Semicon-FarEast.com: http://www.semiconfareast.com/

21   MARINISSEN, E.J., GOEL, S.K., and LOUSBERG, M.: 'Wrapper design for embedded core test'. Proceedings of the IEEE International *Test* Conference (ITC), Atlantic City, NJ, October 2000, pp. 911–920

Chapter 24

High-resolution flash time-to-digital conversion and calibration for system-on-chip testing

Peter M. Levine and Gordon W. Roberts

24.1 Introduction

Temporal uncertainty in modern digital and mixed-signal complementary metal oxide semiconductor (CMOS) integrated circuits (ICs) can have detrimental effects on system-on-chip (SoC) performance. For example, fluctuations in global clock edge placement in microprocessors can cause timing failures along the critical path. In addition, uncertainty in the timing of sampling clock edges in analogue-to-digital converters (ADCs) can reduce the signal-to-noise ratio of the converted signal [1].

Verification of the timing performance of very large scale integration (VLSI) parts in a manufacturing environment is usually carried out using a production tester like that shown in Figure 24.1. With this instrument, the device-under-test (DUT) is attached to a device-interface board (DIB), which in turn sits on the test head. Pin electronics present in the test head excite the DUT with analogue and digital signals and capture its output. Additional hardware located in the tester mainframe supplies power to the DUT, stores test vectors and processes the received output.

Production testers are capable of performing very accurate timing measurement of signals received from the DUT. This is because the electronics in these instruments are often constructed using technologies that have traditionally outperformed the speed and noise performance of CMOS, such as silicon bipolar and gallium arsenide. However, the massive integration and increased performance of CMOS components in recent years due to technology scaling has introduced limitations in the effectiveness of traditional production testing. For example, bandwidth limitations due to the growing electrical distance between the DUT and test head electronics cause attenuation and additional phase delay in high-speed signals. In addition, routing the outputs from deeply buried IC components to the chip boundary for observation by the tester is often impractical and can skew timing results.

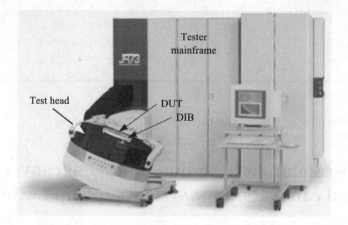

Figure 24.1   A modern mixed-signal VLSI production tester

To address these issues, various paradigms have been established over the years to improve the methods used in evaluating IC timing performance. The first, set forth in Reference 2, is represented in Figure 24.2(a) and involves integrating the timing measurement electronics of a VLSI functional tester on a silicon IC. This was intended to alleviate some of the issues surrounding functional testing, including the rising cost of testers and the worsening electrical problems due to the increasing distance between the tester electronics and DUT.

The second paradigm, proposed in Reference 3, is shown in Figure 24.2(b) and involves placing a single high-resolution timing measurement 'core' directly on the chip being tested. This effectively reduced the electrical distance between tester and DUT even more, making high-frequency timing measurement a reality.

The latest paradigm, developed in Reference 4, sought to improve the testability of integrated intellectual-property (IP) cores in an SoC environment. As is displayed in Figure 24.2(c), each IP core has an associated test system. In addition, each of these systems communicates with the outside world via a digital bus (not shown). One of the main advantages that this test system has over external instrumentation is improved accessibility to the buried IP cores. This is because many cores in an SoC environment interface with each other, rather than the external pins of the IC.

Despite the advantages inherent in these test paradigms, elimination of external testers altogether will not likely occur in the near future. This is mainly due to industrial momentum and resistance to altering test procedures for high-volume production parts [5]. However, it is expected that these techniques will be used to augment the capabilities of pre-existing testers so that high-performance SoCs can still be validated without a complete overhaul of the test infrastructure.

24.1.1   The need for high-resolution on-chip timing measurement

Timing uncertainty in ICs is most often expressed as jitter and skew. The synchronous optical network (SONET) standard defines jitter as the 'short term phase variation

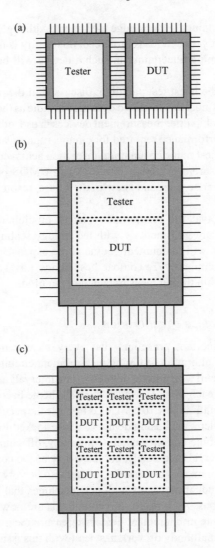

Figure 24.2   VLSI test paradigms. (a) Stand-alone tester IC, (b) integration of a single tester on chip, (c) integration of numerous testers on chip in an SoC environment

of the significant instants of a digital signal from their ideal positions in time' [6]. In this definition, 'significant instants' refer to edges while the system reference clock can be used to define what the 'ideal positions in time' are [7]. Skew, on the other hand, refers to spatial variations in the arrival time of a clock edge at one location in an IC compared to another.

It has been demonstrated experimentally in Reference 8 that it is possible to achieve reliable 10-GHz clock frequency operation in a standard 0.18-μm CMOS technology. At such speeds, a peak-to-peak jitter of only 10 ps translates to a ten per cent uncertainty in clock edge placement. Although processors running at these speeds have not yet been mass produced, this reality is not far off. As a result, verification of the timing uncertainty in such a device will become an issue in the not-so-distant future.

In keeping with the latest test paradigm discussed in the previous section, jitter measurement devices must be integrated onto the same die as the DUT. Unfortunately, the resolutions of on-chip jitter measurement devices in use today are too low to characterise the timing performance required for circuits running at speeds near 10 GHz. For example, the highest reported temporal resolution achieved experimentally from an on-chip time measurement device in a 0.18-μm CMOS process was 19 ps [9]. To accurately characterise jitter and skew in the future, resolutions below 10 ps are required.

While it is true that the resolution of many on-chip measurement systems reported in the literature will increase with technology scaling (like that described in Reference 10), this will undoubtedly come at the expense of greater power consumption, area and complexity. To combat these trends, novel algorithms and circuit architectures for on-chip timing measurement are required.

24.1.2   The importance of calibration

Calibration is an important procedure that every measurement instrument must undergo before use. Calibration of timing measurement circuitry is normally carried out by exciting the circuit with a series of known time intervals and then correlating the output with the input each time. However, such a scheme becomes more difficult as the desired resolution falls below 10 ps. This is because the accuracy of on-chip timing generators (often implemented using delay- or phase-locked loops) is limited by the jitter and mismatch of the circuitry itself. Conversely, off-chip pulse generators can produce such time intervals accurately, but these may be too costly to implement in a production-test environment. Furthermore, it may also be very expensive to upgrade or replace existing production testers with new machines that have higher temporal resolution specifications. As a result, novel methods are now required to calibrate the high-resolution time measurement devices to ensure measurement accuracy. Up to this point, only a small body of work has dealt with this issue [11,12].

24.1.3   Chapter organisation

This chapter is organised as follows. Section 24.2 presents a literature review of electronic time measurement techniques and discusses their applicability to on-chip testing. Section 24.3 describes the development of a high-resolution flash TDC that exploits the temporal offsets present in flip-flops or arbiters for time quantisation. Also, a novel calibration method for this measurement device, based on additive temporal noise, is described and experimentally verified using a programmable-logic device. Implementation details of a custom flash TDC, constructed in the Taiwan Semiconductor Manufacturing Company (TSMC) 0.18-μm CMOS process, are presented in Section 24.4. In addition, experimental results from calibration of the converter, as well as jitter measurement using this device, are also included. The chapter is concluded in Section 24.5.

24.2 Review of time measurement techniques

This section reviews the most popular circuit architectures for performing timing measurement of electronic signals. These can be grouped into three broad categories: homodyne mixing, signal amplitude sampling and time-domain analysis. Each will be discussed and their suitability for SoC timing measurement elucidated.

24.2.1 Homodyne mixing

Homodyne mixing involves using an analogue multiplier, such as a Gilbert cell, followed by a lowpass filter to convert the phase difference between two periodic signals, equal in frequency, to a DC voltage [8] as shown in Figure 24.3. This system is part of a broader class of circuits known as time-to-voltage converters (TVCs).

Given two sinusoids with amplitude A, radial frequency ω, and phases ϕ_1 and ϕ_2, respectively, the output T from the mixer in Figure 24.3 is

$$T = A^2 \cos(\omega t + \phi_1) \cos(\omega t + \phi_2) \tag{24.1}$$

$$= \frac{A^2}{2} (\cos(\phi_1 - \phi_2) + \cos(2\omega t + \phi_1 + \phi_2)) \tag{24.2}$$

The cosine term at 2ω is removed by lowpass filtering, leaving only the DC term which is proportional to the cosine of the phase difference multiplied by a constant.

The DC transfer characteristic of the filtered mixer output is shown in Figure 24.4. At the location of maximum sensitivity, the multiplied signals are in quadrature. The phase difference in seconds can then be determined by first using a calibration technique to find the slope of the graph around the quadrature region. With this information, a measured DC level can be mapped to a corresponding time difference.

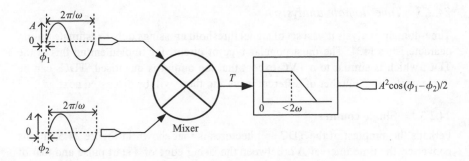

Figure 24.3   Schematic of the homodyne mixing technique

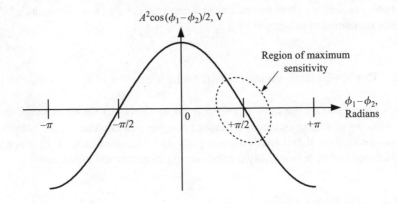

Figure 24.4   Transfer characteristic of the filtered mixer output

Due to the ease of integration of Gilbert-cell mixers and passive filters, this technique is amenable to on-chip time measurement. In addition, it has been used to determine the jitter and skew in a high-frequency clock distribution network [8].

24.2.2   Signal amplitude sampling

Signal amplitude sampling involves sampling a signal at an appropriate rate and using the time of occurrence and amplitude of each sample to obtain phase information about the signal. The eye-diagram method and such frequency-domain techniques as phase noise measurement fall into this category.

This method of time measurement can be implemented on chip by sampling a signal well above the Nyquist rate using high-speed, high-resolution ADCs [13]. However this may be impractical as these devices tend to consume a large die area and can be quite sensitive to process variation. Fortunately, recent research has shown that undersampling techniques can be used to gain sufficient information about a signal using circuits that are more area efficient and robust [14,15].

24.2.3   Time-domain analysis

Time-domain analysis uses a set of signal threshold crossings only to estimate timing characteristics [13]. The most popular type of circuit for implementing this is the TDC, which is similar to a TVC except that the output is quantised. TDCs can be implemented in a number of different ways and these will be discussed next.

24.2.3.1   Single counter

Perhaps the simplest of the TDC architectures is that shown in Figure 24.5. In this converter, the time interval ΔT between the rising edge of a start pulse and that of a stop pulse is measured by a counter running on a high-frequency reference clock. The AND gate ensures that the counter is enabled only when 'Start' and 'Stop' are

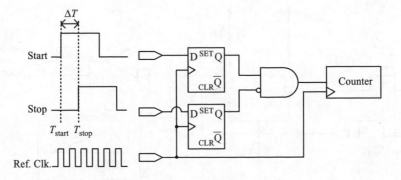

Figure 24.5 TDC implemented using a single counter

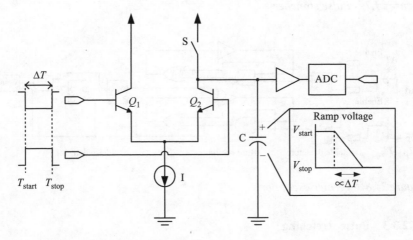

Figure 24.6 Interpolator-based TDC

logically different. The resolution of this device is constrained by the speed of the reference clock and can be no higher than a single clock period. Limitations in the frequency and stability of on-chip clocks makes high-resolution SoC testing difficult using this architecture.

24.2.3.2 Interpolation

The interpolator shown in Figure 24.6 operates by discharging the voltage on a pre-charged capacitor C using the constant current source I [16,17]. The final voltage on the capacitor V_{stop} is proportional to the duration of the input pulse ΔT and is digitised using an ADC. Switch S is used to charge up the capacitor between measurements. This implementation has a wide dynamic range but requires a very-high-resolution ADC to perform picosecond timing measurement. As a result, its integration in an on-chip testing system is challenging.

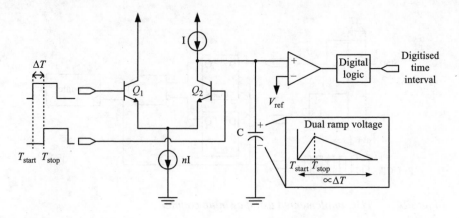

Figure 24.7   Pulse stretcher

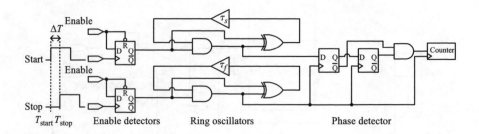

Figure 24.8   Vernier oscillator

24.2.3.3   Pulse stretching

In this technique, two voltage ramps are used to 'stretch' the small time interval ΔT so that a separate low-resolution time measurement device can easily digitise the interval as shown in Figure 24.7 [18]. The charge and discharge rates of the capacitor C are made different using two separate current sources. The ramp voltage is compared to reference voltage V_{ref} and the amount of time that the comparator output is high can then be measured using, e.g. a simple counter.

24.2.3.4   Vernier oscillator

In the Vernier oscillator shown in Figure 24.8, two ring oscillators produce ple-siochronous square waves (set by buffer delays τ_s and τ_f) to quantise a time interval [9]. The 'Start' and 'Stop' pulses enable the oscillators and the time interval is mea-sured by the phase detector and counter. The latter devices are effectively triggered when one oscillator 'catches up' to the other. Due to its small size and relatively high temporal resolution, the Vernier oscillator is amenable for use in on-chip test systems. In addition, the use of the oscillators reduces the matching requirements on the delay buffers used to quantise a time interval. This feature is used to overcome the temporal

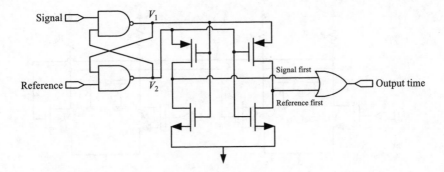

Figure 24.9   MUTEX-based time amplifier

uncertainties caused by component variation in the delay lines of Vernier delay flash TDCs, which will be described later.

One negative aspect of this architecture, however, is that it takes many cycles to complete a single measurement (i.e. has a long dead-time). Compared to flash converters that can make a measurement every cycle, the Vernier oscillator requires a long test time. Such a feature can be cost-prohibitive in a production-test environment.

24.2.3.5   Time amplification

The principle of time amplification involves comparing the phase of two inputs and then producing two outputs that differ in time by a multiple of the input phase difference. The first implementation of a time amplifier was proposed in Reference 19 and uses the mutual exclusion (MUTEX) circuit shown in Figure 24.9. Here, the cross-coupled NAND gates form a bistable while the output transistors switch only when the difference in voltage between nodes V_1 and V_2 reach a certain value [20]. The OR gate connected to the output is used to detect this switching action.

Time amplification is provided by the response time of the bistable when the input time difference is small enough to cause the bistable to exhibit metastability. The difference ΔV in voltages V_1 and V_2 in Figure 24.9 is approximately given by

$$\Delta V \approx \theta \cdot \Delta t \cdot e^{t/\tau}, \qquad (24.3)$$

where θ is the conversion factor from time to initial voltage at nodes V_1 and V_2, Δt is the time difference between the rising edges of the signal and reference and τ is the device time constant. By measuring the time t between the moment that the inputs switch to when the OR gate switches, Δt can be found.

A second implementation of a time amplifier was proposed in Reference 21. This device is displayed in Figure 24.10 and consists of two cross-coupled differential pairs with passive RC loads attached as shown. Upon arrival of the rising edges of ϕ_1 and ϕ_2, the amplifier bias current is steered around the differential pairs and into the passive loads. This causes the voltage at the drains of transistors M_1 and M_2 to be equal at a certain time and that at M_3 and M_4 to coincide a short time later. This

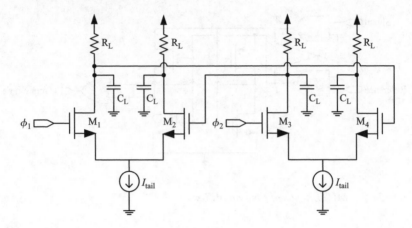

Figure 24.10   Time amplifier constructed with two cross-coupled differential pairs

effectively produces a time interval proportional to the input time difference which can then be detected by a voltage comparator.

Although both of these time amplifiers output an analogue voltage rather than a digital code, they are meant to be used in conjunction with a relatively low-resolution TDC, as was done in the pulse-stretching technique described previously. Such systems may therefore be quite conducive for use in on-chip testing environments.

24.2.3.6   Flash conversion

Flash TDCs are analogous to flash ADCs for voltage amplitude encoding and operate by comparing a signal edge to various reference edges, all displaced in time. The elements which compare the input signal to the reference are usually flip-flops or arbiters (note that an arbiter is a circuit that decides which of the two input signals arrived first). For simplicity, flip-flops and arbiters will be referred to interchangeably herein.

Delay chain flash converter: In the single delay chain flash TDC shown in Figure 24.11, each buffer produces a delay equal to τ [22]. To ensure that τ is known reasonably accurately, the delay chain is often implemented using voltage-controlled buffers stabilised by a delay-locked loop (DLL) [24].

Suppose it is desired to determine the time difference ΔT between the rising edges of pulses P_{start} and P_{stop} using the 8-level delay chain converter in Figure 24.12. Each flip-flop compares the displacement in time of the delayed P_{start} to that of P_{stop}. The thermometer-encoded output indicates the value of ΔT, assuming the flip-flops are given sufficient time to resolve. The drawback to this implementation is that the temporal resolution can be no higher than the delay through a single gate in the semiconductor technology used.

Vernier delay flash converter: To achieve sub-gate temporal resolution, the flash converter can be constructed with a Vernier delay line as shown in Figure 24.13 [24]. This architecture achieves a resolution of $\tau_1 - \tau_2$, where $\tau_1 > \tau_2$.

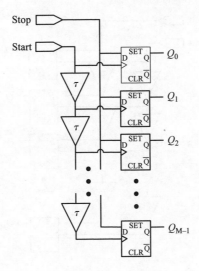

Figure 24.11   Delay chain flash converter

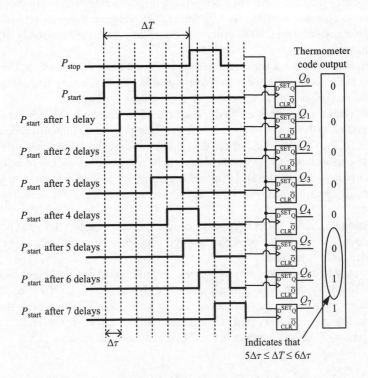

Figure 24.12   Operation of an eight-level delay chain flash converter

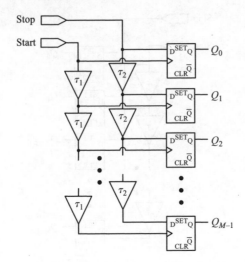

Figure 24.13   Vernier delay flash converter

These converters are well suited for use in on-chip timing measurement systems because they are capable of performing a measurement on every clock cycle and can be operated at relatively high speeds. In addition, they can be easily constructed in any standard CMOS process because they are composed solely of digital components.

However, to obtain a wide dynamic range from a flash TDC, many buffer stages must be built. As a result, these converters tend to consume a very large chip area.[1] Also, the numerous buffers can consume a great deal of power.

The Vernier delay flash converter constructed in Reference 23 was experimentally verified to have a resolution of 30 ps. This particular implementation made use of a DLL to stabilise one of the delay lines. Although scaling such a design to a higher-performance CMOS technology could increase the resolution somewhat, at some point, jitter in the DLL controlling the delay line will cause τ to deviate substantially from its ideal value. This, in turn, severely limits the accuracy of the converter.

In addition, buffer matching is extremely important in achieving good accuracy from the measurement device. This, however, becomes quite difficult as the length of the delay chain increases.

One way to overcome the above issues, and to obtain very high resolution, is to remove the delay buffers completely and use only the temporal offsets on the flip-flops themselves for time quantisation. This type of flash TDC has been shown to have resolutions which vary from a few picoseconds to tens of picoseconds [25]. The development of such a converter is described in the next section.

---

[1] For example, the flash TDC built in Reference 23 occupied a total area of 5.5 mm^2 in a 0.35-μm CMOS technology.

24.3 Development and calibration of a high-resolution flash TDC

24.3.1 Sampling offset TDC

A flash converter that relies solely on flip-flop transistor mismatch can be used to obtain fine time resolution without separate delay buffers. This type of converter is known as a 'sampling offset' TDC (SOTDC) [26] and Figure 24.14 shows how such a converter is related to the basic Vernier delay TDC. As the diagram shows, the Vernier delay flash converter is first represented as a single delay chain in which each buffer has delay $\Delta\tau = \tau_1 - \tau_2$. Note that all flip-flops are assumed to be ideal (i.e. they possess no transistor mismatch). An alternative form of the delay chain flash, in which each buffer has a cumulative delay, can then be drawn. Finally, the latter model can be replaced by the SOTDC, in which each ideal flip-flop has been substituted for one with transistor mismatch. Note that the offsets of the non-ideal flip-flops will be random, and not monotonic or multiples of a fundamental offset as Figure 24.14 might seem to suggest.

Simulations and experiments conducted in References 25 and 20 confirm that mismatches due to process variation can produce time offsets from 30 ps down to 2 ps, depending on the flip-flop architecture and semiconductor technology used. Of course, calibration is required to determine these offsets before the flip-flops can be used for time measurement. Common calibration procedures are described next.

24.3.2 Traditional TDC calibration

The goal of calibration is to determine the time offset t_{os} of each flip-flop or arbiter in the TDC. For purposes of analysis, a flip-flop model is defined first.

24.3.2.1 Flip-flop model

An ideal rising-edge-triggered flip-flop has $t_{os} = 0$ and is free from noise. A non-ideal flip-flop can be modelled as an ideal flip-flop with $t_{os} \neq 0$ as well as a source of thermal noise as shown in Figure 24.15(a) [27]. The noise voltage V_{noise} follows a Gaussian distribution with zero mean and standard deviation σ.

An alternative flip-flop model, which is more appropriate in the context of time measurement, is shown in Figure 24.15(b). Here, the ideal flip-flop takes a time difference ΔT_{eff} as input and produces a '1' if $\Delta T_{eff} > 0$ and a '0' otherwise. The input ΔT is equal to $t_{clock} - t_{data}$, where t_{clock} and t_{data} are the times of the rising edges of ϕ_{clock} and ϕ_{data} in Figure 24.15(a), respectively. In addition, V_{noise} in the original model is expressed as the temporal noise t_{noise}. Assuming a linear relationship between these two variables, t_{noise} follows a Gaussian distribution with zero mean and standard deviation σ_{FF}.

24.3.2.2 Indirect calibration

An indirect calibration technique, involving the use of uncorrelated signals to find the relative offsets of the flip-flops in an SOTDC, was proposed in Reference 25. An implementation of this is displayed in Figure 24.16(a), where ϕ_1 and ϕ_2 are square

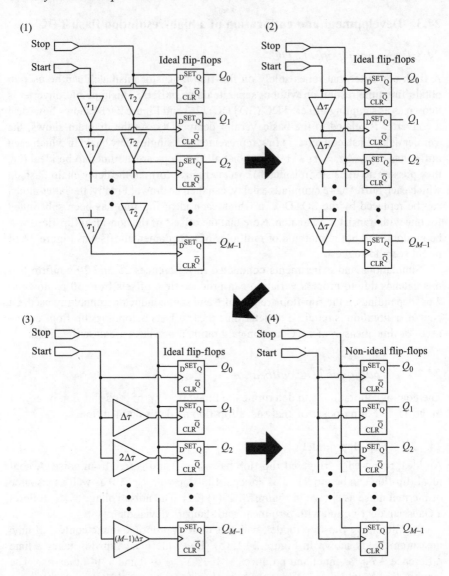

Figure 24.14 Relationship of sampling offset flash TDC to basic Vernier delay TDC. (1) Vernier delay flash converter, (2) single delay chain flash converter, (3) alternative representation of delay chain flash converter, (4) sampling offset flash converter

waves with constant frequencies f_1 and f_2. These are input to two flip-flops having offsets t_{os_1} and t_{os_2}. The relative flip-flop offset is given by $\Delta_{12} = t_{os_1} - t_{os_2}$ and it is assumed that $\Delta_{12} \gg t_{noise}$.

Figure 24.16(b) shows how Δ_{12} can be found empirically. Assuming that f_2 is only slightly greater than f_1, ϕ_1 will appear to 'move past' ϕ_2 in time. This ensures

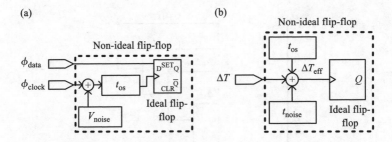

Figure 24.15   Flip-flop models: (a) Voltage model, (b) time model

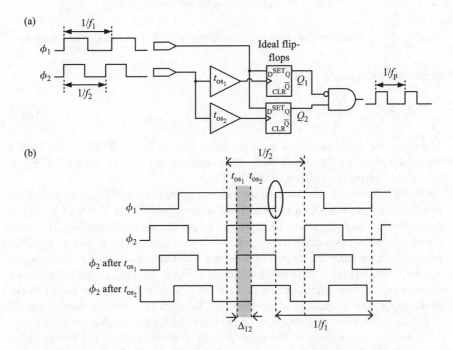

Figure 24.16   Indirect calibration of two flip-flops. (a) System used to determine relative flip-flop offset (noise sources are not shown for simplicity), (b) clocks in the indirect calibration system

that the rising edge of ϕ_1 is uniformly distributed over the interval defined by the period of ϕ_2. Therefore, the probability P_{12} that the circled rising edge of ϕ_1 in Figure 24.16(b) will land in the shaded interval Δ_{12} is given by

$$P_{12} = \frac{\Delta_{12}}{1/f_2} \qquad (24.4)$$

When this occurs, flip-flop output Q_1 in Figure 24.16(a) will be '0' while Q_2 will be '1'. Furthermore, measurements with the circled edge of ϕ_1 residing in Δ_{12} will

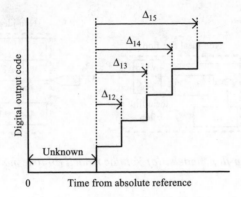

Figure 24.17   Transfer curve of a TDC determined using an indirect calibration

occur with a frequency f_p of

$$f_p = \frac{\Delta_{12}}{1/f_2} f_1 = \Delta_{12} f_1 f_2 \tag{24.5}$$

Therefore, the periodic output from the AND gate in Figure 24.16(a) will have a frequency given by Equation (24.5). This equation can then be solved for Δ_{12} to obtain the relative time offsets of the flip-flops.

Knowledge of the relative flip-flop offsets can be used to obtain a TDC transfer function like that shown for a hypothetical five-level converter in Figure 24.17. In this curve, the flip-flop offsets are all expressed relative to the same flip-flop. However, the time from the absolute reference [which is ϕ_2 in Figure 24.16(a)] to the first offset cannot be surmised from the indirect calibration. Consequently, if it is desired to measure jitter having a Gaussian distribution using an SOTDC calibrated this way, only the standard deviation (or equivalently, the rms value) of the jitter can be found. Furthermore, no information about the mean value of the jitter (i.e. how much the jittery clock deviates, on average, from the reference signal) can be surmised.

However, to acquire both the mean and standard deviation of the timing uncertainty, the absolute values of the offsets must be found. To determine the absolute offsets using the indirect calibration, some information about the offset statistics must be known. Although it could be assumed that the mean offset of a large number of flip-flops constructed on the same die will be zero, this assumption is invalid unless care is taken in the layout to ensure that all flip-flops experience similar process variation. Since this may be difficult to achieve in practice for a large number of devices, the direct calibration technique described next can be used instead.

24.3.2.3   Direct calibration

In a direct calibration, t_{os} of a single flip-flop is found by setting ΔT to M different values in the range $(-\infty, +\infty)$ and recording the flip-flop output each time as shown in Figure 24.18. This process is then repeated N times.

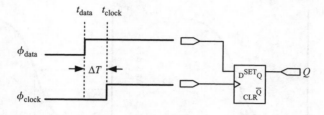

Figure 24.18   Direct calibration of a flip-flop

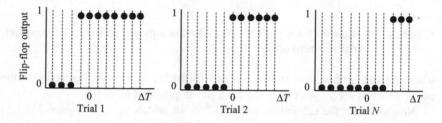

Figure 24.19   Results from each trial of flip-flop calibration

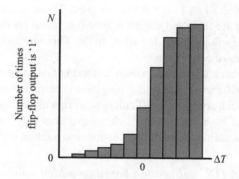

Figure 24.20   Histogram result of flip-flop calibration

For simplicity, assume that a flip-flop under calibration has $t_{os} = 0$. The data collected from each of the N trials may appear as in Figure 24.19. Note that the presence of thermal noise causes the flip-flop output to be different on each trial. By summing the number of times, n, the output from the flip-flop is '1' for each ΔT, the histogram in Figure 24.20 can be plotted. Next, the histogram can be expressed as a cumulative distribution function (cdf) by dividing each n by N and then curve fitting, as shown in Figure 24.21(a). In the cdf, the probability P of the event $\{\Delta \tau \leq \Delta T\}$ is given by

$$P(\Delta \tau \leq \Delta T) = \Phi\left[\frac{\Delta T - (-t_{os})}{\sigma_{FF}}\right] \tag{24.6}$$

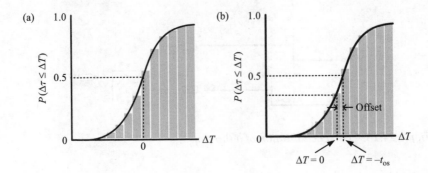

*Figure 24.21   Calibration cdfs of flip-flops having different offsets. (a) Zero offset,
(b) non-zero offset*

where $\Delta\tau$ is a random time and $\Phi(\cdot)$ is the standard Gaussian cdf. Inspection of the cdf reveals that $P(\Delta\tau \leq \Delta T) = 0.5$ for a flip-flop having $t_{os} = 0$.

Now consider the calibration of a flip-flop for which $t_{os} \neq 0$. Figure 24.21(b) shows the cdf of such a device and it is apparent that $P(\Delta\tau \leq 0) \neq 0.5$.

It should be noted that the histogram cdf obtained by calibrating a flip-flop violates the theoretical definition of the cdf, in which $P(\Delta\tau \leq \Delta T) = 0$ only as $\Delta T \rightarrow -\infty$, and where $P(\Delta\tau \leq \Delta T) = 1$ only as $\Delta T \rightarrow +\infty$. These deviations occur because flip-flop inputs that are far enough apart in time (i.e. much greater than $\pm 3\sigma_{FF}$) will cause a constant flip-flop output for all N trials. This violation, however, can be ignored for purposes of calibration.

The main drawback to this calibration method is that to accurately produce the cdf in Figure 24.21(b) and obtain t_{os}, ΔT may have to be set to values of the order of a few picoseconds. Such accuracy is difficult to achieve with on-chip signal generators. Furthermore, use of high-resolution off-chip generators may be too impractical or expensive. Therefore, an improved direct calibration method is described next.

24.3.3   Improved TDC calibration based on added noise

Performing a direct calibration on a flip-flop with a t_{os} of a few picoseconds and a σ_{FF} of a few hundred femtoseconds is difficult because ΔT cannot be made small enough to accurately produce a cdf curve. It would be helpful if σ_{FF} was much larger, say in the order of tens, or even hundreds, of picoseconds, so that points on the cdf could be measured accurately. With this data, t_{os} could be found by curve fitting.

Fortunately, it is possible to increase σ_{FF} by adding a temporal noise to that already present on the flip-flop inputs. A model for this is shown in Figure 24.22, where $t_{noise_{added}}$ is a noise source with zero mean and standard deviation σ_{added} and $t_{noise_{FF}}$ is the thermal noise of the flip-flop. Assuming that the summed random variables are independent, the total noise on the flip-flop will have a mean of zero and a standard deviation that is the square root of the sum of squares of σ_{added} and σ_{FF}. Furthermore, if $\sigma_{added} \gg \sigma_{FF}$, the total noise standard deviation will be very close to σ_{added}.

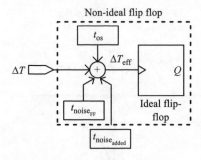

Figure 24.22   Flip-flop model with added noise source

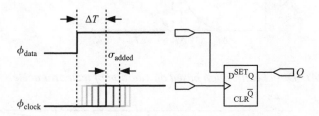

Figure 24.23   Flip-flop input edges with added temporal noise

The sum of the time ΔT and the noise $t_{\text{noise}_{\text{added}}}$ is simply a set of times which follow a Gaussian distribution with mean ΔT and standard deviation σ_{added}. A pictorial representation of this is displayed in Figure 24.23. Calibration using these times is performed just like a traditional direct calibration, however the need to set ΔT to very small values is eliminated. To see why, consider the upper graph in Figure 24.24 where each set of input times is expressed as a probability density function (pdf). As ΔT is increased in discrete steps from $-\infty$ to $+\infty$, a certain number of input times cross the offset threshold of the flip-flop, forcing the output to '1'. If N points are collected from each set of distribution, a histogram can be produced as before. Fitting a Gaussian cdf to the data, the standard deviation of the curve will equal σ_{added}, while the mean will correspond to $-t_{\text{os}}$, as shown in the lower graph in Figure 24.24.

24.3.3.1   Simulation results

Simulations were carried out to demonstrate the viability of the proposed calibration technique. The flip-flop model in Figure 24.22 was built using MATLAB with $\sigma_{\text{FF}} = 0.35$ ps (this was the measured value reported in Reference 25). The value of σ_{added} was set to 250 ps while ΔT was moved from -400 to $+400$ ps in steps of 40 ps. Such time resolutions can be handled with good accuracy by modern pulse generators or on-chip DLLs. Time t_{os} was set to various values in the range $(-40$ ps,$+40$ ps$)$ and $N = 100\,000$.

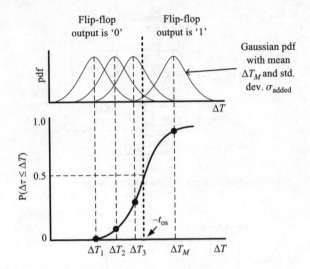

Figure 24.24   Direct calibration based on addition of temporal noise

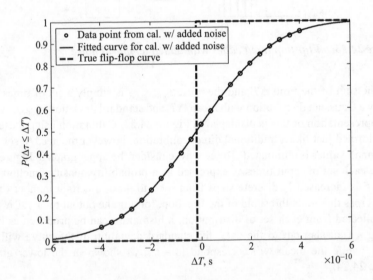

Figure 24.25   Simulated calibration result. Actual t_{os} was 17.4 ps while offset from fitted curve was 17.1 ps

Figure 24.25 shows the result from a single flip-flop calibration. A χ^2 fitting algorithm [28] was used and it is clear that the fitted cdf and the actual flip-flop curve described by Reference 6 nearly coincide at the value of ΔT for which $P = 0.5$.

Table 24.1 compares the actual t_{os} of a flip-flop to that determined through calibration. The results for each calibrated t_{os} are within acceptable percentage error levels. Note that the fitted σ values (not shown) were very close to 250 ps in all cases.

Table 24.1   Comparison of calibrated to actual offset from simulation

Actual offset (ps)	Calibrated offset (ps)	Error (%)
−35.0	−34.6	−1.1
−2.2	−2.3	4.5
5.0	5.3	6.0
17.4	17.1	−1.7

Table 24.2   Comparison of mean of absolute value of percentage errors from calibration of a flip-flop for different values of N

N	Error (%) (mean of 30 runs)
10^6	2.83
10^5	12.2
10^4	34.3
10^3	85.2

24.3.3.2   Calibration time

Calibration of the SOTDC using the method based on added noise depends on the number of values of ΔT used, denoted by M, the number of trials N run for each of these values and the rate f at which the trials are executed. Since all levels of the SOTDC can be characterised simultaneously for each ΔT, the calibration time t_{cal} is given by

$$t_{cal} = MN/f \qquad (24.7)$$

As an example, a calibration with $M = 21$, $N = 10^5$ and $f = 25$ MHz, takes 84.0 ms. Of course, additional time is needed to perform the curve-fitting procedure.

Since the calibration technique is based on a statistical method, decreasing the test time by reducing N will also reduce the accuracy in finding t_{os}. This fact is revealed in Table 24.2, in which the mean of the absolute value of the percentage errors from the simulated calibration of a flip-flop having $t_{os} = 2.0$ ps is determined for N varying from 10^6 down to 10^3.

Of course, test time can also be reduced by decreasing M. The minimum value of M required to accurately fit a Gaussian curve and find t_{os} is 2. In addition, practically speaking, these two points must occur within a $\pm 3\sigma_{added}$ range about the mean. This is because the extremities of the Gaussian cdf will equal exactly 0 or 1 as the absolute

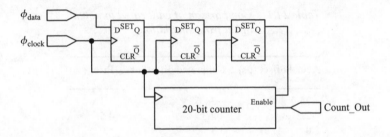

Figure 24.26 Single-level SOTDC synthesised on a CPLD

value of ΔT becomes large. Therefore, having points outside the $\pm 3\sigma_{added}$ region will produce a poor fit.

24.3.3.3 Preliminary experimental results

To gain preliminary experimental evidence in support of the proposed calibration technique, one level of an SOTDC was synthesised using an Altera EPM7128SLC84-7 complex programmable-logic device (CPLD). As shown in the schematic circuit in Figure 24.26, the left-most flip-flop performs the timing measurement by comparing the highly accurate reference ϕ_{clock} with the jittery signal ϕ_{data}. The two additional flip-flops attached in series to the main flip-flop are used to reduce the probability that a metastable value is latched by the counter. The 20-bit counter is enabled when the output from the third flip-flop is high.

To perform the direct calibration discussed in Section 24.3.2.3, the apparatus shown in Figure 24.27(a) was constructed. The CPLD containing the synthesised TDC resided on a multi-layer Altera UP1 development board. Bayonet nut coupling (BNC) connectors and 50-Ω terminating resistors were soldered onto the expansion slots of the board so that it could be interfaced with an Agilent 81334A dual-channel pulse/pattern generator. The latter device was used to produce two calibration clocks and highly control their phase relationship in order to generate the necessary ΔT. It should also be noted that the generator has a delay resolution of 1 ps, a delay accuracy of ± 20 ps and a typical total jitter of 2 ps rms [29] (although the latter specification was experimentally verified to be somewhat lower).

Calibration involving added noise was performed using the experimental setup in Figure 24.27(b). This is similar to that in (a), however an Agilent 33120A function generator was connected to the delay control input of the pulse generator. This allowed the phase of the clock signal to be varied according to the amplitude of a white Gaussian noise signal created by the function generator.

Under both calibration techniques, the clock and data frequencies were set to 20 MHz. The number of cycles considered for each value of ΔT was $N = 1\,048\,575$. In the direct calibration technique, ΔT was adjusted in 1-ps increments in order to find the flip-flop offset. In the technique based on added noise, ΔT was adjusted in increments of 40 ps. Finally, in the proposed technique, σ_{added} was found to be around 250 ps according to an external oscilloscope.

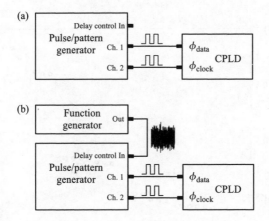

Figure 24.27 Experimental setups to perform TDC calibration. (a) Basic direct calibration technique, (b) calibration technique based on added noise

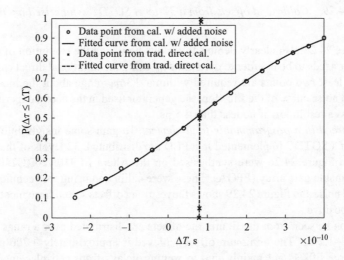

Figure 24.28 Experimental results from calibration of a single-level SOTDC synthesised on a CPLD

Results from the calibration techniques are shown in Figure 24.28. A t_{os} of -81 ps was found using the basic direct calibration while an offset of -71 ps was determined from the technique based on added noise. This results in an error of -12.3 per cent. The larger error compared to the simulation results is likely due to non-linearities in the delay control circuitry of the pulse generator. Such non-linearities can cause the variation in ΔT to not be perfectly Gaussian, which in turn can produce a bias error in the results. The fitted value of σ_{added} was 264 ps, which is within the expected range. In addition, the fitted value of σ_{FF} was approximately 0.75 ps.

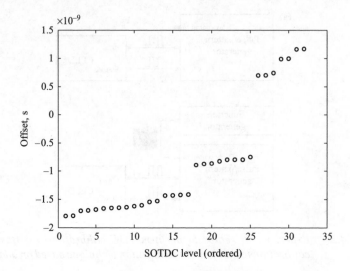

Figure 24.29   Calibrated offsets from a 32-level SOTDC synthesised on an FPGA

Figure 24.28 also clearly demonstrates just how high the resolution of the pulse generator needs to be in order to determine t_{os} using the traditional direct calibration. Since at least two points are required within a $\pm 3\sigma_{FF}$ range about the mean on the Gaussian noise curve of the flip-flop, the generator used in the above experiment had to possess a resolution of no less than 4.5 ps.

Offset spread in a programmable-logic device. To gain some insight into how the offsets of a SOTDC implemented in a PLD are distributed, 32 levels of the SOTDC shown in Figure 24.26 were synthesised on the Altera EPF10K20RC240-4 field-programmable gate array (FPGA). These were calibrated using the technique based on added noise and Figure 24.29 shows the resulting offsets sorted from most negative to most positive.

As can be seen from the figure, the offsets are distributed over a range of about -1.8 to $+1.2$ ns. The minimum offset achieved is approximately $+700$ ps. These rather large offsets are mainly due to routing delays from cell placement within the FPGA. Although modern FPGA software tools allow the designer to have some control over the location of cells within the device, this is nowhere near the control that can be had from a custom silicon implementation. Such an implementation will be described in the next section.

24.4   Custom IC implementation of a sampling offset TDC

To obtain better timing resolution than can be achieved using a PLD, a custom SOTDC was designed and fabricated in a standard TSMC 0.18-μm CMOS process. This chapter provides implementation details of the custom IC and includes

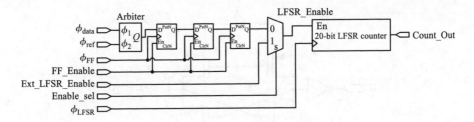

Figure 24.30 *Single level of custom flash SOTDC*

experimental results from calibration of this device. In addition, results from Gaussian jitter measurement are included.

24.4.1 Single level of the custom IC

Each level of the SOTDC was built as shown in Figure 24.30. The arbiter in the figure compares the phase of the highly stable reference clock ϕ_{ref} to that of the signal under test ϕ_{data}. Since the output from the arbiter is not held constant throughout the entire ϕ_{ref} cycle, a flip-flop is required to latch and hold the resulting arbiter decision. Clock ϕ_{FF}, which must be delayed in relation to ϕ_{ref} in order to give the arbiter sufficient time to settle, is used to clock the flip-flops. Note that a serial chain of three flip-flops has been placed after the arbiter in order to reduce the probability of metastable failure.

The output from the flip-flops is used to enable the 20-bit counter. A multiplexer has been placed between the last flip-flop of the serial chain and counter so that the source of the enable signal can be toggled. Signal Ext_LFSR_Enable is driven externally and is used to control the counter when its contents are being shifted out during calibration and measurement modes. The counter is clocked by signal ϕ_{LFSR}, which is in phase with ϕ_{FF}. These two clocks were physically separated on the IC so that greater control of the circuit was possible. More detailed descriptions of the arbiter and counter are described next.

24.4.1.1 Arbiter design

The arbiter displayed in Figure 24.31 [25] is used to compare the rising edge of the highly accurate reference signal ϕ_{ref} to that of ϕ_{data}. If ϕ_{data} arrives before ϕ_{ref}, positive feedback causes the output Q to go to '1'; otherwise it goes to '0'.

Dimensions of the arbiter transistors are shown in Table 24.3. All transistors were designed to have a length of 0.18 μm so that circuit speed was maximised. In addition, because transistor mismatches (and therefore temporal offsets) are exacerbated when smaller devices are used, the widths of the input transistors M_1 and M_2 were kept relatively small.

This particular arbiter architecture was chosen for the application at hand because of its inherent symmetry and simple structure. Other arbiter architectures used in single-ended time measurement systems may require, e.g. both ϕ_{data} and its complement to be present [10,22]. Since an extra inverter in the signal path is required for

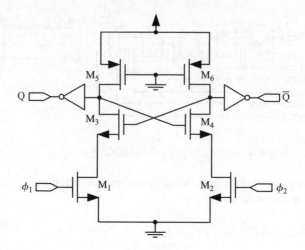

Figure 24.31   Arbiter used in custom SOTDC

Table 24.3   Arbiter transistor sizes

Transistor	Width (μm)
M_1/M_2	1.45
M_3/M_4	1.45
M_5/M_6	0.90

this, instantaneous delay variations in the inverter due to power supply fluctuations could skew the phases of this signal, leading to erroneous measurements even after calibration. In addition, the added inverter may cause the offsets to favour one of the inputs more than the other on average, instead of being distributed evenly around zero offset (assuming that this property is desired).

Effects of process variation: To gain some insight into how process variation affects the arbiter offset, the schematic arbiter was simulated as the width and length of transistor M_1 in Figure 24.31 were varied compared to its nominal values shown in Table 24.3. Table 24.4 displays the offset of the arbiter as the width of M_1 is increased relative to M_2. Note that these simulations were run with the data input connected to ϕ_1 and the reference connected to ϕ_2.

As the width of M_1 becomes larger, relative to M_2, Table 24.4 shows that the ϕ_1 input is favoured. This is expected, as a larger width causes M_1 to have a higher current drive than M_2 for the same input voltage level.

Table 24.5 shows the simulated arbiter offset as the length of M_1 is varied. The offset now favours ϕ_2 because the increased length of M_1 reduces the current drive of this transistor.

Table 24.4   Simulated arbiter offset as the width of M_1 *is varied*

Width of M_1 (μm)	Width of M_2 (μm)	% Variation	t_{os} (ps)
1.49	1.45	2.5	+0.6
1.52	1.45	5	+1.1
1.60	1.45	10	+2.1
1.74	1.45	20	+3.7

Table 24.5   Simulated arbiter offset as the length of M_1 *is varied*

Length of M_1 (μm)	Length of M_2 (μm)	% Variation	t_{os} (ps)
0.1845	0.1800	2.5	−0.5
0.1890	0.1800	5	−1.1
0.1980	0.1800	10	−2.1
0.2160	0.1800	20	−4.3

Although informative, the above simulations cannot fully predict the behaviour of the arbiter under process variation. Obviously, size aberrations of the other arbiter transistors also affect the offset. Therefore, a more extensive Monte Carlo analysis would have to be done to determine the offset contributed from these transistors. In addition, effects on transistor sizing due to such process phenomena as lateral diffusion and overetching [30] can only be fully characterised after a large number of devices have been manufactured and tested.

Arbiter layout: The arbiter was laid out as shown in Figure 24.32. Attempts were made to make the arbiter as symmetric as possible so that the offsets would be distributed around zero. This was done so that phase variation on both sides of ϕ_{ref} could be detected by the arbiter. However, guaranteeing such symmetry was not always possible due to routing constraints. The total area consumed by the arbiter is approximately 250 μm^2.

Because the arbiter layout is not perfectly symmetric about its horizontal axis, some offset will be present due to differences in transistor parasitics. To determine the extent of this offset, the arbiter layout was simulated with extracted capacitive parasitics. Figure 24.33 shows the output from the arbiter as ΔT is swept from 0 to +1 ps in steps of 0.1 ps. Inspection of the figure reveals that the offset is located somewhere between −0.2 and −0.3 ps. Therefore, the intrinsic offset due to layout asymmetries is not substantial.

24.4.1.2   Counter design

The 20-bit counter shown in Figure 24.30 is implemented as a linear feedback shift register (LFSR) [31]. To achieve maximal length (i.e. $2^{20} - 1$ distinct combination

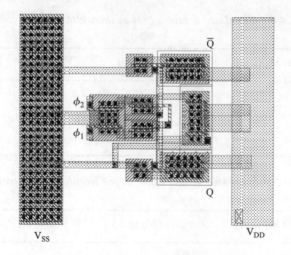

Figure 24.32   Arbiter layout

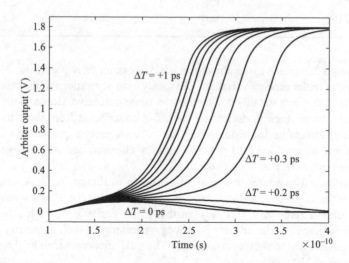

Figure 24.33   Simulation of arbiter layout with extracted capacitive parasitics as input ΔT is swept from 0 to +1 ps in 0.1-ps steps

of 20 bits in this case) using only two feedback taps, the primitive polynomial $x^{19} + x^{16} + 1$ was implemented. This results in the architecture shown in Figure 24.34, in which the output from the 20th and 17th flip-flops are input to the XOR gate in the feedback path.

The multiplexer shown in the feedback path of the LFSR is used to select the source of the feedback signal. During regular operation, this signal comes from the XOR gate. However, when the data must be serially read out from the counter to analyse the measurements, the feedback signal originates from the Count Out in the

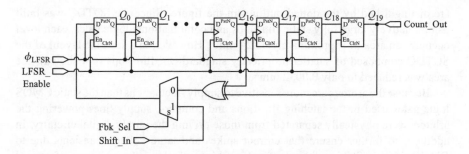

Figure 24.34 Twenty-bit LFSR counter used in custom SOTDC

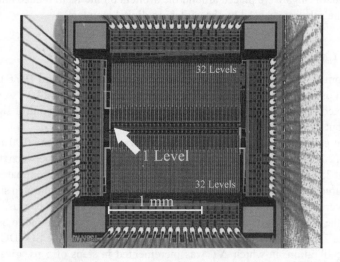

Figure 24.35 A 64-level SOTDC implemented in a TSMC 0.18-μm *CMOS process*

previous level of the SOTDC. This allows all the counters in the SOTDC to be daisy-chained together, forming a serial scan chain which can be read out via a single pin on the IC.

An LFSR was implemented instead of a regular ripple-carry counter because the former employs fewer logic gates and has a simpler routing complexity. In addition, the 20-bit LFSR can be operated at a higher speed than a ripple-carry counter because its critical path consists of only a single XOR gate (neglecting the presence of the feedback multiplexer). The drawback to the LFSR implementation is that the counter output must be decoded using a lookup table containing all LFSR states. Fortunately, for the purposes of production testing, this table can exist on an external computer.

24.4.1.3 Chip layout

The fabricated layout of the custom SOTDC chip is displayed in Figure 24.35. The IC contains two identical sections, each consisting of 32 converter levels. These are distinguished by the dotted outlines in the figure. Since the SOTDC was built using relatively large fully static flip-flops and combinational logic gates, each level occupies an area of approximately 0.024 mm^2. However, in a separate layout of the SOTDC composed of smaller proprietary standard-cell flip-flops (not shown), the area was reduced to only 0.0032 mm^2.

Because the arbiters are more sensitive to supply fluctuations than the static CMOS logic gates used in the latching flip-flops and counters, supply lines powering the arbiters were physically separated from those driving the purely digital circuitry. In addition, to further ensure that current spikes and large voltage transients due to CMOS circuit switching did not affect the arbiters, the latter devices were referenced to 'analogue' ground while the counters were connected to a separate 'digital' ground. Finally, guard rings were placed around the arbiters on the IC to reduce interference through the substrate caused by the switching digital counters.

24.4.2   Calibration of the custom IC

The custom IC was mounted on a two-layer printed-circuit board (PCB) and was calibrated using a test apparatus similar to that shown in Figure 24.27. A Teradyne A567 mixed-signal production tester was used to assert digital control signals and store output from the IC.

For the calibration based on added noise, ΔT was increased from -100 to $+100$ ps in steps of 20 ps. This was done by controlling the phase difference between two clocks running at 25 MHz. The standard deviation of added noise was approximately 29.8 ps (as measured using an external oscilloscope) and $N = 10^5$. Both the step size of ΔT and σ_{FF} were reduced compared to the test setup in Section 24.3.3.3 because the linear range of the delay control circuitry in the current apparatus was found to be smaller.

Figure 24.36 shows the calibration results for one level of the SOTDC. The traditional calibration, in which ΔT was incremented in steps of 5 ps, gives a t_{os} of 9.78 ps while the proposed method produces a t_{os} of 9.80 ps. Best-fit σ using the former method is 9.28 ps and that for the latter is 30.38 ps. Note that the larger σ present in the traditional calibration in the current test setup, compared to the CPLD results, is due to the presence of noise and interference on the PCB. These should be significantly reduced when the IC is tested on a multi-layer board (as in Section 24.3.3.3), in which the signal, supply and ground planes are isolated.

Figure 24.37 displays the calibrated offsets for 32 levels in one section of the custom IC. Results from both calibration methods are included and these are ordered from the smallest to the largest. From the technique based on added noise, the offsets are distributed from about $+3$ to $+16$ ps, giving a dynamic range of 13 ps.

The percentage error between each offset, calibrated in the traditional way and using the proposed technique, is shown in Figure 24.38. The error on only two offsets exceeds 30 per cent and the average of the absolute value of the percentage errors is 14 per cent. Such an error is reasonable considering the fact that direct measurements using a source without picosecond resolution would produce much higher errors.

It would have been desirable to have the arbiter offsets distributed around zero. This is so that deviations in the edge placement of ϕ_{data} on both sides of the ϕ_{ref} edge

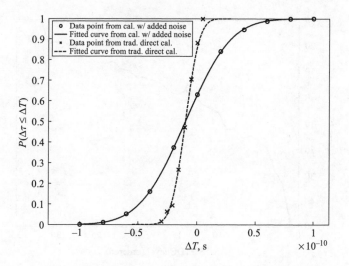

Figure 24.36   *Calibration of a single level of the custom SOTDC*

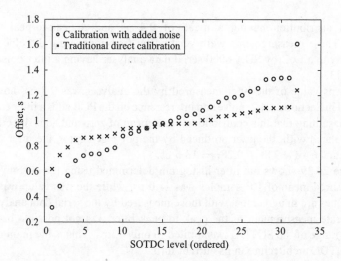

Figure 24.37   *Ordered offsets from one section of the custom IC determined using traditional and proposed calibration methods*

could be detected by the SOTDC. Unfortunately, asymmetries in the original arbiter layout, produced while designing the IC, likely caused all the offsets to favour the same input.

24.4.3   Jitter measurement using the custom IC

The custom SOTDC was used to measure Gaussian jitter. Jitter was produced by varying the phase difference between two 25-MHz clocks according to a

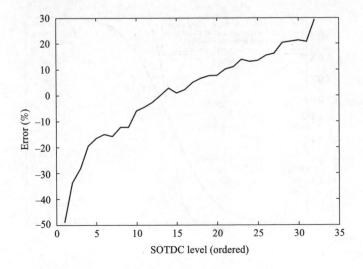

Figure 24.38   Percentage error for each offset

Gaussian distribution having a mean of 9.4 ps and a peak-to-peak value of 79.2 ps. These measurements were verified by first connecting the clock generator directly to a LeCroy SDA 6000 serial data analyser having a jitter noise floor of 1 ps rms [32].

The rms value of the jitter measured by the analyser was 9.8 ps, however, as discussed in Section 24.4.2, noise and interference on the PCB adds a jitter component to this. Assuming that this contribution has a mean of zero and adds directly to, and is uncorrelated with, the jitter produced by the pulse generator, the jitter under test has an actual σ of $\sqrt{9.8^2 + 9.28^2} = 13.5$ ps.

Figure 24.39 shows the jitter histogram determined using the custom SOTDC. The measured mean of 10^5 samples was 11.0 ps, while the rms value was 14.5 ps. These values are in agreement with those measured by the serial data analyser.

Accurate measurement of the peak-to-peak jitter was not possible because the dynamic range of the SOTDC was limited to only 13 ps. This is the main drawback of the SOTDC architecture in its current form.

24.4.3.1   Test time

Test time for jitter measurement is equal to N/f. Therefore, the measurement carried out in Section 24.4.3 took only 4.0 ms. This short test time is characteristic of flash TDCs.

24.4.4   *Performance limitations*

Noise coupled in from the PCB ground planes affects the accuracy of the SOTDC. This is because an increase in such noise raises the value of σ_{FF} of each arbiter. For arbiters that have small differences in t_{os}, the Gaussian curves describing their switching

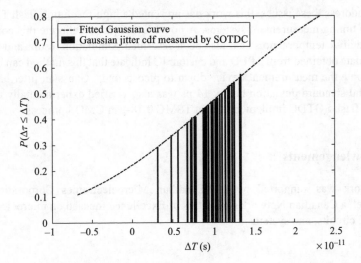

Figure 24.39   Gaussian jitter histogram measured using the custom SOTDC. A Gaussian cdf has been fit to the data

characteristics might start to overlap, causing the converter to output an incorrect code for a particular value of ΔT. Fortunately, averaging over many cycles of ΔT can help reduce the error caused by the added noise. This shows the fundamental trade-off between accuracy and test time inherent in the SOTDC.

The accuracy of the SOTDC is also affected by power supply fluctuations. When these occur, variations in current drive in the arbiter transistors can cause the effective arbiter offset to shift slightly from its nominal value. This is why supply decoupling and the use of regulated power supplies is important when calibrating and using the SOTDC.

Finally, the ability of the SOTDC to measure jitter accurately is also limited by the stability of the reference clock. A clock displaying a high degree of phase variation will undoubtedly introduce errors in the measurement of the signal under test. The availability of a highly controlled reference clock is therefore a fundamental constraint when using the SOTDC.

24.5 Conclusions

On-chip measurement of electrical phenomena using custom IC test cores is an attractive method for overcoming the bandwidth limitations and interconnect timing-related uncertainties common in traditional off-chip test systems. However, to accurately characterise jitter in today's high-performance SoCs, the resolution of on-chip measurement devices must be significantly increased from that of the current state-of-the-art. In addition, new calibration techniques are required to ensure high accuracy in these measurement systems.

To address these issues, this work has presented a high-resolution flash TDC for on-chip timing measurement systems. A novel technique to calibrate this converter using additive temporal noise has also been described. Simulation results and experimental data obtained from a PLD and custom IC indicate that this method can be used to calibrate the measurement device down to picoseconds. Gaussian jitter having a mean and standard deviation below 14 ps was also verified experimentally using a custom flash SOTDC implemented in a TSMC 0.18-μm CMOS process.

Acknowledgements

This work was supported by the Canadian Microelectronics Corporation and Micronet, a Canadian Network of Centres of Excellence focused on microelectronic devices, circuits, and systems.

References

1   WALDEN, R.H.: 'Analog-to-digital converter survey and analysis', *IEEE Journal on Selected Areas in Communications*, 1999, **17**(4), pp. 539–550

2   GASBARRO, J.A., and HOROWITZ, M.A.: 'Integrated pin electronics for VLSI functional testers', *IEEE Journal on Solid-State Circuits*, 1989, **24**(2), pp. 331–337

3   HAFED, M., ABASKHAROUN, N., and ROBERTS, G.W.: 'A stand-alone integrated test core for time and frequency domain measurements'. Proceedings of the international *Test* conference, October 2000, pp. 1031–1040

4   HAFED, M.M., ABASKHAROUN, N., and ROBERTS, G.W.: 'A 4-GHz effective sample rate integrated test core for analog and mixed-signal circuits', *IEEE Journal on Solid-State Circuits*, 2002, **37**(4), pp. 499–514

5   WEINLADER, D.K.: 'Precision CMOS receivers for VLSI testing applications'. Ph.D. dissertation, Stanford University, Palo Alto, CA, USA, November 2001

6   SULLIVAN, D.B., ALLAN, D.W., HOWE, D.A., and WALLS, F.L.: 'Characterization of clocks and oscillators'. NIST, Technical Report, 1337, 1990

7   'Jitter fundamentals: Agilent N4900 Serial BERT Series Jitter injection and analysis capabilities'. Agilent Technologies, July 2003, application note

8   O'MAHONY, F., YUE, C.P., HOROWITZ, M.A., and WONG, S.S.: 'A 10-GHz global clock distribution using coupled standing-wave oscillators', *IEEE Journal on Solid-State Circuits*, 2003, **38**(11), pp. 1813–1820

9   CHAN, A.H., and ROBERTS, G.W.: 'A jitter characterization system using a component-invariant Vernier delay line', *IEEE Transactions on Very Large Scale Integration (VLSI) Systems*, 2004, **12**(1), pp. 79–95

10 WEINLADER, D., HO, R., YANG, C.K.K., and HOROWITZ, M.: 'An eight channel 35 GSample/s CMOS timing analyzer'. IEEE international *Solid-state circuits* conference digest of technical papers, February 2000, pp. 170–171

11 NELSON, B., and SOMA, M.: 'On-chip calibration technique for delay line based BIST jitter measurement'. Proceedings of the IEEE international symposium on *Circuits and systems*, May 2004, pp. 944–947

12 RESTLE, P.J., FRANCH, R.L., JAMES, N.K., *et al.*: 'Timing uncertainty measurements on the Power5 microprocessor'. IEEE international *Solid-state circuits* conference digest of technical papers, February 2004, pp. 354–361

13 TABATABAEI, S., and IVANOV, A.: 'Embedded timing analysis: a SoC infrastructure', *IEEE Design and Test of Computers*, 2002, **19**(3), pp. 22–34

14 HAFED, M.M., and ROBERTS, G.W.: 'A stand-alone integrated excitation/extraction system for analog BIST applications'. Proceedings of the IEEE 2000 *Custom integrated circuits* conference, May 2000, pp. 83–86

15 ZHENG, Y., and SHEPARD, K.L.: 'On-chip oscilloscopes for noninvasive time-domain measurement of waveforms in digital integrated circuits', *IEEE Transactions on Very Large Scale Integration (VLSI) Systems*, 2003, **11**(3), pp. 336–344

16 STEVENS, A.E., VAN BERG R.P., VAN DER SPIEGEL, J., and WILLIAMS, H.H.: 'A time-to-voltage converter and analog memory for colliding beam detectors', *IEEE Journal on Solid-State Circuits*, 1989, **24**(6), pp. 1748–1752

17 RÄISÄNEN-RUOTSALAINEN, E., RAHKONEN, T., and KOSTAMOVAARA, J.: 'A time digitizer with interpolation based on time-to-voltage conversion'. Proceedings of the 40th Midwest symposium on *Circuits and systems*, August 1997, pp. 197–200

18 SUMNER, R.L.: 'Apparatus and method for measuring time intervals with very high resolution'. US Patent 6 137 749, October 24, 2000

19 ABAS, A.M., BYSTROV, A., KINNIMENT, D.J., MAEVSKY, O.V., RUSSELL, G., and YAKOVLEV, A.V.: 'Time difference amplifier', *Electronics Letters*, 2002, **38**(23), pp. 1437–1438

20 KINNIMENT, D.J., BYSTROV, A., and YAKOVLEV, A.V.: 'Synchronization circuit performance', *IEEE Journal on Solid-State Circuits*, 2002, **37**(2), pp. 202–209

21 OULMANE, M., and ROBERTS, G.W.: 'A CMOS time amplifier for femtosecond resolution timing measurement'. Proceedings of the IEEE international symposium on *Circuits and systems*, May 2004, pp. 509–512

22 GRAY, C.T., LIU, W., VAN NOIJE, W.A.M., HUGHES, T.A., JR., and CAVIN, R.K., III: 'A sampling technique and its CMOS implementation with 1 Gb/s bandwidth and 25 ps resolution', *IEEE Journal on Solid-State Circuits*, 1994, **29**(3), pp. 340–349

23 ABASKHAROUN, N., HAFED, M., and ROBERTS, G.W.: 'Strategies for on-chip sub-nanosecond signal capture and timing measurements'. Proceedings of the IEEE international symposium on *Circuits and systems*, May 2001, pp. 174–177

24 DUDEK, P., SZCZEPAŃSKI, S., and HATFIELD, J.: 'A high-resolution CMOS time-to-digital converter utilizing a Vernier delay line', *IEEE Journal on Solid-State Circuits*, 2000, **35**(2), pp. 240–247

25 GUTNIK, V., and CHANDRAKASAN, A.: 'On-chip picosecond time measurement'. Symposium on *VLSI circuits* digest of technical papers, June 2000, pp. 52–53

26 GUTNIK, V.: 'Analysis and characterization of random skew and jitter in a novel clock network'. Ph.D. dissertation, Massachusetts Institute of Technology, Cambridge, MA, USA, 2000

27 LIAN, W., and CHENG, X.S.: 'Analytical model of noise of a switched flip-flop', *Electronics Letters*, 1988, **24**(21), pp. 1317–1318

28 LEON-GARCIA, A.: 'Probability and random processes for electrical engineering' (Addison-Wesley, Reading, MA, 1994, 2nd edn.)

29 Agilent Technologies: 'Agilent Technologies 81133A and 81134A, 3.35 GHz Pulse/Pattern Generators'. [Online] Available: http://cp.literature.agilent.com/litweb/pdf/5988-5549EN.pdf, June 2003

30 JOHNS, D., and MARTIN, K.: 'Analog integrated circuit design' (Wiley, New York, NY, 1997)

31 CLARK, D.W., and WENG, L.-J.: 'Maximal and near-maximal shift register sequences: efficient event counters and easy discrete logarithms', *IEEE Transactions on Computers*, 1994, **43**(5), pp. 560–568

32 LeCroy Corporation: 'LeCroy Serial Data Analyzers Datasheet'. [Online]. Available: http://www.lecroy.com/tm/products/Analyzers/SDA/SDA_Datasheet.pdf, 2003

Chapter 25

Yield and reliability prediction for DSM circuits

Thomas S. Barnett and Adit D. Singh

25.1 Introduction

Yield models for integrated circuits have been in use since the beginning of the semiconductor industry. These models not only allow us to estimate the number of functional integrated circuits before these circuits are manufactured, but also to monitor and improve the manufacturing process. Accurate yield projection is essential in order to ensure that an adequate number of circuits are produced for the customer in a timely fashion. As such, successful yield models play a key role in determining the economic success of semiconductor manufacturers.

From the customer's perspective, it is the continued functionality, or reliability, of the integrated circuit that is of primary importance. Indeed, because integrated circuits are generally built into larger systems, the failure of a single component may result in the failure of the entire system [1–4]. If the individual component that caused the failure cannot be identified and replaced, then the entire system must be replaced. The cost of a reliability failure can therefore be significantly greater than the cost of the individual integrated circuit or circuits that caused the failure. For this reason, companies that produce an unreliable product are not likely to stay in business very long. Integrated circuit manufacturers will therefore go to great lengths to ensure that customer reliability requirements are met.

One way to ensure highly reliable integrated circuits is to subject them to accelerated life tests. For integrated circuits the most common accelerated life test is known as burn-in. Burn-in is a test that attempts to age a device by subjecting it to high-stress conditions such as high temperature and high voltage. Under these conditions, manufacturing defects, too subtle for detection at initial testing, will tend to become more severe and produce circuit failures. Moreover, the population of devices that survive burn-in conditions are unlikely to contain manufacturing defects, and thus, the product shipped to the customer should be of high quality.

Unfortunately, burn-in has become prohibitively expensive for modern integrated circuits. Indeed, the burn-in boards (BIBs) that apply voltage to the IC during burn-in testing are generally in the range of $100 000–$200 000 per set, and burn-in ovens can approach $1 000 000 each. Since several sets of BIBs and several ovens may be required for each product, the overall capital cost can easily run into the millions of dollars, even for relatively low volume products. Moreover, since BIBs cannot generally be used on multiple products, new BIBs often must be purchased with the introduction of each new product. The enormous financial implications of burn-in have left the semiconductor industry aggressively searching for alternatives to burn-in and/or methods to reduce burn-in.

Note that the traditional burn-in methodology treats all die as equal. Thus, once a burn-in duration is determined, all die are subjected to the same burn-in. If, however, the reliability of a die could be identified prior to stress testing, then die that present the greatest reliability risk could be subjected to longer burn-in durations. Conversely, the burn-in duration for the most reliable die could be significantly reduced or eliminated entirely. This approach would significantly reduce overall burn-in durations and therefore dramatically reduce testing costs while maintaining outgoing reliability requirements.

The key to the burn-in methodology described above is in the identification and quantification of a die's reliability. Indeed, while anecdotal evidence has long suggested a connection between yield and reliability, yield and reliability models were never explicitly connected until Huston and Clarke's [5] work in the early 1990s. Therein the authors suggest that those defects that cause wafer probe failures are essentially the same types of defects that cause early-life reliability failures; these defects can be distinguished only by size and location. This has since been verified with large experimental studies by several independent researchers [6–9].

Over the last few years, the modelling for yield and reliability has matured, and now accounts for the clustering of defects over semiconductor wafers [10]. As this chapter will show, incorporating defect clustering resulted in an integrated yield-reliability model that is completely consistent with long-standing yield models, and, at the same time, accurately predicts the early-life reliability of individual die. Applications of the model will be presented that allow one to identify die of varying reliability. When used to reduce burn-in, this information can have very significant financial implications for the semiconductor industry.

25.2   Defect-based yield models

Defect-based yield models for integrated circuits have been used within the semiconductor industry since the early 1960s. These models attempt to describe the distribution of defects over the semiconductor wafer. Once the defect distribution is specified, one can calculate the probability that an individual die contains $0, 1, 2, \ldots$

defects. The yield is then given by the probability of 0 defects.[1] While many different models have been proposed [11–13], only the Poisson model and the negative binomial model are commonly in use today.[2] Of the two models, the Poisson model is the simplest in that it can be described by a single parameter λ, which denotes the average number of defects per chip. The negative binomial distribution, on the other hand, requires the specification of two parameters: the average number of defects per chip, λ, and the clustering parameter α. As the name implies, the clustering parameter describes the degree to which defects over the wafer cluster or group together.

25.2.1 Poisson statistics

The Poisson distribution is used widely to model many different physical phenomena. Examples include radioactive decay, the incoming calls at a telephone exchange and the distribution of bombs dropped on London during the Second World War [14]. The utility of the Poisson distribution lies in its ability to describe the occurence of random events. Here the term random simply denotes the fact that the occurence of one event does not affect the outcome or occurrence of future events. In the radioactive decay of Helium, for example, the number of alpha particles ejected from the nucleus per unit time is random in the sense that the number ejected in one time interval does not depend on the number obtained in previous time intervals. From the point of view of yield statistics, the event of interest is the number of physical defects present on a wafer or an individual die. Thus, if one describes the distribution of these defects with a Poisson distribution, one is assuming the number of defects on one die is independent of the number on any other die.

Mathematically, the Poisson distribution specifies the probability that a die contains a certain number of defects. In particular, if N is a random variable denoting the number of defects present in a die, then the event of exactly q defects in a die, written $N(q)$, has the probability given by

$$P[N(q)] = \frac{\lambda^q}{q!} \exp(-\lambda) \tag{25.1}$$

where λ denotes the average number of defects per chip. From this equation the yield is defined as the probability of 0 defects. If the yield is denoted by Y, then

$$Y = P[N(0)] = \exp(-\lambda) \tag{25.2}$$

Figure 25.1 shows the Poisson distribution for various values of yield. Note that while these curves are shown as continuous, actual defect counts must be discrete.

---

[1] An exception to this occurs for defect tolerant integrated circuits. Such designs can tolerate a certain number of defects and still function. This will be discussed in detail in Section 25.5.

[2] It should be noted that the Poisson model is actually a limiting case of the negative binomial model. However, the yield literature often treats them as separate models, and this approach is taken here.

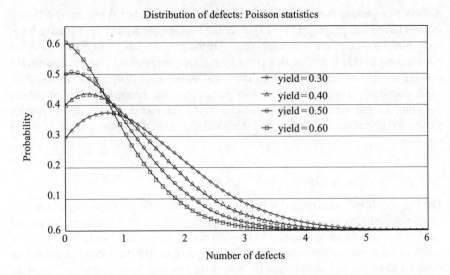

Figure 25.1   Defect distribution when defects follow Poisson statistics. Various values of the yield are shown

Further, each curve corresponds to a different value of λ; the lower the yield, the higher the value of λ. Thus, λ is 1.2, 0.92, 0.69 and 0.51 when the yield is 0.30, 0.40, 0.50 and 0.60, respectively. Moreover, the lower the yield, the more likely it is that a die contains multiple defects. For example, the probability of three defects on a die is 0.087, 0.051, 0.028 and 0.013 when the yield is 0.30, 0.40, 0.50 and 0.60, respectively.

The simplicity of the Poisson equation makes it very easy to use for calculation. Of course, the value of any model rests in its ability to represent experimental data. In this respect, the Poisson model, while simple, has long been known to underestimate the yield. The reason for this lies in the fact that defects over semiconductor wafers are not randomly distributed but have a tendency to cluster. To illustrate the effect of defect clustering, suppose that 100 defects are distributed over a wafer containing 100 die. The average number of defects per chip is simply $\lambda = 100/100 = 1.0$. According to Poisson statistics the yield is $\exp(-1) = 0.37$. If these 100 defects are highly clustered together, however, the yield can be drastically different. For example, if the 100 defects are contained within 10 die and the remaining 90 die are defect-free, then λ is still 1.0, but the yield is now 90 per cent. Defect clustering can therefore significantly impact yield. This fact has led researchers to explore distributions that can account for such behaviour. The most successful of these has been the negative binomial distribution.

25.2.2   Negative binomial statistics

The use of the negative binomial distribution for projecting die yields was introduced in the early 1980s by Charles Stapper of IBM [11]. It was specifically introduced

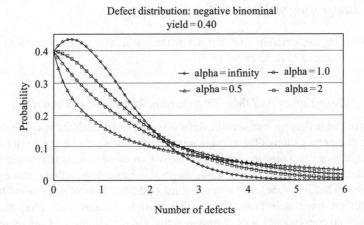

Figure 25.2   Defect distribution when defects follow negative binomial statistics.
Various values of the clustering parameter α are shown. Note that the
yield is 0.40

for its ability to describe the clustering of defects. To describe this effect, however, requires an additional parameter, appropriately called the clustering parameter.

Assuming defects get distributed according to negative binomial statistics, the probability that there are exactly q defects over a die is given by

$$P[N(q)] = \frac{\Gamma(\alpha + q)}{q! \, \Gamma(\alpha)} \frac{(\lambda/\alpha)^q}{(1 + \lambda/\alpha)^{\alpha+q}} \tag{25.3}$$

where $\Gamma(x)$ is the Gamma function, λ is the average number of defects per die and α is the clustering parameter. Values of α typically range from 0.5 to 5 for different fabrication processes; the smaller values indicate increased clustering. As $\alpha \to \infty$ the negative binomial distribution becomes a Poisson distribution, which is characterised by no clustering [11,15–17].

Figure 25.2 shows the negative binomial distribution for the clustering parameters $\alpha = 0.5, 1.0, 2$ and ∞. The yield is 0.40 in each case. $\alpha = 2$ represents a typical value that may be seen in industry today. Notice that the greater the clustering (lower α), the less likely a die is to have a small number of defects. For example, the probability of exactly one defect is 0.17, 0.24, 0.29 and 0.37 for $\alpha = 0.5, 1.0, 2$ and ∞. However, the probability that a die contains a large number of defects increases with a high degree of clustering. Thus, the probability of exactly four defects is 0.054, 0.052, 0.036 and 0.012 for $\alpha = 0.5, 1.0, 2$ and ∞. This is consistent with an intuitive notion of clustering; when defects are highly clustered together, a faulty die is more likely to contain multiple defects, rather than a small number of defects. Conversely, with weak clustering, most die contain about the same number of defects, and few die will have defect counts that differ significantly from the mean.

25.3   Integrating yield-reliability modelling

The modelling presented in this section and the application sections that follow were original developed in References 18–21.

25.3.1   *Defect types and their distribution: killer and latent defects*

As mentioned in the previous section, defect-based yield models for integrated circuits require the specification of the average number of defects per chip, generally denoted by λ. Traditionally, these models have focused on those defects that cause failures detectable at wafer probe testing, while neglecting those defects that cause early-life or reliability failures. One can, however, think of defects to be of three possible types: killer defects, latent defects and defects that cause no failures at all. The latter of the three is of no consequence with regard to actual circuit failures, and can therefore be neglected. Thus, one can write

$$\lambda = \lambda_K + \lambda_L \tag{25.4}$$

where λ_K is the average number of killer defects and λ_L is the average number of latent defects. An example of both defect types is shown in Figure 25.3, where a particle defect obstructs a metal line. If the particle completely obstructs the line, an open circuit results. This can be detected at wafer test. If however, the particle only partially obstructs the line, then the metal wire may function sufficiently at initial wafer probe testing. Under stress conditions, however, current densities in the thinned wire may grow large enough to 'blow' the line, much like an electrical fuse. Figure 25.3 implies that defects that cause failures detectable at wafer probe are fundamentally the same in nature as those which cause early-life reliability failures; size and placement being the primary distinguishing features. This has led researchers to assume that λ_L is

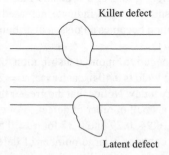

Figure 25.3   *A metal wire is obstructed by two types of defects. The first type, called a killer defect, causes an open circuit. This defect is detectable at wafer probe. The second type, called a latent defect, only partially obstructs the metal line. Such a defect may pass wafer probe but cause an early life failure*

linearly related to λ_K [5–9]. Under this assumption one may write

$$\lambda_L = \gamma \lambda_K \tag{25.5}$$

where γ is a constant. This has recently been shown to agree well with experiments conducted at Intel over a wide range of yield values [8]. In this study, γ was shown to fall within the range of 0.01–0.02. That is, for every 100 killer defects, one expects, on average, 1–2 defects to result in latent faults. While the actual value of γ is expected to be process dependent, these values provide a useful order of magnitude estimate.

Now imagine that an experiment consists of placing a single defect on an integrated circuit. The outcome of this experiment is either a killer or latent defect. If these defects occur with probabilities p_K and p_L, respectively, then a series of q such experiments will follow a binomial distribution. Thus, if $K(m)$ denotes the event of exactly m killer defects and $L(n)$ the event of exactly n latent defects, then, given a total of q defects, the probability of m killer and n latent defects is given by

$$P[K(m), L(n)|N(q)] = \binom{q}{m} p_K^m \, p_L^n \tag{25.6}$$

where $q = m + n$ and $p_K + p_L = 1$. Note that Equation (25.6) implies that the average number of latent defects is $\lambda_L = qp_L$. Similarly, $\lambda_K = qp_K$. Thus, $\lambda_L = (p_L/p_K)\lambda_K$. But from Equation (25.5), $\lambda_L = \gamma \lambda_K$. It follows that $\gamma = p_L/p_K$. Combining this with the equation $p_K + p_L = 1$ relates the probabilities for latent and killer defects to the parameter γ. That is,

$$p_L = \left(\frac{\gamma}{1+\gamma}\right) \quad \text{and} \quad p_K = \left(\frac{1}{1+\gamma}\right) \tag{25.7}$$

Thus, for $\gamma = 0.01$, $p_L \approx 0.0099$ and $p_K \approx 0.9901$.

Equation (25.6) specifies the probability of m killer and n latent defects given q defects. If the value of q is not known, one must specify its probability as well. To do this, and to account for the observed clustering of defects, one assumes that the defects are distributed according to negative binomial statistics [11,15–17,22,23]. $P[N(q)]$ is therefore given by Equation (25.3).

Combining Equations (25.6) and (25.3) gives the probability of exactly m killer and n latent defects. Specifically, with $P[K(m), L(n)] = P[K(m), L(n)|N(q)] \times P[N(q)]$ one can write

$$P[K(m), L(n)] = \frac{\Gamma(\alpha + m + n)}{m!\, n!\, \Gamma(\alpha)} \frac{(\lambda_K/\alpha)^m (\lambda_L/\alpha)^n}{(1 + \lambda/\alpha)^{\alpha + m + n}} \tag{25.8}$$

where $\lambda_K = p_K\lambda$, $\lambda_L = p_L\lambda$ and $\lambda = \lambda_K + \lambda_L$. Thus, the probability that a chip contains zero killer and zero latent defects is given by

$$Y = P[K(0), L(0)] = \left(1 + \frac{\lambda}{\alpha}\right)^{-\alpha} \tag{25.9}$$

This is the fraction of chips that are functional following both wafer probe and stress testing.

25.3.2   Wafer probe yield

Although Equation (25.8) gives the probability of m killer and n latent defects, it is often convenient to consider killer defects separately. Thus, to obtain the probability of exactly m killer defects, $P[K(m)]$, regardless of the number of latent defects, one can sum $P[K(m), L(n)]$ over n. That is,

$$P[K(m)] = \sum_{n=0}^{\infty} P[K(m), L(n)] \tag{25.10}$$

Substituting Equation (25.8) into (25.10) and using the identity $(\Gamma(\beta + n))/(n!\,\Gamma(\beta)) = (-1)^n \binom{-\beta}{n}$ allows one to write the summation as a power series of the form $A \sum_{n=0}^{\infty} \binom{-\beta}{n}(x)^n = A(1-x)^{-\beta}$. The probability of exactly m killer defects can then be written as

$$P[K(m)] = \frac{\Gamma(\alpha + m)}{m!\,\Gamma(\alpha)} \frac{(\lambda_K/\alpha)^m}{(1 + \lambda_K/\alpha)^{\alpha+m}} \tag{25.11}$$

where $\lambda_K = p_K \lambda$. Thus, the number of killer defects follows a negative binomial distribution with parameters (λ_K, α). For $m = 0$ Equation (25.11) gives

$$Y_K = P[K(0)] = \left(1 + \frac{\lambda_K}{\alpha}\right)^{-\alpha} \tag{25.12}$$

This formula is the semiconductor industry standard for predicting wafer probe yields of modern integrated circuits.

25.3.3   Reliability yield

After defining the wafer probe yield as $Y_K = P[K(0)]$, one may be tempted to define the reliability yield similarly as the probability of zero latent defects, $Y_L = P[L(0)]$. This definition, however, is not correct. Indeed, while $P[L(0)]$ does give the probability of zero latent defects, it says nothing about the number of killer defects. Thus, a die containing zero latent defects may still contain one or more killer defects. Such a die will be discarded following wafer testing and not subject to stress testing. To incorporate killer defect information when defining reliability yield, one can calculate the probability of n latent defects given m killer defects, denoted by $P[L(n)|K(m)]$. Using Bayes' Rule $P[K(m), L(n)] = P[L(n)|K(m)]P[K(m)]$ along with Equations (25.8) and (25.11), one can write

$$P[L(n)|K(m)] = \frac{\Gamma(\alpha + m + n)}{n!\,\Gamma(\alpha + m)} \frac{[\lambda_L(m)/(\alpha + m)]^n}{[1 + \lambda_L(m)/(\alpha + m)]^{\alpha+m+n}} \tag{25.13}$$

where $\lambda_L(m) = ((\alpha + m)/\alpha)\gamma\lambda_K/(1 + (\lambda_K/\alpha)) = ((\alpha + m)/\alpha)\lambda_L(0)$ is the average number of latent defects given that there are m killer defects. Thus, $P[L(n)|K(m)]$ follows a negative binomial distribution with parameters $(\lambda_L(m), \alpha + m)$. Defining the reliability yield Y_L as the number of die that are functional following stress divided by the number of die that passed wafer probe testing, one can write $Y_L = P[L(0)|K(0)]$. Thus,

$$Y_L = \left(1 + \frac{\lambda_L(0)}{\alpha}\right)^{-\alpha} \tag{25.14}$$

This is the fraction of die that survive stress testing. Using $\lambda_L = \gamma\lambda_K$ and solving Equation (25.12) for λ_K allows one to write $\lambda_L(0) = \gamma\alpha(1 - Y_K^{1/\alpha})$. Thus, Equation (25.14) may be rewritten as

$$Y_L = [1 + \gamma(1 - Y_K^{1/\alpha})]^{-\alpha} \tag{25.15}$$

Note that Y_K and α are obtained from the results of wafer probe testing, and thus γ is the only unknown parameter in Equation (25.15). γ may be obtained either from the statistical analysis of stress data or from direct calculation. A direct calculation of γ is carried out by considering the details of the circuit layout. This method relies on the calculation of a reliability critical area [5]. In all but the simplest cases, analytical solutions are not possible, and the critical area is obtained through Monte Carlo simulation [13].

   An important limiting case of Equation (25.14) occurs for $\alpha \to \infty$. In this limit $Y_K \to \exp(-\lambda_K)$, $Y_L \to \exp(-\lambda_L(0))$ and $\lambda_L(0) \to \lambda_L = \gamma\lambda_K$. Thus,

$$Y_L \to \exp(-\lambda_L) = \exp(-\gamma\lambda_K) = Y_K^{\gamma} \tag{25.16}$$

This is the expected reliability yield when defects follow a Poisson distribution.

25.4 Numerical results

Figure 25.4 shows the stress failure probability $P_f = (1 - Y_L)$ in per cent as a function of α for various values of Y_K. The parameter $\gamma = 0.015$. Note that clustering can have a significant impact on the probability of failure, particularly for the lower values of Y_K. For example, when $Y_K = 0.30$ and $\alpha = 0.5$, the stress failure probability is 0.676 per cent. Yet when $\alpha = \infty$ (no clustering), Equation (25.16) implies a failure probability of 1.79 per cent. Thus, the number of stress failures predicted by the Poisson model ($\alpha = \infty$) is $1.79/0.676 = 2.65$ times greater than the number predicted with $\alpha = 0.5$ (highly clustered). Note, however, that this ratio decreases as clustering decreases (α increases). Hence, for the more typical value of $\alpha = 2$ and the same wafer probe yield, $Y_K = 0.30$, this ratio falls to $1.79/1.34 = 1.34$.

   The effect of defect clustering is also seen to diminish as Y_K increases. For example, consider the case when $Y_K = 0.70$. For $\alpha = 2$, this gives a stress failure probability of 0.488, while for $\alpha = \infty$, the value is 0.534. The ratio is therefore $0.534/0.488 = 1.09$, and the Poisson model prediction is only slightly different than that of the clustering model.

   It is also of interest to consider how the stress failure probability varies with the parameter γ. This is illustrated in Figure 25.5 for $Y_K = 0.50$ and various values of the clustering parameter α. Note that each curve exhibits nearly linear behaviour with

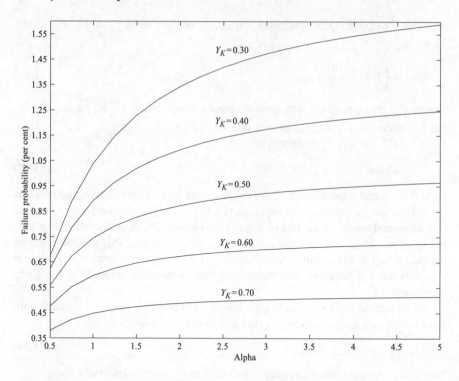

Figure 25.4   Stress failure probability $P_f = (1 - Y_L)$ in per cent as a function of α for various values of wafer probe yield Y_K. The parameter $\gamma = 0.015$

a slope that increases with increasing α. To understand the linearity of the curves in Figure 25.5 one needs to take a closer look at Equation (25.14). In particular, when $\lambda_L(0)/\alpha \ll 1$ this equation can be written as

$$Y_L = \left(1 + \frac{\lambda_L(0)}{\alpha}\right)^{-\alpha}$$

$$\approx 1 - \lambda_L(0) \tag{25.17}$$

The stress failure probability, $P_f = (1 - Y_L)$, is therefore

$$P_f \approx \lambda_L(0)$$

$$= \frac{\gamma \lambda_K}{(1 + \lambda_K/\alpha)}$$

$$= \gamma\alpha(1 - Y_K^{1/\alpha}) \tag{25.18}$$

This is the equation of a line with slope $\alpha(1 - Y_K^{1/\alpha})$ and a vertical intercept of 0. As mentioned above, the accuracy of this approximation is based on the assumption that $\lambda_L(0)/\alpha \ll 1$, where $\lambda_L(0) = \gamma \lambda_K/(1 + (\lambda_K/\alpha))$. With $\lambda_K \sim 0.5\text{–}3$ and $\alpha \sim 1\text{–}4$ for reasonable wafer probe yields, the accuracy of the approximation depends

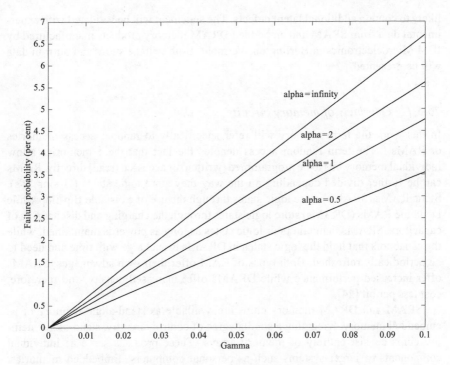

Figure 25.5　*Stress failure probability P_f in per cent as a function of γ for various values of α. The wafer probe yield is $Y_K = 0.50$*

primarily on the value of γ. For the recently reported values of $\gamma \sim 0.01$–0.02 [8], this approximation is very good. For significantly larger values of γ, the accuracy will of course decrease, and the curves will not be so linear in nature.

The integrated yield-reliability model presented thus far forms the underlying framework for the applications presented in sections to come. In the next section the reliability implications for repaired die will be addressed, while in Section 25.6 the influence of local region yield on the reliability of a die is discussed. These sections will show how wafer test information can be used to identify integrated circuits of varying reliability.

25.5　Application to defect tolerant memory circuits

In this section, the integrated yield-reliability model will be extended to estimate the early-life reliability of repaired and unrepaired memory die, and therefore quantify the effect of repairs on early-life reliability. Specifically, the model will be used to calculate the probability that a die with a given number of repairs results in a stress test failure. It will be shown that a die that has been repaired can present a far greater reliability risk than a die with no repairs. The physical reason for this is defect clustering; a die that has been repaired is known a priori to contain defects and is therefore more

likely to contain additional latent defects. The statement will be validated with experimental data from SRAM and embedded DRAM memory products manufactured by
IBM Microelectronics in Burlington, Vermont. Both voltage stress and burn-in data
will be presented.

25.5.1   *Overview of memory circuits*

In this work the term memory will refer specifically to random access memories
or RAMs.[3] The term random access denotes the fact that these memories allow
individual memory bits to be programmed (written) or accessed (read) directly. RAMs
can be further divided depending on the way they store logic states ('1's or '0's).
Static RAMs (SRAMs) store logic states through the use of a bistable flip-flop while
Dynamic RAMs (DRAMs) store logic states through the charging and discharging of
capacitors. SRAMs maintain their logic states as long as power is maintained, while
the capacitors that hold the logic states in DRAMs lose charge with time and need to
be periodically refreshed. Both types of RAM offer their own advantages: SRAMs
offer increased performance while DRAMs offer increased density, and therefore,
cost less per bit [24].

SRAM and DRAM memory chips are available as stand-alone memory or as
embedded memory. Excepting a small degree of control circuitry, stand-alone memory chips consist entirely of memory arrays. These products serve as individual
components of larger systems such as personal computers. Embedded memories,
on the other hand, are built on the same chip with surrounding logic circuits and
are a step in the direction of so-called 'Systems-on-a-chip' or SoCs. The vision of
SoCs is to fully integrate diverse systems into a single integrated circuit. Thus, a
SoC may consist of digital circuitry, analogue circuitry and memory, all on the same
integrated circuit. The close proximity of the various components on an SOC can
result in significant size reduction as well as enhanced performance. Because of these
advantages, individual logic circuits with embedded memory are becoming com-
monplace in applications once employing separate logic and stand-alone memory
units.

Memory circuits, whether SRAM or DRAM, embedded or stand-alone, differ
from standard logic circuits in that they generally offer extensive defect tolerance in
order to boost yield. This simply means that many of the killer defects identified at
wafer probe test can be repaired. While defect tolerance is employed on virtually all
large RAM circuitry, it is not commonly employed for logic designs. This is due to the
fact that memories, in contrast to complex logic, have very simple, regular structures
and architectures. This allows defect tolerance to be easily integrated into memory
designs. The net result is often a significant increase in overall yield. Indeed, it is not
uncommon for repairability to increase memory yield by a factor of ten or more. Of
course, it is the purpose of this section to show that the increase in yield resulting
from repairability comes at the price of decreased early-life reliability.

---

[3] See Reference 24 for a detailed account of semiconductor memories.

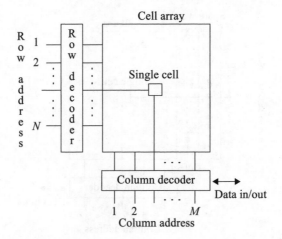

Figure 25.6   A typical memory architecture. An N bit row address allows for 2^N possible word line selections, while M column address bits allows 2^M bit line selections. The total number of single cells is therefore 2^{N+M}

25.5.1.1 Memory architecture

Figure 25.6 shows a simple schematic of a memory array architecture. It consists of an N bit row address and an M bit column address. The row bits get decoded to select 1 of the 2^N possible cell array rows. These are known as word lines. Similarly, the M column bits get decoded to select 1 of the 2^M possible columns. Each column is known as a bit line. The word and bit lines form a grid with a single cell located at each intersection point. Selection of a given word line and bit line therefore allows a single cell to be accessed. Once accessed, the logic state of the cell can be read or a new value loaded through the Data In/Out port. Since there are 2^N word lines and 2^M bit lines, there are a total of 2^{N+M} single cells or bits.

Defect tolerance is generally added to this memory architecture with the addition of spare word lines and bit lines. Such an arrangement is shown in Figure 25.7. The figure shows the replacement of a defective word line with a spare word line. Spare bit lines are also available to replace defective bit lines. In the case of a single cell failure, the cell's word line and a bit line both get replaced. The number of word lines, bit lines or single cells that can be replaced therefore depends on the number of spares available.

It should be noted that defect tolerance for memory circuits is generally limited to the memory cell array. Defects that occur within the control circuitry (e.g. decoders, read/write enable lines), cannot generally be repaired; these regions remain vulnerable to killer defects.

25.5.2   Yield-reliability modelling for redundant memory circuits

This section extends the yield-reliability model to incorporate situations in which the product under consideration contains memory circuits. Such an extension requires

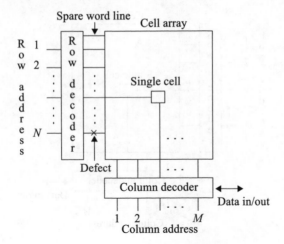

Figure 25.7 *A defective word line is replaced by a spare word line. The defective word line is identified at wafer test. The memory chip is then configured so that the spare word line is activated whenever the defective word line is addressed*

one to compensate for the fact that some of the killer defects can now be repaired. This means that the wafer test yield, defined previously as the probability that a chip contains 0 killer defects, must be altered to include those chips with killer defects that could be repaired. While this is a relatively simple modification, repairability can drastically affect the early-life reliability of a chip. Indeed, it will be shown that chips that have been repaired are more likely to fail stress tests than chips with no repairs. Moreover, the more repairs a chip has, the greater its stress test failure probability.

25.5.2.1 Wafer probe yield with repair

To incorporate repairability one must consider the probability that a killer defect can be repaired. If it is assumed that a given defect is just as likely to land anywhere within the chip's critical area,[4] then the probability that a killer defect lands within the non-repairable critical area, A_{NR}, is given by the ratio $p_{NR} = A_{NR}/A_T$, where $A_T = A_R + A_{NR}$ is the total critical area of the chip. Similarly, the probability that a given defect is repairable is given by $p_R = A_R/A_T$, where A_R is the amount of critical area that is repairable. Note that $p_R + p_{NR} = 1$.

Now, let $G(i)$ be the event that a chip is functional and contains i killer defects. As the chip is functional, the i killer defects must have been repairable. Thus,

$$P[G(i)] = p_R^i P[K(i)] \tag{25.19}$$

---

[4] The critical area denotes the physical area that is susceptible to defects.

where as before, $P[K(i)]$ denotes the probability of exactly i killer defects. The effective wafer probe yield with repair, $Y_{K\text{eff}}$, is therefore

$$Y_{K\text{eff}} = \sum_{i=0}^{\infty} P[G(i)] = \sum_{i=0}^{\infty} p_R^i P[K(i)] \tag{25.20}$$

The summation in Equation (25.20) shows that $Y_{K\text{eff}}$ is simply the generating function for $P[K(i)]$. This generating function, denoted $T(z)$, is given by [16]

$$T(z) = \left[1 + (1-z)\frac{\lambda_K}{\alpha}\right]^{-\alpha} \tag{25.21}$$

$Y_{K\text{eff}}$ can therefore be written as

$$Y_{K\text{eff}} = \left[1 + \frac{\lambda_{K\text{eff}}}{\alpha}\right]^{-\alpha} \tag{25.22}$$

where $\lambda_{K\text{eff}} = (1 - p_R)\lambda_K = p_{NR}\lambda_K$. Thus, repairability has the effect of reducing the average number of killer defects from λ_K to $p_{NR}\lambda_K$. Note that extending the sum to infinity assumes that there is no limit to the number of repairs that can be made. This is justified by the fact that the probability of more than ~25 repairs is negligibly small for most products employing redundancy.

Note that one must now differentiate between $Y_{K\text{eff}}$ and Y_K. To do this, Y_K is often termed the 'perfect' wafer probe yield to distinguish it from the effective yield achievable with repairable or redundant circuits [15,17,23]. Y_K is simply the probability of zero killer defects.

As a numerical example, suppose that 90 per cent of the chip's critical area is repairable. This implies that $p_{NR} = 0.10$. If $\lambda_K = 1$ and $\alpha = 2$, then $Y_{K\text{eff}} = 0.91$. With no repair capabilities, $p_{NR} = 1$, and the yield is $Y_K = 0.44$. Thus, repairability can have a very significant impact on wafer probe yield.

It is also of interest to determine the fraction of functional die with exactly i repaired defects. This can be obtained as follows:

$$f(i) = \frac{P[G(i)]}{\sum_{i=0}^{\infty} P[G(i)]} = \frac{p_R^i P[K(i)]}{Y_{K\text{eff}}} \tag{25.23}$$

This equation, when compared to repair data, can be used to obtain the yield parameters α and λ_K.

25.5.2.2 Reliability yield with repair

Equation (25.13) of Section 25.1 gives an expression for the probability of n latent defects given m killer defects. This was denoted by $P[L(n)|K(m)]$. For logic circuits with no redundancy, one is only interested in the case when no killer defects are present. This corresponds to $m = 0$. For redundant circuits, however, one needs to consider the more general case of $m \geq 0$. Setting $n = 0$ in the expression for

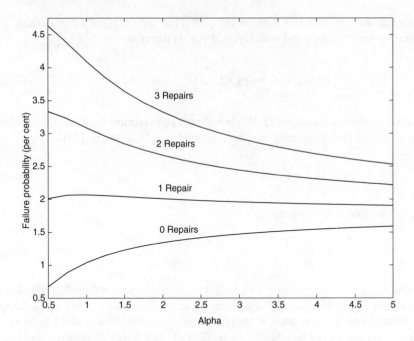

Figure 25.8   *Stress test failure probability $P_f(m)$ in per cent for $m = 0, 1, 2$ and 3 repairs. The value of α ranges from 0.5 to 5. The perfect wafer probe yield is $Y_K = 0.30$, $\gamma = 0.015$ and $p_{NR} = 0.10$*

$P[L(n)|K(m)]$ and defining $Y_L(m) = P[L(0)|G(m)] = P[L(0)|K(m)]$ gives

$$Y_L(m) = \left[1 + \frac{\lambda_L(m)}{\alpha + m}\right]^{-(\alpha+m)} \tag{25.24}$$

This gives the early-life reliability yield of a chip which has been repaired exactly m times. Equation (25.24) is an important result because it shows that $Y_L(m)$ arises from a negative binomial distribution with parameters $(\alpha + m)$ and $\lambda_L(m) = (\alpha + m/\alpha)\lambda_L(0)$. Chips with m repairs therefore contain, on average, $(\alpha + m)$ times as many latent defects as chips with 0 repairs.

25.5.2.3   Numerical results

Figure 25.8 shows the stress test failure probability $P_f(m) = 1 - Y_L(m)$ in per cent as a function of the clustering parameter α. Note that while α can certainly range from 0.5 to 7 in practice, a typical value may be around 2.0. The figure shows four curves corresponding to $m = 0, 1, 2$ and 3 repairs. The perfect wafer probe yield was assumed to be $Y_K = 0.30$, $\gamma = 0.015$ and $p_{NR} = 0.10$. Note also that this implies that the effective wafer probe yield, $Y_{K\mathrm{eff}}$, varies from 0.71 when $\alpha = 0.5$ to 0.88 when $\alpha = 5$.

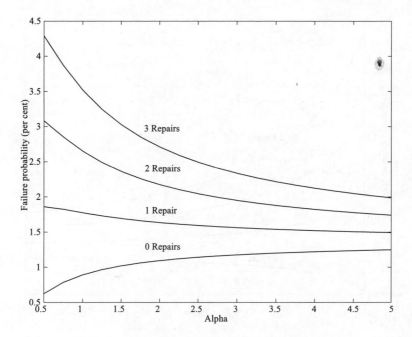

Figure 25.9   *Stress test failure probability $P_f(m)$ in per cent for $m = 0, 1, 2$ and 3 repairs. The value of α ranges from 0.5 to 5. The perfect wafer probe yield is $Y_K = 0.40$, $\gamma = 0.015$ and $p_{NR} = 0.10$*

The figure shows that chips that have been repaired can have a probability of failure that is significantly greater than chips with no repairs. This is particularly apparent when there is a high degree of clustering (low value of α). Indeed, for $\alpha = 0.5$, the probability of failure is 0.68, 2.01, 3.33 and 4.63 per cent for 0, 1, 2 and 3 repairs, respectively. This means that a chip with 1 repair is $2.01/0.68 = 2.96$ times more likely to fail than a chip with no repairs. Furthermore, chips with 2 and 3 repairs are 4.90 and 6.81 times more likely to fail than a chip with no repairs. Note, however, that as α increases, the reliability improvement for chips with no repairs decreases. Thus, for $\alpha = 2$, chips with 1 repair are 1.50 times more likely to fail, while chips with 2 and 3 repairs are 1.99 and 2.48 times more likely to fail than chips with no repairs. This trend continues as α increases. In particular, as $\alpha \to \infty$ (no clustering), the probability of failure becomes independent of the number of repairs. In such a case, repaired memory chips are just as reliable as memory chips with no repairs.

Figures 25.9 and 25.10 show the stress test failure probability as a function of α with 0, 1, 2 and 3 repairs for a perfect wafer probe yield of $Y_K = 0.40$ and $Y_K = 0.50$, respectively. Comparison of Figures 25.8–25.10 indicates that the failure probability decreases as Y_K increases. For example, suppose that $\alpha = 2$ and a chip has been repaired twice. Then the failure probability is 2.67 per cent for $Y_K = 0.30$, 2.18 per cent for $Y_K = 0.40$ and 1.74 per cent for $Y_K = 0.50$. This decrease in failure

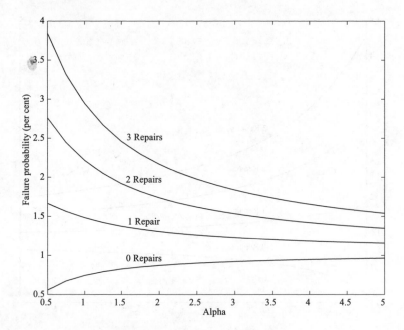

Figure 25.10   *Stress test failure probability $P_f(m)$ in per cent for $m = 0, 1, 2$ and 3 repairs. The value of α ranges from 0.5 to 5. The perfect wafer probe yield is $Y_K = 0.50$, $\gamma = 0.015$ and $p_{NR} = 0.10$*

probability with increasing Y_K follows from the fact that, for a given clustering parameter α, the average number of killer defects decreases as Y_K increases. Since the average number of latent defects, λ_L, is proportional to λ_K, λ_L also decreases as Y_K goes up. The result is a decrease in the number of stress test failures.

Let us now consider more closely how the stress test failure probability depends on the number of repairs and the clustering parameter. This dependence is shown in Figure 25.11, where the stress test failure probability is plotted versus the number of repairs for various values of α. Note that the curves are very linear with a slope that increases with decreasing α. In particular, note that the slope goes to zero when $\alpha = \infty$. This corresponds to a Poisson distribution and implies no clustering.

To understand the linearity of the curves in Figure 25.11 one needs to take a closer look at Equation (25.24). In particular, when $\lambda_L(0)/\alpha << 1$ this equation can be written as

$$Y_L(m) = \left(1 + \frac{\lambda_L(0)}{\alpha}\right)^{-(\alpha+m)}$$

$$\approx 1 - (\alpha + m)\frac{\lambda_L(0)}{\alpha} \qquad (25.25)$$

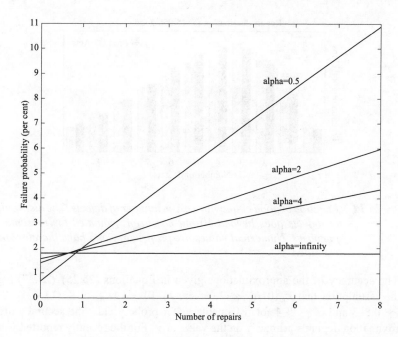

Figure 25.11   *Stress test failure probability $P_f(m)$ in per cent as a function of the number of repairs. Various values of α are shown. The perfect wafer probe yield is $Y_K = 0.30$, $\gamma = 0.015$ and $p_{NR} = 0.10$*

The stress test failure probability for a chip with m repairs, $P_f(m)$, is therefore

$$P_f(m) = 1 - Y_L(m)$$

$$\approx (\alpha + m)\frac{\lambda_L(0)}{\alpha}$$

$$= \frac{\lambda_L(0)}{\alpha}m + \lambda_L(0) \tag{25.26}$$

This is the equation of a line with slope $\lambda_L(0)/\alpha$ and vertical intercept $\lambda_L(0) = P_f(0)$.

As a measure of the stress test failure probability for chips with m repairs as compared with chips with no repairs, one may define the relative failure probability $R_f(m) = P_f(m)/P_f(0)$. Thus, from Equation (25.26) it follows that

$$R_f(m) = \frac{P_f(m)}{P_f(0)} \approx \frac{m}{\alpha} + 1 \tag{25.27}$$

Note that $R_f(m)$ provides a simple way to validate the proposed model. Indeed, according to Equation (25.27), a plot of $R_f(m)$ versus m yields a straight line with slope $1/\alpha$ and a vertical intercept of 1. Further, since Equation (25.27) depends only on the clustering parameter α, one can estimate the relative failure probability for repaired memory chips once the clustering parameter α is known. This is generally known following wafer probe testing.

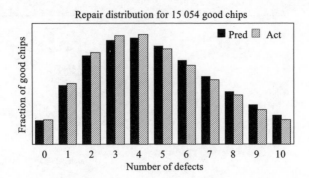

Figure 25.12   Fraction of good chips with a given number of defects. The distribution of defects goes beyond 10 but has been truncated for presentation purposes. Numerical values are proprietary and have therefore been excluded

The accuracy of the approximations given in Equations (25.25)–(25.27) based on the assumption that $\lambda_L(0)/\alpha \ll 1$, where $\lambda_L(0) = \gamma \lambda_K/(1 + (\lambda_K/\alpha))$. With $\lambda_K \sim 0.5$–3 and $\alpha \sim 1$–4 for reasonable wafer probe yields, the accuracy of the approximation depends primarily on the value of γ. For the recently reported values of $\gamma \sim 0.01$–0.02 [8], this approximation is very good. For significantly larger values of γ, the accuracy will of course decrease.

25.5.3   Model verification

This section presents data from a 32 Mbit SRAM product and a product with 8 Mbits of embedded DRAM (eDRAM). In both cases, predictions based on the yield-reliability model are shown to be in excellent agreement with observed data.

25.5.3.1   A 32 Mbit SRAM product with voltage screen

This section presents data from 15 054 SRAM chips from a 0.18 μm CMOS process technology. These chips were functional following wafer test and repair. The repair distribution for these chips is shown in Figure 25.12. Note that this distribution can be easily calculated from Equation (25.23). The parameters for the negative binomial distribution, namely the clustering parameter α and the average number of killer defects per chip, λ_K, can then be determined by fitting Equation (25.23) to the distribution obtained from data. This is done using a standard non-linear least squares method such as that available through the statistical software package SAS [25]. Thus, Figure 25.12 shows that the negative binomial statistics can accurately model the repair distribution.

To test the yield-reliability model one must compare model predictions to stress fail data. Recall that the relative failure probability $R_f(m)$ is just the failure probability for m repairs divided by the failure probability for 0 repairs. According to Equation (25.27), a plot of $R_f(m)$ versus the number of repaired defects m yields a straight line with slope $1/\alpha$ and a vertical intercept of 1. Such a plot is shown

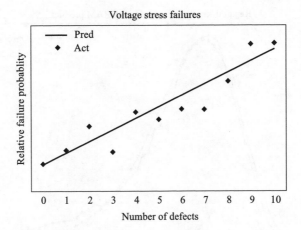

Figure 25.13   *Voltage screen failures as a function of the number of defects repaired following wafer probe testing. Numerical values are proprietary and have therefore been excluded*

in Figure 25.13, where the failures refer to voltage screen failures. Note that the voltage screen relative failure probabilities are in excellent agreement with the predicted line. Indeed, fitting the data with a linear regression model gives a slope nearly identical to the predicted slope of $1/\alpha$. Moreover, the R^2 value of the fitted line was 0.87, indicating excellent agreement between the data and the linear model.

Similar data from other SRAM products manufactured by IBM have also been closely examined for this study. Unfortunately, for proprietary reasons, only a limited amount of data was available for publication. It can be said, however, that the failure probability for SRAM products with voltage screen as well as burn-in shows the same dependence; the failure probability linearly increases with the number of killer defects repaired. In particular, as shown in Equation (25.27), the slope of this line is $1/\alpha$ in both cases.

25.5.3.2   An embedded DRAM product with burn-in data

The previous subsection demonstrates that the yield-reliability model can accurately predict the repair distribution as well as voltage stress fall-out for SRAM products. In this subsection data will be presented that shows the model predictions hold for products with embedded DRAM as well. This will be demonstrated with 2485 burn-in failures from a 0.25 μm CMOS process with 8 Mbits of embedded DRAM.

As with the SRAM product, the clustering parameter α and the average number of killer defects λ_K are determined by fitting Equation (25.23) to the repair distribution. The result of this procedure is shown in Figure 25.14. Once again, the model provides an excellent fit to the data.

Figure 25.15 shows the relative failure probability $R_f(m)$ for 2485 burn-in failures. Note that the curve is again linear. In this case the R^2 value is 0.903, indicating excellent agreement between the model and observed data.

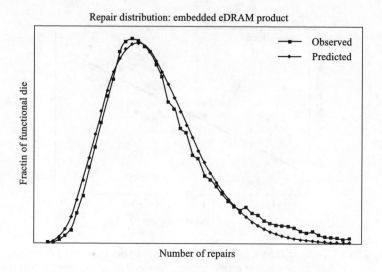

Figure 25.14   *Fraction of good chips with a given number of defects. Numerical values are proprietary and have therefore been excluded*

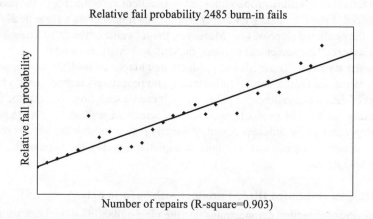

Figure 25.15   *Burn-in failures as a function of the number of defects repaired following wafer probe testing. Numerical values are proprietary and have therefore been excluded*

This section has extended the integrated yield-reliability model to include integrated circuits containing redundant memory circuits. The model has been validated on production data manufactured by IBM Microelectronics in Burlington, Vermont. The model was shown to be valid for SRAM as well as DRAM products. Specifically, it has been shown that the stress test failure probability is a linear function of the number of repairs. Moreover, since the slope of this curve is simply related to the clustering parameter α, early-life reliability estimates can be made following wafer probe testing, when the clustering parameter α is known.

25.6   Local region yield

The previous sections have demonstrated that defect clustering can significantly impact both wafer test yield and early-life reliability. It is the purpose of this section to show that the effects of defect clustering can stretch beyond the boundaries of an individual chip and affect the reliability of neighbouring die. Indeed, it will be shown mathematically and demonstrated experimentally that a chip's local yield can be a strong indicator of chip reliability; chips that test good at wafer probe yet come from regions with many faulty neighbours are greater reliability risks than chips from high yielding regions. Again, this follows from the fact that defects on semiconductor wafers are not uniformly distributed but have a tendency to cluster. Thus, if a die comes from a region where there are many defects, it is more likely to contain a 'smaller' reliability defect than a chip from a high yielding region.

Using the local yield of a chip as an indicator of its quality was proposed in the technical literature as early as 1993 [26], and can easily be extended to incorporate reliability estimation through the use of the integrated yield-reliability model. Although many definitions are possible,[5] a simple definition of local region consists of a central die and its eight adjacent neighbours. This is shown in Figure 25.16. The local yield of a given die can then be determined simply by counting the number of faulty neighbours it has. Die that pass wafer probe testing are then separated into one of nine bins depending on the number of faulty neighbours; die with 0 faulty neighbours go into bin 0, die with 1 faulty neighbour go into bin 1, and so on, until bin 8, where all neighbours were faulty. Since latent defects tend to cluster with killer defects, one expects the die in the higher numbered bins to be more likely to contain a latent defect than those in the lower numbered bins. Thus, the probability that a die fails a stress test should increase as one moves from bin 0 to bin 8. Indeed, as shown in Figure 25.17, this is consistent with actual production data. Figure 25.17 shows the fraction of good die in each bin that eventually fail burn-in. As expected, the general trend is for the burn-in failure probability to increase as one moves up in bin number. Thus, the more faulty neighbours a die has, the more likely it is to fail burn-in. Similar results have been obtained when observing voltage stress data. Moreover, the yield-reliability model allows one to accurately determine the fraction of die and the number of stress failures in each bin. Such calculations have been validated experimentally on several different products manufactured at IBM Microelectronics in Burlington, Vermont. These products had various yields, chip sizes and technologies. Data from one of these products, consisting of 77 000 microprocessor

---

[5] In work done at Intel [8], for example, local region yield is determined as a weighted average of neighbour yields; die closest to the die in question are weighted more heavily than die farther away. The surrounding 24 die have also been considered [9]. Experience at IBM has shown that most of the neighbourhood information is captured with the simple definition given here.

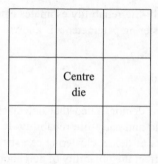

Figure 25.16   *The nine-die neighbourhood. All central die which test good at wafer probe are placed into one of nine bins depending on how many of their neighbours fail wafer probe testing*

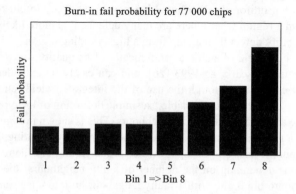

Figure 25.17   *Burn-in failure probability in each bin for 77 000 microprocessor chips that tested good at wafer probe. Note that bin 0 is excluded as no chips had 0 faulty neighbours. This was due to a systematic problem with a reticle, causing a particular reticle location to zero-yield. Numerical values are proprietary and have therefore been excluded*

units that underwent burn-in testing, will be presented here. The results of this study will be described in the following sections of this section, and the burn-in data will be compared to model predictions. It will be shown that the model accurately predicts both the yield and the early-life reliability of the chips in each bin.

25.6.1   Bin distribution calculations

This section provides the mathematical results required to calculate bin distributions. Specifically, the number of good die in each bin and the number of stress test fails in each bin will be calculated. While the calculations presented here should be sufficiently detailed for most purposes, a review of the fundamental mathematical principles (e.g. the principle of inclusion–exclusion) can be found in Reference 14.

25.6.1.1   Chips free of killer defects

Consider a central die and its $N - 1$ adjacent neighbours ($N = 9$ for the 9-die neighbourhood shown in Figure 25.16). Furthermore, let X_K be a random variable denoting the number of chips in the N-die neighbourhood with zero killer defects. According to the inclusion–exclusion principle [14], the probability that exactly k chips are free of killer defects is given by [15]

$$P[X_K(k)] = \binom{N}{k} \sum_{q=0}^{N-k} (-1)^q \binom{N-k}{q} y_{k+q} \qquad (25.28)$$

where

$$y_{k+q} = \left[1 + \frac{(k+q)\lambda_K}{\alpha}\right]^{-\alpha} \qquad (25.29)$$

Now, let C_K denote the event that the central die is free of killer defects. Then the probability that the central die passes wafer probe, given that there are a total of k killer defect-free chips in the N-die neighbourhood, is given by

$$P[C_K | X_K(k)] = \frac{\binom{N-1}{k-1}}{\binom{N}{k}} = \frac{k}{N} \qquad (25.30)$$

Thus, the fraction of good chips with $k - 1$ good neighbours is $P[C_K|X_K(k)]$ $P[X_K(k)]/Y_K$. To write this in terms of the number of faulty neighbours, note that, if the central chip has r good neighbours, then it must have $N - r$ faulty neighbours. The fraction of good die with exactly r bad neighbours is therefore $P[C_K|X_K(N-r)]P[X_K(N-r)]/Y_K$. Using Equation (25.30), this can be written as

$$f(r) = \frac{(1 - (r/N))P[X_K(N-r)]}{Y_K} \qquad (25.31)$$

where $r = 0, 1, \ldots, N - 1$. This equation gives the fraction of good chips in each bin following wafer probe testing.

25.6.1.2   Chips free of both killer and latent defects

When both killer and latent defects are considered one can easily extend the concepts developed in the previous subsection. In particular, one can define $P[X_K(k), X_L(l)]$ as the probability that exactly k chips are free of killer defects and l chips are free of latent defects. This probability can be written as

$$P[X_K(k), X_L(l)] = \sum_{q_1=0}^{(N-k)} \sum_{q_2=0}^{(N-l)} a(q_1, q_2, k, l) y_{k+q_1, l+q_2} \qquad (25.32)$$

where

$$a(q_1, q_2, k, l) = \binom{N}{k}\binom{N}{l}(-1)^{q_1+q_2}\binom{N-k}{q_1}\binom{N-l}{q_2} \qquad (25.33)$$

and

$$y_{n_1,n_2} = \left[1 + \frac{n_1\lambda_K}{\alpha} + \frac{n_2\lambda_L}{\alpha}\right]^{-\alpha} \tag{25.34}$$

Note that while $P[X_K(k), X_L(l)]$ gives the probability of k killer defect-free chips and l latent defect-free chips, it says nothing about which chips are free of both killer and latent defects. Towards this end, let C_K and C_L denote the events that the central die is free of killer and latent defects, respectively. Then $C = C_K \cap C_L$ denotes the compound event that the central die is free of both killer and latent defects. Thus, given k killer defect-free chips and l latent defect-free chips in the N die neighbourhood, the probability that the central die is free of both defect types is

$$P[C|X_K(k), X_L(l)] = \frac{\binom{N-1}{k-1}\binom{N-1}{l-1}}{\binom{N}{k}\binom{N}{l}} = \frac{kl}{N^2} \tag{25.35}$$

Multiplying Equation (25.35) by $P[X_K(k), X_L(l)]$ and summing over l gives $P[C|X_K(k)]P[X_K(k)]$. That is,

$$P[C|X_K(k)]P[X_K(k)] = \sum_{l=0}^{N} P[C|X_K(k), X_L(l)]P[X_K(k), X_L(l)]$$

$$= \frac{k}{N^2} \sum_{l=0}^{N} l P[X_K(k), X_L(l)] \tag{25.36}$$

But from Bayes' Rule $P[X_K(k), X_L(l)] = P[X_L(l)|X_K(k)]P[X_K(k)]$. Therefore the above equation can be written as

$$P[C|X_K(k)] = \frac{k}{N^2} \sum_{l=0}^{N} l P[X_L(l)|X_K(k)] \tag{25.37}$$

where $P[X_L(l)|X_K(k)] = P[X_K(k), X_L(l)]/P[X_K(k)]$ is obtained from Equations (25.28) and (25.32). Equation (25.37) is the probability that a central die passes both wafer probe and early life testing, given that k of the N die in the neighbourhood have passed wafer probe. Since $P[C|X_K(k)] = P[C_L|C_K, X_K(k)]P[C_K|X_K(k)]$, dividing Equation (25.37) by $P[C_K|X_K(k)] = k/N$ gives the probability that a chip passes early life testing, given that it and $(k-1)$ of its neighbours have passed wafer probe. If one defines $Y_L(q)$ as the probability that a good chip with q faulty neighbours passes early life testing, then $Y_L(q) = P[C_L|C_K, X_K(N-q)]$. Hence,

$$Y_L(q) = \frac{1}{N} \sum_{l=0}^{N} l \ P[X_L(l)|X_K(N-q)] \tag{25.38}$$

where $q = 0, 1, \ldots, N-1$. This is the stress test yield (e.g. burn-in yield) for a chip that has passed wafer probe and has q faulty neighbours.

25.6.1.3   Numerical results

Figure 25.18 shows the reliability failure probability $[1 - Y_L(i)]$ for die in each bin for various values of the clustering parameter α, $Y_K = 0.50$ and $\gamma = 0.015$. Recall that a lower value of α indicates increased clustering, while $\alpha = \infty$ implies no clustering. Further, for a value $\gamma = 0.015$, one expects, on average, 1.5 latent defects for every 100 killer defects.

As expected, the figure shows that the probability of failure increases as one moves from the lower numbered bins to the higher numbered bins. An exception to this is the case of $\alpha = \infty$, which corresponds to no clustering. In this case, the probability of failure is constant for each bin number. Thus, binning based on local region yield provides no advantage when defects follow a Poisson distribution.

Consider now the particular case of $\alpha = 0.5$. Note that the probability of failure in the best bin (i.e. bin number 0) is significantly lower than in the other bins. In particular, die from bin 8 have a failure probability of 3.16 per cent compared with 0.08 per cent in bin 0. This means that a die selected from bin 8 is ~39 times more likely to fail burn-in than a die selected from bin 0. Further, compared to the average probability of failure of 0.558 per cent achieved without binning (see Equation (25.38)), bin 0 represents a factor of ~7 improvement. Note, however, that these benefits decrease as the clustering parameter increases. Thus, for $\alpha = 2$ and $\alpha = 4$ the best bin shows a factor of 3.33 and 2.26 improvement over the no binning case, respectively.

Although Figure 25.18 indicates the potential of binning for improved reliability, it is important to realise that the usefulness of this technique depends significantly on the fraction of die in each bin. This is illustrated in Figure 25.19 where the fraction of die in each bin is shown for $\alpha = 0.5, 2.0, 4.0$ and ∞. With $\alpha = 0.5$, most of the defects will be clustered together and there will be many neighbourhoods with few, if any, defects. The result is a large number of die in the lower numbered bins. In particular, bin 0 contains ~40 per cent of the die. When clustering decreases (α increases), however, the defects get distributed more evenly among the neighbourhoods. For the more realistic value of $\alpha = 2.0$, this results in fewer die in the best bin with the maximum number of die in bin 2. For $\alpha = 4$ this effect is accentuated and the higher numbered bins become more heavily populated. Thus, as clustering decreases, fewer die are present in the lower numbered bins. Note that the bin variation for $\alpha = \infty$ is quite irrelevant since the probability of failure is the same in each bin when no clustering is present. Indeed, the bin variation for $\alpha = \infty$ is based solely on the wafer probe yield Y_K. This illustrates the important point that Figures 25.18 and 25.19 must be examined together to accurately evaluate the effectiveness of binning.

Finally, it is important to consider how the above results depend on the wafer probe yield Y_K. For a fixed value of α and γ, low yields imply that, on average, a greater number of defects (both killer and latent) get distributed over each neighbourhood. Thus, as the yield decreases, one expects a higher failure probability in each bin and a lower fraction of die in the lower numbered bins. These effects are illustrated in Figure 25.20 for $\gamma = 0.015, \alpha = 2.0$ and Y_K ranging from 0.10 to 0.90. Note that the bottom curve shows the probability of failure in the best bin divided by the average probability of failure obtained without binning. This ratio indicates the reliability improvement one sees in the best bin as compared to the lot taken as a whole. Note

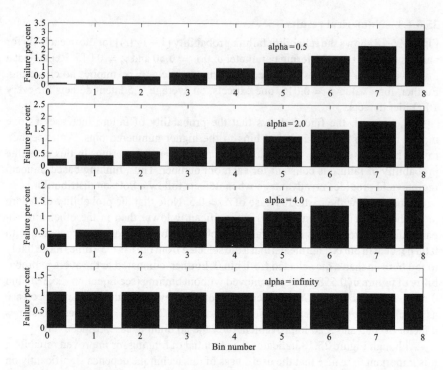

Figure 25.18   Reliability failure probability $[1 - Y_L(i)]$ in each bin in per cent. The value of α ranges from 0.5 to ∞. The wafer probe yield is $Y_K = 0.50$ and $\gamma = 0.015$

that while this ratio is maximum for low yields, the fraction of die present in the best bin under these circumstances is generally quite small.

25.6.2   Burn-in results from 77 000 IBM Microprocessor units

In this section, the predictions of the yield-reliability model will be compared to industry data. It should be noted that, while data from a single product will be presented here, similar results have been observed on many products manufactured by IBM Microelectronics in Burlington, Vermont. This information is proprietary, however, and only a subset of this work was available for publication.

25.6.2.1   Parameter estimation

The use of the negative binomial distribution requires the specification of its two parameters λ and α. This can be accomplished through a technique known as windowing [16]. One begins with the equation

$$Y_K(n) = \left(1 + \frac{n\lambda_K}{\alpha}\right)^{-\alpha} \tag{25.39}$$

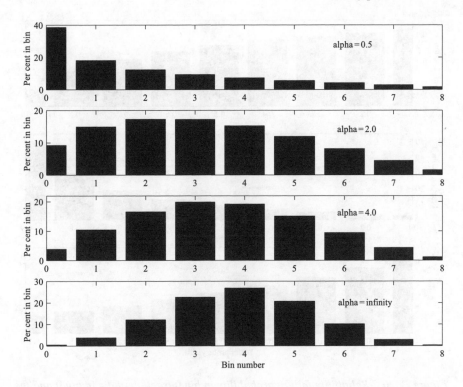

Figure 25.19   Fraction of die in each bin in per cent. Various values of α are shown. The wafer probe yield is $Y_K = 0.50$ and $\gamma = 0.015$

where $Y_K(n)$ is the wafer probe yield for chips taken in groups of n. Thus, $Y_K(1) = Y_K$ is the ordinary wafer probe yield, $Y_K(2)$ is the yield with chips taken in pairs, $Y_K(4)$ is the yield with chips taken in groups of four, and so on. Once data is obtained for several values of n, a non-linear least squares algorithm is performed to obtain the values of α and λ_K. This can be carried out, for example, using the statistical analysis software SAS and its NLIN procedure.

Having obtained values for α and λ_K, one need only determine γ. This can be obtained from stress data. That is, if a sample of chips is subjected to a stress test (e.g. burn-in, voltage stress), Y_L can be determined experimentally. Equation (25.14) can then be inverted to obtain the value of γ.

Note that these parameter will generally change with time. Experience indicates that fluctuations in the clustering parameter, α, are small, while the average number of killer defects per chip, λ_K, decreases as yield learning progresses. For mature products, these parameters are very stable.

25.6.2.2   Model verification

The data presented in this subsection came from 77 000 microprocessor units from a 0.25 μm CMOS process that were subjected to burn-in. Die that tested good at wafer

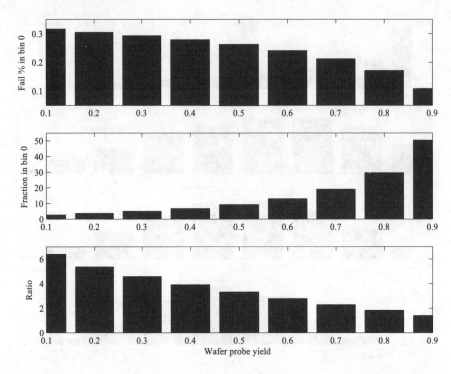

Figure 25.20   Reliability failure probability in bin 0 fraction of die in bin 0 and the
improvement ratio as a function of wafer probe yield Y_K. All values
are given in per cent. Here $\alpha = 2$ and $\gamma = 0.015$

probe were binned based on the number of faulty neighbours they had. The fraction
of good chips in each bin following wafer probe and the fraction of stress failures
in each bin were then obtained from the data and compared to model predictions.
As this section will show, model predictions matched observed data. Indeed, differences between projected and observed reliability were negligible for all practical
purposes.

Figure 25.21 shows the bin distribution of those die that passed wafer probe. Note
that bin 0 has been excluded as no chips were present in this bin. This was due to a
systematic problem; a particular reticle location had zero-yield. The regularity of this
defect made it impossible for a chip to have 0 faulty neighbours. Thus, for calculation
purposes, the nine-die neighbourhood was reduced to eight-die.

Edge chips that lacked complete neighbourhood information were treated in two
different ways. In the data shown in Figure 25.21, missing die were treated as if
they were faulty. A die with three missing neighbours, for example, could never
appear in a bin <3. An alternative method was considered that involved replacing
missing neighbours with die a distance of two die away from the central die rather than
one. In this case, a die with three missing neighbours would choose the next three
closest neighbours with complete wafer test information. In cases where multiple

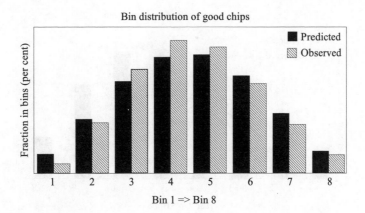

Bin distribution of good chips

Figure 25.21   *Fraction of good die in each bin. Note that bin 0 is excluded as no chips had 0 faulty neighbours. This was due to a systematic problem with a reticle, causing a particular reticle location to zero-yield. Numerical values are proprietary and have therefore been excluded*

replacement possibilities exist, the one with the most faulty die was chosen. Applying this method to the data only minutely changed the results of Figure 25.21. Experience with multiple products indicates that both methods can be used successfully; the latter method being more effective when a large fraction of the good chips are on the edge.

Although the y-axis values are not shown, it is clear that the predicted and observed values track closely. Note that for the purpose of obtaining the bin distribution of good chips, only the yield model is needed. Thus, the main conclusion from this figure is that the negative binomial statistics can accurately describe the bin distribution of good chips.

To test the accuracy of the integrated yield-reliability model, one must compare the fraction of stress failures observed in each bin to model predictions. This is shown in Figure 25.22. Note that the predicted and observed values track closely. Again, the difference between the predicted and observed reliability is small.

Note that the largest fraction of burn-in fails comes from bin 5. This is often a source of confusion, since many expect bin 8 to contribute the largest fraction of fails. However, even though the chips in bin 8 have the greatest failure probability, the number of good chips in bin 8 following wafer probe is relatively small. This emphasises the important point that the stress test failure probability (Figure 25.17) must be considered alongside the bin distribution of good chips (Figure 25.21). Indeed, it is the product of these two figures that gives the fraction of fails in each bin (Figure 25.22). As a numerical example, suppose that the burn-in failure probabilities in bin 0 and 8 are 0.01 and 0.1, respectively. Thus, chips in bin 8 are ten times more likely to fail than chips in bin 0. Suppose further that there are 1000 chips in bin 0 and 100 chips in bin 8 following wafer probe. Then the number of failures from bin 8 is $0.1 \times 100 = 10$ and the number of fails in bin 0 is also $0.01 \times 1000 = 10$.

Fraction of burn-in fails in each bin

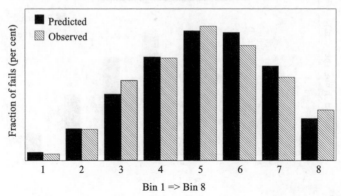

Figure 25.22   *Fraction of failures in each bin. Note that bin 0 is excluded as no chips had 0 faulty neighbours. This was due to a systematic problem with a reticle, causing a particular reticle location to zero-yield. Numerical values are proprietary and have therefore been excluded*

25.7   Conclusions

This chapter has presented a unified approach to yield and reliability modelling. Fundamentally, the successful integration of yield and early-life reliability results from the fact that latent early-life reliability defects, while smaller and placed differently, are essentially the same in nature as those killer defects that cause wafer probe failures. Yield models that accurately describe the distribution of killer defects will therefore also describe the distribution of latent defects.

Any successful defect distribution must account for the fact that defects over semiconductor wafers tend to cluster. The negative binomial distribution, known for two decades to describe killer defect distributions, accurately describes latent defects distributions as well. Defect clustering information can be used to identify the most reliable die, or conversely, those die most likely to contain a latent defect and fail during the early-life period. This information allows the semiconductor manufacturer to optimise stress tests such as burn-in, and can result in significant cost reductions when compared to traditional methods.

References

1   KUO, W., CHIEN, W.-T.K., and KIM, T.: 'Reliability, yield, and stress burn-in' (Kluwer Academic Publishers, Boston, MA, 1998)
2   KUO, W., and KIM, T.: 'An overview of manufacturing yield and reliability modeling for semiconductor products', *Proceedings of the IEEE*, 1999, **87**(8), pp. 1329–1344

3   NELSON, W.: 'Applied life data analysis' (Wiley Series in Probability and Mathematical Statistics, 1982)

4   AJITH AMERASEKERA, E., and NAJM, F.: 'Failure mechanisms in semiconductor devices' (Wiley, New York, 1997, 2nd edn)

5   HUSTON, H.H., and CLARKE, P.C.: 'Reliability defect detection and screening during processing-theory and implementation'. Proceedings of the international *Reliability physics* symposium, 1992, pp. 268–275

6   KUPER, F., VAN DER POL, J., OOMS, E., *et al.*: 'Relation between yield and reliability of integrated circuits: experimental results and application to continuous early failure rate reduction programs'. Proceedings of the international *Reliability physics* symposium, 1996, pp. 17–21

7   VAN DER POL, J., OOMS, E., VAN 'T HOF, T., and KUPER, F.: 'Impact of screening of latent defects at electrical test on the yield-reliability relation and application to burn-in elimination'. Proceeding of the international *Reliability physics* symposium, 1998, pp. 370–377

8   RIORDAN, W., MILLER, R., SHERMAN, J., and HICKS, J.: 'Microprocessor reliability performance as a function of die location for a 0.25μ, five layer metal CMOS logic process'. Proceedings of the international *Reliability physics* symposium, 1999, pp. 1–11

9   MILLER, R., and RIORDAN, W.: 'Unit level predicted yield: a method of identifying high defect density die at wafer sort'. Proceedings of the 2001 international *Test* conference, October 2001, pp. 1118–1127

10 BARNETT, T.S., SINGH, A.D., and NELSON, V.P.: 'Extending integrated circuit models to estimate early-life reliability', *IEEE Transactions on Reliability*, 2003, **52**(3), pp. 296–300

11 STAPPER, C.H., ARMSTRONG, F.M., and SAJI, K.: 'Integrated circuit yield statistics', *Proceedings of IEEE*, 1983, **71**(4), pp. 453–470

12 FERRIS-PRABHU, A.V.: 'Introduction to semiconductor device yield modeling' (Artech House, Boston, MA, 1992)

13 WALKER, D.M.H.: 'Yield simulation for integrated circuits' (Kluwer, Boston, MA, 1987)

14 WILLIAM FELLER: 'An introduction to probability theory and its applications' (Wiley Publications in Statistics, 1957, 2nd edn)

15 KOREN, I., and STAPPER, C.H.: 'Yield models for defect tolerant VLSI circuits: a review', in KOREN, I. (Ed.): 'Defect and fault tolerance in VLSI systems, vol. 1' (Plenum, New York, 1989) pp. 1–21

16 KOREN, I., KOREN, Z., and STAPPER, C.H.: 'A unified negative binomial distribution for yield analysis of defect tolerant circuits', *IEEE Transactions on Computers*, 1993, **42**, pp. 724–437

17 KOREN, I., and KOREN, Z.: 'Defect tolerant VLSI circuits: techniques and yield analysis', *Proceedings of the IEEE*, 1998, **86**, pp. 1817–1836

18 BARNETT, T.S., SINGH, A.D., and NELSON, V.P.: 'Burn-in failures and local region yield: an integrated yield-reliability model'. Proceedings 2001 *VLSI test* symposium, May 2001, pp. 326–332

19   BARNETT, T.S., SINGH, A.D., and NELSON, V.P.: 'Estimating burn-in fall-out for redundant memory'. Proceedings 2001 international *Test* conference, October 2001, pp. 340–347

20   BARNETT, T.S., SINGH, A.D., GRADY, M., and PURDY, K.G.: 'Yield-reliability modeling: experimental verification and application to burn-in reduction'. Proceedings 2002 *VLSI test* symposium, May 2002, pp. 75–80

21   BARNETT, T.S., SINGH, A.D., GRADY, M., and PURDY, K.G.: 'Redundancy implications for product reliability: experimental verification of an integrated yield-reliability model'. Proceedings 2002 international *Test* conference, October 2002, pp. 693–699

22   STAPPER, C.H.: 'Correlation analysis of particle clusters on integrated circuit wafers', *IBM Journal of Research and Development*, **31**(6), 1987, pp. 641–649

23   STAPPER, C.H.: 'Yield model for productivity optimization of VLSI memory chips with redundancy and partially good product', *IBM Journal of Research and Development*, 1980, **24**(3), pp. 398–409

24   SHARMA, A.K.: 'Semiconductor memories' (IEEE Press, Los Alamitos, CA, 1997)

25   SAS Software Package: http://www.sas.com/

26   SINGH, A.D., and KRISHNA, C.M.: 'On optimizing VLSI testing for product quality using die-yield prediction', *IEEE Transactions on CAD*, 1993, **12**(5), pp. 695–709

Index